WOMEN and CRIME

SAGE Text/Reader Series in Criminology and Criminal Justice

Craig Hemmens, Series Editor

1. Walsh and Hemmens: *Introduction to Criminology: A Text/Reader*, 2nd Edition
2. Lawrence and Hemmens: *Juvenile Justice: A Text/Reader*
3. Stohr, Walsh, and Hemmens: *Corrections: A Text/Reader*, 2nd Edition
4. Spohn and Hemmens: *Courts: A Text/Reader*, 2nd Edition
5. Archbold: *Policing: A Text/Reader* (forthcoming)
6. Barton-Belessa and Hanser: *Community Corrections: A Text/Reader*
7. Greene and Gabbidon: *Race and Crime: A Text/Reader*
8. Tibbetts and Hemmens: *Criminological Theory: A Text/Reader*
9. Daigle: *Victimology: A Text/Reader*
10. Mallicoat: *Women and Crime: A Text/Reader*
11. Payne: *White Collar Crime: A Text/Reader*

Other Titles of Related Interest

Chesney-Lind and Pasko: *The Female Offender*, 3rd Edition
Chesney-Lind and Pasko: *Girls, Women and Crime: Selected Readings*
Banks: *Criminal Justice Ethics*, 3rd Edition
Cox, Allen, Hanser, and Conrad: *Juvenile Justice*, 7th Edition
Lawrence and Hesse: *Juvenile Justice: The Essentials*
Scaramella, Cox, and McCamey: *Introduction to Policing*
Stohr and Walsh: *Corrections: The Essentials*
Hanser: *Community Corrections*
Cullen and Jonson: *Correctional Theory*
Pratt: *Addicted to Incarceration*
Hagan: *Introduction to Criminology*, 7th Edition
Lilly, Cullen, and Ball: *Criminological Theory*, 5th Edition
Tibbetts: *Criminological Theory: The Essentials*
Pratt, Gau, and Franklin: *Key Ideas in Criminology and Criminal Justice*
Felson and Boba: *Crime and Everyday Life*, 5th Edition
Boba Santos: *Crime Analysis With Crime Mapping*, 3rd Edition
Gabbidon and Greene: *Race and Crime*, 3rd Edition
Bachman and Schutt: *The Practice of Research in Criminology and Criminal Justice*, 4th Edition
Bachman and Schutt: *Fundamentals of Research in Criminology and Criminal Justice*, 2nd Edition
Gau: *Statistics for Criminal Justice*
Mosher, Miethe, and Hart: *The Mismeasure of Crime*, 2nd Edition
Martin: *Understanding Terrorism*, 4th Edition
Payne: *White Collar Crime: The Essentials*
Howell: *Gangs in America's Communities*
Maguire and Okada: *Critical Issues in Criminology and Criminal Justice*
Hemmens, Brody, and Spohn: *Criminal Courts*
Spohn: *How Do Judges Decide?* 2nd Edition
Lippman: *Criminal Procedure*
Lippman: *Contemporary Criminal Law*, 2nd Edition

WOMEN and CRIME

A Text/Reader

Stacy L. Mallicoat
California State University, Fullerton

Los Angeles | London | New Delhi
Singapore | Washington DC

Los Angeles | London | New Delhi
Singapore | Washington DC

FOR INFORMATION:

SAGE Publications, Inc.
2455 Teller Road
Thousand Oaks, California 91320
E-mail: order@sagepub.com

SAGE Publications Ltd.
1 Oliver's Yard
55 City Road
London EC1Y 1SP
United Kingdom

SAGE Publications India Pvt. Ltd.
B 1/I 1 Mohan Cooperative Industrial Area
Mathura Road, New Delhi 110 044
India

SAGE Publications Asia-Pacific Pte. Ltd.
33 Pekin Street #02-01
Far East Square
Singapore 048763

Acquisitions Editor: Jerry Westby
Editorial Assistant: Erim Sarbuland
Production Editor: Catherine M. Chilton
Copy Editor: Brenda Weight
Typesetter: C&M Digitals (P) Ltd.
Proofreader: Annette R. Van Deusen
Indexer: Jennifer Pairan
Cover Designer: Janet Kiesel
Marketing Manager: Erica DeLuca
Permissions: Karen Ehrmann

Printed in the United States of America

Library of Congress Cataloging-in-Publication Data

Women and crime : a text/reader / editor, Stacy L. Mallicoat.

p. cm.
Includes bibliographical references and index.

ISBN 978-1-4129-8750-9 (pbk.)

1. Feminist criminology. 2. Women—Crimes against. 3. Female offenders. 4. Sex discrimination in criminal justice administration. I. Mallicoat, Stacy L.

HV6030.W66 2012
364.082—dc23 2011029804

This book is printed on acid-free paper.

11 12 13 14 15 10 9 8 7 6 5 4 3 2 1

Brief Contents

Contents

Foreword

You hold in your hands a book that we think is something new. It is billed as a "text/reader." What that means is we have taken the two most commonly used types of books, the textbook and the reader, and blended them in a way that we anticipate will appeal to both students and faculty.

Our experience as teachers and scholars has been that textbooks for the core classes in criminal justice (or any other social science discipline) leave many students and professors cold. The textbooks are huge, crammed with photographs, charts, highlighted material, and all sorts of pedagogical devices intended to increase student interest. Too often, though, these books end up creating a sort of sensory overload for students and suffer from a focus on "bells and whistles," such as fancy graphics, at the expense of coverage of the most current research on the subject matter.

Readers, on the other hand, are typically composed of recent and classic research articles on the subject. They generally suffer, however, from an absence of meaningful explanatory material. Articles are simply lined up and presented to the students, with little or no context or explanation. Students, particularly undergraduate students, are often confused and overwhelmed.

This text/reader represents our attempt to take the best of both the textbook and reader approaches. It is intended to serve either as a supplement to a core textbook or as a stand-alone text. The book includes a combination of previously published articles and textual material introducing these articles and providing some structure and context for the selected readings. The book is divided into a number of sections. The sections of the book track the typical content and structure of a textbook on the subject. Each section of the book has an introduction that serves to introduce, explain, and provide context for the readings that follow. The readings are a selection of the best recent research that has appeared in academic journals, as well as some classic readings. The articles are edited as necessary to make them accessible to students. This variety of research and perspectives will provide the student with a grasp of the development of research as well as an understanding of the current status of research in the subject area. This approach gives the student the opportunity to learn the basics (in the textbook-like introductory portion of each section) and to read some of the most interesting research on the subject.

There is also an introductory chapter explaining the organization and content of the book and providing context for the articles that follow. This introductory chapter provides a framework for the text and articles that follow and introduces relevant themes, issues, and concepts. This will assist the student in understanding the articles.

Each section includes a summary of the material covered and discussion questions following the introductions and each reading. These summaries and discussion questions should facilitate student thought and class discussion of the material.

It is our belief that this method of presenting the material will be more interesting for both students and faculty. We acknowledge that this approach may be viewed by some as more challenging than the traditional textbook. To that we say, "Yes! It is!" But we believe that if we raise the bar, our students will rise to the challenge. Research shows that students and faculty often find textbooks boring to read. It is our belief that many criminology instructors would welcome the opportunity to teach without having to rely on a "standard" textbook that covers only the most basic information and that lacks both depth of coverage and an attention to current research. This book provides an alternative for instructors who want to get more out of the basic criminology courses and curriculum than one can get from a typical textbook that is aimed at the lowest common denominator and filled with flashy but often useless features that merely serve to drive up its cost. This book is intended for instructors who want to go beyond the ordinary, standard coverage provided in textbooks.

We also believe students will find this approach more interesting. They are given the opportunity to read current, cutting-edge research on the subject, while also being provided with background and context for this research.

We hope that this unconventional approach will be more interesting, and thus make learning and teaching more fun, and hopefully more useful as well. Students need not only content knowledge but also an understanding of the academic skills specific to their discipline. Criminology is a fascinating subject, and the topic deserves to be presented in an interesting manner. We hope you will agree.

Craig Hemmens, JD, PhD
Missouri State University

Preface

The purpose of this book is to introduce readers to the issues that face women as they navigate the criminal justice system. Regardless of the role they play in this participation, women have unique experiences that have significant effects on their perspectives of the criminal justice system. In order to effectively understand the criminal justice system, the voices of women must be heard. This book seeks to inform readers on the realities of women's lives as they interact with the criminal justice system. These topics are presented in this book through summary essays highlighting the key terms and research findings and incorporating cutting-edge research from scholars whose works have been published in top journals in criminal justice, criminology, and related fields.

Organization and Contents of the Book

This book is divided into 11 sections, with each section dealing with a different subject related to women and crime. Each section begins with an introduction to the issues raised within each topic and summarizes some of the basic themes related to the subject area. Each introductory essay concludes with a discussion of the policy implications related to each topic and summary points to help guide student review of the material. This discussion is followed by selected readings that focus on research being conducted on critical issues within each topical area. These readings represent some of the best research in the field and are designed to expose students to the discussions of women's issues within contemporary criminal justice. These 11 sections are

- Women and Crime: An Introduction
- Theories on Victimization and Offending
- Women and Victimization: Rape and Sexual Assault
- Women and Victimization: Intimate Partner Abuse
- Women and Victimization: Stalking and Sexual Harassment
- International Issues for Women and Crime
- Girls and Juvenile Delinquency
- Female Offenders and Their Crimes
- Processing and Sentencing of Female Offenders
- The Incarceration of Women
- Women and Work in the Criminal Justice System: Police, Courts, and Corrections

The first section provides an introduction and foundation for the book. This section begins with a discussion on the different data sources that present statistics on women as victims and offenders. In setting the context for the book, this section concludes with a review of the influence of feminism on the study of crime.

The second section begins with a review of the theories of victimization of women. Following a discussion on the fear of victimization as a gendered experience, this section highlights the intersectionality of victimization and offending. The section turns to a review of the historical perspectives of female offending and presents how traditional theories of criminal behavior have failed to adequately explain the offending patterns of women. This section highlights some of the major criminological theories and their applications (or lack thereof) toward understanding the female offender and includes a discussion of contemporary feminist perspectives on offending. The section includes two readings on the development of feminist criminology and one on victimization. The first article in this section, by Joanne Belknap and Kristi Holsinger, discusses the feminist pathways model and how boys' and girls' life events affect their trajectories toward crime and delinquency. The second article, by Meda Chesney-Lind, looks at the role of patriarchy in developing a paradigm of feminist criminology in light of traditional and contemporary theories of crime. Finally, the section concludes with a reading by Bonnie S. Fisher and David May investigating the effects of gender on the fear of victimization by college students.

The third section focuses on the victimization of women by crimes of rape and sexual assault. From historical issues to contemporary standards in the definition of sexual victimization, this section highlights the various forms of sexual assault, the role of the criminal justice system in the reporting and prosecution of these crimes, and the role of victims in the criminal justice system. The introductory essay concludes with a discussion on policy implications related to the issues of rape and sexual assault. The readings in this section highlight some of the critical research on issues related to rape and sexual assault. Beginning with a discussion of sexual assault in the ivory tower, Bonnie S. Fisher, Leah E. Daigle, and Francis T. Cullen investigate the extent of non-rape sexual victimizations on college campuses. We then turn to Renae Franiuk, Jennifer L. Seefelt, Sandy L. Cepress, and Joseph A. Vandello's article on the utilization of rape myths in print journalism and its effects on the public's perception of the offender and victim in a high-profile case of rape accusation. This section concludes with an article by April L. Girard and Charlene Y. Senn and investigates how the voluntary and involuntary uses of drugs and alcohol affect the diffusion of offender responsibility and the assignment of victim blame in cases of sexual assault.

The fourth section presents a discussion of victimization of women in cases of intimate partner abuse. A review of the legal and social research on intimate partner violence addresses a multitude of issues for victims, including the barriers in leaving a battering relationship. This section concludes with a discussion on policy implications of research on intimate partner abuse and focuses on the effects of mandatory arrest versus discretionary arrest policies. The articles in this section address some of the contemporary issues facing victims of intimate partner violence. The readings in this section begin with an essay by Martin D. Schwartz and Walter S. DeKeseredy on the role of patriarchy in the culture of battering and how violence against women can be stopped in a male-dominated society. The second article, authored by Angela M. Moe, draws on narratives by battered women and their interactions with legal and social agencies to highlight the struggles of women in their attempts to leave a violent relationship. This section concludes with a discussion by Edna Erez, Madelaine Adelman, and Carol Gregory on how immigration shapes one's experiences with battering.

The fifth section deals with the topics of stalking and sexual harassment. While these two crimes are different in their own ways, they share the common foundation of an unwanted form of attention that perpetuates fear for the victims. This section discusses the foundations of defining each of these behaviors as criminal acts and discusses the legal and social implications of these crimes for society. The first article, by Heather Melton, uses the words of victims to discuss the experiences of stalking by an intimate partner. The second article, by Karen L. Paullet, Daniel R. Rota, and Thomas T. Swan, focuses on the experiences of cyberstalking among college students. The section concludes with a reading by Kimberly Fairchild and Laurie A. Rudman and focuses on the issues of harassment of women perpetrated by strangers.

The sixth section focuses on international issues for women. Each article centers on an issue of crime and justice facing women of the world. The readings of this section begin with Sujay Patel and Amin Muhammad Gadit's research on honor killings of women in Pakistan. The second article presents an issue that has engaged communities around the world: human trafficking. Padam Simkhada's research on sex trafficking of Nepalese women and girls shares the voices of these victims and presents strategies for anti-trafficking initiatives for international communities. The section concludes with an article by Mark Ensalaco on the systematic rape and murders of women in Mexico, known in international circles as the "dead women of Juárez." Ensalaco follows the stories of these women, the resistance of the Mexican government to provide adequate answers to victims' families, and the rise of grassroots women's and human rights organizations in a search for answers in these cases.

Section VII focuses on girls and the juvenile justice system. Beginning with a discussion on the patterns of female delinquency, this section investigates the historical and contemporary standards for young women in society and how the changing definitions of delinquency have disproportionately and negatively affected young girls. The readings for this section begin with an article by Carla P. Davis that demonstrates how early intervention efforts for girls at risk could dismantle their road to incarceration as adults. The second article is authored by Barry C. Feld and looks at the question of whether girls are becoming more violent, or whether the system is changing how they deal with young girls who "act out." The section concludes with an article by Crystal A. Garcia and Jodi Lane on the needs of girls in the juvenile justice system.

Section VIII deals with women and their crimes, and highlights some of the types of crimes where women are disproportionately represented as offenders. The readings begin with a topic that is at the heart of the dramatic rise of female participation in the criminal justice system: the war on drugs. Vanessa Alleyne presents how the U.S. drug war has not only led to the burgeoning of the prison population and the effects of overcrowding in prisons, but how it has changed the way that the system has responded to addiction, mental health, welfare, and the custody of children. An article by Theresa Porter and Helen Gavin reviews the literature on cases of infanticide and neonaticide, crimes that traditionally have focused on women as the perpetrators. Celia Williamson and Gail Folaron comment on the experiences of women engaged in prostitution and the dangers they confront on the streets. The section concludes with an article by Susan Batchelor, who examines the prevalence and behaviors of girl gangs in the United Kingdom.

The ninth section details historical and contemporary patterns in the processing and sentencing of female offenders, as different from their male counterparts. This section highlights research on how factors such as patriarchy, chivalry, and paternalism within the criminal justice system affect women. Three articles in this section investigate the effects of gender on the processing of offenders: Lori D. Moore and Irene Padavic look at the effects of race and ethnicity on the processing of girls, Candace Kruttschnitt and Jukka Savolainen focus on gender and sentencing from an international perspective in their review of the processing of offenders in Finland, and Tina L. Freiburger and Carly M. Hilinski focus on the effects of race and gender on the pretrial decision stage of the criminal court.

The 10th section examines the incarceration of women. Here, the text and readings focus on the patterns and practices of the incarceration of women. Ranging from historical examples of incarceration to modern-day policies, this section examines how the treatment of women in prison varies from that of their male counterparts and how incarcerated women have unique needs based on their differential pathways to prison. This section concludes with a discussion on the complications of reintegration for women following their sentence. The readings in this section begin with a discussion of best practices for gender-responsive programming and treatment for female offenders by Barbara Bloom, Barbara Owen, and Stephanie Covington. Suzanne Allen, Chris Flaherty, and Gretchen Ely talk with women incarcerated on drug charges to learn how their incarceration has affected their role of mothering. Finally, Mary Dodge and Mark R. Pogrebin present the challenges that women face following incarceration in their quest for a successful transition back to their families and communities.

Section XI concludes this text and brings attention to the women who work within the domain of the criminal justice system and how gender affects their occupational context. This section focuses on women in careers related to policing, courts, and corrections. These readings highlight the levels of stress that women experience in their efforts to work in an environment that has historically been dominated by men. In addition, women face choices and consequences within the context of these career decisions as a result of their gender. Following a discussion of the history of women in these occupations, this section looks at how gender affects the performance of women in these jobs and the personal toll it has on their lives. The first article, by Cara Rabe-Hemp, looks at women in policing and their strategies for surviving in a male-dominated arena. An article by Phyllis Coontz addresses whether gender affects judicial decision-making processes. The section concludes with an article by Marie Griffin and her investigations of whether the stress that correctional officers experience varies by gender.

As you can see, this book provides an in-depth look at the issues facing women in the criminal justice system. From victimization to incarceration to employment, this book takes a unique approach in its presentation by providing a review of the literature on each of these issues followed by some of the key research studies that have been published in academic journals. Each section of this book presents a critical component of the criminal justice system and the role of women in it. As you will soon learn, gender is a pervasive theme that runs deep throughout our system, and how we respond to it has a dramatic effect on the lives of women in society.

Acknowledgments

This book has been such a labor of love for me. Love is an emotion that runs the gamut, from elation, joy, and pride, to frustration and insecurity. Throughout the writing of this book, I have experienced each of these emotions. At the end of the day, this book represents all that I have learned about women and crime from the amazing mentors and scholars in my life. I thank each of you for doing such important work on a population that throughout history has been abused; neglected; forgotten; and made to feel, as my mentor from graduate school, Joanne Belknap, would say, invisible.

I have to give tremendous thanks first to Jerry Westby, acquisitions editor in the Criminology/Criminal Justice division at Sage Publications. One day, I called Jerry to rave about my experience with another book in the text-reader series and asked him, "When will you have a women and crime book in the series?" His response was, "When are you going to write it?" He threw down quite the challenge, and I thank him from the bottom of my heart for not only providing me with such an amazing opportunity, but also for his support and encouragement along the way. I also offer thanks to Craig Hemmens, who first developed this series and has not only developed a resource that brings the best of both worlds in textbooks and readers together into a single volume to enhance the student learning experience, but also provided an opportunity for scholars such as myself to contribute to this collection. Special thanks as well to the staff at Sage who have also helped in breathing life into this book.

Throughout my career, I have been blessed to be surrounded by amazing colleagues and mentors. As a young graduate student at the University of Colorado, Boulder, my mentor and advisor Joanne Belknap introduced me to the study of women and crime and opened my eyes to how gender has significant effects for women as victims and offenders within the criminal justice system. I am forever in her debt for her support throughout my career and for making me into the scholar that I am today. Thanks also to Michael Radelet and Sara Steen, who shared with me important lessons on teaching, mentorship, and scholarship. Dearest thanks to my friends and colleagues—Jill Rosenbaum, Connie Ireland, Allison Cotton, Hillary Potter, Denise Paquette Boots, and the faculty and staff at California State University, Fullerton—who have kept me grounded and laughing, even when I wanted to kick, scream, and cry. Finally, thank you to the amazing network of scholars that is the Division on Women and Crime with the American Society of Criminology, for providing a home filled with love, care, and support for my adventures in the academy.

This book would not have been possible if not for the love and support of my husband, Jeff Randolph. Thank you for putting up with all the blood, stress, sweat, and tears that went into the writing of this book; for holding me high throughout this process; and for tolerating the busy days, nights, and weekends that it took for me to write this book. Thanks to my family and friends for being a sounding board and source of strength throughout this experience—particularly my parents, Gary and Marcia Mallicoat, who provided me with the foundation to succeed in my life, and my extended family of Ann Cherry, Taylor Randolph, Heidi Rodriguez, and Sam Schuman for their love and support. Finally, thank you to my son, Keegan Deane, who literally came into this world during the writing of this book and provided daily opportunities for procrastination through each smile, laugh, and life milestone.

Women and Crime

An Introduction

Section Highlights

- Various data sources that estimate female offending and victimization rates
- Feminist theory, with focus on the emergence of feminism in criminology
- The experiences of women who work in the field of criminal justice

Since the creation of the American criminal justice system, the experiences of women have been reduced to either a cursory glance or have been completely absent. Gendered justice, or rather injustice, has prevailed in every aspect of the system. The unique experiences of women have historically been ignored at every turn—for victims, for offenders, and even for women who have worked within its walls. The criminal justice system is a gendered experience.

Yet the participation of women in the system is growing in every realm. Women make up a majority of the victims for certain types of crimes, particularly when men are the primary offender. These gendered experiences of victimization in crimes such as rape, sexual assault, intimate partner violence, and harassment demonstrate that women suffer disproportionately from these crimes. Yet their cries for help have traditionally been ignored by a system that many in society believe is designed to help victims. Women's needs as offenders are also ignored. From the processing of cases to their incarceration, women in the criminal justice system face a variety of unique circumstances and experiences that are absent from the male offending population. From classical theories of crime that fail to adequately explain female offending behaviors, to the negative treatment women have received by the courts and correctional systems, the conditions of women's lives have gone unnoticed. Rather than look at the experiences of women, most of the understanding about offending was created by male scholars about the lives of men. When the discussion turned to the female offender, most perspectives tended to apply these male principles to women without any consideration or adaptation for women's perspectives. This approach has been criticized by feminist scholars for its failure to understand that the lives and experiences of women provide the foundations for their offending (Belknap, 2007).

Likewise, the employment of women in the criminal justice system has been limited, as women have been traditionally shut out of many of these male-dominated occupations. As women began to enter these occupations, they were faced with a hyper-masculine culture that challenged their gender at every turn. While the participation of women in these traditionally male-dominated fields has grown significantly in modern-day times, women continue to struggle for equality in a world where the effects of the "glass ceiling" continue to pervade a system that presents itself as one interested in the notion of justice (Martin, 1991).

The study of women and crime has seen incredible advances over the past fifty years. The 1960s and 1970s shed light on many significant issues for many groups in society, including women. The momentum of social change as represented by the civil rights and women's movements had significant impacts for society, and the criminal justice system was no stranger in these discussions. Here, the second wave of feminism expanded beyond the focus of the original activists (who were concerned exclusively about women's suffrage and the right to vote) to topics such as sexuality, legal inequalities, and reproductive rights. It was during this time frame that criminology scholars began to talk about women and offending in their discussions. Prior to this time, women were largely forgotten in discussions about crime and criminal behavior. When they were mentioned, they were relegated to a brief footnote or discussed in stereotypical and sexist ways. With the emergence of second-wave feminism, scholars began to talk in earnest about the nature of the female offender and began to ask questions about the lives of women involved in the criminal justice system. Who is she? Why does she engage in crime? And, perhaps most important, how is she different from the male offender and how should the criminal justice system respond to her? Increased attention was also brought to women who were victims of crime. How do women experience victimization? How does the system respond to women who have been victims of a crime? How have criminal justice policies responded to the victimization of women? Feminism also brought a greater participation in the workforce, and the field of criminal justice was no exception. How does gender affect the way in which women work within the police department, correctional agencies, and the legal system as a whole? What issues do women face within the context of these occupations? How has the participation of women in these fields affected the experiences of women who are victims and offenders? Today, scholars in criminology, criminal justice, and related fields explore these issues in depth, in an attempt to shed light on the population of women in the criminal justice system.

▲ **Photo 1.1** The image of Rosie the Riveter became synonymous with the second wave of the women's movement in the United States and the increase of women in the working world.

Official Data Sources on Women as Victims and Offenders

In order to develop an understanding of how often women engage in offending behaviors, it is important to look at how information about crime is gathered. The Uniform Crime Reports (UCR) represents one of the largest datasets on crime in the United States. The Federal Bureau of Investigation (FBI) is charged with collecting and publishing the arrest data from over 17,000 police agencies in the United States. These statistics are published annually and present the rates and volume of crime by offense type based on arrests made by police. The dataset includes a number of demographic variables to evaluate these crime statistics, including age, gender, race/ethnicity, location (state), and region (metropolitan, suburban, rural).[1] UCR data give us a general understanding of the extent of crime in the United States and are often viewed as the most accurate assessment of crime. In addition, the UCR data allow us to compare how crime changes over time, as they allow for the comparison of arrest data for a variety of crimes over a specific time frame (e.g., 1990–2000) or from one year to the next. Generally speaking, it is data from the UCR findings that are typically reported to the greater society through news media outlets and form the basis for headline stories that proclaim the rising and falling rates of crime.

However, the reporting of this information as the "true" extent of crime is flawed. While the UCR represents crime statistics from almost 95% of the population, it is important to take several issues into consideration. First, the UCR data represent statistics on only those crimes that are reported to the police. This means that unreported crimes are not recognized in these statistics. Sadly, many of the victimization experiences of women, such as intimate partner violence and sexual assault, are significantly underreported. Second, the UCR only collects data on certain types of crime, versus all forms of crime. The classification of crime is organized into two different types of crime: Part 1 offenses and Part 2 offenses. Part 1 offenses, known as index crimes, include eight different offenses: aggravated assault, forcible rape, murder, robbery, arson, burglary, larceny-theft, and motor vehicle theft. Note that the data do not include information on attempted rapes, only completed rapes. Part 2 includes information on several other forms of crime. These crimes are viewed as less serious than Part 1 offenses.[2] Third, the reporting of the crimes to the UCR is incomplete, as only the most serious crime is reported in cases where multiple crimes are committed during a single criminal event. These findings skew the realities of crime where several different offenses may occur within the context of a single crime incident. For example, a crime involving physical battery, forcible rape, and murder is reported to the UCR as the most serious crime, murder. Fourth, the participation by agencies in reporting to the UCR has fluctuated over time. While there are no federal laws requiring agencies to report their crime data, many states today have laws that direct law enforcement agencies to comply with UCR data collection. However, this means that analysis of crime trends over time needs to take into consideration the number of agencies involved in the reporting of crime data. Failure to do so could result in a flawed reporting of crime rates over time.

In an effort to develop a better understanding of the extent of offending, the National Incident Based Reporting System (NIBRS) was implemented in 1988. Rather than compile monthly summary reports on crime data in their jurisdictions, agencies now forward data to the FBI for every crime incident. The NIBRS

[1] Up-to-date statistical reports on crime data from the Uniform Crime Reports can be accessed at http://www.fbi.gov/ucr/ucr.htm

[2] Part 2 offenses are defined as simple assault, curfew offenses and loitering, embezzlement, forgery and counterfeiting, disorderly conduct, driving under the influence, drug offenses, fraud, gambling, liquor offenses, offenses against the family, prostitution, public drunkenness, runaways, sex offenses, stolen property, vandalism, vagrancy, and weapons offenses.

catalog involves data on 22 offenses categories and includes 46 specific crimes known as Group A offenses. Data on 11 lesser offenses (Group B offenses) are also collected.[3] The NIBRS allows for a more comprehensive understanding of crime in the United States. However, the transition of agencies to the NIBRS has been slow, as only 23 states have been certified by the FBI as of 2003. The reporting from these states only comprises 20% of law enforcement agencies in the United States. Additionally, this system still carries over a fatal flaw from the UCR, in that the NIBRS also only accounts for reported crimes to the police. In spite of this, it is hoped that the improvements in official crime data collection will allow an increased understanding of the extent of female crime, and offending patterns in general, compared to the UCR system.

While the UCR and NIBRS data provide a wealth of statistics about crime, the results present an incomplete understanding about crime in society, as the findings suffer from issues in reporting data as well as sampling parameters that limit the types of data collected. In an effort to develop a better understanding of how crime affects society, scholars have created additional ways to gather data on these issues. Self-report studies ask women about their experiences of crime, both as victims and offenders. Self-report studies can include both quantitative and qualitative measures and have yielded tremendous amounts of data on all aspects of the criminal justice system that is not reflected by official measures of crime. Not only do these data give us an expanded picture about the true extent of crime, we also benefit from the additional details about crime offending and victimization that are provided by these studies and are absent from the UCR/NIBRS data.

Studies on offenders allow us to investigate the lives of women involved in crime. This research is conducted at every stage of the criminal justice process and beyond. Some self-report studies ask women about their participation in crimes in an effort to understand the risk factors for women to engage in criminal behavior. Others seek to understand how incarceration affects the lives of women and their families. These data help inform scholars not only about some of the flaws of criminal justice policies when it comes to women, but also about what considerations criminal justice policies should make when it comes to women offenders.

While the UCR includes limited data on victims of crime, the National Crime Victimization Survey (NCVS) is the largest victimization study conducted in the United States. Its greatest achievement lies in its attempt to fill the gap between reported and unreported crime, often described as the dark figure of crime. The NCVS gathers additional data about crimes committed and gives criminologists a greater understanding of the types of crimes committed and victim characteristics. In 2007, the NCVS interviewed 73,600 individuals age 12 and older in 41,500 households. Based on their survey findings, generalizations are made to the population regarding the prevalence of victimization in the United States. However, the NCVS is not error-free, as victims may not necessarily remember the details of their crime or the offender. Additionally, the NCVS is a structured questionnaire, which limits the type of data that are collected. However, the NCVS is not the only investigation of victimization. Many research studies address victims of crime and how the justice system responded to their victimization. These studies provide valuable data in understanding the experiences of women as victims that is not reflected by the NCVS or UCR data.

In summary, official crime statistics offer only one perspective on the extent of crime in society. Through the use of self-report studies and victimization surveys, scholars can investigate issues of gender and crime that are not reflected in official crime measures. While each source of data has its strengths and weaknesses in terms of the types of data that are collected and the methods that are utilized, together they provide a wealth of information that is invaluable in understanding the complex nature of gender and crime.

---

[3]Information on the National Incident Based Reporting System can be found at http://www.fbi.gov/ucr/faqs.htm.

Women as Victims of Violence

The experience of victimization is something that too many women are intimately familiar with. Women comprise the majority of victims of certain forms of violent crime, particularly intimate partner violence and sexual assault. Women are most likely to be victimized by someone they know. They often do not report these crimes to the police. In many cases when they do seek help from the criminal justice system, charges are not always filed or are often reduced through plea bargains, resulting in offenders receiving limited, if any, sanctions for their criminal behavior. In many cases, victims find their own lives put on trial to be criticized by the criminal justice system and society as a whole. Based on these circumstances, it is no surprise that many women have had little faith in the criminal justice system.

In 1994, the Bureau of Justice statistics estimated that five million women were victimized by violent crimes, three million of whom had acts perpetrated on them by people they knew (Craven, 1997). While the rates of violent victimization of women have significantly declined from a historical high in 1994 (43.0 per 1,000), women in 2007 continued to experience crimes of violence at a rate of 18.9 per 1,000 persons, or an estimated 2.4 million cases (Rand, 2008). Based on NCVS data, the Bureau of Justice Statistics estimates that there were 236,980 cases of rape and sexual assault in 2007, a 25% increase compared to the 176,540 cases in 2005. Likewise, cases of intimate partner violence increased 42% with 554,260 cases in 2007 compared to 389,100 cases in 2005 (Rand, 2008; Catalano, 2006). However, it is difficult to determine a true understanding of the levels of crime perpetrated against women, as many do not report their victimization—only 47.3% of women surveyed by the NCVS in 2007 reported their crimes to the police (Rand, 2008).

The Extent of Female Offending

A review of arrest data from the Uniform Crime Reports (UCR) indicates that the overall levels of crime for women increased 11.6% in 2008 compared to rates in 1999. For the same time period, the number of arrests for men declined 3.1%. Have the crime rates of women skyrocketed while the rates of men have fallen? A true understanding of this question requires a deeper look. Table 1.1 illustrates the UCR data on arrest trends for men and women for 1999 and 2008. In 2008, the UCR indicates that 8,068,607 arrests were made, and women accounted for 24.6% of these arrests (1,985,133). Numerically, the number of crimes committed by men decreased (comparing 2008 data to 1999 data) while the number of crimes committed by women increased. Since men represent a majority of arrests, any decrease in the number of male arrests results in a small impact to male arrest rates. Likewise, because women represent a minority of arrests, increases in the raw number of incidences have a significantly greater effect on female arrest rates. So, an increase of 206,632 arrests of women leads to an increase of 11.6% in female arrests, comparing 1999 to 2008 data. In comparison, a decrease of 195,645 arrests for men is only a 3.1% reduction. These numbers illustrate the effects of population size in determining the proportion of male and female involvement in arrests. A closer look at the data finds that men make up the majority of all offending types, except for the crimes of embezzlement and prostitution.

When assessing overall crime trends, it is important to consider the time period of evaluation. While the 10-year arrest trends demonstrate an increase for females and a decrease for males, the data for 2008 actually represent a decrease from the overall crime rates for both men and women compared to 2007. Table 1.2 demonstrates the arrest trends for these two years. The crime rate for men fell 1.9%, while the crime rate for women decreased 0.3%. In addition, the types of crimes affect how these data are interpreted. A deeper look shows that violent crime fell 0.7% for men and 0.2% for women between 2007 and 2008, while property

Table 1.1 10-Year UCR Arrest Trends

	Men			Women		
	1999	2008	%	1999	2008	%
All arrests	6,279,139	6,083,494	−3.1	1,778,501	1,985,133	+11.6
Violent crime	305,271	286,255	−6.2	63,085	63,943	+1.4
Homicide	6,636	6,292	−5.2	831	780	−6.1
Forcible rape	15,452	12,474	−19.3	179	148	−17.3
Robbery	54,658	64,844	+18.6	6,261	8,615	+37.6
Aggravated assault	228,525	202,645	−11.3	55,814	54,400	−2.5
Property crime	680,364	639,470	−6.0	290,439	347,701	+19.7
Burglary	149,875	157,341	+5.0	23,106	29,055	+25.7
Larceny-theft	461,632	431,212	−6.6	254,629	308,011	+21.0
Motor vehicle theft	60,540	43,801	−27.6	11,331	9,344	−17.5
Arson	8,317	7,116	−14.4	1,373	1,291	−6.0

NOTE: 7,892 agencies reporting.

crime rates rose 3.2% for men and 10.3% for women. A look at the individual crimes presents an interesting lesson about the rise and fall of crime rates. While one violent crime showed a significant increase in the percentage of females involved, it is important to look at the raw data (numbers) versus the percentage increase between years. In 2007, 167 women were arrested for the crime of forcible rape. In 2008, this number jumped to 208, an increase of 39 cases. Yet when you interpret these numbers as a percentage increase, we find that the arrests of women in cases of forcible rape increased 24.6% compared to the previous year. Due to the small number of women who engaged in this behavior, any relatively marginal increase in the number of women arrested has a significant effect on the overall percentage change. Another example of the effects of small numbers of women's participation in crime and its effects on the year-to-year comparisons can be found by looking at the crime of arson. In 2007, 1,628 women were arrested for arson compared to 1,655 in 2008. This increase of 27 cases amounts to an increase of 1.7%. Therefore, when reviewing the increases or decreases of women's (and men's) participation in crime, it is important to note both the raw numbers as well as the rates of crime relative to the population.

Given that the NCVS documents both reported and unreported crime, the findings generally document projected crime rates that are higher than the rates of crime reported by UCR data. But what does this mean about the true extent of crime? While women comprise a small proportion of most offending categories, are they becoming increasingly more violent? Or have police department arrest policies and practices changed

Table 1.2 1-Year UCR Arrest Trends

	Men			Women		
	2007	2008	%	2007	2008	%
All arrests	7,718,150	7,572,871	−1.9	2,498,353	2,489,838	−0.3
Violent crime	348,490	345,946	−0.7	77,764	77,614	−0.2
Homicide	8,149	7,855	−3.6	940	968	+3.0
Forcible rape	15,985	16,141	+1.0	167	208	+24.6
Robbery	79,082	81,108	+2.6	10,495	10,838	+3.3
Aggravated assault	245,274	240,842	−1.8	66,162	65,600	−0.8
Property crime	770,438	795,093	+3.2	389,656	429,672	+10.3
Burglary	185,254	189,959	+2.5	32,186	32,966	+2.4
Larceny-theft	507,878	540,161	+6.4	340,809	383,260	+12.5
Motor vehicle theft	68,139	56,201	−17.5	15,033	11,791	−21.6
Arson	9,167	8,772	−4.3	1,628	1,655	+1.7

NOTE: 11,056 agencies reporting.

toward how women are treated in cases of violent crime offending? A comparison of UCR and NCVS data between 1980 and 2003 finds that the difference between male and female participation in crimes of violence has remained relatively stable. According to UCR data, there has been little change in the gender gap for homicide arrests. Additionally, NCVS data indicate no significant increases in women's participation in assaultive behaviors (Steffensmeier, Zhong, Ackerman, Schwartz, & Agha 2006). Indeed, women are no more likely today to seek out participation in violent crimes that have historically been dominated by men, such as robbery, gangs, and organized crimes (Steffensmeier & Ulmer, 2005). Instead, it appears that policy changes are to blame for much of the "increase" in female crime. For example, crimes that were once dealt with on an informal basis, such as disorderly conduct, minor harassment, and resisting arrest, are now considered cases of simple assault. This process of overcharging has created a new category of "violent" women in cases where little change in behavior actually occurred. This process of net widening has expanded the number of women in the criminal justice system, particularly in low-level cases. Finally, the perception that women are engaging in increasingly higher levels of crime has translated to an increased reporting of female assailants for crimes, even though the rates of victimization by women has remained relatively constant. This trend illustrates a shift in the processing of female offenders and reflects a reduction in the chivalrous treatment of women by the criminal justice system (Steffensmeier et al., 2006).

The comparison of UCR and NCVS data can also provide evidence that the gender gap in offending between men and women is narrowing. Certainly, men engage in crime at a higher rate than women, and men are more likely to be involved in violent crimes than women. With the increased attention in the media on the "violent girl," the question arises of whether women are indeed becoming more violent in their offending practices. A review of data from both the UCR and NCVS between 1973 and 2005 indicates that female rates of violent offending have actually decreased over time, not increased as many might assume. In contrast to research by Steffensmeier et al. (2006), which held that the closing of the gender gap was due to changes in police practices (and an increased likelihood to process female offenders using official channels), Lauritsen, Heimer, and Lynch (2009) suggest that the reduction in the gender gap is a result of how the rates of both male and female offending decreased. Here, they suggest that the gender gap between male and female offending has narrowed because the proportion of male offending has decreased at a greater rate than the proportion of female offending.

Women and Work in the Criminal Justice System

Just as the social movements of the 1960s and 1970s increased the attention on female offenders and victims of crime, the access to opportunities to work within the walls of criminal justice expanded for women. Prior to this era of social change, few women were granted access to work as police officers, correctional officers, lawyers, or judges. Even when women were present, their duties were significantly limited compared to those of their male counterparts, and their opportunities for advancement were essentially nonexistent. In addition, these majority-male workforces resented the presence of women in their masculine environment.

While many scholars have reported that the experience of women in these male-dominated fields has significantly improved in recent years, women continue to face a number of challenges directly related to their status as women, such as on-the-job sexual harassment, the work-family balance, and maternity and motherhood. In addition, research reflects on how women manage the roles, duties, and responsibilities of their positions and the effects of gender on this process. Indeed, research indicates that the experience of womanhood can impact the work environment, both personally and culturally.

The Influence of Feminism on Studies of Women and Crime

One of the criticisms of traditional research on gender and crime is that it fails to recognize the intricate details of what it means to be a woman in society. The feminist movement has had a significant effect on how we understand women and their relationships with crime by making gender an integral part of the research experience. By incorporating a feminist perspective to the research environment, scholars are able to present a deeper understanding of the realities of women's lives by placing women and women's issues at the center of the research process. The concept of giving women a voice, particularly in situations where they have been historically silenced, is a strong influence of feminist research methods. From the conceptualization of the research question to a discussion of which methods of data collection will be utilized, and how the data will be analyzed, feminist methods engage in practices that are contrary to the traditional research paradigms. While the scientific method focuses on objectivity and the collection of data is detached from the human condition, the use of feminist methods requires a paradigm shift from what is traditionally known as research. While many of the researchers who first engaged in research through a feminist lens were women, feminist methodology does not dictate that the gender of the research participant or researcher be a woman. Rather, the philosophy of this method refers to the types of data a researcher is seeking and the process by which data are obtained (Westervelt & Cook, 2007). Feminist methods are largely qualitative in nature and allow for emotions and values to be present as part of the research process. While some feminist methodologists have criticized the process by which data are often quantified, as it does not allow for the intricate nature and quality of women's lives to be easily documented, others argue that quantitative data have a role to play within a feminist context. Regardless of the approach, the influence of feminism allows for researchers to collect data from a subject that are theoretically important for their research, versus data that are easily categorized (Hessy-Biber, 2004; Reinharz, 1992).

▲ **Photo 1.2** The icon of Lady Justice represents many of the ideal goals of the justice system, including fairness, justice, and equality.

There is no single method of research that is identified as the "feminist method." Rather, the concept of feminist methodology refers to the process by which data are gathered and the relationship between the researcher and the subject. This process involves five basic principles: (1) acknowledging the influence of gender in society as a whole (and inclusive of the research process), (2) challenging the traditional relationship between the researcher and the subject and its link to "scientific research" and the validity of findings, (3) engaging in consciousness raising about the realities of women's lives as part of the methodological process, (4) empowering women within a patriarchal society through their participation in research, and (5) an awareness by the researcher of the ethical costs of the research process and a need to protect their subjects (Cook & Fonow, 1986).

The traditional relationship between the researcher and subject is one of separation, objectivity, and distance. In contrast, feminist research methodology seeks connectivity between the researcher and the subject. Many feminist researchers discuss the need for a continuum of care for the people who participate in their research. This principle of care involves concern for the person as a whole person, not just as a part of the research process and for the information that a subject can provide (Westervelt & Cook, 2007). This ethic of care involves a greater connectivity between the researchers and the respondent. Indeed, this leads to a pathway of empowerment for those participating in the research studies. "By defining the research as collaborative and the 'subjects' as active participants, the feminist method allows participants agency and an opportunity to own their stories and be heard and accepted in a way often denied to them otherwise" (Westervelt & Cook, 2007, p. 34).

For many researchers who study women in the criminal justice system, the use of feminist methodologies is particularly beneficial. Not only does it allow for researchers to explore in depth the issues that women face as victims and offenders, but it provides the opportunity for the researcher to delve into their topics in a way that traditional methods fail to explore, such as the context of women's lives and their experiences in offending and victimization. For example, a simple survey question might inquire about whether an incarcerated woman has ever been victimized. We know that scholarship on incarcerated women has consistently documented the relationship between early life victimizations and participation in crime in their adolescent and adult lives. Yet, traditional methods may underestimate the extent and nature of the victimization. Feminist methodologies not only allow for the exploration of these issues at a deeper level, but they allow for scholars to develop an understanding of the multifaceted effects of these experiences.

Conclusion

The feminist movement has had a significant effect on the experience of women in the criminal justice system—from victims to offenders to workers. Today, the efforts of some of the pioneer women of feminist criminology have led to an increased understanding of what leads a woman to engage in crime and the effects of her life experiences on her offending patterns, as well as the challenges in her return to the community. In addition, the victim experience has changed for many women, such that their voices are beginning to be heard by a system that either blamed them for their victimization or ignored them entirely in years past. The feminist movement has also shed light on what it means to be a woman working within the criminal justice system and the challenges that she faces every day as a woman in this field. While women have experienced significant progress over the last century, there are still many challenges that they continue to face as offenders, victims, and workers within the world of criminal justice.

Summary

- The feminist movement of the 1960s and 1970s (known as second-wave feminism) has had a significant impact on discussions about female participation in crime.
- Data from the Uniform Crime Reports (UCR) and National Incident Based Reporting System (NIBRS) often fail to identify much of female victimization, as crimes of rape, sexual assault, and intimate partner abuse go largely underreported.
- Victimization studies, such as the National Crime Victimization Survey (NCVS) and self-report survey data, help illuminate the "dark figure of crime" by collecting data on crimes that are not reported to police.
- A comparison of UCR and NCVS data finds that policy changes are to blame for the "perceived increase" in female offending.
- Feminist research methods give women a voice in the research process and influence how data on gender are collected.

KEY TERMS

Uniform Crime Reports (UCR)

National Incident Based Reporting System (NIBRS)

National Crime Victimization Survey

Dark figure of crime

Feminism

Gendered justice

Feminist methodology

DISCUSSION QUESTIONS

1. Discuss how the Uniform Crime Reports and the National Incident Based Reporting System represent the measure of female crime and victimization in society.
2. How does the National Crime Victimization Survey investigate the "dark figure of crime"?
3. How do feminist research methods inform studies on women and crime?

WEB RESOURCES

Uniform Crime Reports: http://www.fbi.gov/about-us/cjis/ucr/ucr

National Incident Based Reporting System: http://www.icpsr.umich.edu/icpsrweb/NACJD/NIBRS/

National Crime Victimization Survey: http://www.icpsr.umich.edu/icpsrweb/NACJD/NCVS/

Theories on Victimization and Offending

Section Highlights

- Theories of female victimization
- Gender and fear of victimization
- The intersectionality of criminal victimization and offending
- Feminist criminology

This section is divided into two topics: victimization and offending. The section begins with a review of the victim experience in the criminal justice system and the practices of help seeking and victim blaming. The section then turns to the discussion of victimization, and how theories of victimization seek to understand the victim experience and place it within the larger context of the criminal justice system and society in general. The discussion on victimization concludes with a summary of how the fear of being victimized is a gendered experience. The section then turns to a review of the theoretical perspectives on offending and the failures of mainstream criminology to provide adequate explanations for women who offend. In tracing the evolution of a feminist perspective to explain crime, this section highlights how feminist researchers have sought out new perspectives to describe the nature of the female offender and her social world. The essay concludes with a discussion on feminist criminology and how the offending patterns of women are often intertwined with their experiences of victimization.

Victims and the Criminal Justice System

Why do victims seek out the criminal justice system? Do they desire justice? What does "justice" mean for victims of crime? Is it retribution? Reparation? Something else? Victims play an important role in the criminal justice process—indeed, without a victim, many cases would fail to progress through the system. However, many victims who seek out the criminal justice system for support following their victimization are often sadly disappointed in their experiences. In many cases, victims are reduced to a tool of the system or a piece of evidence in a criminal case. As a result, many of these victims express frustration over a system that seems to do little to represent their needs and concerns.

As a result of increased pressures to support the needs of victims throughout the criminal justice process, many prosecutor offices began to establish victim-assistance programs during the mid-1970s to provide support to victims as their cases moved through the criminal justice process. In some jurisdictions, nonprofit agencies for particular crimes, such as domestic violence and rape crisis, also began to provide support for victims during this time. Community agencies such as rape crisis centers developed in response to the perceived need for sexual assault prevention efforts, a desire for increased community awareness, and a wish to ameliorate the pain that the victims of crime experience. In response to a backlash against the rights of criminal defendants as guaranteed by the U.S. Constitution, citizens and legislatures have increased their efforts toward establishing rights for victims in the criminal justice process. Several pieces of federal legislation have passed in reference to victims' rights in particular crimes, such as sexual offenders and domestic violence. While attempts to include an amendment to the U.S. Constitution on victims' rights have currently been unsuccessful, many state constitutions have been amended to reflect the rights of victims in criminal cases. Table 2.1 illustrates some of the core rights of victims found in many state constitutions.

Much of what we know about victims comes from official crime datasets or research studies on samples of victimized populations. A comparison between official crime data (arrest rates) and victimization data indicates that many victims do not report their crime to law enforcement, which affects society's understanding regarding the realities of crime. According to the National Victimization Survey, only 46% of victims surveyed report their victimization to law enforcement. There are many reasons why a victim might choose not to report their victimization to the police. Some victims believe that the crime was not serious enough to make "a big deal over it," while others may be embarrassed by the crime or believe it is a personal matter. Still others may decide not to report a crime to the police out of the belief that nothing could be done.

Table 2.1 Core Rights of Victims

1. Protection and safety from the offender
2. Confidentiality of victim information in reports and documents
3. Information about available victim services
4. Notice of rights and proceedings
5. Right to be present and heard in criminal justice proceedings
6. Information about the status and location of the offender during incarceration
7. Restitution and compensation

Reporting behaviors also vary with the type of offense—robbery was the most likely crime reported (66%), followed by aggravated assault (57%). While women are generally more likely to report crimes to law enforcement compared to men, cases of personal violence are significantly underreported among female victims. For example, the NCVS indicates that only 42% of rapes and sexual assaults are reported, and the Chicago Women's Health Risk Study found that only 43% of women who experience violent acts from a current or former intimate partner contacted the police (Davies, Block, & Campbell, 2007; Rand, 2008). Certainly the relationship between the victim and offender is a strong predictor in reporting rates, as women who are victimized by someone known to them are less likely to report compared to women who are victimized by a stranger (Resnick et al., 2000). However, a failure to report does not mean that women do not seek out assistance for issues related to their victimization experience. Several studies on sexual assault and intimate partner abuse indicate that victims are more likely to seek help from resources outside of law enforcement such as family and friends. Women are also more likely to seek out mental health services following a victimization experience (Kaukinen, 2004). While many victims may be reluctant to engage in formal help seeking, research suggests that victims who receive positive support from informal social networks such as friends and family are more likely to seek out formal services such as law enforcement and therapeutic resources. In these cases, informal networks act as a support system to seek professional help and to make an official crime report (Davies et al., 2007; Starzynski, Ullman, Townsend, Long, & Long, 2007). The literature on barriers to help seeking indicates that fears of retaliation can affect a victim's decision to make a report to the police. This is particularly true for victims of intimate partner violence where research documents that violence can increase following police intervention. The presence of children in domestic violence situations also affects reporting rates, as many victims may incorrectly believe that they will lose their children as a result of intervention from social service agents.

The presence of victim blaming has also been linked to the low reporting rates of crime, particularly for cases such as intimate partner abuse and sexual assault. In many cases like these, victims reach out to law enforcement, community agencies, and family/peer networks in search of support and assistance and are met with blame and refusals to help. These experiences have a negative effect on the recovery of crime victims. Consider the reporting practices by the news media in cases involving high-profile defendants. In the case of the sexual assault charge against Kobe Bryant, a famous basketball player for the Los Angeles Lakers, hundreds of articles were written about the case in the months before the trial. A review of these articles demonstrates that they included positive statements about the defendant's character as well as his talent as an athlete. In contrast, cases involving an ordinary citizen would be unlikely to receive such treatment. In this case, these articles also contained negative statements about the victim, many of which reflected an acceptance of rape myths such as "she's lying" or "she asked for it" (Franiuk, Seefelt, Cepress, & Vandello, 2008). Based on the experiences of this one victim, it is not surprising that many victims may be fearful of reporting sexual assault if they believe that their experiences will be invalidated and if they are concerned that they will be emotionally attacked by outsiders who give credence to rape myths.

The practice of victim blaming is an example of secondary victimization. Secondary victimization refers to the process whereby victims of crime feel traumatized as a result of not only their victimization experience, but by the official responses to their victimization by the criminal justice system. For those cases that progress beyond the law enforcement investigative process, few are chosen to have charges filed by prosecutors, and only in the rare case is a conviction secured. Indeed, the "ideal" case for the criminal justice system is one that represents stereotypical notions of what rape looks like, rather than the realities of this crime. Given the nature of the criminal justice process, the acceptance of rape myths by jurors can ultimately affect the decision-making process. In an attempt to limit victim blaming in cases of sexual assault, many jurisdictions developed rape shield laws, which are used to limit the types of evidence that can be admitted regarding a victim's background and history. Despite their existence, many of these laws allow for judges to make decisions about when and how these laws will be applied. The experience of secondary victimization indicates that many rape victims would not have reported the crime if they had known what was in store for them (Logan, Evans, Stevenson, & Jordan, 2005).

Why do we blame the victim? The process of victim blaming is linked to a belief in a just world. The concept of a just world posits that society has a need to believe that people deserve whatever comes to them—bad things happen to bad people and good things happen to good people (Lerner, 1980). Under these assumptions, if a bad thing happens to someone, then that person must be at fault for the victimization because of who he or she is and what he or she does. A just world outlook gives a sense of peace to many individuals. Imagining a world where crime victims must have done something foolish, dangerous, or careless allows members of society to distinguish themselves from this identity of victimhood—"I would never do that, so therefore I must be safe from harm"—and allows individuals to shield themselves from feelings of vulnerability and powerlessness when it comes to potential acts of victimization. There are several negative consequences stemming from this condition: (1) victim blaming assumes that people are able to change the environment in which they live, (2) victim blaming assumes that only "innocent" victims are true victims, and (3) victim blaming creates a false sense of security about the risks of crime.

Given the nature of victimization patterns in society, few meet the criteria of a culturally ideal victim. This process of victim blaming allows society to diffuse the responsibility of crime between the victim and the offender. For example, the battered woman is asked "why do you stay?"; the rape victim is asked "what were you wearing?"; the assault victim is asked "why didn't you fight back?"; the burglary victim is asked "why didn't you lock the door?"; and the woman who puts herself in harm's way is asked "what were you thinking?" Each of these scenarios shifts the blame away from the perpetrator and assigns responsibility to the victim. Victim blaming enables people to make sense of the victimization. In many cases, the process of victim blaming allows people to separate themselves from those who have been victimized—"I would never have put myself in that situation"—and this belief allows people to feel safe in the world.

Fear of Victimization

The majority of Americans have limited direct experience with the criminal justice system. Most are left with images of crime that are generated by the portrayal of victims and offenders in mass media outlets (Dowler, 2003). These images present a distorted view of the system, with a generalized understanding that "if it bleeds, it leads." This leads to the over-exaggeration of violent crime in society (Maguire, 1988). Research indicates that as individuals increase their consumption of local and national television news, their fears about crime increase, regardless of crime rates or a history of victimization (Chiricos, Padgett, & Gertz, 2000). In addition to the portrayal of crime within the news, stories of crime, criminals, and criminal justice have been a major staple of television entertainment programming. These images, too, present a distorted view of the reality of crime, as they generally present crime as graphic, random, and violent incidents (Gerbner & Gross, 1980).

> Imagine yourself walking across a parking lot toward your car. It's late and the parking lot is poorly lit. You are alone. Standing near your car is a man who is watching you. Are you afraid?

When this scenario is presented to groups, we find that men and women respond to this situation differently. When asked who is afraid, it is primarily women who raise their hands. Rarely do men respond to this situation with emotions of fear. This simple illustration demonstrates the fear of victimization that women experience in their daily lives. "Fear of crime is, in other words, partly a result of feelings of personal discomfort and uncertainty, which are projected onto the threat of crime and victimization" (De Groof, 2008, p. 281). Indeed, research demonstrates that girls are more likely to indicate levels of fear of victimization due to situations such as poorly lit parking lots and sidewalks, overgrown shrubbery, and groups loitering in public spaces (Fisher & May, 2009).

How women develop a fear of victimization is inherently a gendered experience, rooted in how men and women are socialized differently. From a young age, young girls are taught about fear, as parents are more likely to demonstrate concern for the safety of their daughters, compared to their sons. This fear results in the lack of freedom for girls, in addition to an increase in the parental supervision of girls. These practices, which are designed to protect young women, can significantly affect their confidence levels regarding the world around them. The worry that parents fear for their daughters continues as they transition from adolescence to adulthood (de Vaus & Wise, 1996). Additionally, this sense of fear can be transferred from the parent to the young female adult as a result of the gendered socialization that she has experienced throughout her life.

Research indicates that the fear of crime for women is not necessarily related to the actual levels of crime that they personally experience. Women are less likely to be victimized than men, yet they report higher levels of fear of crime (Fattah & Sacco, 1989). These high levels of overall fear of victimization may be perpetuated by a specific fear of crime for women—rape and sexual assault. Indeed, rape is the crime that generates the highest levels of fear for women. These levels of fear are validated, as women make up the majority of victims for sexually based crimes (Warr, 1984, 1985). However, research indicates that this fear of sexual victimization extends to all crimes. The "shadow of sexual assault" thesis suggests that women experience a greater fear of crime in general, because they believe that any crime could ultimately become a sexually based victimization (Fisher & Sloan, 2003). For women, their fear of victimization is related to feelings of vulnerability. This is particularly true for women with prior histories of victimization (Young, 1992). Yet even when women engage in measures to keep themselves safe, their fear of sexual assault appears to increase, rather than decrease (Lane, Gover, & Dahod, 2009). This sense of vulnerability is portrayed by "movie of the week" outlets that showcase storylines of women being victimized by strange men who lurk in dark alleys and behind bushes. Unfortunately, these popular culture references toward rape and sexual assault paint a false picture of the realities of rape, as most women are victimized not by strangers, as these films would indicate, but instead by people known to them (Rand, 2008).

The fear of crime and victimization has several negative consequences. Women who are fearful of crime, particularly violent or sexual crimes, are more likely to isolate themselves from society in general. This fear reflects not only the concern of potential victimization, but also a threat regarding the potential loss of control that a victim experiences as a result of being victimized. Fear of crime can also be damaging toward one's feelings of self-worth and self-esteem. Here, potential victims experience feelings of vulnerability and increased anxiety. The effects of fears of victimization are reflected both in individual and societal actions. Agents of criminal justice can respond to a community's fear of crime by increasing police patrols, while district attorneys pursue tough-on-crime stances in their prosecution of criminal cases. Politicians respond to community concerns about violent crime by creating and implementing tough-on-crime legislation such as habitual sentencing laws like "three strikes" and targeting perceived crimes of danger such as "the war on drugs." While the public's concern about crime may be very real, it can also be encouraged by inaccurate data on crime rates or a misunderstanding about the community supervision of offenders and recidivism rates.

Theories on Victimization

In an effort to understand the victim experience, social science researchers began to investigate the characteristics of crime victims and the response by society to these victims. While criminology focuses more on the study of crime as a social phenomena and the nature of offenders, the field of victimology places the victim at the center of the discussion. Early perspectives on victimology focused on how victims, either knowingly or unconsciously, can be at fault for their victim experience, based on their personal life events and decision-making processes. One of the early scholars in this field, Benjamin Mendelsohn (1963), developed a typology of victimization that distinguished different types of victims based on the responsibility of the victim and the degree to which victims have the power to make decisions that can alter their likelihood of victimization. As a result of his work, the study of victimology began to emerge as its own distinct field of study.

Mendelsohn's theory of victimology is based on six categories of victims. The first category is the innocent victim. This distinction is unique in Mendelsohn's typology as it is the only classification that does not have any responsibility for the crime attributed to the victim. An innocent victim is someone who is victimized by a random crime, such as a school shooting or a natural disaster. According to Mendelsohn's classification, the victims in the fall of the twin towers in New York City on September 11, 2011 would be considered innocent victims. The second category is the victim with minor guilt. In this case, victimization occurs as a result of one's carelessness or ignorance. The victim with minor guilt is someone who, if they had given better thought or care to their safety, would not have been a victim of a crime. For instance, someone who was in the wrong place at the wrong time, or someone who potentially places himself or herself in dangerous areas where he or she is at risk for potential victimization is characterized as a victim with minor guilt. The third category is a victim who is equally as guilty as the offender. This victim is someone who shares the responsibility of the crime with the offender and deliberately placed himself or herself in harm's way. An example of this classification is the individual who seeks out the services of a sex worker, only to contract a sexually transmitted infection as a result of their interaction. The fourth category represents the case whereby the victim is deemed by society as "more guilty" than the offender. This is a "victim" who is provoked by others to engage in criminal activity. An example of this category is one who kills a current or former intimate partner following a history of abuse. The fifth category is a victim who is solely responsible for the harm that comes to him or her. These individuals are considered to be the "most guilty" of victims, as they engaged in an act that was likely to lead to injury on their part. Examples of the "most guilty" victim include a suicide bomber who engages in an act that results in his or her death. The final category is the imaginary victim. This is an individual who, as a result of some mental disease or defect, believes that he or she has been victimized by someone or something, when in reality this person has not.

While Mendelsohn focused on the influence of guilt and responsibility of victims, von Hentig's (1948) typology of victims looked at how personal factors, such as biological, psychological, and social factors, influence risk factors for victimization. Table 2.2 lists the 13 categories of von Hentig's typology of crime victims. While his theory was applied to developing an understanding of victims in general, several of his categories can be applied to explain the victimization of women specifically. For instance, young girls who run away from home are easy targets for pimps who "save" girls from the dangers of the streets and "protect" them from harm. The youth of these girls places them at a higher risk for violence and prostitution activities under this "guise" of protection. While von Hentig's category of mentally defective was designed to capture the vulnerability of the mentally ill victim, he also referenced the intoxicated individual within this context. Under this category, women who engage in either consensual acts of intoxication, or who are subjected to substances unknown to them, can be at risk for drug-facilitated sexual assault. Immigration can also play a

Table 2.2 Hans von Hentig's Typology of Crime Victims

Category	Description
Young	Youth are seen as vulnerable to victimization as a result of their age.
Female	Women are at risk as they are physically weaker compared to men.
Old	Less physical strength due to age, and greater financial resources.
Mentally defective	Mentally ill or intoxicated individuals are easy prey for attackers.
Immigrants	A new environment (culture, language, acculturation) may place someone at risk for victimization.
Minorities	Prejudice and discrimination can affect the victim experience.
Dull normals	A vulnerable individual who is easily exploited. Typically have lower than average IQ.
Depressed	Depressed individuals can be more submissive and less likely to fight off an attacker.
Acquisitive	One who is at risk for being taken advantage of due to a desire for financial advantage.
Wanton	One who is particularly vulnerable to stressors at various stages during the life cycle.
Lonesome/heartbroken	One who is taken advantage of due to a fear of being alone.
Tormentors	The victim becomes the perpetrator of a crime, following years of abuse.
Blocked	One who is unable to help or defend himself or herself out of a negative situation.

key role for women victims. Many abusers use a woman's immigration status as a threat to ensure compliance. In these cases, women may be forced to endure violence in their lives out of fear of deportation or mistreatment by government officials. Von Hentig also discusses how race and ethnicity can affect the victim experience, and significant research has demonstrated how these factors affect the criminal justice system at every stage.

While early theories of victimization provided a foundation to understand the victim experience, modern victimization theories expand from these concepts to investigate the role of society on victimization and to address how personal choices affect the victim experience. One of the most influential perspectives in modern victimology is Cohen and Felson's (1979) routine activities theory. Routine activities theory suggests that the likelihood of a criminal act (and in turn, the likelihood of victimization) occurs when you have someone who is interested in pursuing a criminal action (offender), a potential victim (target) "available" to be taken advantage of, and the absence of someone or something (guardian) that would deter the offender from making contact with the available victim. The name of the theory is derived from a belief that victims and guardians exist within the normal, everyday patterns of life. They posit that lifestyle changes during the second half of the 20th century created additional opportunities for the victim and offender to come into contact with each other as a result of changes to daily lifestyle and activities. Cohen and Felson's theory was created to discuss the risk of victimization in property crimes. Here, if individuals were at work, or out enjoying events in the community, they were less likely to be at home to guard their property against potential victimization. Routine activities theory has been criticized by feminist criminologists, who disagree with the theory's original premise that men are more vulnerable to the risks of victimization than women. Indeed, the "guardians" that Cohen and Felson suggest can protect victims from crime, may instead be the ones most likely to victimize women, particularly in cases of intimate partner abuse and sexual assault. Research by Schwartz, DeKeseredy, Tait, and Alvi (2001) indicate that women who engage in recreational substance use (such as alcohol or drugs) are considered to be a suitable target by men who are motivated to engage in offending patterns. The attempts by administrators to increase safety on college campuses by implementing protections such as escort patrols, lighted paths, and emergency beacons (modern-day "guardians") may have little effect on sexual assault rates on campus, given that many of these incidents take place behind closed doors in college dormitories and student apartments. In addition, the concept of self-protective factors (or self-guardians) may not be able to ward off a potential attacker, given that the overwhelming majority of sexual assaults on college campuses are perpetrated by someone known to the victim (Mustaine & Tewksbury, 2002).

Like routine activities theory, lifestyle theory seeks to relate the patterns of one's everyday activities to the potential for victimization. While routine activities theory was initially designed to explain victimization from property crimes, lifestyle theory was developed to explore the risks of victimization from personal crimes. Research by Hindelang, Gottfredson, and Garafalo (1978) suggests that people who engage in risky lifestyle choices place themselves at risk for victimization. Based on one's lifestyle, one may increase the risk for criminal opportunity and victimization through increased exposure to criminal activity and an increased exposure to motivated offenders. Given the similarities between the foundations of lifestyle theory and routine activities theory, many researchers today combine the tenets of these two perspectives to investigate victimization risks in general. These perspectives have been used to explain the risks of sexual assault of women on college campuses. Research by Fisher, Daigle, and Cullen (2010) illustrates that young women in the university setting who engage in risky lifestyle decision-making processes (such as use of alcohol) and have routine activity patterns (such as living alone or frequenting establishments such as bars and clubs where men are present and alcohol is readily available) are at an increased risk for sexual victimization. In addition, women who are at risk for a single incident remain at risk for recurrent victimizations if their behavior patterns remain the same.

Theoretical Perspectives on Female Criminality

Since the late 19th century, researchers have been investigating the relationship between gender, crime, and punishment. "Female lawbreakers historically (and to some degree today) have been viewed as abnormal and as worse than male lawbreakers—not only for breaking the law but also for stepping outside of prescribed gender roles of femininity and passivity" (Belknap, 2007, p. 34). In addition, scholars grew increasingly concerned with the limited explanations of female criminality. A review of criminological theories throughout the early 20th century indicates that most theories failed to include women in their studies. Those that did utilized limited samples of girls in their study of crime and developed perspectives that were subjected to gross gendered stereotypes about women and girls. Arguments regarding the nature of female criminality range from the aggressive, violent female to the passive, helpless female who is in need of protection. Consequently, theories on the etiology of female offending have reflected both of these perspectives. It is important to debate the tenets of early theories of female crime, as they provide a foundation to build a greater understanding of female offending.

Cesare Lombroso and William Ferrero represent the first attempts by criminologists to investigate the nature of the female offender. Expanding on his earlier work *The Criminal Man*, Lombroso joined with Ferrero in 1895 to publish *The Female Offender*. Lombroso's basic idea was that criminals are biological throwbacks to a primitive breed of man and can be recognized by various "atavistic" degenerative physical characteristics. To test this theory for female offenders, Lombroso and Ferrero went to women's prisons where they measured body parts and noted physical differences of the incarcerated women. To the female criminal, they attributed a number of unique features, including occipital irregularities, narrow foreheads, prominent cheekbones, and a "virile" type of face. While they found that the female offenders had fewer degenerative characteristics compared to male offenders, they explained these differences by suggesting that women, in general, are biologically more primitive and less evolved than males. They also suggested that the "evil tendencies" of female offenders "are more numerous and more varied than men's" (Lombroso & Ferrero, 1895, p. 151). Female criminals were believed to be more like men than women, both in terms of their mental and physical qualities. They suggested that female offenders were more likely to experience suppressed "maternal instincts" and "ladylike" qualities. They were convinced that women who engage in crime would be less sensitive to pain, less compassionate, generally jealous, and full of revenge—in short, the criminal women possessed all of the worst characteristics of the female gender while embodying the criminal tendencies of the male. The methods and findings of Lombroso and Ferrero have been harshly criticized, mostly due to their small sample size and the lack of heterogeneity of their sample demographics. They also failed to control for additional environmental and structural variables, which might explain criminal behavior regardless of gender. Finally, their key assumptions about women had no scientific basis. Their claim that the female offender was more ruthless and unmerciful had more to do with the fact that she violated sex-role and gender-role expectations than the nature of her actual offending behaviors.

More than a half-century passed between the publication of Lombroso and Ferrero's *The Female Offender* and the publication of *The Criminality of Women* by Otto Pollak (1961). Pollak believed that crime data sources failed to reflect the true extent of female crime. His assertion was that since the majority of the crimes in which women engage are petty in nature, many victims do not report these crimes, particularly when the victim is a male. Additionally, he suggested that many police officers exercise discretion when confronted with female crime and may only issue informal warnings in these cases. His data also indicated that women were more likely to be acquitted compared to their male counterparts. All together, he concluded that the crimes of women are underreported, and when reported, these women benefit from preferential treatment by the systems of criminal justice. His discussion of the "masked criminality" of women suggested that women gain power by deceiving men through sexual playacting, faked sexual responses, and menstruation. This power allowed female criminality to go undetected by society. Likewise, Pollak believed that the traditional female role status of homemaker, caretaker, and domestic worker gave women an avenue to engage in crimes against vulnerable populations, such as children and the elderly.

Theories based in biological determinism led to early psychological theories of female criminality. While biological determinism dealt with physical traits, early psychological theories drew from the physiological traits of women, such as their reproductive abilities, in interpreting psychological reactions. Crime was seen as a rejection of, or rebellion against, the socially prescribed traditional notions of femininity. Regardless of the individual or personal reasons for their behaviors, scholars during this time interpreted that the root for all causes of crime was related to a woman's sexuality. By defining women's behaviors as inherently sexual, the position of women as reproductive and domestic workers was reinforced. The use of a sexual explanation allowed for the criminologists to reinforce the concept of a patriarchal rule in society.

The emergence of works by Freda Adler and Rita Simon in the 1970s marked a significant shift in the study of female criminality. The works of Adler and Simon were inspired by the emancipation of women that was occurring during the 1960s and 1970s in the United States and the effects of the second wave of the feminist movement. Both authors highlighted how the liberation of women would lead to an increased participation of women in criminal activities. While Adler (1975) suggested that women's rates of violent crime would increase, Simon (1975) hypothesized that women would commit a greater proportion of property crimes as a result of their "liberation" from traditional gender roles and restrictions. While both authors broke new ground in the discussion of female crime, their research has also been heavily criticized. Analysis on crime statistics between 1960 and 1975 indicates that while female crime rates for violent crimes skyrocketed during this time period, so did the rates of male violent crime. In addition, one must consider the reference point of these statistics. True, more women engaged in crime. However, given the low number of female offenders in general, small increases in the number of crimes can create a large percentage increase, which can be misinterpreted and over-exaggerated. For example, if women are involved in 100 burglaries in one year, and this number increases to 150 burglaries in the next year, this reflects a 50% increase from one year to the next. If, however, males participated in 1,000 burglaries in one year, and in 1,250 during the next, this is only a 25% increase, even though the actual numerical increase is greater for males than females. Another criticism of Adler and Simon's works focuses on their overreliance on the effects of the women's liberation movement. While it is true that the emancipation of women created increased opportunities and freedoms to engage in crime, this may not indicate that women were more compelled to engage in crime. Changing policies in policing and the processing of female offenders may reflect an increase of women in the system as a result of changes in the response by the criminal justice system to crimes involving women.

In addition to Adler and Simon's focus on the emancipation of women as an explanation for increasing crime rates, the 1960s and '70s saw an increased interest in developing an understanding of female criminality and in explaining crime in general. Here, theories shifted from an individual pathology focus to one that referenced social processes and the greater social environment. These theorists believed that women and girls are socialized in different ways than men and boys. While boys are granted greater freedoms than girls and are encouraged to be more aggressive, ambitious, and outgoing, girls are expected to be nonviolent. As a result of this socialization process, girls have fewer opportunities to engage in crime, which is reflected in the lower rates of crime for women. When girls did engage in crime, they did so in very stereotypical ways—they went from shoppers to shoplifters. Even for women who engaged in violent acts, they did so in very gendered ways. In cases of violence and homicide, a woman's victims were more likely to be a relative, a lover, or at least someone known to her, a pattern that remains true today. They generally used a kitchen knife as a weapon, a tool that further reflects a gendered experience. Acknowledging these differences in gender roles and gender role socialization became a launching point for modern-day feminist discussions of crime.

With a shift away from the biological and psychological explanations for female criminality, the 1980s built upon the themes of gender roles and socialization to explain patterns of female offending. Here, early tenets of feminist criminology began with a discussion of the backgrounds of female offenders in an effort to assess who she is, where she comes from, and why she engages in crime. Feminist criminology reflects several key themes in its development. Frances Heidensohn (1985, p. 61) suggested that "feminist criminology began with the awareness that women were invisible in conventional studies in the discipline . . . feminist criminology began as a reaction . . . against an old established male chauvinism in the academic discipline." While some criminologists suggested that traditional theories of crime could account for explanations in female offending, others argue that in order to accurately theorize about the criminal actions of women, a new approach to the study of crime needs to be developed.

> Theoretical criminology was constructed by men, about men. It is simply not up to the analytical task of explaining female patterns of crime . . . thus, something quite different will be needed to explain women and crime . . . existing theories are frequently so inconsistent with female realities that specific explanations of female patterns of crime will probably have to precede the development of an all-inclusive theory. (Leonard, 1982, pp. xi–xii)

One of the defining features of feminist criminology is its focus on the lives of women and girls as a context for the causes of offending behavior. Sampson and Laub's (1993) life course theory suggests that the events of one's life (from birth to death) can provide insight as to why one might engage in crime. Life course theory highlights the importance of adolescence as a crucial time in the development of youthful (and ultimately adult) offending behaviors. While not specifically a feminist theory of crime, life course theory does allow for a gender-neutral review of how the different developmental milestones in one's life can explain criminal behavior. However, few scholars have relied on this perspective to understand female offending (see Belknap, 2007, for a review of these studies). In particular, life course theory can be limiting in its definitions and applications of what constitutes a traumatic life event and its potential relationship to criminality. An expansion of this category could benefit not only the understanding of female criminality, but male criminality as well (Belknap & Holsinger, 2006).

Perhaps some of the most influential research to date on female offending is the feminist pathways approach. Feminist pathways research seeks to use the historical context of women's and girls' lives to relate how life events (and traumas) affect the likelihood to engage in crime. Researchers have identified a cycle of violence for female offenders that begins with their own victimization and results with their involvement in offending behavior. In their study, Belknap and Holsinger (1998) identify that the entry to delinquency for boys and girls is a gendered experience. Their research suggests that understanding the role of victimization in the histories of incarcerated women may be one of the most significant contributions for feminist criminology, as female offenders report substantially high occurrence of physical, emotional, and sexual abuse throughout their lifetimes.

While such an explanation does not fit all female offenders (and also fits some male offenders), the recognition of these risks appears to be essential for understanding the etiology of offending for many girls and women. Yet this link between victimization and offending has largely been invisible or deemed inconsequential by the powers that be in criminology theory building and by those responsible for responding to women's and girls' victimizations and offenses (Belknap & Holsinger, 1998, p. 32).

Indeed, the feminist pathways approach may provide some of the best understanding of how women find themselves stuck in a cycle that begins with victimization and leads to offending. Research on women's and girls' pathways to offending provides substantial evidence for the link between victimization and offending, as incarcerated girls are three to four times more likely to be abused than their male counterparts. A review of case files of delinquent girls in California indicates that 92% of delinquent girls in California reported having been subjected to at least one form of abuse (emotional = 88%; physical = 81%; sexual 56%; Acoca & Dedel, 1997). For female offenders, research indicates that a history of abuse leads to a propensity to engage in certain types of delinquency offending. However, the majority of research points to offenses such as running away and school failures, rather than acts of violence. The effects of sexual assault are also related to drug and alcohol addiction and mental health traumas, such as posttraumatic stress disorder and a negative self-identity (Raphael, 2004). In a cycle of victimization to offending, young girls often run away from home in an attempt to escape from an abusive situation. In many cases, girls were forced to return home by public agencies such as the police, courts, and social services—agencies designed to "help" victims of abuse. Girls who refused to remain in these abusive situations were often incarcerated and labeled as "out of parental control." In a review of case files of girls who had been committed to the California Youth Authority during the 1960s, most girls were incarcerated for status offenses (a legal charge during that time frame). Many of these girls were committed to the Youth Authority for running away from home where significant levels of alcoholism, mental illness, sexual abuse, violence, and other acts of crime were present. Unfortunately, in their attempt to escape from an abusive situation, girls often fall into criminal behaviors as a mechanism of survival. Running away from home places girls at risk for crimes of survival such as prostitution, where the level of violence they experience is significant, and includes behaviors such as robbery, assault, and rape. These early offenses led these girls to spend significant portions of their adolescence behind bars. As adults, these same girls who had committed no crimes in the traditional sense later were convicted for a wide variety of criminal offenses, including serious felonies (Rosenbaum, 1989). Gaarder and Belknap (2002) characterize the pathway to delinquency as one of "blurred boundaries," as the categories of victim and offender are not separate and distinct. Rather, girls move between the categories throughout their lives, as their victimization does not stop once they become offenders. In addition to the victimization they experience as a result of their survival strategies, many continue to be victimized by the system through its failure to provide adequate services for girls and women (Gaarder & Belknap, 2002).

Developments in feminist research have addressed the significant relationship between victimization and offending. A history of abuse is not only highly correlated with the propensity to engage in criminal behaviors, but it often dictates the types of behaviors in which young girls engage. Often, these behaviors are methods of surviving their abuse, yet the criminal nature of these behaviors brings these girls to the attention of the justice system. The success of a feminist perspective is dependent on a theoretical structure that not only has to answer questions about crime and delinquency, but also address issues such as sex-role expectations and patriarchal structures within society. The inclusion of feminist criminology has important policy implications for the justice system.

> The ramifications of the traditionally male-centered approaches to understanding delinquency not only involve ignorance about what causes girls' delinquency but also threaten the appropriateness of systemic intervention with and treatment responses for girls. (Belknap & Holsinger, 2006, pp. 48–49)

Summary

- Not all victims report their crimes to the police but may seek out support from other sources.
- Victim blaming has been linked to low reporting rates.
- Mendelsohn's typology of victimization distinguishes different categories of victims based on the responsibility of the victim and the degree to which victims have the power to make decisions that can alter their likelihood of victimization.
- Von Hentig's typology of victimization focuses on how personal factors such as biological, social, and psychological characteristics influence risk factors for victimization.
- Routine activities theory and lifestyle theory have been used to investigate the risk of sexual assault of women.
- Women experience higher rates of fear of crime compared to males.
- Early biological studies of female criminality were based on gross assumptions of femininity and had limited scientific validity.
- Classical theories of crime saw women as doubly deviant—not only did women break the law, but they violated traditional gender role assumptions.

- Theories of female criminality during the 1960s and 1970s focused on the effects of the emancipation of women, gendered assumptions about female offending, and the differential socialization of girls and boys.
- The feminist pathways approach has identified a cycle for women and girls that begins with their own victimization and leads to their offending.

KEY TERMS

Victim blaming

Secondary victimization

Just world hypothesis

Benjamin Mendelsohn's six categories of victims

Hans von Hentig's 13 categories of victims

Routine activities theory

Lifestyle theory

Fear of victimization

Shadow of sexual assault

Cesare Lombroso and William Ferrero

Otto Pollak

Biological theories of crime

Masked criminality of women

Emancipation/liberation theory

Freda Adler

Rita Simon

Feminist criminology

Feminist pathways perspective

Cycle of victimization and offending

QUESTIONS FOR DISCUSSION

1. How do early theories of victimization distinguish between different types of victims? How might the criminal justice system use these typologies in making decisions about which cases to pursue?
2. What type of help-seeking behaviors do female crime victims engage in? How are these practices related to the reporting of crimes to law enforcement?
3. What effects does the practice of victim blaming have for future potential crime victims and the criminal justice system?
4. How might feminist criminologists critique modern-day victimization theories such as routine activities theory and lifestyle theory?
5. How have historical theories on female offending failed to understand the nature of female offending?
6. What contributions has feminist criminology made in understanding the relationship between gender and offending?

WEB RESOURCES

Bureau of Justice Statistics: http://bjs.ojp.usdoj.gov

The National Center for Victims of Crime: http://www.ncvc.org

Office of Victims of Crime: http://www.ojp.usdoj.gov/

Feminist Criminology: http://fcx.sagepub.com

How to Read a Research Article

As a student of criminology or criminal justice, you may have learned about the types of research that scholars engage in. In many cases, researchers publish the findings of their studies as articles in academic journals. In this section, you will learn how to read these types of articles and how to understand what researchers are saying about issues related to criminology and criminal justice.

In the majority of cases, a research article that is published in an academic journal includes five basic elements: (1) an introduction, (2) a review of the literature related to the current study, (3) the methods used by the researcher to conduct the study, (4) the findings or results of the research, and (5) a discussion of the results and/or conclusion.

Research in the social sciences generally comes in two basic forms: quantitative research and qualitative research. Quantitative research often involves surveys of groups of people, and the results are reported using numbers and statistics. Qualitative research can involve interviews, focus groups, and case studies and relies on words and quotes to tell a story. In this book, you will find examples of both of these types of research studies.

In the introduction section of the article, the author will typically describe the nature of the study and present a hypothesis. A hypothesis frames the intent of the research study. In many cases, the author(s) will state the hypothesis directly. For example, a research study in criminology or criminal justice might pose the following hypothesis: As the number of arrests increase, the length of the incarceration sentence will increase. Here, the author is investigating whether a relationship exists between a defendant's prior criminal record and sentence length. Similar to a hypothesis is the research question. Whereas a hypothesis follows an "if X happens, then Y will occur" format, research questions provide a path of inquiry for the research study. For example, a research question in criminology might ask, "what are the effects of a criminal record on the likelihood of incarceration?" While the presentation of a hypothesis and the presentation of a research question differ from each other, their intent is the same as each sets out a direction for the research study and may reference the expected results of the study. It is then left up to the researcher(s) and their data findings to determine whether they prove or disprove their hypothesis or if the results of their study provide an answer to their research question.

The next section of the article is the literature review. In this section, the author provides a review of the previous research conducted on this issue and the results of these studies. The purpose of the literature review is to set the stage for the current research and provide the foundation for why the current study is important to the field of criminology and criminal justice. Some articles will separate the literature review into its own section, while others will include this summary within the introductory section. Using the example from our earlier sample hypothesis, a literature review will consider what other scholars have said about the relationship between criminal history and incarceration and how their findings relate to the current research study. It may also point out how the current study differs from the research that has previously been conducted.

In the methods section, the researcher presents the type of data that will be used in the current study. As I mentioned earlier, research can be either quantitative or qualitative (and some studies may have both types of data within the same research project). In the methods section, the researcher will discuss who the participants of the study were; how the data were collected (interview, survey, observation, etc.); when, where, and how long the study took place; and how the data were processed. Each of these stages represents a key part of the research experience, and it is important for researchers to carefully document and report on this process.

The results section details the findings of the study. In quantitative studies, the researchers use statistics (often accompanied by tables, charts, or graphs) to explain whether the results of the study support or reject the hypothesis/research question. In qualitative studies, the researchers look for themes in the narrative data.

The research article concludes with a discussion and summary of the findings. The findings are often discussed within the context of the hypothesis or research question and relate the findings of this research study to related research in the field. Often, scholars will highlight their findings in light of the methods used in the study or the limitations of the study. The section concludes with recommendations for future research or may discuss the policy implications of the research findings.

Now that you've learned a bit about the different components of a research article, let's apply these concepts to one of the articles here in your book.

The Gendered Nature of Risk Factors for Delinquency

1. What is the thesis or main idea from this article?
   - The main idea of this article references that delinquent boys and girls have different pathways to delinquency that are often not acknowledged by traditional mainstream theories of crime. The failure by criminological theories to address these gender differences has a significant impact when developing and implementing policy decisions about gender-specific programming.
2. What is the hypothesis or research question?
   - The research question for this study is located at the end of the first paragraph. The authors of this study seek to understand how variables such as family, school, peers, victimization, and mental health affect the pathways to delinquency for girls and boys.
3. Is there any prior literature related to the hypothesis?
   - Yes. The authors list three different criminological theories to investigate how scholars have examined gender to explain differences in male and female delinquency. These three theories are general strain theory (GST), the life course perspective, and the feminist pathways and cycle of violence perspectives. Within these explanations, the authors provide a review of the existing literature on how each perspective has been used to explore gender differences.
4. What methods are used to support the hypothesis/research question?
   - This study contains data from a larger study on the gender-specific needs of girls. For this article, the authors surveyed 163 girls and 281 boys who were incarcerated in juvenile facilities in the state of Ohio in 1998. The data in this report rely on self-reported data on childhood abuse, family relationships, school experiences, and mental health needs. The youth surveyed in this study were from three different facilities in the state. Given the gender differences in the number of youth who were incarcerated, all of the girls were eligible to participate the study

and a random sample of boys (1/10) were eligible to participate. The questions on the survey were derived from information that was obtained from focus group interviews that were conducted during an earlier phase of the study.

5. Is this a quantitative or qualitative study?
    - In looking at the results section, you can determine that this is a quantitative study through its use of numbers and statistics.
6. What are the results, and how do(es) the author(s) present the results?
    - The authors present the data in four different tables and look at gender differences in terms of victimization histories, their relationships with their families, their experiences with school and peers, and their mental health histories. The authors find that girls experience significantly higher levels of all types of abuse (emotional, physical, and sexual) compared to boys. In terms of family relationships, 56% of girls (compared to 45% of boys) reported that they had been abandoned by one of their parents. Several other youth indicated that at least one of their parents was deceased, incarcerated, or institutionalized. The lack of positive family relationships appeared to affect the lives of both boys and girls in this study. Girls were significantly more likely to report greater mental health needs, including histories of self-mutilation, suicidal thoughts, and attempts of suicide. Girls were also more likely to have negative feelings about themselves. Finally, girls are more likely than boys to drop out of school. In addition to gender differences for each of these measures, several of the findings also indicated differences by race and sexual orientation.

      Comparing these research findings to the research question, the authors suggest that these results are important for the development of criminological theory on gender and delinquency. For example, they suggest that the feminist pathways perspective makes important contributions to the literature by emphasizing the importance of a history of abuse for both female and male delinquency. While they indicate that GST and life course theories do provide value when understanding the issues of gender and delinquency, the feminist pathways approach provides the most comprehensive explanation for understanding the link between victimization and delinquency. In addition, the authors assert that the data from this study highlight how the understanding of a history of victimization is important when developing and implementing policy and programming for youth offenders and youth at risk.
7. Do you believe the author(s) provided a persuasive argument? Why or why not?
    - While the assessment of whether the authors provided a persuasive argument is ultimately up to the reader, the data in this study do provide a compelling link between gender and delinquency, particularly in the case of a history of victimization. Given the limited inclusion of gender into traditional theories of crime, the research in this article provides an important contribution to the field.
8. Who is the intended audience of this article?
    - In thinking about the intended audience of the article, it may be useful to ask yourself "who will benefit from reading this article?" This article appeared in an academic journal, which is typically read by students, professors, scholars, and justice officials. Here, the information in their

article can not only add to the classroom experience for students and professors who study this issue, but it can also ultimately influence policy decisions about how cases are handled in the juvenile court.

9. What does the article add to your knowledge of the subject?
   - The answer to this question will vary for each student, as it asks students to reflect about what they learned from this research and how it relates to their previous experience with the topic. An example from this article might be the understanding that boys and girls have different life experiences that affect their risk factors and pathways to delinquency.

10. What are the implications for criminal justice policy that can be derived from this article?
    - While not always the case, this article in particular provides a specific link between the data findings of the study and the implications for criminal justice policy. In this article, the authors suggest that by developing an understanding of the pathways of delinquent girls, particularly in reference to how a history of victimization can place girls at risk for future offending behaviors, criminal justice policy makers can consider these issues when designing prevention and intervention programming for at-risk and offending youth.

Now that you've applied these concepts to an article, continue this practice as you go through each reading in your text. Most articles will have the same basic contents (introduction, literature review, methods, results, and conclusion). Some articles will be easier to understand while others will be more challenging. You can refer back to this example if you need help with additional articles in the book.

READING

In the section introduction, you learned about pathways theory and the importance of considering the impact of victimization on offending behaviors. This reading looks at how pathways theory might help explain girls' delinquency by looking at a number of variables, including demographics, abuse histories, family, school and peer relationships. and feelings of self-esteem. The authors suggest the importance of recognizing the link between victimization and criminality not only for developing criminological theories, but also in developing prevention and intervention strategies.

The Gendered Nature of Risk Factors for Delinquency

Joanne Belknap and Kristi Holsinger

One of the greatest limitations of existing criminological research is the low priority given to the role of gender in the etiology of offending. In 1982, Eileen Leonard wrote, "Despite the endless volumes written to account for it, sex, the most powerful variable regarding crime, has been virtually ignored" (p. xi). Early theories on delinquency and crime either fail to include girls (and women) or if included, theorize about them in sexist and stereotypical ways (Burman, Batchelor, & Brown, 2001). It is notable that mainstream criminology still tends to ignore the importance of gender and how events perceived as risks for offending, such as poor school and family experiences, may be gendered. The ramifications of the traditionally male-centered approaches to understanding delinquency not only involve ignorance about what causes girls' delinquency but also threaten the appropriateness of systemic intervention with and treatment responses to girls. Since the early 1990s, government agencies have increasingly called for "gender-specific" approaches in intervening with and treating delinquent girls. The purpose of this article is to merge what has been learned from applications of various theoretical approaches to understand how characteristics and experiences often associated with gender may or may not be gendered among a delinquent population. More specifically, this study is based on the most comprehensive data including girl and boy delinquents' childhood traumas. We surveyed 163 girls and 281 boys incarcerated in Ohio in 1998 about their family, school and peers, victimization, and mental health histories and experiences to examine how risk factors may be gendered and the implications for gender-specific services.

Theoretical Perspectives Explaining Delinquency

In 1979, Cernkovich and Giordano wrote, "The development of accurate theory continues to suffer because some scholars rush to print with the causes of delinquency

SOURCE: Belknap, J., & Holsinger, K. (2006). The gendered nature of risk factors for delinquency. *Feminist Criminology, 1*(1), 48–71.

before they know what it is they should be explaining" (p. 144). It is notable that "malestream" criminological theories have questionable applicability to girls' offending largely because they were developed to understand boys' delinquency and even then, almost always fail to explain the role of gender in boys' lives (e.g., masculinity). This work has a long legacy of affecting how girls' offending is both explained and officially processed. Three more recent perspectives (generalized strain theories [GSTs], the life-course perspective, and feminist pathways and cycle of violence theories), outlined in this section, were chosen, as each offers more promising avenues to accurately theorize about the etiology of crime and delinquency, how it may be gendered, and how responses to delinquent youth might be improved.

General Strain Theory (GST)

GST was designed to explain how frustration because of unequal legitimate access to culturally agreed on goals is an important explanation for offending (e.g., Cohen, 1955; Merton, 1949). The focus was on economic strains, and the applications were almost exclusively on middle- and working-class boys. Faith (1993) critiqued this theory from a gender perspective, pointing out that relative to boys and men, girls and women commit far less crime and they "constitute the most impoverished group of every Western Society" (p. 107). GST has also been criticized for omitting some of the major strains in youths' lives, such as abuse, sexism, racism, and other traumas (Belknap, 2001).

Some scholars have argued that the gender gap in offending is best explained by Agnew's (1992) revised GST, which acknowledges a wider variety of strain sources and allows for diverse adaptations to strains (Broidy & Agnew, 1997; Hoffman & Su, 1997). Rather than a narrow focus on the failure of the lower class to achieve monetary success, GST outlines three major social-psychological sources of strain—the failure to achieve positively valued goals, the loss of positively valued stimuli, and the presentation of negative stimuli. Another compelling aspect of GST is that it allows individuals' goals to be varied based on gender, racial, and class differences (Broidy, 2001). Thus, strains can be examined as artifacts of a gendered society and allow for gendered socialization to shape both the strains experienced and individuals' responses to them. Broidy and Agnew (1997) concluded, among other findings, that girls are more likely than boys to be the targets of sexual, emotional, and physical abuse. Girls are more concerned than boys about establishing and maintaining close relationships, whereas boys are more strained than girls about external achievements, particularly material success. Although two studies find few gender differences regarding the impact of stressful life events on delinquency (Hoffman & Su, 1997; Mazerolle, 1998), it is remarkable that they do not include any measure of abuse victimization among the stressful life events. Sharp and her colleagues' (Sharp, Brewster, & Love, 2005; Sharp, Terling-Watt, Atkins, & Gilliam, 2001) tests of GST suggest a complex interaction of variables in understanding the gendered nature of delinquent (and deviant) responses to strain. A study of college students by Sharp et al. (2005) reports that although both women and men respond to strain by being angry, women are more likely than men to also respond to strain with additional, more internalized negative emotions, which might mediate their likelihood of subsequent delinquent behaviors. Sharp et al. suggested that this complex gendered difference in responding to strain potentially explains the gendered gap in offending.

The Life-Course Perspective

The life-course perspective examines human development and views problem behaviors as age associated because of patterns in developmental stages (Laub & Lauritsen, 1993; Loeber & Le Blanc, 1990; Sampson & Laub, 1990). This perspective advocates the need to study important transitions in an individual's life, as these transitions affect offending, and to identify causal factors that occur before and, thus, may influence behavioral development (Loeber & Le Blanc, 1990; Sampson & Laub, 1992). As noted by Pollock (1999) and Belknap (2001), however, the bulk of the life-course studies focus on males (e.g., Laub & Sampson, 1993; Loeber, 1996; Moffitt, 1990, 1993; Nagin, Farrington, & Moffitt, 1995; Piquero, Brame,

Mazerolle, & Haapanen, 2003; Piquero, MacDonald, & Parker, 2002; Stattin & Magnusson, 1991). One longitudinal study including girls that compares the separate and combined effects of parental psychiatric disorders and supportive parent-child communications finds that low parental support has a greater effect on boys' deviance than girls' (R. Johnson, Su, Gerstein, Shin, & Hoffman, 1995).

Feminist Pathways and Cycle of Violence Perspectives

Feminist research claims not only that to truly understand delinquency, the differences between girls' and boys' experiences and "realities" must be examined (Holsinger, 2000) but also that patriarchy must be central to the study of causes of delinquency (Daly & Chesney-Lind, 1988). Within this perspective, the variables leading to problem behavior may be attributed to a variety of sources—socialized gender roles, structural oppression, vulnerability to abuse from males, and female responses to male domination. In other words, girls' and boys' trajectories into delinquency may be partially gender specific—with gender differences in developmental processes, resulting problem behaviors, and social and official responses to troubling (and other) behaviors. Unlike the GST or life-course perspective, the feminist pathways approach to understanding the causes of illegal behavior emphasizes childhood abuses as significant risks for subsequent delinquency. Studies on delinquent girls and incarcerated women report abuse victimizations much higher than the general population of women and girls (e.g., Arnold, 1990; Browne, Miller, & Maguin, 1999; Chesney-Lind & Rodriguez, 1983; Daly, 1992; Gaarder & Belknap, 2002; Gilfus, 1992). Moreover, the research conducting gender comparisons of childhood abuse and neglect routinely reports abuse and neglect are more common, start earlier, and last longer for girls (e.g., Artz & Riecken, 1994; Chesney-Lind, 1989; Dembo, Williams, Wothke, Schmeidler, & Brown, 1992; McClellan, Farabee, & Crouch, 1997; Miller, Trapani, Fejes-Mendoza, Eggleston, & Dwiggins, 1995). Moreover, girls' reactions to abuse are relevant. For example, girls' running away from a sexually abusive home is often officially tagged as offending behavior (Gilfus, 1992).

Regarding mental health, a review of literature on adolescent female development reports that girls experience greater depression, more suicide attempts, and a decrease in self-concept, whereas boys report improved self-concept and self-esteem (Miller et al., 1995). The delinquent girls interviewed in one study reported very few, if any, positive attributes associated with being female, accepting as inevitable the routine sex discrimination, lack of respect, sexual double standards, sexual harassment, and abuse they experienced (Artz, 1998). Although these girls described themselves as "tough," they conveyed a strong sense of personal worthlessness that they hoped to change through male approval (i.e., in their relationships with boys and men).

Consistent with the feminist pathways research is Cathy Spatz Widom and her colleagues' work on the cycle of violence. Widom (1989) and Rivera and Widom (1990) compared subsequent delinquency and adult offending histories of individuals with and without formal records of childhood abuse and neglect victimizations. Overall, childhood victimizations, including neglect, placed individuals at risk for both delinquency and adult offending (Widom, 1989). It is notable that abused and neglected girls were more likely than non-abused/neglected girls to have violent delinquency offense arrests, whereas abused and neglected boys were no more likely to have violent delinquency offense arrests than non-abused/neglected boys (Rivera & Widom, 1990).

Method

As part of the Office of Juvenile Justice and Delinquency Prevention Act of 1992 to assess the gender-specific needs of girls, the authors conducted two phases of research in Ohio. Phase I was a statewide focus group study of girls and professionals who work with them (Belknap, Holsinger, & Dunn, 1997). Phase II, the source of the data for this article, was a survey of incarcerated girls and boys. The primary goal of this study was to broaden the scope of youthful risks hypothesized in both mainstream and feminist criminology to increase the likelihood of delinquent behavior and determine whether these risk factors were gendered among delinquent youth. Unlike the vast majority of tests of GST, life-course, and feminist pathways perspectives, which include only one sex in the sample, or the cycle of violence studies that rely on officially reported childhood victimizations, this study included both girls and boys and drew on self-reported childhood abuses (a more valid measure than official reports), as well as the youths' reports on their families, schools, and mental health.

Sample

In Ohio, institutionalized delinquent girls are held in 1 of 2 Department of Youth Services (DYS) girls' facilities (the majority in Scioto Village and a small population in The Freedom Center). All the girls housed in these 2 institutions were included in the sample ($n = 163$) and all completed usable surveys. Adjudicated delinquent boys are held in 1 of 10 DYS boys' facilities. Given the lack of feasibility in surveying all the boys, the Department of Research in DYS drew a random sample of 350 boys from the most recent list of admissions for DYS institutions. Of the boys randomly selected, 83% were provided with the opportunity to take the survey, and 89% of these were completed and usable. Results are based on 281 surveys completed by incarcerated boys (or 80% of those sampled) and 163 girls (100% response rate).

Measurement Instrument and Data Collection

Unlike much of the research on delinquent youth drawn from official reports, this was a "youth-centered" design (see Morrill, Yalda, Adelman, Musheno, & Bejarano, 2000), permitting the youth to directly report their own experiences (and pilot-tested on youth who gave their feedback). The 15-page measurement instrument was the result of what was learned in the focus groups in Phase I and an extensive review of research on delinquency etiology, processing, and treatment. We used the qualitative data gleaned from the focus groups to make survey items. For example, in our focus groups, the girls reported many childhood traumas as life events leading to their delinquent behavior that were not in the literature (e.g., a parent's death, abandonment by a parent, and witnessing abuse). The self-esteem items were adopted from Rosenberg (1989).[1]

A release form was used to verify voluntary participation from the youth and to convey the goals of the survey, assuring them of confidentiality and anonymity. The youth were informed that the survey would require approximately 60 minutes to complete. The cooperation of the youth was impressive, particularly given the survey length. Notably, the boys had more questions, required more assistance, and took longer to complete the survey than the girls, whereas the girls were more likely than the boys to seem to "enjoy" taking the survey and thank the research staff for conducting the study. (Typically, the girls took about 45 minutes and the boys took close to 90 minutes to complete the survey.) Finally, we structured the data collection so that youth who did not want to take the survey could sit and "doodle" on their surveys while the rest of the youth completed them, and everyone was given their own blank manila envelope to put the surveys in when they were completed. These were collected directly by the research staff (no institutional staff members were involved in data collection). Thus, we believe the environment was one in which it was safe for youth to write honest answers, including self-disclosure about abuse, family, school and peer, mental health, and offending histories.

Limitations

The major shortcoming of this study is that there is no comparison between delinquent and nondelinquent youth—we sampled only institutionalized delinquents. The state would not allow us a comparison group of nondelinquents. However, given the frequent assumptions that boys' primary childhood stressors are economic in nature and that girls and boys have unique stressors, these data allow for the much needed examination of the gendered nature of childhood experiences with family, abuse, school and peer, and mental health among delinquent youth. This approach is also valid for examining the gender-specific needs of delinquent youth.

Another concern is temporal order. Given that these data were not collected prospectively, it is difficult to tell the sequencing of events. For example, did the abuse precede or follow the delinquency? This is why we directly asked the youth if they believe their abuse was related to their subsequent delinquency. Another concern is whether the youth told the truth, particularly about victimization. In this regard, we believe the data collection method (as described above) allowed a safe environment to complete the survey anonymously and confidentially. Finally, we did not have reliable scale items for many of our items, given that we were basing a significant portion of the survey items on the qualitative findings from our focus groups and the qualitative interview research from incarcerated women.

Findings

The findings are divided into five sections. The first section describes the sample and the following four reflect areas identified by theories as potentially influential to gender and delinquency: abuse victimization, family, self-esteem/mental health, and school experiences. Tables 2.3 through 2.7 report (bivariate) gender comparisons. In addition, to account for whether the gender effects were "direct" or changed when controlling for race, age, and sexual identity, multivariate logistic regression analyses were conducted. The appendix presents the 10 cases where the gender became nonsignificant when controlling for race, age, and sexual identity, as well as the 2 cases where gender went from being nonsignificant to significant when controlling for these variables.

Demographic Characteristics

The sample, reported in Table 2.3, was almost half White (46%) and about two fifths (41%) African American, with the remaining 13% of the sample describing themselves as Native American, Latino/a, Puerto Rican, Spanish, Asian American, South African, or Biracial. Although youth of color are disproportionately represented in this sample of incarcerated youth by a large margin, there were no significant gender differences in the racial representation in the sample. The sample ranged in age from 12 to 20 years old with a mean age of 16.35 years. Girls, however, were significantly younger than the boys; approximately one third of the girls and one fifth of the boys were 15 years old or younger ($t = 4.77, p \leq .001$). This is consistent with a significant amount of research claiming younger girls receive harsher sentences because of the paternalistic nature of the courts (Belknap, 2001).

The life-course perspective purports that adolescent development and transitions are crucial to identify to understand youth behaviors. The existing empirical applications of life-course, GST, and cycle of violence approaches never mention sexual identity, and when addressed in the pathways research, it is done so only in passing. C. Johnson and K. Johnson's (2000) careful review of high-risk behaviors among lesbian/gay adolescents provides compelling arguments for including sexual identity in research on delinquency. They emphasized the isolation from family, peers, and mainstream culture often experienced by youth coming to terms with a lesbian/gay identity. Moreover, lesbian, gay, and bisexual youth appear to be at an increased risk of sexual and physical abuse compared to their straight counterparts (D'Augelli & Dark, 1994). In the current study, there were strong significant differences between girls and boys regarding self-reported sexual identity (see Table 2.3). Heterosexual identities were reported by 95% of the boys and 73% of the girls. Girls (22.4%) were 6 times as likely as boys (3.6%) to identify as bisexual and 3 times (4.6%) as likely as boys (1.6%) to identify as lesbian/gay ($\chi^2 = 39.85, p \leq .001$). With these data, it is difficult to determine whether boys are less likely to report gay or bisexual identities or if it is an identity that places girls, but not boys, at increased risk of marginalization and delinquency. Perhaps lesbian and bisexual girls are more stigmatized as "masculine" and, thus, "delinquent" (Robson, 1992) relative to their gay and bisexual male counterparts and heterosexual female counterparts.

Finally, Table 2.3 reports on the intimate/romantic relationship status, children, and pregnancies as

Table 2.3 Youths' Demographic Characteristics

Variable	*n*	Girls		Boys		Total		Test Statistic
		Percentage	(*n*)	Percentage	(*n*)	Percentage	(*n*)	
Sex	444	36.7	(163)	63.3	(281)			
Race	441							3.05
African American		36.2	(59)	43.9	(122)	41.0	(181)	
White		47.9	(78)	44.2	(123)	45.6	(201)	
Other[a]		16.0	(26)	11.9	(33)	13.4	(59)	
Age[b]	444							4.77***
12 to 13		6.1	(10)	1.8	(5)	3.4	(15)	
14 to 15		26.4	(43)	17.8	(50)	20.9	(93)	
16 to 17		55.2	(90)	57.7	(162)	56.8	(252)	
18 to 20		12.3	(20)	22.8	(64)	18.9	(84)	
Sexual identity	404							39.85***
Heterosexual		73.0	(111)	94.8	(239)	86.6	(350)	
Lesbian/Gay		4.6	(7)	1.6	(4)	2.7	(11)	
Bisexual		22.4	(34)	3.6	(9)	10.6	(43)	
Relationship status	436							4.57
Single		20.4	(33)	20.8	(57)	20.6	(90)	
Boy/Girlfriend		74.1	(120)	77.4	(332)	76.1	(332)	
Married/Common-Law		5.6	(9)	1.8	(5)	3.2	(14)	
Children	412							12.78***
Yes		13.8	(22)	29.0	(73)	23.1	(95)	
No		86.3	(138)	71.0	(179)	76.9	(317)	
Pregnancy history[c]	158							
Never		64.5	(102)					
Once		20.9	(33)					
Twice		8.9	(14)					
Three or more		5.7	(9)					

NOTE: Chi-square tests were used for all variables except age, in which case a *t* test was used.

a. This category includes Native American, Latino/a, Puerto Rican, Spanish, Asian, South African, and Biracial.

b. The girls' mean age was 15.94, and the boys' mean age was 16.59 years old.

c. This question was asked only of the girls. Of the 158 girls who reported a pregnancy, 27.8% ($n = 44$) reported experiencing a miscarriage and 5.7% ($n = 9$) reported having an abortion.

***$p \leq .001$.

reported by the youth. Although girls and boys were indistinguishable in terms of relationship status (whether they were single, dating, or married), girls were significantly more likely to be involved with partners older than themselves compared to boys (not reported in the tables). The extent to which relationships with older boys (and in some cases, men) should also be examined as a gendered pathway for girls' involvement in crime and delinquency needs to be addressed in future research. Boys reported themselves as the father to more children than girls reported giving birth to, but girls, not surprisingly given society's gender roles, reported more actual parenting/custody than the boys. The girls were asked to report on their pregnancy histories. Almost two thirds (64.5%) had never been pregnant, one fifth (20.9%) had been pregnant once, almost one tenth (8.9%) had been pregnant twice, and about one twentieth (5.7%) had been pregnant three or more times. Of the girls who reported a pregnancy, more than one quarter (27.8%) reported experiencing a miscarriage and about one twentieth (5.7%) reported having an abortion.

Abuse Histories

Table 2.4 summarizes the direct effects of gender on self-reported abuse victimizations. There are two overwhelming patterns in this table. First, for virtually every abuse variable, girls reported experiencing significantly greater amounts of abuse. Second, although lower, boys' rates of reported abuse are still extremely high. Regarding verbal abuse, two thirds of girls and more than half of the boys reported experiencing verbal abuse from a family member ($\chi^2 = 5.60, p \leq .05$). More than half the girls and one third of the boys reported verbal abuse from someone outside of the family ($\chi^2 = 20.83, p \leq .001$). Physical abuse included a wide range of behaviors: spanking or slapping, pushing or grabbing, having something thrown at you, kicking or hitting, beating, choking, burning, or having weapons used or threatened to be used against you. Three quarters of the girls and almost two thirds of the boys reported physical abuse perpetrated by a family member ($\chi^2 = 6.23, p \leq .05$). Two thirds of the girls and more than one third of the boys reported physical abuse by a non-family member ($\chi^2 = 34.20, p \leq .001$). More than three fifths of the girls and more than two fifths of the boys reported physical abuse repeated with time ($\chi^2 = 14.45, p \leq .001$).

"Unwanted sexual contact" was used to measure all types of sexual abuse that involved physical contact. Almost three fifths of the girls and almost one fifth of the boys reported sexual abuse by anyone (family or non-family member; $\chi^2 = 75.73, p \leq .001$). Close to one quarter of the girls and about 1 in 12 boys reported sexual abuse by a family member ($\chi^2 = 17.45, p \leq .001$). More than half of the girls and about 1 in 7 boys reported sexual abuse from a non-family member ($\chi^2 = 77.10, p \leq .001$). Close to half of the girls and about 1 in 7 boys reported sexual abuse that was repeated with time ($\chi^2 = 42.70, p \leq .001$). Regarding the number of sexual abusers, almost half of the girls and about one sixth of the boys reported one or two different sexual abusers. Almost one eighth of the girls and about 3% of the boys reported having three or more different sexual abusers ($t = -8.55, p \leq .001$). Stated alternatively regarding the sexual abuse data, of the almost three fifths of the girls who reported any sexual abuse, almost two fifths experienced sexual abuse by a family member, about nine tenths experienced sexual abuse perpetrated by a non-family member, and many of these girls had more than one abuser. In a similar manner, of the almost one fifth of boys who reported any sexual abuse, almost half reported family member-perpetrated sexual abuse and three quarters reported sexual abuse from a non-family member.

Habermas and Bluck (2000) reported that most adolescents are adept at assessing the causal ordering of important events in their lives. In the focus groups of Phase I, we asked the girls whether they could identify events that "caused" or led to their offenses. Many of the girls reported abusive and other traumatic experiences. When probed further, they explained an understanding of the sequencing or causal nature of the trauma-to-offending link. Although a fair amount of research indicates a high correlation between

Table 2.4 Delinquent Youths' Abuse History

Variable	*n*	Girls		Boys		Total		Test Statistic
		Percentage	(*n*)	Percentage	(*n*)	Percentage	(*n*)	
Verbal abuse from family[a]	444	66.3	(154)	54.8	(262)	59.0	(262)	5.60*
Verbal abuse from others[b]	444	55.2	(90)	33.1	(93)	41.2	(183)	20.83***
Physical abuse from family	444	74.8	(122)	63.3	(178)	67.6	(300)	6.23*
Physical abuse from others	444	65.0	(106)	36.3	(102)	46.8	(208)	34.20***
Physical abuse repeated with time	379	62.9	(101)	42.8	(101)	50.4	(191)	14.45***
Sexual abuse from anyone[c]	444	58.9	(96)	18.5	(52)	33.3	(148)	75.73***
Sexual abuse from family member	444	22.7	(37)	8.5	(24)	13.7	(61)	17.45***
Sexual abuse from others	444	52.8	(86)	13.9	(39)	28.2	(125)	77.10***
Sexual abuse repeated with time	343	45.8	(65)	13.9	(28)	27.1	(93)	42.70***
Total number of sexual abusers[d]	444							–8.55***
None		41.1	(67)	81.5	(229)	66.7	(296)	
1 to 2		47.2	(77)	16.0	(45)	27.5	(122)	
3+		11.7	(19)	2.5	(7)	5.8	(26)	
Abuse led to getting in trouble	310	55.8	(78)	40.1	(73)	48.7	(151)	13.05***
Witnessed verbal abuse	444	49.1	(91)	42.0	(118)	47.1	(209)	7.93**
Witnessed physical abuse	444	49.1	(80)	30.6	(86)	37.4	(166)	15.04***
Witnessed sexual abuse	444	12.3	(20)	6.0	(17)	8.3	(37)	5.23*

a. This category includes father, stepfather, mother, stepmother, brother, or sister.

b. This category includes boyfriend, girlfriend, spouse, friend, stranger, or anyone else.

c. This category includes sexual abuse from family or others.

d. Where interval-level data are used and means are reported, significance is based on nests. The mean number of sexual abusers for girls was 1.14 and for boys was 0.29.

*$p \leq .05$. **$p \leq .01$. ***$p \leq .001$.

female delinquency and a history of physical and sexual abuse, research comparing males and females on the link between abuse and subsequent delinquency is underdeveloped. To our knowledge, no one has asked this question of youth in previous quantitative research, but given the pathways research suggestions of this link and the manner it was verbalized in our focus groups, we included this item on the survey. When asked about whether they viewed the abuses they experienced as leading to their trouble with offending, more than half of the girls and two fifths of the boys believed their victimizations had influenced their subsequent offending ($\chi^2 = 13.05, p \leq .001$; see Table 2.4).

Turning to a rarely asked form of abuse, the youth were asked if they had witnessed any family members being verbally, physically, or sexually abused by another family member. Half of the girls and more than two fifths of the boys witnessed verbal abuse ($\chi^2 = 7.93, p \leq .01$), half of the girls and almost one third of the boys witnessed physical abuse ($\chi^2 = 15.04, p \leq .001$), and girls (12%) were twice as likely as boys (6%) to report witnessing someone else's sexual abuse ($\chi^2 = 5.23, p \leq .05$).

When controlling for race, age, and sexual identity, four of the gender differences in abuse victimizations became nonsignificant: verbal and physical abuse from family and witnessing verbal and physical abuse (see the appendix). However, physical abuse from family remained borderline significant for gender ($p \leq .054$). In the case of witnessing sexual abuse, none of the independent variables were significantly related. For the remaining variables, race was significant: White youth were more likely than youth of color to report experiencing family verbal abuse, family physical abuse, and witnessing family verbal abuse.

Relationships With Parents and Family History

All of the theoretical perspectives included in this article suggest youths' experiences with their parents/guardians and family are important in examining delinquency risks. It is notable that more than 10% of the sample reported that at least one of their parents had died, about two thirds reported that at least one parent had been incarcerated, and about 7% reported a parent institutionalized in a mental hospital. Although there were no gender differences in these variables (a deceased parent, an incarcerated parent, and a parent in a mental hospital), girls (56%) were significantly more likely than boys (45%) to report desertion or abandonment by a parent ($\chi^2 = 11.33, p \leq .001$). In some ways this may be more troubling to a child than a parent who dies or is incarcerated or institutionalized in a mental hospital, because the deserting parent is choosing to not be with her or his child. It also suggests that parenting daughters is viewed as less important, less of a responsibility, than parenting sons. In support of this interpretation, studies find that sons provide greater marital stability than daughters as fathers may be more involved and invested in their sons' (than daughters') lives (Dahl & Moretti, 2003; Morgan, Lye, & Condran, 1988).

There were no gender differences when the youth were asked whether they were raised by at least one of their parents; almost nine tenths of the youth reported this (see Table 2.5). When asked to report people other than parents who had helped raise them, girls (47%) were more likely than boys (36%) to report that others had participated in raising them. There were no significant differences in girls' (15%) and boys' (19%) reports that they were the first person in their family incarcerated. However, girls (14%) were significantly more likely than boys (9%) to report that they would rather be living in the delinquent institution than living at home ($\chi^2 = 8.68, p \leq .01$). This suggests worse home experiences for delinquent girls relative to delinquent boys, as suggested in previous research (e.g., Moore, 1999), or even that girls could commit offenses as a means to be taken out of their homes.

Two items were asked to assess general relationships with fathers and mothers. There were no gender differences in reports on the youths' general relationships with their fathers, with about one quarter of both boys and girls reporting they did not have a relationship and another one quarter of both boys and girls reporting the relationship with their father was "great." About 13% of both girls and boys reported "poor" relationships with their fathers, and almost 3 in 10 reported their relationships with their fathers were "OK." On the other hand, these delinquent youth reported gendered evaluations of their general relationships with their mothers. Boys and girls did not differ much in the frequency of reporting relationships with their mothers as "poor" or nonexistent, but boys (59%) were more likely than girls (41%) to describe the maternal relationship as "great" ($\chi^2 = 12.55, p \leq .01$).

When multivariate analyses were conducted, the only family variables where previously gendered relationships became nonsignificant were for whether the youth was raised by others and the individual's relationship with her or his mother (see the appendix). In the multivariate analysis, none of the independent variables (gender, race, age, and sexual identity) were significantly related to whether the youth was raised by others. It is notable that lesbian/gay/bisexual youth reported worse relationships with their mothers than straight youth ($p \leq .01$).

Mental Health and Self-Esteem Histories

Table 2.6 presents a summary of the items on mental health and self-esteem variables, areas of frequent gender differences in research on youth. In all four measures of mental health, girls reported a significantly higher likelihood of mental health problems. Girls (54%) were more likely than boys (46%) to report purposefully hurting or harming themselves ($\chi^2 = 19.55, p \leq .001$). When asked to elaborate on how they hurt themselves, girls (43%) were more than twice as likely as boys (19%) to indicate that they had cut or burned their bodies ($\chi^2 = 30.92, p < .001$). Finally, not only were girls (52%) more likely than boys (28.5%) to report thinking about committing suicide (suicide ideation; $\chi^2 = 22.34$, $p \leq .001$) but also girls (46%) were more than twice as likely as boys (19%) to report that they had tried committing suicide ($\chi^2 = 31.19, p \leq .001$).

The measures of self-esteem included 10 statements written with Likert-type responses, 4 of which indicated no gender differences (see Table 2.6). The youth were very likely to agree with positive self-esteem statements about "having good qualities" (93%), "doing things as well as most people" (90%), "feeling OK about myself" (87%), being "a person of worth" (86%), and being "satisfied with myself" (77%). Yet a sizeable majority of the youth also agreed with the negative self-esteem statement "I wish I could have more self-respect" (62%), almost half (45%) agreed that they "felt useless at times," and more than one third (35%) reported "at times, I think I am no good at all." Almost one quarter (23%) agreed with the statement "I do not have much to be proud of" and 15% agreed with the statement "I am a failure."

With the exception of the fairly moderate self-esteem question about feeling "OK about myself," all the gender differences in reported self-esteem had to do with endorsement of the negative self-esteem items. Thus, there were no significant gender differences for any of the statements about high self-esteem, which were highly endorsed by all youth (ranging from 77% to 99% agreement). However, for the one "mediocre" self-esteem item, "I feel OK about myself," boys (89%) were more likely than girls (81.5%) to agree ($\chi^2 = 5.84$, $p \leq .05$), and for all of the low self-esteem statements, the girls were significantly more likely to agree with these than the boys. More specific, girls were more likely than boys to agree with the items "I wish I could have more self-respect" (72% of girls and 56% of boys, $\chi^2 = 11.00, p \leq .001$), "I feel useless at times" (51% of girls and 41% of boys, $\chi^2 = 4.22, p \leq .05$), "at times, I think I'm no good at all" (45% of girls and 29% of boys, $\chi^2 = 11.94, p \leq .001$), and "I am a failure" (21% of girls and 12% of boys, $\chi^2 = 5.93, p \leq .05$).

Regarding the multivariate analyses, when gender, race, age, and sexual identity were controlled, the effect of gender became nonsignificant in two analyses on self-esteem. Although gender was no longer a predictor of "not much to be proud of" and "feel useless at times," sexual identity was related to both, and in both cases lesbian/gay/bisexual youth were more likely to report agreeing with these negative self-esteems ($p \leq .01$ in both cases). In addition, race was related to reports of feeling useless, with White youth more likely to report agreeing with this item ($p \leq .001$).

Table 2.5 Youths' Parent and Family History

Variable	*n*	Girls		Boys		Total		Test Statistic
		Percentage	(*n*)	Percentage	(*n*)	Percentage	(*n*)	
Deserted by a parent	411	55.7	(83)	38.2	(100)	44.5	(183)	11.83***
Parent in a mental hospital	436	7.5	(12)	6.5	(18)	6.9	(30)	0.15
Death of a parent	444	12.3	(20)	10.3	(29)	11.0	(49)	0.40
Raised by parents[a]	437	86.4	(247)	89.8	(247)	88.6	(387)	1.16
Raised by others[b]	437	46.9	(76)	30.2	(83)	36.4	(159)	12.33***
Rather be here than at home	438	14.1	(23)	5.8	(16)	8.9	(39)	8.68**
First person in family incarcerated	425	15.7	(25)	21.8	(58)	19.5	(83)	2.34
Parent incarcerated	425	69.2	(110)	62.4	(166)	64.9	(276)	2.01
Relationship with dad	427							4.89
Don't have one		25.5	(41)	30.1	(80)	28.3	(121)	
Poor		16.1	(26)	11.7	(31)	13.3	(57)	
OK		32.9	(53)	26.7	(71)	29.0	(124)	
Great		25.5	(41)	31.6	(84)	29.3	(125)	
Relationship with mom	429							12.55**
Don't have one		9.4	(15)	7.1	(19)	7.9	(34)	
Poor		6.9	(11)	5.6	(15)	6.1	(26)	
OK		42.5	(68)	28.6	(77)	33.8	(145)	
Great		41.3	(66)	58.7	(158)	52.2	(224)	

a. Youth being raised by parents includes youth being raised by at least one parent or their mother and father living together.

b. This category includes all others that were reported as raising the respondent. The most frequently mentioned others were grandparents, siblings, aunts, uncles, foster parents, and group homes.

$^{**}p \leq .01$. $^{***}p \leq .001$.

Table 2.6 Youths' Mental Health and Self-Esteem

Variable	*n*	Girls		Boys		Total		Test Statistic
		Percentage	(*n*)	Percentage	(*n*)	Percentage	(*n*)	
Mental health								
Purposely harmed self	425	54.1	(86)	32.3	(86)	40.5	(172)	19.55**
Cut or burned self	444	42.9	(70)	18.5	(52)	27.5	(122)	30.92***
Thought about suicide	428	51.9	(82)	28.5	(77)	37.1	(159)	23.34***
Tried suicide	428	46.3	(74)	19.0	(51)	29.1	(125)	31.19***
Self-esteem[a]								
I am a person of worth	435	84.0	(136)	87.2	(238)	86.0	(374)	0.88
I have good qualities	440	90.2	(147)	94.9	(263)	93.2	(410)	3.66
I do things as well as most people	443	88.3	(144)	91.1	(255)	90.1	(399)	0.86
I am satisfied with myself	431	72.2	(117)	79.6	(214)	76.8	(331)	3.05
I am a failure	439	21.0	(34)	12.3	(34)	15.5	(68)	5.93*
I do not have much to be proud of	440	28.4	(46)	20.1	(56)	23.2	(102)	3.91*
I feel OK about myself	441	81.5	(132)	89.6	(250)	86.6	(382)	5.84*
I wish I could have more self-respect	437	71.8	(117)	55.8	(153)	61.8	(270)	11.00***
I feel useless at times	435	51.3	(82)	41.1	(113)	44.8	(195)	4.22*
At times, I think I am no good at all	438	45.4	(74)	29.1	(80)	35.2	(154)	11.94***

a. This is the percentage of the sample who agree or strongly agree with this statement. Respondents could also mark disagree or strongly disagree.

*$p \leq .05$. **$p \leq .01$. ***$p \leq .001$.

School and Peer Experiences

The life-course perspective emphasizes school experiences and delinquency as potentially influencing each other (e.g., Laub & Lauritsen, 1993), and some of the feminist pathways research interviewing incarcerated women reports extremely alienating childhood school experiences as risk factors for offending (e.g., Arnold,

1990; Gilfus, 1992). Other scholars have noted the potential or real significance of peer relationships regarding youths' risks of offending; some in terms of the life course (e.g., Laub & Lauritsen, 1993) and some addressing whether, and perhaps how, peer relationships and their effects on offending might be gendered (e.g., Bottcher, 2001; Heimer & DeCoster, 1999; Messerschmidt, 2000; Nagasawa, Qian, & Wong, 2000). In the current study, there were few gender differences in youths' self-reported school experiences and no differences in their self-reported peer experiences (see Table 2.7). Evaluations of the overall educational experience were quite positive and not gendered. More than half the sample (54%) reported the school experience as "good," one third (34%) as "adequate," and about one eighth (12%) as "poor." There were also no gender differences in the (high) rates in which they reported getting in trouble at school (almost half reported "usually" or "always" getting in trouble) or in their reports of being suspended (24%), expelled (22%), or repeating a grade (66%). Regarding the youths' self-reported peer experiences, there were no gender differences, with 93% of youth reporting friends use alcohol and or drugs, 84% reporting friends are involved in crime, 67% reporting that friends stay out of trouble, and 24% reporting they are currently gang members (see Table 2.7).

The only significant gender differences in the youths' reports of their school and peer experiences are related to whether and why they dropped out of or quit school. Girls (41%) were significantly more likely than boys (31%) to report ever dropping out of or quitting school ($\chi^2 = 4.35$, $p \leq .05$). Girls were more than twice as likely as boys to report they dropped out of or quit school because they "could not keep up" at school (42% of girls and 19% of boys, $\chi^2 = 9.32$, $p \leq .01$) and because they had "left home" (48% of girls and 23% of boys, $\chi^2 = 11.80$, $p \leq .001$). These findings are notable for a number of reasons, including research indicating the devaluing of girls' education and the high rates of girls running away from home because of sexual abuse. In addition, in this study, girls were more than 3 times as likely as boys to report that they dropped out of or quit school because "no one cared if I learned or attended" (12% of girls and 3% of boys, $\chi^2 = 4.20$, $p \leq .05$) and "nobody liked me at school" (12% of girls and 3% of boys, $\chi^2 = 4.20$, $p \leq .05$).

Without accounting for gender differences, the top four reasons youth reported for dropping out of school were "trouble with the law" (67%), "I left home" (33%), "I could not keep up at school" (29%), and "conflict with teachers" (25%). That "trouble with the law" is by far the most frequently given reason by both boys and girls for quitting or dropping out of school is consistent with the life-course perspective, which contends that schools are major institutions in children's lives (see Farrington, 1994), that a child being expelled or skipping school affects delinquency, and that delinquency can affect going to school. Some of the less commonly reported reasons for dropping out of or quitting school, in order of frequency, which were not gendered, were "family moved a lot" (9%), "pregnancy-related reasons" (for boys because their girlfriends were pregnant) (8%), "I had to work to help my family earn money" (7%), "transportation problems" (3%), and "health problems" (2%). It is interesting that there were no significant differences between boys' and girls' rates of listing pregnancy as a reason for dropping out of or quitting school.

Although gender became nonsignificant for two reasons listed for leaving school (no one cared and nobody liked me) when race, age, and sexual identity were controlled, two reasons not gendered in the direct (bivariate) analyses became gendered. More specific, there were no direct gender effects in reporting leaving school because of pregnancy or trouble with the law until race, age, and sexual identity were controlled: In the multivariate analyses, girls were more likely than boys to report leaving school because of a pregnancy and because of trouble with the law ($p \leq .01$ in both cases). Age was also related to both of these reasons, and in both cases older youth were more likely to report leaving because of pregnancies and because of trouble with the law ($p \leq .01$ in both cases). Given that the peer variables were never significantly gendered (as reported in Table 2.7) and that when the peer variables were added to the multivariate logistic regressions they were never gendered, they were not included in the final multivariate analyses presented in the appendix. Perhaps that peer variables were never gendered is consistent with Bottcher's (2001) claim that youths' daily activities are all manifestations of male dominance. For example, Bottcher found "that high-risk families usher boys out of their homes and into crime. Conversely, the most troubled of these families almost force girls out" (p. 922). One could speculate that although girls' and boys' self-reports of peer experiences are not gendered in terms of rates, boys' peer experiences are more likely than girls' peer experiences to result in the type of offending more likely to "land" them in a delinquent institution.

Discussion

This study highlights the need to broaden the identification and inclusion of strains in youths' lives that may be linked to delinquency. The findings are important for not only theory but also prevention of and responses to delinquency. Furthermore, the findings reported herein attest to not only important gender issues but also how gender intersects with race, sexual identity, and sometimes age to explain delinquency risk factors. The most overwhelming patterns reported in these data concern abuse victimization. Although girls reported higher rates of abuse victimization, boys' reported rates were also alarmingly high. Moreover, almost half of the youth believed the abuse was related to their subsequent delinquency, and girls were significantly more likely than boys to report this. Prior to controlling for race, age, and sexual identity, girls reported higher rates of all abuses than boys. The gender differences were particularly strong for the items measuring sexual abuse.

There are four implications of the abuse victimization findings. First, mainstream theories should incorporate what has been learned from feminist pathways theory in assessing the role of abuse events as delinquency risks for both girls and boys. More specifically, the pathways approach to delinquency risk factors appears to provide the most support for determining not only girls' risks but also boys' risks. Unlike the GST and life-course approaches, the pathways approach specifically advances the need to identify childhood traumas as precursors to delinquency (and adult offending). These findings emphasize the need to expand our definitions of childhood traumas (e.g., broadening definitions of abuse victimization, including desertion by a parent, a parent in a mental hospital) not only for girls but for boys as well. In sum, the findings from this study are more consistent with the pathways approach than the GST or life-course approaches as they are currently articulated and most commonly tested.

The second implication from the victimization findings is that improved intervention responses to abused children (girls and boys) are not only the morally appropriate actions but also crucial in deterring delinquency. Sadly, the devaluation of the lives of both juvenile delinquents and nondelinquent abused children from more economically disadvantaged homes appears to have become status quo in the United States. The ramifications of underfunded or nonexistent governmental programs means not only a serious threat to children's well-being but also the likelihood of an increase in delinquency.

In a similar manner, the third implication from the high victimization findings among these delinquent youth is that abuse treatment programs must be high in quality and availability in both girls' and boys' delinquency facilities. Thus, although the second implication reported better intervention for childhood abuses prior to delinquency, we must also attend to better programming for youth already caught up in the system. This provides the most promise for these youth to lead nonoffending lives when leaving the institutions.

Finally, the victimization findings emphasize the importance of accounting for intersecting oppressions and identities when assessing gender differences. In this study, when the combined impacts of gender, race, age, and sexual identity were examined, race appeared to mediate the impact of gender for experiencing verbal abuse, physical abuse, and witnessing verbal abuse in delinquent youths' families: These three abuses were all more common in White than other homes.

The other set of variables, which were most consistently gendered, included those measuring mental health in terms of harming oneself. Girls always reported higher levels than boys of physically harming themselves and contemplating and attempting suicide. Five of the nine items measuring the youths' self-esteem were also gendered, with girls reporting lower self-esteem on all five. However, in the two cases

Table 2.7 Youths' School and Peer Experiences

Variable	*n*	Girls		Boys		Total		Test Statistic
		Percentage	(*n*)	Percentage	(*n*)	Percentage	(*n*)	
Educational experience	442							0.18
Poor		12.9	(21)	12.2	(34)	12.4	(55)	
Adequate		32.5	(53)	34.4	(96)	33.7	(149)	
Good		54.6	(89)	53.4	(149)	53.8	(238)	
Got in trouble at school	444							1.21
Never		9.8	(16)	8.9	(25)	9.2	(41)	
Sometimes		42.3	(69)	45.6	(128)	44.4	(197)	
Usually		22.1	(36)	23.8	(67)	23.2	(103)	
Always		25.8	(42)	21.7	(61)	23.2	(103)	
Ever suspended	389	27.8	(52)	21.8	(52)	24.2	(94)	1.79
Ever expelled	389	25.2	(38)	20.2	(48)	22.1	(86)	1.34
Repeated a grade	429	63.9	(101)	67.2	(182)	66.0	(283)	0.47
Ever dropped out or quit	444	41.1	(67)	31.3	(88)	34.9	(155)	4.35*
Reasons dropped out or quit[a]	155							
Pregnancy-related reason		9.0	(6)	6.8	(6)	7.7	(12)	0.24
Trouble with the law		71.6	(48)	63.6	(56)	67.1	(104)	1.10
Could not keep up at school		41.8	(28)	19.3	(17)	29.0	(45)	9.32**
Family moved a lot		7.5	(5)	10.2	(9)	9.0	(14)	0.35
I left home		47.8	(32)	21.6	(19)	32.9	(51)	11.80***
Conflict with teachers		23.9	(16)	25.0	(22)	24.5	(38)	0.03

Variable	*n*	Girls		Boys		Total		Test Statistic
		Percentage	(*n*)	Percentage	(*n*)	Percentage	(*n*)	
No one cared if I learned or attended		11.9	(8)	3.4	(3)	7.1	(11)	4.20*
Nobody liked me at school		11.9	(8)	3.4	(3)	7.1	(11)	4.20*
I had to work to help my family earn money		6.0	(4)	8.0	(7)	7.1	(11)	0.23
Transportation problems		4.5	(3)	1.1	(1)	2.6	(4)	1.69
Health problems		1.5	(1)	2.3	(2)	1.9	(3)	0.12
Peer variables								
Friends use drugs/alcohol	431	93.7	(149)	93.0	(253)	93.3	(402)	0.08
Friends involved in crime	430	81.1	(129)	85.6	(232)	84.0	(361)	1.49
Friends stay out of trouble	429	62.9	(100)	69.6	(188)	67.1	(288)	2.06
Currently a gang member	427	22.6	(35)	24.6	(67)	23.9	(102)	0.23

a.These analyses were conducted only on the 155 youths who reported dropping out of or quitting school. Respondents were to report as many/all reasons they dropped out of school; thus an individual could report more than one reason and the reasons do not total 100.0%.

*$p \leq .05$. **$p \leq .01$. ***$p \leq .001$.

where gender became nonsignificant when controlling for other demographic characteristics, a lesbian/gay/bisexual identity became the predominant predictor of low self-esteem (feeling useless and having nothing to be proud of). Thus, although the literature on gender-specific programming for delinquent girls frequently addresses low self-esteem as a girl problem, we found that aspects of self-esteem are also prevalent among boys and are particularly acute for lesbian/gay/bisexual youth. Research and programs must address and understand why and how these variables are gendered and related to sexual identity. For example, delinquent girls' self-esteem programs might need to be quite different than those for delinquent boys, and both should include a healthy and nonheterosexist approach to sexuality.

This study is consistent with the GST, life-course, and pathways research concerning findings about nonabuse family history, although the nonabuse family history variables were rarely gendered (see Table 2.5). Of the youth, 7% reported a parent in a mental hospital and 11% reported that at least one of their parents was dead. Almost two thirds reported that at least one of their parents had been incarcerated, whereas only one fifth reported they were the first person in their family incarcerated. The two gender differences regarding families were telling: Girls were far more likely than boys to report abandonment by a parent and twice as likely as boys to report that they would rather be in the juvenile delinquency facility than at home. The former indicates parents valuing sons over daughters. It is notable that two family variables that were gendered before controlling for race, age, and sexual identity were nonsignificant in this multivariate analysis. In one case, being raised by someone other than a parent was not significantly related to any of these control variables. In the other case, the only predictor of delinquent youths' relationships with their mothers was sexual identity: Lesbian/gay/bisexual youth reported significantly worse relationships with their mothers (and inclusion of this variable relegated gender to a nonsignificant predictor).

In summary, this study emphasizes the importance of both broadening our definitions of risk factors for delinquent youth and including girls and boys in these studies to truly assess gender-specific needs. Overall, the findings are supportive of the GST, life-course, pathways/feminist, and cycle of violence perspectives but emphasize the need to broaden what is included under strains, traumas, and significant life events. For GSTs to have relevance, it is clear that abusive experiences, mental health reports, and sexual identity must be included, along with better indicators of school experiences, particularly reasons for dropping out. Although the GST and life-course perspectives broaden the range of the delinquency risk factors identified in more traditional theories, the findings herein are most consistent with the feminist pathways perspective and suggest the importance of including "life events" reported in pathways research on offending women and girls in analyses of boys (and men). The feminist pathways approach not only offers a better understanding of offending girls' and women's risk factors and needs for intervention and treatment but also provides a better understanding for boys' and men's risk factors and intervention needs. The findings also stress the necessity of investigating intersecting oppressions and unique to this study, verify the need to examine sexual identity in future research on delinquency.

Note

1. The survey items used for this article, although strongly influenced by existing qualitative and quantitative studies' findings, were most heavily influenced by the findings from our focus groups in the Phase I study (Belknap, Holsinger, & Dunn, 1997). Thus, we developed the survey items used in this article, with the exception of the self-esteem items that are from Rosenburg (1989).

References

Agnew, R. (1992). *Foundation for a general theory of crime and delinquency.* Criminology, *30,* 47–87.

Arnold, R. A. (1990). Women of color: Processes of victimization and criminalization of Black women. *Social Justice, 17,* 153–166.

Artz, S. (1998). *Sex, power, and the violent school girl.* Toronto, Ontario, Canada: Trifolium Books.

Artz, S., & Riecken, T. (1994). *The survey of student life: In a study of violence among adolescent female students in a suburban school district.* Unpublished report, British Columbia Ministry of Education, Education Research Unit, Canada.

Belknap, J. (2001). *The invisible woman: Gender, crime, and justice* (2nd ed.). Cincinnati, OH: Wadsworth.

Belknap, J., Holsinger, K., & Dunn, M. (1997). Understanding incarcerated girls: The results of a focus group study. *The Prison Journal, 77,* 381–404.

Bottcher, J. (2001). Social practices of gender: How gender relates to delinquency in the everyday lives of high-risk youth. *Criminology, 39,* 893–931.

Broidy, L. M. (2001). A test of general strain theory. *Criminology, 39,* 9–36.

Broidy, L. M., & Agnew, R. (1997). Gender and crime: A general strain theory perspective. *Journal of Research in Crime and Delinquency, 34,* 275–306.

Browne, A., Miller, B., & Maguin, E. (1999). Prevalence and severity of lifetime physical and sexual victimization among incarcerated women. *International Journal of Law and Psychiatry, 22,* 301–322.

Burman, M. J., Batchelor, S. A., & Brown, J. A. (2001). Researching girls and violence. *British Journal of Criminology, 41,* 443–459.

Cernkovich, S., & Giordano, P. (1979). A comparative analysis of male and female delinquency. *Sociological Quarterly, 20,* 131–145.

Chesney-Lind, M. (1989). Girls' crime and woman's place: Toward a feminist model of female delinquency. *Crime & Delinquency, 35,* 5–29.

Chesney-Lind, M., & Rodriguez, N. (1983). Women under lock and key. *The Prison Journal, 3,* 47–65.

Cohen, A. K. (1955). *Delinquent boys: The culture of the gang.* New York: Free Press.

Dahl, G., & Moretti, E. (2003). *The demand for sons: Evidence from divorce, fertility and shotgun marriage.* Cambridge, MA: National Bureau of Economic Research. Retrieved from http://gsbwww.uchicago.edu/ labor/dahl.pdf

Daly, K. (1992). Women's pathway to felony court. *Review of Law and Women's Studies, 2,* 11–52.

Daly, K., & Chesney-Lind, M. (1988). Feminism and criminology. *Justice Quarterly, 5,* 497–535.

D'Augelli, A. R., & Dark, L. J. (1994). Lesbian, gay, and bisexual youths. In L. D. Eron, J. H. Gentry, & P. Schlegal (Eds.), *Reason to hope: A*

Appendix Table Logistic Regression Models Where the Significance of Sex/Gender Changed When Controlling for the Impact of Race, Age, and Sexual Identity

	Sex/Gender		Race		Age		Sexual Identity		Model
Dependent Variable	**Coefficient**	**SE**	**Coefficient**	**SE**	**Coefficient**	**SE**	**Coefficient**	**SE**	**Chi-Square**
Abuse									
Verbal abuse from family	.325	.236	.621**	.214	.016	.078	.489	.351	14.479**
Physical abuse from family	.427	.254	.899***	.231	.048	.083	.614	.393	24.004***
Witness verbal abuse	.377	.228	.792***	.208	–.012	.077	.588	.328	24.257***
Witness sexual abuse	.587	.386	.604	.365	–.004	.154	.524	.464	8.021
Parent history									
Raised by others	–1.902	1.198	–1.869	1.172	–.606	.439	.335	1.199	7.104
Relationship with mom	–.148	.331	.220	.302	–.155	.113	–.835*	.386	9.046
Self-esteem									
Not much to be proud of	–.329	.269	–.256	.247	–.035	.091	–.946**	.334	14.108**
Feel useless at times	–.135	.234	–.842***	.213	–.089	.079	–1.060**	.340	29.525***
(Why left) school									
Pregnancy related	2.014*	.810	–.415	.746	.895**	.294	–1.351	1.136	14.833**
Trouble with law	.581*	.273	–.323	.251	.318**	.097	.441	.342	20.985***
No one cared	1.609	.834	.498	.667	–.190	.237	.512	.746	8.444
Nobody liked me	1.401	.777	2.536*	1.075	.398	.277	.543	.772	15.209**

NOTE: Regarding the independent variables, gender was coded 0 = boys and 1 = girls, age was coded as a ratio variable, race was coded 0 = youth of color and 1 = White, and sexual identity was coded 0 = heterosexual/straight and 1 = gay/lesbian/bisexual. Regarding the dependent variables, except for the self-esteem items, these were measured 0 = no and 1 = yes. The self-esteem items were collapsed into agree and disagree. For the items that are negatively worded (I don't have much to be proud of, I feel useless at times), 0 indicates agreement with the statements or low self-esteem and 1 indicates disagreement or higher self-esteem. For self-esteem questions that are positively worded (I have good qualities, I am satisfied with myself, etc.), 1 = agree, 2 = disagree. Thus, for the self-esteem items, a high score represents higher self-esteem. The independent variables were checked for multicollinearity problems; none existed.

$^{*}p \leq .05$. $^{**}p \leq .01$. $^{***}p \leq .001$.

psychological perspective on violence and youth (pp. 177–196). Washington DC: American Psychological Association.

Dembo, R., Williams, L., Wothke, W., Schmeidler, J., & Brown, C. H. (1992). The role of family factors, physical abuse, and sexual victimization experiences in high-risk youths' alcohol and other drug use and delinquency: A longitudinal model. *Violence and Victims, 7,* 245–266.

Faith, K. (1993). *Unruly women: The politics of confinement and resistance.* Vancouver, British Columbia, Canada: Press Gang.

Farrington, D. P. (1994). Human development and criminal careers. In M. Maguire, R. Morgan, & R. Reiner (Eds.), *The Oxford handbook of criminology* (pp. 511–584). New York: Oxford University Press.

Gaarder, E., & Belknap, J. (2002). Tenuous borders: Girls transferred to adult court. *Criminology, 40*(3), 481–517.

Gilfus, M. E. (1992). From victims to survivors to offenders: Women's routes of entry and immersion into street crime. *Women & Criminal Justice, 4,* 63–90.

Habermas, T., & Bluck, S. (2000). Getting a life: The emergence of the life story in adolescence. *Psychological Bulletin, 126,* 748–769.

Heimer, K., & DeCoster, S. (1999). The gendering of violent delinquency. *Criminology, 37,* 277–318.

Hoffman, J. P., & Su, S. S. (1997). The conditional effects of stress on delinquency and drug use: A strain theory assessment of sex differences. *Journal of Research in Crime and Delinquency, 34,* 46–78.

Holsinger, K. (2000). Feminist perspectives on female offending: Examining real girls' lives. *Women & Criminal Justice, 12*(1), 23–51.

Johnson, C. C., & Johnson, K. A. (2000). High-risk behavior among gay adolescents: Implications for treatment and support. *Adolescence, 35,* 619–637.

Johnson, R. A., Su, S. S., Gerstein, D. R., Shin, H.-C., & Hoffmann, J. P. (1995). Parental influences on deviant behavior in early adolescence: A longitudinal response analysis of age- and ender-differentiated effects. *Journal of Quantitative Criminology, 11,* 167–193.

Laub, J. H., & Lauritsen, J. L. (1993). Violent criminal behavior over the life course: A review of the longitudinal and comparative research. *Violence and Victims, 8,* 235–252.

Laub, J. H., & Sampson, R. J. (1993). Turning points in the life course: Why change matters to the study of crime. *Criminology, 31,* 301–325.

Leonard, E. (1982). *Women, crime and society.* New York: Longman.

Loeber, R. (1996). Developmental continuity, change, and pathways in male juvenile problem behavior. In J. D. Hawkins (Ed.), *Delinquency and crime* (pp. 1–28). New York: Cambridge University Press.

Loeber, R., & Le Blanc, M. (1990). Toward a developmental criminology. In M. Tonry & N. Morris (Eds.), *Crime and justice: A review of the research* (pp. 375–473). Chicago: University of Chicago Press.

Mazerolle, P. (1998). Gender, general strain, and delinquency: An empirical examination. *Justice Quarterly, 15,* 65–91.

McClellan, D. S., Farabee, D., & Crouch, B. M. (1997). Early victimization, drug use and criminality: A comparison of male and female prisoners. *Criminal Justice and Behavior, 24,* 455–476.

Merton, R. K. (1949). *Social theory and social structure.* Glencoe, IL: Free Press.

Messerschmidt, J. W. (2000). *Nine lives: Adolescent masculinities, the body, and violence.* Boulder, CO: Westview.

Miller, D., Trapani, C., Fejes-Mendoza, K., Eggleston, C., & Dwiggins, D. (1995). Adolescent female offenders: Unique considerations. *Adolescence, 30,* 429–435.

Moffitt, T. E. (1990). Juvenile delinquency and attention deficit disorder: Boys' development trajectories from age 3 to age 15. *Child Development, 61,* 893–910.

Moffitt, T. E. (1993). Adolescence-limited and life-course persistent antisocial behavior: A developmental taxonomy. *Psychological Review, 100,* 674–701.

Moore, J. W. (1999). Gang members' families. In M. Chesney-Lind & J. M. Hagedorn (Eds.), *Female gangs in America: Essays on girls, gangs and gender* (pp. 159–176). Chicago: Lakeview Press.

Morgan, S. P., Lye, D. N., & Condran, G. A. (1988). Sons, daughters and the risk of marital disruption. *American Journal of Sociology, 94,* 110–129.

Morrill, C., Yalda, C., Adelman, M., Musheno, M., & Bejarano, C. (2000). Telling tales in school: Youth culture and conflict narratives. *Law & Society Review, 34,* 521–566.

Nagasawa, R., Qian, Z., & Wong, P. (2000). Social control theory as a theory of conformity: The case of Asian/Pacific drug and alcohol nonuse. *Sociological Perspectives, 43,* 581–603.

Nagin, D., Farrington, D. P., & Moffitt, T. E. (1995). Life-course trajectories of different types of offenders. *Criminology, 33,* 111–138.

Office of Juvenile Justice and Delinquency Prevention Act of 1992, Pub. L. No. 102–586, § 8(c)(1), 106 Stat. 5036.

Piquero, A. R., Brame, R., Mazerolle, P., & Haapanen, R. (2003). Crime in emerging adulthood. *Criminology, 40,* 137–169.

Piquero, A. R., MacDonald, J. M., & Parker, K. F. (2002). Race, local life circumstances, and criminal activity. *Social Science Quarterly, 83,* 654–670.

Pollock, J. M. (1999). *Criminal women.* Cincinnati, OH: Anderson.

Rivera, B., & Widom, C. S. (1990). Childhood victimization and violent offending. *Violence and Victims, 5,* 19–35.

Robson, R. (1992). *Lesbian (out)law: Survival under the rule of law.* Ithaca, NY: Firebrand.

Rosenberg, M. (1989). *Society and the adolescent self-image* (Rev. ed.). Middletown, CT: Wesleyan University Press.

Sampson, R. J., & Laub, J. H. (1990). Crime and deviance over the life course: The salience of adult social bonds. *American Sociological Review, 55,* 609–627.

Sampson, R. J., & Laub, J. H. (1992). Crime and deviance in the life course. *Annual Review of Sociology, 18,* 63–84.

Sharp, S. F., Brewster, D. R., & Love, S. R. (2005). Disentangling strain, personal attributes, affective response and deviance: A gendered analysis. *Deviant Behavior, 26,* 122–157.

Sharp, S. F., Terling-Watt, T. L., Atkins, L. A., & Gilliam, J. T. (2001). Purging behavior in a sample of college females: A research note on general strain theory and female deviance. *Deviant Behavior, 22,* 171–188.

Stattin, H., & Magnusson, D. (1991). Stability and change in criminal behavior up to age 30. *The British Journal of Criminology, 31*(4), 327–346.

Widom, C. S. (1989). The cycle of violence. *Science, 244,* 160–166.

DISCUSSION QUESTIONS

1. How do the perspectives of general strain theory, the life course perspective, and feminist pathways and the cycle of violence theory provide an increased understanding of the relationship between gender and crime?
2. What effect does physical, sexual, and emotional abuse have on the offending patterns of delinquent youth?
3. Which factors best explain the causes of delinquency for girls? For boys?

READING

As you learned in the section introduction, feminist criminology not only challenged the male-dominated views of criminality, but it provided a new perspective to understand the offending behaviors of women. Here, Dr. Chesney-Lind argues for the importance of examining the interrelationship of race, gender, and crime in order to effectively address issues of racism and sexism within the political agenda of crime control policies. She provides three examples where feminist criminology has been effective in fighting against the right-wing "tough on crime" philosophy: the use of the media in creating the image of the "bad" girl, the criminalization of victimization, and the negative effects of "equality" in the institutionalization of female offenders.

Patriarchy, Crime, and Justice

Feminist Criminology in an Era of Backlash

Meda Chesney-Lind

A product of the second wave of the women's movement, feminist criminology has been in existence now for more than three decades. Although any starting point is arbitrary, certainly one could point to the publication of key journal issues and books in the 1970s,[1] and it is clear that a signal event was the founding of the Women and Crime Division of the American Society of Criminology in 1982 (Rafter, 2000, p. 9). Since that time, the field has grown exponentially, which makes it increasingly impossible to do justice to all its dimensions in the space of an article. This article, instead, focuses on the challenges facing our important field as we enter a millennium characterized by a deepening backlash against feminism and other progressive movements and perspectives.

Feminist Criminology in the 20th Century: Looking Backward, Looking Forward

The feminist criminology of the 20th century clearly challenged the overall masculinist nature of theories of crime, deviance, and social control by calling attention to the repeated omission and misrepresentation of women in criminological theory and research (Belknap, 2001; Cain, 1990; Daly & Chesney-Lind, 1988). Turning back the clock, one can recall that prior to path-breaking feminist works on sexual assault, sexual harassment, and wife abuse, these forms of gender violence were ignored, minimized, and trivialized. Likewise, girls and women in conflict with the law were overlooked or excluded in mainstream works while demonized, masculinized, and sexualized in the marginalized literature that brooded on their venality. Stunning gender discrimination, such as the failure of most law schools to admit women, the routine exclusion of women from juries, and the practice of giving male and female "offenders" different sentences for the same crimes went largely unchallenged (see Rafter, 2000, for a good overview of the history in each of these areas).

The enormity of girls' and women's victimization meant that the silence on the role of violence in women's lives was the first to attract the attention of feminist activists and scholars. Because of this, excellent work exists on the problem of women's victimization—especially in the areas of sexual assault, sexual harassment, sexual abuse, and wife battery (see, e.g., Buzawa & Buzawa, 1990; Dziech & Weiner, 1984; Estrich, 1987; D. Martin, 1977; Rush, 1980; Russell, 1986; Schechter, 1982; Scully, 1990).

In retrospect, the naming of the types and dimensions of female victimization had a significant impact on public policy, and it is arguably the most tangible accomplishment of both feminist criminology and grassroots feminists concerned about gender, crime, and justice. The impact on the field of criminology and particularly criminological theory was mixed, however, in part because these offenses did not initially seem to challenge andocentric criminology per se. Instead, the concepts of "domestic violence" and "victimology," although pivotal in the development of feminist criminology, also supplied mainstream criminologists and some criminal justice practitioners with a new area in which to publish, "new" crimes to study (and opportunities to secure funding), and new men to jail (particularly men of color and other marginalized men). More recent, the field of domestic violence has even been home to a number of scholars who have argued that women are as violent as men (for critical reviews, see DeKeseredy, Sanders, Schwartz, & Alvi, 1997; DeKeseredy & Schwartz, 1998; S. Miller, 2005). In part because of these trends, the focus on girls' and women's victimization has produced a range of challenges for feminist criminology and for feminist activists that have become even more urgent as we move into the new century.

Compared to the wealth of literature on women's victimization, interest in girls and women who are labeled, tried, and jailed as "delinquent" or "criminal" was slower to fully develop[2] in part because scholars of "criminalized" women and girls had to contend early on with the masculinization (or "emancipation") hypothesis of women's crime, which argues in part that "in the same way that women are demanding equal opportunity in the fields of legitimate endeavor, a similar number of determined women are forcing their way into the world of major crimes" (Adler, 1975, p. 3; see also Simon, 1975). Feminist criminologists, as well as mainstream criminologists, debated the nature of that relationship for the next decade and ultimately concluded it was not correct (Chesney-Lind, 1989; Steffensmeier, 1980; Weis, 1976), but this was a costly

SOURCE: Chesney-Lind, M. (2006). Patriarchy, crime, and justice: Feminist criminology in an era of backlash. *Feminist Criminology, 1*(1), 6–26.

intellectual detour (and also a harbinger of things to come, as it turned out).

The 1980s and 1990s, however, would see breakthrough research on the lives of criminalized girls and women. Rich documentation of girls' participation in gangs, as an example, challenges earlier gang research that focuses almost exclusively on boys (see Chesney-Lind & Hagedorn, 1999; Moore, 1991). Important work on the role of sexual and physical victimization in girls' and women's pathways into women's crime (see Arnold, 1995; Chesney-Lind & Rodriquez, 1983; Chesney-Lind & Shelden, 1992; Gilfus, 1992) began to appear, along with work that suggests unique ways in which gender and race create unique pathways for girl and women offenders into criminal behavior, particularly in communities ravaged by drugs and overincarceration (Bourgois & Dunlap, 1993; Joe, 1995; Maher & Curtis, 1992; Richie, 1996). Needless to say, the focus on girls' and women's gender also highlights the fact that masculinity and crime need to be both theorized and researched (Bowker, 1998; Messerschmidt, 2000).

Instead of the "add women and stir" (Chesney-Lind, 1988) approach to crime theorizing of the past century (which often introduces gender solely as a "variable" if at all), new important work on the gender/crime nexus *theorizes gender*. This means, for example, drawing extensively on sociological notions of "doing gender" (West & Zimmerman, 1987) and examining the role of "gender regimes" (Williams, 2002) in the production of girls' and women's behavior. Contemporary approaches to gender and crime (see Messerschmidt, 2000; J. Miller, 2001) tend to avoid the problems of reductionism and determinism that characterize early discussions of gender and gender relations, stressing instead the complexity, tentativeness, and variability with which individuals, particularly youth, negotiate (and resist) gender identity (see Kelly, 1993; Thorne, 1993). J. Miller and Mullins (2005), in particular, have argued for the crafting of "theories of the middle range" that recognize that although society and social life are patterned on the basis of gender, it is also the case that the *gender order* (Connell, 2002) is "complex and shifting" (J. Miller & Mullins, 2005, p. 7).

Feminist Criminology and the Backlash

Feminist criminology in the 21st century, particularly in the United States, finds itself in a political and social milieu that is heavily affected by the backlash politics of a sophisticated and energized right wing—a context quite different from the field's early years when the initial intellectual agenda of the field evolved. Political backlash eras have long been a fixture of American public life, from reconstruction after the Civil War to the McCarthy era of the 1950s. Most of these have certain common characteristics, including a repression of dissent, imperialistic adventure, a grim record of racism, and "resistance to extending full rights to women" (Hardisty, 2000, p. 10).

The current backlash era, however, uses crime and criminal justice policies as central rather than facilitating elements of political agenda—a pattern clearly of relevance to feminist criminology. The right-wing intent to use the "crime problem" became evident very early. Consider Barry Goldwater's 1964 unsuccessful presidential campaign where he repeatedly used phrases such as *civil disorder* and *violence in the streets* in a "covertly racist campaign" to attack the civil rights movement (Chambliss, 1999, p. 14). Both Richard Nixon and Ronald Reagan refined the approach as the crime problem became a centerpiece of the Republican Party's efforts to wrest electoral control of southern states away from the Democratic Party. Nixon's emphasis on *law and order* and Reagan's *war on drugs* were both built on "white fear of black street crime" (Chambliss, 1999, p. 19). With time, crime would come to be understood as a code word for race in U.S. political life, and it became a staple in the Republican attacks on Democratic rivals. When Reagan's former vice president, George Bush Sr., ran for office, he successfully used the Willie Horton incident (where an African American on a prison furlough raped and murdered a woman) to derail the candidacy of Michael Dukakis in 1988 (Chambliss, 1999).

His son, George W. Bush, would gain the presidency as a direct result of backlash criminal justice policies, because felony disenfranchisement of largely African American voters in Florida was crucial to his political strategy in that state (Lantigua, 2001). In Bush's second election campaign, however, another feature would be added to the Republican mix: an appeal to "moral values." Included in the moral values agenda, designed to appeal to right-wing Christians, is the rolling back of the gains of the women's movement of the past century, including the recriminalization of abortion and the denial of civil rights to gay and lesbian Americans. Bush's nominee to the Supreme Court, John Roberts, has even questioned "whether encouraging homemakers to become lawyers contributes to the public good" (Goldstein, Smith, & Becker, 2005).

The centrality of both crime and gender in the current backlash politics means that feminist criminology is uniquely positioned to challenge right-wing initiatives. To do this effectively, however, the field must put an even greater priority on *theorizing patriarchy and crime*, which means focusing on the ways in which the definition of the crime problem and criminal justice practices support patriarchal practices and worldviews.

To briefly review, patriarchy is a sex/gender system[3] in which men dominate women and what is considered masculine is more highly valued than what is considered feminine. Patriarchy is a system of social stratification, which means that it uses a wide array of social control policies and practices to ratify male power and to keep girls and women subordinate to men (Renzetti & Curran, 1999, p. 3). Often, the systems of control that women experience are explicitly or implicitly focused on controlling female sexuality (such as the sexual double standard; Renzetti & Curran, 1999, p. 3). Not infrequently, patriarchal interests overlap with systems that also reinforce class and race privilege, hence, the unique need for feminist criminology to maintain the focus on intersectionality that characterizes recent research and theorizing on gender and race in particular (see Crenshaw, 1994).

Again, in this era of backlash, the formal system of social control (the law and criminal justice policies) play key roles in eroding the rights of both women and people of color, particularly African Americans but increasingly, other ethnic groups as well. Feminist criminology is, again, uniquely positioned to both document and respond to these efforts. To theorize patriarchy effectively means that we have done cutting-edge research on the interface between patriarchal and criminal justice systems of control and that we are strategic about how to get our findings out to the widest audience possible, issues to which this article now turns.

Race, Gender, and Crime

If feminist criminology is to fully understand the interface between patriarchal control mechanisms and criminal justice practices in the United States, we must center our analysis on the race/gender/punishment nexus. Specifically, America's long and sordid history of racism and its equally disturbing enthusiasm for imprisonment must be understood as intertwined, and both of these have had a dramatic effect on African American women in particular (Bush-Baskette, 1998; Horton & Horton, 2005; Johnson, 2003; Mauer, 1999).

More than a century ago, W. E. B. Du Bois saw the linkage between the criminal justice system and race-based systems of social control very clearly. Commenting on the dismal failure of "reconstruction," he concluded,

> Despite compromise, war, and struggle, the Negro is not free. In well-nigh the whole rural South the black farmers are peons, bound by law and custom to an economic slavery from which the only escape is death or the penitentiary. (as quoted in Johnson, 2003, p. 284)

Although the role of race and penal policy has received increased attention in recent years, virtually all of the public discussion of the issues has focused on African American males (see, e.g., Human Rights Watch, 2000). More recent, the significant impact of mass incarceration on African American and Hispanic women has received the attention it deserves. Current data show that African American women account for "almost half (48 percent)" of all the women we incarcerate (Johnson, 2003, p. 34). Mauer and Huling's (1995) earlier research adds an important perspective here; they noted that the imprisonment of African

American women grew by more than 828% between 1986 and 1991, whereas that of White women grew by 241% and of Black men by 429% (see also Bush-Baskette, 1998; Gilbert, 2001). Something is going on, and it is not just about race or gender; it is about both—a sinister synergy that clearly needs to be carefully documented and challenged.

Feminist criminologist Paula Johnson (2003), among others, advocated a "Black Feminist analysis of the criminal justice system." An examination of Black women's history from slavery through the Civil War and the postwar period certainly justifies a clear focus on the role that the criminal justice system played in the oppression of African American women and the role of prison in that system (Rafter, 1990). And the focus is certainly still relevant because although women sometimes appear to be the unintended victims of the war on drugs, this "war" is so heavily racialized that the result can hardly be viewed as accidental. African American women have always been seen through the "distorted lens of otherness," constructed as "subservient, inept, oversexed and undeserving" (Johnson, 2003, pp. 9-10), in short, just the "sort" of women that belong in jail and prison. Hence, any good work on criminalized women must also examine the ways in which misogyny and racism have long been intertwined themes in the control of women of color (as well as other women) in the United States, as the next section demonstrates.

Media Demonization and the Masculinization of Female Offenders

As noted earlier, the second wave of feminism had, by the 1980s, triggered an array of conservative political, policy, and media responses. In her book *Backlash: The Undeclared War Against American Women*, Susan Faludi (1991), a journalist, was quick to see that the media in particular were central, not peripheral, to the process of discrediting and dismissing feminism and feminist gains. She focused specific attention on mainstream journalism's efforts to locate and publicize those "female trends" of the 1980s that would undermine and indict the feminist agenda. Stories about "the failure to get husbands, get pregnant, or properly bond with their children" were suddenly everywhere, as were the very first stories on "bad girls"; Faludi noted that "NBC, for instance, devoted an entire evening news special to the pseudo trend of 'bad girls' yet ignored the real trends of bad boys: the crime rate among boys was climbing twice as fast as for girls" (p. 80).

Faludi's (1991) recognition of the media fascination with bad girls was prescient. The 1990s would produce a steady stream of media stories about violent and bad girls that continues unabated in the new millennium. Although the focus would shift from the "gansta girl," to the "violent girl," to the "mean girl" (Chesney-Lind & Irwin, 2004), the message is the same: Girls are bad in ways that they never used to be. As an example, the Scelfo (2005) article titled "Bad Girls Go Wild," published in the June 13, 2005, issue of *Newsweek*, describes "the significant rise in violent behavior among girls" as a "burgeoning national crisis" (p. 1).

Media-driven constructions such as these generally rely on commonsense notions that girls are becoming more like boys on both the soccer field and the killing fields.[4] Implicit in what might be called the "masculinization" theory (Chesney-Lind & Eliason, in press; Pollock, 1999) of women's violence is the idea that contemporary theories of violence (and crime more broadly) need not attend to gender but can, again, simply add women and stir. The theory assumes that the same forces that propel men into violence will increasingly produce violence in girls and women once they are freed from the constraints of their gender. The masculinization framework also lays the foundation for simplistic notions of "good" and "bad" femininity, standards that will permit the demonization of some girls and women if they stray from the path of "true" (passive, controlled, and constrained) womanhood.

Ever since the first wave of feminism, there has been no shortage of scholars and political commentators issuing dire warnings that women's demand for equality would result in a dramatic change in the character and frequency of women's crime (Pollak, 1950; Pollock, 1999; Smart, 1976). As noted earlier, in the 1970s, the notion that the women's movement was causing changes in women's crime was the subject of extensive media and scholarly attention (Adler, 1975; Chesney-Lind, 1989; Simon, 1975). Again, although this perspective was definitely refuted by the feminist criminology of the era (see Gora, 1982; Steffensmeier & Steffensmeier, 1980; Weis, 1976), media enthusiasm about the idea that feminism encourages women to become more like men and, hence, their "equals" in crime, remains undiminished (see Chesney-Lind & Eliason, in press).

As examples, *Boston Globe Magazine* proclaimed in an article as well as on the issue's cover, over the words *BAD GIRLS* in huge red letters, that "girls are moving into the world of violence that once belonged to boys" (Ford, 1998). And from *San Jose Mercury News* came a story titled "In a New Twist on Equality, Girls' Crimes Resemble Boys'" that features an opening paragraph arguing that

> juvenile crime experts have spotted a disturbing nationwide pattern of teenage girls becoming more sophisticated and independent criminals. In the past, girls would almost always commit crimes with boys or men. But now, more than ever, they're calling the shots. (Guido, 1998, p. 1B)

In virtually all the stories on this topic (including the Scelfo, 2005, article appearing in *Newsweek*), the issue is framed as follows. A specific and egregious example of female violence is described, usually with considerable, graphic detail about the injury suffered by the victim—a pattern that has been dubbed "forensic journalism" (Websdale & Alvarez, 1997, p. 123). In the *Mercury News* article, for example, the reader hears how a 17-year-old girl, Linna Adams, "lured" the victim into a car where her boyfriend "pointed a .357 magnum revolver at him, and the gun went off. Rodrigues was shot in the cheek, and according to coroner's reports, the bullet exited the back of his head" (Guido, 1998, p. 1B). Websdale and Alvarez (1997) noted that this narrative style, while compelling and even lurid reading, actually gives the reader "more and more information about less and less" and stresses "individualistic explanations that ignore or de-emphasize the importance of wider social structural patterns of disadvantage" (p. 125).

These forensic details are then followed by a quick review of the Federal Bureau of Investigation's arrest statistics showing what appear to be large increases in the number of girls arrested for violent offenses. Finally, there are quotes from "experts," usually police officers, teachers, or other social service workers, but occasionally criminologists, interpreting the narrative in ways consistent with the desired outcome: to stress that girls, particularly African American and Hispanic girls whose pictures often illustrate these stories, are getting more and more like their already demonized male counterparts and, hence, becoming more violent (Chesney-Lind & Irwin, 2005).

There are two problems with this now familiar frame: One, there are considerable reasons to suspect that it is demonstrably wrong (i.e., that girls' violence is not increasing) and two, it has created a "self-fulfilling prophecy" that has had dramatic and racialized effects on girls' arrests, detentions, and referrals to juvenile courts across our country.

Although arrest data consistently show dramatic increases in girls' arrests for "violent" crimes (e.g., arrests of girls for assault climbed an astonishing 40.9%, whereas boys' arrests climbed by only 4.3% in the past decade; Federal Bureau of Investigation, 2004), other data sets, particularly those relying on self-reported delinquency, show no such trend (indeed they show a decline; Chesney-Lind, 2004; Chesney-Lind & Belknap, 2004; Steffensmeier, Schwartz, Zhong, & Ackerman, 2005). It seems increasingly clear that forces other than changes in girls' behavior have caused shifts in girls' arrests (including such forces as zero-tolerance policies in schools and mandatory arrests for domestic violence; Chesney-Lind & Belknap, 2004). There are also indications that although the hype about bad girls seems to encompass all girls, the effects of enforcement policies aimed at reducing "youth violence" weigh heaviest on girls of color whose families lack the resources to challenge such policies (Chesney-Lind & Irwin, 2005).

Take juvenile detention, a focus of three decades of deinstitutionalization efforts. Between 1989 and 1998, girls' detentions increased by 56% compared to a 20%

increase seen in boy's detentions, and the "large increase was tied to the growth in the number of delinquency cases involving females charged with person offenses (157%)" (Harms, 2002, p. 1). At least one study of girls in detention suggests that "nearly half" the girls in detention are African American girls, and Latinas constitute 13%; Caucasian girls, who constitute 65% of the girl population, account for only 35% of those in detention (American Bar Association & National Bar Association, 2001, pp. 20-21).

It is clear that two decades of the media demonization of girls, complete with often racialized images of girls seemingly embracing the violent street culture of their male counterparts (see Chesney-Lind & Irwin, 2004), coupled with increased concerns about youth violence and images of "girls gone wild," have entered the self-fulfilling prophecy stage. It is essential that feminist criminology understand that in a world governed by those who self-consciously manipulate corporate media for their own purposes, newspapers and television may have moved from simply covering the police beat to constructing crime "stories" that serve as a "non conspiratorial source of dominant ideology" (Websdale & Alvarez, 1997, p. 125). Feminist criminology's agenda must consciously challenge these backlash media narratives, as well as engage in "newsmaking criminology" (Barak, 1988), particularly with regard to constructions of girl and women offenders. The question of how to do this is one that must also engage the field. As a start on such a discussion, consider the advice of Bertold Brecht (1966):

> One must have the courage to write the truth when the truth is everywhere opposed; the keenness to recognize it, and although it is everywhere concealed; the skill to manipulate it as a weapon; the judgment to select in whose hands it will be effective, and the cunning to spread the truth among such persons. (p. 133)

The advocacy work coupled with excellent research that one sees with reports issued by The Sentencing Project and the Center for Juvenile and Criminal Justice provide models of how this work might be done. It certainly requires that we work more closely with progressive journalists than many academics are used to, but given the success of these agencies in doing just that, feminist criminologists should consider this a priority and use our national and regional meetings, as a start, to develop strategies toward this end.

Criminalizing Victimization

Many feminist criminologists have approached the issue of mandatory arrest in incidents of domestic assault with considerable ambivalence (see Ferraro, 2001). On one hand, as noted earlier, the criminalization of sexual assault and domestic violence was in one sense a huge symbolic victory for feminist activists and criminologists alike. After centuries of ignoring the private victimizations of women, police and courts were called to account by those who founded rape crisis centers and shelters for battered women and those whose path-breaking research laid the foundation for major policy and legal changes in the area of violence against women (see Schecter, 1982).

On the other hand, the insistence that violence against women be handled as a criminal matter threw victim advocates into an uneasy alliance with police and prosecutors—professions that feminists had long distrusted and with good reason (see Buzawa & Buzawa, 1990; Heidensohn, 1995; S. Martin, 1980). The criminal justice approach, however, was bolstered in the mid-1980s by what appeared to be overwhelming evidence that arrest decreased violence against women (Sherman & Berk, 1984). Although subsequent research would find that arrest was far less effective than originally thought (see Ferraro, 2001; Maxwell, Garner, & Fagan, 2002), for the policy world, the dramatic early research results seemed to ratify the wisdom of a law enforcement–centered approach to the problem of domestic violence. Ultimately, the combined effects of the early scientific evidence; political pressure from the attorney general of the United States, the American Bar Association, and others; and the threat of lawsuits against departments who failed to protect women from batterers "produced nearly unanimous agreement that arrest was the best policy for domestic violence" (Ferraro, 2001, p. 146).

As the academic debate about the effectiveness of arrest in domestic violence situations continued unabated, the policy of "mandatory arrest" became routinized into normal policing and quite quickly, other unanticipated effects began to emerge. When arrests of adult women for assault increased by 30.8% in the past decade (1994 to 2003), whereas male arrests for this offense fell by about 5.8% (Federal Bureau of Investigation, 2004, p. 275), just about everybody from the research community to the general public began to wonder what was happening. Although some, such as criminologist Kenneth Land, quoted in a story titled "Women Making Gains in Dubious Profession: Crime," attributed the increase to "role change over the past decades" that presumably created more females as "motivated offenders" (Anderson, 2003, p. 1), others were not so sure. Even the Bureau of Justice Statistics looked at a similar trend (increasing numbers of women convicted in state courts for "aggravated assault") and suggested the numbers might be "reflecting increased prosecution of women for domestic violence" (Greenfeld & Snell, 1999, pp. 5–6).

Much like the increases seen in girls' assaults, this trend requires critical review, a process that takes the reader through the looking glass to a place that the feminists who worked hard to force the criminal justice system and the general public to take wife battery seriously could never have imagined. In this world, as in California recently, the female share of domestic violence arrests tripled (from 5% in 1987 to 17% in 1999; S. Miller, 2005, p. 21); and as it turned out, it was not just a California phenomenon.

Despite the power of the stereotypical scenario of the violent husband and the victimized wife, the reality of mandatory arrest practices has always been more complicated. Early on, the problem of "mutual" arrests—the practice of arresting both the man and the woman in a domestic violence incident if it is not clear who is the "primary" aggressor—surfaced as a concern (Buzawa & Buzawa, 1990). Nor has the problem gone away, despite efforts to clarify procedures (Bible, 1998; Brown, 1997); indeed, many jurisdictions report similar figures. In Wichita, Kansas, for example, women were 27% of those arrested for domestic violence in 2001 (Wichita Police Department, 2002). Prince William County, Maryland, saw the number of women arrested for domestic violence triple in a 3-year period, with women going from 12.9% of those arrested in 1992 to 21% in 1996 (Smith, 1996). In Sacramento, California, even greater increases were observed; there the number of women arrested for domestic violence rose by 91% between 1991 and 1996, whereas arrests of men fell 7% (Brown, 1997).

A Canadian study (Comack, Chopyk, & Wood, 2000) provides an even closer look at the impact of mandatory arrest on arrests of women for crimes of violence. Examining the gender dynamics in a random sample of 1,002 cases (501 men and 501 women) involving charges filed by the Winnipeg, Manitoba, police services for violent crimes during the period 1991 through 1995, the researchers found that the "zero-tolerance" policy implemented by the police force in 1993 had a dramatic effect on women's arrest patterns. Although the policy resulted in more arrests of both men and women for domestic violence, the impact on women's arrests was most dramatic. In 1991, domestic violence charges represented 23% of all charges of violence against women; by 1995, 58% of all violent crime charges against women were for partner violence (Comack et al., 2000, p. ii). Most significant, the researchers found that in 35% of the domestic violence cases involving women, the accused woman had actually called the police for help (only 5% of male cases showed this pattern; Comack et al., 2000, p. 15).

Susan Miller's (2005) study of mandatory arrest practices in the state of Delaware adds an important dimension to this discussion. Based on data from police ridealongs, interviews with criminal justice practitioners, and observations of groups run for women who were arrested as offenders in domestic violence situations, Miller's study comes to some important conclusions about the effects of mandatory arrest on women.

According to beat officers S. Miller (2005) and her students rode with, in Delaware, they do not have a "pro-arrest policy, we have a pro-paper policy" (p. 100) developed in large part to avoid lawsuits. What initially surfaces as a seemingly minor, and familiar, lament begins to take on far more meaning. It emerges that at least in Delaware (but one suspects elsewhere), police departments, often in response to threatened or real lawsuits, have developed an "expansive definition" (S. Miller, 2005, p. 89) of domestic violence, including a wide range of family disturbances. As a consequence, although the officers "did not believe there was an increase in women's use of violence" (S. Miller, 2005, p. 105), "her fighting back now gets attention too" (S. Miller, 2005, p. 107) because of this sort of broad interpretation of what constitutes domestic violence.

Another significant theme in police comments reflects male batterers' increased skill in deploying the criminal justice system to further intimidate and control their wives. Officers reported that men are now more willing to report violence committed by their wives and more willing to use "cross-filings" in securing protective orders against their wives and girlfriends. Police particularly resent what they regard as "bogus" violations of protective orders that are actually just harassment (S. Miller, 2005, p. 90).

None of the social service providers and criminal justice professionals S. Miller (2005) spoke with felt women had become more aggressively violent; instead, they routinely called the women "victims." They noted that at least in Delaware, as the "legislation aged," the name of the game began to be "get to the phone first" (S. Miller, 2005, p. 127). Social service workers noted that male batterers tended to use their knowledge of the criminal justice system and process as a way to threaten their wives with the loss of the children, particularly if they had managed to get the woman placed on probation for abuse. Workers echoed the police complaints about paperwork, noting it takes 8 hours to do the paperwork if an arrest is made, but then they made a crucial link to the arrest of women, noting that police, weary of being told they were the problem, have channeled at least some of their resentment into making arrests of women who act out violently (regardless of context or injury) because "according to police policy," they have to make an arrest.

Essentially, it appears that many mandatory arrest policies have been interpreted on the ground to make an arrest if any violent "incident" occurs, rather than considering the context within which the incident occurs (Bible, 1998). Like problematic measures of violence that simply count violent events without providing information on the meaning and motivation, this definition of *domestic violence* fails to distinguish between aggressive and instigating violence from self-defensive and retaliatory violence. According to S. Miller (2005) and other critics of this approach, these methods tend to produce results showing "intimate violence is committed by women at an equal or higher rate than by men" (p. 35). Although these findings ignited a firestorm of media attention about the "problem" of "battered men" in the United States (Ferraro, 2001, p. 137), the larger question of how to define *domestic violence* in the context of patriarchy is vital. Specifically, much feminist research of the sort showcased here is needed on routine police and justice practices concerning girls' and women's "violence." In particularly short supply are studies of girls' arrests, particularly those of girls of color (who are often detained for "assault"), although indirect evidence certainly suggests this is happening (see Chesney-Lind & Irwin, 2005). The evidence to date suggests the distinct possibility that in addition to the well-documented race and class problems, with draconian criminal justice approaches to domestic violence (S. Miller, 1989; Richie, 1996), we have a gender issue: Are these policies criminalizing women's (and girls') attempts to protect themselves?

Women's Imprisonment and the Emergence of Vengeful Equity

When the United States embarked on a policy that might well be described as mass incarceration (Mauer & Chesney-Lind, 2002), few considered the impact that this correctional course change would have on women. Yet the number of women in jail and prison continues to soar (outstripping male increases for most of the past decade), completely untethered from women's crime rate, which has not increased by nearly the same amount (Bloom & Chesney-Lind,

2003). The dimensions of this shift are staggering: For most of the 20th century, we imprisoned about 5,000 to 10,000 women. At the turn of the new century, we now have more than 100,000 women doing time in U.S. prisons (Harrison & Beck, 2004, p. 1). Women's incarceration in the United States not only grew during the past century but also increased tenfold; and virtually all of that increase occurred in the final two decades of the century.

The number of women sentenced to jail and prison began to soar at precisely the same time that prison systems in the United States moved into an era that abandoned any pretense of rehabilitation in favor of punishment. As noted earlier, decades of efforts by conservative politicians to fashion a crime policy that would challenge the gains of the civil rights movement as well as other progressive movements in the 1960s and 1970s had, by the 1980s, born fruit (Chambliss, 1999). Exploiting the public fear of crime, particularly crime committed by "the poor, mostly nonwhite, young, male inner-city dwellers" (Irwin, 2005, p. 8), all manner of mean-spirited crime policies were adopted. The end of the past century saw the war on drugs and a host of other "get tough" sentencing policies, all of which fueled mass imprisonment (see Mauer, 1999). The period also saw the development of what Irwin (2005) has called "warehouse prisons," a correctional regime that focuses on a physical plant designed to control (not reform), rigid enforcement of extensive rules, and easy transfer of unruly prisoners to even more draconian settings.

Although feminist legal scholars can and do debate whether equality under the law is necessarily good for women (see Chesney-Lind & Pollock-Byrne, 1995), a careful look at what has happened to women in U.S. prisons might serve as a disturbing case study of how correctional equity is implemented in practice. Such a critical review is particularly vital in an era where decontextualized notions of gender and race "discrimination" are increasingly and successfully deployed against the achievements of both the civil rights and women's movements (Pincus, 2001/2002).

Consider the account of Martha Sierra's experience of childbirth. As she writhed in pain at a Riverside hospital, laboring to push her baby into the world, Sierra faced a challenge not covered in the childbirth books: her wrists were shackled to the bed. Unable to roll on her side or even sit straight up, Sierra managed as best she could. The reward was fleeting . . . she watched as her daughter, hollering and flapping her arms, was taken from the room. (Warren, 2005, p. A1)

As difficult as it was to talk about giving birth while serving time in prison, Sierra was particularly "distressed and puzzled" by her medical treatment: "Did they think I was going to get up and run away?" asked the 28-year-old California prisoner (Warren, 2005, p. A1).

Sierra's story is unfortunately all too familiar to anyone who examines gender themes in modern correctional responses to women inmates. In fact, her experience is less horrific than the case of Michelle T., a former prisoner from Michigan who told Human Rights Watch (1996) that she was accompanied by two male correctional officers into the delivery room:

> According to Michelle T., the officers handcuffed her to the bed while she was in labor and positioned themselves where they could view her genital area while giving birth. She told [Human Rights Watch] they made derogatory comments about her throughout the delivery. (p. 249)

Basically, male prisoners have long used visits to hospitals as opportunities to escape, so correctional regimes have generated extensive security precautions to assure that escapes do not occur, including shackling prisoners to hospital beds (Amnesty International, 1999, p. 63). This is the dark side of the equity or parity model of justice—one which emphasizes treating women offenders as though they were men, particularly when the outcome is punitive, in the name of equal justice—a pattern that could be called vengeful equity.

Vengeful equity could have no better spokesperson than Sheriff Joe Arpaio who, when he defended his controversial chain gang for women in Maricopa County, Arizona, proclaimed himself an "equal opportunity incarcerator" and went on to explain his controversial move by saying,

> If women can fight for their country, and bless them for that, if they can walk a beat, if they can protect the people and arrest violators of the law, then they should have no problem with picking up trash in 120 degrees. (Kim, 1996, p. A1)

Other examples of vengeful equity can be found in the creation of women's boot camps, often modeled on the gender regimes found in military basic training. These regimes, complete with uniforms, shorn hair, humiliation, exhausting physical training, and harsh punishment for even minor misconduct have been traditionally devised to "make men out of boys." As such, feminist researchers who have examined them contended, they "have more to do with the rites of manhood" than the needs of the typical woman in prison (Morash & Rucker, 1990).

Although these examples might be seen as extreme, legal readings by correctional administrators and others that define any attention to legitimate gender differences as "illegal" have clearly produced troubling outcomes. It is obviously misguided to treat women as if they were men with reference to cross-gender supervision, strip searches, and other correctional regimes while ignoring the ways in which women's imprisonment has unique features (such as pregnancy and vulnerability to sexual assault). Recently, this approach has been correctly identified by Human Rights Watch (1996) as a major contributing factor to the sexual abuse of women inmates.

Reviewing the situation of women incarcerated in five states (California, Georgia, Michigan, Illinois, and New York) and the District of Columbia, Human Rights Watch (1996) concluded,

> Our findings indicate that being a woman prisoner in U.S. state prisons can be a terrifying experience. If you are sexually abused, you cannot escape from your abuser. Grievance or investigatory procedures where they exist, are often ineffectual, and correctional employees continue to engage in abuse because they believe they will rarely be held accountable, administratively or criminally. Few people outside the prison walls know what is going on or care if they do know. Fewer still do anything to address the problem. (p. 1)

Human Rights Watch also noted that their investigators were "concerned that states' adherence to U.S. anti-discrimination laws, in the absence of strong safeguards against custodial sexual misconduct, has often come at the fundamental rights of prisoners" (p. 2).

Institutional subcultures in women's prisons, which encourage correctional officers to "cover" for each other, coupled with inadequate protection accorded women who file complaints, make it unlikely that many women prisoners will formally complain about abuse. In addition, the public stereotype of women in prison as bad makes it difficult for a woman inmate to support her case against a correctional officer in court. Finally, what little progress has been made is now threatened by recent legislation that curtails the ability of prisoners and advocates to commence a legal action concerning prison conditions (Stein, 1996, p. 24).

Finally, it appears that women in prison today are also recipients of some of the worst of the more traditional, separate-spheres approach to women offenders (which tends to emphasize gender difference and the need to focus on "saving" women by policing even minor behaviors, particularly sexual behaviors; Rafter, 1990). Correctional officers often count on the fact that women prisoners will complain, not riot, and as a result, often punish women inmates for offenses that would be ignored in male prisons. McClellan (1994) found this pattern quite clearly in her examined disciplinary practices in Texas prisons. Following up two cohorts of prisoners (one male and one female), she found most men in her sample (63.5%) but only a handful of women (17.1%) had no citation or only one citation for a rule violation. McClellan found that women prisoners not only received numerous citations but also were charged with different infractions than men. Most frequent, women were cited for "violating posted rules," whereas males were cited most often for "refusing to work" (McClellan, 1994, p. 77). Women were more likely than men to receive the most severe sanctions.

McClellan (1994) noted that the wardens of the women's prisons in her study stated quite frankly that they demand total compliance with every rule on the books and punish violations through official mechanisms. McClellan concluded by observing that there exists

> two distinct institutional forms of surveillance and control operating at the male and female facilities.... This policy not only imposes extreme constraints on adult women but also costs the people of the State of Texas a great deal of money. (p. 87)

Much good, early feminist criminology focuses on the conditions of girls and women in training schools, jails, and prisons (see Burkhart, 1976; Carlen, 1983; Faith, 1993; Freedman, 1981). Unfortunately, that work is now made much harder by a savvy correctional system that is extremely reluctant to admit researchers, unless the focus of the research is clearly the woman prisoner and not the institution. That said, there is much more need for this sort of criminology in the era of mass punishment, and the work that is being done in this vein (see Bloom, 2003; Owen, 1998) points to the need for much of the same. Huge numbers of imprisoned girls and women are targeted by male-based systems of "risk" and "classification" (Hannah-Moffat & Shaw, 2003) and then subjected to male-based interventions such as "cognitive behaviorism" to address their "criminal" thinking as though they were men (Kendall & Pollack, 2003). Good work has also been done on the overuse of "chemical restraints" with women offenders (Auerhahn & Leonard, 2000; Leonard, 2002). In short, as difficult as it might be to do, in this era of mass imprisonment, feminist criminology needs to find creative ways to continue to engage core issues in girls' and women's carceral control as a central part of our intellectual and activist agenda. As Adrian Howe (1994) put it, "Academics must not let 'theoretical rectitude' deter them from committing themselves as *academics* and *feminists* to campaigns on behalf of women lawbreakers" (p. 214).

Theorizing Patriarchy: Concluding Thoughts

In 1899, Jane Addams was asked to address the American Academy of Political and Social Science. She took the occasion to reflect on the role of the social science of her day:

> As the college changed from teaching theology to teaching secular knowledge the test of its success should have shifted from the power to save men's souls to the power to adjust them in healthful relations to nature and their fellow men. But the college failed to do this, and made the test of its success the mere collecting and dissemination of knowledge, elevating the means unto an end and falling in love with its own achievement. (Addams, 1899, pp. 339–340)

Recall that when Addams worked in Chicago, criminology as a discipline was taking shape at the University of Chicago, whose researchers often relied heavily on contacts made at Hull House while systematically excluding women from its faculty ranks and distancing themselves from the female-dominated field of social work (see Deegan, 1988).

How do we avoid the pitfalls Addams (1899) observed in the male-dominated criminology of her day? This article argues that although feminist criminology has made a clear contribution to what might be described as the criminological project, it is positioned to play an even more central role in the era of political backlash. Certainly, we, as feminist scholars, shoulder many burdens, but perhaps the most daunting is the one articulated by Liz Kelly (as quoted in Heidensohn, 1995): "Feminist research investigates aspects of women's oppression while seeking at the same time to be a part of the struggle against it" (p. 71).

For feminist criminology to remain true to its progressive origins in very difficult times, we must seek ways to blend activism with our scholarship (and senior scholars, in particular, need to make the academy safe for their junior colleagues to do just that by redefining tenure criteria to make this work a part of "scholarship"). We must discuss the many tensions and difficulties with this work, again in an era of backlash when the right is actively patrolling faculty behavior (Horowitz & Collier, 1994), and be honest about the many challenges ahead. We must create venues for feminist criminology, including peer-reviewed journals (such as *Critical Criminology*, *Feminist Criminology*, and *Women & Criminal Justice*) while also writing for broader audiences, particularly practitioners and policy makers (see *Women, Girls & Criminal Justice*). We must engage in continued activism on the part of girls and women who are the victims of crimes and whose very experiences are being trivialized by well-funded backlash research (Hoff Sommers, 1994) while also documenting the problems those same girls and women have when they take their cases to court (see Estrich, 1987; Matoesian, 1993). It means close attention and continued vigilance about the situations of women working in various aspects of the criminal justice system, particularly as the right wing cynically appropriates concepts such as discrimination. Finally, and most important, it means activism on behalf of criminalized girls and women, the least powerful and most marginalized of all those we study.

Again, given the focus of the backlash, this article argues that feminist criminology is uniquely positioned to do important work to challenge the current political backlash. To do so effectively, however, it is vital that in addition to documenting that gender matters in the lives of criminalized women, we engage in exploration of the interface between systems of oppression based on gender, race, and class. This work will allow us to make sense of current crime-control practices, particularly in an era of mass incarceration, so that we can explain the consequences to a society that might well be ready to hear other perspectives on crime control if given them (consider the success of drug courts and some initiatives that encourage alternatives to incarceration; Mauer, 2002). Researching as well as theorizing both patriarchy and gender is crucial to feminist criminology so that we can craft work, as the right wing does so effectively, that speaks to backlash initiatives in smart, media-savvy ways. To do this well means foregrounding the role of race and class in our work on gender and crime, as the work showcased here makes clear. There is simply no other way to make sense of key trends in both the media construction of women offenders and the criminal justice response that increasingly awaits them, particularly once they arrive in prison.

Finally, we must also do work that will document and challenge the policy and research backlash aimed at the hard fought and vitally important feminist and civil rights victories of the past century. To do any less would be unthinkable to those who fought so long to get us where we are today, and so it must be for us.

Notes

1. One might cite the appearance of the classic special issue on women and crime of *Issues in Criminology*, edited by Dorie Klein and June Kress (1973); two important books on the topic of women and crime by Rita Simon (1975) and Freda Adler (1975); and the publication of Del Martin's (1977) *Battered Wives* and Carol Smart's (1976) *Women, Crime and Criminology*.

2. Early but important exceptions to this generalization are Klein and Kress (1973), Smart (1976), Crites (1976), Bowker, Chesney-Lind, and Pollock (1978), Chapman (1980), and Jones (1980). There has also been an encouraging outpouring of more recent work on women offenders in the past decade. See Belknap (2001), Chesney-Lind and Pasko (2004), and DeKeseredy (1999) for reviews of this recent work.

3. Sex-gender systems include the following elements: (a) the social construction of gender categories on the basis of biological sex, (b) a sexual division of labor in which specific tasks are allocated on the basis of sex, and (c) the social regulation of sexuality, in which particular forms of sexual expression are positively and negatively sanctioned (Renzetti & Curran, 1999, p. 3).

4. I owe this analogy to Frank Zimring who, in response to a question from a reporter, quipped, "Women's liberation didn't turn girls into boys—violence is still particularly male. There has been much more diversification of gender roles on the soccer field than the killing field" (Ryan, 2003, p. 2).

References

Addams, J. (1899). A function of the social settlement. *Annals of the Academy of Political and Social Science, 13*, 323–345.

Adler, F. (1975). *Sisters in crime*. New York: McGraw-Hill.

American Bar Association and the National Bar Association. (2001, May 1). *Justice by gender: The lack of appropriate prevention, diversion and treatment alternatives for girls in the justice system.* Retrieved from http://www.abanet.org/crimjust/juvjus/justicebygenderweb.pdf

Amnesty International. (1999). *Not part of my sentence: Violations of the human rights of women in custody*. Washington, DC: Author.

Anderson, C. (2003, October 28). Women making gains in dubious profession: Crime. *Arizona Star*, p. A1.

Arnold, R. (1995). Processes of criminalization: From girlhood to womanhood. In M. B. Zinn & B. T. Dill (Eds.), *Women of color in American society* (pp. 136-146). Philadelphia: Temple University Press.

Auerhahn, K., & Leonard, E. (2000). Docile bodies? Chemical restraints and the female inmate. *The Journal of Criminal Law and Criminology, 90*(2), 599–634.

Barak, G. (1988). Newsmaking criminology: Reflections on the media, intellectuals, and crime. *Justice Quarterly, 5*(4), 565–587.

Belknap, J. (2001). *The invisible woman: Gender, crime and justice* (2nd ed.). Belmont, CA: Wadsworth.

Bible, A. (1998). When battered women are charged with assault. *Double-Time, 6*(1/2), 8–10.

Bloom, B. (Ed.). (2003). *Gendered justice*. Durham, NC: Carolina Academic Press.

Bloom, B., & Chesney-Lind, M. (2003). Women in prison: Vengeful equity. In R. Muraskin (Ed.), *It's a crime: Women and the criminal justice system* (pp. 175–195). New Jersey: Prentice Hall.

Bourgois, P., & Dunlap, E. (1993). Exorcising sex—for crack: An ethnographic perspective from Harlem. In M. Ratner (Ed.), *The crack pipe as pimp* (pp. 97–132). New York: Lexington Books.

Bowker, L. (Ed.). (1998). *Masculinities and violence*. Thousand Oaks, CA: Sage.

Bowker, L., Chesney-Lind, M., & Pollock, J. (1978). *Women, crime, and the criminal justice system*. Lexington, MA: D. C. Heath.

Brecht, B. (1966). *Galileo* (E. Bentley, Ed., C. Laughton, Trans.). New York: Grove.

Brown, M. (1997, December 7). Arrests of women soar in domestic assault cases. *Sacramento Bee*. Retrieved July 31, 2005, from http://www.sacbee.com/static/archive/news/projects/violence/part12.html

Burkhart, K. W. (1976). *Women in prison*. New York: Popular Library.

Bush-Baskette, S. (1998). The war on drugs as a war against Black women. In S. L. Miller (Ed.), *Crime control and women* (pp. 113–129). Thousand Oaks, CA: Sage.

Buzawa, E., & Buzawa, C. G. (1990). *Domestic violence: The criminal justice response*. Newbury Park, CA: Sage.

Cain, M. (1990). Realist philosophy and standpoint epistemologies or feminist criminology as a successor science. In L. Gelsthorpe & A. Morris (Eds.), *Feminist perspectives in criminology* (pp. 124–140). Buckingham, UK: Open University Press.

Carlen, P. (1983). *Women's imprisonment: A study in social control.* London: Routledge.

Chambliss, W. (1999). *Power, politics and crime*. Boulder, CO: Westview.

Chapman, J. R. (1980). *Economic realities and the female offender*. Lexington: Lexington Books.

Chesney-Lind, M. (1988, July-August). Doing feminist criminology. *The Criminologist, 13*, 16–17.

Chesney-Lind, M. (1989). Girls' crime and woman's place: Toward a feminist model of female delinquency. *Crime & Delinquency, 35*(10), 5–29.

Chesney-Lind, M. (2004, August). Girls and violence: Is the gender gap closing? *National Electronic Network on Violence Against Women*. Retrieved from http://www.vawnet.org/DomesticViolence/Research/VAWnetDocs/ARGirlsViolence.php

Chesney-Lind, M., & Belknap, J. (2004). Trends in delinquent girls' aggression and violent behavior: A review of the evidence. In M. Putallaz & P. Bierman (Eds.), *Aggression, antisocial behavior and violence among girls: A development perspective* (pp. 203–222). New York: Guilford.

Chesney-Lind, M., & Eliason, M. (in press). From invisible to incorrigible: The demonization of marginalized women and girls. *Crime, Media, Culture: An International Journal.*

Chesney-Lind, M., & Hagedorn, J. M. (Eds.). (1999). *Female gangs in America: Essays on gender and gangs*. Chicago: Lakeview Press.

Chesney-Lind, M., & Irwin, K. (2004). From badness to meanness: Popular constructions of contemporary girlhood. In A. Harris (Ed.), *All about the girl: Culture, power, and identity* (pp. 45–56). New York: Routledge.

Chesney-Lind, M., & Irwin, K. (2005). Still "the best place to conquer girls": Gender and juvenile justice. In J. Pollock-Byrne & A. Merlo (Eds.), *Women, law, and social control* (pp. 271–291). Boston: Allyn & Bacon.

Chesney-Lind, M., & Pasko, L. (2004). *The female offender*. Thousand Oaks, CA: Sage.

Chesney-Lind, M., & Pollock-Byrne, J. (1995). Women's prisons: Equality with a vengeance. In J. Pollock-Byrne & A. Merlo (Eds.), *Women, law, and social control* (pp. 155–176). Boston: Allyn & Bacon.

Chesney-Lind, M., & Rodriguez, N. (1983). Women under lock and key. *The Prison Journal, 63*, 47–65.

Chesney-Lind, M., & Shelden, R. G. (1992). *Girls, delinquency and juvenile justice*. Belmont, CA: Wadsworth.

Comack, E., Chopyk, V., & Wood, L. (2000). *Mean streets? The social locations, gender dynamics, and patterns of violent crime in Winnipeg.* Ottawa, Ontario: Canadian Centre for Policy Alternatives.

Connell, R. W. (2002). *Gender*. Cambridge, UK: Polity.

Crenshaw, H. (1994). Mapping the margins: Intersectionality, identity politics, and violence against women of color. In M. A. Fineman & R. Mykitiuk (Eds.), *The public nature of private violence* (pp. 93–118). New York: Routledge.

Crites, L. (Ed.). (1976). *The female offender*. Lexington, MA: Lexington Books.

Daly, K., & Chesney-Lind, M. (1988). Feminism and criminology. *Justice Quarterly, 5*(4), 497–538.

Deegan, M. J. (1988). *Jane Addams and the men of the Chicago School, 1892–1918*. New Brunswick, NJ: Transaction Books.

DeKeseredy, W. (1999). *Women, crime, and the Canadian criminal justice system*. Cincinnati, OH: Anderson.

DeKeseredy, W., Sanders, D., Schwartz, M., & Alvi, S. (1997). The meanings and motives for women's use of violence in Canadian college dating relationships. *Sociological Spectrum, 17*, 199–222.

DeKeseredy, W., & Schwartz, M. (1998, February). *Measuring the extent of woman abuse in intimate heterosexual relationships: A critique of the conflict tactics scales*. Retrieved from VAWnet Web site: http://www.vawnet.org/DomesticViolence/Research/VAWnetDocs/AR_ctscrit.php

Dziech, B. W., & Weiner, L. (1984). *The lecherous professor*. Boston: Beacon.

Estrich, S. (1987). *Real rape*. Cambridge, MA: Harvard University Press.

Faith, K. (1993). *Unruly women: The politics of confinement & resistance*. Vancouver, British Columbia, Canada: Press Gang.

Faludi, S. (1991). *Backlash: The undeclared war against American women*. New York: Anchor Doubleday.

Federal Bureau of Investigation. (2004). *Crime in the United States, 2003*. Washington, DC: Government Printing Office.

Ferraro, K. (2001). Women battering: More than a family problem. In C. Renzetti & L. Goodstein (Eds.), *Women, crime and criminal justice* (pp. 135–153). Los Angeles: Roxbury.

Ford, R. (1998, May 24). The razor's edge. *Boston Globe Magazine*, pp. 3, 22–28.

Freedman, E. (1981). *Their sisters' keepers: Women and prison reform, 1830–1930*. Ann Arbor: University of Michigan Press.

Gilbert, E. (2001). Women, race, and criminal justice processing. In C. Renzetti & L. Goodstein (Eds.), *Women, crime and criminal justice* (pp. 222–231). Los Angeles: Roxbury.

Gilfus, M. (1992). From victims to survivors to offenders: Women's routes of entry into street crime. *Women & Criminal Justice, 4*(1), 63–89.

Goldstein, A., Smith, J., & Becker, J. (2005, August 19). Roberts resisted women's rights. *Washington Post*, p. A1.

Gora, J. (1982). *The new female criminal: Empirical reality or social myth*. New York: Praeger.

Greenfeld, A., & Snell, T. (1999). *Women offenders: Bureau of Justice Statistics, special report*. Washington, DC: U.S. Department of Justice.

Guido, M. (1998, June 4). In a new twist on equality, girls' crimes resemble boys'. *San Jose Mercury News*, p. 1B–4B.

Hannah-Moffat, K., & Shaw, M. (2003). The meaning of "risk" in women's prisons: A critique. In B. Bloom (Ed.), *Gendered justice* (pp. 25–44). Durham, NC: Carolina Academic Press.

Hardisty, J. V. (2000). *Mobilizing resentment*. Boston: Beacon.

Harms, P. (2002, January). *Detention in delinquency cases, 1989–1998* (OJJDP Fact Sheet No. 1). Washington, DC: U.S. Department of Justice.

Harrison, P. M., & Beck, A. J. (2004). *Prisoners in 2003*. Washington, DC: U.S. Department of Justice, Bureau of Justice Statistics.

Heidensohn, F. (1995). Feminist perspectives and their impact on criminology and criminal justice in Britain. In N. H. Rafter & F. Heidensohn (Eds.), *International feminist perspectives in criminology* (pp. 63–85). Buckingham, UK: Open University Press.

Hoff Sommers, C. (1994). *Who stole feminism? How women have betrayed women*. New York: Simon & Schuster.

Horowitz, D., & Collier, P. (1994). *The heterodoxy handbook: How to survive the PC campus*. Lanham, MD: National Book Network.

Horton, J. O., & Horton, L. (2005). *Slavery and the making of America*. Oxford, UK: Oxford University Press.

Howe, A. (1994). *Punish and critique: Towards a feminist analysis of penality*. London: Routledge.

Human Rights Watch. (1996). *All too familiar: Sexual abuse of women in U.S. state prisons*. New York: Human Rights Watch.

Human Rights Watch. (2000, May). Punishment and prejudice: Racial disparities in the war on drugs. *Human Rights Watch Reports, 12*(2). Available from http://www.hrw.org/reports/2000/usa/

Irwin, J. (2005). *The warehouse prison*. Los Angeles: Roxbury.

Joe, K. (1995). Ice is strong enough for a man but made for a woman: A social cultural analysis of methamphetamine use among Asian Pacific Americans. *Crime, Law and Social Change, 22*, 269–289.

Johnson, P. (2003). *Inner lives: Voices of African American women in prison*. New York: New York University Press.

Jones, A. (1980). *Women who kill*. New York: Fawcett Columbine.

Kelly, D. M. (1993). *Last chance high: How girls and boys drop in and out of alternative schools*. New Haven, CT: Yale University Press.

Kendall, K., & Pollack, S. (2003). Cognitive behaviorism in women's prisons. In B. Bloom (Ed.), *Gendered justice* (pp. 69–96). Durham, NC: Carolina Academic Press.

Kim, E.-K. (1996, August 26). Sheriff says he'll have chain gangs for women. *Tuscaloosa News*, p. A1.

Klein, D., & Kress, J. (Eds.). (1973). Women, crime and criminology [Special issue]. *Issues in Criminology, 8*(3).

Lantigua, J. (2001, April 30). How the GOP gamed the system in Florida. *The Nation*, pp. 1–8.

Leonard, E. (2002). *Convicted survivors: The imprisonment of battered women*. New York: New York University Press.

Maher, L., & Curtis, R. (1992). Women on the edge: Crack cocaine and the changing contexts of street-level sex work in New York City. *Crime, Law and Social Change, 18*, 221–258.

Martin, D. (1977). *Battered wives*. New York: Pocket Books.

Martin, S. (1980). *Breaking and entering: Police women on patrol*. Berkeley: University of California Press.

Matoesian, G. (1993). *Reproducing rape domination through talk in the courtroom*. Chicago: University of Chicago Press.

Mauer, M. (1999). *Race to incarcerate*. New York: New Press.

Mauer, M. (2002). State sentencing reforms: Is the "get tough" era coming to a close. *Federal Sentencing Reporter, 15*, 50–52.

Mauer, M., & Chesney-Lind, M. (Eds.). (2002). *Invisible punishment: The collateral consequences of mass imprisonment*. New York: New Press.

Mauer, M., & Huling, T. (1995). *Young Black Americans and the criminal justice system: Five years later*. Washington, DC: The Sentencing Project.

Maxwell, C., Garner, J. H., & Fagan, J. A. (2002). The preventive effects of arrest on intimate partner violence: Research, policy and theory. *Criminology & Public Policy, 2*(1), 51–80.

McClellan, D. S. (1994). Disparity in the discipline of male and female inmates in Texas prisons. *Women & Criminal Justice, 5*(20), 71–97.

Messerschmidt, J. W. (2000). *Nine lives: Adolescent masculinities, the body, and violence*. Boulder, CO: Westview.

Miller, J. (2001). *One of the guys: Girls, gangs, and gender*. New York: Oxford University Press.

Miller, J., & Mullins, C. (2005). *Taking stock: The status of feminist theories in criminology*. Unpublished manuscript.

Miller, S. (1989). Unintended side effects of pro-arrest policies and their race and class implications for battered women: A cautionary note. *Criminal Justice Policy Review, 3*, 299–317.

Miller, S. (2005). *Victims as offenders: Women's use of violence in relationships*. New Brunswick, NJ: Rutgers University Press.

Moore, J. (1991). *Going down to the barrio: Homeboys and homegirls in change*. Philadelphia: Temple University Press.

Morash, M., & Rucker, L. (1990). A critical look at the idea of boot camp as a correctional reform. *Crime & Delinquency, 36*(2), 204–222.

Owen, B. (1998). *"In the mix": Struggle and survival in a women's prison*. Albany: State University of New York Press.

Pincus, F. (2001/2002). The social construction of reverse discrimination: The impact of affirmative action on Whites. *Journal of Inter-Group Relations, 38*(4), 33–44.

Pollak, O. (1950). *The criminality of women*. Philadelphia: University of Pennsylvania Press.

Pollock, J. (1999). *Criminal women*. Cincinnati, OH: Anderson.

Rafter, N. H. (1990). *Partial justice: Women, prisons and social control*. New Brunswick, NJ: Transaction Books.

Rafter, N. H. (Ed.). (2000). *Encyclopedia of women and crime*. Phoenix, AZ: Oryx Press.

Renzetti, C., & Curran, D. J. (1999). *Women, men and society*. Boston: Allyn & Bacon.

Richie, B. (1996). *Compelled to crime: The gender entrapment of battered Black women*. New York: Routledge.

Rush, F. (1980). *The best kept secret: Sexual abuse of children*. New York: McGraw-Hill.

Russell, D. (1986). *The secret trauma*. New York: Basic Books.

Ryan, J. (2003, September 5). Girl gang stirs up false gender issue: Data show no surge in female violence. *San Francisco Chronicle*, p. 2.

Scelfo, J. (2005, June 13). Bad girls go wild. *Newsweek*. Retrieved July 31, 2005, from http://www .msnbcnsn.com/id.8101517/site/newsweek/page/2/

Schecter, S. (1982). *Women and male violence: The visions and struggles of the battered women's movement*. Boston: South End.

Scully, D. (1990). *Understanding sexual violence*. Boston: Unwin Hyman.

Sherman, L. W., & Berk, R. A. (1984). The specific deterrent effects of arrest for domestic assault. *American Sociological Review, 49*(1), 261–272.

Simon, R. (1975). *Women and crime*. Lexington, MA: Lexington Books.

Smart, C. (1976). *Women, crime and criminology: A feminist critique*. London: Routledge Kegan Paul.

Smith, L. (1996, November 18). Increasingly, abuse shows a female side: More women accused of domestic violence. *Washington Post*, p. B1.

Steffensmeier, D. J. (1980). Sex differences in patterns of adult crime, 1965–1977. *Social Forces, 58*, 1080–1108.

Steffensmeier, D. J., Schwartz, J., Zhong, H., & Ackerman, J. (2005).An assessment of recent trends in girls' violence using diverse longitudinal sources. *Criminology, 43*, 355–406.

Steffensmeier, D. J., & Steffensmeier, R. H. (1980). Trends in female delinquency: An examination of arrest, juvenile court, self-report, and field data. *Criminology, 18*, 62–85.

Stein, B. (1996, July). Life in prison: Sexual abuse. *The Progressive*, 23–24.

Thorne, B. (1993). *Gender play*. New Brunswick, NJ: Rutgers University Press.

Warren, J. (2005, June 19). Rethinking treatment of female prisoners. *Los Angeles Times*, p. A1.

Websdale, N., & Alvarez, A. (1997). Forensic journalism as patriarchal ideology: The newspaper construction of homicide-suicide. In D. Hale & F. Bailey (Eds.), *Popular culture, crime and justice* (pp. 123–141). Belmont, CA: Wadsworth.

Weis, J. G. (1976). "Liberation and crime": The invention of the new female criminal. *Crime and Social Justice, 6*, 17–27.

West, C., & Zimmerman, D. H. (1987). Doing gender. *Gender & Society, 1*, 125–151.

Wichita Police Department. (2002). *Domestic violence statistics: 2001*. Retrieved from http://wichitapolice .com/DV/DV_statistics.htm

Williams, L. S. (2002). Trying on gender, gender regimes, and the process of becoming women. *Gender & Society, 16*, 29–52.

DISCUSSION QUESTIONS

1. How have attempts of "vengeful equity" failed to meet the needs of women offenders?
2. How has the experience of patriarchy influenced the criminal justice system?
3. How has feminist criminology shed light on the race/gender/punishment nexus? Provide examples from the reading to support your answer.

READING

In the section introduction, you learned that many of the fears of crime are gendered. Women experience different levels of fear about crime than men do. This reading highlights how gender affects how someone experiences fear about being victimized on a college campus for certain types of crimes, but that no differences exist between males and females for other types of crimes. In addition, the authors expose what factors lead to increased levels of fear for both men and women.

College Students' Crime-Related Fears on Campus

Are Fear-Provoking Cues Gendered?

Bonnie S. Fisher and David May

The notion that gender plays a central role in determining crime-related fear levels is so tightly woven into thinking about fear that it is by and large no longer subject to question. Decades of empirical scrutiny by sociologists, victimologists, psychologists, planners, and geographers have established that there are gender-based differences in fear levels across crime types and in certain types of environments, such as public places (Day, 1994; Fisher & Sloan, 2003; Klodawsky & Lundy, 1994; Lane & Meeker, 2003; May, 2001a; Nasar & Fisher, 1993; Reid & Konrad, 2004).

Despite these widely accepted gendered findings, much of the crime-related research has focused almost exclusively on why women are fearful (Madriz, 1997; Pain, 1997; Starkweather, 2007). A small, but growing, number of researchers have turned their attention to men as victims of fear and why they are fearful (Brownlow, 2005; Day, Stump, & Carreon, 2003). Only recently a few comparative research pieces have been published that identify and explain which factors, if any, differentiate crime-related fear between women and men (Lane & Meeker, 2003; May, 2001b; May & Dunaway, 2000; May, Vartanian, & Virgo, 2002; Reid & Konrad, 2004; Schafer, Huebner, & Bynum, 2006; Smith & Tortenson, 1997; Wallace & May, 2005).

Avenues of research grounded in both environmental and spatial cognition and psychological theories have been used to identify what cues in the immediate environment provoke fear of crime and constrain behavior (Kitchin, 1996; Pain, 2001; Valentine, 1990). Researchers have shown that stimuli that generate fear among individuals (hereafter referred to as "fear-provoking cues") vary from specific physical features of the built environment to the presence of others to the visibility of police officers whose duty is to provide surveillance and protection (Cordner, 1986; Fisher &

SOURCE: Fisher, B. S., & May, D. (2009). College students' crime-related fears on campus: Are fear-provoking cues gendered? *Journal of Contemporary Criminal Justice, 25*(3), 300–321.

Nasar, 1992; Herzog & Chernik, 2000; Herzog & Miller, 1998; Loewen, Steel, & Suedfeld, 1993; Nasar & Fisher, 1993; Warr, 1990, 2000; Winkel, 1986). Much evidence supportive of the positive relationship between individuals' assessment of these cues and their crime-related fears has amassed.

Researchers have offered a plausible explanation as to why these types of cues elicit an emotional fear response. They reason that persons infer from certain cues in the environment that this could be a situation in which possible impending physical danger or harm to oneself or property awaits them (see Fisher & Nasar, 1992; Herzog & Chernik, 2000; Nasar & Fisher, 1993). Fear-provoking cues speak to the amygdale—the brain's emotion region—not the neocortex—the logic and reasoning region (Begley, 2007).

Despite researchers' efforts and explanations, gaps in understanding the relationship between fear-provoking cues and subsequent fear of crime are evident. The current research takes several important steps in a long overdue effort to close the gaps about what is known about which, if any, fear-provoking cues differ across gender and which ones, if any, influence males' and females' fear of specific types of crime. We employ survey data from a large sample of undergraduate students at a 4-year public university to address three research questions about the possible gendered nature of cues and crime-related fear that have not been previously addressed. First, are fear-provoking cues gendered? Simply stated, do males and females evaluate cues known to provoke fear of crime the same way or differently? Second, which of these cues, if any, predict males' and females' fear of specific types of crimes? That is, are fear-provoking cues offense specific? For example, do certain fear-provoking cues predict fear of theft but not fear of violent crime? Third, do these cues equally predict specific crime-related fear across males and females? Simply put, do fear-provoking cues equally predict the same types of fear for males and females alike? Addressing these questions is among the first attempts to enhance our understanding about whether the relationship between fear-provoking cues and crime-related fears is a gendered one among college students while on campus.

Cognitive Mapping and Fear-Provoking Cues

From an evolutionary perspective, individuals use a cognitive map as an efficient mechanism for managing spatial and temporal information about the physical and social nature of their environment to guide behavioral decisions (Kitchin, 1996). Cognitive maps are important to individuals' safety; they protect people from harm. Environmental psychologists have suggested that these maps give individuals "a selective advantage in a difficult and dangerous world that is necessary for survival" (Kitchin, 1994, p. 2). Sighted individuals scan their immediate environment for cues of danger, physical threat, or harm that would make themselves, others, or their property vulnerable to attack. Merry (1981, pp. 11–12) described this process when she explained that

> [C]ues are structured into spatial, temporal, and personal cognitive maps that define places, times, and categories of persons who are likely to be safe or dangerous. The decision that a situation is (is not) dangerous depends on the intersection of these maps. To understand fear of crime, it is much less useful to ask how afraid an individual feels than it is to explore the content of his or her cognitive maps and the frequency with which he or she encounters situations these maps define as dangerous.

Ultimately, these cognitive maps shape an individual's sense of potential criminal victimization, and it is from these maps that individuals draw inferences about their fear levels. Van der Wurff, Van Staalduinen, and Stringer (1989, p. 145) noted that as individuals venture into a specific place, they immediately heed the "criminalizability" of that space.

Given the gendered focus of this article, the nexus between cognitive maps and crime-related fear gives rise to an issue directed at possible differences between males and females in their perceptions of fear-provoking cues. At the core of this issue are three quite simple, yet unanswered questions: (a) Do males and females differ with respect to their assessment of fear-provoking cues? (b) Do the same or different fear-provoking cues predict fear of different types of crimes, namely property and violent ones? (c) Are these fear-provoking cues the same for males and females? The answers to these questions, however, are not so obvious.

Fear-Provoking Cues

There is ample evidence cutting across a variety of academic disciplines that supports a significant association between specific features of the immediate physical environment and crime-related fear (Brownlow, 2005; Fisher & Nasar, 1992; Merry, 1981; Nasar & Fisher, 1993; Warr, 1990, 2000). However, the findings from fear-provoking cues research suggests that there is not one cue that influences fear but rather a constellation of cues that include specific features of the physical environment to the presence of others to the visibility of police officers whose duty is to provide surveillance and protection.

Are the Cues That Predict Crime-Related Fear Gendered?

As we have discussed, cognitive maps are helpful in understanding how sighted individuals assess their fear of crime level. The growing body of research on fear-provoking cues also provides insights into what cues influence individuals' fear of crime. Bringing these bodies of research together raises issues as to whether the cues that predict crime-related fear are gendered. For example, does poor lighting influence fear of crime for both males and females, or does poor lighting only influence females' fear of crime? Does police presence predict males and females being fearful or only predict males being fearful?

It is somewhat surprising that researchers have largely neglected the integration of these bodies of research to examine if the relationship between cues and crime-related fear are similar or different across males and females. In part, the lack of attention may lie in the fact that researchers' focus has primarily highlighted females' experiences of crime-related fear (Madriz, 1997; Pain, 1997; Starkweather, 2007; for exception see Brownlow, 2005; Day et al., 2003; Lane & Meeker, 2003; May, 2001b; May & Dunaway, 2000).

Among the very few published studies to offer some guidance into addressing our gendered-based questions about the fear-provoking cues relationship is Brownlow's (2005) study of fear among young men and women in Philadelphia's Cobb Creek Park. From the focus group discussions with these youth and their rating of slides from the park, he concluded that "clear differences distinguish how the young men and women of the study negotiated their fears in public spaces" (p. 589). He reported that unlike their female counterparts, males do not judge an environment safe based on the presence or absence of environmental cues. Brownlow found that males judge an environment based on their sense and perceptions of their negotiation of an environment, namely whether they see themselves as being able to flee a risky or uncertain situation. Males consider their youth, physical strength, and speed to be a key in managing dangerous situations. These results suggest that environmental cues to crime-related fear differ across sexes. His conclusions provide starting points to unpacking the gendered nature, if it exists, of cues predicting students' crime-related fear while on campus.

Lighting and Gender

Previous research has reported that lighting affects sighted individuals' ability to see a potentially dangerous environment. Research has also shown that lighting is a significant correlate and predictor of fear of crime, in part, because poor lighting does not offer adequate illumination to observe environmental cues to danger such as being physically attacked or having property stolen. Poor lighting in certain areas, such as parking garages that have perceptional tendency to be isolated, may have more pronounced effects on predicting fear than poor lighting in more public spaces such as sidewalks. Regardless of the exact place of the lighting, poor lighting on campus might have different effects on whether males and females are fearful. In line with Brownlow's conclusions, we would expect that poor lighting might influence whether females are fearful but not whether male counterparts are fearful. Research has shown that most females are physically and sexually vulnerable to attack and are physically challenged to thwart off such an attack. Most males, however, would have physical strength and ability to thwart off such an attack but, in line with Brownlow's work, if they cannot see how to escape when confronted this situation might make them fearful. So, poor lighting might equally influence both sexes' fearfulness.

Foliage and Gender

Researchers have shown that foliage influences individuals' fear of crime because it provides refuge or hiding places for a predator who can surprise attack a victim or even walk from inside or behind the greenery. On one hand, foliage such as overgrown shrubbery might have a positive effect on fear for women because of their sexual and physical vulnerability and physical ability to thwart an attack. Overgrown foliage on campus might not influence males' fear because of their physical confidence to thwart attack but could also present an element of confrontation that might heighten their fear. Hence, the effect of foliage on crime-related fear may be the same for males and females.

Youth Loitering and Gender

The presence of others, especially youth, congregating or loitering has been shown to heighten fear of victimization. Researchers attribute the elevated fear in these types of situations to individuals' perceiving a breakdown in social control, suggesting that if confronted by these youth, the infraction would go unchallenged by others. Much research has shown that this cue results in a lack of a sense of social control in that others may not effectively respond to the situation at hand (Skogan, 1990; Warr, 2000). For males, this might be especially so in light of research that shows that for many men public places and situations that challenged their gender identity, in particular their masculinism, generated fear (Day et al., 2003). Supportive of Day et al.'s (2003) results are those reported by Brownlow (2005), who reported that males felt less safety and security in situations where they lack the ability to flee a risky or uncertain situation. Groups loitering around campus may well predict males' fear but not females' fear. Another plausible speculation is that both sexes might sense a lack of social control in this situation in that others may not effectively respond to the situation at hand or that they would be unable to escape attack since they are outnumbered. The fear that confrontation would increase risk of being victimized might loom equally for both sexes.

Police Visibility and Gender

The relationship between police visibility and fear of crime for the sexes is less clear. The relationship appears to be contingent on the type of activities the police are engaged in during the time they are visible. As such, it is quite possible that the impact of police visibility on fear of crime will vary by gender as well. Increased visibility of police might reduce fear of crime among women because of their vulnerabilities discussed above yet may have no effect on male fear because they lack those same vulnerabilities.

Drawing from the research examining cognitive mapping, fear-provoking cues, and fear of crime it is plausible that fear-provoking cues have different effects on whether or not males and females are fearful of being victimized. But it may be equally likely that there are no cues that significantly predict whether males and females are fearful, and hence there are no fear-provoking cue differences predicting fear across the sexes. Because we are not certain whether cues that predict crime-related fear are gendered or not, we turn to our empirical analyses to explore this overlooked relationship.

Method

Data Collection

In March 2008, we asked for and received a list of all the general education courses offered during the current spring term on campus at a large public institution in the south. We randomly selected 25 of those courses and e-mailed each professor who was listed as the instructor for the course, requesting permission to administer the survey at one of his or her class meetings.

At the mutually agreed on time, a research team member visited the classroom and read a protocol that (a) described the process through which their course was selected, (b) asked the students for their cooperation, (c) ensured them that their responses were voluntary and anonymous, (d) asked for their assistance with the data collection effort, and (e) advised them that if they had already completed the survey, to inform the research team member. The surveys were then distributed to the students, who took approximately 10 minutes to complete them.

Across the 24 participating courses, there were 904 students enrolled on February 17, 2008, the day the sample was randomly selected. None of the classes that we visited had the same number of students in attendance as were enrolled for that course; as such, data from students not attending were not obtained. In addition, a small number of students (approximately 2% of those contacted) were registered for more than one of the 24 classes we visited and thus completed the survey only in the first course we visited. Furthermore, eight students either declined to participate or submitted a blank survey at the end of the data collection period. Finally, one student indicated that they were a graduate student; this person was subsequently deleted prior to data analysis to insure that only undergraduate students had participated in the study. Our final sample consisted of 607 students, resulting in a response rate of 67.1%.

Dependent Variables

The dependent variable is fear of criminal victimization while on campus. We included a number of questions that asked the student about being afraid of being a victim of different types of crime while on campus. Students were asked to indicate their level of agreement on a 4-point Likert-type scale with the following statements:

> While on campus at (name of school):
>
> I am afraid of being attacked by someone with a weapon.
>
> I am afraid of having my money or possessions taken from me.
>
> I am afraid of being beaten up.
>
> I am afraid of being sexually assaulted.

For the purpose of this study, the first item will serve as an indicator of fear of aggravated assault, followed by fear of larceny-theft, fear of simple assault, and fear of sexual assault.

In the original instrument, students were asked to indicate the relative strength with which they agreed with the above statements (e.g., *strongly agree*, *somewhat agree*, *somewhat disagree*, *strongly disagree*). With the exception of one variable (fear of larceny-theft, where 7% strongly agreed), less than 5% of the respondents strongly agreed that they were fearful of that situation. As such, we created a dichotomous measure of each of the four types of fear, with those who strongly agreed or agreed that they were fearful coded as 1 and those who strongly disagreed or disagreed coded as 0.

Independent Variables

The survey data allowed us a unique opportunity to examine the relationship between different fear-provoking cues and types of fear of victimization because included in the survey were five cue-specific fear of crime measures. Students were asked to indicate their level of agreement on a 4-point Likert-type scale with the following[2]:

> Since the beginning of this school year, I have been fearful of crime victimization on campus because of . . .
>
> poorly lit parking lots
>
> poorly lit sidewalks and common areas
>
> overgrown or excess shrubbery
>
> groups congregating or loitering
>
> visibility of public safety officials

In the original question, students were asked to indicate the relative strength with which they agreed with the above statements (e.g., *strongly agree*, *somewhat agree*, *somewhat disagree*, *strongly disagree*). Due to the skewed nature of the distribution of each variable, we created a dichotomous variable from each of the five cue-specific fear of crime measures, with those who strongly agreed or agreed that they were fearful coded as 1 and those who strongly disagreed or disagreed were coded as 0.

Each of these cue-specific variables measures a different factor that past research has found to be associated with high levels of fear of crime. The two poor lighting variables measure students' ability to see if a threatening or dangerous situation is in view (e.g., to observe if predator is close). The overgrown or excess shrubbery variable captures the notion of possible hiding places for would-be offenders. Groups congregating or loitering is an indication of some level of social incivility that could create an impression about the concentration of possible motivated offenders. Visibility of public safety officials is a measure of police presence that provides formal guardianship.

Control Variables

Two sets of control variables were used in our analyses. First, due to their association with fear of criminal victimization reported in the past research, we also included measures of age, student's current academic classification (freshman, sophomore, junior, and senior), course load status (full- or part-time student), and residence status (on- or off-campus). Summary statistics for the control variables are presented in Table 2.8.

The results presented in Table 2.8 show that over half (55.9%) of the sample were females. Almost two in three were freshmen or sophomores (66.1%), freshmen being the largest academic group across all categories (37.1%). Most were full-time students (96.4%) and were between the ages of 18 and 24 (86.9%), with the mean age of the sample being nearly 21 (20.88) years of age. Approximately half of the respondents lived on-campus (52.6%).

Given that the emphasis of this research is on gender differences in fear of criminal victimization, we examined how the distribution on the aforementioned variables varied by sex. These results, presented in Table 2.8, demonstrate that the distribution of student classification ($p = .002$) and residence status ($p = .078$) were significantly different between females and males.

Second, given previous research, the importance of perceived risk of victimization as a significant predictor of fear of crime cannot be overlooked in any analysis. In light of the consistent positive effect of perceived risk on fear of crime, we included perceived risk as a control variable. Perceived risk of victimization was defined as the chance that a specific type of crime would happen to the student while on campus during the coming year. Students were asked to rate their perceived risk of specific types of crime on a 10-point scale from 1 meaning that *it is not at all likely to happen* to 10 meaning *it is very likely to happen.*

Perceived risk of four specific types of crimes was used as control variables: larceny-theft and aggravated, simple, and sexual assault. Each specific type of risk was used as a control variable for the specific type of fear. For example, perceived risk of larceny-theft was used as a control variable only for predicting fear of larceny-theft.

Much of the past research has shown that females, in particular college women, have higher perceived risk of different types of victimization than males (Fisher & Sloan, 2003; see May, 2001a, for review; for exception see Lane and Meeker, 2003). As shown in Table 2.8, our student sample follows this previously reported college student risk pattern reported by Fisher and Sloan (2003): females reported being more at risk of victimization than males. In other words, females' perceived risk mean for each type of crime was significantly higher than the respective males' mean.

Results

Are Fear of Criminal Victimization-Provoking Cues Gendered?

The first step in examining whether fear-provoking cues are gendered was to explore the proportion of females and males who reported that a specific cue provoked their fear of victimization while on campus. As presented in Table 2.9, at first glance it appears that

Table 2.8 Sample Characteristics ($N = 607$)

		Sex		
	Total Sample	Females	Males	
Characteristic	% (*n*)	% (*n*)	% (*n*)	*p* Value
Sex				
Female	55.9 (335)			
Male	44.1 (264)			
Current academic classification				
Freshman	37.1 (221)	43.8 (145)	28.8 (76)	.002
Sophomore	29.0 (173)	26.3 (87)	32.6 (86)	
Junior	20.1 (120)	16.3 (54)	24.6 (65)	
Senior	13.8 (82)	13.6 (45)	14.0 (37)	
Type of student				
Traditional[a]	87.1 (520)	88.9 (296)	84.8 (223)	.145
Nontraditional/exchange	12.9 (77)	11.1 (37)	15.2 (40)	
Current course load				
Full time	96.4 (563)	96.0 (313)	96.9 (249)	.145
Part time	3.6 (21)	4.0 (13)	3.1 (8)	
Residence status				
On campus[b]	52.6 (314)	56.0 (187)	48.5 (127)	.078
Off campus	47.4 (283)	44.0 (147)	51.5 (135)	
	M (SD)	**M (SD)**	**M (SD)**	
Age in years	20.88 (4.53)	20.75 (4.90)	21.06 (3.98)	.418
Perceived risk				
Larceny-theft	4.58 (2.76)	4.86 (2.84)	4.22 (2.58)	.005
Aggravated assault	2.52 (1.90)	2.81 (1.98)	2.13 (1.67)	.000
Simple assault	2.86 (2.03)	3.29 (2.12)	2.26 (1.67)	.000
Sexual assault	2.44 (2.10)	3.13 (2.24)	1.55 (1.48)	.000

a. Traditional students are those who are between the ages of 18 and 24 years old. Nontraditional students are those 25 years and older. Less than 1% (0.5%, $n = 3$) of the sample are an exchange student.

b. On campus includes on-campus dormitories (50.9%, $n = 304$) and on-campus apartments and family housing (1.7%, $n = 10$).

fear-provoking cues might be gendered. There are statistically significant differences in the proportion of female students who agreed that specific cues provoked fear of crime victimization while on campus when compared to their male counterparts. For example, 65% of females reported that poorly lit parking lots provoked their on-campus fear of victimization compared to 34% of males, a 30 percentage point difference. About a third of females (32%) reported that overgrown or excessive shrubbery provoked their fear, whereas 19% of males reported feeling fear, a 13 percentage point difference.

These results, however, may be a bit misleading since research has consistently shown that females in general are more fearful than males, and our sample also shows this pattern as female students are more fearful than males for each of our four crime-related fears (larceny-theft, aggravated assault, simple assault, and sexual assault).

Another way to examine these fear-provoking cues results is to look at the ordering of the cues between females and males by rank ordering females' and males' proportion from largest proportion agreeing that the cue provoked them to be fearful of victimization while on campus to smallest proportion who agreed. From this lens, the rank ordering can be seen as an indicator of the relative magnitude of the order of fear-provoking cues between females and males from largest to smallest percentage agreeing. As can be seen in Table 2.9, the Spearman's rank order correlation of their ranking is quite strong, but it is not statistically significant ($p = .19$). There appears to be no significant difference between females and males in the rank ordering of the fear-provoking cues, suggesting that these cues do not vary by gender in their ranking and, therefore, might not be gendered.

Are the Cues That Predict Crime-Related Fear Gendered?

The second step of our analyses examined which cues, if any, predict crime-related fear for females and males and which cues, if any, are different across female and male students. Findings from Table 2.10 indicate that different fear-provoking cues are evident for females and males. Across fear of larceny-theft, aggravated assault, simple assault, and sexual assault, the visibility of public safety officials increased females' fearfulness. Overgrown or excessive shrubbery also increased women's fearfulness of larceny-theft and aggravated and sexual assault. Poor lighting on sidewalks and common areas also increased their fear of larceny-theft and aggravated assault. Groups loitering only increased females' fear of simple assault. Poorly lit parking lots did not predict fearfulness of any type of crime for females. For males, only two cues were significant predictors of fearfulness. Overgrown or excessive shrubbery increased their fear of aggravated assault. Groups congregating or loitering increased their fearfulness of larceny-theft.

Turning to possible gendered effects of cues on crime-related fear, the results from the equality of coefficients test indicate that none of the cues had significantly different effects across females and males. None of the fear-provoking cues had a stronger effect for either sex compared to the other, thus suggesting that fear-provoking cues are not gendered.

Discussion

Among the major goals of this exploratory study was to begin to close the gaps about what is known about fear-provoking cues among females and males and to examine if these cues were gendered. To these ends, the results reported are among the first steps to unpacking the relationship among different fear-provoking cues and crime-related offense specific fear among females and males and provide informative findings for future research.

The wide range in proportions (19%–65%) of both females and males who indicated that specific cues provoked them to be fearful of criminal victimization while on campus gives credence to the past research findings that individuals see and distinguish cues in their immediate environment as fear generating. Interestingly, despite the relative difference in these proportions between females and males, there was not a statistically

Table 2.9 Type of Fear-Provoking Cue by Sex of Respondent

Type of Cue	Proportion Agreeing Cue Provoked Them to Be Fearful of Victimization While on Campus		Proportions Test	Rank Order	
	Females	Males			
	% (*n*)	% (*n*)	z Score (p Value)	Females	Males
Poorly lit parking lots	64.5 (213)	34.0 (88)	7.37 (.000)	1	2
Poorly lit sidewalks and common areas	62.1 (205)	30.4 (79)	7.66 (.000)	2	3
Groups congregating or loitering	53.0 (175)	37.2 (97)	3.84 (.0001)	3	1
Visibility of public safety officials	35.0 (114)	22.7 (58)	3.24 (.0001)	4	4
Overgrown or excessive shrubbery	32.1 (105)	18.9 (49)	3.63 (.0003)	5	5

Spearman's rank order correlation = .70, $p = .19$.

significant relationship between the rank orders of these proportions, suggesting that fear-provoking cues, at least in at the bivariate level, are not gendered. In addition, when considering the multivariate results, there were no significant differences across gender in the impact of the cues on either fear of larceny-theft, aggravated assault, or simple assault. As such, it appears that the fear-provoking cues under study are not gendered as Brownlow's research suggests.

There are a number of plausible explanations why our results suggest that fear-provoking cues are not gendered. With the exception of the limited number of lighting and foliage studies, most of the fear-provoking cues research has not been done within a university setting. The unique nature of the university setting, especially being relatively safe and secure, may contribute to the lack of associations found in our study. It might be that the unique setting of a campus, relatively open to all yet populated with young studious adults, faculty, and staff on a daily basis, might influence the type of persons who loiter around the grounds. Many college students may find comfort (and therefore be less fearful) in seeing members of the college community congregating on campus.[4]

Another explanation for the lack of "gendered findings" revolves around the fact that the questions used to measure fear-specific cues were not as detailed as they could have been. As discussed earlier, these questions did not incorporate the element of time of day, which may have reduced the impact of the fear-specific cue on the fear of the respondent. For example, it is possible that groups congregating or loitering and visibility of public safety officials at night might have different impact on fear of crime among females (or, conversely, males) than these cues during the day. Future research should carefully word these measures to distinguish between daytime and nighttime cues to further explore this effect.

Despite the fact that female students are more fearful than their male counterparts for each offense-specific fear, the relative safety and security of the university setting may also reduce the impact of gender on these relationships, as neither male or female students were generally fearful on campus. With the exception of fear of larceny-theft, where two in five respondents (40.0%) agreed that they were fearful of victimization on campus, the levels of fear among these respondents were relatively low (18.4% agreed that they were at least somewhat fearful of aggravated assault and only 12.2%

Table 2.10 Fear of Type of Crime Logit Models Results

Independent Variable: Specific Fear-Provoking Cue	Larceny-Theft			Aggravated Assault			Simple Assault			Sexual Assault
	Females	Males	Equality of Coefficient Test z Score	Females	Males	Equality of Coefficient Test z Score	Females	Males	Equality of Coefficient Test z Score	Females
	b (SE)	b (SE)		b (SE)	b (SE)		b (SE)	b (SE)		b (SE)
Poorly lit parking lots	.12 (.37)	.13 (.53)	−.02	.20 (.44)	.71 (1.01)	−.46	.83 (.56)	.47 (.91)	.34	.49 (.42)
Poorly lit sidewalks and common areas	.86 (.36)**	.47 (.56)	.59	1.03 (.44)***	1.28 (1.02)	−.23	−.05 (.48)	1.32 (.93)	−1.31	.58 (.15)
Overgrown or excessive shrubbery	.64 (.33)**	.32 (.44)	.58	.61 (.35)*	1.86 (.70)***	−1.60	.58 (.39)	.55 (.63)	.04	.70 (.35)**
Groups congregating or loitering	.43 (.29)	.99 (.40)***	−1.13	.27 (.33)	.99 (.83)	−.81	.84 (.39)**	.84 (.70)	.00	.42 (.33)
Visibility of public safety officials	.54 (.31)*	.06 (.47)	.85	.59 (.33)*	.51 (.75)	.10	.61 (.37)*	−.15 (.82)	.84	.95 (.34)***
Constant	−3.14 (1.09)***	−1.71 (1.50)		−4.80 (1.21)****	−4.22 (2.90)		−5.0 (1.38)****	−3.73 (2.07)*		−4.33 (1.17)***
Model chi-square (df)	88.62 (10)	51.75 (10)		76.52 (10)	48.29 (10)		38.11 (10)	34.30 (10)		129.67 (10)
Significance	.000	.000		.000	.000		.000	.000		.000

NOTE: The respective perceived risk, age and current residence status, academic classification, and course load were used as control variables.

$^{*}p < .1$, $^{**}p < .05$, $^{***}p < .01$, $^{****}p < .001$.

agreed that they were fearful of simple assault). In light of these findings, future research should attempt to replicate and build from our current study in non-university settings, such as residential communities or even computer-generated settings that vary characteristics by known fear-generating cues, to determine if the relative safety and security of the university setting masks any impact that fear-provoking cues might have on fear of criminal victimization. Equally important to future research is examining whether this relationship is gendered. The past fear-provoking research provides ample evidence to suggest that there is quite a strong association between fear-provoking cues and fear of crime but the question about this relationship conditioned on gender remains ripe for inquiry.

Despite the lack of gender differences in the association between fear-provoking cues and offense-specific fear, the reported results inform the research community about gender differences in fear of crime on a number of dimensions. First, the multivariate results offer some support for the "shadow of powerlessness" that has been used to explain differences in fear of crime among adolescent males (May, 2001b).

As May (2001b) suggests, males who feel that they have less power in a situation are likely to be more fearful of that situation. Both Day et al.'s (2003) and Brownlow's (2005) research are supportive of May's shadow of powerlessness suggestion. Their research jointly suggests that males are fearful in environments in which they experience a loss of control because, for some males, their masculine gender identity (e.g., aggression, physical strength) is challenged. For males, there were only two significant associations between fear-provoking cues and fear of crime found in the current study. For males, fear of crime because of groups congregating or loitering had a statistically significant association with fear of larceny-theft and fear of crime because of overgrown or excessive shrubbery had a statistically significant association with fear of aggravated assault. In both of these situations, this relationship might be explained by the challenges to their gender identity some males feel in these types of environments.

Males may feel that their odds of resisting larceny-theft are reduced in a group setting where they are surrounded by a number of young adult males and females (the demographic most likely to loiter and congregate on a university campus); as such, the powerlessness they feel to overcome these odds may be responsible for the significant association between fear caused by groups loitering and fear of larceny-theft. These feelings of powerlessness may also explain male fear of aggravated assault in this sample as well. Although males may think that they can evade a person who wants to commit aggravated assault against them in a poorly lit parking lot or sidewalk or when public safety officials are not present, they may think they are less likely to be able to evade an assailant who confronts them in an area with overgrown shrubbery. As such, those males most fearful because of the overgrown shrubbery cue are significantly more fearful of aggravated assault than their male counterparts who are not as fearful because of that cue. This evidence of the impact of the shadow of powerlessness related to gender identity suggests that this line of thinking is a potentially rich area of exploration for the continued research into the possible gendered relationship between fear-provoking cues and offense-specific fear.

A second interesting gender-specific finding concerns the relationship between the visibility of police and crime-related fear for females. Females (but not males) who were most fearful of crime because of the visibility of public safety officials were significantly more likely than their counterparts to be fearful of every crime under consideration. Given that over 90% of both male and female respondents felt that the university public safety officials were either somewhat or very visible, this finding would appear to indicate that the visibility of police increases fear of crime for females but not for males. Nevertheless, analysis of a follow-up question reveals that this may not be the case. For males, one in three (38.1%) respondents agreed that they would "feel safer if public safety officials were more visible than they currently are"; two in three (67.9%) females agreed with that statement. As such, the presence of police may be more relevant for decreasing fear of crime among females than males. Given that this finding has not been uncovered in any study of which we are aware, this provides another particularly rich area of research that could inform the study of fear of crime.

In much the same way that the shadow of powerlessness may partially explain fear of victimization among the males in this sample, there is some evidence to suggest that the shadow of sexual assault (see May, 2001a, for review; Fisher & Sloan, 2003) may partially explain fear of victimization among the female students in this sample as well. Females who were most fearful because of overgrown or excessive shrubbery and visibility of public safety officials (but none of the other specific fear-provoking cues) were significantly more likely to be fearful of sexual assault than their counterparts were. This finding would suggest that certain cues, in this case, overgrown shrubbery and low police visibility, are relevant to increasing women's fear of sexual assault. This result can also be seen through the lens of several studies that have found that women are primarily fearful of being sexually assaulted, especially in public places at night because they are afraid of being attacked by a stranger (see Fisher & Sloan, 2003; Merry, 1981; Pain, 2001; Valentine, 1990). Again, this line of thinking provides another rich area of exploration in the area of gendered fear-provoking cues and fear of crime, in particular fear of sexual assault.

Although we have uncovered a number of interesting findings, this study is not without limitations. First, and most importantly, future researchers should develop measures of fear-provoking cues that incorporate richer descriptions of a specific cue. For example, although our survey question asked students about fear of groups congregating or loitering, the question was not specific about the demographic or nonstudent status composition of the group, the location, the activity of the group who was loitering, or the time of day the group was loitering. Anecdotal evidence suggests that those groups loitering on the campus under study here were mostly male college students loitering outside of one or more dormitories on campus who routinely verbally harass other students (particularly female students). The one measure included in the survey used to collect group loitering information did not allow us to fully examine these relationships which Warr's (2000) work suggests influences fear of crime. In addition, as alluded to above, the measure of visibility of public safety officials could be improved by following the example of Salmi et al. (2004) reviewed earlier but including even more types of police visibility (e.g., foot patrol, bicycle patrol, face-to-face interaction) to better unpack the police presence and crime-related fear relationship, especially to see if this relationship is gendered. The day-night distinction with respect to fear-provoking cues and fear of crime is also another measurement issue that was not fully addressed in the current research. It could well be that certain fear-provoking cues, for example poor lighting, only influence certain offense specific fears during the night time but not during daylight. We could not address such issues in our work but leave this issue to future researchers to address.

Whether fear-provoking cues are gendered is clearly an issue deserving more scholarly attention. The current study is an important first exploration to informing an agenda for future researchers to examine the possible gendered nature of fear-provoking cues.

Like we have done in the current research, we would encourage future researchers to draw from the variety of disciplines that has examined different aspects of crime-related fear and integrate their theoretical approaches and findings to more fully comprehend which, if any, fear-provoking cues are gendered and their effects on offense-specific fears. Hopefully, in the next decade, a better understanding of the possible gendered relationship between fear-provoking cues and crime-related fear will mature and provide practical means to address fear-provoking cues and thereby reduce crime-related fears among both females and males.

References

Begley, S. (2007, December 24). The roots of fear. *Newsweek*, pp. 37–40.

Brownlow, A. (2005). A geography of men's fear. *Geoforum, 36*, 581–592.

Cordner, G. (1986). Fear of crime and the police—An evaluation of a fear-reduction strategy. *Journal of Police Science and Administration, 14,* 223–233.

Day, K. (1994). Conceptualizing women's fear of sexual assault on campus. *Environment and Behavior, 26,* 742–767.

Day, K., Stump, C., & Carreon, D. (2003). Confrontation and loss of control: Masculinity and men's fear of public spaces. *Journal of Environmental Psychology, 23,* 311–322.

Ferguson, K., & Mindel, C. H. (2007). Modeling fear of crime in Dallas neighborhoods: A test of social capital theory. *Crime & Delinquency, 53,* 322–349.

Fisher, B. S., & Nasar, J. L. (1992). Fear of crime in relation to three exterior site features: Prospect, refuge, and escape. *Environment and Behavior, 24,* 35–65.

Fisher, B. S., & Sloan, J. J. (2003). Unraveling the fear of sexual victimization among college women: Is the "shadow of sexual assault" hypothesis supported? *Justice Quarterly, 20,* 633–659.

Herzog, T. R., & Chernick, K. K. (2000). Tranquility and danger in urban and natural settings. *Journal of Environmental Psychology, 20,* 28–39.

Herzog, T. R., & Miller, E. J. (1998). The role of mystery in perceived danger and environmental preference. *Environment and Behavior, 30,* 429–449.

Holmberg, L. (2002). Personalized policing: Results from a series of experiments with proximity policing in Denmark. *Policing: An International Journal of Police Strategies and Management, 25,* 32–47.

Kirk, K. L. (1988). Factors affecting perceptions of safety in a campus environment. *Environmental Design Research Association, 3,* 215–221.

Kitchin, R. M. (1994). Cognitive maps: What are they and why study them? *Journal of Environmental Psychology, 14,* 1–19.

Kitchin, R. M. (1996). Are there sex differences in geographic knowledge and understanding? *Geographical Journal, 162,* 273–286.

Klodawsky, F., & Lundy, C. (1994). Women's safety in the university environment. *Journal of Architectural and Planning Research, 11,* 128–331.

Kuo, F. E., Bacaicao, M., & Sullivan, W. (1998). Transforming inner-city landscapes: Trees, sense of safety, and preference. *Environment and Behavior, 30,* 28–59.

Lane, J., & Meeker, J. W. (2003). Women's and men's fear of gang crimes: Sexual and nonsexual assault as perceptually contemporaneous offenses. *Justice Quarterly, 20,* 337–371.

Loewen, L. J., Steel, G. D., & Suedfeld, P. (1993). Perceived safety from crime in the urban environment. *Journal of Environmental Psychology, 13,* 323–331.

Madriz, E. (1997). *Nothing bad happens to good girls.* Berkeley: University of California Press.

May, D. C. (2001a). *Adolescent fear of crime, perceptions of risk, and defensive behaviors: An alternate explanation of violent delinquency.* Lewiston, NY: Edwin Mellen Press.

May, D. C. (2001b). The effect of fear of sexual victimization on adolescent fear of crime. *Sociological Spectrum, 21,* 141–174.

May, D. C., & Dunaway, R. G. (2000). Predictors of adolescent fear of crime. *Sociological Spectrum, 20,* 149–168.

May, D. C., Vartanian, L. R., & Virgo, K. (2002). The impact of parental attachment and supervision on fear of crime among adolescent males. *Adolescence, 37,* 267–287.

Merry, S. E. (1981). *Urban danger: Life in a neighborhood of strangers.* Philadelphia: Temple University Press.

Nasar, J. L., & Fisher, B. S. (1993). 'Hot spots' of fear and crime: A multi-method investigation. *Journal of Environmental Psychology, 13,* 187–206.

Pain, R. (1997). Social geographies of women's fear of crime. *Transactions of the Institute of British Geographies, New Series, 22,* 231–244.

Pain, R. (2001). Gender, race, age and fear in the city. *Urban Studies, 38,* 899–913.

Reid, L. W., & Konrad, M. (2004). The gender gap in fear: Assessing the interactive effects of gender and perceived risk on fear of crime. *Sociological Spectrum, 24,* 399–425.

Salmi, S., Gronroos, M., & Keskinen, E. (2004). The role of police visibility in fear of crime in Finland. *Policing: An International Journal of Police Strategies and Management, 2,* 573–591.

Schafer, J. A., Huebner, B. M., & Bynum, T. S. (2006). Fear of crime and criminal victimization gender-based contrasts. *Journal of Criminal Justice, 34,* 285–301.

Shaffer, G. S., & Anderson, L. M. (1983). Perceptions of the security and attractiveness of urban parking lots. *Journal of Environmental Psychology, 5,* 311–323.

Skogan, W. G. (1990). *Disorder and decline: Crime and the spiral of decay in American neighborhoods.* New York: Free Press.

Skogan, W. G., & Hartnett, S. (1997). *Community policing, Chicago style.* New York: Oxford University Press.

Smith, W. R., & Torstenson, M. (1997). Gender differences in risk perception and neutralizing fear of crime. *British Journal of Criminology, 37,* 608–634.

Stamps, A. E., III. (2005). Enclosure and safety in urbanscapes. *Environment and Behavior, 37,* 102–133.

Starkweather, S. (2007). Gender, perceptions of safety and strategic responses among Ohio university students. *Gender, Place, and Culture, 14,* 355–370.

Valentine, G. (1990). Women's fear and the design of public space. *Built Environment, 16,* 279–287.

Van der Wurff, A., Van Staalduinen, L., & Stringer, P. (1989). Fear of crime in residential environments: Testing a social psychological model. *Journal of Social Psychology, 129,* 141–160.

Wallace, L. H., & May, D. C. (2005). The impact of relationship with parents and commitment to school on adolescent fear of crime at school. *Adolescence, 40,* 458–474.

Warr, M. (1990). Dangerous situations: Social context and fear of victimization. *Social Forces, 68,* 891–907.

Warr, M. (2000). Fear of crime in the United States: Avenues for research and policy. In D. Duffee (Ed.), *Measurement and analysis of crime and justice: Crime justice* (Vol. 4, pp. 451–490). Washington, DC: Department of Justice.

Winkel, F. W. (1986). Reducing fear of crime through police visibility: A field experiment. *Criminal Justice Policy Review, 1,* 381–398.

DISCUSSION QUESTIONS

1. How is the fear of victimization similar for men and women? How is it different?
2. Is a fear of crime related to specific offenses? Does this fear vary by gender?
3. How can university administrators use the findings of this study to increase safety on college campuses?

Women and Victimization

Rape and Sexual Assault

Section Highlights

- History of sexual victimization of women
- Contemporary paradigms for sexual victimization
- Categories of sexual assault
- Criminal justice treatment and processing of female sexual assault victims
- Policy implications for female sexual assault victims

When is sex a criminal act? Consider the following scenario. An adult male takes a teenage girl to his friend's home as part of a modeling photography session. She is given alcohol and drugs. Despite her pleas for him to return her to her family home, he proceeds to engage in oral, vaginal, and anal intercourse against her will. Under most circumstances, most would argue that these circumstances equate to the crimes of rape and sexual assault. However, this was no ordinary case. The year was 1977 and the adult male was famed Hollywood director Roman Polanski. He was 44, while his victim was only 13 years old. Initially charged with multiple crimes, including rape by the use of drugs and lewd and lascivious acts upon a child under 14, Polanski pled guilty to a lesser charge as part of a deal with prosecutors. Prior to his sentencing, Polanski fled the United States and remained a fugitive for 32 years. During this time, he lived as a free man in France, though he never returned to the United States out of fear that he would face the criminal sentencing that he had evaded. Throughout the years, most seemed to forget that he had pled guilty to the crime of unlawful sexual intercourse and instead focused on his artistic accolades, which included an Academy Award for Best Director in 2003. In 2009, he was arrested in

Switzerland and faced extradition back to California to be sentenced to a punishment that he had long avoided. At the time of his arrest, Hollywood debated whether or not Polanski should face punishment for the crimes he committed so many years ago. Would the same be said for someone not of his notoriety?

Historical Perspectives

Rape is one of the oldest crimes in society. It has existed in every historical and contemporary society around the world. Yet, we are still attempting to understand and respond to this crime. Images of rape have appeared in historical works of art, in literature, and throughout contemporary popular culture. Laws prohibiting the act of rape have existed for almost four thousand years. One of the first laws on rape as a crime can be found in the Code of Hammurabi from Babylon in 1900 CE. Ancient Greek, Roman, and Judaic societies also criminalized the act of rape under various circumstances. Some laws distinguished between the rape of a married versus an unmarried woman, and the punishments for these crimes varied based on the status of the victim. Others viewed rape not as a violent sexual offense, but as a property crime (Burgess-Jackson, 1999). Laws during these times did little to protect women. The acknowledgment of a rape was an admission of sexual activity. In many cases of forcible sexual assault, women were often blamed for tempting offenders into immoral behaviors. During court hearings, a woman's sexual history was put on display in an attempt to discredit her in front of a jury. The courts did not request similar information about a man's sexual history, as it would be considered prejudicial in the eyes of the jury (Odem, 1995). Punishment in early society ranged from fines to the execution of the offender. In addition, early laws provided the first distinctions of marital rape as exempt from criminal prosecution, as well as class-based definitions of crime and punishment (Burgess-Jackson, 1999).

Until the 20th century, early American statutes on rape limited the definition to a narrow view of sexual assault. Consider the following definition of rape that was included in the Model Penal Code in 1955:

Section 213.1: Rape and Related Offenses

1. Rape. A male who has sexual intercourse with a female not his wife is guilty of rape if:
   a. he compels her to submit by force or by threat of imminent death, serious bodily injury, extreme pain or kidnapping, to be inflicted on anyone; or
   b. he has substantially impaired her power to appraise or control her conduct by administering or employing without her knowledge drugs, intoxicants or other means for the purpose of preventing resistance; or
   c. the female is unconscious; or
   d. the female is less than 10 years old.

What is wrong with this definition? First, it reduces the definition of rape to the act of intercourse. Second, it limits the victim-offender relationship to a man and a woman. Third, it implies that force, or the threat of force, must be used in order for an act to qualify as rape and focuses on violence and brutality as proof of the crime. Fourth, it suggests that it is not a crime for a husband to rape his wife. Finally, it fails to acknowledge attempted rapes as a crime and the traumatic effects of these "near misses" of victimization. However, we do see two positive signs from this early definition that have affected modern-day laws on rape. First, the Model Penal Code acknowledges that the impairment of drugs or alcohol, as well as a state of unconsciousness, prohibits a victim from being able to consent to sexual activity. Second, it acknowledges that rape of a child is a crime, even if it limits this definition to the rape of female children. In reality, the issues of rape expand far beyond what this early definition provided.

While contemporary definitions of rape vary from state to state, many include similar provisions. Today, most laws broadly define sexual victimization as sexual behaviors that are unwanted and harmful to a person. Most emphasize the force or coercion that is displayed by the offender rather than focusing on the response of the victim. This is not to say that the actions of the victim are not debated by defense counsel or by members of the jury, but the law itself does not require victims to demonstrate their levels of resistance. Another development in contemporary rape laws involves the abolishment of the marital-rape exemption clause, as every state now has laws on the books that identify rape within the context of marriage as a criminal act. In an effort to resolve some of the limitations with the word "rape," the term "sexual assault" is used to identify forms of sexual victimization that are not included under the definition of rape. These laws expanded the definitions of sexual assault beyond penile-vaginal penetration and include sodomy, forced oral copulation, and unwanted touching of a sexual nature. Cases of child sexual assault are treated differently in many jurisdictions, and age of consent laws have led to the development of statutory rape laws. Finally, sex offender registration laws such as Megan's Law and Jessica's Law require the community notification of sexual offenders and place residential, community, and supervision restrictions on offenders.

Defining Sexual Victimization

Many women who experience acts that are consistent with a legal definition of rape or sexual assault may not label their experience as such. These unacknowledged victims do not see themselves as victims and therefore do not report these crimes to the police, nor do they seek out services. Research indicates that in many of these cases, women who experience these acts do not define themselves as victims because their experience differs from their personal definitions of what rape and sexual assault look like. For many of these women, rape and sexual assault involves a stranger who attacks them in their home or on a dark sidewalk at night. These incidents involve high levels of violence by people unknown to the victim.

The lack of an understanding of a definition of rape and sexual assault affects offenders, as well. Many do not define their own actions as rape or sexual assault. One of the most frequently cited studies on rape and sexual assault surveyed 2,971 college men regarding whether they had engaged in conduct that met the definitions of rape, attempted rape, sexual coercion, and unwanted sexual contact. The results indicated that 1,525 acts of sexual assault had occurred, including 187 acts of rape. Of those whose acts met the legal definition of rape, 84% believed that their acts did not constitute rape (Warshaw, 1994).

Prevalence of Rape and Sexual Assault

Despite the acknowledgment that rape and sexual assault are among the most underreported types of crimes, the known data indicate that these crimes pervade our society. According to the Rape, Abuse and Incest National Network, a rape, attempted rape, or sexual assault occurs approximately once every 2 minutes. This figure represents the 248,300 victims of these crimes that are documented by the U.S. Department of Justice's National Crime Victimization Survey.[1] While the U.S. Department of Justice (2003) found that 40% of victims report their crime to the police, other research has placed this number significantly lower, at 16% for adult women (Kilpatrick, Resnick, Ruggiero, Conoscenti, & McCauley, 2007) and only 2% for college women (Fisher et al., 2003).Given the stigmatizing nature of this crime, it is not surprising that rape, attempted rape, and sexual assault are some of the most underreported crimes, making it difficult to determine the extent of this

[1]The National Crime Victimization Survey does not include victims under the age of 12.

problem. While researchers attempt to estimate the prevalence of sexual assault, they are faced with their own set of challenges, including differences in defining sexual assault, the emphasis on different sample populations (adolescents, college-age adults, adults, etc.), or different forms of data (arrest data vs. surveys). Regardless of these issues and the data it yields, it appears that sexual assault affects most individuals in some way (either personally, or through someone they know) at some point in their lifetime.

Prevalence studies report a wide range of data results on the pervasiveness of rape and sexual assault in the United States. A national study on rape published in 2007 indicated that 18% of women in America have experienced rape at some point in their lifetime, with an additional 3% of women experiencing an attempted rape. A comparison of these findings to the Violence Against Women survey in 1996 indicates that little change has occurred in the prevalence of this crime over time (15% of women). Indeed, these results demonstrate an increase in the number of rape cases, which is contrary to the belief that rape has declined significantly in recent times. Rates of sexual assault appear to be higher on college campuses, where it is estimated that between 20 and 25% of women will experience a completed or attempted rape at some point during their collegiate career (Fisher, Cullen, & Turner, 2000). The collegiate experience contains many variables that may increase the risk for sexual assault—campus environments that facilitate a "party" atmosphere, easy access to alcohol and drugs, increases in freedom, and limited supervision by older adults (Sampson, 2006). Findings from studies such as these have led researchers, rape-crisis organizations, and policy makers to posit that one in four American women has experienced rape or attempted rape.

Rape Myths

Rape myths are defined as "attitudes and beliefs that are generally false but are widely and persistently held, and that serve to deny and justify male sexual aggression against women" (Lonsway & Fitzgerald, 1994, p. 134). Table 3.1 highlights some of the most commonly perpetuated myths about rape. The acceptance of rape myths by society is a contributing factor in the practice of victim blaming. First, the presence of rape myths allows society to shift the blame of rape from the offender to the victim. By doing so, we can avoid confronting the realities of rape and sexual assault in society. This denial serves as a vicious cycle: as we fail to acknowledge the severity of rape and sexual assault, which leads to victims not reporting their crime to authorities, this results in a larger acceptance that the crime is not taken seriously by society as a whole. Second, the presence of rape myths lends support to the notion of a just world hypothesis, which suggests that only good things happen to good people and bad things happen to those who deserve it. Rape myths such as "she asked for it" serve to perpetuate the notion of the just world in action (Lonsway & Fitzgerald, 1994).

Offenders often use rape myths to excuse or justify their actions. Excuses occur when offenders admit that their behavior was wrong, but blame their actions on external circumstances outside of their control. In these instances, offenders deny responsibility for their actions. Statements such as "I was drunk" or "I don't know what came over me" are examples of excuses. In comparison, justifications occur when offenders admit responsibility for their actions, but argue that their behavior was acceptable under the circumstances. Examples of justifications include "she asked for it" or "nothing really happened." Miscommunication appears to play a significant role for men, as well, who ask "when does no mean no, or when does no mean yes?" By suggesting that men "misunderstand" their victim's refusal for sexual activity, the responsibility of rape is transferred back to the woman.

Some victims accept excuses or justifications for their assault that minimize or deny the responsibility of their offender. In cases where the offender "got carried away," victims accept the actions of the offender

Table 3.1 Rape Myths

- A woman who gets raped usually deserves it, especially if she has agreed to go to a man's house or park with him.
- If a woman agrees to allow a man to pay for dinner, then it means she owes him sex.
- Acquaintance rape is committed by men who are easy to identify as rapists.
- Only women can be raped or sexually assaulted by men.
- Women who don't fight back haven't been raped.
- Once a man reaches a certain point of arousal, sex is inevitable and he can't help forcing himself upon a woman.
- Most women lie about acquaintance rape because they have regrets after consensual sex.
- Women who say "No" really mean "Yes."
- Certain behaviors such as drinking or dressing in a sexually appealing way make rape a woman's responsibility.
- If she had sex with me before, she has consented to have sex with me again.
- A man can't rape his wife.
- Only "bad" women get raped.
- Women secretly enjoy being raped.

as a natural consequence of male sexuality. In these cases, victims feel that they deserve their victimization as a result of their own actions. Many victims argue that "they should have known better" or that "they didn't try hard enough to stop it." In these cases, victims believe that they put themselves at risk as a result of their own decision-making process.

The prevalence and acceptance of rape myths in society does a significant disservice for both victims and society in general in terms of understanding the realities of rape. These myths permit us to believe that stranger rape is "real" rape, whereas acquaintance rape is less serious, less significant, and less harmful. Rape myths perpetuate the belief that women should be more fearful of the stranger who lurks in the alley or hides in the bushes and surprises the victim. Rape myths suggest that in order for a woman to be raped, she needs to fight back against her attacker and leave the scene with bruises and injuries related to her efforts to thwart the assault. This myth also suggests that "real" rape victims always report their attackers and have evidence collected, and that an offender is identified who is then arrested, prosecuted, and sentenced to the fullest extent under the law. In a "real" rape, the victim is an innocent victim who holds no responsibility for causing her own rape. In reality, however, the majority of rapes in society possess few, if any, of these characteristics. Rape myths serve to limit the definition of rape and the understanding of society about the realities of rape.

Acquaintance Versus Stranger Assault

Contrary to the belief perpetuated by popular culture, stranger rapes are not the most prevalent type of rape and sexual assault. Young women are socialized to be wary of walking alone at night, to be afraid that a scary man will jump out of the bushes and attack them. Many prevention efforts focus on what women can do to keep themselves safe from sexual assault and focus on situations that are more likely to protect women from stranger victimizations such as "lock your doors" and "don't walk alone at night." While these tools are certainly valuable in enhancing women's safety, they fail to acknowledge the reality of sexual assault, as the majority of rape and sexual assault cases involve a perpetrator known to the victim. Acquaintance rape accounts for 90% of all rapes of college women (Sampson, 2006). Additionally, 60% of all rape and sexual assault incidents occur either at the victim's home, or at the home of a friend, neighbor,

▲ **Photo 3.1** Kobe Bryant enters a courtroom in Eagle, Colorado, where he faces allegations of sexually assaulting a 19-year-old woman in a Colorado hotel room in June 2003. The case was ultimately dismissed against Bryant after the victim refused to participate in the criminal process. Throughout the preliminary stages of the case, the victim experienced significant acts of victim blaming in the media and personal threats by members of the public.

or relative (Greenfeld, 1997). Cases of acquaintance rape and sexual assault tend to entail less force by the offender and involve less resistance by the victim (Littleton, Breitkopf, & Berenson, 2008).

It is difficult to assess how many sexual assault victims disclose their victimization to police. Research conducted by Millar, Stermac, and Addison (2002) documented that 61% of acquaintance rapes are not reported to the police. In comparison, Rickert, Wiemann, and Vaughan (2005) found that only one of 86 study participants made a report to law enforcement authorities, and an additional four victims sought services from a mental health professional. While these findings demonstrate a dramatic range of reporting rates, it is safe to conclude that acquaintance rape is significantly underreported. Society tends to discount the validity of acquaintance rape, suggesting that it is less serious than stranger rape ("real rape"). Yet research demonstrates that victims of acquaintance rape also suffer significant mental health trauma as a result of their victimization. However, this trauma is often exacerbated by the fact than many victims of acquaintance rape tend to blame themselves for their own victimization. In many cases, these victims are less likely to seek assistance from rape crisis or counseling services.

Drug-Facilitated Sexual Assault

A drug-facilitated rape is defined as an unwanted sexual act following the deliberate intoxication of a victim. In comparison, an incapacitated rape is an unwanted sexual act that occurs after a victim voluntarily consumes drugs or alcohol. In both cases, the victim is too intoxicated to be aware of her behavior and is unable to consent. Kilpatrick et al. (2007) found that 5% of women experience drug-facilitated or incapacitated rape.

Recent research has discussed a rise in drug-facilitated rapes. The terms "date rape drug" and "drug-facilitated sexual assault" have been used to identify how the involuntary consumption of substances have been used in sexual assault cases. Table 3.2 provides a description of the different types of substances that are commonly used in cases of drug-facilitated sexual assault. In many cases, these substances are generally colorless, odorless, and/or tasteless when dissolved in a drink and result in a rapid intoxication that renders a potential rape victim unconscious and unable to recall events that occurred while she was under the influence. One research study identified that less than 2% of sexual assault incidents were directly attributed to the deliberate covert drugging of the victim (Scott-Ham & Burton, 2005). However, these findings document reported cases of sexual assault, and it is reasonable to conclude that many cases of drug-facilitated sexual assault go unreported, as victims may be reluctant to report a crime for which they have little recollection. Research indicates that in cases where victims are deliberately intoxicated, they are less likely to be judged as responsible for their victimization (Girard & Senn, 2008).

With the exception of alcohol, the majority of the substances that are used in cases of drug-facilitated sexual assault (such as GHB, ketamine, and Rohypnol) are labeled as controlled substances, and the

Table 3.2 Substances Commonly Used in Drug-Facilitated Sexual Assaults

- GHB (Gamma-Hydroxybutyric acid)
  - GHB comes in a few forms—a liquid that contains no odor or color, a white powder, and a pill. GHB has not been approved by the FDA since 1990, so it is considered illegal to possess or sell. GHB can take effect in as little as 15 minutes and can last for 3 to 4 hours. GHB is considered a schedule 1 drug under the Controlled Substances Act. GHB leaves the body within 10–12 hours, making it very difficult to detect.
- Ketamine
  - Ketamine is an anesthetic that is generally used to sedate animals in a veterinarian's office. Ketamine can be particularly dangerous when used in combination with other drugs and alcohol. It is very fast-acting and can cause individuals to feel as if they are disassociated from their body and be unaware of their circumstances. It can cause memory loss, affecting the ability of a victim to recall details of the assault.
- Rohypnol (Flunitrazepam)
  - Rohypnol is a dissolvable pill of various sizes and colors (round, white, oval, green-gray). Rohypnol is not approved for medical use in the United States and much of the supply comes from Mexico. However, the manufacturer of this drug recently changed the chemistry of the pill such that if it is inserted into a clear liquid, it will change the color of the drink to a bright blue color, allowing for potential victims to increase the chance that they could identify whether their drink has been altered. Rohypnol effects can be noticeable within 30 minutes of being ingested; the individual appears overly intoxicated, and the drug affects their balance, stability, and speech patterns. Like many other substances, Rohypnol leaves the body in a rapid fashion, generally between 36–72 hours of ingestion.
- Alcohol
  - Alcohol is one of the most common "date rape" drugs. Here, victims drink to excess, placing themselves at risk for sexual assault. Not only do victims willingly consume alcohol, it is (generally, based on the age of the individual) legal and easily obtained. The consumption of alcohol impairs judgment, lowers inhibition, and affects a victim's ability to recognize potentially dangerous situations.

possession of these drugs is considered a federal offense under the Controlled Substances Act of 1970. In addition, the Drug-Induced Rape Prevention and Punishment Act of 1996 provides penalties for up to 20 years for the involuntary drugging of an individual in cases of violence (National Drug Intelligence Center, n.d.). Many states have created laws that provide specific sanctions in cases of drug-facilitated sexual assault. For example, California penal code 261(a)(3) provides the following definition for the crime of Rape by Intoxication: (1) a male and female engaged in an act of sexual intercourse who are not married; (2) the victim was prevented from resisting by an intoxicating substance; and (3) the victim's condition was known, or reasonably should have been known by the accused. Here, state law provides an assessment of a victim's ability to consent to sexual relations and holds that the level of intoxication, combined with the resulting mental impairment of the individual, must affect the victim's ability to exercise reasonable judgment. Under this law, convicted offenders can be punished for either 3, 6, or 8 years. In most cases, offenders are sentenced to 6 years and can receive a reduced sentence (3 years) if mitigating factors are present or an enhanced sentence (8 years) if aggravated factors are present.

While there has been increased attention to sexual assault due to involuntary intoxication, this is not the primary form of drug-facilitated sexual assault. Rather, cases where the victim is sexually assaulted following a voluntary intoxication make up the majority of drug-facilitated sexual assaults. In rape cases of college-aged women, alcohol was involved in 79% of cases of nonforcible rape (Kilpatrick et al., 2007). The use of drugs and alcohol places women at a greater risk for sexual assault. Not only may women be less aware of the risk for sexual assault and labeled as a target for potential offenders due to a reduction of their inhibitions, but they may struggle to resist their attackers due to their incapacitated state. Additionally, intoxicated individuals are legally incapable to give consent for sexual activity (Beynon, McVeigh, McVeigh, Leavey, & Bellis, 2008). However, these victims are held as the most responsible of all sexual assault victims, since they chose to use intoxicating substances recreationally. In addition, the actions of these perpetrators are most likely to be excused or diminished (Girard & Senn, 2008).

Spousal Rape

Throughout history, the rape of a wife by her husband was not considered to be a crime. The marital rape exception argued that women automatically consent to sex with their husbands as part of their marriage contract. Women were considered not as equal partners, but as an item of property that men were free to do with as they wished. The relationship between a man and wife was viewed as a private manner, and not one for public scrutiny. This belief system permitted the criminal justice system to maintain a "hands-off" policy when it came to spousal rape. As existing rape laws began to change throughout the 1970s and 1980s, increased attention was brought to the marital rape exception. In 1978, only 5 states defined marital rape as a crime. By 1993, all 50 states had at least one statute prohibiting rape within the context of marriage.

Perpetrators use a variety of different tactics to coerce sex from their victims. The majority of cases of marital rape involve cases of emotional coercion, rather than physical force. Examples of emotional coercion include inferences that it is a "wife's duty" to engage in sex with her husband (referred to as social coercion) or the use of power by a husband to exert sexual favors from his wife (referred to as interpersonal coercion). A third form of emotional coercion involves cases where a wife engages in sex for fear of unknown threats or damages that may occur if she refuses. Many of these occurrences are related to cases of domestic violence, where the possibility of violence exists. Cases of marital rape by the use of physical force are referred to as battering rape. The physical effects of marital rape are generally greater compared to cases of stranger and acquaintance rape. In cases of battering rape, the sexual assault is an extension of the physical and emotional violence that occurs within the context of the relationship (Martin, Taft, & Resick, 2007).

Contrary to popular belief, marital rape is as prevalent as other forms of rape. Much of this victimization is hidden from public view. Results from randomized studies find that 7–14% of women experienced completed or attempted rape within the context of marriage, cohabitating, or intimate relationship (Bennice & Resick, 2003). Community samples tend to yield significantly higher rates of marital rape—however, they tend to draw from shelters or therapeutic settings, which offer skewed results. These studies find that 10% to 34% of women studied experienced rape within the context of marriage (Martin et al., 2007).

Despite the criminalization of spousal rape, the cultural acceptance of marital rape still fails to identify these women as victims. By leaving them with the belief that their experiences are not considered as "real" rape, these women are less likely to seek assistance for their victimization.

Same-Sex Sexual Violence

Much of the existing research on rape involves a male offender and a female victim. Many of the theories to explain rape involve the use of violence by men to exert power and control over women. This explanation is rooted in a heterosexist ideology. Indeed, our laws, which in many states identify the crime of rape as the unlawful penetration of a penis into a vagina, do not allow for us to legally identify same-sex cases as rape (though most have additional statutes of sexual assault that would be inclusive of same-sex acts of sexual violation).

Much of the discussion about same-sex rape is limited to male-on-male sexual assault, and many of these studies are conducted within an incarcerated setting. Research on woman-to-woman sexual violence is limited. One study that compared levels of violence experienced by lesbian and heterosexual women found that women involved in same-sex relationships experienced significantly higher levels of nonsexual physical violence (51%) compared to heterosexual women (33%; Bernhard, 2000).

Women (and men) who report same-sex sexual violence are often confronted with a system where agents of the criminal justice system may reflect homophobic views. Such perspectives can potentially silence victims and prevent them from seeking legal remedies and social services. Indeed, advocacy services have been slow in responding to the unique needs of this population. Some community service providers express a fear that offering services to the LGBT population could potentially restrict their donations from government or socially conservative individuals and organizations. These conflicts limit the opportunities to identify same-sex sexual assault as a social problem (Girshick, 2002).

The Role of Victims in Sexual Assault Cases

Many women do not identify themselves as victims. According to a national survey of college women, 48.8% of women who were victimized did not consider the incident to be rape. In many cases, victims may not understand the legal definition of rape. Others may be embarrassed and not want others to know. In some cases, women may not want to identify their attacker as a rapist (Fisher, Cullen, & Turner, 2000).

Several factors increase the likelihood that a victim will report the crime to the police, including injury, concern over contracting HIV, and if they identified the crime as rape. However, victims are less likely to report the crime if the offender is a friend or if they were intoxicated (Kilpatrick et al., 2007). For college-age women, less than 5% of completed and attempted rapes were reported to the police. While women do not report these crimes to law enforcement or school officials, they do not necessarily stay silent, as over two thirds of victims confided in a friend about their attack. The decision by victims to not report their assault to the police stems from a belief that the incident was not harmful or important enough to report. For these women, it may be that they did not believe that they had been victims of a crime or did not want family members or others to know about the attack. Others had little faith in the criminal justice system, as they were concerned that the criminal justice system would not see the event as a serious incident or that there would be insufficient proof that a crime had occurred (Fisher, Cullen, & Turner, 2000).

For those victims who decide to report their crime, the most common reason to report was to prevent the crime from happening to others (Kilpatrick et al., 2007). Key findings from the National Violence Against Women Survey documented that only 43% of reported rapes resulted in an arrest of the offender. Of those reported, only 37% of these cases were prosecuted. Fewer than half (46.2%) of those prosecuted were convicted, and 76% of those convicted were sentenced to jail or prison. Taking unreported rapes into consideration, only 2.2% of all rapists are incarcerated. Of those who reported their rape, less than half of victims indicated that they were satisfied with the way their case was handled by the authorities: 47.7% of victims were satisfied with how their case was handled by the police, while 48.6% of female victims were pleased with the outcome of their case by the courts (Tjaden & Thoennes, 2006).

Victims who seek out medical treatment are more likely to report their sexual assaults (or attempted assaults) compared to those who do not seek medical treatment, perhaps because medical treatment for their injuries occurs in conjunction with the collection of evidence during a sexual assault examination.

▲ Photo 3.2 a and b The Clothesline Project is a collection of more than 500 projects around the world that uses t-shirts created by victims and survivors to represent women's stories of abuse and violence. Each color of shirt represents a different experience of victimization, ranging from rape, to child abuse, to domestic violence. The Clothesline travels around communities to educate society about violence against women.

In a study of women who sought medical assistance following a sexual assault, 52% had some form of injury to their body, while 20% had direct injuries to their genitals as a result of the sexual assault (Sugar, Fine, & Eckert, 2004). Experiences of rape and sexual assault also place victims at risk for developing long-term mental health concerns. Over half of the victims of sexual assault experience symptoms of posttraumatic stress disorder (PTSD) at some point during their lifetime. Symptoms of PTSD can appear months or even years following the assault. The levels of emotional trauma that victims experience lead to significant mental health effects such as depression, low self-esteem, anxiety, and fear for personal safety. Women with a history of sexual assault are more likely to have seriously considered attempting suicide and are more likely to engage in behaviors that put them at risk, including risky sexual behaviors with multiple partners, extreme weight loss measures, and substance abuse involving alcohol and illegal drugs (Gidycz, Orchowski, King, & Rich, 2008; Kaukinen & DeMaris, 2009). Women who are victimized by strangers may experience anxiety and fear about their surroundings, particularly if the assault occurred in a public setting. For women who were assaulted by a family member, acquaintance, or date, they may experience issues with trusting people (Illinois Coalition Against Sexual Assault, n.d.).

Victims of rape and sexual assault have both immediate and long-term physical and emotional health needs. Rape-crisis services play an important role in victim advocacy. The current rape crisis movement developed in response to the perceived need for prevention, community awareness, and amelioration of victims' pain. However, even the best community services are limited and lack adequate resources to effectively combat all the needs for victims of sexual assault. While attempts to help survivors of sexual assault involve friends, family members, community agencies, and criminal justice personnel, efforts in help seeking may actually enhance the trauma that victims experience due to lack of support, judgment, and blame by support networks. Additionally, victims may experience further trauma by being forced to relive their trauma as part of the official processing of the assault as a crime (Kaukinen & DeMaris, 2009). Due to these negative experiences in disclosure, many victims choose to keep their assault a secret.

Ultimately, cases of rape and sexual assault can be very difficult to prove in a court of law. Convictions are rare, and many cases are plea bargained to a lesser charge, many of which carry little to no jail time. Alas, the acceptance of rape myths by police, prosecutors, judges, and juries limits the punishment of offenders in cases of sexual assault. Figure 3.1 highlights how each stage of the criminal justice system reduces the likelihood that offenders will be arrested, charged, and punished for these cases. The effects of these practices can further discourage victims from reporting these crimes, believing that little can be done by criminal justice officials.

Figure 3.1 Punishment and Rape

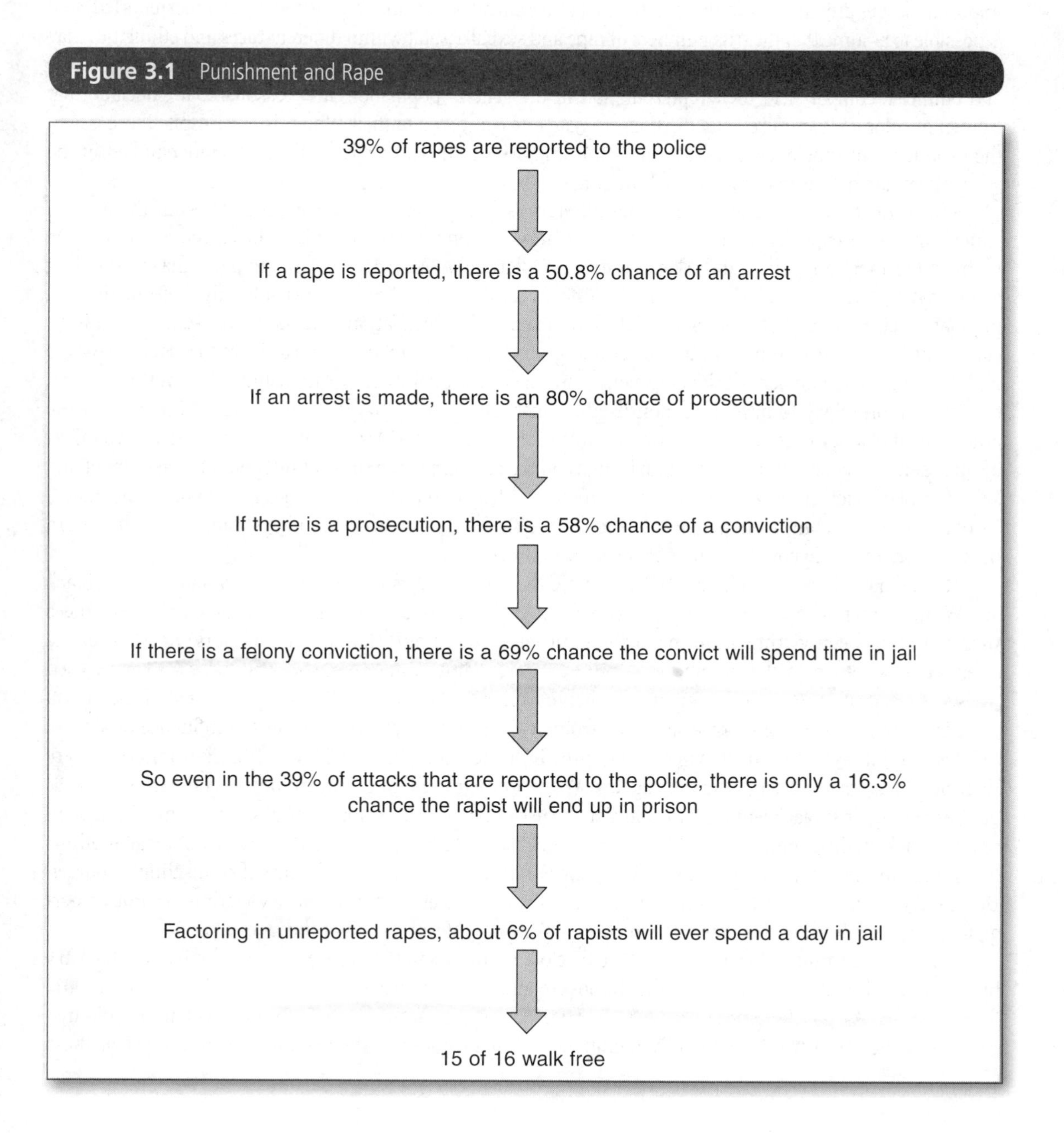

Racial Differences in Sexual Assault

Research suggests that women of color have different experiences of sexual assault, compared to Caucasian women. These differences can be seen in prevalence rates, reporting behaviors, disclosure practices, help-seeking behaviors, and responses by the justice system. For example, research indicates that 18% of White women, compared to 19% of Black women, 34% of American Indian/Alaska Native women, and 24% of women who identify as mixed race report a rape or sexual assault during the course of their lifetime (Tjaden & Thoennes, 2006). Two important issues are raised with these statistics: (1) We already know that rape generally is underreported, so it is possible to assume that the true numbers of rape and sexual assault within different races and ethnicities may be significantly higher than these data indicate; and (2) given the unequal distribution of these statistics by race and ethnicity, compared to their representation in the general population, it is reasonable to conclude that women of color are victimized at a disproportionate rate compared to their White sisters. Despite these issues, the experience of rape and sexual assault within minority communities is significantly understudied in the scholarly research. How does race and ethnicity affect the experience of rape and sexual assault?

While much of the literature on racial differences in rape and sexual assault focuses on the African American female experience, statistics by Tjaden and Thoennes (2006) highlight the extreme rates of rape within the American Indian and Alaska Native population (AIAN). These data are particularly troubling given that the AIAN population is a small minority in the population, comprising only 1.5% of the U.S. population (U.S. Bureau of the Census, 2000). Research using the National Crime Victimization data indicates that compared to other racial and ethnic groups, AIAN women are most likely to experience rape within an intimate partner relationship, versus stranger or acquaintance relationships. Within this context, they were more likely to have a weapon used against them and to be physically assaulted as part of the attack. Alcohol and drugs also play a stronger role in the attacks of AIAN women, with more than two thirds of offenders under the influence of intoxicants, compared to only one third of offenders in cases involving White or Black victims. While AIAN victims are more likely to report these crimes to the police, the majority of these reports come from people on behalf of the victim (family, officials, others) rather than the victim herself (Bachman, Zaykowski, Lanier, Poteyva, & Kallmyer, 2010).

Research by Boykins et al. (2010) investigates the different experiences of sexual assault among Black and White women who sought emergency care following their attack. While no racial and ethnic differences were found between victims in terms of the location of the assault (home, car, outdoors) or whether the offender was known to the victim, Black women were significantly more likely to have a weapon used against them during the attack compared to White women (42% vs. 16.7%). The intoxication of the victim (and offender) also varied by race, as White women were more likely to be under the influence of alcohol (47.2% of White women reported being under the influence, compared to 23.8% of Black women), as were their perpetrators (47.2% of offenders against White women were under the influence, compared to 23.8% of offenders against Black women). In contrast, the use of illicit drugs prior to the assault was more common among Black victims compared to White victims (28.7% vs. 12.5%). However, there were no racial or ethnic differences in the reporting of the assault to police or of the offering or acceptance of counseling resources. Despite the importance of these findings, it is important to keep in mind that few victims seek out emergency services following their assault, which may skew the interpretation of these results.

Not only are women of color less likely to disclose sexual assault, there are a number of factors that vary by race and ethnicity that can affect the disclosure and recovery process. Research by Washington (2001) found that less than half of the women interviewed had disclosed their victimization—when they did disclose, they did so to friends or family members within 24 hours of the assault. However, most of these women experienced incidents of victim blaming as a result of their disclosure. As a result of historical personal and cultural experiences with law enforcement, the majority of the women did not seek out the police to make an official report of their attack. In addition, many of the Black women talked about not reporting as a cultural expectation of keeping their business to themselves. They also mentioned not wanting to perpetuate additional racist views against members of the African American community, particularly if their assailant was also Black. For example, one woman reported to Washington (2001):

> We have this element in our community that it's the White man or the White race that causes most, if not all, of the problems we have in our communities. If we begin to point out the Black male for specific problems, we tend to get heat . . . and even from some women because we as women have been socialized as well. And it's "Don't bring the Black man down. . . . He's already going to jail, dying, rumored to be an endangered species; so why should we as Black women bring our wrath against him?" (p. 1269)

Likewise, cultural expectations also can inhibit the official reporting practices of women within the Asian American and Pacific Islander population (AAPI). Like the African American community, there is a high level of distrust of public officials (often due to negative experiences either in the United States or in the cases of immigrant and refugee individuals, in their home country) as well as a cultural expectation to keep personal issues in the private sphere. In addition, concerns over immigration status, lack of knowledge about the criminal justice system, and language barriers affect both reporting practices as well as the utilization of mental health resources. In addition, research has highlighted that many AAPI women fail to understand the definitions of rape and sexual assault, which further limits the likelihood that such incidents will be reported (Bryant-Davis, Chung, & Tillman, 2009). The same factors that limit the reporting rates of crimes such as rape and sexual assault also affect the use of therapeutic resources. Indeed, AAPIs have the lowest utilization of mental health services of any racial or ethnic minority group (Abe-Kim et al., 2007).

Within the Hispanic community, Latina women have the highest rates of attempted sexual assault of all ethnic groups. The experience of an attempted rape/sexual assault carries with it many of the same psychological traumas as a completed rape. Stereotypes of Latina women as passionate and sexual women can affect the fears of rape victims that they could have contributed to the assault, and therefore limit the likelihood that they will report (or that their reports will be taken seriously), making it important for agencies in Hispanic/Latino communities to reach out to the population and dismantle some of the stereotypes and attitudes that can inhibit reporting and help-seeking behaviors (Bryant-Davis et al., 2009). Like the AAPI community, some Latina women are confronted by concerns over their immigration status and language barriers that can affect reporting rates of completed and attempted sexual assaults.

Culture shapes the manner in which people represent themselves, make sense of their lives, and relate to others in the social world. Indeed, the experience of trauma is no different, and we find that women of color are less likely to engage in help-seeking behaviors from traditional models of assistance. While many women of color believe that agencies such as rape-crisis centers can provide valuable resources to victims of sexual assault, they may be hesitant to call upon these organizations for fear that these organizations would be unable to understand their experiences as a woman of color. Instead, victims may turn to sympathetic leaders and women within their own communities. In order to increase the accessibility of these services to women of color, victims and scholars argue that services need to be culturally sensitive and address the unique considerations that women of various racial and ethnic identities face as victims of sexual assault (Tillman, Bryant-Davis, Smith, & Marks, 2010).

CASE STUDY

The Duke Lacrosse Case

On March 13, 2006, players from the Duke lacrosse team hosted a party off campus where they hired some entertainment for the evening. Crystal Mangum, a 27-year-old African American female exotic dancer, was one of the women who showed up to the party. She accused three of the players (Reade Seligmann, Collin Finnerty, and David Evans), all of whom were White, of raping her in the bathroom of the house. She reported the incident to the police, and the lead prosecutor, Mike Nifong, announced that a rape had indeed occurred. Indictments were handed down for first degree forcible rape, first degree sexual offense, and kidnapping against the three college students.

The racial dynamics of the case set off fireworks within the community. Duke is one of the most affluent universities in the nation, and many were concerned that privilege would prevail over the interest of justice. The case became more about race and class, while issues of gender and sexual assault were secondary issues. The accused players were suspended from the university and the lacrosse team forfeited their season. All of the White lacrosse players on the team were forced to submit their DNA for testing. However, significant skepticism began to arise in the case, as the details of Mangum's story changed several times. None of the players' DNA matched the evidence found on Mangum. In addition, it was revealed that the lead prosecutor in the case, Mike Nifong, deliberately withheld exculpatory evidence from the defense. He was criticized for his inflammatory and prejudicial comments to the media and stepped down from the case. The charges against Seligmann, Finnerty, and Evans were ultimately dismissed and apologies were made for Nifong's "tragic rush to accuse." Nifong was disbarred by the North Carolina State Bar for his handling of the case and was found guilty of criminal contempt. Seligmann, Finnerty, and Evans have filed a civil suit requesting $30 million in damages and reforms to the legal process.

While Mangum was never charged with filing false charges in the case, she has had several interactions with the law in the years since. In 2010, she was convicted on misdemeanor charges for setting a fire in her home that almost killed her three children. In April 2011, she was indicted for first-degree murder for stabbing her boyfriend, who later died from his injuries.

Policy Implications

Research on rape and sexual assault indicates a number of areas where the criminal justice system and other social institutions can improve prevention and intervention efforts. Given that adolescents and young adults have higher rates of acquaintance rape and sexual assault, much of the prevention efforts have been targeted toward college campuses. Evidence indicates that the prevention efforts may be most effective if targeted at the beginning of the college career, versus later in life. While college campuses have increased their educational activities aimed toward preventing rape on campuses in recent times, these efforts may still be inadequate given the number of assaults that occur on campuses around the nation each year. However, the age of victimization appears to be decreasing, indicating a need for education efforts focused on high school students.

Victims indicate that an increase in public education about acquaintance rape and increased services for counseling would encourage more victims to report their crimes (Kilpatrick et al., 2007). Programs focusing on rape and sexual assault prevention should provide accurate definitions of sexual assault behaviors, the use of realistic examples, discussions about alcohol use and sexual assault, and an understanding of what it means to consent to sexual activity. By tailoring education efforts toward combating myths about rape, these efforts can help reduce the levels of shame that victims can experience as a result of their victimization and encourage them to seek help following a sexual assault. Services need to be made available and known to students, both in terms of services and outreach on campus and information available online.

Summary

- Rape is one of the most underreported crimes of victimization.
- The risk of rape and sexual assault appears to be higher on college campuses.
- The acceptance of rape myths by society contributes to the practice of victim blaming.
- Many victims of rape and sexual assault fail to identify their experiences as a criminal act.
- Excuses and justifications allow perpetrators of rape and sexual assault to deny or minimize levels of blame and injury toward their victims.
- The majority of rapes and sexual assaults involve individuals who are known to the victim prior to the assault.
- The terms "date rape drugs" have been used to identify a group of drugs, such as GHB, Rohypnol, and Ketamine, that have been used to facilitate a sexual assault.
- Marital (spousal) rape is as prevalent as other forms of rape, though it is significantly underreported and hidden from public view.
- Victims of rape and sexual assault are at risk for long-term physical and emotional health concerns.

KEY TERMS

Rape

Sexual assault

Rape myths

Victim blaming

Stranger rape

Acquaintance rape

Drug-facilitated rape

Incapacitated rape

Spousal rape

Same-sex sexual assault

DISCUSSION QUESTIONS

1. How has the definition of rape evolved over time?
2. Why do many victims of rape and sexual assault choose not to report their crimes to the police?
3. What impact do rape myths play in victim blaming and the denial of offender culpability?

4. Why do many victims of rape and sexual assault fail to identify themselves as victims of a crime?
5. Why are acquaintance rape cases not viewed as "real" rape?
6. What tactics do perpetrators use to coerce sex from their victims?
7. How can prevention efforts be used to educate women and men about the realities of rape and sexual assault?

WEB RESOURCES

Bureau of Justice Statistics: http://bjs.ojp.usdoj.gov

Rape, Incest and Abuse National Network: http://www.rainn.org

National Clearinghouse on Marital and Date Rape: http://ncmdr.org/

The National Center for Victims of Crime: http://www.ncvc.org

Office of Victims of Crime: http://www.ojp.usdoj.gov/

READING

Much of the literature on sexual assault focuses specifically on the crime of rape, and similarly related sexual offenses (forced oral copulation, forced sodomy, etc.). This article reviews how there is more to sexual victimization than just rape. In understanding acts outside of rape, the authors of this article investigate three different dimensions: (1) type of contact, (2) degree of coercion, and (3) degree of action. They highlight how cases of sexual coercion and unwanted and coerced sexual contact are important acts for consideration in the spectrum of sexual assault and suggest that communities need to acknowledge these acts of victimization in prevention and intervention efforts.

Beyond Rape

The Pervasiveness of Sexual Victimization

Bonnie S. Fisher, Leah E. Daigle, and Francis T. Cullen

The criminal status of rape, coupled with grassroots efforts by campus safety and women's advocacy groups, have combined to keep rape prominent on the policy agendas of federal and state policymakers and of campus administrators. Indeed, rape on campus has been among the most extensively addressed issues by state-level and congressional statutes (Carter & Bath, 2007; Sloan & Shoemaker, 2007). A key requirement of the federal The Jeanne Clery Disclosure of Campus Security Policy and Campus Crime Statistics Act (20 USC § 1092(f)) (hereafter the Clery Act), for example, is that postsecondary institutions publish and distribute an annual report of seven major categories of crime statistics. Included in this crime reporting requirement is sex offenses. Consistent with the FBI's National Incident-Based Reporting System, sex offenses must be divided into and reported in two categories: (1) forcible offenses, including rape as defined as the carnal knowledge of a female forcibly and against her will, forcible sodomy, sexual assault with an object, and forcible fondling, and (2) nonforcible offenses such as incest or statutory rape (Ward & Lee, 2005).

These deeply rooted scholarly and legislative interests in rape among college students often overshadow other types of sexual victimization experienced by female students during their college years. In a way, rape is the "tip of the iceberg"—the part of the problem that receives the most attention but that sits atop a larger foundation of victimization. Indeed, a growing body of research documents that, beyond rape, college women experience much unwanted and coerced sexual contact.

Our exploration of this issue involves three main discussions. First, we examine why most attention has been devoted to rape and, in turn, why it is now important to understand the diverse ways in which college women are sexually victimized. Some of these are fairly

SOURCE: Fisher, B. S., Daigle, L. E., & Cullen, F. T. (2010). *Unsafe in the ivory tower: The sexual victimization of college women.* Thousand Oaks, CA: Sage.

serious; others are limited. But taken as a whole, they impose a cost on many female students that detracts from the quality of their college experience—a burden not confronted by most of their male counterparts. Second, we furnish a conceptual framework for understanding types of sexual victimization other than rape. In this regard, we categorize the dimensions of sexual victimization as a means of constructing a simple, but hopefully illuminating, framework. Third, the remainder of the chapter presents an overview of what researchers know about the extent of different types of sexual victimization beyond rape—namely (1) sexual coercion, (2) unwanted sexual contact, and (3) noncontact sexual abuse.

Moving Beyond the Study of Rape

Women experience a wide range of different types of sexual victimization that, although not legally rape, are not consensual and not wanted. These "non-rapes" are part of a broader constellation of different types of unwanted and coerced sexual contact and abuses that many women, especially young college women, experience. They are often neglected, we suspect, because of the belief that unwanted sex compelled in the absence of force, threat of force, or incapacitation, is either not important or not that "serious." Unlike the construct of rape, these experiences lack any name in the nation's lexicon that intuitively captures their distinctive nature. As Basile (1999, p. 1039) observes, they exist in a "grey area"—not legally qualifying as a rape but not acts that can be dismissed as inconsequential.

Non-rapes are sexual victimizations that do not meet the legal criteria for completed or attempted rape. A variety of different types of victimization fall under this conceptual umbrella. Some types of sexual victimization, which would be included in the category of sexual coercion, involve sexual intercourse or penetration obtained by verbal coercion, such as continual arguments or misuse of authority. Other types of sexual victimization involve unwanted sexual contact that does not involve penetration of the body. This would include the perpetrator intentionally touching the victim's body in a sexual manner, such as fondling breasts or genitals, touching the thighs or neck, kissing lips or other body parts, or rubbing against a person's body. The perpetrator can seek sexual contact through a variety of forms of verbal coercion, misuse of authority, or threatened or actual use of force. *Noncontact sexual abuses* are non-penetrative or nontouching acts. Examples include the perpetrator making sexist comments, jeering or taunting, or asking questions about one's sex life. Unwanted exposure to pornography, indecent exposure, exhibition, and voyeurism are also examples of noncontact sexual abuses (see Basile & Saltzman, 2002).

There are four main reasons that it is important to focus on sexual victimization other than rape. First, a substantial, if not alarming, proportion of college women experience non-rape victimization. This does not appear to be a new occurrence. Recall the research by Kirkpatrick and Kanin (1957), who first measured "male sexual aggression on a university campus" in the 1950s. In their investigation, 27% of the college women reported themselves offended at least once during the school year by a level of "erotic intimacy" that did not involve forceful attempts at "sex intercourse." Of the 1,022 offensive erotic intimacy episodes that college women reported happening during the academic year, 73% involved "necking and petting above the waist" and 19% involved "petting below the waist."

Most recent studies of non-rape sexual victimization suggest that little has changed since the 1950s with respect to the extent of these types of victimizations committed against college women. In their study during the 1980s, Koss and her colleagues reported that a large proportion of college women had been sexually victimized at least once during the past year. About 40% of their sample had experienced at least one non-rape experience during the previous year. Fifty-four percent of the incidents involved unwanted touching—fondling, kissing, or petting but not intercourse—and 22% of the incidents involved sexual intercourse that was coerced but not forced through the perpetrator's threat or physical action.

Results from two studies during the 1990s parallel these earlier findings. First, White, Smith, and Humphrey

(2001) conducted a 4-year longitudinal study in the early 1990s on a single campus (see also White & Smith, 2001). Their analysis revealed that 46% of women experienced at least one non-rape—coerced sexual intercourse or unwanted touching—during their four college years. This is five times higher than the 9% of college women who reported having been raped at least once during their college years. Second, our National College Women Sexual Victimization Study found that 15% of college women in the sample were sexually victimized, with the vast majority of these women, 13%, experiencing at least one non-rape during the academic year. Two thirds of the sexual incidents involved unwanted sexual touching with or without the perpetrator threatening or using force. One-fifth of the incidents involved coerced sexual intercourse without threat or use of physical force.

Second, there is mounting evidence that the proportion of women who experience non-rape sexual victimizations far exceeds the proportion who experience rape (e.g., Abbey et al., 2004; Fisher et al., 2000; Koss et al., 1987). Take, for example, the results of a study by the American College Health Association, which used a large probability sample of college students from over hundreds of postsecondary institutions nationally. This research reported that 2% of college women were raped, on average, each year. Twice as many college women, 4%, during this time experienced an attempted rape. Most striking, during this 9-year period, a considerably larger proportion of college women, 11% annually, experienced sexual touching against their will. An annual average of 4% of college women were verbally threatened for sex against their will (American College Health Association, 2008b).

White and her research team's work is similarly instructive (Smith et al., 2003; White et al., 2001). Figure 3.2 shows that the sample of women on the campus studied were far more likely to experience a non-rape than rape victimization. Thus, on average annually, 21.5% of these college women experienced a non-rape victimization compared to the 7.1% who were raped. This finding also was true year by year. During each collegiate year from first to fourth, a much larger proportion of women experienced at least one non-rape victimization than experienced a rape (see Figure 3.2). For example, during the first year of college, 29% of women reported experiencing a non-rape, whereas 11% reported being raped. The difference between those who experienced a non-rape and those who were raped across each year of college is always 15 percentage points or higher. The fourth year of college, in fact, shows the largest difference between these two groups of victimized college women. There is a 21 percentage point difference between women who experienced a non-rape (24%) and a rape (3%).

Our National College Women Sexual Victimization (NCWSV) Study is further relevant. We found that within an academic year, 2.8% of college women reported being raped at least once. More college women, 3.7%, were coerced (but not by the perpetrator's threatened or use of force) into having intercourse at least once during an academic year. An even much larger proportion of the national sample, 11%, experienced at least one incident of unwanted sexual contact during the school year. Noncontact sexual abuses, such as someone "flashing" his penis at her or having sexually tinged catcalls or sexually charged sounds or remarks directed at her, were experienced by the largest proportion of college women. Seventy-seven percent of the respondents experienced at least one form of noncontact sexual abuse during the school term.

Third, non-rape and rape victimization may have unique characteristics and correlates, which can only be discerned through careful investigation. In a study of a community sample of women, Testa and Derman (1999) revealed that the tactics used by perpetrators to obtain sexual contact is different in rape and other types of sexual victimization, in particular sexual coercion. Our NCWSV Study discovered that students' lifestyles and routines differentially affected the risk of college women experiencing rape and non-rape (Fisher & Cullen, 1999a). These findings suggest that whether they are researchers, policymakers, or campus officials, those who seek to unravel what places college women at risk will need to focus on the similar and unique risk factors associated with rape and different types of non-rape sexual victimization.

Figure 3.2 The Percentage of Women Victimized During Each Collegiate Year by Rape and Non-rape Victimization

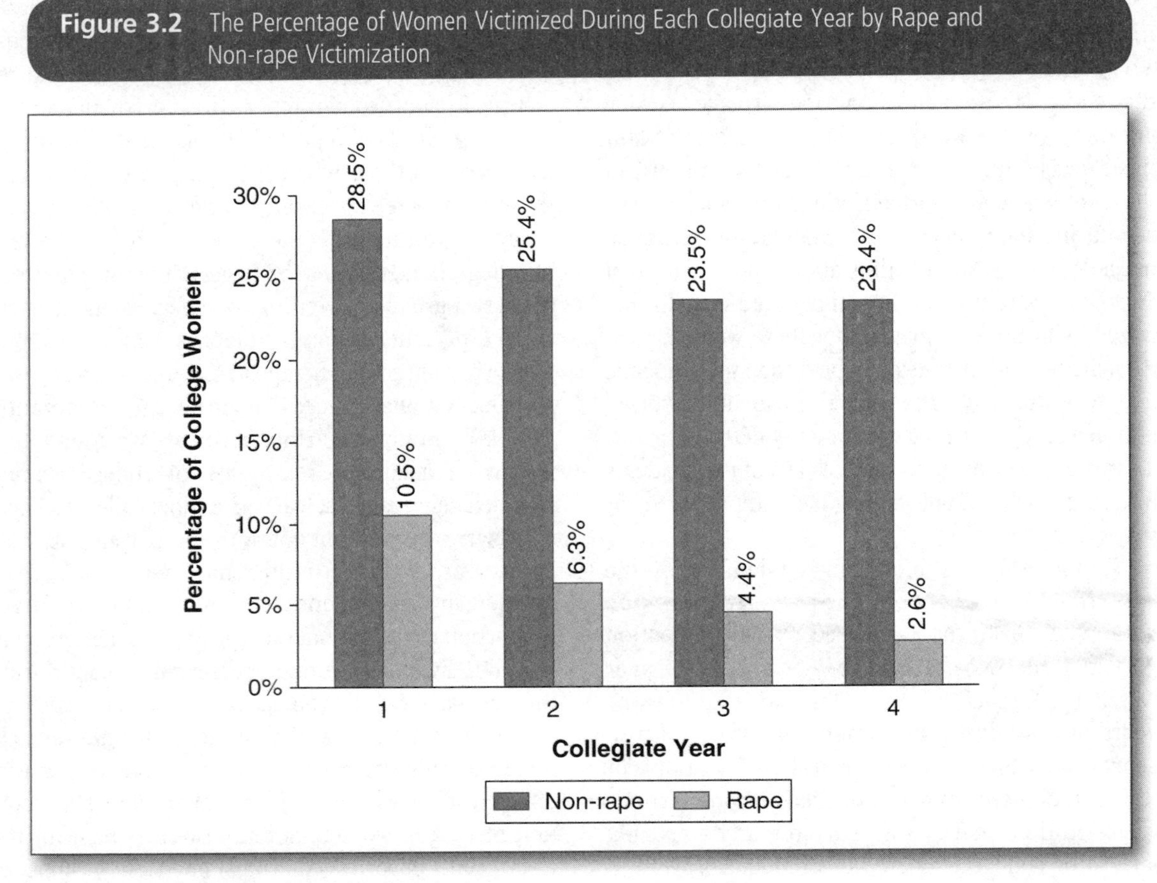

SOURCE: Adapted from White and Smith (2001).

Fourth, for the allocation of prevention and intervention resources on campuses, it is important for campus administrators to understand the extent to which students are experiencing diverse types of sexual victimization, other than rape. This knowledge is needed to formulate a more informed campus policy and programmatic response that supplements the Clery Act crime statistics. Note that the Clery Act's sole focus is on a limited number of types of sexual offenses as defined by the Federal Bureau of Investigation. An exclusive focus on these offenses risks turning a blind eye to other forms of sexual victimization that are more commonly experienced and diminish the quality of college women's lives.

Categorizing Sexual Victimization

The categorizing of a sexual victimization as a rape depends on the presence of penetration or intercourse, the use or threatened use of force by the perpetrator, and the lack of consent by the victim. Not all sexual victimizations are characterized by all of these three dimensions—intercourse or penetration, threat of force or use of force, and lack of consent. However, this is not to say that those who experience unwanted sexual touching—for example, fondling of genitals or breasts—have not been sexually victimized. It is only to say that they have not been raped.

Non-rape sexual victimizations involve the assailant having unwanted physical contact with the victim. These contacts or acts are characterized by varying degrees of coercion, but they are not characterized by threats or use of actual physical force. Similar to rape, they also can be either completed or attempted acts.

To understand more fully different types of sexual victimization, it is useful to think about these behaviors falling along three dimensions. These dimensions can be used to categorize an act as a particular type of sexual victimization— whether rape or some form of non-rape. The three dimensions that characterize sexual victimization are as follows:

Type of contact

Degree of coercion

Degree of action

It is useful to present each dimension along a continuum that captures the range of the behaviors included in each dimension. By doing so, different types of sexual experiences can be identified and then measured as distinct types of sexual victimizations. It is important to distinguish different types of sexual victimization beyond rape in terms of both naming the unwanted sexual contact and describing its dimensions.

Type of Contact

The first dimension of sexual victimization categorizes the type of sexual contact that the perpetrator engages in to obtain physical interaction with the victim. As shown in Figure 3.3, the type of contact varies along a contact continuum. This continuum ranges from the perpetrator (1) having no sexual contact with any part of the victim's body to (2) using sexual advances to intentionally touch any part of the victim's body to (3) penetrating or engaging in intercourse with the victim's vagina, anus, or mouth. Having physical contact with any part of the victim's body can happen either underneath or on top of her clothing.

On one end of the continuum shown in Figure 3.3, the perpetrator has no sexual contact with any part of the victim's body. The perpetrator may direct a variety of behaviors or actions at the victim. Examples of no sexual contact include the perpetrator making general sexist remarks or comments to the victim. Also included are gendered behaviors such as undue attention (e.g., being too eager to please or help), body language (e.g., leering or standing too close), verbal advances (e.g., expressions of sexual attraction), or invitations for dates (Belknap & Erez, 2007). Other acts may involve no sexual contact yet be sexually abusive to the victim. These acts include the perpetrator exposing the victim to pornography or voyeurism (Basile & Saltzman, 2002).

In the middle of the type of contact continuum are sexual advances to intentionally physically touch any part of the victim's body. This would include physical advances such as kissing, touching, or rubbing with any part of the body, especially erogenous zones such as the head, lips, neck, hands, breasts, thighs, groin, or buttocks. Also included is the fondling of breasts, nipples, or a genital or sex organ (e.g., vagina, clitoris, vulva). Examples of sexual advances to physically touch someone's body would include a man intentionally touching a woman's breasts or buttocks or trying to kiss her lips or cheeks when she does not want him to do so.

Figure 3.3 Continuum of Type of Contact

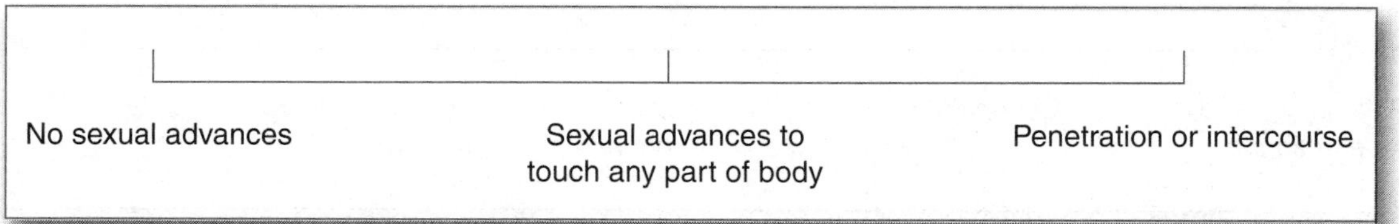

No sexual advances	Sexual advances to touch any part of body	Penetration or intercourse

On the farther end of the continuum is penetration or intercourse. This includes the insertion of the perpetrator's body part, such as his penis, mouth, tongue, digit (e.g., fingers), or an object (e.g., dildo, bottle) into any opening of the victim's body, including vagina, anus, and mouth. A man inserting his penis into the vagina or anus of a woman is an example of sexual intercourse or penetration. A woman performing oral sex on a man's penis (i.e., fellatio) or a man performing oral sex on a woman's vulva or clitoris (i.e., cunnilingus) are examples of sexual intercourse or penetration.

Degree of Coercion

The second dimension, degree of coercion, refers to the characteristics of physical actions or tactics used by the perpetrator with the intent of having sexual contact with the victim. Various scholars have identified different levels or types of coercion to describe the diverse tactics through which the perpetrator uses coercion to obtain sexual contact with the victim—that is, to victimize her (Basile & Saltzman, 2002; Thompson, Basile, Hertz, & Sitterle, 2006). Finkelhor and Yllo (1985) were among the first researchers to conceptualize the meaning of sexual coercion. They identified four types of coercion: (1) social (e.g., institutional or cultural expectations as to the role of women in a relationship), (2) interpersonal (e.g., threatening behavior on the part of perpetrator), (3) threatened physical (e.g., threats to use physical force), and (4) physical (e.g., use of physical force).

Other researchers built on this theoretical foundation in an attempt to clarify and expand the conceptualization of the level of coercion. Thus, Hamby and Koss (2003) argue that in assessing different types of coerced sexual victimizations other than rape it is important to specify the type of force involved (see also Koss, 1996). For example, they use the term *psychological coercion* to include situations in which the perpetrator uses continual nagging, false promises, or similar strategies that "are not desirable but at the same time not crimes" (p. 244). Further, in his review of 120 studies of sexual victimization, Spitzberg (1999) identified five major categories of sexual coercion: (1) pressure and persistence, (2) deception, (3) threat, (4) physical restraint, and (5) physical force or injury.

The continuum shown in Figure 3.4 takes into account Finkelhor and Yllo's types of coercion, Hamby and Koss's degrees of coercion, and Spitzberg's five categories of sexual coercion. The continuum of degree of coercion has five discrete categories, ranging from the perpetrator engaging in psychological or emotional pressure to using physical force to obtain sexual contact with the victim. Across this continuum, the perpetrator is using varying degrees of intimidation or force. The perpetrator uses coercion to get the victim to engage in sexual contact in which she does not want to engage with the perpetrator. The victim's compliance is not voluntary, but rather it is the result of the perpetrator's level of coercion.

The first category on the degree of coercion continuum is psychological or emotional pressure. This category includes the assailant using verbal or emotional persuasion to obtain sexual contact with the victim's body. These persuasive tactics include continual requesting, nagging, pleading, urging, pressuring, pouting, or misusing authority to obtain sexual contact. Verbal aggression, not including threats of violence,

Figure 3.4 Continuum of Degree of Coercion

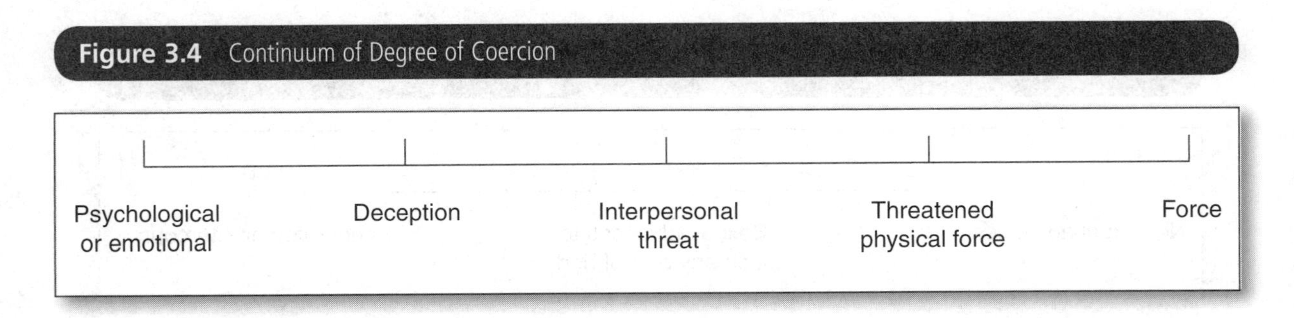

could also be included, such as the perpetrator belittling or swearing at the victim. The perpetrator uses the power of persuasion to manipulate the woman into having sexual contact with him by eliciting feelings of guilt or obligation.

An example of verbal pressure is a boyfriend who pesters his girlfriend about having sexual contact—intimate touching, penetration, or intercourse—and continually mentions in virtually every communication with her that they should have more sex more often. She has repeatedly told him she is not interested, but he continues to verbally pressure her to have sexual contact with him. An example of emotional pressure could be a man making the victim feel sorry for him. She then has sex with him because of feelings of guilt or sympathy that he intentionally evoked.

The second category on this continuum, deception, includes the perpetrator communicating emotions, feelings, or promises that he does not really mean or intend to fulfill. These communications could include falsely professing some level of romantic emotion (e.g., love) to the victim, telling her lies, or making false promises. The perpetrator assures his victim that he will fulfill obligations once she gives in to having sexual contact with him. Examples of deception include a professor telling a student he will not pass her in the course if she does not let him fondle her breasts and genitals. Another example is a classmate telling the victim that he will give her answers to the exam if she has oral sex with him.

The third category on the degree of coercion continuum, interpersonal threat, is another form of coercion used by a perpetrator to obtain sexual contact from his victim. Interpersonal threat involves the perpetrator threatening to terminate the relationship with the victim or to seek another person for sexual contact if the victim does not agree to have sexual contact with him. This type of coercion does not involve the perpetrator being physically violent toward the victim. Interpersonal threat entails the perpetrator threatening to carry through with unkind, callous, cruel, or mean actions directed at harming or terminating the relationship if the victim does not provide sexual contact to him. Also, the threat does not have to be communicated solely by word of mouth; it can be communicated via writing through a variety of mediums, such as a handwritten note, text messaging, an e-mail message, or a post on a social networking site such as Facebook.

Examples of an interpersonal threat include a man who tells his date or even a platonic friend that he will not drive her home unless she has intercourse with him. Another example is a man who pressures his girlfriend into having anal sex with him by telling her that, if she does not, he will end their relationship. He communicates his threat via text messaging to ensure his girlfriend receives the message.

Moving along the continuum toward physical force, the level of coercion shifts to include the perpetrator threatening to use, or using, physical means to obtain the victim's compliance to sexual contact. The fourth category on the level of force continuum is what Finkelhor and Yllo (1985) refer to as "threatened physical coercion," which specifically implies the use of physical force if sexual contact is not obtained. Threats of physical force involve the perpetrator threatening to use his body, such as his arms and hands, mouth, or feet, to inflict harm or pain on his victim or to threaten to use his strength to restrain her. He can also threaten to use a weapon, such as a knife or gun, so as to make her comply with his demands for sexual contact.

Examples of threatened physical coercion include a perpetrator threatening to hit, slap, or physically harm the victim if she does not have sexual contact with him. A husband who threatens to pull his wife's hair or to hit her if she does not perform oral sex on him is another example of threatened physical force.

The last category on the level of coercion continuum is physical force. Finkelhor and Yllo (1985) refer to physical force as physical coercion. As the term *physical force* suggests, the perpetrator actually uses physical force to obtain sexual contact with his victim. Physical force involves the perpetrator using his body, such as his arms and hands, mouth, or feet to inflict harm or pain on his victim or to use his strength to restrain her. He can also employ a weapon to force his victim to have sexual contact with him.

Examples of the use of physical force would include the perpetrator using his hands to slap the victim, holding her body down against her will, biting her, or pulling her hair in order to have sexual contact with her. Spitzberg (1999) uses a similar definition as that used by Finkelhor and Yllo. Spitzberg, however, distinguishes between physical restraint and physical force. Physical restraint is the perpetrator holding her arm down or her legs apart. Physical force is the perpetrator hitting or slapping his victim or using a weapon to get her to comply with his sexual demands.

It is important to understand and distinguish use of coercion from threatened use or use of physical force as one of the dimensions of a sexual victimization. It is also important when distinguishing pertinent characteristics of sexual victimization other than rape to name or label the experience as a victimization and not merely as a "less serious" experience.

Degree of Action

As shown in Figure 3.5, the third dimension of a sexual victimization is the degree of action. The continuum ranges from a perpetrator threatened to have sexual contact with the victim, to attempted to but was unsuccessful in having sexual contact with the victim, to successfully obtaining sexual contact.

Figure 3.5 Continuum of Degree of Action

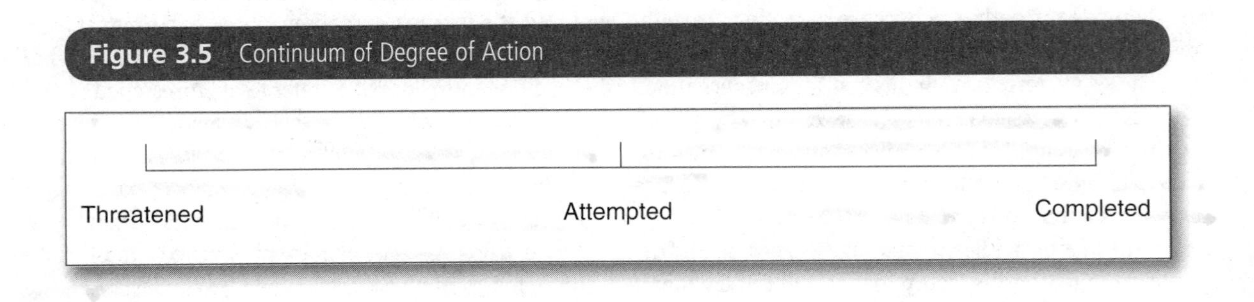

A perpetrator can threaten to have sexual contact with a victim by verbally stating his intent in person or over the telephone. He also can threaten in writing that he will obtain sexual contact with the victim by sending her a text message or e-mail message.

If the perpetrator's action to obtain sexual contact with the victim was interrupted or thwarted, this could be considered an attempted sexual victimization. An example of an attempted sexual contact would be a perpetrator who reaches out to touch the breasts of a woman but she thwarts him before he does so.

A completed sexual contact would result if the perpetrator achieved sexual contact with the victim. A completed sexual contact includes actually having sexual contact with any part of the victim's body, such as her lips, neck, thighs, genitals, or buttocks. An example of a completed sexual contact would be when the perpetrator uses his fingers, hand, or penis to touch a woman's vaginal or anal areas.

A Framework for Categorizing Sexual Victimization

The three dimensions of sexual victimization discussed thus far provide a framework for defining and naming different types of sexual victimization beyond rape; that is, non-rape victimizations. These different types of sexual victimization go beyond the legal, yet narrowly focused, view of sexual victimizations as only including rape. The view of sexual victimizations defined as rape is limited. It only includes completed or attempted unwanted sexual intercourse or penetration that occurs as a result of the assailant using or threatening to use force or the victim's inability to consent.

There are many categories on each of the three continuums that can be used to distinguish rape from non-rape victimizations. Characteristics from each of the three continuums—type of contact, level of coercion, and degree of action—can be employed to categorize these experiences as distinct types of sexual victimizations beyond rape. The analysis to follow focuses on three main categories: (1) sexual coercion (2) unwanted sexual contact.

The next sections highlight what is known from the research about the extent of different types of sexual victimization other than rape that college women experience before and during their higher education tenure. The extent of a range of sexual victimizations from sexual coercion to unwanted sexual contact are discussed below.

Sexual Coercion

Acts of intercourse or penetration are not always performed by the perpetrator with the threatened or the actual use of force. Sexual coercion is one such type of sexual victimization for which it is true that threats of force or force are not used by the perpetrator to obtain intercourse or penetration.

Sexual coercion is like rape on one dimension—type of contact. Like rape, sexual coercion is defined by vaginal, anal, or oral intercourse or penetration with the perpetrator's penis, mouth, tongue, digit (e.g., fingers), or with an object (e.g., dildo, bottle). Sexual coercion, like rape, is a matter of degree. Both rape and sexual coercion can be attempted or completed acts. Sexual coercion is unlike rape on the degree of coercion continuum. The type of contact that defines a sexual coercion is subsequent to the perpetrator's use of a variety of psychological or emotional tactics. Unlike with rape, the perpetrator does not engage in threats or use of force to obtain intercourse or penetration. In a sexual coercion, the perpetrator's purpose for using these tactics is to persuade or manipulate the victim into having sex she otherwise does not want to engage in.

In our NCWSV Study, we defined completed and attempted sexual coercion in the following way (Fisher et al., 2000, p. 8):

> Unwanted completed or attempted penetration with the threat of non-physical punishment, promise of reward, or pestering/verbal pressure. Penetration includes: penile-vaginal, mouth on your genitals, mouth on someone else's genitals, penile-anal, digital-vaginal, digital-anal, and object-anal.

Tactics

Livingston, Buddie, Testa, and VanZile-Tamsen (2004) identified four types of tactics from women's descriptions of their verbal sexual coercion experiences that men use to obtain contact with the victim's body. These four tactics are (1) verbal persuasion, (2) persistence, (3) physical persuasion, and (4) gaining access.

They identified three types of verbal persuasion. First, negative sexual persuasion includes behaviors that were manipulative and emotionally hurtful to the victim, such as threats to end the relationship or go elsewhere for sex and belittling her or making her feel sorry for him. Second, positive sexual persuasion tactics were also used as a tactic by perpetrators to manipulate women into having sex. These tactics typically included the perpetrator sweet-talking or offering compliments to his victim, such as making promises to deceptively allure her into having sex. Third, neutral verbal persuasion was neither negative nor positive. Neutral verbal persuasion included the man continually nagging or pleading for sex without using emotionally charged statements. The man's intent is that his persistence will eventually wear the woman's resistance down and she will give in to having sex.

Persistence included the continual use of tactics to persuade the woman to comply with his sexual advances, such as sweet talk or sexual contact. Physical persuasion tactics involved sexual contact (e.g., kissing, sexual touching) or physical aggression (e.g., being held down). Gaining access strategies, such as isolating the woman or using false pretences to be alone with her, typically were employed by the perpetrator as a function of being in the relationship or by sexual precedence.

Using only those incidents involving verbal sexual coercion, Livingston et al. reported that verbal persuasion (81.6% of the incidents) and persistence (61.4%) were among the two most common tactics used by those who commit sexual coercion, followed by physical persuasion (48.2%) and gaining access (10.5%). In the incidents in which the perpetrator used verbal persuasion, negative verbal persuasion was most common (49.1% of the incidents), followed by positive verbal persuasion (19.3%) and neutral/nagging (18.4%). Nearly half of the women reported that the perpetrator used physical persuasion tactics in conjunction with the use of verbal tactics.

Measurement of Sexual Coercion

Koss and her colleagues were among the first researchers to develop and incorporate the systematic measurement of sexual coercion into the measure of the extent of different types of sexual victimization that college women had experienced. They defined sexual coercion as experiences of "sexual intercourse subsequent to the use of menacing verbal pressure or the misuse of authority" (Koss et al., 1987, p. 166).

Within the Sexual Experiences Survey (SES), there are questions that distinguish the type of coercion experienced. In Table 3.3, the questions that Koss et al. used to measure the extent of sexual coercion among college women are presented again. The SES captures three different degrees of coercion: verbally coerced, misuse of authority, and threat or use of force.

Table 3.3 Questions Used to Measure Sexual Coercion

Form of Coercion	Survey Question
Koss et al.: Sexual Experiences Survey (SES) Have you ever. . . .	
Verbally coerced	given in to sexual intercourse when you did not want to because you were overwhelmed by a man's continual arguments or pressure?
Misuse of authority	had sexual intercourse when you didn't want to because a man used his position of authority (boss, teacher, camp counselor, supervisor) to make you?
Threatened or used force	had sexual intercourse when you didn't want to because a man threatened or used some degree of physical force (twisting your arm, holding you down, etc.) to make you?
Broach and Petretic: Modified Sexual Experiences Survey Been in a situation where a man became so sexually aroused that you. . . .	
Guilty/obligation	had sexual intercourse with him, even though you did not want to, because you felt responsible for "leading him on"?
Felt unable to physically/ Useless to try to stop him	would not be able to physically stop him/useless to try to stop him, so you had sexual intercourse with him, even though you did not want to?
Threatened to end relationship	had sexual intercourse with a man when you didn't really want to because he threatened to end your relationship otherwise?
Fisher et al.: National College Women Sexual Victimization Study (NCWSV) (screen questions) Since school began, has anyone made or tried to make you have sexual intercourse when you did not want to by. . . .	
Threats of nonphysical punishment	making threats of nonphysical punishment such as lowering a grade, being demoted or fired from a job, damaging your reputation or being excluded from a group for failure to comply with requests for any type of sexual activity?
Promises of rewards	making promises of rewards such as raising a grade, being hired or promoted, being given a ride or class notes, or getting help with course work from a fellow student if you complied sexually?
Verbally coerced	simply being overwhelmed by someone's continual pestering and verbal pressure?

SOURCES: Broach and Petretic (2006); Fisher, Cullen, and Turner (1999); Koss, Gidycz, and Wisniewski (1987).

Several researchers have used Koss's original questions to estimate the extent of sexual coercion (see Spitzberg, 1999; Thompson et al., 2006). Other researchers have modified these questions to take into account additional forms of sexually coercive tactics (see Jordan, Wilcox, & Pritchard, 2007; Thompson et al., 2006). One example of the modified version of the SES is by Broach and Petretic (2006). As shown in Table 3.3, their questions provide a description of three additional forms of coerced sexual intercourse to measure the degree of coercion used by the perpetrator. The first two forms, the victim's feeling of guilt or obligation to have intercourse and victim's feeling unable or useless to stop the perpetrator, are examples of psychological coercion. The third form, the perpetrator threatening to end the relationship, is an example of an interpersonal threat.

As also shown in Table 3.3, our NCWSV Study measured verbally coerced sex. In addition to measuring verbal coercion like the SES, this study included screen questions about two additional types of coercion that occur within a campus context: threats of nonphysical punishment and promises of reward. An example of nonphysical punishment includes a professor giving a student a bad grade for not having sexual intercourse with him. On the degree of coercion continuum, this could fall under interpersonal threat. Examples of a promise of rewards for having sexual intercourse with the perpetrator is a male student promising to give the answers to a test to his victim if she gives him oral sex or a teaching assistant giving the student a better grade for sex. On the degree of coercion continuum, these are examples of deception.

Extent of Sexual Coercion

The extent that college women experience sexual coercion is a relatively new, but growing, field of study within the sexual victimization research. This escalating interest may well be fueled by research reporting that the college years are among the vulnerable ones for experiencing sexual coercion. Thus, Abbey, McAuslan, Ross, and Zawacki (1999) report that of those women who had been sexually coerced, 61% reported that the "most serious" sexual coercion happened when they were between 18 and 21 years old, 14% reported that they were between 22 and 25 years old, and 7% reported they were between 26 and 28 years old. The first two age groups encompass the years when a large number of women are enrolled pursuing undergraduate and graduate college degrees.

Single-Campus Studies. Some of this emerging research has been conducted by researchers who undertook a survey on a single campus. Obviously, single-campus studies are limited in their generalizability. However, when taken together, they show common findings that lend credence to the assessment that their individual results are not idiosyncratic. In this case, it appears that the risk of sexual coercion is relatively high even over a short period of time and increases the longer a woman is a college student.

This conclusion can be drawn from studies conducted on single campuses since 2000 that employ different reference periods.

- Over a 3-month period, Gidycz, Orchowski, King, and Rich (2008) reported that nearly 4% of women on a single campus were sexually coerced.
- Using a reference period of 8 months, Messman-Moore, Coates, Gaffey, and Johnson (2008) reported that almost 12% of women were sexually coerced into having intercourse.
- Within a 2-year time frame, Kalof (2000) found that 20% of college women were sexually coerced.
- Using a reference period of "since enrolling in college," Gross, Winslett, Roberts, and Gohm (2006) reported that 9.1% of the female students answered "yes" to having been physically unable, or that it was useless to try, to stop their male assailant from having sexual intercourse.
- White et al.'s data (2001) show that 30% of women experienced at least one form of sexual coercion over the course of an undergraduate career. Similar to other studies, this one found that more women were verbally coerced, 30%, compared to the 1.9% who were coerced by misuse of authority.

White et al.'s (2001) study is further relevant because as a longitudinal study, it was able to show that a sizable number of women are sexually coerced *each* of their four college years. These researchers followed two cohorts of freshmen at a single campus to examine their annual experiences with sexual coercion. Their results as to the proportion of women sexually coerced during each of their four college years are presented in Table 3.4. As can be seen, in their first year on campus, nearly 16% of the respondents reported being sexually coerced. Despite age and maturing, virtually the same proportion of the sample in their fourth year reported being victimized by sexual coercion. What is noteworthy about these results is that the proportion of sexually coerced college women does not change dramatically across college years. In fact, the results suggest that this proportion remains quite steady at about 15% annually.

National-Level Studies. Only three national-level studies have been published that report the extent of sexual coercion among women during their collegiate years. First, each year since 2000, the American College Health Association has been tracking the proportion of college women who have experienced a range of sexual victimizations, including sexual coercion. Between 2000 and 2007, the annual average of college women who experienced verbal threats for sex against their will has been 4%. There has been little year-to-year variation in this annual estimate across these 8 years. Every year, slightly more or less than 4% of college women are sexually coerced (American College Health Association, 2000a, 2000b, 2001a, 2001b, 2002a, 2002b, 2003a, 2003b, 2004a, 2004b, 2005a, 2005b, 2006a, 2006b, 2007a, 2007b, 2008a, 2008b).

Second, Koss and her colleagues (1987) found that within the past year, 11.5% of college women reported being sexually coerced into having intercourse. Eleven percent of women were verbally coerced. Much fewer, less than 1% of these women, were coerced through the misuse of authority.

Third, as noted above, our National College Women Sexual Victimization Study reported the extent of sexual coercion among college women during an academic year.

Recall that we used a two-step process: screen questions followed by an incident report to confirm what had occurred. This is perhaps one reason why the prevalence of sexual coercion is lower in the NCWSV Study than in investigations not employing a follow-up incident report. Even so, we found that over a 6.9-month reference period during the school year, 3.7% of college women experienced at least one completed or attempted sexual coercion.

Further, the use of the incident report allowed us to measure in the NCWSV Study what kind of coercion the perpetrator employed. Close to 4% of women experienced verbal coercion. Compared to other kinds of coercion, such as threatened with nonphysical force experienced by 0.5% of women or promise of reward experienced by 0.2%, verbal coercion was by far the dominant form of sexual coercion used against college women.

Table 3.4 White, Smith, and Humphrey's Longitudinal Study of Sexual Coercion Against College Women

Sexual Coercion	Year in College			
	Year 1 Percentage (*n*)	Year 2 Percentage (*n*)	Year 3 Percentage (*n*)	Year 4 Percentage (*n*)
	15.9	15.2	14.9	14.5
	(222)	(179)	(142)	(108)

SOURCES: Adapted from White and Smith (2001); White, Smith, and Humphrey (2001).

NOTE: Percentages are based on total sample size that varied by year (Year 1, $n = 1{,}378$; Year 2, $n = 1{,}178$; Year 3, $n = 954$; and Year 4, $n = 747$).

This finding that verbal coercion was the most likely form of coercion mirrors the result by Koss et al. (1987). Of course, some commentators might argue that verbally coerced sexual intercourse is simply effective foreplay by males. But recall that the participants in the NCWSV Study were asked about being made to have sexual intercourse "when you did not want by simply being *overwhelmed* by someone's continual pestering and verbal pressure." It is when sexual advances reach this stage—unwanted and achieved through overwhelming verbal coercion—that a sexual victimization can be said to have occurred.

Unwanted Sexual Contact

Unwanted and coerced sexual contact includes the perpetrator intentionally using different coercive tactics to make sexual advances to touch any part of victim's body. This type of sexual contact is not penetrative but does involve physical touching. The perpetrator makes sexual advances to any part of the victim's body—lips, breasts, genitals, buttocks, or any erogenous zone—either over or underneath her clothing. Sexual contact includes such acts as kissing, groping or fondling, rubbing, petting, licking or sucking, or any other form of unwanted sexual contact perpetrated by the assailant.

The perpetrator engages in tactics along the continuum of coercion to obtain sexual contact with the victim's body. The degree of coercion used by the perpetrator ranges from using no force to using psychological or emotional coercion, deception, interpersonal threats, to using threats or force in his attempt to obtain sexual contact with the victim's body. The perpetrator's action to have contact with any of the victim's body can be an attempted or completed act.

Measurement of Unwanted Sexual Contact

There is no standard way of measuring unwanted sexual contact. However, to illuminate how this form of sexual victimization is assessed, we discuss two prominent measurement strategies. First, the most widely used measure of unwanted sexual contact is with items on the SES (Koss et al., 2007). As shown in Table 3.5, the three items from the SES that measure unwanted sexual contact describe the perpetrator's behavior in detail and his use of different forms of coercion to engage in "sex play." Recall that this does not include sexual penetration, which would qualify an act as a rape or sexual coercion. The three types of coercion represented in these SES items are (1) verbal coercion, (2) misuse of authority, and (3) threatened use of force or use of force. Verbal coercion is a form of emotional or psychological coercion. Under certain circumstances, misuse of authority can be an example of deception. Threatened use of force or use of force is the more severe form of physical coercion on the continuum of degree of coercion presented previously in Figure 3.4.

Second, our NCWSV Study offers a related but different way of measuring this phenomenon. Recall, we used a two-stage measurement process with a series of screen questions and an incident report to confirm which type of unwanted sexual contact, if any, had occurred during the course of each incident. We included three behaviorally specific screen questions that describe the forms of coercive behaviors used by the perpetrator to obtain sexual contact. As shown in Table3.5, similar to the Koss et al.'s SES, we incorporated a question about the use of verbal coercion. Two screen questions asked about other additional forms of coercion: threats of nonphysical punishment and promises of rewards. A threat of nonphysical punishment can be seen as an example of an interpersonal threat, whereas a promise of rewards is an example of deception.

Again, unlike the SES and most other investigations, the NCWSV Study measured rape, sexual coercion, and unwanted sexual contact through a two-step process. Thus, female students were categorized as experiencing unwanted sexual contact based on their responses to questions about type of physical contact and type of coercion in the incident report. They could be categorized in this way regardless of whether they screened into the incident report from the questions meant to capture unwanted sexual contact or from questions meant to capture other forms of sexual victimization (e.g., rape).

Table 3.5 Sexual Experiences Survey Items of Unwanted Sexual Contact

Form of Coercion or Force	Survey Question
Koss et al.: Sexual Experiences Survey (SES) Have you ever. . . .	
Verbally coerced	given in to sex play (fondling, kissing, or petting, but not intercourse) when you didn't want to because you were overwhelmed by a man's continual arguments and pressure?
Misuse of authority	had sex play (fondling, kissing, or petting, but not intercourse) when you didn't want to because a man used his position of authority (boss, teacher, camp counselor, supervisor) to make you?
Threatened use or use of physical force	had sex play (fondling, kissing, or petting, but not intercourse) when you didn't want to because a man threatened or used some degree of physical force (twisting your arm, holding you down, etc.) to make you?
Fisher et al.: National College Women Sexual Victimization Study (NCWSV)(screen questions) Since school began, has anyone made or tried to make you have sexual contact when you did not want to by. . . .	
Threats of nonphysical punishment	making threats such as lowering a grade, being demoted or fired from a job, damaging your reputation or being excluded from a group for failure to comply with requests for any type of sexual activity?
Promises of rewards	making promises such as raising a grade, being hired or promoted, being given a ride or class notes, or getting help with course work from a fellow student if you complied sexually?
Verbally coerced	simply being overwhelmed by someone's continual pestering and verbal pressure?

SOURCES: Fisher, Cullen, and Turner (1999); Koss, Gidycz, and Wisniewski (1987).

Extent of Unwanted Sexual Contact

The existing research shows that a large proportion of college women have experienced unwanted sexual contact. Single-campus and national-level studies are similarly revealing.

Single-Campus Studies. Recent single-campus studies have explored the extent of this type of sexual victimization. What is evident looking across these studies is that unwanted sexual contact is a persistent issue confronting college women. A substantial percentage of college women experience unwanted sexual contact during their college tenure. The percentage of women who experience unwanted sexual contact increased as the reference period lengthened—from 7.9% over 3 months (Gidycz et al., 2008) to 19.6% over about 6 months (Banyard, Ward, et al., 2007), to 29.8% among first-year and second-year undergraduate students during their time at the university (Flack et al., 2008). Gross et al. (2006) limited their research to those experiencing sexual contact through force or threat of force; 13.3% indicated being victimized.

Data from White and her colleagues' (2001) study of one campus over 4 years also reveals that a very large proportion of college women will experience unwanted sexual contact. Over 45% of the female students in the sample experienced at least one incident of unwanted sexual contact while in college. As seen in Table 3.6, there was a tendency for victimization to decline as the students' college careers progressed. Thus, the highest total victimization occurred in the first year of college, with 27.0% of the sample experiencing an unwanted sexual contact. Still, by year 4 in college, fully 22.1% of the female students were victimized by unwanted sexual contact.

Table 3.6 White, Smith, and Humphrey's Longitudinal Study of Unwanted Sexual Contact Against College Women

Type of Unwanted Sexual Contact	Year in College			
	Year 1 Percentage	Year 2 Percentage	Year 3 Percentage	Year 4 Percentage
Unwanted sexual contact without force	26.2	23.4	22.0	22.0
Unwanted sexual contact	4.7	3.1	2.6	1.6
Total	27.0	23.9	22.3	22.1

SOURCES: Adapted from White and Smith (2001); White, Smith, and Humphrey (2001).

NOTE: Percentages are based on those who have at least one valid value across all 4 years of data ($n = 1407$).

National-Level Studies. National data are similarly revealing. As discussed above, the American College Health Association has been tracking sexual victimization over time. Between 2000 and 2008, it found that, on average annually, 12% of college women experienced sexual touching against their will. This statistic has remained relatively stable during this time, with the proportion of women sexually touched against their will ranging from a low of 10.6% to a high of 12.4% (American College Health Association, 2000a, 2000b, 2001a, 2001b, 2002a, 2002b, 2003a, 2003b, 2004a, 2004b, 2005a, 2005b, 2006a, 2006b, 2007a, 2007b, 2008a, 2008b).

Again, we consult the findings of Koss et al.'s (1987) classic studies. Their research revealed that 27.8% of the college women in their sample experienced unwanted sexual contact during the past year. Verbal coercion was the most prevalent form, but sexual contact also was accomplished in a minority of cases through threatened use or use of force (3.5%) and misuse of authority (1.6%).

Finally, we can report the findings from our NCWSV Study. We found that 10.9% of college women experienced at least one completed or attempted unwanted sexual contact incident since the start of the current academic year. This overall pattern included the following:

- Seven percent experienced verbally coerced contact.
- Slightly fewer, 5%, were threatened or physical force was used.
- Less than 1% of college women experienced an unwanted sexual contact through the perpetrator's promise of rewards (0.7%) or use of nonphysical punishment (0.5%).

Conclusion

Scholars' traditional focus on rape is understandable. After all, rape is a serious crime; acquaintance rape on college campuses and elsewhere was long hidden from view; and the consequences of being raped are potentially substantial for the victim. Even so, as a growing number of researchers have realized, the focus on rape should not obscure the extent to which college women face other forms of sexual victimization. These include the use of sexual coercion for purposes of intercourse, being groped or physically touched when the attention is uninvited, and having men express sexist remarks, whistle, make obscene comments in person or over the telephone, or involuntarily expose women to degrading sexual images.

We can anticipate that some critics will respond that our analysis is just an exercise in political correctness. Men, after all, "will be men"; it is just nature taking its course. In the pursuit of sexual gratification, young men will be persistent, clumsy, inarticulate, and crude; and some might commit criminal acts. For Katie Roiphe, the author of *The Morning After*, women simply have to get a backbone and "deal with it." Roiphe admits to having her "bad nights." The solution is not to whine and assume a victim's role but to take responsibility and show self-efficacy. More broadly, claims Roiphe, this concern with supposed sexual victimization deprives sexual relations of essential human qualities:

> People pressure and manipulate and cajole each other into all sorts of things all the time. . . . No human interactions are free from pressure, and the idea that sex is, or can be, makes it what Sontag calls a "special case," vulnerable to inconsistent expectations of double standard . . . these feminists are endorsing their own utopian vision of sexual relations: sex without struggle, sex without power, sex without persuasion, sex without pursuit. (1993, pp. 79–80)

In the end, according to Roiphe, the attempt to place a protective veil over women robs them of their human agency and of their obligation to meet life as it presents itself. Under the supposed protection of feminists, college students are reduced to passive victims, who live in a world in which they feel "pinched, leered at, assaulted daily by sexual advances, encroached upon, kept down, bruised by harsh reality" (1993, p. 172).

Of course, when any new problem is discovered—in this case, sexual victimization—there is the risk of hyperbole, of trying to illuminate the problem's seriousness so that it will no longer remain hidden and ignored. So Roiphe has a point—but only to a degree. The sexual victimization reported in this chapter is not a mere fantasy but rather an empirical reality. Study after study—whether conducted 20 years ago or 2 years ago, whether conducted on a single campus or across the nation, whether asking one set of questions or another—all reach the same general conclusion: college women often reside in a sexualized environment where their bodies and personal space are violated in unwanted and potentially damaging ways.

In Roiphe's world, females are to manifest individual human agency but are expected to be passive about the gender relations as they find them. Their role is to cope with, but not try to change, this world—as the feminists she criticizes have argued should be done. More than this, the men in Roiphe's world lack human agency or responsibility. Implicitly, her argument suggests that beyond "real rape," males are free to use any form of coercion short of physical force to gratify their sexual desires. They are free to leer and whistle and grope and make sexist remarks. It is just men being men, or at least some men being who they are. This perspective, however, robs men of their human agency, of their responsibility to monitor their own behavior and to avoid ethical lapses. It offers male students a ready-made excuse for their sexual misconduct: "I was only being a guy." It is this thinking that can be transformed into "rape myths"—the techniques of neutralization that weaken normative restraint and permit potential assailants to traverse moral boundaries.

More broadly, Roiphe's attempt to correct what she sees as feminist excess ignores the fact that women on college campuses must endure sexual victimization that rarely touches men. This is a form of social inequality that imposes an unfair cost on female students not borne by their male counterparts. Roiphe's attempt to normalize this inequality only reinforces feminists' claims that patriarchy is hegemonic. In such a social context, male sexual aggression is given normative status, and it is up to women to cope with the "natural order of things." Efforts to complain about—or even to study—women's sexual victimization are in turn dismissed as political correctness run amok.

In the end, political debate is not the issue. Rather, what matter are the lives of college women that, each day, are affected in small and sometimes large ways by the sexual victimization they endure. This hidden cost of being a female student should not be dismissed as making a mountain out of a molehill. Rather, further research should be undertaken to confirm the conclusions of existing studies and to map out the ways in which diverse types of sexual victimization potentially diminish the quality of life experienced by college women.

DISCUSSION QUESTIONS

1. Why is it important for researchers to focus on sexual victimizations other than rape on college campuses?
2. Discuss the three dimensions that characterize sexual victimization.
3. How do we measure sexual victimizations such as sexual coercion and unwanted sexual contact? What is the extent of these crimes on college campuses?

READING

The case of the alleged sexual assault by Los Angeles Lakers basketball star Kobe Bryant in 2003 made national headlines. Many of these headlines commented less on the accused actions of Mr. Bryant and more on his athletic prowess or engaged in victim-blaming strategies against his accuser. This article highlights not only how rape myths affected the reporting of this case by agents of the press, but also the implications of these rape myths on whether people believed that Bryant could be guilty of sexual assault or whether this was a case in which the victim lied in order to gain publicity and attention.

Prevalence and Effects of Rape Myths in Print Journalism

The Kobe Bryant Case

Renae Franiuk, Jennifer L. Seefelt, Sandy L. Cepress, and Joseph A. Vandello

Helping the media shape the perception of a case is the single most important thing a lawyer can do.

—Alan Dershowitz (quoted in Chancer, 2005, p. 133)

On July 1, 2003, a woman went to authorities in Eagle County, Colorado, and reported that Los Angeles Lakers basketball player Kobe Bryant sexually assaulted her the night before. Kobe Bryant acknowledged having sexual intercourse with this woman but said that the sex was consensual. Two weeks later, Mark Hurltbert, Eagle County district attorney, decided there was enough evidence to proceed to trial, and Kobe Bryant was formally charged with one count of felony sexual assault. In the 14 months that preceded

SOURCE: Franiuk, R., Seefelt, J. L., Cepress, S. L., & Vandello, J. A. (2008). Prevalence and effects of rape myths in print journalism: The Kobe Bryant case. *Violence Against Women, 14*(3), 287–309.

the trial, hundreds of articles were published in newspapers around the country and on the Internet about this case. More than 70 articles were written for the *Denver Post* alone before the trial was set to begin. On September 1, 2004, the Eagle County DA dropped the charges filed against Kobe Bryant mainly because of the alleged victim's decision not to testify. Leading up to the trial, several errors on the part of the court led to confidential material about the trial being leaked to the press. The alleged victim believed that she could not get a fair trial because of how the case had been discussed in the media in the year preceding the trial.

For most people, sexual assault is conceptualized as a brutal crime that occurs between strangers and deserves swift and harsh punishment for the offender. However, in actuality, most sexual assaults are committed by acquaintances of the victim, go unreported, and, when reported, typically go unpunished (U.S. Department of Justice, 2003). To explain this disconnect between people's images of sexual assault and the reality of sexual assault, Martha Burt (1980) outlined several "rape myths" that highlight the distinction between sexual assaults that actually occur and ones that we (prefer to) believe occur. Research supports that people are more likely to label a situation as sexual assault when it fits the prototype—for example, when a "no" is explicit, when it is not a dating couple (Check & Malamuth, 1983; Goodchilds, Zellman, Johnson, & Giarrusso, 1988; Sawyer, Pinciaro, & Jessell, 1998). When sexual assault does not fit this image of what Estrich (1987) labels "real rape," people are likely to employ one or more rape myths to explain away the assault. In the present research, we explored the prevalence of rape myths in print journalism and the effects of exposure to rape myths on people's beliefs about sexual assault. The recent charges brought against Kobe Bryant presented a unique opportunity to study rape myths in the media coverage of a high-profile case (Study 1). Furthermore, this case allowed for a real-world test of media exposure to rape myths on people's attitudes and beliefs about the case (Study 2). In reality, only Kobe Bryant and his alleged victim can know exactly what happened that night in June 2003. It is the goal of this article not to suggest otherwise but rather to show the extent to which stereotypes and misconceptions are still used when discussing sexual assault and the impact that these myths can have on beliefs about sexual assault cases. Because of the uniquely high profile of Kobe Bryant and the subsequent media saturation of the case, this story had great potential to shape public opinion about sexual assault in general.

Rape Myths

Rape myths are generalized and widely held beliefs about sexual assault that serve to trivialize the sexual assault or suggest that a sexual assault did not actually occur. Brownmiller (1975) was one of the first to discuss the long history of myths and misconceptions about sexual assault. A few years later, Burt (1980; Burt & Albin, 1981) developed a measure of several rape myths that reflect common responses to sexual assaults that do not fit the prototype described above. Burt (1980) described myths about the victim, the perpetrator, and the nature of sexual assault. Myths about the victim suggest that she is lying and has ulterior motives,[1] was "asking for it" (e.g., by going to the perpetrator's apartment for a drink), is not the type of woman who gets raped (i.e., it only happens to promiscuous women), or changed her story after the fact (i.e., she wanted it at the time). Myths about the perpetrator excuse his behavior (i.e., he didn't mean to) or paint a narrow picture of those who commit sexual assault (i.e., sex-crazed psychopaths).[2] People also hold the false belief that rape is trivial (i.e., she wasn't really hurt) or natural (i.e., men have a biological predisposition to get sex through force). Although it is possible that for any specific case the above beliefs may not actually be myths (i.e., the "she is lying" allegation is accurate if a woman has made a false report), these are "myths" in the sense that data do not generally support these popular beliefs about sexual assault (for a review, see Lonsway & Fitzgerald, 1994).[3]

Although endorsing rape myths may seem malicious and cruel, this is usually not the explicit motivation of those maintaining these beliefs. First, Brinson (1992) noted that sexual assault contradicts our culture's values of personal integrity and justice. As a culture, we pride ourselves on respecting one's personal integrity and in punishing those who violate such integrity. Sexual assault is a serious violation of the victim's personal integrity, and consistency demands that we severely punish those who violate this cultural norm. However, the majority of sexual assaults go unreported (Koss, 1992), and the majority of those reported go unpunished (Bureau of Justice Statistics, 1998; U.S. Department of Justice, 2003). The employment of rape myths may explain why judges and juries are not harshly punishing this crime that they would otherwise view as very serious (Brinson, 1992; Burt, 1991). By using rape myths to explain away the majority of sexual assaults that occur, a culture maintains that sexual assault is a serious violation that should be punished harshly (in the rare instances) when it does occur.

A related explanation for the widespread employment of rape myths is the pervasive motivation to believe the world is just. Lerner (1980) argued that the belief in a just world allows people to give order to and make sense out of troubling events. A belief in a just world encourages the attribution that good things happen to good people and bad things happen to bad people. Therefore, when a negative event such as sexual assault occurs, people search for a way to make sense of it. It is threatening to accept that a sexual assault could have happened under less prototypical (and, therefore, less predictable) circumstances, so people have a tendency to use just-world explanations for the event (Cowan & Curtis, 1994; McCaul, Veltum, Boyechko, & Crawford, 1990; Wyer, Bodenhausen, & Gorman, 1985). The thinking goes, "If this woman who is not promiscuous, who was not dressed provocatively, who clearly did say 'no' and was with her boyfriend was sexually assaulted, what's to prevent *me* from getting sexually assaulted too?" People have a powerful incentive to maintain rape myths as a way of bringing predictability and control to otherwise random events. Furthermore, internalizing rape myths may protect us from disturbing thoughts that we have been victims of or have committed sexual assault (Lonsway & Fitzgerald, 1995).

Several studies have shown that rape myths are endorsed by a significant portion of the population and that men are almost always more accepting of rape myths than are women (for a review, see Lonsway & Fitzgerald, 1994). Rape myths, no matter how strongly endorsed by an individual, have serious consequences for sexual assault victims. People who endorse rape myths are less likely to label a scenario as sexual assault, even when it meets the legal criteria (Muehlenhard & MacNaughton, 1988; Norris & Cubbins, 1992). Endorsement of rape myths leads people to be less likely to blame the man for an assault (Check & Malamuth, 1985; Linz, Donnerstein, & Adams, 1989; Muehlenhard & MacNaughton, 1988). Muehlenhard and MacNaughton (1988) showed that women who endorsed rape myths were 3 times more likely to be victims of coerced sex than were those who did not strongly endorse rape myths (though Koss and Dinero, 1989, did not find this distinction). Priming men's rape myth acceptance increased their self-reported likelihood of sexually assaulting a woman (Bohner et al., 1998; Bohner, Jarvis, Eyssel, & Siebler, 2005). Furthermore, research has shown associations between the endorsement of rape myths and hostility toward women, endorsement of stereotypical attitudes and sex roles for women, and negative evaluations of rape survivors (for a review, see Lonsway & Fitzgerald, 1994). Finally, rape myth acceptance has been shown to lead to greater victim blame, lower conviction rates for accused rapists, and shorter sentences for convicted rapists by juries in mock trials (Finch & Munro, 2005; also see Lonsway & Fitzgerald, 1994). It follows that rape myths may lead others to advise a sexual assault victim away from pressing charges, may lead law enforcement to doubt the legitimacy of a woman's claim, and may lead lawmakers away from enacting appropriate legislation.

As suggested above, rape myths serve to indirectly perpetuate sexual violence through creating beliefs and attitudes about sexual assault that distort the definition of sexual assault and shift the blame to the victim. Rape myths may also directly contribute to sexual violence by leading to a greater likelihood to commit sexual assault. Several studies have shown correlations between endorsement of rape myths and sexual aggression (e.g., Koss, Leonard, Beezley, & Oros, 1985), whereas other studies have shown causal associations between endorsement of rape myths and aggressive behavior against women (but not men) in the laboratory (e.g., Donnerstein & Malamuth, 1997). Finally, Lanier (2001), in a longitudinal study of 851 young men, found that rape myth supportive attitudes predicted sexually aggressive behavior, but sexual aggression did not predict rape myth attitudes. Despite these findings, it is admittedly difficult to go beyond correlational data or measures of aggression in a controlled setting to empirically establish a causal association between rape myth endorsement and actual sexual aggression.

Media Exposure and Views of Sexual Assault

Rape myths are part of transmitted culture. They get passed from person to person through many channels. Popular media are one such channel. Although we usually view movies and television shows as fictional accounts of events and news media as factual, the similarities between the two in promulgating stereotypical views of sexual assault are striking. The media's treatment of sexual assault not only serves to prime and reinforce rape myths in those who already hold them but also may construct these thoughts for those who do not already have them.

Evaluating sexual assault's treatment in the media, researchers have primarily focused on television shows and print journalism. Cuklanz (2000), evaluating prime-time television shows depicting sexual assaults from 1976 to 1990, found that acquaintance rapes became more prevalent in the late 1980s and that cases of false accusations were overrepresented on TV. Brinson (1992), reviewing 26 episodes involving sexual assault from various television shows in the 1980s, found that each episode averaged just more than five uses of rape myths, with the most commonly endorsed myths being "she asked for it" (46% of the episodes) and "she wanted it" (42% of the episodes). However, Cuklanz (1996, 2000) found fictional depictions on TV to be more sympathetic to sexual assault victims and issues than mass media depictions of actual sexual assault cases during the same period. This is consistent with findings that reality "crime-solving" police programs tend to engage in victim blame by focusing on victims instead of perpetrators to sensationalize crimes for the purpose of garnering viewers (Dobash, Schlesinger, Dobash, & Weaver, 1998). Cuklanz (1996) suggested that the "fragmented nature of news" (p. 50) perpetuates traditional, stereotypical views of sexual assault by discussing many elements out of context. First, victim-blame themes are common in newspaper accounts of sexual assault cases (Korn & Efrat, 2004; Los & Chamard, 1997; Smart & Smart, 1978). Los and Chamard (1997) reviewed several hundred cases of sexual assault covered in Canadian newspapers in the early 1980s. They found that although stranger rapes were reported more frequently during the 5-year period, acquaintance rape cases received more attention (i.e., more articles on the one case), and the reputation of the victim was usually the focus. Second, news media accounts of sexual assault cases seem to focus on the stereotypical stranger rape, unusual cases, and rare cases in which the accusation had been falsified (Caringella-MacDonald, 1998; Gavey & Gow, 2001; Los & Chamard, 1997; Soothill & Walby, 1991; Surette, 1992). Labeling a sexual assault claim as "false" may have more to do with law enforcement adhering to rape myths than with the actual dishonesty of the alleged victim, but Gavey and Gow (2001) found that allegedly false claims are taken as indisputable fact.

Although the vast majority of sexual assaults never get any publicity (Meyers, 1997), the ones that do get publicity serve an important role in shaping and maintaining our perceptions of sexual assault. The above research discusses rape myths on TV and in the print media, but quantitative research in this area remains scant, particularly with regard to the American press. The first goal of this research is to add to the existing literature by assessing the prevalence of rape myths in print journalism surrounding a highly publicized case of acquaintance rape. Although previous researchers have offered speculation about the effects of exposure to such media on attitudes about sexual assault, they have not empirically tested these effects. Therefore, the second goal of the present research is to assess the impact of the depiction of sexual assault in print journalism on people's opinions about a sexual assault case.

Overview

The two studies presented here address two related issues: (a) How do the media present information about sexual assault cases? and (b) How does this presentation affect its audience? The recent case involving Kobe Bryant allowed us to investigate the print news media's treatment of a high-profile sexual assault case. This research adds to the literature by empirically assessing the endorsement of rape myths in the American print media and then assessing the effects of exposure to these rape myths. Study 1 is a content analysis of more than 150 news articles for their endorsement of rape myths and other information that may have influenced readers about the case. In Study 2, we designed an experiment to assess the causal impact of exposure to rape myths in news articles on people's attitudes and beliefs about the case.

Study 1

Method

Sample. A total of 156 unique articles was gathered from 76 different online sources (major newspapers and news sources, such as CNN, ESPN). Based on U.S. Census Bureau groupings, 18 (11.5%) articles were from Northeastern newspapers, 16 (10.2%) from Midwestern newspapers, 20 (12.8%) from Southern newspapers, 44 (28.2%) from Western newspapers, and 58 (37.2%) from national papers or Web sites. Nine of the 10 most highly circulated newspapers (with the exception of *The Wall Street Journal*) were included in the sample, along with 30 other newspapers among the top 150 most widely circulated (Audit Bureau of Circulation, 2004). Given the large number of articles written about this trial, a sample of articles that was geographically diverse was analyzed for this article. Although only one article was chosen from most newspapers, some sources composed a greater percentage of the sample. The most articles from single sources came from *The Denver Post* (13 articles) and espn.com (12 articles).

Procedure. Collection of articles started when the media first broke the story on July 6, 2003, and stopped when the charges were dropped on September 1, 2004. The focus of this study was pretrial media. The articles were chosen using search engines (e.g., Google) and the keywords *Kobe Bryant sexual assault.*

Two raters coded the articles for endorsement of seven rape myths: (a) she's lying, (b) she asked for it, (c) she wanted it, (d) rape is trivial, (e) he didn't mean to, (f) he's not the kind of guy who would do this, and (g) it only happens to "certain" women (Burt, 1980). The articles were also coded for endorsement of any myths suggesting Kobe Bryant was guilty (e.g., because he is the type to cheat on his wife, he is probably also guilty of sexual assault), positive statements about Kobe Bryant, positive statements about the alleged victim, negative statements about Kobe Bryant, and mention of race (of the alleged victim or the alleged perpetrator). Myths were coded as present in an article only if the article endorsed those myths.[4] If an article mentioned rape myths by countering them, the myths were not counted in the present analysis. Only 13 of the 156 articles included statements countering rape myths. The two coders were trained together about the seven myths and common examples of each. They were each given a sample article to rate before being given the remainder of the articles. The two coders independently reached consensus on the myths present in this sample article. Each rater coded approximately half of the articles and coded 15 redundant articles to check for interrater reliability. The intraclass correlation to assess interrater reliability for this sample was high at $r_1 = .88$ (for an explanation of using intraclass correlations, see Shrout & Fleiss, 1979).

Results

Rape myths in articles. On average, there were 1.66 myth-endorsing statements per article, with 65.4% of the articles ($n = 102$) having at least one myth-endorsing statement. The number of myth-endorsing statements per article ranged from 0 statements (in 34.6% of the articles; $n = 54$) to 15 statements (in 0.6% of the articles; $n = 1$), as some myths were occasionally represented more than once in the same article. Also, on average, there was one distinct myth mentioned per article. As seen in Table 3.7, the most frequently endorsed myths

Table 3.7 Percentage of Articles Endorsing Each Rape Myth

Rape Myth	Articles Endorsing (%)	Articles (*n*)
She's lying	42.3	66
She asked for it	8.3	13
She wanted it	31.4	49
Rape is trivial	1.3	2
He didn't mean to	1.3	2
He's not the type	17.9	28
It only happens to "certain" women	1.9	3

NOTE: $N = 156$ articles.

were that the victim was lying (mentioned in 42.3% of all articles) and that she wanted it (mentioned in 31.4% of all articles). A contrast of proportions indicated that the articles were significantly more likely to endorse Rape Myth 1 (she's lying) than all other rape myths combined ($z = 7.90, p < .001$). Articles were significantly more likely to endorse Rape Myth 3 (she wanted it) than all other rape myths, excluding Rape Myth 1 ($z = 6.61, p < .001$). Articles were significantly more likely to endorse Rape Myth 6 (he's not the type) than all other rape myths, excluding Rape Myths 1 and 3 ($z = 4.67, p < .001$). Articles were more likely to endorse Rape Myth 2 (she asked for it) than all other rape myths, excluding Rape Myths 1, 3, and 6 ($z = 4.86, p < .001$). Endorsement of Rape Myths 4 (rape is trivial), 5 (he didn't mean to), and 7 (it only happens to certain women) did not significantly differ in this sample.

Other coded variables in articles. In addition to rape myths, we also coded the articles for other statements that may have influenced readers' opinions about the sexual assault case (Table 3.8). We found that 24.4% of the articles had at least one positive comment about Kobe Bryant as an athlete (e.g., "he is one of the best in the league" and "one of the greatest superstars ever to step foot on the court"—Kilson-Anderson, 2003), and 21.2% of the articles had at least one positive comment about Bryant as a person (e.g., "the boy next door"—Dilbeck, 2003; "a squeaky-clean image, a devoted husband and father"—Wilson, 2003). Admittedly, 67.0% ($n = 22$) of the articles with positive comments about Bryant as a person also had positive comments about Bryant as an athlete. In all, 41 (26.3%) unique articles had positive comments about Bryant as an athlete and/or person. In contrast, only 5.1% of the articles ($n = 8$) had positive comments about the victim as a person ($z = 5.47, p < .001$; e.g., "'A good kid,' reporters were told"—Eagan, 2003). It is important to note that the above analyses did not include other articles that discussed Bryant's performance during the 2003–2004 NBA season. Because the analysis was focused on articles about the sexual assault case, this research does not address the additional press Bryant was receiving outside of the case.

Although most of the information presented in the articles was likely to bias the audience in favor of Bryant's position (i.e., that a sexual assault did not occur), we also coded the articles for information that might bias the audience against Bryant's position. Only 2.6% of the articles included negative comments about Bryant as an athlete, and 14.1% of the articles included negative comments about Bryant as a person. Furthermore, 7.7% of the articles included statements that we coded as "myths about Kobe" because they drew unsubstantiated correlations between events to suggest Bryant's guilt (e.g., "People will say Kobe bought the verdict . . . call it O.J.'s legacy"—Reynolds, 2003). Therefore, myths questioning the alleged victim's honesty (found in 42.3% of articles) were much more common than were myths questioning Bryant's ($z = 7.62, p < .001$). Finally, in 23.5% of the articles ($n = 37$), Bryant and/or the alleged victim's race was mentioned (Bryant is Black and the alleged victim is White). Although Black men are not convicted at a higher rate than are White men in sexual assault cases, Black men convicted of sexual assault against a White woman get the longest prison sentences of all defendants (Wortman, 1985). This may stem from

Table 3.8 Percentage of Articles Including Other Potentially Influential Statements

Statement Type	Articles Including (%)	Articles (*n*)
Positive comments about Bryant as an athlete	24.4	38
Positive comments about Bryant as a person	21.2	33
Positive comments about victim as a person	5.1	8
Negative comments about Bryant as an athlete	2.6	4
Negative comments about Bryant as a person	14.1	22
Statements suggesting Bryant's dishonesty	7.7	12
Statements suggesting victim's dishonesty	42.3	66

NOTE: $N = 156$ articles.

a long-standing myth about the commonality of Black men's sexually assaulting White women (Brownmiller, 1975; Epstein & Langenbaum, 1994) and general stereotypes of Black men as violent (Peffley & Hurwitz, 1998). In addition, Knight, Giuliano, and Sanchez-Ross (2001) found that people were more likely to give harsher punishments to Black celebrities than Black non-celebrities in a hypothetical sexual assault case. Therefore, it is reasonable to think that the mention of race in these articles might have worked against Bryant's position.

Finally, given Kobe Bryant's celebrity status, it is likely that more information in general was written about Kobe Bryant than the alleged victim (who was not publicly known prior to this case). Therefore, a significant difference between the amount of information (positive and negative) written about Kobe Bryant and that written about the alleged victim is expected. A better test of a bias in the presentation of information about Bryant and the alleged victim is a within-person comparison of the amount of positive and negative information written about each. Of the articles, 24.0% had positive statements about Bryant as an athlete, and 2.6% of the articles contained negative statements about Bryant as an athlete. Articles were significantly more likely to include positive information about Bryant's athletic performance ($z = 5.91$, $p < .001$). Although one can present relatively objective data to determine whether or not Bryant deserved this positive assessment of his athletic skills, this information may still function to support Rape Myth 6. However, comparing statements about Bryant's character also shows that journalists were more likely to write positive than negative statements about Bryant as a person ($z = 1.65$, $p < .05$). In contrast, articles were much more likely to contain negative statements (i.e., she's lying, she wanted it, or she asked for it) than positive statements about the alleged victim's character ($z = 21.7$, $p < .001$).

Discussion

Analyzing articles from around the country spanning the 14 months from the point charges were filed until charges were dropped against Kobe Bryant, we found that a high percentage of articles include rape myth–endorsing statements. Rape myths negate the experience of the assault victim and perpetuate our misperceptions about sexual assault. The articles in Study 1 were most likely to endorse the myths that the alleged victim was lying and that a sexual assault therefore did not occur, that the alleged victim's actions indicated that she actually wanted the sex (that she is claiming was an assault), implying that the accused cannot be held responsible for interpreting her actions as such, and that the accused is not the type of man who would commit such a heinous act. The alarming frequency with which these myths are perpetuated in the media is highlighted in this content analysis, and the potential impact of these myths in this particular sexual assault case is the focus of Study 2.

Study 2

Study 1 demonstrated the prevalence of rape myths surrounding the Kobe Bryant case in articles from widely circulated newspapers. The purpose of Study 2 is to assess the causal impact of exposure to these myths on people's beliefs about this particular case by giving participants bogus articles about the case. In addition, Study 2 allows for a test of the impact of exposure to media breaking rape myths. It was expected that participants would be more likely to believe that Kobe Bryant was not guilty after reading an article endorsing rape myths than they did before they read the article. Participants should be more likely to believe that Kobe Bryant is guilty after reading an article challenging rape myths. Finally, participants should be more likely to hold beliefs in favor of Bryant's position after reading an article endorsing rape myths than after reading an article challenging rape myths.

Method

Participants. Participants were 62 undergraduate students (18 male, 44 female) at a Midwestern university. Their ages ranged from 18 to 49 years, with a mean age of 23.9 years ($SD = 5.3$) and a median age of 21.5 years. Of the sample, 87% were White, 3% Native American, 3% Asian, 2% Latino, and 5% Other or Missing.

Procedure. This study was conducted in the summer of 2004, approximately 1 year after news about the Kobe Bryant case broke. All data were collected before the charges were dropped on September 1, 2004. Participants were first asked five questions about their existing knowledge of the Kobe Bryant case: (a) if they knew who Kobe Bryant was, (b) if they knew that Kobe Bryant had been charged with sexual assault, (c) to rate how informed they believed they were on a 1 (*not at all informed*) to 7 (*extremely well informed*) scale about "the case in general," (d) to rate how well informed they believed they were about "physical evidence that may be used in court against Kobe Bryant," and (e) to rate how informed they believed they were about "the alleged victim's history." In addition, participants were asked to rate the extent to which they believed Kobe Bryant was guilty of the charges brought against him on a 1 (*definitely not guilty*) to 7 (*definitely guilty*) scale. Participants were then randomly assigned to read one of two fictitious articles about the case. One article was rape myth endorsing (RME) and the other was rape myth challenging (RMC).[5] The two articles were of approximately the same length (around 1,000 words), with the RME article including 11 statements endorsing rape myths and the RMC article including 9 statements countering rape myths. The RME article was fashioned after many of the actual articles that had been printed in the media (mainly focusing on Rape Myths 1, 3, and 6) but was a more extreme version of these articles in the sense that it included more RME statements than most of the articles from Study 1. The RME article included statements such as "We also know that the woman had planned to see Kobe that evening, expected him to make a move on her, was flirtatious with him, and admitted to willingly kissing him." The RMC article took a position cautioning readers against employing rape myths and gave reasons to "explain away" the myths that the media had been presenting (again focusing on Myths 1, 3, and 6). The RMC article included statements to counter rape myths such as

> Reports have stated that the accuser knew she would be seeing Bryant that night and that she expected him to make a move on her. These statements, though, do not imply that she indeed wanted *sex* or that she didn't change her mind once alone with Bryant.

After reading the article, participants were asked two questions about their opinions about the Kobe

Bryant case. As they were before the manipulation, participants were asked to rate the extent to which they believed that Kobe Bryant was guilty of the charges brought against him, and participants were asked to rate the extent to which they believed that the alleged victim was lying on a 1 (*definitely not lying*) to 7 (*definitely lying*) scale. Participants were asked about the victim's honesty post manipulation only to minimize bias and suspicion prior to reading the stimulus article. Also, following the manipulation, participants were again asked Questions 3 to 5 above regarding how informed they felt they were about the case. Participants then filled out a short demographics form including questions about gender, year in school, age, ethnicity, religiosity, and relationship status. Participants were thanked, debriefed, and given a list of community and campus resources for sexual assault support.

Results

Preexisting knowledge of the case. All but 2 participants said that they knew who Kobe Bryant was, and all but 1 of the remaining participants said that they knew that Kobe Bryant had been charged with sexual assault. The participants who knew about the sexual assault case rated their knowledge slightly below the midpoint on the 7-point scale ($M = 3.57$, $SD = 1.56$). Therefore, it follows that participants felt even less informed about the alleged victim's history ($M = 3.02$, $SD = 1.74$) and the physical evidence in the case ($M = 2.63$, $SD = 1.64$). That participants believed they knew more about the alleged victim's history than the physical evidence against Bryant is practically and statistically significant, $t(58) = -2.65$, $p < .01$. This is consistent with findings from Study 1 on the prevalence of rape myths in the media surrounding this case. There were no preexisting differences between experimental conditions on any of these variables, t values < 1.34. After reading the articles about the case, all participants felt more informed about the case in general, more informed about the victim's sexual history, and more informed about the physical evidence in the case (t values > -2.84, p values $< .01$) than they did prior to reading the article. Furthermore, there were no preexisting differences between experimental conditions on participants' ratings of Bryant's guilt, $t(57) = -0.75$, *ns*.

Beliefs about the case. Reading the articles shifted participants' beliefs about Bryant in the predicted directions. After reading the RME article, participants were more likely to believe that Bryant was not guilty, $t(25) = 3.49$, $p < .01$. After reading the RMC article, participants were less likely to believe that Bryant was not guilty, $t(29) = -3.10$, $p < .01$. In addition, participants who read the RME article ($M = 4.48$, $SD = 1.34$) were significantly more likely to believe that the alleged victim was lying than were those who read the RMC article ($M = 3.79$, $SD = 1.0$), $t(57) = 2.25$, $p < .05$.

Discussion

Study 2 highlights the causal effects of exposure to articles endorsing and challenging rape myths. This study demonstrated how exposure to articles endorsing rape myths leads participants to be more likely to side with the defendant in a sexual assault case than prior to exposure. Furthermore, exposure to articles challenging rape myths leads participants to be more likely to believe an alleged victim's claim of sexual assault than prior to exposure. Given the widespread endorsement of rape myths in the media (as supported by Study 1), Study 2 suggests the effects that such media exposure could have had on the Kobe Bryant sexual assault case and on sexual assault cases in general.

General Discussion

Taken together, findings from the current studies show the media's role in perpetuating rape myths and reinforcing beliefs about men and women who support sexual assault in American culture. Study 1 demonstrated the extent to which rape myths are endorsed in print journalism. More than 65% of articles discussing the Kobe Bryant sexual assault case included at least one statement endorsing popular rape myths. Finding that "she's lying" and "she wanted it" were the most commonly perpetuated myths was consistent with past research on rape myth endorsement in the media (Caringella-MacDonald, 1998; Los & Chamard, 1997). Study 2 allowed for a test of the effects of these myths in this particular case. Participants were much more likely to think that the defendant was not guilty after reading an RME article compared to pre-exposure beliefs, and participants were much more likely to think the defendant was guilty after reading an RMC article. Finally, participants were more likely to think that the victim was lying after reading an RME article than after reading an RMC article. Study 2 is an important demonstration of the potential devastating effects of the saturation of media coverage of sexual assault cases with rape myths.

In Study 1, not only did a high percentage of articles contain RME statements, but the articles also often contained other irrelevant information about Kobe Bryant and the alleged victim that might have swayed readers' opinions of the case. For example, numerous articles mentioned Bryant's (good) performance as an athlete during past and present NBA seasons. In addition, many articles discussed general sentiment by the public and other NBA players about Bryant's (good) character. The alleged victim in the case did not receive such additional positive editorial comments in articles about the case. And although it is important to acknowledge that there was irrelevant information presented in the articles that may have biased readers *against* Bryant (e.g., his race), this information was much less likely to be presented than information that led readers to believe a sexual assault did not occur.

The findings from Study 2 are consistent with findings from other studies investigating the effects of rape myths in the media. Exposure to rape myths reinforces people's prototypical representations of sexual assault, making them more likely to dismiss or explain away claims of sexual assault that do not fit their narrow definitions (for a review, see Lonsway & Fitzgerald, 1994). Franiuk, Seefelt, and Vandello (in press) found that men were less sympathetic to sexual assault victims after reading newspaper headlines endorsing rape myths. Furthermore, exposure to rape myths may either lead victims of sexual assault to dismiss their own experiences or scare them away from reporting sexual assault (Peterson & Muehlenhard, 2004; Pitts & Schwartz, 1997). According to survey evidence, only 10% to 40% of sexual assaults (and possibly far fewer) are reported to police (Kilpatrick, Edmunds, & Seymour, 1992; Koss, 1992; U.S. Department of Justice, 2003). Finally, exposure to rape myths can lead men to excuse or dismiss their own sexually assaulting behavior (Lonsway & Fitzgerald, 1995; Sinclair & Bourne, 1998).

Rape Myths and the Media

The Kobe Bryant case is unusual in that it deals with an acquaintance rape, unlike most sexual assault stories in the news (Los & Chamard, 1997). Therefore, the prevalence of rape myths used in this case is not surprising given that this case does not meet the criteria for the prototypical "stranger" rape. In other words, because this case does not meet the stereotypical criteria for a sexual assault, people may be particularly inclined to dismiss it as a sexual assault. Researchers have suggested that coverage of acquaintance rape cases often perpetuates rape myths by focusing on the misinterpretations and misunderstandings between the victim and the accused (Los & Chamard, 1997; Smart & Smart, 1978). Results from the present research support this assertion given the prevalence of rape myths found in Study 1.

Rape myths serve to not only perpetuate misinformation about sexual assault but also prevent communication of accurate information about sexual assault. Some of the information used against the alleged victim in this case (e.g., emotional instability, promiscuity) could have been used to discuss her *heightened* vulnerability to sexual assault (Gold, Sinclair, & Balge, 1999). Suggesting that a woman's promiscuity makes it more likely that she "wanted" the sexual assault is mutually exclusive of suggestions that sexual promiscuity may put a woman in more sexual situations, thereby increasing her chances of being assaulted (Koss, 1985; Koss & Dinero, 1989). Furthermore, Rape Myth 3 (suggesting she is promiscuous) and Rape Myth 7 (suggesting it only happens to promiscuous women) contradict one another. Suggesting that a woman is promiscuous implies that she is the type to want sex (and, therefore, cannot be assaulted), which runs contrary to the myth that sexual assault happens *only* to promiscuous women.

That people endorse these contradictory myths suggests that people employ not all rape myths at once, just the ones that assist in dismissing the current sexual assault. Although endorsement of rape myths is to be expected from a defendant's attorney, it seems that the court of public opinion, fueled by media reports, often tries a case before it actually makes it to the courtroom (Chancer, 2005). By her own admission, the alleged victim in this case was no longer willing to testify in the criminal trial after a year of being vilified by the press ("Experts Were to Testify," 2004). Results from Study 2 (and past research on rape myths) support her fears that she would not have been able to receive a fair trial from an unbiased judge and jury (for a review, see Lonsway & Fitzgerald, 1994).

Given what we know about people's perceptions of sexual assault, it is no surprise that we continue to see the media flooded with rape myths when charges of sexual assault make the headlines. However, given the prevalence of sexual assaults that do not fit the prototype, it may seem surprising that people have not changed their views of sexual assault. As social cognitive research on motivated reasoning and perseverance biases has repeatedly demonstrated, though, people will often maintain erroneous beliefs in the face of contradictory evidence (e.g., Anderson, Lepper, & Ross, 1980; Kunda, 1990). It takes mental effort to change existing beliefs, especially when this change may be threatening and cause personal distress. Although it is difficult to acknowledge truths about sexual assault, it is ironically more harmful to ourselves and others not to do so.

When confronted with data from studies demonstrating the negative influence of the media, many people defensively say that they are able to separate "truth" from fiction in television shows and movies. Although the effects of the media are well documented (e.g., Bryant & Zillman, 1994; Emmers-Sommer & Allen, 1999), there are likely certain audiences that give less credibility to the messages they receive from television and movies than do other audiences. Messages received through television and print journalism, however, may not be filtered with the same skepticism as other media. Although some viewers may dismiss rape myths in movies and television shows as distortions of reality, these same viewers may look at television and print news as unbiased presentations of fact (Gaziano, 1988; Robinson & Kohut, 1988; Surette, 1992). Any form of media that transmits rape myths is clearly problematic, but news media may have a greater impact on audiences' (false) beliefs about sexual assault given our almost blind faith that their reports are impartial. Rape myths in the news may contribute to the development of rape myths, and, more likely, they may prime rape myths already held by the audience and make people more likely to use them in the future (Malamuth & Check, 1985).

Of interest, sexual assault cases seem to present a counterexample to the common finding that pretrial publicity (PTP) usually biases potential jurors *against* defendants (for a review, see Devine, Clayton, Dunford, Seying, & Pryce, 2001); however, much PTP related to sexual assaults is anti prosecution and/or prodefendant.[6] Past researchers have found that men, in particular, are less likely to display an antidefendant bias after exposure to PTP for sexual assault cases (Hoiberg & Stires, 1973; Mullin, Imrich, & Linz, 1996). The current studies suggest a possible reason for PTP biases favoring defendants in sexual assault cases—namely, journalists' employment of rape myths.

It is important to note that journalists' motives behind endorsing rape myths in coverage of sexual assault cases are not elucidated by the present research. First, according to Websdale and Alvarez (1998), journalists' use of "forensic journalism" causes them to give many details of a crime without discussing these details within the context of the greater social issues related to that crime. This journalistic strategy also involves getting information quickly, regardless of the source, and is readily capitalized on by defense attorneys in sexual assault cases (Chancer, 2005; Websdale & Alvarez, 1998). Second, it is possible that journalists consciously employ rape myths in their writing to sensationalize a story and increase newspaper sales. It is also possible that the use of rape myths in print journalism is less a reflection of malicious intent by an author and more a reflection of that author's internalization of our culture's beliefs about sexual assault. Journalists may believe that they are merely presenting reasonable alternatives to a sexual assault claim. Although it would be interesting to assess journalists' personal endorsement of rape myths, bringing the current research to the attention of journalists is important for reducing rape myths in the print media regardless of journalists' motives.

In Study 1, we found that nearly 35% of articles did not mention rape myths at all. Clearly, bias-free journalism is possible, if not probable. The media have great potential for positive effects, too, as past research has shown the positive impact of prosocial messages on television (Fisch, Truglio, & Cole, 1999; McAlister, 2000; Mussen & Eisenberg-Berg, 1977). In addition to showing the negative impact of media coverage of sexual assault cases, Study 2 suggests the potentially powerful effects of *countering* rape myths when discussing sexual assault cases. RMC messages in the media can break myths (at least temporarily). It is important to note that in Study 2 it is impossible to determine whether reading an RMC article leads people to espouse an antidefendant stance unfairly or helps people evaluate the case on the evidence and accurate information about sexual assault. Future research to delineate the short- and long-term effects of both RME messages and RMC messages is imperative.

On September 1, 2004, the district attorney's office in Eagle, Colorado, dropped the charges against Kobe Bryant primarily because of the alleged victim's decision not to testify. Many have speculated about this turn of events, but the alleged victim has said little publicly about her decision. Her attorneys said that she believed she could not get a fair trial after all of the leaks and errors in this case ("Experts Were to Testify," 2004). Furthermore, her attorneys cited the alleged victim's fear of how she was going to be the one put on trial through cross-examination (which would undoubtedly employ rape myths; "Experts Were to Testify," 2004). We will never know what specific role the media's saturation with rape myths played in the alleged victim's decision, but given the research presented here, we can fairly confidently cite negative repercussions. And more important, we will never know the full impact that this case will have on future sexual assault victims and perpetrators. Research has shown that men are more likely to accept rape myths after a not-guilty verdict in a sexual assault case (Sinclair & Bourne, 1998). At least for men, not-guilty verdicts strengthen their beliefs that excuse men as perpetrators of sexual assault. The rape myths surrounding the Bryant case likely played a large role in preventing the alleged victim from believing she could receive a fair trial; and, consequently, her decision to not testify in the criminal trial resulted in an effectively "not-guilty" verdict for Bryant that further validates rape myths.

Summary

One of the biggest barriers that remains for reducing sexual assault is people's inability or refusal to recognize it when it occurs. When a sexual assault does not meet the criteria for a prototypical sexual assault (and it often does not), we are likely to use rape myths. Unfortunately, the employment of rape myths creates a vicious cycle that makes it increasingly harder for sexual assault victims to report the crime. In the unlikely event that a sexual assault victim actually reports the crime, rape myths are self-reinforcing when they influence the way the victim is treated along the entire chain of the criminal justice process (e.g., intake by hospital personnel, questioning and investigation by law enforcement, jury verdicts, judge's sentencing). For example, the myth that women lie about sexual assault may contribute to law enforcement personnel's not accepting a woman's sexual assault claim and deeming it a false report or "unfounded" (Estrich, 1987). This boosts the belief that women often make false claims of sexual assault and leads law enforcement to be more skeptical the next time a woman claims assault, thereby fueling the "she's lying" myth. Unfortunately, though, sexual assaults that *do* conform to the prototype reinforce rape myths as well. Although these women are more likely to be believed and these cases more likely to be prosecuted, giving legitimacy to the prototypical assault and dismissing the atypical assault reinforces the prototype and the rape myths that support it. In the present research, we discuss one way that rape myths are reinforced in our culture. The more rape myths are used in the media, the more accessible they are to those responding to sexual assault victims and the harder it is to eliminate sexual assault.

Notes

1. It is difficult to get true estimates on false accusation rates in sexual assault cases. Lonsway and Fitzgerald (1994), in a review of the literature on rape myths, found numbers ranging from 2% to 9% in data estimating falsely reported cases. With any crime, a certain level of subjectivity and uncertainty may be involved in labeling a report a "false report" when allegations are not backed up by sufficient evidence. However, sexual assault cases may be particularly susceptible to being labeled false reports given the tendencies for law enforcement to employ rape myths and to reduce sexual assault cases to "he said–she said" situations. In the present analysis, because no one besides Bryant and his alleged victim can know who is lying in this case, statements questioning Bryant's and the alleged victim's honesty should be equally presented in newspaper articles.

2. It should be noted that recent research has identified some groups that are at a high risk to commit rape, namely, fraternity members (Boswell & Spade, 1996; Humphrey & Kahn, 2000) and athletes (for a review, see Benedict, 1998). Therefore, it is possible that a popular athlete such as Kobe Bryant fits people's prototypical notions of someone who would be likely to commit sexual assault. However, results from Study 1 indicate otherwise.

3. For this article, feminine gender pronouns will be used to refer to the victim and masculine gender pronouns to refer to the perpetrator. Although men are victims of sexual assault (in approximately 10% of reported cases) and women are perpetrators of sexual assault (in approximately 2% of reported cases), the overwhelming majority of sexual assaults are committed by men on women (Bureau of Justice Statistics, 1999; U.S. Department of Justice, 2003).

4. The authors of this article consider "endorsement" and "presence" of a rape myth to be equivalent. In the absence of any qualifying statements to admonish the particular rape myth, "use" of a rape myth in these articles (e.g., suggesting that a woman "asked for it" by her actions) is implicit endorsement. We are not suggesting that any one author's purpose is explicit endorsement of rape myths; it is our assumption that most journalists believe that they are reporting unbiased "facts."

5. Both articles are available from the first author on request.

6. In addition, Brown, Duane, and Fraser (1997) have suggested that pretrial publicity may generate public sympathy for celebrity defendants they are motivated to like. They found that media exposure was correlated with greater beliefs in O.J. Simpson's innocence, regardless of race and gender of the respondent.

7. Less than 5% of participants answered a 1 (*not at all informed*) when asked to rate their self-knowledge about the case. Of participants, 24% felt they were "not at all informed" about the alleged victim's history, and 30% of participants felt that they were "not at all informed" about the physical evidence against Bryant. That participants felt least informed about the alleged evidence against Kobe Bryant is consistent with the data presented in Study 1.

References

Anderson, C. A., Lepper, M. R., & Ross, L. (1980). Perseverance of social theories: The role of explanation in the persistence of discredited information. *Journal of Personality and Social Psychology, 39*, 1037–1049.

Audit Bureau of Circulation. (2004, September 30). *FAS-FAX*. Retrieved April 9, 2005, from http://www .accessabc.com

Benedict, J. R. (1998). *Athletes and acquaintance rape*. Thousand Oaks, CA: Sage.

Bohner, G., Jarvis, C. I., Eyssel, F., & Siebler, F. (2005). The causal impact of rape myth acceptance on men's rape proclivity: Comparing sexually coercive and noncoercive men. *European Journal of Social Psychology, 35*, 819–828.

Bohner, G., Reinhard, M.-A., Rutz, S., Sturm, S., Kerschbaum, B., & Effler, D. (1998). Rape myths as neutralizing cognitions: Evidence for a causal impact of anti-victim attitudes on men's self-reported likelihood of raping. *European Journal of Social Psychology, 28*, 257–268.

Boswell, A. A., & Spade, J. Z. (1996). Fraternities and collegiate rape culture: Why are some fraternities more dangerous places for women? *Gender & Society, 10*, 133–147.

Brinson, S. L. (1992). The use and opposition of rape myths in prime-time television dramas. *Sex Roles, 27*, 359–375.

Brown, W. J., Duane, J. J., & Fraser, B. P. (1997). Media coverage and public opinion of the O.J. Simpson trial: Implications for the criminal justice system. *Communication Law and Policy, 2*, 261–287.

Brownmiller, S. (1975). *Against our will: Men, women and rape.* New York: Simon & Schuster.

Bryant, J., & Zillman, D. (1994). *Media effects: Advances in theory and research*. Hillsdale, NJ: Lawrence Erlbaum.

Bureau of Justice Statistics. (1998). *Crime in the United States and in England and Wales, 1981–1996*. Washington, DC: U.S. Department of Justice.

Bureau of Justice Statistics. (1999). *Women offenders*. Washington, DC: U.S. Department of Justice.

Burt, M. R. (1980). Cultural myths and support for rape. *Journal of Personality and Social Psychology, 38*, 217–230.

Burt, M. R. (1991). Rape myths and acquaintance rape. In A. Parrot & L. Bechhofer (Eds.), *Acquaintance rape: The hidden crime* (pp. 26–40). New York: John Wiley.

Burt, M. R., & Albin, R. S. (1981). Rape myths, rape definitions, and probability of conviction. *Journal of Applied Social Psychology, 11*, 212–230.

Caringella-MacDonald, S. (1998). The relative visibility of rape cases in national popular magazines. *Violence Against Women, 4*, 62–80.

Chancer, L. S. (2005). *High-profile crimes: When legal cases become social causes*. Chicago: University of Chicago Press.

Check, J. V. P., & Malamuth, N. M. (1983). Sex role stereotyping and reactions to depictions of stranger versus acquaintance rape. *Journal of Personality and Social Psychology, 45*, 344–356.

Check, J. V. P., & Malamuth, N. M. (1985). An empirical assessment of some feminist hypotheses about rape. *International Journal of Women's Studies, 8*, 414–423.

Cowan, G., & Curtis, S. R. (1994). Predictors of rape occurrence and victim blame in the William Kennedy Smith Case. *Journal of Applied Social Psychology, 24*, 12–20.

Cuklanz, L. M. (1996). *Rape on trial: How the mass media construct legal reform and social change*. Philadelphia: University of Pennsylvania Press.

Cuklanz, L. M. (2000). *Rape on prime time: Television, masculinity, and sexual behavior*. Philadelphia: University of Pennsylvania Press.

Devine, D. J., Clayton, L. D., Dunford, B. B., Seying, R., & Pryce, J. (2001). Jury decision making: 45 years of empirical research on deliberating groups. *Psychology, Public Policy, and Law, 7*, 622–727.

Dilbeck, S. (2003, July 19). Beyond belief even in age of athletes' bad behavior, charge against Kobe shocking. *Daily News*, p. S1.

Dobash, R. E., Schlesinger, P., Dobash, R., & Weaver, C. K. (1998). "Crimewatch UK": Women's interpretation of televised violence. In M. Fishman & G. Cavender (Eds.), *Entertaining crime: Television reality programs* (pp. 37–58). Hawthorne, NY: Aldine de Gruyter.

Donnerstein, E., & Malamuth, N. M. (1997). Pornography: Its consequences on the observer. In L. B. Schlesinger & E. Revitch (Eds.), *Sexual dynamics of anti-social behavior* (2nd ed., pp. 30–49). Springfield, IL: Charles C Thomas.

Eagan, M. (2003, July 20). Stereotypes define Kobe case. *Boston Herald*, p. 16.

Emmers-Sommer, T. M., & Allen, M. (1999). Surveying the effect of media effects: A meta-analytic summary of the media effects in Human Communication Research. *Human Communication Research, 25*, 478–497.

Epstein, J., & Langenbaum, S. (1994). *The criminal justice and community response to rape*. Washington, DC: U.S. Department of Justice.

Estrich, S. (1987). *Real rape*. Cambridge, MA: Harvard University Press.

Experts were to testify about "battering ram" injuries. (2004, September 8). Retrieved January 29, 2005, from http://www.espncom

Finch, E., & Munro, V. E. (2005). Juror stereotypes and blame attribution in rape cases involving intoxicants. *British Journal of Criminology, 45*, 25–38.

Fisch, S., Truglio, R. T., & Cole, C. F. (1999). The impact of Sesame Street on preschool children: A review and synthesis of 30 years' research. *Media Psychology, 1*, 165–190.

Fischer, G. J. (1987). Hispanic and majority student attitudes toward forcible date rape as a function of differences in attitudes toward women. *Sex Roles, 17*, 93–101.

Foley, L. A., Evancic, C., Karnik, K., King, J., & Parks, A. (1995). Date rape: Effects of race of assailant and victim and gender of subjects on perceptions. *Journal of Black Psychology, 21*, 6–18.

Franiuk, R., Seefelt, J. L., & Vandello, J. A. (in press). Prevalence of rape myths in headlines and their effects on attitudes toward rape. *Sex Roles*.

Freeman, J. (1993). The disciplinary function of rape's representation: Lessons from the Kennedy Smith and Tyson trials. *Law and Social Inquiry, 18*, 517–546.

Gavey, N., & Gow, V. (2001). "Cry wolf," cried the wolf: Constructing the issue of false rape allegations in New Zealand media texts. *Feminism and Psychology, 11*, 341–360.

Gaziano, C. (1988). How credible is the credibility crisis? *Journalism Quarterly, 65*, 267–279.

Gold, S. R., Sinclair, B. B., & Balge, K. A. (1999). Risk of sexual revictimization: A theoretical model. *Aggression and Violent Behavior, 4*, 457–470.

Goodchilds, J. D., Zellman, G. L., Johnson, P. B., & Giarrusso, R. (1988). Adolescents and their perceptions of sexual interactions. In A. W. Burgess (Ed.), *Rape and sexual assault* (Vol. 2, pp. 245–270). New York: Garland.

Hoiberg, B. C., & Stires, L. K. (1973). The effect of several types of pretrial publicity on the guilt attributions of simulated jurors. *Journal of Applied Social Psychology, 3*, 267–275.

Humphrey, S. E., & Kahn, A. S. (2000). Fraternities, athletic teams, and rape: Importance of identification with a risky group. *Journal of Interpersonal Violence, 15*, 1313–1322.

Jimenez, J. A., & Abreu, J. M. (2003). Race and sex effects on attitudinal perceptions of acquaintance rape. *Journal of Counseling Psychology, 50*, 252–256.

Kilpatrick, D. G., Edmunds, C. N., & Seymour, A. K. (1992). *Rape in America: A report to the nation*. Arlington, VA: National Victim Center and Medical University of South Carolina.

Kilson-Anderson, K. (2003, October 31). Kobe: Damned if he did, damned if he didn't. *Portland State University Vanguard*. Retrieved October 1, 2004, from http://www.dailyvanguard.com/vnews/display.v/ART/2003/10/31/3fa20e7f58dd8

Knight, J. L., Giuliano, T. A., & Sanchez-Ross, M. G. (2001). Famous or infamous? The influence of celebrity status and race on perceptions of responsibility for rape. *Basic and Applied Social Psychology, 23*, 183–190.

Korn, A., & Efrat, S. (2004). The coverage of rape in the Israeli popular press. *Violence Against Women, 10*, 1056–1074.

Koss, M. P. (1985). The hidden rape victim: Personality, attitudinal, and situational characteristics. *Psychology of Women Quarterly, 9*, 193–212.

Koss, M. P. (1992). The underdetection of rape: Methodological choices influence incidence estimates. *Journal of Social Issues, 48*, 61–75.

Koss, M. P., & Dinero, T. E. (1989). Discriminant analysis of risk factors for sexual victimization among a national sample of college women. *Journal of Consulting and Clinical Psychology, 57*, 242–250.

Koss, M. P., Leonard, K. E., Beezley, D. A., & Oros, C. J. (1985). Nonstranger sexual aggression: A discriminant analysis of the psychological characteristics of undetected offenders. *Sex Roles, 12*, 981–992.

Kunda, Z. (1990). The case for motivated reasoning. *Psychological Bulletin, 108*, 480–498.

Lanier, C. A. (2001). Rape-accepting attitudes: Precursors to or consequences of forced sex. *Violence Against Women, 7*, 876–885.

Lerner, M. J. (1980). *The belief in a just world: A fundamental delusion*. New York: Plenum.

Linz, D. G. (1989). Exposure to sexually explicit materials and attitudes toward rape: A comparison of study results. *Journal of Sex Research, 26*, 50–84.

Linz, D. G., Donnerstein, E., & Adams, S. M. (1989). Physiological desensitization and judgments about female victims of violence. *Human Communication Research, 15,* 509–522.

Lonsway, K. A., & Fitzgerald, L. F. (1994). Rape myths: In review. *Psychology of Women Quarterly, 18,* 133–164.

Lonsway, K. A., & Fitzgerald, L. F. (1995). Attitudinal antecedents of rape myth acceptance: A theoretical and empirical reexamination. *Journal of Personality and Social Psychology, 68,* 704–711.

Los, M., & Chamard, S. E. (1997). Selling newspapers or educating the public? Sexual violence in the media. *Canadian Journal of Criminology, 39,* 293–328.

Malamuth, N. M., & Check, J. V. P. (1985). The effects of aggressive pornography on beliefs in rape myths: Individual differences. *Journal of Research in Personality, 19,* 299–320.

McAlister, A. (2000). Action-oriented mass communication: Theory and application illustrated. In J. Rappaport & E. Seidman (Eds.), *Handbook of community psychology* (pp. 379–396). New York: Academic Press.

McCaul, K. D., Veltum, L. G., Boyechko, V., & Crawford, J. J. (1990). Understanding attributions of victim blame for rape: Sex, violence, and foreseeability. *Journal of Applied Social Psychology, 20,* 1–26.

Meyers, M. (1997). *News coverage of violence against women: Engendering blame.* Thousand Oaks, CA: Sage.

Mixon, K. D., Foley, L. A., & Orme, K. (1995). The influence of racial similarity on the O.J. Simpson trial. *Journal of Social Behavior & Personality, 10,* 481–490.

Muehlenhard, C. L., & MacNaughton, J. S. (1988). Women's beliefs about women who "lead men on." *Journal of Social and Clinical Psychology, 7,* 65–79.

Mullin, C., Imrich, D. J., & Linz, D. (1996). The impact of acquaintance rape stories and case-specific pretrial publicity on juror decision making. *Communication Research, 23,* 100–135.

Mussen, P., & Eisenberg-Berg, N. (1977). *Roots of caring, sharing, and helping: The development of pro-social behavior children.* Oxford, UK: W. H. Freeman.

Norris, J., & Cubbins, L. A. (1992). Dating, drinking, and rape: Effects of victim's and assailant's alcohol consumption on judgments of their behavior and traits. *Psychology of Women Quarterly, 16,* 179–191.

Peffley, M., & Hurwitz, J. (1998). Whites' stereotypes of Blacks: Sources and political consequences. In J. Hurwitz & M. Peffley (Eds.), *Perception & prejudice* (pp. 58–99). New Haven, CT: Yale University Press.

Peterson, Z. D., & Muehlenhard, C. L. (2004). Was it rape? The function of women's rape myth acceptance and definitions of sex on labeling their own experiences. *Sex Roles, 51,* 129–144.

Pitts, V. L., & Schwartz, M. D. (1997). Self-blame in hidden rape cases. In M. D. Schwartz (Ed.), *Researching sexual violence against women: Methodological and personal perspectives* (pp. 65–70). Thousand Oaks, CA: Sage.

Reynolds, B. (2003, August 7). Must-see TV, sadly, is Kobe on trial. *Providence Journal,* p. D01.

Robinson, M., & Kohut, A. (1988). Believability and the press. *Public Opinion Quarterly, 52,* 174–189.

Sawyer, R. G., Pinciaro, P. J., & Jessell, J. K. (1998). Effects of coercion and verbal consent on university students' perception of date rape. *American Journal of Health Behavior, 22,* 46–53.

Shrout, P. E., & Fleiss, J. L. (1979). Intraclass correlations: Uses in assessing rater reliability. *Psychological Bulletin, 86,* 420–428.

Sinclair, H. C., & Bourne, L. E. (1998). Cycle of blame or just world: Effects of legal verdicts on gender patterns in rape-myth acceptance and victim empathy. *Psychology of Women Quarterly, 22,* 575–588.

Smart, C., & Smart, B. (1978). *Women, sexuality and social control.* London: Routledge & Kegan Paul.

Soothill, K., & Walby, S. (1991). *Sex crime in the news.* London: Routledge.

Surette, R. (1992). *Media, crime and criminal justice: Images and realities.* Pacific Grove, CA: Brooks/Cole.

U.S. Department of Justice. (2003). *Criminal victimization, 2003.* Washington, DC: Author.

Varelas, N., & Foley, L. A. (1998). Blacks' and Whites' perceptions of interracial and intraracial date rape. *Journal of Social Psychology, 138,* 392–400.

Websdale, N., & Alvarez, A. (1998). Forensic journalism as patriarchal ideology: The newspaper construction of homicide-suicide. In F. Bailey & N. Dale (Eds.), *Popular culture, crime & justice* (pp. 123–141). Belmont, CA: Wadsworth.

Wilson, A. (2003, July 20). No matter the outcome of his trial, Kobe will pay a steep price. *Buffalo News,* p. C2.

Wortman, M. S. (1985). *Women in American law.* New York: Holmes & Meier.

Wyer, R. S., Bodenhausen, G. V., & Gorman, T. F. (1985). Cognitive mediators of reactions to rape. *Journal of Personality and Social Psychology, 48,* 324–338.

DISCUSSION QUESTIONS

1. What role do rape myths play in the reporting of sexual assault cases in the media?
2. How does the use of rape myths affect the culpability of alleged victims and perpetrators of sexual assault?
3. What impact does the reporting of high-profile rape cases have for the everyday victim of these crimes?

READING

As you learned in the section introduction, a variety of different substances are used in cases of drug-facilitated sexual assault. While we tend to think these types of assaults involve intoxicating individuals without their knowledge or consent, there are also cases where sexual assault occurs after someone voluntarily becomes intoxicated. While states limit the ability to consent to sexual activity if one is under the influence of drugs or alcohol (thereby labeling these cases as rape and sexual assault), this study highlights the differences in the assignment of blame and responsibility of the victim and offender based on whether someone was voluntary or involuntarily intoxicated.

The Role of the New "Date Rape Drugs" in Attributions About Date Rape

April L. Girard and Charlene Y. Senn

Sexual assault, including date and acquaintance rape, has been discussed quite thoroughly since the 1980s. As adapted from the Criminal Code of Canada, sexual assault can be defined as any sexual contact without voluntary consent (Criminal Code, 1985, s. 271). This definition of sexual assault has been broadened by case law to include forced kissing, fondling, anal intercourse, and oral sex. Under the U.S. Federal Criminal Code, sexual abuse (including rape) is similarly defined (Sexual Abuse Act, 1986). Although there are many ways to define sexual assault, it is ultimately society's attitudes and attributions that determine how it is labeled (Burt, 1980).

Studies have determined that the rate of sexual assault in North America is very high, especially among college-age women. In the United States, 15% of college women report an experience that fits the legal definition of rape (Brener, McMahon, Warren, & Douglas, 1999; Koss, Gidycz, & Wisniewski, 1987). Dekeseredy, Schwartz, and Tait (1993) discovered that 32.8% of Canadian female undergraduates surveyed had been victims of some form of sexual assault.

On college campuses, many sexual assaults involve the use of drugs or alcohol, often within the context of a dating relationship. A total of 13% of Canadian female students studied said that when they were drunk or high, a man attempted unwanted sexual intercourse (Dekeseredy et al., 1993). If more general questions are asked about whether alcohol was consumed by either perpetrator or victim preceding sexual coercion or assault, even higher rates are reported in the United States (Abbey, Ross, McDuffie, & McAuslan, 1996). Although intoxication or belief that consent has been given are not defenses to sexual assault under section 273.2 of the Canadian Criminal Code (1992) and deliberate intoxication is an aggravating factor under U.S. law since 1986 (aggravated sexual abuse), men may be more likely to misinterpret friendly cues as sexual invitations,

SOURCE: Girard, A. L. & Senn, C. Y. (2008). The role of the new "date rape drugs" in attributions about date rape. *Journal of Interpersonal Violence, 23*(1), 3–20.

and women are more at risk of having diminished coping responses and being unable to ward off a potential attack when under the influence of alcohol (Abbey, Zawacki, Buck, Clinton, & McAuslan, 2004; Adams-Curtis & Forbes, 2004; Dudley, 2005; Rickert & Weinmann, 1998). This may begin to explain why the problem of alcohol- and drug-involved rapes is so rampant.

Alcohol is the most common drug associated with allegations of drug-facilitated sexual assaults, and many studies contribute to our knowledge about the relationship between date rape and alcohol. The Canadian Federation of Students (2001) reported that 90% of the sexual assaults reportedly experienced by Canadian female students involved alcohol. Yet, the relationship between sexual assault and other drugs is virtually unknown, despite the growing concern about the use of other drugs, such as Rohypnol, GHB, and Ketamine, in sexual assaults since the late 1990s (Hensley, 2002; Weir, 2001). Any substance that is administered to lower sexual inhibition and enhance the possibility of unwanted sexual intercourse is potentially a date rape drug (Weir, 2001). However, the term *date rape drugs* has been coined by the media to label a few specific drugs (i.e., Rohypnol, GHB, and Ketamine) because of the frequency with which they are used by men to facilitate rape (Hindmarch & Brinkmann, 1999). The use of these drugs for the purpose of inducing amnesia and rapid sedation of the victim is becoming more common (Pope & Shouldice, 2001; Hensley, 2002).

The majority of research conducted on date rape drugs focuses on their effects on individuals. Rohypnol, GHB, and Ketamine have many depressant effects that resemble the effects of alcohol. Therefore, neither the victim nor others around them are likely to be aware of the drugging. The victim may experience confusion, dizziness, nausea, visual disturbances, physical and/or motor impairment, reduced inhibition, drowsiness, impaired judgment, slurred speech, amnesia, and an inability to stop the offender because they are incapable of resisting. The amnesia is a particularly troubling symptom as this makes it much more difficult for the victim to accurately report the crime (Canadian Federation of Students, 2001).

The symptoms associated with the administration of these drugs compound the problem for rape victims because they affect many of the domains on which society makes its judgments of rape and rape victims. A common belief in North American society is that women are responsible for the regulation of sexual interactions between men and women (Burt, 1980). This belief is held widely despite the reality that factors such as who drove, the location, and the activities on the date, all contribute to the increased risk of date rape, but are all traditionally controlled by men (Rickert & Weinmann, 1998). The belief in women's responsibility could explain why people are less likely to perceive a forced sexual encounter as rape if the man invested time and money into the situation (Jenkins & Dambrot, 1987). People in North America also tend to perceive date rape as more permissible than stranger rape (Rickert & Weinmann, 1998). In fact, the greater the acquaintanceship between a man and woman, the less likely people are to judge an instance of forced sexual contact as sexual assault (Bridges, 1991).

Research has demonstrated that attributions vary widely when alcohol is involved in sexual assaults (Adams-Curtis & Forbes, 2004; Cameron & Stritzke, 2003; Dudley, 2005; Finch & Munro, 2005; Rickert & Weinmann, 1998; Schuller & Wall, 1998; Stormo, Lang, & Stritzke, 1997). More blame and responsibility may be attributed to the perpetrator for taking advantage of the victim. However, sometimes more blame and responsibility are attributed to the victim for "putting themselves in that situation" by getting drunk. Yet, because alcohol is legal and has been observed quite frequently in sexual assaults, it may be viewed differently than drug-related sexual assaults. Alcohol use is almost uniformly socially acceptable whereas drug use is not. Furthermore, the individuals involved may be perceived differently. If a female voluntarily drinks alcohol and is sexually assaulted, she may be viewed differently than a female who voluntarily takes drugs and is sexually assaulted.

There is a lack of research on attitudes and attributions toward victims and perpetrators in drug-related date rapes. Therefore, the goal of this study was to explore people's attributions when date rape drugs are involved in a rape. This study was designed to extend the work of Stormo et al. (1997) on attributions toward perpetrators and victims related to alcohol use. Their scenarios describing a clear date rape situation between a male and female (John and Cathy) were used as the control (both individuals sober) and the alcohol (both individuals drunk) scenarios respectively. The scenario was modified in the first study to include a third and fourth condition. These represent a voluntary drug situation (both individuals take drugs), and an involuntary drug situation (Cathy is drugged without her knowledge). GHB is a date rape drug that is also sometimes taken recreationally; therefore, it was an appropriate drug for these scenarios. Our second study was identical to the first except we added a fifth condition (involuntary drunk) to provide a better comparison for our involuntary drugging scenario. In this scenario, John purposely mixes much stronger drinks for Cathy than for himself, without her knowledge.

Based on the literature that contributes to our understanding of the features of rape situations that change attributions about victims and perpetrators, it was expected that purposeful drugging of women would be seen as more serious than the usual acquaintance rape scenario with no drug or alcohol involvement. Therefore, it was hypothesized that the perpetrator would be held more responsible and blameworthy when he purposefully drugged the victim than in all other situations presented. Similarly, it was hypothesized that people would attribute less blame to the involuntarily drugged victim than they would to the drunk or sober victim. Furthermore, it was expected that more blame would be attributed to a victim that willingly ingested a drug or drank alcohol than a victim who had no control over the situation or one who remained sober. It was also hypothesized that those who were more accepting of rape myths would blame the victim more and the perpetrator less across all situations than those who have lower rape-myth acceptance. A gender difference was expected to emerge with women blaming the victim less than men, regardless of her behavior.

Method

Participants

The participants were 280 undergraduates (143 men and 137 women), 160 participants in Study 1 (80 men and 80 women) and 120 in Study 2 (63 men and 57 women). Participants were chosen at random with gender-balance criteria from the participant pool in the Department of Psychology. For the two studies combined, the age of the participants ranged from 17 to 53, with a mean age of 21.75 ($SD = 4.90$). The majority of the participants were heterosexual (94%) and White (75%). The remaining 25% of non-White students were Latin American (1%), Middle Eastern (5%), African (2%), and Far Eastern (17%).

Date rape scenarios (DRS). The scenarios used in this study were adapted from the two rape scenarios in Stormo et al.'s (1997) study, one in which both individuals were drinking the same thing, and the other in which both individuals were sober. A third and fourth scenario were added to include the use of date rape drugs, one in which both John and Cathy take the drug voluntarily while also drinking and one in which John makes a regular drink for himself but slips a drug into Cathy's drink without her knowledge. In the fifth scenario, John purposely mixes Cathy's drinks much stronger (triple strength) than his without her knowledge.

Results

Coercive Sexual Experiences

A total of 75% of the participants had engaged in consensual sexual activity that included intercourse. 56% of female participants had never been victimized, 27% had been sexually coerced, 6% had been sexually abused, and 11% sexually assaulted (i.e., completed rape). Of the males in the sample, 83% reported never having engaged in coercion of any type, 13% admitted to being sexually coercive, 1% to being sexually abusive, and 3% to being sexually assaultive.

Victimization. Nine women (6.6%) reported that sexual activity against their wishes was attempted or completed by a man who gave them alcohol or drugs. Eight of these women had experienced attempted sexual assault involving drugs or alcohol with two thirds ($n = 5$) having more than one such incident. The majority of these experiences (75%, $n = 6$) involved alcohol rather than drugs (25%, $n = 2$). Seven women reported completed sexual assault involving drugs or alcohol,[1] with 43% of those women ($n = 3$) reporting more than one such incident. The majority of these completed sexual assault experiences (71%, $n = 5$) involved alcohol. However, there were two drug-related experiences (29%, $n = 2$).

Beyond the experiences participants were sure of, a number had symptoms on at least one occasion that made them believe that they had been drugged. A total of 20% of the women in the sample ($n = 17$) suspected that they had been drugged due to their feelings of intoxication after having consumed only minimal amounts of alcohol. A small percentage of women (8%, $n = 11$) reported having woken up at least once not remembering the night before, and a majority of these ($n = 7$, 5% of total sample) believed that their memory loss was drug related rather than due to the quantity of alcohol consumed.

Perpetration. A total of 3% of men ($n = 4$) admitted to attempting or completing intercourse with a woman against her will by giving her alcohol or drugs. Of the men who admitted using drugs and alcohol to induce coercion, all had done so more than once and with more than one woman. Half of these men (1.4% of total sample, $n = 2$) admitted to rape using drugs. One man reported surreptitious drugging of a woman. Attempted sexual assault was evenly divided between the use of alcohol and drugs.

Dating-Related Alcohol and Drug Use

When asked how often they consumed alcohol on a date, a sizeable proportion reported that if they do drink alcohol on a date it does not occur that often (83.8%). A minority of participants reported drinking fairly often or always (11.2%), and 5% did not date. Participants were asked how much alcohol they consume on an average date. Their responses ranged from 0 to 8 with the average being 2 drinks ($M = 1.64$, $SD = 1.49$).

Most participants did not report voluntary drug use under any circumstances (95%). Almost 5% reported some drug use combined with dating. Furthermore, 8.1% have thought that on one occasion, someone had given them a drug without their knowledge.

Assessment of Realism of Scenarios and Levels of Perpetrator and Victim Blame and Responsibility

Participants indicated how likely they believed the story in the scenario was on a scale from 1 to 5 where 5 was *very likely*. Participants believed that the situation depicted in the scenario was quite likely to occur in the world outside the laboratory ($M = 4.04$; $SD = .88$).

As would be expected in this unambiguous date-rape situation, the perpetrator was held more responsible ($M = 86.41$; $SD = 16.18$; range 20-100%) and blameworthy ($M = 89.41$; $SD = 16.52$; range = 5-100%) than the victim ($Ms = 14.96$ and 11.16; $SDs = 17.82$ and 16.75; ranges = 0-100% and 0-95%, respectively). However, as can be seen from the ranges and standard deviations, there was a fair amount of variability on these judgments.

Effect of Scenario Content (and Rape Myths) on Attributions

Responsibility. Univariate tests showed that rape-myth beliefs predicted attributions of responsibility for both John ($F(1,269) = 63.84$, $p < .001$, $\eta^2 = .19$) and Cathy ($F(1,269) = 66.79$, $p < .001$, $\eta^2 = .20$), with individuals with higher rape-myth belief scores attributing more responsibility to Cathy and less to John than people with lower rape-myth belief scores across all scenarios. With rape-myth beliefs controlled, the main effect of scenario was significant for both John and Cathy, $Fs\ (4, 269) = 4.73$ and 4.32 respectively, $\eta^2 s = .07$ and .06, $ps < .01$. Post-hoc pairwise comparisons (see Table 3.9 for means) demonstrated that participants assigned less responsibility to John and more to Cathy in the voluntary drug condition than in the control (sober) condition. Furthermore, less responsibility was assigned to John in the voluntary drug than in the involuntary drug and the involuntary drunk conditions. In other words, attributions of responsibility for

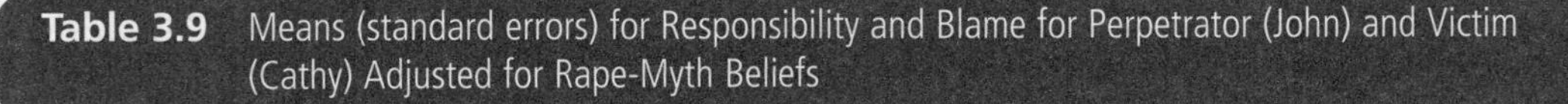

Table 3.9 Means (standard errors) for Responsibility and Blame for Perpetrator (John) and Victim (Cathy) Adjusted for Rape-Myth Beliefs

Variable	Control/Sober (n = 64)	Voluntary Alcohol (n = 64)	Voluntary Drug (n = 64)	Involuntary Drug (n = 64)	Involuntary Drunk (n = 24)
Responsibility					
John	88.84[a] (1.81)	86.69[abc] (1.81)	79.95[b] (1.80)	91.34[ac] (2.94)	88.32[ac] (1.80)
Cathy	11.01[a] (1.99)	14.16[ab] (1.99)	21.68[b] (1.98)	11.48[ab] (3.24)	14.32[ab] (1.98)
Blame					
John	91.40[a] (1.84)	89.95[a] (1.84)	82.21[b] (1.83)	94.52[a] (2.99)	92.17[a] (1.83)
Cathy	9.13[a] (1.85)	10.30[ab] (1.86)	16.95[b] (1.85)	8.75[a] (3.03)	9.15[ab] (1.85)

NOTE: Superscripts that differ within rows indicate significant differences, $p < .05$.

the perpetrator were unique (with a single exception) in the situation where the victim took GHB recreationally, with lower levels of responsibility attributed to him than in any other circumstances (except when she drank enough to get drunk). Attributions of responsibility for the victim followed the same trends across the board (all $ps < .10$), but the victim who used GHB voluntarily received significantly higher attributions of responsibility from participants than the situation where the victim was sober.

Blame. Significant multivariate effects for the covariate, rape-myth beliefs, $F(2,268) = 35.55, p < .001$, and scenario were found, $F(8,536) = 2.90, p < .05$. Univariate tests showed that rape-myth beliefs predicted attributions of blame for both John ($F(1,269) = 61.00, p < .001, \eta^2 = .19$) and Cathy ($F(1,269) = 69.36, p < .001, \eta^2 = .21$), with individuals with higher rape-myth belief scores attributing more blame to Cathy and less to John than people with lower rape-myth belief scores across all scenarios. With rape-myth beliefs controlled, the main effect of scenario was significant for both John and Cathy, $Fs\ (4,269) = 5.52$ and 3.34, respectively, η^2 s = .08 and .05, $ps < .01$. Post-hoc simple contrasts (see Table 3.9 for means) showed that less blame was assigned to John and more to Cathy in the voluntary drug than in the control (sober) and the involuntary drug conditions. Furthermore, less blame was assigned to John in the voluntary drug than in the drunk and the involuntary drunk conditions. All other mean differences were not significant. In other words, the amount of blame attributed to the perpetrator in an acquaintance rape situation was significantly lower when the victim had engaged in recreational GHB use than in any other circumstance. For blame assigned to the victim, again the amount of blame was the highest in the situation of voluntary drug use but was only significantly higher than the blame attributed to sober or drugged victims.

Discussion

In this study, five unambiguous date rape scenarios were used to evaluate participants' attributions for the sexual assault to the perpetrator and victim under varying types of drug and alcohol involvement. University students' judgments of responsibility and blame for the sexual assault showed that when women are drugged or deliberately provided with large amounts of alcohol without their knowledge, perpetrators are held more responsible and blameworthy as when no alcohol or drugs are involved in the situation. Similarly, the victim of the sexual assault is judged to have very low levels of responsibility and blame for what was done to her. The surreptitiously drugged or drunk victim seemed to fit participants' view of a "real rape" with the lowest victim blame and highest perpetrator responsibility. Furthermore, individual differences in rape-myth acceptance did not play a role in these judgments as they were statistically controlled here. This finding suggests that individuals may be more punitive to offenders and blame victims less in sexual assaults where drugging, whether with alcohol or other drugs, is used as a weapon, than in the more usual university acquaintance rape situation where both the victim and the perpetrator have been drinking.

Previous studies have varied alcohol use in rape scenarios but none that we are aware of have investigated reactions to these other drugs. As shown by participants' responses, use of date rape drugs (in this case GHB) by a man to facilitate a sexual assault also appears to place him clearly in the position of other perpetrators who use force or threats with no alcohol or drug involvement. This finding could mean that the victims of these types of crimes would receive more positive treatment by the justice system to their reports of sexual assault (Finch & Munro, 2005). Yet, in the case of GHB and other similar drugs, there would still be problems with victims' memory, which may hinder prosecutions. The implications of this finding are further complicated by the reality that most victims are unlikely to know "for sure" that they have been drugged or mixed stronger drinks. It would be important for future research to examine whether the claim of surreptitious drugging alone would modify attributions or whether, as in our scenario, direct knowledge or proof of this fact is necessary.

In contrast, women who engage in recreational drug use are harshly judged by university students and the perpetrators of sexual assault against them are marginally excused. As we hypothesized, the female victim was held most responsible and blameworthy when she took GHB recreationally and significantly more so than when she was sober or was slipped GHB. There were no significant differences found between the situations where Cathy took GHB or drank alcohol and got drunk, although there was a trend toward drug use being responsible for higher blame and responsibility attributions for the victim ($p < .10$). Similarly, the perpetrator of an unambiguous sexual assault was viewed as least blameworthy (although still above 80%) when the woman had taken GHB voluntarily that evening. He was also viewed as less responsible in this situation than in situations where no alcohol or drugs were present or where he used alcohol or drugs to facilitate sexual assault. His responsibility was marginally lower when the woman took GHB than when she got drunk ($p < .10$).

Drugs, perhaps even to a greater extent than alcohol, seem to be seen as a separate class of circumstances in which "voluntary" use by a woman decreases their worthiness as a "victim." This finding extends the work of Norris and Cubbins (1992), who demonstrated that victim and assailant drinking diminished the view that a rape had occurred to the domain of drug use. This could have serious ramifications for victims (see also Schuller & Wall, 1998). Norris and Cubbins (1992) have suggested that if a woman reports that she has been raped after drinking with her date, her report may not be taken as seriously as in other circumstances, and the impact of the psychological trauma may be underestimated. This would have even worse consequences in a "date rape" drug situation because many drug-induced rapes leave the victim without a memory of the incident (K. McIntosh, personal communication, November 12, 2003). When a woman cannot remember, or adequately report the incident (Canadian Federation of Students, 2001), this can have a negative impact on how others view the assault. The amnesia is a characteristic of the drugs and is not dependent on whether the person is aware that they have consumed it. Although it seems unusual that people would take "date rape" drugs voluntarily, there is evidence that GHB and other drugs are quite routinely taken recreationally (Jones, 2001; Weir, 2001).

The Role of Rape-Myth Beliefs and Gender in Attributions

There was a significant difference between our two samples on the attitudinal variable, rape-myth beliefs, with students recruited during the spring/summer semester holding significantly higher rape-myth beliefs than those students who participated in the winter semester. As a result, we were compelled to control rape-myth acceptance in all analyses. However, rape-myth beliefs were associated with higher levels of victim blame and responsibility and lower levels of perpetrator blame and responsibility across all date-rape situations. Furthermore, although there were no gender differences in attributions when rape-myth beliefs were controlled, men had higher levels of these erroneous beliefs than did the women in the sample. It is therefore likely that without statistical control of attitudes, the gender difference shown in attributions in other studies would have been present (e.g., Jenkins & Dambrot, 1987). These are therefore still important factors in explaining attributions and should continue to be included in future research.

Burt (1980) and others have theorized that socialization fosters the acquisition of false beliefs about rape. It would also follow that socialization would be responsible for the acquisition of beliefs about drug use. This study found that the man who sexually assaulted a woman was perceived to be less blameworthy when she had taken GHB recreationally than in any other circumstance we measured. In North American society, drinking is something that is socially acceptable, although drunkenness by women is not (Clark & Lewis, 1977). The findings of the current study, along with other research done previously (Finch & Munro, 2005; Norris & Cubbins, 1992; Stormo et al., 1997), have shown that when a woman was drinking and was raped, she was blamed more for what was done to her than if she had been sober. Drug use appears to have marginally stronger effects on people's judgments than alcohol, with voluntary drug use being used to excuse the perpetrator and blame the victim even more. This could be in part because drugs are currently illegal or that there are greater social taboos for women who use drugs. Future research using qualitative methods could explore people's reasoning further to assess the underlying causes of these problematic judgments.

Research conducted on factors contributing to rape point out that circumstances such as who pays for the date, who drove, and who drank, all account for increased risks of rape and higher levels of blame and/or responsibility attributed to the victim (Burt, 1980; Dekeseredy et al., 1993; Jenkins & Dambrot, 1987; Rickert & Weinmann, 1998; Schuller & Wall, 1998; Stormo et al., 1997). The current study revealed another factor that is related to increased responsibility and blame attributed to the victim. This factor is whether the victim voluntarily or involuntarily took a drug or drank alcohol.

Given that our scenarios (adapted from those originally written by Stormo et al., 1997) clearly and unambiguously described a deliberate, forceful, sexual assault, these effects could be expected to be even stronger where the situation is less clear in people's minds. Education efforts to counter these victim-blaming tendencies when voluntary alcohol or drug use is involved may be helpful to improve the social conditions for victims, disclosing sexual assault under these fairly common circumstances.

Are Alcohol- and Drug-Related Sexual Coercion Really a Problem on Campus?

The findings regarding sexual coercion and assault involving alcohol or drugs, suggest that it is a problem among undergraduate students on this campus. Although the absolute numbers of victims and perpetrators are relatively small in this sample, across an entire university campus this would translate into a significant minority of students who have had these experiences. A total of 7% of women reported having men attempting to engage in or engaging in sexual intercourse against their will by giving them drugs or alcohol. A total of 3% of men admit to using these tactics to induce women to have sex against their will. Most of the women had more than one of these experiences and all of the men who claimed these tactics used them on more than one woman. Alcohol was used more commonly than drugs in these coercive episodes, but at least one quarter or more were drug-related incidents. These rates are conservative because they do not include all sexual assaults where alcohol or drugs were involved but rather only those where the coercive tactic was the drug or alcohol. Furthermore, if suspected rather than known drugging is included in these figures, the rates of date-rape-drug-involved coercion may be much higher. Although not all of these suspected drugging situations reported here by women (5–20%) were known to include sexual assaults, they do indicate that rates of date rape drugging of women on campus may be higher than it at first appears. One woman wrote, "I felt dizzy and numb, I remembered flashbacks, and I was overly tired to the point that I couldn't get out of bed" (043).[2] Another woman commented that "I was very dizzy and ended up in different clothes" (068). For the minority of students (12%) who often or always drink or take drugs on dates, the risk may be even higher.

The current study contributes to our understanding of a fairly new social phenomenon, drug-facilitated sexual assault. The study utilized and adapted scenarios from those used previously by researchers (Stormo et al., 1997) to explore attributions related to sexual assault, which strengthens the generalizations that can be made across these studies. Attributions about date rape are affected by the involvement of date rape drugs. These attributions are not simply a replication of the attributions about alcohol use as a coercive tactic but rather appear to be exaggerated perhaps by the illegal status of GHB. If a man surreptitiously drugs a woman with GHB, he is held almost completely responsible whereas she is seen as blameless. On the other hand, if a woman voluntarily takes GHB prior to being attacked, the male perpetrator of an unambiguous sexual assault is assigned less blame than in any other situation tested. Victim blaming is an obstacle to effective treatment for victims, legal action against perpetrators, and the search for justice. Although only 5% of students in this Canadian sample report voluntary drug use during dating, the social climate for reporting sexual assaults that do happen under these circumstances are bleak. More research on this issue is desperately needed.

Notes

1. A total of 78% (7/9) of women had both attempted and completed sexual assaults committed against them.
2. Identified by participant number.

References

Abbey, A., Ross, L. T., McDuffie, D., & McAuslan, P. (1996). Alcohol and dating risk factors for sexual assault among college women. *Psychology of Women Quarterly, 20*, 147–169.

Abbey, A., Zawacki, T., Buck, P. O., Clinton, A. M., & McAuslan, P. (2004). Sexual assault and alcohol consumption: What do we know about their relationship and what types of research are still needed? *Aggression and Violent Behaviour, 9*, 271–303.

Adams-Curtis, L. E., & Forbes, G. B. (2004). College women's experiences of sexual coercion: A review of cultural, perpetrator, victim, and situational variables. *Trauma, Violence, & Abuse, 5*, 91–122.

Brener, N. D., McMahon, P. M., Warren, C. W., & Douglas, K. A. (1999). Forced sexual intercourse and associated health risk behaviours among female college students in the U.S. *Journal of Counseling and Clinical Psychology, 67*(2), 252–259.

Bridges, J. S. (1991). Perceptions of date and stranger rape: A difference in sex role expectations and rape supportive beliefs. *Sex Roles, 24*, 291–307.

Burt, M. R. (1980). Cultural myths and supports for rape. *Journal of Personality and Social Psychology, 38*, 217–230.

Cameron, C. A., & Stritzke, W. G. K. (2003). Alcohol and acquaintance rape in Australia: Testing the presupposition model of attributions about responsibility and blame. *Journal of Applied Social Psychology, 33*, 983–1008.

Canadian Federation of Students. (2001). *No means no campaign against sexual assault.* Retrieved October 10, 2003, from http://www.city.vancouver.bc.ca/police/InvesServDiv/sos-dvach/soscfs.htm

Carr, J. L., & VanDeusen, K. M. (2004). Risk factors for male sexual aggression on college campuses. *Journal of Family Violence, 19*, 279–289.

Clark, L. M. G., & Lewis, D. J. (1977). *Rape: The price of coercive sexuality.* Toronto: The Women's Press.

Criminal Code, R.S.C., 1985, c. C-46, s. 271.

Criminal Code, R.S.C., 1992, C. 38, s. 273.

Dekeseredy, W. S., Schwartz, M. D., & Tait, K. (1993). Sexual assault and stranger aggression on a Canadian university campus. *Sex Roles, 28*, 263–277.

Dudley, C. A. (2005). Alcohol, sexual arousal, and sexually aggressive decision-making: Preventative strategies and forensic psychology implications. *Journal of Forensic Psychology Practice, 5*, 1–34.

Finch, E., & Munro, V. E. (2005). Juror stereotypes and blame attribution in rape cases involving intoxicants. *British Journal of Criminology, 45*, 25–38.

Gaither, G. A., Sellbom, M., & Meier, B. P. (2003). The effect of stimulus content on volunteering for sexual interest research among college students. *The Journal of Sex Research, 40*(3), 240–248.

Hensley, L. G. (2002). Drug-facilitated sexual assault on campus: Challenges and interventions. *Journal of College Counseling, 5*, 175–181.

Hindmarch, I., & Brinkmann, R. (1999). Trends in the use of alcohol and other drugs in cases of sexual assault. *Human Psychopharmacology, 14*, 225–231.

Jenkins, M. J., & Dambrot, F. H. (1987). The attribution of date rape: Observer's attitudes and sexual experiences and the dating situation. *Journal of Applied Social Psychology, 17*, 875–895.

Jones, C. (2001). Suspicious death related to gamma-hydroxybutyrate (GHB) toxicity. *Journal of Clinical Forensic Medicine, 8*, 74–76.

Koss, M. P., & Gidycz, C. A. (1985). Sexual Experiences Survey: Reliability and validity. *Journal of Consulting and Clinical Psychology, 53*, 422–423.

Koss, M. P., Gidycz, C. A., & Wisniewski, N. (1987). The scope of rape: Incidence and prevalence of sexual aggression and victimization in a national sample of higher education students. *Journal of Consulting and Clinical Psychology, 55*, 162–170.

Koss, M. P., & Oros, C. J. (1982). Sexual Experiences Survey: A research instrument investigating sexual aggression and victimization. *Journal of Consulting and Clinical Psychology, 50*, 455–457.

Monson, C. M., Langhinrichsen-Rohling, J., & Binderup, T. (2000). Does "no" really mean "no" after you say "yes"? *Journal of Interpersonal Violence, 15*, 1156–1177.

Norris, J., & Cubbins, L. A. (1992). Dating, drinking and rape: Effects of victim and assailant's alcohol consumption on judgments of their behaviour and traits. *Psychology of Women Quarterly, 16*, 179–191.

Payne, D., Lonsway, K., & Fitzgerald, L. (1999). Rape myth acceptance: Exploration of its structure and its measurement using the Illinois Rape Myth Acceptance Scale. *Journal of Research in Personality, 33*, 27–68.

Pope, E., & Shouldice, M. (2001). Drugs and sexual assault: A review. *Trauma Violence & Abuse, 2*, 51–55.

Rickert, V. J., & Weinmann, C. M. (1998). Date rape among adolescents and young adults. *Journal of Pediatric and Adolescent Gynecology, 11*, 167–175.

Saunders, D. M., Fisher, W. A., Hewitt, E. C., & Clayton, J. P. (1985). A method for empirically assessing volunteer selection effects: Recruitment procedures and responses to erotica. *Journal of Personality and Social Psychology, 49*(6), 1703–1712.

Schuller, R. A., & Wall, A. M. (1998). The effects of defendant and complainant intoxication on mock jurors' judgments of sexual assault. *Psychology of Women Quarterly, 22*, 555–573.

Sexual Abuse Act of 1986, 18 USCS § 2241–1143 (LexisNexis 2005).

Stormo, K. J., Lang, A. R., & Stritzke, W.G.K. (1997). Attributions about acquaintance rape: The role of alcohol and individual differences. *Journal of Applied Social Psychology, 27*, 279–305.

Weir, E. (2001). Drug-facilitated date rape. *Canadian Medical Association Journal, 165*, 80.

DISCUSSION QUESTIONS

1. How does the voluntary nature of alcohol and/or drugs affect the assignment of responsibility and blame to the perpetrator in acquaintance sexual assault cases?

2. How is responsibility and blame assigned to victims of sexual assault in voluntary consumption of drugs and/or alcohol? How does this change in cases of involuntary consumption of drugs and/or alcohol?

3. How is the responsibility and blame of sexual assault attributed to the victim and offender when drugs and/or alcohol are not involved in the situation?

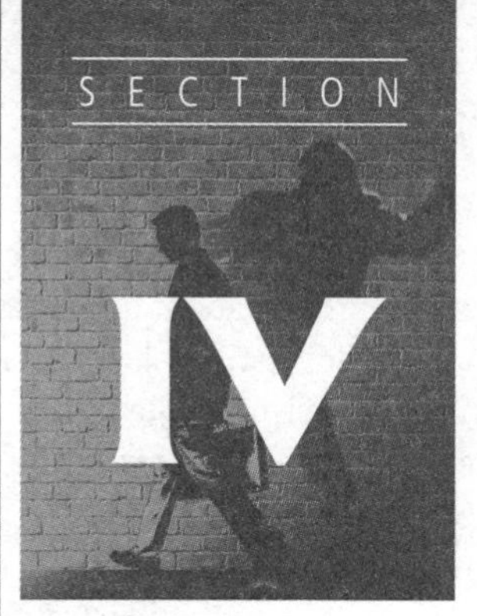

Women and Victimization

Intimate Partner Abuse

Section Highlights

- Historical overview of intimate partner abuse (IPA)
- Contemporary issues in IPA
- Barriers to leaving abusive relationships
- Legal remedies and policy implications for IPA

Much of history has documented the presence of violence within relationships. Throughout history, women were considered the property of the men in their life. Wife beating was a legal and accepted form of discipline of women by their husbands. During ancient Roman times, men were allowed to beat their wives with "a rod or switch as long as its circumference is no greater than the girth of the base of the man's right thumb" (Stevenson & Love, 1999). The "rule of thumb" continued as a guiding principle of legalized wife beating throughout early European history and appeared in English common-law practices, which influenced the legal structures of the early settlers in America. While small movements against wife beating appear in the United States throughout the 18th and 19th century, it wasn't until 1871 that Alabama and Massachusetts became the first states to take away the legal right of men to beat their wives. However, significant resistance still existed in many states on the grounds that the government should not interfere in the family environment. It wasn't until 1882 when wife beating became a crime in the state of Maryland that the act received criminal consequences. However, the enforcement of the act as a crime was limited, and husbands rarely received significant penalties for their actions.

The rise of the feminist movement in the late 1960s and early 1970s gave a foundation for the battered women's movement. Shelters and counseling programs began to appear throughout the United States during the 1970s; however, these efforts were small in scale and the need for assistance significantly outweighed the availability of services. While police officers across the nation began to receive training about domestic violence calls for service, most departments had a non-arrest policy toward cases of domestic violence, as many officers saw their role as a peacemaker or interventionist, rather than as an agent of criminal justice. In these cases, homicide rates continued to increase due to the murders of women at the hands of their intimate partners, and more officers were dying in the line of duty responding to domestic violence calls.

The grassroots battered women's movement of the 1970s led to systemic changes in how the police and courts handled cases of domestic violence. The Minneapolis Domestic Violence Experiment illustrated that when an arrest was made in a misdemeanor domestic violence incident, recidivism rates were significantly lower compared to cases in which police simply "counseled" the aggressor (Sherman & Berk, 1984). However, replication studies did not produce similar results and instead indicated that arresting the offender led to increases in violence.

Throughout the 1980s, state and nonprofit task forces assembled to discuss the issues of intimate partner abuse. By 1989, the United States had over 1,200 programs for battered women and provided shelter housing to over 300,000 women and children each year (Dobash & Dobash, 1992, Stevenson & Love, 1999). In 1994, Congress passed the Violence Against Women Act (VAWA) as part of the federal Crime Victims Act. The VAWA provided funding for battered women's shelters and outreach education, as well as funding for domestic violence training for police and court personnel. It also provided the opportunity for victims to sue for civil damages as a result of violent acts perpetrated against them. In 1995, the Office on Violence Against Women (OVW) was created within the U.S. Department of Justice and today is charged with administering grant programs aimed at research and community programming toward eradicating intimate domestic and intimate partner abuse in our communities (OVW, n.d.).

Defining and Identifying Intimate Partner Abuse

A number of different terms have been used to identify acts of violence against women. Many of these descriptions fall short in capturing the multifaceted nature of these abusive acts. The term "wife battering" fails to identify cases of violence outside of marriage, such as violent relationships between cohabitating individuals, dating violence, or even victims who were previously married to their batterer. Excluding these individuals from the official definition of "battered" often denies these victims any legal protections or services. The most common term used in recent history is "domestic violence." However, this term combines the crime of woman battering with other contexts of abuse found within a home environment, such as the abuse of children or grandparents. Today, many scholars and community activists prefer the term "intimate partner abuse" (IPA) as it captures any form of abuse between individuals who currently have, or have previously had, an intimate relationship (Belknap, 2007).

According to the National Crime Victimization Survey, cases of intimate partner abuse have steadily declined over the past 12 years. However, these rates remain high and indicate that intimate partner abuse remains a significant issue in society. In the majority of cases, men are the aggressor and women are the victim (85%)[1] with an estimated 1.3 million women physically victimized each year (CDC, 2003). A review of state laws across the nation reveals that most crimes of domestic violence are considered a misdemeanor offense,

[1]Given that the majority of data finds men as the perpetrator and women as the victim, this text generally uses to term "he" to refer to the abuser and the term "she" as the victim. The use of these terms is not meant to ignore male victims of violence or abuse within same-sex relationships, but only to characterize the majority of cases of intimate partner abuse.

even for repeat offenders. Most of the time, prosecutors charge offenders with the crime of simple assault (77.9% of cases), which carries with it a penalty of no more than 1 year in jail (Klein, 2004, Smith & Farole, 2009).

Violence between intimates is often a difficult crime for researchers to measure. Much of the abuse occurs behind closed doors and is not visible to the community. Many victims are reluctant to report cases of abuse to anyone (police, friends, or family members) due to the high levels of shame that they feel as a result of the battering. Many victims also do not disclose their victimization out of fear that it will happen again or because they believe that the police will not do anything to help. Research demonstrates that women have both positive and negative experiences in reporting crimes of intimate partner abuse. Some women indicated that the police scolded them for not following through on previous court cases. For others, they were either blamed for causing the violence or were told to fix the relationship with the offender (Fleury-Steiner, Bybee, Sullivan, Belknap, & Melton, 2006).

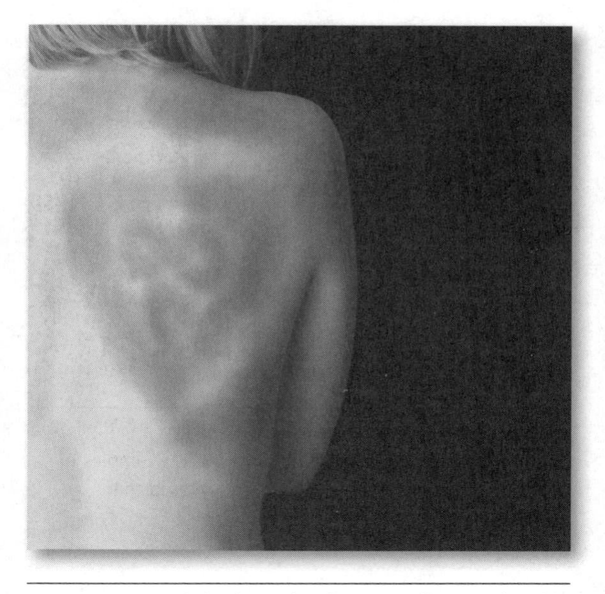

▲ **Photo 4.1** While physical violence, as illustrated in this picture, is often a significant component of intimate partner violence, it is not the only form of abuse perpetrated against victims.

For a small number of women, physical violence in an intimate relationship escalates to murder. While intimate partner homicide is on the decline, women are more likely to die at the hands of a loved one than a stranger. In 2004, 1,159 women were killed by their partners (Bureau of Justice Statistics, 2006). For these women, death was the culmination of a relationship that had been violent over time, and in many cases, the violence occurred on a frequent basis. The presence of a weapon significantly increases the risk of homicide, as women who were threatened or assaulted with a gun or other weapon are twenty times more likely to be killed (Campbell et al., 2003). Three fourths of intimate partner homicide victims had tried to leave their abusers, refuting the common question of "why doesn't she leave." While many of these women had previously sought help and protection from their batterers, their efforts failed (Block, 2003).

Most people think of physical battering/abuse as the major component of intimate partner abuse. However, abuse between intimates runs much deeper than physical violence. Perhaps one of the most common (and some would argue the most damaging in terms of long-term abuse and healing) is emotional battering/abuse. Those who batter their partner emotionally may call them derogatory names, prevent them from working or attending school, or limit access to family members and friends. An abuser may control the finances and limit access and information regarding money, which in turn makes the victim dependent on the perpetrator. Emotional abuse is a way in which perpetrators seek to control their victims, whether it be in telling them what to wear, where to go, or what to do. They may act jealous or possessive of their partner. In many cases, emotional abuse turns violent toward the victim, child(ren), or pet(s). Following acts of physical or sexual violence, the emotional abuse continues when a batterer blames the victim for the violent behavior by suggesting that "she made him do it" or by telling the victim that "you deserve it." Emotional abuse is particularly damaging because it robs the victim of her self-esteem and self-confidence. In many cases, victims fail to identify that they are victims of intimate partner abuse if they do not experience

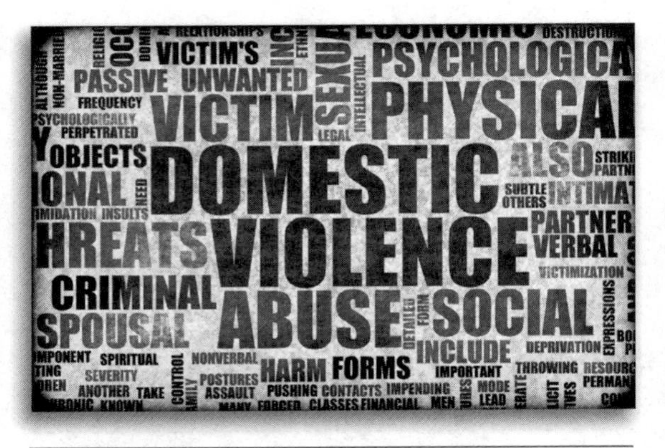

▲ **Photo 4.2** Domestic violence can involve a variety of different forms. This image highlights a few examples of behaviors and victim categories of intimate partner violence.

physical violence. Yet the scars left by emotional abuse are significant and long lasting. Unfortunately, few laws characterize the acts of emotional abuse as a criminal offense.

Victims of Intimate Partner Abuse

Dating Violence

While initial laws on intimate partner abuse only recognized physical violence between married couples, recent laws have been changed to reflect the variety of relationship types where intimate partner abuse can occur. One such example is dating violence. Even though two people are unmarried and not living together, such relationships are not immune from violence. Prevalence rates of dating violence on college campuses indicate that 32% of students report a history of dating violence in a previous relationship, and 21% indicate that they currently experience violence in their dating relationship (Sellers & Bromley, 1996). Teens, in particular, are at high risk for dating violence as a result of their inexperience in relationships and their heightened views of "romantic love," combined with a desire to be independent from their parents (Alabama Coalition Against Domestic Violence [ACADV], n.d.). Rates of intimate partner violence indicate that teens may be at a greater risk for abuse by a significant other than adults (Silverman, Raj, Mucci, & Hathaway, 2001). Given the severity of this issue, it is concerning that few parents believe that dating violence is a significant issue for their children (Women's Health, 2004). Research estimates that one third of youth experience dating violence during adolescence. Unfortunately, few states allow for legal action in teen dating violence, such as protective orders. Adolescent girls who experience physical and sexual dating violence are at an increased risk for a variety of health issues, including (1) use of alcohol, tobacco, and cocaine; (2) poor eating habits and dangerous weight management methods; (3) risky sexual health behaviors, including unprotected sex, multiple partners, and risk of pregnancy; and (4) suicidal ideation and suicide attempts. The early onset of violence and abuse in a relationship carries on for victims into adulthood, as adolescent victims often find themselves in a pattern of abusive relationships as adults (Silverman et al., 2001).

Children of Intimate Partner Abuse

Intimate partner abuse not only affects the victim, but her children as well. Research indicates that 68–87% of incidents involving intimate partner abuse occur while children are present (Raphael, 2000). Children are significantly affected by violence within the home environment, even if they are not the direct victims of the abuse. Despite attempts by mothers to hide their abuse from their children, children are affected. One battered woman spoke of the effects this victimization has on children: "Our kids have problems dealing with us. When we argue and fight in front of them, when they see our husbands humiliating, beating, and cursing us, they will get affected. They will learn everything they see" (Sullivan, Senturia, Negash, Shiu-Thornton, & Giday, 2005, p. 928).

Children who reside in a home where violence is present tend to suffer from a variety of negative mental health outcomes such as feelings of low self-worth, depression, and anxiety. Affected children often suffer in academic settings and have higher rates of aggressive behavior (Goddard & Bedi, 2010). Additionally, many children exposed to violence at a young age continue the cycle of violence into adulthood, as they often find themselves in violent relationships of their own. In an effort to respond to families in need, many agencies that advocate for victims of intimate partner violence are connecting with child welfare agencies to provide a continuum of care for children and their families.

Same-Sex Intimate Partner Abuse

While the majority of intimate partner abuse involves a female victim and a male offender, data indicate that battering also occurs in same-sex relationships. The National Crime Victimization survey found that 3% of females who experienced IPA were victimized by another woman, while 16% of male victims were abused by their male counterpart (Catalano, 2007). Like heterosexual victims of intimate partner abuse, many same-sex victims are reluctant to report their abuse. A review of 16 urban counties found that only 4% of cases handled by the courts involved same-sex offender and victim relationships (Smith & Farole, 2009). Research indicates that female victims of same-sex intimate partner violence face many of the same risk factors for violence as heterosexual battering relationships, including jealousy and controlling behaviors, substance use, a history of violent behaviors, and attempts to leave the relationship. However, additional factors, such as homophobia and discrimination at individual and societal levels, also act as risk factors for female same-sex intimate partner violence (FSSIPV). Gender-role stereotyping has a significant effect on the perceptions of FSSIPV. Research by Hassouneh and Glass (2008) identified four themes where gender-role stereotypes affect women's experience of violence within a same-sex battering relationship. Each of these themes has a significant impact on the denial of harm and victimization. The first theme, "girls don't hit other girls," illustrated that many of the women involved in same-sex battering relationships saw their abuse as an indicator of relationship problems where they were to blame, rather than a relationship where abuse occurred. The second theme, "the myth of lesbian utopia," suggested that the absence of patriarchy meant that there is no oppression or violence within a lesbian relationship. The third theme, labeled "cat fight," discussed how many women thought that violence within their relationship was less significant than the levels of violence that occur in a male-female domestic violence situation. The fourth theme, of "playing the feminine victim," made it difficult for outsiders to identify cases of intimate partner abuse, particularly when agents of law enforcement were involved. Playing the victim allowed offenders to avoid arrest, as law enforcement would rely on traditional gender-role stereotypes to identify who was the victim and who was the perpetrator. Given that victims of FSSIPV are in the minority, few programs and services exist to meet the unique needs of this population. Effective programming needs to address the use of gender-role stereotypes when developing education and intervention efforts for the community.

Effects of Race and Ethnicity on Intimate Partner Abuse

Issues of race and ethnicity add an additional lens through which one can view issues of intimate partner violence. While much of the early research on intimate partner violence focused exclusively on the relationships of gender inequality as a cause of abuse, the inclusion of race and ethnicity (and socioeconomic status) adds additional issues for consideration. For women of color, issues of gender inequality become secondary in the discussion of what it means to be a battered woman. Here, scholars acknowledge the role of cultural differences and structural inequality in understanding the experiences of IPV in ethnically diverse communities (Sokoloff, 2004). When investigating issues of violence among women of color, it is important that scholars not limit their discussions to race and ethnicity. Rather, research needs to reflect on the collision of a number of different factors. "Age, employment status, residence, poverty, social embeddedness, and isolation combine to explain higher rates of abuse within black communities—not race or culture per se" (Sokoloff, 2004, p. 141).

As a population, African American women are at an increased risk to be victimized in cases of intimate partner violence. Scholars are quick to point out that it is not race that affects whether one is more likely to be abused by a partner. Rather, research highlights how economic and social marginalization can place women of color at an increased risk for victimization (West, 2004). Research by Potter (2007b) highlights how interracial abuse among Black women and men is related to feelings of being "devalued" by social stereotypes about "the Black man."

Racial and ethnic identity also affects how victims deal with the abuse perpetuated by an intimate partner. Within the African American community, research highlights how the role of religion and spirituality can serve as methods through which victims are able to cope with the violence in their lives. While many of the women were not active members of a church at the time of their violence, many drew upon spiritual beliefs or connections for support in both enduring the abuse as well as leaving their batterers. For the women who did have a relationship with a Christian congregation and sought out religious leaders for advice and support, many were discouraged by the response they received from their clergy, as they were encouraged to try to stay and work things out. While the women did not generally waiver in their personal faith and spiritual connections with God, these experiences led many to leave either their current congregation or to abandon organized religion in general. In contrast, women who associated with the Islamic (Muslim) faith received greater levels of support from religious leaders and citizens within their community, as they were more likely to condone the violence against women (Potter, 2007a).

Women experiencing IPV may be faced with a multitude of physical and psychological issues, and race and ethnicity can affect whether a victim will seek out support from social service agencies, such as therapeutic and shelter resources. While there were no racial/ethnic differences in the use of services provided by domestic violence agencies, Black women were significantly more likely to use emergency hospital services, police assistance, and housing assistance, compared to Caucasian and Hispanic/Latina women. For example, 65.4% of Black IPV females indicated that they used housing assistance during the past year, compared to only 26.9% of White IPV women and 7.7% of Hispanic/Latina women (Lipsky, Caetano, Field, & Larkin, 2006).

Unique Issues for Immigrant Victims of Intimate Partner Abuse

While intimate partner abuse is a considerable issue for any community, the effects are particularly significant for immigrant communities. Research indicates that men in these communities often batter their partner as a way to regain control and power in their lives, particularly when their immigrant status has deprived them of this social standing. Battering becomes a way in which these men regain their sense of masculinity. For many of these men, their education and training in their home countries does not transfer equally upon their arrival to the United States. "Vietnamese immigrant men have lost power after immigrating to the U.S. Many felt bad because they lack language and occupational skills and could not support their families" (Bui & Morash, 2008, p. 202). Faced with their husband's inability to find a job to support the family, many immigrant women are faced with the need to work, which many immigrant men find to be in opposition to traditional cultural roles and a threat to their status within the family. This strain against traditional roles leads to violence. Many men blame the American culture for the gender clash occurring in their relationships. However, many women accept the violence as part of the relationship, as such behavior is considered normative for their culture. For example, violence is accepted behavior in Vietnamese traditional cultures, wherein men are seen as aggressive warriors and women are seen as passive and meek. Research on intimate partner violence in the Vietnamese immigrant community reveals high levels of verbal (75%), physical (63%), and sexual abuse (46%), with 37% experiencing both physical and sexual abuse (Bui & Morash, 2008).

For Ethiopian-immigrant women, the violent behavior of men is also accepted within the community, making it difficult for women to develop a community understanding that battering is a crime and to seek out services. Help seeking is seen as a complaint by women, and in such cases, members of the community turn to support the perpetrator, not the victim (Sullivan et al., 2005). Intimate partner abuse is also discussed as a normal part of relationships for Russian-immigrant women. One woman stated that domestic violence "is part of the destiny, and you have to tolerate it" (Crandall, Senturia, Sullivan, & Shiu-Thornton, 2005, p. 945).

Cultural expectations may inhibit women from seeking out assistance, as it would bring shame onto the victim and her family, both immediate and extended. Strict gender-role expectations may lead women to believe that they do not have the right to disobey their partner, which legitimizes the abuse. One woman who emigrated from Russia describes the cultural silence that prohibits women from talking of their abuse: "We were raised differently. I do not know, maybe this is a very developed country, and maybe they think it is best if they tell everyone what is going on in their families, their lives, and everything. We are not used to that. We were ashamed of that. But here it is all different" (Crandall et al., 2005, p. 945). Latina victims of intimate partner abuse are less likely to leave an abusive relationship, and in many cases stay with their batterer. For these women, a desire to maintain the family unit, fear of losing their children in a custody battle, and a hope that the batterer will change his behavior all contribute to their decision to remain in a violent relationship (Dutton, Orloff, & Hass, 2000).

Many perpetrators use the fear of deportation to prevent victims from leaving an abusive relationship. Indeed, Latina women are likely to remain in a battering relationship for a longer period of time due to fear surrounding their undocumented immigration status. In addition, Latina immigrants are less likely to seek out help for intimate partner abuse compared to Latina non-immigrants (Ingram, 2007). While the 2005 reauthorization of the Violence Against Women Act increased the protection of immigrant women who are victims of a crime (including domestic violence), it is unclear how many immigrant women are aware of these protections.

Perpetrators often build upon a negative experience of law enforcement from their home country in an effort to create a sense of distrust of the U.S. legal system. For many Vietnamese women, a call to the police for help was a last resort and often done not to facilitate an arrest, but rather to improve the relationship between the perpetrator and the victim by stopping the violence. Most victims did not want to have their partner arrested or prosecuted for domestic violence, but rather wanted to send a message that the abuse was wrong. However, many were reluctant to seek police intervention as they feared losing control over the process and expressed concern and fear over any civil implications that a criminal record would bring, particularly in jurisdictions with mandatory arrest policies (Bui, 2007).

Language barriers may also affect a victim's ability to seek help, as they may not be able to communicate with law enforcement and court personnel, particularly when resources for translators may be significantly limited (National Coalition Against Domestic Violence, n.d.). In an effort to expand access to the courts in domestic violence cases, California amended its domestic violence laws in 2001 to ensure that legal documents in domestic violence cases would be made available in multiple languages. Today, paperwork to request a restraining order and other related documents are available in five different languages: English, Chinese, Spanish, Vietnamese, and Korean.[2] Language skills, combined with a lack of understanding for the American legal system, also can prevent an immigrant/refugee woman from leaving her violent relationship. Not only may a victim not know what services are available, she may not understand how to navigate social systems such as welfare and housing and educational opportunities that are necessary in order to achieve economic independence from her batterer (Sullivan et al., 2005).

Immigrant victims are often unlikely to seek out traditional domestic violence services due to cultural norms. In order to provide assistance to these victims, training and education on intimate partner violence should be made available to other service providers likely to come into contact with these women, such as immigration lawyers and health services personnel. Additionally, public service announcements on the laws against intimate partner abuse and the availability of social services should be made available to all women of an immigrant community regardless of whether they are a victim. This ensures that even if a victim is unlikely to report her abuse to the police, she may tell a friend or family member, who could then direct her to available services (Dutton et al., 2000).

The Cycle of Violence

In explaining why women stay in abusive relationships, Lenore Walker (1979) conceptualized the cycle of violence to explain how perpetrators maintain control over their victims over time. The cycle of violence is made up of three distinct time frames. The first is referred to as "tension building," where a batterer increases control over a victim. As anger begins to build for the perpetrator, the victim tries to keep her partner calm. She also minimizes problems in the relationship. During this time, the victim may feel as though she is "walking on eggshells" because the tension between her and her partner is high. The tension-building phase is characterized by poor communication skills between the partners. It is during the second time frame, referred to as the "abusive incident," where the major incident of battering occurs. During this period, the batterer is highly abusive, and engages in physical and/or sexual violence to control his victim. Following the abusive event, the perpetrator moves to stage three, known as the "honeymoon" period. During this stage, the offender is apologetic to the victim for causing harm. He often is loving and attentive and promises to change his behavior. While this stage is filled with manipulation of the victim's feelings by the perpetrator, he is viewed as sincere and in many cases is forgiven by the victim. Unfortunately, the honeymoon phase doesn't last forever, and in many cases of intimate partner abuse, the cycle begins again, tensions increase and additional acts of violence occur. Over time, the honeymoon stage may disappear entirely. While Walker's (1979) cycle of violence does not explain all relationships where intimate partner abuse occurs, it does provide a framework to understand the cyclical nature of battering.

[2]Each state has different policies on the availability of legal documents in languages other than English. Forms for the State of California are located at http:www.courtinfo.ca.gov.

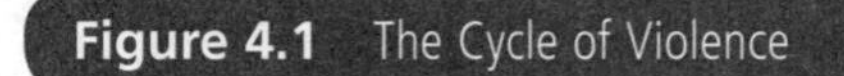

Figure 4.1 The Cycle of Violence

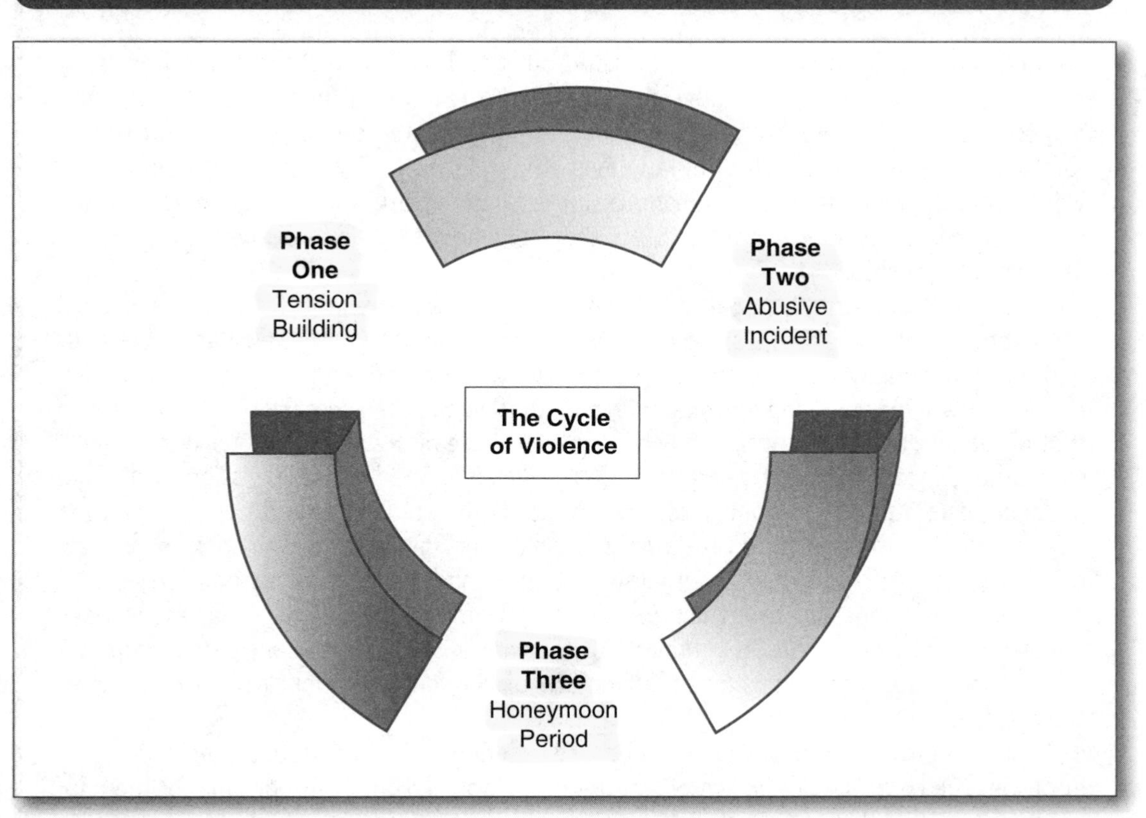

CASE STUDY

Molly Midyette

Molly Midyette is in many ways a classic case of intimate partner violence, yet her story is very different. Like many battered women, she ignored the concerns of her friends about a new relationship with a man named Alex. She suppressed her own apprehensions and before long was engaged to be married. Alex's charming personality soon gave way to a different persona, and he began to express episodes of anger that were often directed toward Molly. He would yell at her, call her names, and throw things at her—abusive acts that define the nature of intimate partner violence. Yet Molly would overlook her husband's behaviors. Even the news that she was expecting did little to stem the violence by Alex toward Molly.

Their son, Jason Jay Midyette, was born on December 17, 2005. However, his life was cut short a mere two months following his birth. On February 24, 2006, Molly and Alex

(Continued)

(Continued)

learned that their son had multiple broken bones and a skull fracture. Jason later died from his injuries. With only Molly and Alex as the caregivers for their son, they soon came under scrutiny. Her father-in-law hired a high-powered defense team of attorneys for both his son and daughter-in-law. Even though she had her own attorney, Molly would defer her comments and questions to the family attorney. Here, her father-in-law's power over the situation would add another layer to her experience with living in an abusive relationship.

Despite her faith that the legal system would accept that there must have been a medical explanation for Jason's death, she grew increasingly concerned with her husband's behavior and her father-in-law's control of their legal strategy. Her friends shared that Alex had continued his partying ways and Molly later learned that he had even cheated on her. Concerned that his anger and use of cocaine may have played a role in the death of her son, she finally shared her concerns with her own attorney. Alex was infuriated with Molly's "betrayal," and his anger escalated to the point where he physically threatened her and her parents. Even her father-in-law threatened to take away everything she valued if she ever spoke negatively of her husband in the future. Even Alex's attorney made threats about potential consequences if she suggested that Alex was responsible for Jason's death. As a result, Molly remained silent on her concerns throughout the trial and kept important details from her attorney. She also refrained from expressing any of her concerns about her husband when she testified in her own trial.

Molly was ultimately convicted of child abuse resulting in death and sentenced to 16 years in prison. The court didn't need to prove that Molly was personally responsible for the injuries to Jason, only that she didn't do anything to protect her son from harm. Yet, Molly says that behind bars she finally felt free of the abuse and control that she had experienced at the hands of her husband and father-in-law. While in prison, Molly divorced Alex, who was later convicted of a lesser charge and was also sentenced to 16 years behind bars. Today, she is currently appealing her sentence and fights for her freedom. Her case is but one example where her experience of intimate partner violence ultimately led to tragedy and her own incarceration.

Women Who Kill Their Abusers

For a small group of women, their abuse ends when they kill their batterer. These women typically sought out the help of the criminal justice system numerous times in their attempts to leave their violent relationships. The very system from which women unsuccessfully sought assistance now treats these once-victims as criminals and works to punish them for their crimes, despite their documented history of abuse, outreach for assistance, and lack of a criminal history of violent behaviors. Their status as a victim is trumped by their conviction for manslaughter or murder sentences (Leonard, 2001).

Developed by Lenore Walker (1979), the battered woman syndrome has become the most recognized explanation of the consequences of intimate partner abuse for victims and has been introduced as evidence to explain the actions of women on trial for killing their batterers. The goal of introducing evidence of abuse is to provide an understanding to juries regarding why women in these extreme cases of intimate partner abuse believed that their lives were in danger and believed that violence was the only option to ensure their own safety. However, juries and judges generally show little sympathy for women who kill their abusers, and many of these women receive life sentences (Leonard, 2001). Indeed, the use of the battered woman syndrome as a theory of self-defense is often negated by the defendant's own actions to defend herself from her batterer, as research suggests that women who fight back or act aggressively at any time during their relationship are less likely to prove to a jury that killing their abuser was an act of self-defense (Schuller & Rzepa, 2002).

Barriers to Leaving an Abusive Relationship

When hearing of cases of domestic violence, many members of the public ask the question "why doesn't she just leave?" Leaving a relationship where intimate partner abuse is present is a difficult and complex process. There are many issues that a victim must face, including housing relocation and safety concerns, as well as the needs of children and family pets. One of the greatest barriers in leaving a battering relationship is the financial limitations that victims face. Women who lack economic self-sufficiency are less likely to report intimate partner abuse and less likely to leave the relationship, as they depend on their abuser for financial support.

Inherent in the question of "why doesn't she just leave?" is the question of "why does she stay?" This question places the responsibility on the victim for staying with a violent partner, rather than focusing on why her partner chooses to be violent. The reality is that many women do leave their batterers. The average battered woman leaves seven to eight times before she is successful in leaving for good (ACADV, n.d.). Violence doesn't always end when women report their crimes or leave their abuse. For some women, the levels of violence increase; women who were separated from their batterers reported higher rates of violence, compared to women who were married or divorced from their batterer (Catalano, 2007). These acts of violence can involve not only the initial victim, but can spread out, placing children, friends, and extended family members of the woman at risk. Concerns regarding these potential increases in violence may influence these women to remain in the relationship out of concern for their loved ones.

For some women, their children become the deciding factor in leaving intimate partner abuse. Some mothers believe that it is their responsibility to keep the family together, and despite the violence they endure, do not want to take the children away from their father. While many women were more likely to tolerate the abuse when it only occurred to them, they were less likely to continue the relationship once their children were negatively affected. In these cases, the decision to leave was based on either a child's request for safety from the violent parent or the mother's conclusion that her child's physical and emotional needs overruled any question of remaining in the relationship (Moe, 2009).

In their search for support, some women may turn toward religious institutions for assistance in leaving a relationship characterized by intimate partner abuse. For many women, their faith gave them strength to leave (Wang, Horne, Levitt, & Klesges, 2009), even if they were not regularly attending a congregation

(Potter, 2007a). Unfortunately for some of these women, their spirituality may hinder their abilities to leave. Cultural scripts of some religious doctrines may encourage women to try to resolve the struggles of their relationship, as divorce and separation are not viewed as acceptable under the eyes of the church. Here, congregations encourage women to forgive the violence that their partners display (Potter, 2007a). Additionally, clergy may be ill equipped to deal with the issue of intimate partner abuse within their congregations, due to a lack of understanding of the realities of the problem and limited training on service and support needs (Shannon-Lewy & Dull, 2005).

Many women struggle with their decision to leave an abusive relationship. Some women may still love their partner, despite the violence that exists within the relationship. Others may hope that their partner will change and believe the promises made by their loved one for a different life. In some multicultural communities, there is a greater pressure outside of the family unit to return to one's batterer. Members of these communities often place significant pressures on victims to reunite with their batterer (Sullivan et al., 2005). For many women, they fear what their lives will be like without their partner. These fears may include how they will support themselves (and their children), the possibility that future relationships will have similar results, and even fear of loneliness. A key to successfully leaving an abusive relationship is the victim's belief that she will be better off without her batterer and have the confidence to make a new life free from violence.

Policy Implications

Restraining Orders

With the increased attention on criminal prosecutions of intimate partner violence, several jurisdictions began to issue civil protection orders or restraining orders as a way for victims to receive legal protection from their batterers. Today, domestic protection orders are available in every jurisdiction in the United States. Protection orders are designed to provide the victim with the opportunity to separate herself from her abuser and generally prohibit the perpetrator from contacting the victim. As a legal document, violations of the protection order are subject to sanctions by the judiciary. For some victims, the restraining order gives them a sense of safety from the batterer.

In some jurisdictions, temporary restraining orders are issued by a police officer when they are called to a domestic violence incident and an arrest is made. By virtue of their name, temporary restraining orders are indeed temporary and are usually only valid for a specific period of time. In many cases, temporary restraining orders expire within a few days, or at the time of the first court appearance by the offender. Following the expiration of a temporary restraining order, victims must apply for an order that will remain in effect for a longer period of time. Not all victims of intimate partner violence seek out a restraining order against their batterer. Victims who believe that their partner will change his behavior are less likely to seek out a civil protection order against their loved one. Those who do apply for protective orders generally do so following a long history of violence and abuse. In order for protective orders to be an effective tool in combating intimate partner abuse, they must be enforced. Federal and state sentencing guidelines require punishment for violations of restraining orders. However, a review of these processes demonstrates that the "full enforcement of the law" is rarely enacted for restraining order violations. Sentencing guidelines in Utah mandate that violators shall be sentenced to batterer intervention programs and jail time, and be ordered to pay restitution as punishment for violating a protective order. However, only 24.1% of offenders were sentenced to batterer intervention programming, while 48.9% were sentenced to jail time and 39.1% were ordered to pay a fine. In addition, the surrender of firearms was requested in only 4.5% of cases, even though laws in all 50 states require that offenders in domestic violence cases be required to surrender their guns to the police. This finding is particularly disturbing given that the leading cause of death in intimate partner homicides involves firearms (Diviney, Parekh, & Olson, 2009). By not sentencing offenders to the fullest extent under the law, we send a message to perpetrators of intimate partner violence that following the guidelines of a protective order is optional and that protecting the victim is not a primary concern for the criminal justice system.

Mandatory Versus Discretionary Arrest Policies

Drawing from criticisms regarding the discretionary arrest policies of many police departments, mandatory arrest or pro-arrest policies began to surface in police departments across the nation during the 1980s and 1990s. The intent behind these laws was to stop domestic violence by deterring offenders. The movement toward mandatory arrest clarified roles of officers when dealing with domestic violence calls for service. It also removed the responsibility of arrest from the victim's decision and onto the shoulders of police personnel. For many women, they believed that a mandatory arrest policy would make officers understand that domestic violence is a serious issue and that it would legitimize their victimization. Here, women believed that an arrest would decrease levels of violence and send a message to the offender that battering is a crime and he would be punished. However, they acknowledged that the decrease in violence was only a temporary measure and that there existed a possibility of increased violence after an offender returned to the family home following an arrest or court proceedings (Barata & Schneider, 2004). In contrast, research by Sokoloff (2004) reflects that many women call the police simply to stop the abuse, not to facilitate an arrest of their partner. While one study found that the majority of women supported the application of mandatory arrest policies in a theoretical sense, they did not believe that such laws would benefit them directly (Smith, 2000). This belief that mandatory arrest policies would not be applicable to their lives is an example of Walker's theory of "learned helplessness," which suggests that a victim may believe that her batterer is exempt from laws against battering and that her status as a victim is unworthy (Walker, 1979). While mandatory arrest policies removed the victim's responsibility for instituting formal charges against an offender, there were some unintentional consequences. In many cases, a victim's call to the police for help resulted in her own arrest, as officers responding to the scene were often unable or unwilling to determine who was the victim and who was the offender. Other victims may be less likely to call for intervention knowing that their batterer (or themselves) would be arrested (Gormley, 2007; Miller & Peterson, 2007). For many women experiencing intimate partner abuse, they supported the concept of mandatory arrest in general, and for other victims, but were less likely to agree that it was necessary in their own lives (Barata & Schneider, 2004).

In response to many mandatory arrest policies, many jurisdictions instituted "no-drop" policies. Rather than force a victim to participate against her will, these jurisdictions developed evidence-based practices that would allow the prosecutor to present a case based on the evidence collected at the scene of the crime, regardless of any testimony by the victim (Gormley, 2007). Such policies were developed in response to a victim's lack of participation in the prosecution of her batterer. These policies may actually work against victims. When victims feel that their voice is not being heard by the criminal justice system, they may be less likely to report incidents of intimate partner violence. While no-drop policies were designed to prevent victims from dismissing charges against their batterer, they instead led to disempowering victims.

Programming Concerns for Victims of Intimate Partner Abuse

Not only are programs needed to address the needs of victims, but it is important to consider the role of battering prevention programs for men. Over the past three decades, batterer intervention programming has become one of the most popular options when sentencing offenders in cases of intimate partner violence. Given the high correlation between substance use and intimate partner violence, most programs include substance abuse treatment as a part of their curriculum. The majority of these programs offer group therapy, which is popular not only for its cost effectiveness, but scholars suggest that the group environment can serve as an opportunity for program participants to support and mentor one another. One criticism of battering intervention programs is that they generally assume that all batterers are alike. This approach does not offer the opportunity for programs to tailor their curriculum to address the differences among men who abuse (Rosenbaum, 2009). In addition, victims of domestic violence voice their dissatisfaction with many of these types of programs, arguing that they are ineffective in dealing with the issues that the men in their lives face (Gillum, 2008).

One of the major themes highlighted by the research findings is the need for services and programming that reflect the unique needs of women. Intimate partner violence attacks every community, age, religion, race, class, and sexual identity. Like rape crisis, programs that provide services for victims of battering are acknowledging the need for options that are culturally diverse and reflect the unique issues within different racial and ethnic communities. The need for culturally relevant programming also extends to shelter programs for victims of domestic violence. In particular, women noted the absence of women of color (particularly African American women) within the administration and staff, even in environments where the majority of the clientele was Black (Gillum, 2008). Additionally, scholars have noted the need for such programs to be based within the targeted community to ensure participation from the community residents—if programs are difficult to access geographically, women are less likely to seek out services as a result of time, money (work and child-care responsibilities), and transportation limitations. Research has also highlighted the need for increased public service information in communities, particularly in neighborhoods where women of color and immigrant women reside. Victims of violence also discuss the need to be proactive and engage in prevention efforts with young women and men in the community (Bent-Goodley, 2004).

However, as Sokoloff (2004) points out, culturally diverse programs are not enough to combat issues of violence between intimate partners. Rather, intervention efforts need to attack the systems that create social inequalities—racism, sexism, classism, and so on. In addition, the legal system and program providers need to understand how these issues are interrelated and not dominated by a single demographic factor. Regardless of their individual effects on a single person, many of these interventions have the potential to fail at the macro level, as long as the social culture of accepting male violence against women remains (Schwartz & DeKeseredy, 2008).

Summary

- Intimate partner abuse is difficult to identify, as much of the abuse occurs behind closed doors and victims are reluctant to report cases of abuse.
- The Violence Against Women Act of 1994 provided funding for battered women shelters, outreach education, and training on domestic violence for police and court personnel.
- Women are more likely to be killed by someone close to them, compared to a stranger.

- Children who are exposed to violence in the home are at risk for negative mental health outcomes. Additionally, many children continue the cycle of violence as adults.
- Gender-role stereotypes and homophobic views have a significant effect on identifying and receiving assistance for victims of same-sex intimate partner abuse.
- Immigrant victims of domestic violence face a variety of unique issues, such as cultural norms regarding violence, gender-role expectations, and a fear of deportation, that affect their experience with battering.
- Walker's Cycle of Violence (1979) helps explain how perpetrators maintain control within a battering relationship.
- Women are confronted with a variety of barriers in their attempts to leave a relationship where intimate partner abuse is present.
- Most women make multiple attempts to leave a violent relationship before they are successful.
- Restraining orders (also known as protection orders) are designed to provide victims of intimate partner abuse the opportunity to separate themselves from their abuser, and generally prohibit the perpetrator from contacting the victim.
- For many women, mandatory arrest policies have resulted in only a temporary decrease in the violence in their lives, with the potential of increased violence in the future.
- In response to mandatory arrest policies, many jurisdictions instituted no-drop policies, which allow prosecutors to file charges without the consent or participation of the victim.

KEY TERMS

Battered women's movement

Minneapolis Domestic Violence Experiment

Violence Against Women Act

Intimate partner abuse

Physical battering/abuse

Emotional battering/abuse

Dating violence

Immigrant victims of intimate partner abuse

The cycle of violence

Tension building

Abusive incident

Honeymoon period

Battered woman syndrome

Restraining orders

Mandatory arrest

Discretionary arrest

Learned helplessness

No-drop policies

DISCUSSION QUESTIONS

1. How have mandatory arrest policies improved the lives of women involved in cases of intimate partner abuse? How have these policies negatively affected victims?
2. What unique issues do immigrant victims of intimate partner abuse face?
3. Describe the different forms of violence that can occur within an intimate partner abusive relationship.
4. Explain how the cycle of violence attempts to explain incidents of intimate partner battering.
5. How has the identification of battered woman syndrome been used in cases of women who kill their batterers?

6. What effects have no-drop policies had for victim rights?

7. What barriers exist for women in their attempts to leave a battering relationship?

WEB RESOURCES

Bureau of Justice Statistics: http://bjs.ojp.usdoj.gov

The National Center for Victims of Crime: http://www.ncvc.org

Office of Victims of Crime: http://www.ojp.usdoj.gov/

The National Domestic Violence Hotline: http://www.ndvh.org/

National Coalition Against Domestic Violence: http://www.ncadv.org/

Office on Violence Against Women: http://www.ovw.usdoj.gov/

READING

In this reading, Martin Schwartz and Walter S. DeKeseredy discuss the need for a shift in the way we think and respond to acts of violence against women. They argue that while efforts at the individual level may provide assistance to victims, these efforts do little to solve the problems of violence at the societal level. Indeed, they suggest that men can be the strongest tools in fighting a culture that is accepting of violence against women.

Interpersonal Violence Against Women

The Role of Men

Martin D. Schwartz and Walter S. DeKeseredy

Many years ago there was a story told so often it became a cliché. Because we have not heard it for a while, it might be useful to drag it out again in a new context. In any of the story's variants, a group of people were being tested for mental health, common sense, or intelligence. They were told that it was essential to keep as much water off the floor as possible to prevent damage. They were issued mops, and a faucet was turned on. The winners in this exercise were not the ones who devoted their lives to mopping as long and hard as they possibly could, but the ones who went over and turned off the faucet.

In many ways this can be applied to the problem of interpersonal violence against women. The authors here have cumulatively put in more than 20 years work in the shelter house movement, and have only the greatest respect for those who are devoted to the sometimes dangerous and always difficult cause of protecting and sheltering battered women from their intimate partners. Unfortunately, such aid sometimes does not solve the problem. It may ameliorate various pieces of the damage caused by violent men, although it may also make things worse, possibly even leading to a male backlash that results in the death of the women (Dugan, Nagin, & Rosenfeld, 2003). Shelters have been called "Band-aids" to the problem, but sometimes they might not even be that.

Generally, the first call of social scientists is for funds to study the problem more. At least on the level of discovering how much interpersonal violence against women exists, we have and have had for many years ample evidence that a phenomenal amount of such violence is committed in North America every day (not to mention the rest of the world). It is not that we do not have enough data, although the issue is sometimes purposely confused by men's rights groups claiming that minor or self-protection violence by women must be counted as equal to extreme or injury-causing violence by men. The problem is that our policies do not reflect the extraordinary amount of information already in our possession.

To speak directly to the issue of programming, there are several major problems with programming

SOURCE: Schwartz, M. D., & DeKeseredy, W. S. (2008). Interpersonal violence against women: The role of men. *Journal of Contemporary Criminal Justice, 24*(2), 178–185.

over interpersonal violence against women. In a short article we will center our comments on three issues, although they will not be given equal attention.

Programming—Attention and Money

The first problem is one of attention and money. In a badly divided country where politics and media-induced moral panics too often overrule logic, money tends to flow to the issues *du jour,* rather than the most important problems. Barry Glassner (1999) asks whether we as a people are afraid of the wrong things. The American media and the populace following behind are afraid of whatever is being newly hyped: terrorist attacks, road rage attackers, methamphetamine, rape drugs, school shootings, and other events that are statistically relatively rare. Meanwhile, statistically more likely events are ignored: homelessness; the lack of proper medical care, particularly among pregnant women (leading to a truly embarrassingly large infant mortality rate) and children; malnourishment of children; and the most extraordinarily low literacy rate in the Western World. A large percentage of our population is highly organized to protest against abortion, for example, but once the child is born there seems to be much less interest in helping to keep the child alive, or later to educate the child. And, of course, various studies have provided statistics that show that as many as one in four college women are the victims of some sort of sexual assault and more than 10% of all women are physically abused.

In the United States today, in addition to the War in Iraq, the popular place to spend money is on the prevention of terrorist attacks, even in places where it boggles the mind to imagine a terrorist attack ever occurring; and fighting wars against either more minor outbreaks of drug use, or relatively harmless drugs. There is very little call to spend more money, for example, on preventing stalking, a crime that absolutely terrorizes many women and even men. National Institute of Justice-sponsored studies, for example, estimate that more than 13% of college women were stalked in one school year, most often by what they characterize as an intimate partner (Office on Violence Against Women, 2007). Looking at all Americans, the Centers for Disease Control and Prevention found that a smaller but still amazingly large number of women have been stalked in their lives (1 in 12 to 1 in 14), and a smaller but still significant number of men (1 in 50) (Basile, Swahn, Chen, & Saltzman, 2006). Not to misrepresent the situation, there has been an enormous reaction to these statistics, resulting in most states passing laws on stalking, and the federal government beefing up similar laws. Stalking incidents are now counted so that reports can be written. There does not, however, seem to have been much imaginative interest in developing programs beyond increasing penalties for people we don't often convict, and keeping track of reports. Programs to actually stop stalking are not very popular.

The same CDC basic study also looked at forced sex, and found that in a national study that victimization rates have remained constant since the 1990s, and that most victims (female or male) were 17 or younger at the time of the first forced sex (Balile, Chen, Black, & Saltzman, 2007). In other words, we have been fairly ineffective not only in preventing *physical* abuse of women and children, but also in preventing *sexual* abuse of intimate partners and their children.

One of the most popular pieces of U.S. legislation is the Violence Against Women Act, but most American programs dealing with the results of such violence operate on shoestring budgets. Worse, for a long and complicated set of reasons, those who try to provide services for victimized women find that to maintain funded facilities they must conform to governmental requirements. To get money from county mental health budgets their clients often must have diagnoses and prognoses. Services must be aimed at the individual problems of the client. Child Protection Services often are required in the first instance to try to maintain the family, even if one member is a batterer or child sexual abuser. Services, money, and programs do not deal with broader social forces in America. Miller and Iovanni (2007) make it clear: "These concessions have shifted the discourse and action away from challenging the root causes of battering—including issues related to power and privilege—and away from prevention efforts" (p. 294).

Most important of all, in a climate where most violence against women consists of men harming women, it is not the women who will stop violence by changing. Rather, men will need to change if there is to be a reduction in the amount of violence against women in North American society. The main place this has been recognized is with the development of batterer intervention programs for men. Although a variety of programs have been tried, and the political popularity of "doing something" has made them the darlings of judges, for the most part they have not been very successful (Jackson, Feder, Davis, Maxwell, & Taylor, 2003; Saunders & Hamill, 2003). One or two hours now and then of counseling, perhaps an emphasis on anger management, and the lack of a system that motivates men to attend and enforces their attendance, all have created a flawed system in most of the country. Most recently attention has been centered on sophisticated programs of coordinated community response plans among courts, probation, shelters, and other community agencies. The main thrust of these programs has been to sweep offenders off the street and lock them up to prevent them from repeating their offenses. These programs have been subject to the most detailed evaluation of any batterer intervention programs, but unfortunately what seems to be the finding is that any changes that come from judicial oversight demonstration projects have come in incapacitation, not in changes in attitudes or deterrence (Harrett, Schaffer, DeStefano, & Castro, 2006). In other words, once the men are let go, they pose the same danger that they posed before.

What all of these programs have in common are two things. First, they deal with men one at a time. To incapacitate all of the spousal assaulters in North America would require a hard-to-imagine further dramatic expansion of our already extraordinarily overloaded penal system. Yet, the problem only gets worse. If men leave batterer treatment programs or batterer incapacitation jail cells and return immediately to their patriarchal families, patriarchal places of work, and patriarchal places of leisure, it is hard to imagine that there will ever be any change in their attitudes, and eventually in their behavior. And, of course, this is exactly what we have been finding.

Few programs have dealt with the problems that started this article: turning off the faucet. If we live in a patriarchal society that encourages male violence against women, we must deal with that society, not only with men one by one. To take an example, a tremendous amount of outrage was unleashed on Atlanta Falcons quarterback Michael Vick in 2007 when he was alleged to have taken part in the killing of two dogs. It was not only a campaign by the radical People for the Ethical Treatment of Animals, but a broad national sense of outrage. People do not like to see dogs harmed. Imagine a Hollywood movie that featured the torture and death of a dog. It will not happen. A few hundred men can be killed in a movie, often very graphically, and rape scenes are no problem at all. In real life, a full-time scorekeeper would be required just to keep track of the number of college and professional athletes in the United States and Canada who have been accused or convicted of beating or raping women, let alone assaulting and/or killing men. It would be quite understandable if Michael Vick were sitting in prison wondering why murderers, rapists, and vicious assaulters of women were playing sports without penalty today. The outrage and economic pressure (e.g., losing lucrative endorsements) just is not there in America for people who harm women. Just dogs.

As we shall see later, the first step thus in programming for the end of interpersonal violence is to actually program for it.

Violence Against Women as a Cause of Crime

Today there has slowly been a growing recognition that being a witness to woman abuse as a child is dangerous to healthy development. We have known for quite some time that many adult criminals grew up in homes marked by domestic terrorism, where they were forced to witness and sometimes experience woman abuse on a regular basis. We know that although they were still children, these witnesses to violence against women act out in serious problematic ways, and suffer from important stress and strain that can lead to drug and alcohol use as time goes on (Emery, 2006).

What has not been commonly recognized has been the relationship between the two. If we have a large number of adult criminals with this background, then we can make the direct connection that growing up in a home marked by extensive violence against women can be seen as a cause of some unknown but certainly large amount of the juvenile and adult delinquency in America. This has been mostly studied in terms of whether there is an intergenerational effect, where children grow up to beat their wives. What has not been studied is the extent to which children who live in terroristic households grow up to join gangs, commit armed robberies, use and sell illegal drugs, commit burglaries, and generally become what society calls street criminals (Schwartz, 1989).

Thus, one important area for study in the future is the extent to which ending interpersonal violence against women can be seen as a strategy for reducing adolescent and adult criminal behavior in later years.

Male Peer Support

Some years ago we proposed a male peer support model of woman abuse, which has been tested many times on both college and community populations, including a national representative sample (Schwartz & DeKeseredy, 1997; Sinclair, 2002). This complex model starts with the proposition that the ultimate cause of woman abuse is societal patriarchy, and provides a corollary that more patriarchal men are more likely to be batterers. What is different about this model is that it suggests that the focus of our attention should not be on women's behavior, but on men's behavior. Unfortunately, and this is very difficult to say, services for women are good to ameliorate many kinds of pain. Counseling and therapy can be very important for dealing with the individual suffering of women, and can help them look at their lives to see if they wish to make changes. Shelters can provide short-term protection, and under some circumstances longer term protection, if they can separate the woman some physical distance from the abuser. However, shelters may not solve the problem. Shelter house directors are fond of saying that under the best of circumstances—if this particular woman is put into a permanent protective environment—batterers will just go on to their next victim; it is hard to see how this can solve any problem except for one particular woman one particular time.

Solving Problems

The most obvious beginning in most introduction to sociology books is the distinction made by C. Wright Mills: the difference between private and public troubles. Private troubles are terrible. One may have cancer or gangrene, be unable to find housing or a sufficient amount of food to eat, or any of a host of other problems. For most of us, sleeping under a bridge in the winter while in pain would be a terrible thing, but it is a private problem; it is *our* problem. To be a sociologist is to look at public problems. If it is not one person who is homeless, but a large percentage of the population, then there is a confluence of social forces here that causes a broad amount of pain. The same applies to interpersonal violence against women. Centering attention on counseling, batterer intervention, protection orders, shelter houses, and the like will not end the problem of male violence, although it may ameliorate the private troubles of some smaller group of women.

There is an emerging number of men who believe that if men are the problem here, then men have to be part of the solution. Jackson Katz (2006) in particular has written on this subject, decrying the fact that so few institutions that affect young males (e.g., schools) actively program to try to reduce misogynist and violent attitudes. Meanwhile, these same institutions may through sports, games, role models, films, and other devices work hard to reinforce the notion that men have an entitlement to be in charge, and to force their way if women resist. Katz has found that there are many men willing to listen, if not actively participate in bystander intervention, having been silenced all their lives for fear of not being manly. He developed his extremely popular MVP program with athletes, the Marines, and others, not because these men are the most difficult or dangerous, but because they make effective leaders: If football linemen can speak out against violence against women, others may feel similarly enabled. Others (e.g., Banyard,

Moynihan, & Plante, 2007) have found that bystander education can be effective and long lasting.

What these programs point out is something that we have long known in dealing in crime, which is that informal social control is more effective than formal social control. For more than a decade men have been recommending a variety of informal social controls. Ron Thorne-Finch (1992), for example, has suggested a variety of one-on-one confrontations that men can make to convince their colleagues not to engage in abuse or sexist jokes. Rus Ervin Funk (1992) argued that men could reduce violence against women by engaging in extensive efforts at what he calls "educational activism." DeKeseredy and Schwartz (1996) argued that men can work in community and local political forums to develop political and informational campaigns. None of these ideas are likely to have an immediate dramatic impact, but all might begin to have a smaller impact, chipping away at the problem.

Rather, what is needed is a major national effort to end interpersonal violence against women. The Michael Vick example may be a good one. There are many similar ones, of course, such as when filmmakers portrayed the death of an animal, and did not make it clear enough that it was not a real animal. Why do people get so upset by the death of animals, but not women? In Pittsburgh, a sports radio personality pointed out that Michael Vick would never have gotten into as much trouble if he had limited himself to raping women. He got into trouble, and was removed from the air, but the fact remains that he was right. Why do most athletes accused of battering or rape end up with the charges dismissed and the woman complainant vilified (Benedict, 1997)?

Many of the activists cited here recommend individual-level patterns of confrontation and struggle to let people know that such behavior and the attitudes that facilitate it is not acceptable (Banyard et al., 2007; DeKeseredy & Schwartz, 1996; Katz, 2006). However, this is not enough. Although there are education programs across the country in this area, there must be significantly more. What is needed is a national-level discussion of programming sufficient to change people's overall attitudes, to where, who knows, maybe raping a woman will come to be seen as bad as killing a pit bull.

Conclusion

The main argument in this essay has been that interpersonal violence against women will not be ended by ameliorative efforts aimed at women. These may be necessary, important, and useful for the women involved, but they will not stop the flow of violence. It is just as unlikely that individual programs such as batterer intervention programs will have much effect, especially if they remain short interventions that have little effect on men's overall environment. Rather, what is needed is major intervention aimed directly at the patriarchal attitudes that facilitate interpersonal violence against women in the United States, and that allow men who commit such crimes to get away with them.

References

Banyard, V. L., Moynihan, M. M., & Plante, E. G. (2007). Sexual violence prevention through bystander education: An experimental evaluation. *Journal of Community Psychology, 35,* 463–481.

Basile, K. C., Chen, J., Black, M. C., & Saltzman, L. E. (2007). Prevalence and characteristics of sexual violence victimization among U.S. adults, 2001-2003. *Violence and Victims, 22,* 437–448.

Basile, K. C., Swahn, M. H., Chen, J., & Saltzman, L. E. (2006). Stalking in the United States: Recent national prevalence estimates. *American Journal of Preventive Medicine, 31,* 172–175.

Benedict, J. (1997). *Public heroes, private felons: Athletes and crimes against women.* Boston: Northeastern University Press.

DeKeseredy, W. S., & Schwartz, M. D. (1996). *Contemporary criminology.* Belmont, CA: Wadsworth.

Dugan, L., Nagin, D. S., & Rosenfeld, R. (2003, November). Do domestic violence services save lives? *National Institute of Justice Journal, 250,* 20–25.

Emery, C. R. (2006). *Consequences of childhood exposure to intimate partner violence.* Washington, DC: National Institute of Justice.

Funk, R. E. (1992). *Stopping rape: A challenge for men.* Philadelphia: New Society.

Glassner, B. (1999). *The culture of fear: Why Americans are afraid of the wrong things.* New York: Basic Books.

Harrett, A., Schaffer, M., DeStefano, C., & Castro, J. (2006). *The evaluation of Milwaukee's judicial oversight demonstration.* Washington, DC: Urban Institute.

Jackson, S., Feder, L., Davis, R. C., Maxwell, C., & Taylor, B. G. (2003). *Do batterer intervention programs work?* Washington, DC: National Institute of Justice.

Katz, J. (2006). *The macho paradox: Why some men hurt women and how all men can help.* Naperville, IL: Sourcebooks.

Miller, S., & Iovanni, L. (2007). Domestic violence policy in the United States. In L. L. O'Toole, J. R. Schiffman, & M. L. Kiter Edwards (Eds.),

Gender violence: Interdisciplinary perspectives (pp. 287–296). New York: New York University Press.

Office on Violence Against Women. (2007). *Report to Congress on stalking and domestic violence, 2005–2006*. Available from the U.S. Department of Justice Web site, http://www.usdoj.gov

Saunders, D. G., & Hamill, R. M. (2003). *Violence against women: Synthesis of research on offender interventions.* Washington, DC: National Institute of Justice.

Schwartz, M. D. (1989). Family violence as a cause of crime: Rethinking our priorities. *Criminal Justice Policy Review, 3*, 115–132.

Schwartz, M. D., & DeKeseredy, W. S. (1997). *Sexual assault on the college campus: The role of male peer support*. Thousand Oaks, CA: Sage.

Sinclair, R. L. (2002). *Male peer support and male-to-female dating abuse committed by socially displaced male youth: An exploratory study.* Unpublished doctoral dissertation, Carleton University, Ottawa, Ontario, Canada.

Thorne-Finch, R. (1992). *Ending the silence: The origins and treatment of male violence against women.* Toronto, Ontario, Canada: University of Toronto Press.

DISCUSSION QUESTIONS

1. How is intimate partner violence a "male" issue?
2. How does a patriarchal society perpetuate violence against women?
3. What type of programming and interventions are necessary to stop violence against women?

READING

In this article, students will learn about the various help-seeking methods that women engage in as part of their attempt to leave a battering relationship. These include seeking help from friends, relatives, the legal or justice system, victim services, and social services. Women also face a number of barriers in their help-seeking efforts, including their criminal history, homelessness, language and immigration barriers, and racism.

Silenced Voices and Structured Survival

Battered Women's Help Seeking

Angela M. Moe

Conceptual Framework

Ptacek's (1999) notion of *social entrapment* provides a contextual framework for understanding the ways in which battered women survive and resist violence. According to Ptacek, the combination of coercive control tactics by abusers, and social and institutional failures to adequately address battering, are largely responsible for the social entrapment of battered women. Abusers, backed by a patriarchal, racist, classist, and homophobic society, effectively stunt women's efforts to safely live. When deserted by community support networks that

SOURCE: Moe, A. M. (2007). Silenced voices and structured survival: Battered women's help seeking. *Violence Against Women, 13*(7), 676–699.

could help them resist victimization, including friends, family, neighbors, schools, workplaces, and various victim services such as shelters, hotlines, and advocacy centers, women may feel as if there is little they may do to stop their victimization. Failures of social institutions, such as the criminal justice system, social services, and health care, in appropriately responding to women also contribute to their entrapment.

Social entrapment is similar in conceptualization to Gondolf and Fisher's (1988) *survivor hypothesis*, which as a response to earlier theories on learned helplessness (Walker, 1984), held that women continually resist their victimization through help-seeking efforts that are largely unsuccessful because of institutional failures. Gondolf and Fisher argued that if women have sufficient resources and social support, they will leave abusers and live independent lives. This hypothesis has been supported by research by Websdale and Johnson (1997) on the effects of appropriate services and support on women's abilities to end abusive relationships, live productive lives, and avoid reassault.

It is from this framework that I examine the success and failure of help seeking from the perspectives of battering survivors living in a domestic violence shelter. Through qualitative, semistructured interviews, women in this study described a myriad of help seeking, including calling the police, obtaining orders of protection, asking friends and family for support, utilizing victim and social services, and seeking medical attention. Although some successful and empowering experiences were reported, most efforts were fraught with failure and disappointment. With respect to their standpoints, I argue that women's resistance to intimate partner abuse and success at utilizing avenues of help seeking are shaped by structural inequalities predicated on patriarchy, poverty, and racism or ethnic bias. Indeed, although most battered women are likely to face some amount of disbelief, discreditation, or even outright hostility in their efforts to seek help with their victimization, in the long run they may fare better or worse within the social welfare, criminal justice, or political-legal system depending on the intersection of various sociostructural inequalities in their lives. To begin, I will review the literature on the social, legal, and institutional outlets from which battered women may seek help.

Literature Review

Justice System

Police have historically neglected battered women's requests for protection. Although there has been widespread implementation of proarrest and mandatory arrest policies during the past two decades, in large part because of Sherman and Berk's (1984) policing experiment, such efforts have been critiqued as to their appropriateness in specific situations and their effectiveness at deterring future violence. Specifically, studies suggest that situational factors, such as the officer's beliefs about the likelihood of future violence, seriousness of the assault, victim's willingness to cooperate with the criminal justice system, suspect's demeanor, and whether the suspect was present at the time the police arrived at the scene, have been correlated with arrest rates (Feder, 1996, 1999; Kane, 1999; McKean & Hendricks, 1997). Further research has found that the race of the perpetrator and victim affects how officers interpret these factors (Robinson & Chandek, 2000) as well as the characteristics and beliefs of responding officers (Feder, 1997; Saunders, 1995; Stith, 1990). Departmental policies and political affiliations have also been correlated with police responses (Belknap & McCall, 1994).

Assuming an arrest is made, further problems abound because of high recidivism (Dunford, 1992; Sherman, 1992), which may also be influenced by court and correctional responses to domestic violence. Several laws have been passed during the past 30 years to help ensure that batterers are criminally processed. However, many of the earliest laws relied on the cooperation of victims, who had to raise criminal complaints against their batterers and testify in court about their experiences. Many victims were unwilling to cooperate with police investigations or prosecutorial efforts out of concern for their safety and economic stability or because they were discouraged by the ways in which the justice system had previously failed to address their victimization (Erez & Belknap, 1998). To make these laws more effective, prosecutors have been given the ability to continue in their efforts with or without victim cooperation. Victims may even be held in contempt of court for failing to appear to testify against their batterers when subpoenaed. These reforms prohibit women from withdrawing criminal complaints against their abusers and strip them of the ability to control the processing of criminal cases regardless of their personal or financial positions (Schechter, 1982). Such practices contradict studies that find that criminal justice practitioners may best protect victims of domestic violence and lessen the risk of batterer recidivism by respecting abused women's wishes with regard to prosecutorial efforts (Fleury, 2002; Ford & Regoli, 1993).

One way that women have been able to seek legal protection aside from criminal prosecution is through a court injunction (i.e., restraining order or protective order). These orders allow women to initiate cost-effective legal actions against their abusers, which may be processed more quickly and with lower standards of proof than criminal proceedings (Chaudhuri & Daly, 1992). There are also psychological benefits for women who feel a sense of empowerment by initiating legal proceedings over which they have control (Fischer, 1993). However, restraining orders have been criticized because of their ineffectiveness in preventing future violence (Davis & Smith, 1995). This is partly because protective orders are limited in scope with regard to the types of situations to which they apply. They also require an immediate and total separation between abusers and victims, which is difficult for women with few economic and social resources (Horton, Simonidis, & Simonidis, 1987). Moreover, police have been inconsistent in their enforcement of these orders (Erez & Belknap, 1998; Rigakos, 1997).

Victim Services

Battered women may also seek help from various services unrelated to the justice system. The most popular of these has been shelters, which provide temporary housing, food, counseling, and support services. However, shelter practices have been found to be inconsistent and infantilizing, wherein traditional notions of individualism, self-sufficiency, and financial independence are emphasized, without an appreciation for the benefits of social support (Ferraro, 1983). With the continual struggle for scarce resources, a more recent study (Donnelly, Cook, & Wilson, 2004) found that agency administrators and staff often make choices about which women they will serve based on whom they see as most appropriate or legitimate. The study found that such decisions are made using long-standing stereotypes about women of color, lesbians, middle-class women, homeless women, rurally located women, mentally ill women, and elderly women.

Moreover, some shelters seem to be moving toward a more traditional, social-service type of agency, adhering to increasingly rigid bureaucratic expectations of self-sufficiency, similar to the philosophy of welfare reform (Chang, 1992; Schillinger, 1988). Such changes lessen the autonomy and control shelter residents have over their lives, slow their development of self-confidence and self-esteem, and impede their creation of supportive networks. Thus, these shelters are, in effect, fostering women's psychological, economic, and legal dependence on the state (Schillinger, 1988).

In contrast, shelter programs and victim services that are flexible in providing women assistance specific to their needs, along with continuous social support, report higher success rates in terms of recipient satisfaction and long-term independence (R. Campbell, Sullivan, & Davidson, 1995; Chang, 1992; Sullivan & Rumptz, 1994; Websdale & Johnson, 1997). Within such programs, victim advocates can provide a helpful link to resources. For the most part, advocates work within the legal arena, assisting women with obtaining legal documents, filing petitions, and negotiating the court system. Although little research has been conducted on advocacy, one study (Weisz, 1999) found that legal advocates provided an empathic presence during court proceedings and were instrumental in disseminating valuable information to women. As a result, many who had received legal assistance from an advocate felt empowered to seek further legal action against their abusers. Other studies (McDermott & Garofalo, 2004; Moe Wan, 2000) have confirmed these findings, while also pointing out the ways in which such intervention may be undesirable and contrary to the goal of empowering women, depending on how services are provided. In a more recent study, Allen, Bybee, and Sullivan (2004) called attention to the need for advocacy in a more comprehensive manner. Based on the needs of a sample of 278 past shelter residents, approximately half of whom were selected to receive advocacy services, the researchers found support for community-based advocacy that would address a myriad of women's needs, including housing, education, employment, and legal issues.

Social Services

Child protection services (CPS) and welfare agencies are also very likely to have contact with battered women and are well positioned to offer victims assistance with safety planning and referrals to appropriate victim-based agencies. CPS often becomes involved with battered women after being alerted to allegations of child maltreatment within their families, as the co-occurrence of woman battering and child abuse or neglect is common (Appel & Holden, 1998; Jones, Gross, & Becker, 2002). The involvement of CPS places battered women in precarious positions as they struggle to survive their own victimization and comply with state interventions on behalf of their children. Court hearings, mediations, mandated counseling, and home visits can be risky for women whose abuse has not yet been publicly disclosed. The risk that a batterer will retaliate against a woman or manipulate her into lying to authorities is heightened in this context, forcing many to negotiate their own and their children's safety. Subsequently, women may appear uncooperative, subversive, and resistant to intervention during court appearances and communications with CPS caseworkers. Moreover, battered women are increasingly being charged with neglect for failing to protect their children from the abuse of their partners (Jacobs, 1998; Jones et al., 2002).

For women who flee abuse, homelessness and poverty are often inevitable consequences (Zorza, 1991). According to Browne and Bassuk's (1997) longitudinal study of low-income and homeless women, more than 83% suffered intimate partner battering. Baker, Cook, and Norris (2003) found similar results among a sample of 110 battered women, 38% of whom had experienced homelessness after separation. A myriad of problems plagued the women who were able to eventually find housing, including being late in paying rent (40%), skipping meals (32%), or neglecting other bills to save for rent (50%), and leaving their homes because of financial problems or continued harassment by their abusers (25%). This study, in particular, illustrated the interconnectedness of various institutions in their responses to domestic violence, finding that women's homelessness was mitigated through positive contact with social service and law enforcement agencies.

Under such circumstances, welfare subsidies often become necessary for economic survival. However, the 1996 Personal Responsibility and Work Opportunity Reconciliation Act (PRWORA) substantially eroded welfare benefits for the poor (Brandwein, 1999). The law encourages marriage, mandates the establishment of paternity in exchange for assistance, and makes it increasingly difficult for women to receive assistance for any length of time (Brush, 2000; Raphael, 1996). Although PRWORA provides a stipulation for domestic violence victims whereby they may be exempted from the time limits placed on recipients for receiving benefits, these exemptions are rarely made available to abused women, and many do not seem to know that they may request such exemptions (Brush, 2000; Busch & Wolfer, 2002). State-by-state policies for implementing the Family Violence Option widely vary as welfare case workers are not always required or encouraged to offer the exemption and have been found to question the legitimacy of women's claims of abuse and require documentation through police or medical records (Brandwein & Filiano, 2000; Levin, 2001). Thus, asking for help through social service agencies, for many women, is inviting heightened scrutiny over their personal lives.

Other Means of Help Seeking

Battered women may also seek medical services or the support of friends, relatives, or clergy as means of dealing with their victimization. Unfortunately, medical and mental health care professionals, as well as religious leaders, too often have no training on domestic violence. The predominant thinking that the family unit ought to be preserved at whatever cost (Goolkasian, 1986) has infiltrated the teachings of various religious institutions (Fortune, 1993). Inconsistent responses to domestic violence, if not outright support for patriarchy and violence against women, has been documented in studies of Christian (Cooper-White, 1996; Dixon, 1995), Muslim (Ayyub, 2000), and Judaic (Cwik, 1997; Graetz, 1998) traditions. Such was evidenced in Fowler and Hill's (2004) study of battered African American women's coping mechanisms, wherein depression and posttraumatic stress disorder (PTSD) remained significantly related to abuse despite the women's reliance on spirituality.

Of specific concern are health care professionals because they are well positioned to intervene in domestic violence situations when women come to emergency rooms or family practitioners for treatment of injuries or routine checkups (Kernic, Wolf, & Holt, 2000). However, this intervention, let alone recognition of the obvious cause of the injuries, has been slow and intermittent (Gerbert, Johnston, Caspers, & Bleecker, 1996; Warshaw, 1993). Women report concealing their abuse out of fear of retaliation from their partners because they encountered health care providers who were not interested in or sympathetic toward their needs and/or because they felt that the health care system was not structured in a way that allowed doctors and nurses sufficient time to deal with issues beyond the treatment of immediate injuries (Gerbert et al., 1996; Hathaway, Willis, & Zimmer, 2002).

Fortunately, progress has been made in recent years (Sheridan, 1998). In particular, administrators of emergency rooms have developed protocols for recognizing, documenting, and intervening in domestic violence. Specific attention is paid to assessing the dangerousness of relationships for battered women through these protocols (J. C. Campbell, 1995). Women who have sought medical care by such providers report feelings of validation, relief, and comfort (Gerbert, Abercrombie, Caspers, Love, & Bronstone, 1999; Hathaway et al., 2002).

With regard to relatives, studies have found that reaching out to family members is one of the most common, and often one of the first, ways in which women seek help (Goodkind, Gillum, Bybee, & Sullivan, 2003; Gordon, 1997; Lockhart & White, 1989). Unfortunately, it has also been ranked as among the least helpful, as women have too often felt that their disclosures were met with judgment and a lack of empathy. Specifically, Goodkind et al. (2003) found that the responses of family and friends depended on several factors, such as the nature of the woman's relationship with her abuser, the number of times she had tried to leave her partner, how many children were involved, and whether friends and family had also been threatened. Such responses affected women's well-being when they were negative in nature or when they included offers of tangible support. Lempert's (1997) findings shed further light on the way in which women's well-being may be affected by the negative responses of friends and family. In this study, women who told their stories felt that the complexity of their situations were ignored; the focus instead turned to the violence without attention to other aspects of the relationship. As a result, the women felt that they were placed in a subordinate position, similar to the one they occupied with their abusers, in that they needed to accept someone else's definition of the situation to receive support and assistance. The mental health consequences of such responses can be devastating. Kocot and Goodman (2003) found that women's coping mechanisms were worsened and statistically associated with PTSD and depression when they received mixed advice or advice to stay with their partners from their closest friends and relatives.

Method

In the present study, qualitative, semistructured interviews were conducted under the auspices of epistemic privilege and standpoint feminist theory with the goal of understanding the help-seeking efforts of a sample of shelter residents. Epistemic privilege holds that members of marginalized groups are better positioned than members of socially dominant groups to describe the ways in which the world is organized according to the oppressions they experience (Collins, 1989; Hartsock, 1987; Smith, 1987). Thus, I approached this research with the assumption that battered women serve as experts of their own lives. This view coincides with standpoint feminism, a central tenet of which supports privileging the experiences and voices of women who participate in our projects over other available discourses (Harding, 1987; Hartsock, 1983, 1985; Smith, 1974, 1989).

In this research, the focus was on battered women's perspectives about their help-seeking efforts. My purpose was to emphasize their accounts over other, more commonly accepted explanations for their victimization offered, for example, by medical professionals, therapists, police officers, politicians, and even batterers themselves. Thus, the *survivor speech* (Alcoff & Gray, 1993) included here challenges hegemonic discourses on women's victimization. Although I do not mean to suggest that all of the women in this study were completely cognizant or critical of the ways in which various social structures affected their life experiences or world views, I do argue that their voices are at least as, or perhaps more, legitimate sources on the realities that battered women in our culture face.

The interview process was facilitated through my position as a fill-in staff member at Tami's Place (pseudonym), an emergency shelter for victims of domestic violence in the Phoenix metropolitan area of Arizona. After obtaining permission from the administrators of the shelter and human subjects approval from my university, I discussed this project with the shelter residents and conducted 19 confidential interviews in the summer and fall of 2000. With each participant's permission, the interviews were audiotaped and transcribed. The interviews were conducted at the discretion of the women in private rooms of the shelter and lasted an average of 55 minutes. Each woman was given the opportunity to provide her own pseudonym for identification purposes, a remuneration of $10 cash, and access to her transcript. The transcripts were then coded for emergent and recurring themes and analyzed using a grounded approach (Glaser & Strauss, 1967). As many narratives as possible were included in this article to illustrate these themes. However, in the interest of brevity, for areas where multiple narratives spoke to a similar theme, the most succinct were chosen.

Although working in the shelter may have hindered the level of trust the women had with me, they actually exhibited a great deal of comfort through their honesty and candor. I asked my interviewees to describe what had brought them to the shelter, the ways in which they had sought help, and responses by social entities to their efforts. All of my questions were intentionally open ended and semistructured so as to provide the women with opportunities to shape the flow and content of their contributions (Reinharz, 1992). This approach yielded a wealth of information on various aspects of the women's lives while simultaneously allowing them to stay within their personal boundaries of comfort and safety. Though this method may be criticized because of its lack of reliability and generalizability, it may also be credited for producing a richly detailed and descriptive set of narratives that would not have been possible through alternative means of data collection (Kvale, 1996).

Profile of Participants

The women were diverse in terms of race, ethnicity, age, educational attainment, and socioeconomic class. Of the women, 9 (47%) identified themselves as White, 4 (21%) as African American, 2 (11%) as American Indian, 2 (11%) as Latina, and 2 (11%) as biracial (American Indian and White, African American and White). Five (26%) women were between the ages of 18 and 25, 10 (53%) were between 26 and 35, and 4 (21%) were between 36 and 45. Eight (42%) women had obtained less than a high school education, 5 (26%) had either graduated from high school or obtained a GED, and 6 (32%) had completed at least some college. Of the women, 5 (26%) reported being consistently poor and/or homeless prior to their stay at the shelter, 10 (53%) reported being lower or working class, and 4 (21%) described themselves as middle class.

The women were quite homogenous in terms of motherhood and their experiences of abuse. Sixteen (84%) had children, and all but one of these children were minors. The majority of the children were living at the shelter. Four (21%) women were also pregnant at the time of the interviews. All had suffered severe and multiple forms of battery: 18 (95%) stated that they had been physically assaulted, 7 (37%) reported having been sexually assaulted, 16 (84%) described instances of emotional or psychological abuse, and 13 (68%) recounted experiences of financial and/or property abuse.

Findings

Attempting to Leave

Most of the women's help-seeking efforts occurred during their abusive relationships as they attempted to prevent further assaults and salvage their relationships. However, such efforts were not divorced from their attempts to leave and terminate their relationships. Indeed, at the time of the interviews, all of the women were separated from their abusers by virtue of residing in the shelter. Prior to this, 6 (32%) women had moved within the state to escape their abusive partners and 4 (21%) had moved out of state. All of them returned to their abusers for reasons such as having no money; being threatened, stalked, sabotaged, or harassed by their abusers; being encouraged by their families to reconcile; feeling guilty; being lonely; and still loving their partners. Thus, the women's help-seeking efforts occurred amid the realization that ending their abusive relationships was not as simple as leaving a physical residence. Despite this realization, however, few exemplified learned helplessness, as suggested by early research (Walker, 1984). On the contrary, they were very active help seekers who felt entrapped in their relationships, largely because of the failures of various agencies to adequately assist them. Indeed, they expressed more of a learned hopelessness than helplessness (LaViolette & Barnett, 2000), which the following sections illustrate.

Relying on Friends and Relatives

Seeking help or support from friends and relatives was among the most common and initial help-seeking strategies employed by the women, which substantiates previous findings on the use of relatives and friends along with or in lieu of formal sources of help (Gordon, 1997; Lempert, 1997; Lockhart & White, 1989) and the importance of social support for women in domestic violence situations (Tan, Basta, Sullivan, & Davidson, 1995). Of the women, 13 (68%) reported that they had told at least one friend or family member about their abuse and, in many cases, had asked for emotional or financial support from them. Responses by friends and relatives varied greatly with some women reporting that they had received much assistance, whereas others reported being abandoned by those close to them. Unfortunately, the latter appeared to be the most common. As Michelle recalled, "My mom calls me and says, 'Well what did you do? What did you do to deserve that?'" After such reactions by family members, feelings of guilt and self-blame emerged. As Terri explained,

> My older sister goes, "I don't understand you not leaving him before." It makes you feel bad when your sister says that. You kind of question who you are and what you're about . . . what really happened. It makes you want to say, "It was really my responsibility. How did I cause him to do that to me?"

Although some women had little or no support from their families or friends, others were too ashamed to tell their relatives about the abuse. Anna Marie, an undocumented immigrant, was particularly concerned about her parents' reaction:

> Now that I went through all this abuse, I'll try to do the best I can to go forward and not go back to Mexico defeated. They don't know what happened between my husband and me. I don't want to tell them anything.

Similar concerns have been noted by immigrant Mexican women in other research (Acevedo, 2000) and by Vietnamese American women (Bui, 2003). In this way, cultural and ethnic norms seem to play into the ways in which women view and approach sources of help.

For those who did have family or friends on whom they could rely, help sometimes came with an ultimatum. Women described instances in which friends or relatives agreed to help them on the condition that they were never to have contact with their partners again. Such scenarios did little to help the women, as Michelle described: "My dad flew out. . . . He told me I needed to leave this man before he killed me. And he said if I ever go back to this man he would disown me, and I did go back to him."

Fortunately, some women did receive unconditional support from their friends and relatives. In these cases, such assistance helped women escape abuse. As Michelle stated, "I went to a friend's house who was a domestic violence counselor and she gave me the list for the hotline and all of the domestic violence shelters." Mothers, fathers, aunts, uncles, grandparents, siblings, best friends, and small cliques of friends all were named as people who offered this type of support. Even women who left their abusers a number of times reported going to these people for assistance over and over again. It was very important to them to know that someone they trusted would unconditionally help them, without blame or judgment.

Involving the Legal or Justice System

Another popular means of help seeking were efforts to involve the legal and justice system. Of the women, 13 (68%) reported taking such actions as calling the police, requesting that their partner be arrested, and cooperating with investigators and district attorneys to bring criminal charges against their abusers, and 11 women (58%) reported filing for restraining orders, seeking divorces, and attempting to maintain full custody of their children on separation. These rates are comparable to previous studies (Gondolf, 1998; Gondolf, Fisher, & McFerron, 1990) and to the National Crime Victimization Survey, which found that 58% of women called the police after being assaulted by their partners (Rennison & Welchans, 2000).

It was clear from the women's narratives that more would have called the police had they been able to. This type of help seeking was extremely dangerous:

> I tried the last time to call the police and he ripped both the phones out of the walls. . . . That time he sat on my upper body and had his thumbs in my eyes and he was just squeezing. He was going, "I'll gouge your eyes out. I'll break every bone in your body. Even if they do find you alive, you won't know to tell them who did it to you because you'll be in intensive care for so long you'll forget." (Terri)

In a few situations, taking the risk proved somewhat worthwhile when batterers were arrested. Women reported that abuse sometimes ceased for some time after an arrest; however, no women reported that their abuse completely stopped after their partners were arrested. None of them stated that their abusers had spent substantial time incarcerated either. In one of the most brutal assaults described, the sentence levied was just more than 1 month in jail and 3 years on house arrest. Rachel described the incident that nearly killed her:

> I don't remember a lot of it. I remember being hit in the head with a can opener, a couple of slashes, a fist to my left side, and a dragging feeling out of my hair. Then 3 days later I woke up in the hospital. He broke four bones in my face, broke my nose, broke a rib so bad that it cut my lung and my kidney, bruised my collarbone, and he stabbed me 47 times. Four or five times I went to court trying to keep him in jail. They sentenced him to 37 days in jail and 3 years house arrest.

Although this attack occurred more than one year before her interview, Rachel was back in the shelter because she had come home one evening to find this man sitting on her porch. This case suggests that the historical complacency toward violence against women and violence within our homes continues to affect the processing of domestic violence cases.

Despite the failure to deter further assaults, arrest did serve other purposes. Having their partners arrested gave some women time to move out of a shared residence. Nina described an incident during which police intervention enabled her to flee the state with her children: "We drove right into a police station and he followed, cut us off right in front of the police department. They said to get as far from Arizona as we could and just held him for 3 or 4 hours." Although her boyfriend did eventually find her, arrest did at least provide Nina with a temporary respite from abuse.

In the majority of cases, batterers were not arrested, and many women reported that calling the police did nothing to improve their situation. Cynthia's experience typified many others: "He'd be gone and they'd say, 'Well he's not here now.' They didn't show up for 2 hours. I called the police all the time at first . . . and then I just said, 'Forget it. They're not helping me.'"

Although many women complained that the police did nothing, others complained of being arrested right along with their abusers. As Amanda explained,

> He assaulted me and my mom at the front office of the apartment complex where we were at. Called the police and the police took us both to jail. He said that I hit him, but there was no marks but there were marks on me, so they took us both to jail for 24 hours. They then let me go and let him go without any charges. After that I told my mom, "I'm never going to call the cops again. If I'm gonna get arrested, I might as well just stay here."

The more women called for police protection to no avail, and the longer the court system failed to respond, the more batterers seemed to learn that they could continue doing what they were doing without repercussions. As Patsy explained,

> I called the police. My manager at the apartments where we were moving out of called the police. The manager of the apartments that we moved into, she called the police. He's went to jail. He's got out. I was scared to press charges on him because he used to tell me some crazy stuff . . . like telling me he would chop me up. He told me I was never going to leave him. They would let him out, and he would be like pounding on the door. I had an order of protection on him. He just kept on coming around. He ripped up the paper.

Such experiences with protective orders were common for the women who obtained them, as were complications with obtaining orders of protection in different states. Markeelie explained how the court system responded to her when she tried to obtain an injunction prior to fleeing the state in which her abuser resided: "I tried to get an order of protection but they won't do it because we're in two different states." In Markeelie's case, it appeared as if she had been given inaccurate information because at the time she filed for the injunction, the Full Faith and Credit provision of the Violence Against Women Act would have required that the order be enforced out of state (National Coalition Against Domestic Violence, 2000). For those who were able to obtain restraining orders *and* whose abusers had not violated the injunctions, the experience of going to court was incredibly empowering. As Terri described, "When the judge gave me the order of protection for me and my kids it was so validating . . . he didn't even question it." Unfortunately, such stories were few and far between.

In addition, diversion is amply available for first-time domestic violence offenders in Arizona. According to the state's domestic violence statute (Arizona Revised Statutes, 1980), first-time offenders may complete a treatment program in exchange for a suspended sentence. A guilty plea is not always required of offenders before entering treatment, and, under certain circumstances, all criminal charges may be dropped on successful completion of the program. The success of batterer intervention seems to be contingent on various factors, including the curriculum and the theoretical underpinnings, format, and duration of a program (Gondolf, 1999; Tolman & Edleson, 1995). Debate has ensued as to the appropriateness of the anger-management curriculum, which has been generally discredited as misdirected and short sighted (Hollenhorst, 1998; Orme, Dominelli, & Mullender, 2000). Despite these findings, women reported that the intervention programs their batterers attended were indeed called "anger management." Moreover, these classes were sometimes not even pertinent to domestic violence. In some cases, alcohol and drug abuse counseling was the focus of the program. Follow-up for noncompliance seemed to be lacking as well:

> He said it seemed like it was an alcohol and drug treatment program rather than anger management. He refused to go. The deal was if he didn't go to that he was going to get arrested. Like 5 years later they caught him and arrested him. Because we're not together anymore, the judge just threw it out. (Nina)

Given such experiences with the legal and justice system, many of the women increasingly turned toward services specifically tailored to victims of domestic violence.

Utilizing Victim Services

Prior to their current shelter stay, 15 (79%) of the women had relied on victim-based services such as shelters, hotlines, support groups, and advocacy centers. This percentage is higher than in an earlier study, which found that, of a sample of 6,612 shelter residents, 63% had previously contacted a shelter or obtained legal advice and that 14% had previously resided in a shelter (Gondolf et al., 1990). This discrepancy is perhaps indicative of the increased number of victim-based services available today and the heightened recognition of domestic violence. However, the women still encountered problems.

For some, finding shelter space had been problematic, either because of their recent drug or alcohol use (many shelters have policies against admitting substance users) or because of a lack of space. During fiscal year 1999–2000, 16,600 women and children requested bed space from the eight domestic violence shelters in Maricopa County (Phoenix metropolitan area). At the time of request, 14,164 women and children were denied shelter, primarily because of a lack of availability (Arizona Department of Economic Security and Department of Health Services, 2001).

Finding bed space was even more problematic for women with children. The shelters in Maricopa County are consistently filled to capacity and usually do not have several beds available at one time to accommodate large families. Moreover, many shelters will not accept children, particularly boys, older than a certain age; only one of the eight shelters in the Phoenix area accepts boys older than 13. As Nina recalled, "My kids are 9, 11, 13, and 14. Most of the shelters don't take kids over 13 and most of the ones that do separate them overnight."

Finding and being admitted to shelter were just the first obstacles women faced; remaining at the facility was another challenge. A woman's stay at the shelter was usually contingent on her (and her children's) compliance with numerous policies. Many of the women found these rules to be restrictive and counterproductive, particularly with regard to the time limits placed on their stay. Tazia's comments were indicative:

> They give you 30 days to do what you have to do. . . . Not many women that have kids and don't have a car are able to hurry up and get a job within 30 days. Hurry up and get a place in 30 days. It's a lot of pressure and deadlines living here. I'm thinking that a shelter is a place to gain self-esteem, to get out of that controlling situation, and get the will power to be self-sufficient and independent, and really there's more stress here with all the rules and extensions and groups and time limits.

The pressure women faced in trying to reestablish their lives in less than a month was enormous. Although the shelter did provide extensions, looming deadlines were always imminent. The most feasible option for most women under these circumstances was to enter a transitional housing program, which provided counseling and subsidized apartments for up to two years. Like shelters, a woman's stay in such a program was contingent on her compliance with numerous policies.

Despite their concerns, those who were able to abide by shelter policies found the experience to be generally positive. Many were grateful to have a safe place to sleep, food for their children, and access to clothing and personal supplies. As Cynthia stated,

> It's a place to stay, they give you food, they give you clothes, they give you the counseling you need, they try to make available to you services that you can get . . . assistance with housing, and everything. They go out of their way to try to make it a good experience and try to make it as much like home as possible.

For a few women, shelter stays played a pivotal role in their efforts to escape abusive men, and many women who resided at a shelter were able to secure subsidized transitional housing. However, others were not as fortunate, particularly those with criminal records. As Lee, who had a felony on her record for drug possession, described,

> There's supposed to be all this help out there. You get in here, you get safe and that's wonderful, but now I find out that they can't help me. I have all this ambition. I was going to be able to go to college. I was going to be able to go to a decent job . . . now to be just slapped in the face. "I'm sorry there's nowhere for you to go." Nobody has an answer. It seems like the system is working more against me than for me.

Beyond shelter, women utilized numerous other victim services, including crisis hotlines and legal advocacy services. As has been documented in previous research (Gondolf et al., 1990), residing in the shelter seemed to open doors to a myriad of other services. Although the women had been quite active help seekers prior to their shelter stay, they became even more active during their stay.

Few complaints were issued with regard to non-shelter victim services. However, Anna Marie, who was a monolingual Spanish-speaking woman, reported that she had difficulty reaching any of the agencies that offered services in Spanish to which she had been initially referred: "A lot of those times I would get someone in Spanish but it was an answering machine so I didn't get the immediate help that I needed." Fortunately, Anna Marie eventually found a bilingual legal advocate who helped her file for a divorce and an order of protection. She also found transitional housing with bilingual support services. Again, all of these services were only made available after she had been admitted to the shelter. Her experiences were consistent with other research, which has found that preference or necessity for services in Spanish are a significant cultural barrier to help seeking for Latinas (West, Kantor, & Jasinski, 1998).

Resorting to Social Services

Women also turned to social service agencies for assistance, including the welfare system, CPS, the mental health care system, and child support enforcement. Eleven (58%) women reported such help-seeking efforts, which was a much higher figure than that reported in an earlier study (Gondolf et al., 1990), which found that only 11% of shelter residents had contact with social service agencies. Social service agencies were seen as powerful extensions of the state's social control mechanisms yet also potentially helpful outlets for those with relatively few alternative resources. The women relied on such help-seeking outlets with hopeful ambivalence, knowing that cooperation could yield many benefits (e.g., counseling, financial support) but that revictimization was also possible (e.g., having their children removed for failing to protect them from abuse).

The women in this study reported that they received food stamps, subsidized health insurance, and cash assistance through Arizona's Department of Economic Security. The main concern expressed about welfare had to do with the recently revised employment requirements. Some of the women were forced to leave their children to work when they felt their kids were too young to go to day care. The sacrifice hardly seemed worthwhile to those who could only find short-term and low-paying employment with few, if any, fringe benefits. Without resources to improve their education, the women found the employment requirements shortsighted. As Amanda explained,

> I asked about schooling 'cause I don't have a GED. They can't help you. People that are undereducated, the only way that they are

> going to find good jobs to support themselves and their families is if they have a good education. If they can't help with that then there's no point in going out there and lookin' for a job.

Another concern expressed about the welfare system had to do with transfer of benefits and eligibility renewals. Women who had to move to escape their abusers found that transferring their cases to another office sometimes took months, during which time they did not receive benefits. Moreover, they faced constant challenges to their eligibility, which required continual submission of renewal applications. Given these difficulties, few women used the benefits for extended periods of time or in excess, and several expressed guilt for relying on government subsidies: "I feel like less of a woman sometimes because I can't take care of my son" (Amy). They often opted for the minimal amount of assistance possible.

Although some social service agencies seemed willing to help the women, sometimes these agencies were unable to do so because of a lack of resources. Women were subsequently placed in a double bind in that the agencies on which they depended for assistance harmed them as well. In some cases, these unfulfilled promises of assistance were caused by nonsensical bureaucratic policies. For example, Markeelie looked into applying for child support benefits from her ex-boyfriend, only to find out that doing so would mandate her to disclose her whereabouts. Her ex-boyfriend had previously shown no interest in parenting the child and was actually legally barred from visitation because of his criminal record. Despite this, Markeelie was told that she could not collect child support without disclosing where she was living, thus endangering her safety:

> I'm supposed to be receiving child support from my child's father but they won't help me because I don't want to tell him where I'm at. If he pays his child support, he has the right to know where his son is. They're asking me to choose between my life and receiving child support benefits for my son.

Such contradictions have been noted elsewhere (Varcoe & Irwin, 2004) in terms of custody arrangements, whereby women are expected to preserve their children's relationships with their fathers while simultaneously protecting them from their fathers. Given these scenarios, it seemed relying on government-sponsored social service agencies provided marginal assistance to the women, which is disheartening given that such public services are often essential to the livelihood of people in difficult circumstances.

Further Sociostructural Impediments to Help Seeking

The experiences of the women in this study lend support to Ptacek's (1999) notion of social entrapment and to Gondolf and Fisher's (1988) survivor hypothesis. Women who received unconditional and empathetic institutional and/or social support in response to their help-seeking efforts felt legitimated in their survivor status. They were empowered to continue resisting the coercive control tactics of their partners and continued to pursue safe, productive, and independent lives. Alternatively, those whose abuse had been ignored or downplayed when they reached out for help felt as if they had been deserted, silenced, and blamed for their victimization. These women seemed more likely to internalize their hurt (often through depression, self-mutilation, or suicide attempts), blame themselves, and return to their batterers.

Examining the ways in which social institutions respond to battered women must involve a critical analysis of gender relations and the maintenance of institutionalized forms of patriarchy. Indeed, some of the women's narratives exemplified the patriarchal underpinnings within and outside the law that disqualify women's voices, concerns, and interests, effectively disempowering many of those who try to use social systems for protection (Smart, 1995). However, it is not as simple as labeling the entire social and legal system misogynist, as the process of gendered exclusion and marginalization is more complex. Of foremost concern are the sociostructural barriers that contribute to battered women's failed help seeking (Zweig, Schlichter, & Burt, 2002).

For example, 5 (26%) of the women in this study had some sort of criminal record that excluded them from many of the services from which they could benefit. Decent-paying jobs and transitional housing were largely unavailable to these women. Even the decision to call for police protection was a risky one for women with outstanding warrants. Although such exclusion may be a matter of agency policy and prioritization processes (e.g., selecting candidates for sparse transitional housing), legal barriers may also affect women's help-seeking efforts. As Hirsch (2001) found in a study of abused women with drug convictions, those with felony drug records face lifetime bans on public subsidies such as cash assistance and food stamps. Given that women's use of drugs has been correlated with battering and that the majority of women in the criminal justice system have experienced intimate partner battering, it may be particularly damaging to be excluded from public assistance because of such criminalization (Hirsch, 2001; Moe, 2004).

Another structural barrier was homelessness. Four (21%) of the women reported having been homeless intermittently throughout their adulthood, often as a consequence of fleeing abuse. These women were told by police officers, social service providers, and victim service providers to obtain orders of protection against their abusers. However, to obtain these injunctions, petitions usually require a street address for both parties. In this way, homeless women, and women whose abusers are homeless, are blocked from obtaining a form of legal protection. Even for those with homes, financial difficulties prevent other means of legal intervention, such as divorce, that could provide some protection. So although the legal system may not always be helpful to women, it is not even available to others.

A third structural barrier to help seeking relates to race and ethnicity. Services geared toward specific populations, such as the Phoenix Indian Hospital, proved extremely helpful for women of color, such as Patsy. However, being monolingual in Spanish and undocumented produced obstacles for Anna Marie. Although she eventually found the help she wanted, many other women who are either undocumented or under the auspices of immigration law are not as fortunate (Acevedo, 2000; Bui & Morash, 1999; Dasgupta, 2000; Davis & Erez, 1998). It is likely that race and ethnicity influenced the women's interactions with criminal justice and social service personnel as well, based on prior literature on such responses to women of color (McGillivray & Comaskey, 1999; Rasche, 1995; Razack, 1998). However, subtle and institutionalized forms of racism are not easily detectable, particularly by persons in the middle of a crisis, as were the women in this study. Hence, it is likely that their help-seeking efforts were affected in ways beyond those explicitly identified.

Discussion and Conclusion

The women in this study reported horrific abuses at the hands of their partners. Their experiences signify the way in which our most intimate of spaces may be infused with violence. Given the extent of their victimization, the women's strength and perseverance in resisting the abuse was compelling. They sought help multiple times from various outlets and, in some cases, continued to do so despite compounding failures. Constructive critiques of the ways in which our societal structures and institutions are responding to battered women are an important first step toward eradicating violence against women.

Indeed, being socially marginalized in ways beyond gender plays an important role in one's accessibility to institutional assistance. As Gondolf and Fisher (1988) argued, women are active help seekers. Given the psychological consequences of battering, combined with the social isolation and emotional degradation caused by batterers, it is quite remarkable that abused women are such active help seekers. Of course, selection bias affected the findings of my research, as the women in this sample were active and, to some extent, successful help seekers because of their ability to secure temporary housing in the shelter in which I conducted interviews. Women who have never told anyone or reached out to social service, victim-based, or criminal justice agencies for help are certainly a much more difficult population to study. In addition, it would be hard to discuss help seeking with a woman who either does not label herself *battered* or who is so deeply controlled and isolated that nobody can gain access to her. Thus, my conclusions are necessarily limited to the population of battered women who do identify their experiences as abuse, have reached out for help, and have to some extent been successful at obtaining it.

Among these women and, more specifically, the sample of women in my study, diligence was an overriding theme in their narratives. Beyond the success of obtaining shelter, the women remained diligent in their efforts to stay safe and removed from their partners, obtain legal protection and intervention, maintain custody of their children, be good mothers, and build a support system of family and friends. In instances in which one or several help-seeking mechanisms came through for them, the women's stories illustrated positive outcomes. Unfortunately, such stories were too few and far between, confirming Ptacek's (1999) social entrapment thesis. Most of the women interviewed for this study seemed to be cumulatively affected by their partners' abuse tactics and the failed or inadequate responses by social and institutional outlets. The main conclusion to be drawn from this is that every little bit matters. One helpful response may spur further help-seeking efforts. It may also legitimize a woman's claims to other agencies. Just as failed help seeking may be cumulative in effect, so too might successful help seeking.

In terms of pragmatic policy change and advocacy, many of the problems women face during the course of seeking help could be addressed without a complete overhaul of the social structure and, in some cases, without substantial increases in funding. Such changes are a matter of increased integrity and empathy and perhaps a bit of ingenuity and legal reform. However, such changes are not the obligation solely of those working in programs that have contact with battered women. As Gondolf and Fisher (1988) acknowledge, service providers often feel overwhelmed in workload and limited in the resources they may direct toward any particular individual, despite sincere feelings of empathy and compassion. The fallout of working in such environments has been deemed a kind of learned helplessness in itself, in that as service providers are faced with restricted resources and indefinite need, less effort may be devoted to any particular case. Thus, coordinated community-response protocols that provide support and collaboration within and between various social service, victim, and criminal justice agencies could go a long way toward efficient use of resources (Uekert, 2003).

References

Acevedo, M. J. (2000). Battered immigrant Mexican women's perspectives regarding abuse and help-seeking. *Journal of Multicultural Social Work, 8,* 243–282.

Alcoff, L., & Gray, L. (1993). Survivor discourse: Transgression or recuperation? *Signs, 18,* 260–290.

Allen, N. E., Bybee, D. I., & Sullivan, C. M. (2004). Battered women's multitude of needs: Evidence supporting the need for comprehensive advocacy. *Violence Against Women, 10,* 1015–1035.

Appel, A. E., & Holden, G. W. (1998). The co-occurrence of spouse and physical child abuse: A review and appraisal. *Journal of Family Psychology, 12,* 578–599.

Arizona Department of Economic Security and Department of Health Services. (2001). *Domestic violence shelter services in Maricopa County.* Phoenix: Author.

Arizona Revised Statutes. (1980). *Domestic violence* (13-3601). Retrieved March 1, 2004, from http://www.azleg.state.az.us

Ayyub, R. (2000). Domestic violence in the South Asian Muslim immigrant population in the United States. *Journal of Social Distress and the Homeless, 9,* 237–248.

Baker, C. K., Cook, S. L., & Norris, F. H. (2003). Domestic violence and housing problems: A contextual analysis of women's help-seeking, received informal support, and formal system response. *Violence Against Women, 9,* 754–783.

Belknap, J., & McCall, K. D. (1994). Woman battering and police referrals. *Journal of Criminal Justice, 22,* 223–236.

Brandwein, R. A. (1999). *Battered women, children, and welfare reform: The ties that bind.* Thousand Oaks, CA: Sage.

Brandwein, R. A., & Filiano, D. M. (2000). Toward real welfare reform: The voices of battered women. *Affilia, 25,* 224–243.

Browne, A., & Bassuk, S. S. (1997). Intimate violence in the lives of homeless and poor housed women: Prevalence and patterns in an ethnically diverse sample. *American Journal of Orthopsychiatry, 67,* 261–278.

Brush, L. D. (2000). Battering, traumatic stress, and welfare-to-work transition. *Violence Against Women, 6,* 1039–1065.

Bui, H. (2003). Help-seeking behavior among abused immigrant women: A case of Vietnamese American women. *Violence Against Women, 9*, 207–239.

Bui, H., & Morash, M. (1999). Domestic violence in the Vietnamese immigrant community. *Violence Against Women, 5*, 769–795.

Busch, N. B., & Wolfer, T. A. (2002). Battered women speak out: Welfare reform and their decisions to disclose. *Violence Against Women, 8*, 566–584.

Campbell, J. C. (Ed.). (1995). *Assessing dangerousness: Violence by sexual offenders, batterers, and child abusers.* Thousand Oaks, CA: Sage.

Campbell, R., Sullivan, C. M., & Davidson, W. S. (1995). Women who use domestic violence shelters: Changes in depression over time. *Psychology of Women Quarterly, 19*, 237–255.

Chang, D. B. K. (1992). A domestic violence shelter: A symbolic bureaucracy. *Social Process in Hawaii, 34*, 37–52.

Chaudhuri, M., & Daly, K. (1992). Do restraining orders help? Battered women's experiences with male violence and legal process. In E. S. Buzawa & C. G. Buzawa (Eds.), *Domestic violence: The changing criminal justice response* (pp. 227–252). Westport, CT: Auburn House.

Collins, P. H. (1989). The social construction of Black feminist thought. *Signs, 14*, 745–773.

Cooper-White, P. (1996). An emperor without clothes: The church's views about treatment of domestic violence. *Pastoral Psychology, 45*, 3–20.

Cwik, M. S. (1997). Peace in the home? The response of rabbis to wife abuse within American Jewish congregations—Part 2. *Journal of Psychology and Judaism, 21*, 5–81.

Dasgupta, S. D. (2000). Charting the course: An overview of domestic violence in the South Asian community in the United States. *Journal of Social Distress and the Homeless, 9*, 173–185.

Davis, R. C., & Erez, E. (1998). *Immigrant populations as victims: Toward a multicultural criminal justice system.* Washington, DC: U.S. Department of Justice.

Davis, R. C., & Smith, B. (1995). Domestic violence reforms: Empty promises or fulfilled expectations? *Crime & Delinquency, 41*, 541–552.

Dixon, C. K. (1995). Violence in families: The development of a program to enable clergy to provide support. *Journal of Family Studies, 1*, 14–23.

Donnelly, D. A., Cook, K. J., & Wilson, L. A. (2004). Provision and exclusion: The dual face of services to battered women in three deep south states. *Violence Against Women, 10*, 1015–1035.

Dunford, F. W. (1992). The measurement of recidivism in cases of spouse assault. *Journal of Criminal Law and Criminology, 83*, 120–136.

Erez, E., & Belknap, J. (1998). In their own words: Battered women's assessment of systemic responses. *Violence and Victims, 13*, 3–20.

Feder, L. (1996). Police handling of domestic calls: The importance of offender's presence in the arrest decision. *Journal of Criminal Justice, 24*, 481–490.

Feder, L. (1997). Domestic violence and police response in a pro-arrest jurisdiction. *Women and Criminal Justice, 8*, 79–98.

Feder, L. (1999). Police handling of domestic violence calls: An overview and further investigation. *Women and Criminal Justice, 10*, 49–68.

Ferraro, K. J. (1983). Negotiating trouble in a battered women's shelter. *Urban Life, 12*, 287–307.

Fischer, K. (1993). The psychological impact and meaning of court orders of protection for battered women. *Dissertation Abstracts International, 53*, 6612–6613.

Fleury, R. E. (2002). Missing voices: Patterns of battered women's satisfaction with the criminal justice system. *Violence Against Women, 8*, 181–205.

Ford, D. A., & Regoli, M. J. (1993). The criminal prosecution of wife assaulters: Process, problems, and effects. In N. Z. Hilton (Ed.), *Legal responses to wife assault: Current trends and evaluation* (pp. 127–164). Newbury Park, CA: Sage.

Fortune, M. M. (1993). The nature of abuse. *Pastoral Psychology, 41*, 275–288.

Fowler, D. N., & Hill, H. M. (2004). Social support and spirituality as culturally relevant factors in coping among African American women survivors of partner abuse. *Violence Against Women, 10*, 1267–1282.

Gerbert, B., Abercrombie, P., Caspers, N., Love, C., & Bronstone, A. (1999). How health care providers help battered women: The survivor's perspective. *Women and Health, 29*, 115–135.

Gerbert, B., Johnston, K., Caspers, N., & Bleecker, T. (1996). Experiences of battered women in health care settings: A qualitative study. *Women and Health, 24*, 1–18.

Glaser, B., & Strauss, A. L. (1967). *The discovery of grounded theory: Strategies for qualitative research.* Chicago: Aldine.

Gondolf, E. W. (1998). The victims of court-ordered batterers. *Violence Against Women, 4*, 659–676.

Gondolf, E. W. (1999). A comparison of four batterer intervention systems: Do court referral, program length, and services matter? *Journal of Interpersonal Violence, 14*, 41–61.

Gondolf, E. W., & Fisher, E. R. (1988). *Battered women as survivors.* New York: Lexington.

Gondolf, E. W., Fisher, E., & McFerron, J. R. (1990). The help-seeking behavior of battered women: An analysis of 6,000 shelter interviews. In E. C. Viano (Ed.), *The victimology handbook: Research findings, treatment, and public policy* (pp. 113–127). New York: Garland.

Goodkind, J. R., Gillum, T. L., Bybee, D. I., & Sullivan, C. M. (2003). The impact of family and friends' reactions on the well-being of women with abusive partners. *Violence Against Women, 9*, 347–373.

Goolkasian, G. A. (1986). *Confronting domestic violence: The role of criminal court judges.* Washington, DC: U.S. Department of Justice, National Institute of Justice.

Gordon, J. S. (1997). Effectiveness of community, medical, and mental health services for abused women. *Dissertation Abstracts International, 57*, 7225.

Graetz, N. (1998). *Silence is deadly: Judaism confronts wifebeating.* Northvale, NJ: Jason Aronson.

Harding, S. (1987). Is there a feminist method? In S. Harding (Ed.), *Feminism and methodology: Social science issues* (pp. 1–14). Bloomington: Indiana University.

Hartsock, N. (1983). The feminist standpoint: Developing the ground for a specifically feminist historical materialism. In S. Harding & M. B. Hintikka (Eds.), *Discovering reality: Feminist perspectives on epistemology, metaphysics, methodology, and philosophy of science* (pp. 283–310). Dordrecht, the Netherlands: D. Reidel.

Hartsock, N. (1985). *Money, sex and power: Towards a feminist historical materialism.* Boston: Northeastern University Press.

Hartsock, N. C. M. (1987). The feminist standpoint: Developing a ground for a specifically feminist historical materialism. In S. Harding (Ed.), *Feminism and methodology* (pp. 157–180). Milton Keynes, UK: Open University Press.

Hathaway, J. E., Willis, G., & Zimmer, B. (2002). Listening to survivors' voices: Addressing partner abuse in the health care setting. *Violence Against Women, 8,* 687–719.

Hirsch, A. E. (2001). "The world was never a safe place for them." Abuse, welfare reform, and women with drug convictions. *Violence Against Women, 7,* 159–175.

Hollenhorst, P. S. (1998). What do we know about anger management programs in corrections? *Federal Probation, 62,* 52–64.

Horton, A. L., Simonidis, K. M., & Simonidis, L. L. (1987). Legal remedies for spousal abuse: Victim characteristics, expectations, and satisfaction. *Journal of Family Violence, 2,* 265–278.

Jacobs, M. S. (1998). Requiring battered women die: Murder liability for mothers under failure to protect statutes. *Journal of Criminal Law and Criminology, 88,* 579–660.

Jones, L. P., Gross, E., & Becker, I. (2002). The characteristics of domestic violence victims in a child protective service caseload. *Families in Society, 83,* 405–415.

Kane, R. J. (1999). Patterns of arrest in domestic violence encounters: Identifying a police decision-making model. *Journal of Criminal Justice, 27,* 65–79.

Kernic, M. A., Wolf, M. E., & Holt, V. L. (2000). Rates and relative risk of hospital admission among women in violent intimate partner relationships. *American Journal of Public Health, 90,* 1416–1420.

Kocot, T., & Goodman, L. (2003). The roles of coping and social support in battered women's mental health. *Violence Against Women, 9,* 323–346.

Kvale, S. (1996). *InterViews: An introduction to qualitative research interviewing.* Thousand Oaks, CA: Sage.

LaViolette, A. D., & Barnett, O. W. (2000). *It could happen to anyone: Why battered women stay.* Thousand Oaks, CA: Sage.

Lempert, L. B. (1997). The other side of help: Negative effects in the help-seeking processes of abused women. *Qualitative Sociology, 20,* 289–309.

Levin, R. (2001). Less than ideal: The reality of implementing a welfare-to-work program for domestic violence victims and survivors in collaboration with the TANF department. *Violence Against Women, 7,* 211–221.

Lockhart, L., & White, B. W. (1989). Understanding marital violence in the Black community. *Journal of Interpersonal Violence, 4,* 421–436.

McDermott, M. J., & Garofalo, J. (2004). When advocacy for domestic violence victims backfires: Types and sources of victim disempowerment. *Violence Against Women, 10,* 1245–1266.

McGillivray, A., & Comaskey, B. (1999). *Black eyes all of the time: Intimate violence, aboriginal women, and the justice system.* Toronto, Canada: University of Toronto Press.

McKean, J., & Hendricks, J. E. (1997). The role of crisis intervention in the police response to domestic disturbances. *Criminal Justice Policy Review, 8,* 269–294.

Misra, D. (2002). *The women's health data book: A profile of women's health in the United States.* Menlo Park, CA: Henry J. Kaiser Family Foundation and Jacobs Institute of Women's Health.

Moe, A. M. (2004). Blurring the boundaries: Women's criminality in the context of abuse. *Women's Studies Quarterly, 32,* 116–138.

Moe Wan, A. (2000). Battered women in the restraining order process: Observations on a court advocacy program. *Violence Against Women, 6,* 606–632.

National Coalition Against Domestic Violence. (2000). *Violence Against Women Act of 2000 as passed by the Senate and the House of Representatives.* Available from Alabama Coalition Against Domestic Violence Web site, http://www.acadv.org/ publicpolicy/ vawapassed.htm

Orme, J., Dominelli, L., & Mullender, A. (2000). Working with violent men from a feminist social work perspective. *International Social Work, 43,* 89–105.

Ptacek, J. (1999). *Battered women in the courtroom: The power of judicial responses.* Boston: Northeastern University Press.

Raphael, J. (1996). *Prisoners of abuse: Domestic violence and welfare receipt.* Chicago: Taylor Institute.

Rasche, C. E. (1995). Minority women and domestic violence: The unique dilemmas of battered women of color. In B. R. Price & N. J. Sokoloff (Eds.), *The criminal justice system and women: Offenders, victims, and workers* (pp. 246–261). New York: McGraw-Hill.

Razack, S. (1998). What is to be gained by looking White people in the eye? Culture, race, and gender in cases of sexual violence. In K. Daly & L. Maher (Eds.), *Criminology at the crossroads: Feminist readings in crime and justice* (pp. 225–245). New York: Oxford University Press.

Reinharz, S. (1992). *Feminist methods in social research.* New York: Oxford University Press.

Rennison, C. M., & Welchans, S. (2000). *Intimate partner violence.* Washington, DC: U.S. Department of Justice, National Institute of Justice.

Rigakos, G. S. (1997). Situational determinants of police responses to civil and criminal injunctions for battered women. *Violence Against Women, 3,* 204–216.

Robinson, A. L., & Chandek, M. S. (2000). Differential police response to Black battered women. *Women and Criminal Justice, 12,* 29–61.

Saunders, D. G. (1995). The tendency to arrest victims of domestic violence: A preliminary analysis of officer characteristics. *Journal of Interpersonal Violence, 10,* 147–158.

Schechter, S. (1982). *Women and male violence: The visions and struggles of the battered women's movement.* Boston: South End.

Schillinger, E. (1988). Dependency, control, and isolation: Battered women and the welfare system. *Journal of Contemporary Ethnography, 16,* 469–490.

Sheridan, D. J. (1998). Health care-based programs for domestic violence survivors. In J. C. Campbell (Ed.), *Empowering survivors of abuse: Health care for battered women and their children* (pp. 23–31). Thousand Oaks, CA: Sage.

Sherman, L. W. (1992). *Policing domestic violence: Experiments and dilemmas.* New York: Free Press.

Sherman, L. W., & Berk, R. A. (1984). The specific deterrent effects of arrest for domestic assault. *American Sociological Review, 49,* 261–272.

Smart, C. (1995). *Law, crime and sexuality: Essays in feminism.* London: Sage.

Smith, D. E. (1974). Women's perspective as a radical critique of sociology. *Sociological Inquiry, 4,* 1–13.

Smith, D. E. (1987). *The everyday world as problematic: A feminist sociology.* Toronto, Canada: University of Toronto Press.

Smith, D. E. (1989). Sociological theory: Methods of writing patriarchy. In R. A. Wallace (Ed.), *Feminism and sociological theory* (pp. 34–64). Newbury Park, CA: Sage.

Stith, S. M. (1990). Police response to domestic violence: The influence of individual and familial factors. *Violence and Victims, 5,* 37–49.

Sullivan, C. M., & Rumptz, M. H. (1994). Adjustment and needs of African-American women who utilized a domestic violence shelter. *Violence and Victims, 9,* 275–286.

Tan, C., Basta, J., Sullivan, C. M., & Davidson, W. S., II. (1995). The role of social support in the lives of women exiting domestic violence shelters: An experimental study. *Journal of Interpersonal Violence, 10,* 437–451.

Tolman, R. M., & Edleson, J. L. (1995). Intervention for men who batter: A review of research. In S. R. Stith & M. A. Straus (Eds.), *Understanding partner violence: Prevalence, causes, consequences and solutions* (pp. 262–273). Minneapolis: National Council on Family Relations.

Uekert, B. K. (2003). The value of coordinated community responses. *Criminology and Public Policy, 3,* 133–135.

Varcoe, C., & Irwin, L. G. (2004). "If I killed you, I'd get the kids": Women's survival and protection work with child custody and access in the context of woman abuse. *Qualitative Sociology, 27,* 77–99.

Walker, L. E. (1984). *The battered woman syndrome.* New York: Springer.

Warshaw, C. (1993). Limitations of the medical model in the care of battered women. In P. B. Bart & E. G. Moran (Eds.), *Violence against women: The bloody footprints* (pp. 134–146). Newbury Park, CA: Sage.

Websdale, N., & Johnson, B. (1997). Reducing woman battering: The role of structural approaches. *Social Justice, 24,* 54–81.

Weisz, A. N. (1999). Legal advocacy for domestic violence survivors: The power of an informative relationship. *Families in Society, 80*(2), 138–147.

West, C. M., Kantor, G. K., & Jasinski, J. L. (1998). Sociodemographic predictors and cultural barriers to help-seeking behavior by Latina and Anglo American battered women. *Violence and Victims, 13,* 361–375.

Zorza, J. (1991). Woman battering: A major cause of homelessness. *Clearinghouse Review, 25,* 421.

Zweig, J. M., Schlichter, K. A., & Burt, M. R. (2002). Assisting women victims of violence who experience multiple barriers to services. *Violence Against Women, 8,* 162–180.

DISCUSSION QUESTIONS

1. What methods of help seeking did women use in their efforts to exit their battering relationships? How successful were these efforts?

2. What barriers did women face in an attempt to leave their abusers?

3. How did women's involvements with police, courts, and social services place them at risk for continued victimizations?

4. How do criminal justice and social welfare policies act as barriers in building a new life following a violent relationship?

READING

In the section introduction, you were exposed to a number of different conditions that can alter the experience of being victimized by a significant other. This article highlights how immigration status can alter women's abilities to seek help in a domestic violence setting. By listening to the voices of women involved in these types of situations, you will learn not only how their access of resources and responses to domestic violence are limited, but that many women fail to identify themselves as victims. Here, immigration provides a new lens through which we can learn about the experience of battering by an intimate partner.

Intersections of Immigration and Domestic Violence

Voices of Battered Immigrant Women

Edna Erez, Madelaine Adelman, and Carol Gregory

Over the past 30 years, feminist academics and practitioners have revealed the extent and variety of gender violence, ranging from street-level sexual harassment (Stanko, 1985) to woman battering (Dobash & Dobash, 1979). According to Chesney-Lind (2006), "naming of the types and dimensions of female victimization had a significant impact on public policy, and it is arguably the most tangible accomplishment of both feminist criminology and grassroots feminists concerned about gender, crime, and justice" (p. 7). Indeed, feminist criminological research was part of the battered woman's movement's hard-won efforts to criminalize domestic violence (Adelman & Morgan, 2006). Feminist criminologists, their cross-disciplinary associates, and others also have been part of the growing critique of the limits or unintended effects of the criminalization of domestic violence (Britton, 2000; Chesney-Lind, 2006; Coker, 2001; Snider, 1998). Together, scholars and activists have identified harms induced by the criminal justice system not only on battered women, and poor battered women of color in particular, but also on men who batter, and in particular poor men of color who batter (Merry, 2000).

Noting the interconnection between racist violence, violence against women, and the institutionalization of the battered woman's movement within U.S. social service and criminal justice systems, feminist criminologists and others have called for antiracist, multicultural feminist analyses of gender violence and other forms of crime (Burgess-Proctor, 2006; Potter, 2006; see Baca Zinn & Thornton Dill, 1996, and

SOURCE: Erez, E., Adelman, M., & Gregory, C. (2009). Intersections of immigration and domestic violence: Voices of battered immigrant women. *Feminist Criminology, 4*(1), 32–56.

Crenshaw, 1991, for foundational elaborations on intersectionality). Much of this analysis has looked at immigrant status as part of one's racial location in the social hierarchy (e.g., Crenshaw, 1991; Scales-Trent, 1999). Here, we build on the history of feminist criminology with an integrated feminist analysis of immigration and domestic violence. Rather than consider immigration as a variable or static category within race, we consider immigration as part of the multiple grounds of identity shaping the domestic violence experience. It is part of the interactive dynamic processes that, along with race, gender, sexual orientation, and class, inform women's experiences of and responses to domestic violence. We do so by analyzing one-on-one interviews with immigrant battered women from a variety of countries, revealing common experiences among immigrants in an effort to highlight *immigrant* as a separate and multiplicative aspect of identity, violence, and oppression.

We situate our study within the literature on gender, immigration, and domestic violence, noting the scholarly focus on discrete groups of immigrants (e.g., by ethnicity or national origin) rather than the commonalities experienced by various immigrant groups. We then outline our research methods and sample, followed by an analysis of the data that focus on commonalities across immigrant battered women's experiences. Specifically, we suggest that although significant investment has been made by federal and state governments, and local community-based organizations, to improve the criminal justice system response to immigrant battered women in terms of legal reform, law enforcement training, and increased services, immigrant battered women continue to face considerable structural barriers to safety. These barriers exist prior to immigration (e.g., social pressure to marry) and as a result of immigration (e.g., economic disadvantage that has gendered consequences). In turn, immigration law and women's perceptions of law enforcement inform their attitudes toward reporting intimate partner violence. We conclude with a discussion of our research findings and their implications for theory and practice, expressing concern with the level of awareness of existing legal options for battered immigrant women and the growing anti-immigrant trend across the United States to devolve enforcement of federal immigration law to local authorities.

Feminist Theory of Intersectionality

Feminist discourse on intersectionality has developed over the past two decades. Although there are some differences in interpretation and application, intersectionality theory considers the ways that hierarchies of power exist along multiple socially defined categories such as race, class, and gender. These categories mutually construct each other via structural inequalities and social interaction, creating a matrix of intersecting hierarchies that is not merely additive but multiplicative in terms of unearned privilege, domination, and oppression (Baca Zinn & Thornton Dill, 1996; Collins, 1991/2000; Crenshaw, 1991; Higginbotham, 1997; Steinbugler, Press, & Johnson Dias, 2006). In this way, both opportunities (including social and material benefit) and oppressions may be simultaneously created by intersecting forms of domination (Baca Zinn & Thornton Dill, 1996; Steinbugler et al., 2006). Thus, for instance, "a gay Black man may experience privilege vis-à-vis his maleness but be marginalized for his race and sexuality" (p. 808). Angela Harris (1990), along with other critical race feminism legal scholars, refers to this notion of intersecting, indivisible identities as "multiple consciousness." Theories of intersectionality have inspired scholars across many disciplines to notice how various forms of privilege and oppression operate simultaneously as well as to reveal those forms of social identities that go unnoticed.

Writings on intersectionality use country of origin as an example of how racial and ethnic identities result in domination or oppression. Crenshaw (1991) specifically refers to immigrant status as an example of how race affects violent victimization in the United States. In this article, we show how the experiences of legal and undocumented immigrants are different from those of U.S. citizens and yet similar to one another, regardless of country of origin. Notwithstanding the racialized politics associated with immigration in the

United States, and recognizing the racism that many immigrants face, our effort here is to build on the substantial literature on intersectionality to reveal the intersection of immigration and domestic violence. We do so to highlight the salience of immigration for battered women in terms of how immigration affects the level and types of intimate partner violence women experience and shapes marital dynamics and women's helpseeking opportunities. We also examine how immigration and the policing of immigration may compromise women's safety. Thus, although we attend to the racialized category of immigration and the racist anti-immigrant sentiment aimed at immigrants, analytically, we have separated immigrant status from race/ethnicity as a category of intersectionality.

Immigration

Twenty-first century migration across international borders is a significant global phenomenon (Sassen, 1998). Motivated by a combination of push and pull factors such as impoverishment and economic opportunities, political instability and the opening of previously closed borders, and the loss or gain of family ties, large numbers of people enter key receiving countries such as the United States each year. The United States is considered "a nation of immigrants." Nevertheless, who is allowed to legally immigrate has varied over time. U.S. immigration and naturalization laws have shaped the resulting immigrant pool in terms of gender, race or nationality, sexual orientation, and marital status. These social identities have been central to U.S. immigration law, ranging from the exclusion of Chinese prostitutes in the 1870s to the men-only Bracero Program instituted in 1942 (Calavita, 1992). Subsequent changes in immigration policy, including an amnesty initiative in the mid-1980s, led to heterosexual family reunification and an increase in the numbers of women and children who migrated to the United States. Such gendered, racialized, and sexualized patterns reflect how immigration and naturalization law serves to police the purported moral as well as political boundaries of the nation (Gardner, 2005). These immigration laws affect why, when, how, and with whom women immigrate and their experiences of domestic violence subsequent to arrival in the United States.

One factor among many that motivates emigration from southern toward northern tier states is immigration policies that focus on family reunification. Other factors include the intensification of economic globalization under neo-liberal policies and relative ease of movement between political borders. Together, these factors are responsible for women making up an ever-increasing proportion of immigrants to the United States. Indeed, by the turn of the century, "close to 60 percent of immigrants from Mexico, China, the Philippines and Vietnam were female"; a similar percentage of female immigrants were between ages 15 and 44, significantly younger than their native-born counterparts (Zhou, 2002, p. 26). This young age cohort requires of female immigrants a long-term commitment to domestic and workplace labor in their new country of residence. In addition to their unpaid domestic and paid workplace labor, female immigrants also frequently contribute financially to the economy of their countries of origin via remittances home. In areas other than age and labor, however, female immigrants, as a whole, are a diverse group: migrating alone or with children and family; undocumented and/or dependent on male kin who sponsor their immigration. Some women arrive as highly skilled workers and successfully secure well-paid jobs. Other women, regardless of their skill sets, become among the lowest paid in the U.S. workforce. Still, female immigrants share the gendered effects of their border crossing.

As research on the gendered nature of immigration has emerged in terms of changing patterns over time of migration, identity formation and transformation, education, fertility, health care, and employment (Gabaccia, 1992; Hondagneu-Sotelo, 2003; Pessar, 1999; Strum & Tarantolo, 2002), so too has insight into the so-called domestic lives of immigrants. Ethnographers, for example, have analyzed how the meaning of marriage, along with women's and men's expectations of intimate relationships, may change as a result of migration patterns, access to education, and women's economic opportunities (Hirsch, 2003). These studies of immigrant domestic life help trace continuities and disruptions of the construction of gender across the migration process. For our purposes, one of the most critical links lies between the transformation of gender across the migration process and domestic violence.

Immigration and Domestic Violence

Violence against women is one of the most common victimizations experienced by immigrants (Davis & Erez, 1998; see also Erez, 2000, 2002; Raj & Silverman, 2002). Working together, battered immigrant women, activists, and scholars have documented how immigration intensifies domestic violence and creates vulnerabilities that impair immigrant women's management of domestic violence, preventing them from successfully challenging men's violence, from securing decreases in rates or types of men's violence, or from leaving their intimate partners. According to domestic violence scholars, "immigrant women arrive with disadvantages in social status and basic human capital resources relative to immigrant men" (Bui & Morash, 1999, p. 774) or cannot participate as actively in networks as male counterparts do (Abraham, 2000). As a result, barriers to safety for immigrant women include a lack of resources for battered women, social isolation or lack of local natal kin, economic instability, and perceptions that disclosure of battering to outsiders sullies community status. Criminal justice agencies that lack translation services and/or knowledge of immigration law, lack of trust in law enforcement and/or government authorities, and immigration law that dictates legal and sometimes economic dependency on the batterer, who may be undocumented or lacking legal immigrant status, also pose significant barriers (Bui, 2004; Dasgupta, 2000; Wachholz & Miedema, 2000).[1]

U.S. immigration law endangers battered immigrant women by giving near total control over the women's legal status to the sponsoring spouses, replicating the doctrine of coverture, under which "a wife could not make a contract with her husband or with others" (Calvo, 1997, p. 381). Coverture, in effect, identifies the married couple as a single legal entity, within which the husband has control over the property and body of the wife and their children. Similarly, women who immigrate as wives of U.S. citizens, legal permanent residents, diplomats, students, or workers are legally dependent on others to sponsor, pursue, and complete their visa petitions. This legal dependency intensifies gendered inequality, creates new ways for men to abuse and control their intimate partners, and entraps battered women (Erez, 2002; Salcido & Adelman, 2004). As part of the Violence Against Women Act (VAWA), legal reforms have been instituted to relieve some of the legal and economic dependencies imposed on battered immigrant women. These reforms include self-petition, which lets an abused spouse apply for a green card on his or her own; cancellation of removal, which lets an abused spouse who has already been subjected to removal proceedings request to remain in the United States; the U-visa, which lets a victim of crime (including domestic violence) who has been helpful to its investigation or prosecution apply for a nonimmigrant visa and work permit; and access to public benefits such as food stamps (Orloff, 2002; see also Wood, 2004). Obstacles to these well-intentioned legal reforms for immigrant battered women remain, in particular due to the complex nature of legal qualifications, including who is eligible to apply for which form of legal relief, and meeting the threshold required to demonstrate having been subjected to battery or extreme cruelty. The rise in anti-immigrant public sentiment has resulted both in the exclusion of some immigrants from access to education and medical care and in increased local law enforcement of federal immigration law. When coupled with post-9/11 delays in processing visa applications, the consequences of anti-immigrant sentiment further complicate the implementation of legal reforms for immigrant battered women.

Knowledge of immigrants' experiences with domestic violence is largely culled from case studies of discrete communities. Due in large part to the depth of social and cultural capital required to conduct sensitive research with members of marginalized immigrant communities, researchers tend to focus on small, local samples of battered women from specific immigrant communities (but see Menjivar & Salcido, 2002). Thus, we have insightful contributions based on the experiences of domestic violence by immigrant women to the United States from, for example, Bosnia (Muftic & Bouffard, 2008), Cambodia (Bhuyan, Mell, Senturia, Sullivan, & Shiu-Thornton, 2005), Mexico (Salcido & Adelman, 2004), Russia (Crandall, Senturia, Sullivan, & Shiu-Thornton, 2005), South Asia (Abraham, 2000), and Vietnam (Bui & Morash, 1999). These studies generate critical albeit partial knowledge with regard to immigration and domestic violence. In addition, until now, much of the holistic knowledge on immigrant battered women has been (rightly) directed toward services and policy-based interventions.

In this study, we take a different approach. We offer a detailed analysis situated within a theoretical framework of intersectionality, using *immigrant* as a positioned identity within the social structure as well as within interactions. This approach highlights the commonalities experienced by battered immigrant women, regardless of their ethnic or national group membership or countries of origin. Aware of the specific and unique contextual elements affecting domestic violence in each immigrant group, and the heterogeneity of domestic violence experiences that immigrant women from different cultures or ethnic groups endure, in focusing on the commonalities experienced rather than the unique elements of violence against immigrant women, we expect to highlight the theoretical value of the findings as well as draw public policy implications.

Research Methods

As previously noted, extant case studies of immigrant battered women typically consist of small, local samples derived from within one discrete community group. Our goal was to create a relatively large sample of diverse participants to be interviewed about their experiences with immigration and domestic violence. Diversity of participants in this study is based on each participant's language, ethnicity, nationality, cultural groupings, and country of origin. The sampling frame originated in states with large numbers of recent immigrants, with diverse immigrant communities, and with communities residing in both urban and rural areas: California, New York, Florida, Texas, Michigan, Wisconsin, and Iowa were selected as research sites.

Major immigration legal assistance organizations in these states helped to identify relevant social service agencies that provide direct services to immigrants. The directors of the social service agencies were contacted by phone about possible participation. In addition, members of social service agencies from other parts of the country who attended various regional and national meetings related to training or discussions about battered immigrant women and other issues concerning domestic violence and immigration were also approached for possible participation. Representatives from several agencies in New Jersey, Ohio, and Washington who expressed interest in participation were added to the list of participating agencies. Altogether, 17 agencies participated in the study, conducting interviews.

The interviews also addressed contacts with the criminal justice system, which some immigrant women may be unwilling to discuss with strangers. In light of the sensitive nature of the interview content and common reluctance among immigrant battered women to disclose detailed accounts of victimization and criminal justice experiences to outsiders, each participating agency instructed its bilingual social service provider to initiate contact with battered immigrant women with whom the provider had previously established rapport and a helping relationship of trust. As with much feminist research, one considers the positionality of the research subject in devising the methodology and conducting the research. The providers' relationship with the immigrant women was an integral component of the data collection phase because the providers were not only familiar with interviewees' strengths, concerns, and needs but also shared their language and, commonly, their culture. Therefore, the provider asked each woman if she was willing to be interviewed, explained the purpose of the research project, and, once the woman gave her consent, conducted the interview.

We recognize that where a power differential existed between the social service agency staff and the helpseeking interviewees, it may have compromised the validity of those data pertaining directly to access to or quality of social services. However, as noted below, many of the social service agents were battered immigrant women turned advocates, where the power differential was minimal. Furthermore, given the logistical barriers (e.g., training and sending interviewers to agencies in multiple states) and skill-based challenges (e.g., language competency) involved in collecting sensitive data from such a diverse sample, on balance we determined that access to a range of immigrant battered women, secured in large part due to the relationship of trust they had established with the agency staff and the linguistic comfort afforded to participants, overrode this limited, albeit important, methodological concern.

The bilingual social service providers who conducted the interviews ($N = 20$), were employees or volunteers who either had training in social services or, in some cases, were themselves survivors of domestic violence who had become battered women advocates. Each was given sets of questionnaires and instructions concerning the interviews (e.g., ethical standards such as confidentiality and interview techniques such as probing questions). The questionnaires, originally written in English, were sent ahead of time to the agencies so that the interviewers could become familiar with their content and be prepared, if necessary, to conduct simultaneous translations.[2] The social service providers/interviewers most often conducted interviews in the immigrant women's native language (i.e., in about two thirds of the cases).

The interviewees ($n = 137$) were immigrant women who sought help related to their immigration and/or domestic violence problems. As such, they are not necessarily representative of all battered immigrant women but represent a subsample of this population: those who have overcome barriers to reveal abuse or seek help, and those whose battering came to the attention of social services, often due to the gravity of their victimization. Furthermore, they are not representative of the subgroup of immigrant women seeking help, as they have been recruited through requests for interviews by agencies that agreed to participate in the study. There were several organizations that for practical or resource reasons did not elect to participate ($N = 8$). Some could not afford the time to conduct lengthy interviews; others were not successful in identifying battered immigrant women who were willing to participate. The sample, therefore, is not a random representation of the universe of battered immigrant women in the United States. The value of the data reported in this study, however, lies in providing accounts of the dynamics of the interaction between domestic violence and immigration from a diverse sample of women who vary by language, ethnicity, nationality, and country of origin.

Most interviews were conducted in the first (non-English) language of the interviewees, as reported by the interviewers.[3] English also was used in some interviews in part or throughout the interview, if the woman being interviewed was well versed and expressed comfort in speaking English. The interviews lasted between 45 minutes and 2½ hours and included closed- and open-ended questions about the women's demographic characteristics, circumstances of their arrival in this country, experiences with abuse and violence in their home countries and in the United States, and their attempts to seek criminal justice and/or social services to ameliorate their situations.

Interviewees were offered a modest stipend ($20) for their time, regardless of whether they completed the interview. Interviews were completed most commonly in one session, but a few were completed during a second session. Any requests to skip a certain question because an interviewee was uncomfortable about describing issues she considered private were honored. Despite an extensive list of interview questions, most women responded to our questions in great detail. Translation problems invalidated some of the responses or resulted in partial responses.[4] For these reasons, the results for a small number of items in the interview schedule present only the range of responses rather than a quantified version of the responses.

Quantitative data were calibrated and the open-ended questions transcripts were analyzed through coding techniques described by Glaser (1992). As we read each response, we searched for and identified patterns and variations in participants' experiences and we reached a set of conceptual categories or propositions. The analysis was conducted by applying the logic of analytic induction, which entails the search for "negative cases" and progressively refining empirically based conditional statements (Katz, 1983). When negative cases were encountered, we revised our propositions until the data were saturated, making the patterns identified and the propositions offered consistent throughout the data. Once no new conceptual categories could be added, or propositions had to be reformulated, it was assumed that saturation had been reached.

 Research Sample Profile

Female immigrants to the United States in the final research sample ($n = 137$) came from 35 countries.[5] They self-identified with a variety of religions: Christian (58%, of which 36% identified as Catholic),[6] Muslim (22%), Hindu (5%), and Jewish (1%). The age of the women ranged from 19 to 56 years, with a mean age of 32.5 and median age of 31.

In terms of marital status, approximately the same percentage of women were married in their home countries (45%) or were never married (i.e., single and/or living apart from an intimate partner) before coming to the United States (43%). The rest of the sample were either divorced (4%), separated (2%), or living with someone (2%) in their home countries prior to immigrating to the United States. At some point after immigrating to the United States, most single women got involved with an intimate partner. The percentage of "never married" decreased from 43% to 6% and those living with someone increased from 2% to 18%. Although the percentage of women in the sample who were married during the interview was the same as those who were married in their home countries prior to the move to the United States (45%), the percentage of women who stated their marital status was "divorced" at the time of the interview increased from 4% to 18%, and the percentage of women who were separated from their spouses rose from 2% to 23% of the sample.

The range of years the women have lived in the United States was from 1 to 30 years, with a mean of 8.7 years and a median of 6. The length of time they lived with the abuser was between 1 and 30 years, with a mean of 7.6 and median of 6 years.

In terms of family size, the overwhelming majority of interviewees had children (86%). The mean number of children was 2.4, and the median was 2. The educational level of the interviewees ranged from 5 to 16 years of education, with a mean of 11.6 and a median of 11 years of education (where 12 refers to high school graduate), excluding one woman who stated she had no education at all.

A quarter (25%) of the women in this sample had no ability to speak English, whereas 48% had some ability and another 26% were fluent English speakers. Thus, the use of interviewers skilled in the participants' native language was imperative. Only 27% were fluent readers of the English language, whereas 25% were fluent writers. The vast majority of women sampled had only some or no ability to read (46% some ability; 27% no ability) or write (37% some ability; 38% no ability) in English. The English proficiency of the sample as reported by interviewees is detailed in Table 4.1.

Immigration status varied among interviewees and between interviewees and their intimate partners at the time of the interview (see Table 4.2). Immigration status was divided into the following categories: U.S.-born citizens, naturalized citizens, lawful permanent residents (LPRs), VAWA self-petition, work visa, undocumented, and temporary visa. Consistent with the definition of immigrant, none of the women in this sample were U.S.-born citizens, whereas 11% of partners were natural-born citizens. Two categories described the largest percentage of female participants: LPR and undocumented. Thirty-four percent of participants were LPRs whereas 36% of their partners were LPRs, and 24% of participants were undocumented immigrants whereas only 15% of partners were undocumented. Naturalized citizens were 19% of our sample of women and 34% of partners. Nine percent of participants and 4% of partners had temporary visas, 9% were VAWA self-petitions, and 5% had work visas. No partners in this study had work visas or were VAWA

Table 4.1 English Proficiency

English Language Literacy	Fluent	Some Ability	No Ability	Total
Reading	27%	46%	27%	100%
Writing	25%	37%	38%	100%
Speaking	26%	48%	25%	99%

self-petitions. In general, male partners occupied a citizenship status with greater rights and privileges than did the female victims in this study.

More than half of the women (58%) were employed at the time of the interview. Most often, employment involved unskilled work, and domestic labor was the most common type of work reported (15%) by those employed. Almost half of the women (42%) had no gainful employment. More than three quarters of the husbands or partners (78%) were employed, most often in menial, service, unskilled, or skilled labor. About one quarter of both men (27%) and women (26%) sent money remittances to family in their home countries. More than one third of the women (39%) either used or planned to use public benefits.

Women reported being subjected to a lengthy period of abuse, ranging from 6 months to 25 years, with a mean of 5.5 years and median of 4 years of mistreatment, which included physical, mental, and sexual abuse, as well as verbal assaults. Women were also subjected to threats of being reported to Immigration and Naturalization Services (INS, now referred to as

Table 4.2 Immigration Status

Immigration Status	Female Immigrants	Intimate Partners
U.S.-born citizen	—	11%
Naturalized citizen	19%	34%
Lawful permanent resident (LPR)	34%	32% 4% amnesty LPR[a]
VAWA self-petition	9%	—
Work visa	5%	0%
Undocumented	24%	15%
Temporary visa[b]	9%	4%
Total	100%	100%

NOTE: VAWA = Violence Against Women Act.

a. Previously undocumented, but secured LPR as part of 1986 Immigration Reform and Control Act.

b. Temporary visas included tourist, student, and work visas.

Immigration and Customs Enforcement [ICE]), being deported, or having their children taken away. The abuse also included tactics of isolating the woman to perpetuate her dependency on the abuser (e.g., she was not allowed to go to English classes, to go to school, to have employment, to be in touch with friends or family members, etc.).

Becoming an Immigrant Battered Woman

Women reported various reasons for coming to the United States. One third (34%) followed their spouses, and one eighth (13%) married U.S. citizens, most of whom (n = 10) were military men.[7] About one fifth (16%) came for family reunification. A substantial proportion of the women immigrated for economic reasons: 29% came to improve their economic status and 12% to work. Another significant proportion fled violence in the home country (18%) or political repression (10%).

In the United States,[8] most of the women (87%) reported that the gendered division of labor was clear-cut; women focused on being a wife and mother and were solely responsible for housework and child care. In a minority of cases (17%), women were responsible for grocery or child-related shopping. Most often, they did not have access to a car or did not have a driver's license (60%). Men were responsible for gainful employment and money transactions related to the family, and only in a minority of cases (13%), the women stated that their men helped with work around the house.

According to female interviewees, the abuse resulted in severe mental and physical harm, including depression, withdrawal, numbness, and anxiety. About one third of the women (34%) required hospitalization to treat the injuries that resulted from the battering. Almost half of the women (46%) reported being battered while they were pregnant, with the abuser often trying to hit, kick, or otherwise interfere with the pregnancy. This abuse took place in all parts of the house, in particular in the bedroom or kitchen. Contrary to popular myths concerning domestic violence, it also occurred in public areas such as medical clinics, cars, and various social service offices, in front of family, children, neighbors, and other community members. Members of the husband's family often participated in the abuse.[9] Victimization in the presence of others is indicative of a perception that the abuse is justified or that it will garner no consequence to the perpetrator. The former suggests that the offender's actions are condoned by friends, family, and the community. The latter raises questions about institutional responses to publicly displayed abuse and how the immigrant status of the victim affects the perceptions and reactions of medical and social service workers.

Immigrant women have an added risk of victimization due to relocation. For women who immigrated with a spouse or partner, the move seemed to have an adverse effect on men's level of violence and control tactics. Following their arrival to the United States, for half of these women, the level of violence increased, and almost one quarter (22%) stated that the violence began after arrival: "It has gotten worse. Now he takes out all the frustration on me." For one fifth (20%) of the women, the level of violence stayed the same, for 6% it decreased, and for 2% it stopped. The escalation of abuse was particularly difficult for immigrant women who had left their natal families behind: "I don't have family here, so he tells me that I don't have another choice but to stay with him." Another woman argued that "if he were in Syria, he would take into consideration my parents and would not act abusively as in U.S." Lacking natal family and an extended kin network led to a high rate of social isolation and a deep sense of vulnerability for immigrant women.

Immigration affected husbands and wives differently. For example, some women reported that immigration removed what they understood as constraints against domestic violence, which were rooted in their home countries. "If I want to compare it to Iraq and the U.S., of course the move has affected us. In Iraq we have family, parents, relatives. Here there is drinking and open society, especially for men." Women explained that men acquired new interests, such as alcohol, drugs, gambling, and women, which often accompanied the abusive behaviors of the spouse.

In addition to marital arguments to which men who batter often respond with abuse (e.g., jealousy, infidelity, drinking, money issues, child discipline, or education issues), there also were distinct issues created by the move to the United States that caused tension in the marriage and exacerbated the abuse. For example, many of the women reported that remittances they or their husbands made (i.e., sending money to family members in country of origin) often precipitated arguments or fights. Other issues included the husband's inability to provide for the family in the new country or his insistence that the wife, although now in the United States, continue to be a "traditional woman and never ask him about anything" or that she remain "a very traditional Latina wife, waiting on him hand and foot and never raising my voice on him." Women often explained the reasons for their battering as "my being a bad wife and mother" or "I needed to do what he told me to do, when he told me to do it."

According to women who took advantage of economic opportunities opened for them in the new country, this change provoked their spouses and led to abuse: "In the U.S. he suffered jealousy attacks and saw me prosper—he did not like that."

Economic Challenges

Economic challenges are not unique to immigrant families, but finding suitable employment or any job at all presents major difficulties for most immigrant families. The difficulty of securing employment that matches one's skills is a significant source of conflict between husbands and wives (e.g., being an engineer but working at a gas station). One woman attributed domestic violence to her husband's unemployment and resultant idleness: "He did not work, stayed home, which made him crazy." Another suggested that unemployment, per se, was not the problem. Rather,

> the dissatisfaction, failure, disappointment, not being able to meet one's economic expectations in life switches the burden on the wife. She becomes the reason of his failures. She is blamed all the time. She consistently tries to please him; it doesn't work. She gets all the frustration and all kinds of abuses.

At the same time, battered immigrant women also are deprived of supportive community, extended family, or a social network that could help them during such difficulties.

> If a spouse did not have work in home country, family or relatives would extend him money and help him. Here in U.S., there are many bills to pay; there is no one to give you a hand. One gets embarrassed.

On the other hand, for women working outside the home, their absence is often seen by men as a threat to the gender hierarchy. Women reported that although they worked outside the home, they controlled little to none of the money they earned and were subject to abuse and domination by their husbands.

> It was really good in the beginning, and then he lost his first job and things started getting really bad. It has not been very happy at work, and that is why he would take things out on me. We used to be happy. He would always keep the money and occasionally would demand a lot of sex, but then after a few years, he really started beating me up . . . [in particular] when I had to file his immigration papers.

Immigrant Status

Some women reported that the increase in emotional, sexual, and physical abuse coincided with immigration-specific activities such as entering the country, filing immigration papers, or accessing social welfare systems. The majority of women who came with their spouses reported that the transition and move to the United States altered the dynamics of the relationship: "He has had more power to manipulate in the U.S. because I am illegal and depended on him and I didn't have any rights here." An immigrant woman's dependency on her male partner elevates his position of dominance over her. At the same time, legal dependency represents a macro-structural vulnerability that systematically marginalizes immigrant women by limiting their access to goods and resources, such as work, social services, protection under the law, and so on. Although law is not intentionally gender biased, one that creates a status-marriage dependency, such as immigration law, makes immigrant women more vulnerable to the domestic violence power dynamic (Erez, 2000; Menjivar & Salcido, 2002).

Husbands became increasingly abusive, and the physical and emotional battering became more conspicuous and severe. One woman explained that "the relationship had gotten bad in Mexico and continued the same in the U.S. The abuse changed from verbal to physical." Another woman agreed that the violence worsened after immigration: "I believe when I came to the U.S. my husband treated me more like a kid. I do not have control over my life." Still another woman explained how "he has become more abusive. He knows the system; I don't. He speaks English; I don't. I don't have family support or someone living with me, so he can lie about me." Even one woman who had divorced her husband still was being threatened by him with regard to her immigration status: "He's going to call INS, because I lied that I was single instead of divorced. [From California] he stalks me, contacts me at home, at work in Michigan."

The overwhelming majority of women (75%) described how men used immigrant status to force them into compliance. "He used my immigration status against me. He would tell me that without him, I was nothing in this country." Men threatened women in a number of ways with regard to immigration including that they would call ICE officials and report their immigration status (40%); get them deported (15%); withdraw their petition to immigrate or otherwise interfere with the naturalization process (10%); take away the children or deny their custodial rights (5%); and, more generally, use immigration status to humiliate or degrade them (5%). One undocumented woman succinctly stated, "He makes threats to report me to the INS if I don't do what he wants."

Women also illustrated the connection between immigration and domestic violence being particularly painful for mothers. "He would tell me I did not have any rights in this country. He threatened to take our children—and he finally did!" In another instance, a woman was forced to trade custody of her children for an adjustment of her immigration status. In addition, mothers feared that their children would be deprived of opportunities for a brighter future that, in the minds of the women, the United States can provide. One woman was concerned about "employment for my older children and their immigration status. [My] son wants to be a U.S. citizen, to attend school and work here." Women did not want to jeopardize their children's immigration status and thought that divorce or leaving the United States would have negative consequences for their children.

Many battered immigrant women who do not have lawful permanent residency believe that divorce means losing their right to work or stay in this country. "If ever I challenge him to stay here, he will divorce me; I will lose my green card and will not be able to financially survive." This translates to jeopardizing her ability to sustain herself financially. Although the VAWA (1994) and its subsequent reauthorization (2000) Public Law 103-322, Violence Crime Control and Law Enforcement Act of 1994 Public Law 106-386, Victims of Trafficking and Violence Protection Act of 2000 provided battered immigrant women a self-petition option, most immigrant women are not aware of it. A husband uses the woman's lack of knowledge, dependency, and immigration status as a weapon to threaten and demand compliance. A man can easily manipulate his control over the relationship and the family because of an immigrant woman's actual or perceived legal dependency: "What prevents me from leaving is the immigration status. I need my green card." Abusers commonly convinced immigrant women that they have no rights (or that they are not entitled to any rights in this country) or that the abusers have the power to cancel their status at any time. Some threatened to withdraw the petitions already filed on the women's behalf or to tell ICE officials that the women married for the sole purpose of legal residency. Most of the women reported enduring abuse for long periods of time because of their desire to remain in the United States, in hopes that their husbands would change their immigration status to legal.

Culture and Community

The majority of the women (65%) reported abuse-tolerant perspectives in their home countries where, they explained, domestic violence is not considered a crime. On one end of the abuse tolerant–intolerant continuum, a woman stated that "my national community doesn't believe that domestic violence exists." Another woman described another position along the continuum: "In Armenian culture, it is okay for a husband to hit his wife, and she should accept it. In America, it is considered a crime." Other women also drew a sharp contrast between their home countries, where domestic violence is a normal part of the marriage, and the United States: "There's a difference because here it's a crime. In Nicaragua if the couple makes up, then it's okay." Overall, women reported being raised in households where fathers and husbands were considered authoritarian decision makers with the right to wield violence as needed to secure women's compliance and that their communities expected them to reproduce such marital arrangements.

The man is the center of authority. He is the supreme decision maker. He is the breadwinner; without him, in general, it is very hard to survive financially, especially if you are unskilled or uneducated.

> I was raised in a Hindu household . . . to be obedient and considerate of your elders.
>
> Tradition [says] that you stay with the person you married no matter what he does. Women stay home, to be housewife and put up with domestic violence. Here divorce is acceptable more so domestic violence not accepted.
>
> Women in Latin America and Mexico are supposed to suffer a lot with their husbands.
>
> We have to listen to men more than the American women. We have to stay home most of the time when we get married. We have to be more responsible for children and husbands.

These general comments were reinforced by more individualized lessons:

> My mother and father told me to go back and be a better wife. Otherwise I would be shaming them.
>
> My mother told me to bear it, since it was my decision to marry him.
>
> At first they were sad, told me to be patient. God will solve it.

Family members warned that divorce would negatively affect their children's welfare or chances for a good marriage or would decrease their younger sisters' prospects to marry. They used fear of shame, gossip, and guilt to convince their daughters to stay with their abusive husbands. In addition, some women also expressed fears, based on their respective husbands' threats, that leaving would lead to serious injury or even death. Despite their fear and familial admonitions to "put up with domestic violence" and "listen to men," the majority (85%) of women made one or more attempts to leave the abusers. Many of the women tried from 1 to 15 times to extricate themselves from the violence. Some women stated that they attempted to leave hundreds of times.

Reporting Abuse

Women reflected on the expectation that "everything stays in family. Sometimes we don't even tell our families, only after many years of problems." According to their immigrant communities, marital strife was to be kept private and should not be disclosed:

> A man can do anything; he is the head of the family, and a woman should always sacrifice to make things work. The expectations for men and women are different. Our culture does not welcome outside intervention. We don't involve outsiders in family issues. We do not consider

> domestic violence as a crime; police do not get involved. We don't go to shelters. Legal system does not get involved.
>
> They don't like [public intervention], because they want to have the liberty of committing family violence at will.
>
> In this town, it will label the woman. It will make it harder on the woman. [Public intervention is] not a good idea.

In the face of abuse-tolerant and privacy-affirmative perspectives, more than half (54%) of the women stated that they did not report the abuse because of their culture or religion. Nearly half the women did end up dealing with the criminal justice system as a result of the abuse (46%); however, in one third of these cases (35%), it was because someone other than the victim called the police (neighbor, family member, friend, or hospital staff).

Given the public pressure to keep domestic violence private, women struggled to maintain their social identity and status within their immigrant communities as they struggled to obtain safety for their children and themselves. "I will be ostracized and then where will I go?" Women reflected on distinctions between "home" and "here" attitudes toward criminal justice and other public interventions into domestic violence: "Here the police will help you. In El Salvador, they won't." Unaccustomed to involving outsiders or reporting domestic violence to the police at home ("I'm from Haiti; there is no such law to protect women against domestic violence"), women discussed the tension here in immigrant communities about disclosing abuse to family members and law enforcement.

Female interviewees "became aware of domestic violence in this country, because we know that many people can help us with our problem," including law enforcement, who "are very responsive here and very helpful." Immigrant women "now . . . think [domestic violence] is a crime here," and "Americans treat it like a crime, because that's what it is." Moreover, "here in U.S., a woman demands her rights. The Arab woman does not have a say in Arab countries." As a whole, women identified that "in the U.S. there is more support and protection for the victims, more services" and that "a woman in U.S. has her say, can make her own decisions. The government helps her to have the kids. In our country, no welfare benefits." One woman was impressed that "the clergy here in U.S. encourage you to report [domestic violence] to authorities."

Overall, women felt empowered by having at least the option to mobilize the justice system for help. It provided them a "big relief," or they found it "positive" or "helpful." In some cases, individual women's growing awareness was matched by communal acceptance of domestic violence as a behavior that deserves intervention, in particular when abuse resulted in serious injury. Women distinguished between those who shared ethnic or national identities in the home country and those in the United States.

> The Armenians from Armenia think police intervention is bad but Armenians in the U.S. generally do not think police intervention is a bad thing.
>
> In Mexico, they do not interfere until the woman is sent to the hospital; in the U.S. they interfere at an early stage, before there is need to send women to hospital.

Women also distinguished between known cases of domestic violence and those that remain hidden from sight, due to either literally or figuratively closed doors:

> It depends. When cases are really bad, like publicly seen abuse, the community 100% supports. When cases happen behind closed doors, the community is hesitant.
>
> It depends from case to case. If you or your family has a social standing.

However, they were well aware that their communities, or segments thereof, did not view favorably intervention by outsiders, in particular law enforcement.

> The community is accepting the outside intervention, except the religious leaders. Still even

> if the spouse is very abusive, they do not give religious divorce to victims. The batterer immediately remarries while the victim is helpless. Also, the community is not very supportive to a divorced woman.

In light of these mixed messages, "it makes you hesitate. Even if you know it is the right thing to do, you postpone the outside intervention."

Some immigrant women had negative experiences (either in the home country or here) with the justice system. Ambiguous messages about and ambivalent attitudes toward law enforcement when coupled with a persistent lack of material resources made many battered women reluctant to seek such intervention. These immigrant community views affected women's responses to the abuse, prolonged their marriages, or prevented them from seeking outside help. Still, individual women prevailed with assistance from immigrant community organizations to secure a semblance of physical security, social standing, and legal stability: "My children and the family unit is what keeps me in the relationship. However, he has promised to stop hitting me. I used to fear deportation, not anymore—I filed my own papers. I also wanted to protect my children."

Conclusion

Battered women in general face a number of interrelated and intricate barriers that complicate their pursuit of safety. Women struggle with, among other factors, embarrassment and shame about disclosing abuse and seeking help from social service or criminal justice agencies; emotional connection to and economic dependency on batterers; reluctance to break up families; and fear of myriad forms of violence, control, and retaliation by abusers and their communities. Although heterosexual men who batter are found in all social groups and at all economic levels, regardless of ethnicity, religion, national origin, cultural affiliation, or immigration status (Volpp, 2001), we have demonstrated that men who batter immigrant women, the majority of whom are immigrants themselves, have access to unique forms of domination and control, some of which are facilitated or even sanctioned by federal immigration law.

In our analysis of 137 battered women who had immigrated to the United States from 35 countries across the globe, we found that the general difficulties that battered women face coexist with challenges they experience as immigrants. Battered immigrant women face a range of legal, economic, and social challenges to safety. Legal challenges include lack of familiarity with or access to social service or criminal justice systems that possess limited immigrant-related cultural and linguistic competencies; legal dependency on batterers; and lack of legal knowledge. In terms of economic barriers, immigrant battered women report that their communities' economic marginalization combined with the continued responsibility for sending remittances home figures large in batterers' justification for abuse. The social implications of battering are no less central to immigrant battered women than legal and economic barriers. Internal to the community, individual women are limited by a deep fear of losing social status in and the support of their immigrant communities—often the only communities they know—and a fear of various forms of violence, control, and retaliation by the husband and his family, often the only kin they have in the new country. Among other social complications external to the community, immigrant battered women face racist anti-immigrant public sentiment that exacerbates their desire to keep violence private in order to transmit an untarnished and positive image of the immigrant community. These patterns persist, despite any differences among the sample.

The interaction of domestic violence and immigration informs not only the level and type of abuse men perpetrate but also individual and community-based responses to the abuse. We found that, over time, immigration shaped the meaning that battered women gave to the controlling behaviors and violence perpetrated against them by their intimate partners. For the most part, women distinguished between attitudes and practices related to domestic violence "here" and "there." That is, they labeled their home countries as abuse tolerant and their adopted country as abuse intolerant. Moreover, despite existing antiracist critiques of the institutionalization of the criminalization of domestic violence, and mixed messages from their own communities as to the appropriateness of reporting domestic violence to the authorities, immigrant battered women seemed to appreciate that domestic violence was considered a crime in the United States and perceived that law enforcement officers were willing to assist as they sought safety for themselves and their children.

However, although at least some immigrant battered women feel empowered to mobilize the criminal justice system, few seem to be familiar with new policies promulgated to protect battered immigrant woman, such as the VAWA self-petition option. And even for those who obtain relevant information and meet legal criteria, pursuit of such remedies may be limited by lack of access to legal assistance or fear of turning to legal authorities, including the criminal justice system. Undocumented immigrants, as well as those in the midst of applying for legal status, or even legal immigrants may avoid engagement with the criminal justice system, in particular if they are part of a "mixed-status" immigrant family or in order to prevent law enforcement from entering an immigrant-majority neighborhood.

The commonalities among immigrants from across such a wide range of countries of origin raise two additional concerns related to immigrant battered women and the criminal justice system. First, over the past decade, the criminalization of immigrants has escalated in the United States, where immigrants are perceived of as criminals-in-the-making who make "real" Americans vulnerable to uninsured drivers, lower wages, unemployment, and property crimes as well as drug, gang, and trafficking-related violent crime. It is "immigrant" on "American" visible forms of crime that populate public discourse. Rarely mentioned is the less visible crime of intimate partner violence. When referenced, intimate partner violence among immigrants is either naturalized (i.e., that's just the way they are) or culturalized (i.e., that's how they treat their women). Naturalization and culturalization of immigrant domestic violence blame intimate partner violence on membership within the group, minimize the effect of intimate partner violence on its victims, and dismiss victims' claims for justice. Moreover, it erases intimate partner violence among so-called "assimilated" and/or native-born members of U.S. society. As such, although we acknowledge that meanings and patterns of domestic violence vary across cultures, we write against the tendency to stereotype domestic violence as an inherent part of "other" cultures (Razack, 1998; Volpp, 1996, 2001). Such views reinforce the notion that gender-based violence does not warrant state intervention because it is part of the "way of life" (Ferraro, 1989), is the "mentality," or is "part of the culture" (Adelman, Erez, & Shalhoub-Kevorkian, 2003) of certain religious, ethnic, or national groups. This perception also precludes examinations of how structural inequalities and systemic responses (e.g., criminal justice system) may sometimes diminish the material conditions and safety options for individual immigrant women and their families. Dismissing domestic violence as an immigrant or cultural problem also precludes serious considerations of how to ameliorate commonly experienced structural inequalities or how to work with battered immigrant women to identify helpful systemic responses.

Second, in the post-9/11 era, the trend in cities, counties, and states is to enter "287(g) agreements" with the federal government to enforce immigration law as proxies for ICE (Versanyi, 2008). This means that local law enforcement officers, those charged with protecting battered women, are now responsible for enforcing the civil matters of federal immigration law as well. Undocumented immigrants, as well as legal immigrants who face criminal charges, are at risk for deportation, with or without their children. As a result, immigrants, in general, and immigrant women, in particular, regardless of legal status, may go further underground with their need for domestic violence services, thereby rejecting the investment made into the criminal justice system for victims of domestic violence. Further complicating immigrant battered women's pursuit of safety is the recent move by local governments to bar undocumented immigrants from education and social services. These developments make ambiguous which government agencies, including the criminal justice system and members of law enforcement, immigrants and their families have the right to approach—and whom to trust. Individual immigrant women, who commonly shoulder the responsibility for their children's welfare, face the structurally produced hardship of choosing between their safety and a stable, brighter future for their children. Designing social and legal policies that do not further entrap battered immigrant women will continue to challenge feminist criminologists.

Notes

1. Collaborative efforts among battered immigrant women, activists, and researchers also have resulted in the identification of strategies productively used by immigrant women. For examples of barriers and safety strategies, see online materials available at www.immigrantwomennetwork.org, produced by the National Network to End Violence Against Immigrant Women. Many of these issues shaped the legislation addressing the plight of battered immigrant women in the Violence Against Women Act of 1994 and its subsequent revisions.

2. Due to confidentiality requirements, it was not possible to conduct quality control of the translation. However, agencies did not report translation of the questions as a problem.

3. Primary languages included Arabic, Armenian, Bengali, Farsi, French, Haitian, Hindi, Japanese, Malaysian, Portuguese, Russian, Spanish, and Turkish.

4. Most questions invalidated due to translation pertained to criminal justice procedural issues associated with the events described during the interviews.

5. These countries are Armenia, Bahrain, Bangladesh, Brazil, Colombia, Costa Rica, Egypt, El Salvador, Former Yugoslavia, Albania, Germany, Great Britain, Guatemala, Guyana, Haiti, Honduras, India, Iran, Iraq, Israel, Palestine, Japan, Latvia, Lebanon, Mexico, Morocco, New Zealand, Nicaragua, Peru, Syria, Trinidad, Turkey, Venezuela, Vietnam (South), and Yemen.

6. Christians described themselves as Adventist, Armenian Apostolic, Assyrian Christian, Baptist, Jehovah's Witness, Lutheran, Mormon, Pentecostal, Protestant, or Roman Catholic.

7. The circumstances and experiences of these "military brides" are described in Erez and Bach (2003).

8. This clear-cut division of labor was also the case in the home country. We focus on the U.S. responses to examine whether division of labor changed as a result of immigration to the United States.

9. Those who have family members in the United States can immigrate due to family unification laws. Thus, men who immigrate have family members in the United States whereas women who follow their husbands leave their own families behind.

References

Abraham, M. (2000). *Speaking the unspeakable: Marital violence among South Asian immigrants in the United States.* New Brunswick, NJ: Rutgers University Press.

Adelman, M., Erez, E., & Shalhoub-Kevorkian, N. (2003). Policing violence against women in a multicultural society: Gender, minority status and the politics of exclusion. *Police and Society, 7,* 103–131.

Adelman, M., & Morgan, P. (2006). Law enforcement versus battered women. *Afflia: Journal of Women and Social Work, 21*(1), 28–45.

Baca Zinn, M., & Thornton Dill, B. (1996). Theorizing difference from multiracial feminism. *Feminist Studies, 22*(2), 321–331.

Bhuyan, R., Mell, M., Senturia, K., Sullivan, M., & Shiu-Thornton, S. (2005). "Women must endure according to their karma": Cambodian immigrant women talk about domestic violence. *Journal of Interpersonal Violence, 20*(8), 902–921.

Britton, D. M. (2000). Feminism in criminology: Engendering the outlaw. *Annals of the American Academy of Political and Social Science, 571*(1), 57–76.

Bui, H. (2004). *In the adopted land: Abused immigrant women and the criminal justice system.* Westport, CT: Praeger.

Bui, H., & Morash, M. (1999). Domestic violence in the Vietnamese community: An exploratory study. *Violence Against Women, 5,* 769–795.

Burgess-Proctor, A. (2006). Intersections of race, class, gender and crime: Future directions for feminist criminology. *Feminist Criminology, 1*(1), 27–47.

Calavita, K. (1992). *Inside the state: The Bracero Program, immigration and the INS.* New York: Routledge.

Calvo, J. (1997). Spouse-based immigration law: The legacy of coverture. In A. Wing (Ed.), *Critical race feminism: A reader* (pp. 380–386). New York: New York University Press.

Chesney-Lind, M. (2006). Patriarchy, crime and justice: Feminist criminology in an age of backlash. *Feminist Criminology, 1*(1), 6–26.

Coker, D. (2001). Crime control and feminist law reform in domestic violence law: A critical review. *Buffalo Criminal Law Review, 4*(2), 801–860.

Collins, P. H. (2000). *Black feminist thought: Knowledge, consciousness and the politics of empowerment.* New York: Routledge. (Original work published 1991)

Crandall, M., Senturia, K., Sullivan, M., & Shiu-Thornton, S. (2005). "No way out": Russian-speaking women's experiences with domestic violence. *Journal of Interpersonal Violence, 20*(8), 941–958.

Crenshaw, K. (1991). Mapping the margins: Intersectionality, identity politics and violence against women. *Stanford Law Review, 41,* 1241–1298.

Dasgupta, S. (2000). Charting the course: An overview of domestic violence in the South Asian community in the United States. *Journal of Social Distress and the Homeless, 9*(3), 173–185.

Davis, R. C., & Erez, E. (1998). *Immigrant population as victims: Toward a multicultural criminal justice system* [Research in brief]. Washington, DC: National Institute of Justice.

Dobash, R. E., & Dobash, R. (1979). *Violence against wives: A case against the patriarchy.* New York: Free Press.

Erez, E. (2000). Immigration, culture conflict and domestic violence/woman battering. *Crime Prevention and Community Safety: An International Journal, 2,* 27–36.

Erez, E. (2002). Migration/immigration, domestic violence and the justice system. *International Journal of Comparative and Applied Criminal Justice, 26*(2), 277–299.

Erez, E., & Bach, S. (2003). Immigration, domestic violence and the military: The case of "military brides." *Violence Against Women, 9*(9), 1093–1117.

Ferraro, K. J. (1989). Policing woman battering. *Social Problems, 36*(1), 61–74.

Gabaccia, D. (Ed.). (1992). *Seeking common ground: Multidisciplinary studies of immigrant women in the U.S.* Westport, CT: Greenwood Press.

Gardner, M. (2005). *The qualities of a citizen: Women, immigration and citizenship, 1870–1965.* Princeton, NJ: Princeton University Press.

Glaser, B. G. (1992). *Basics of grounded theory analysis.* Mill Valley, CA: Sociology Press.

Harris, A. (1990). Race and essentialism in feminist legal theory. *Stanford Law Review, 42,* 581–616.

Higginbotham, E. (1997). Introduction. In E. Higginbotham & M. Romero (Eds.), *Women and work: Exploring race, ethnicity, and class* (pp. xv–xxxii). Thousand Oaks, CA: Sage.

Hirsch, J. S. (2003). *A courtship after marriage: Sexuality and love in Mexican transnational families.* Berkeley: University of California Press.

Hondagneu-Sotelo, P. (Ed.). (2003). *Gender and U.S. immigration: Contemporary trends.* Berkeley: University of California Press.

Katz, J. (1983). A theory of qualitative methodology. In R. Emerson (Ed.), *Contemporary field research* (pp. 127–148). Boston: Little, Brown.

Menjivar, C., & Salcido, O. (2002). Immigrant women and domestic violence: Common experiences in different countries. *Gender & Society, 6*(6), 898–920.

Merry, S. (2000). *Colonizing Hawai'i: The cultural power of law.* Princeton, NJ: Princeton University Press.

Muftic, L. R., & Bouffard, L. A. (2008). Bosnian women and intimate partner violence: Differences in experiences and attitudes for refugee and nonrefugee women. *Feminist Criminology, 3,* 173–190.

Orloff, L. (2002). Women immigrants and domestic violence. In P. Strum & D. Tarantolo (Eds.), *Women immigrants in the United States* (pp. 49–57). Washington, DC: Woodrow Wilson International Center for Scholars and the Migration Policy Institute.

Pessar, P. (1999). Engendering migration studies: The case of new immigrants in the United States. *American Behavioral Sciences, 42,* 577–600.

Potter, H. (2006). An argument for Black feminist criminology: Understanding African American women's experiences of intimate partner violence using an integrated approach. *Feminist Criminology, 1*(2), 106–124.

Raj, A., & Silverman, J. (2002). Violence against immigrant women: The roles of culture, context, and legal immigrant status on intimate partner violence. *Violence Against Women, 8,* 367–398.

Razack, S. (1998). *Looking White people in the eye: Gender, race, and culture in courtrooms and classrooms.* Toronto, CA: University of Toronto Press.

Salcido, O., & Adelman, M. (2004). "He has me tied with the blessed and damned papers": Undocumented-immigrant battered women in Phoenix, Arizona. *Human Organization, 63*(2), 162–173.

Sassen, S. (1998). *Globalization and its discontents.* New York: New Press.

Scales-Trent, J. (1999). African women in France: Immigration, family and work. *Brooklyn Journal of International Law, 24,* 705–737.

Snider, L. (1998). Toward safer societies: Punishment, masculinities and violence against women. *British Journal of Criminology, 38*(1), 1–39.

Stanko, E. (1985). *Intimate intrusions.* New York: HarperCollins.

Steinbugler, A. C., Press, J. E., & Johnson Dias, J. (2006). Gender, race and affirmative action operationalizing intersectionality in survey research. *Gender & Society, 20*(6), 805–825.

Strum, P., & Tarantolo, D. (Eds.). (2002). *Women immigrants in the United States.* Washington, DC: Woodrow Wilson International Center for Scholars and the Migration Policy Institute.

Versanyi, M. (2008, April 20). Should cops be la migra? *Los Angeles Times.* Available April 22, 2008, from www.latimes.com

Volpp, L. (1996). Talking "culture": Gender, race, nation, and the politics of multiculturalism. *Columbia Law Review, 96*(6), 1573–1617.

Volpp, L. (2001). Feminism versus multiculturalism. *Columbia Law Review, 101*(5), 1181-1218.

Wachholz, S., & Miedema, B. (2000). Risk, fear, harm: Immigrant women's perceptions of the "policing" solution to women abuse. *Crime, Law and Social Change, 34*(3), 301–317.

Wood, S. (2004). VAWA's unfinished business: The immigrant women who fall through the cracks. *Duke Journal of Gender, Law & Policy, 11,* 141–155.

Zhou, M. (2002). Contemporary female immigration to the United States: A demographic profile. In P. Strum & D. Tarantolo (Eds.), *Women immigrants in the United States* (pp. 23–34). Washington, DC: Woodrow Wilson International Center for Scholars and the Migration Policy Institute.

DISCUSSION QUESTIONS

mmigration status affect the levels of violence experienced by the women in this study?

2. How does a culture of violence in a woman's home country affect her experience of domestic violence in the United States?
3. What implications do cultural values about domestic violence have for help-seeking strategies for victims of intimate partner abuse?

◈

Women and Victimization
Stalking and Sexual Harassment

Section Highlights

- The crime of stalking, its various forms, and its impact on women
- Sexual harassment of women
- Legal definitions and responses to stalking and sexual harassment
- Policy implications for stalking and sexual harassment

The crimes of stalking and sexual harassment are very different in many ways. A stalker can be someone known (like a significant or ex-significant other) or unknown (stranger) to the victim, whereas sexual harassment is generally perpetrated by someone known to the offender, such as a coworker, boss, or landlord. However, they share a number of similar characteristics. Both crimes involve unwanted attention from individuals who engage in behaviors that harass, threaten, and intimidate their victims. In turn, these experiences can often cause victims to be afraid of their assailant. In addition, both events are rarely a single act, but involve a number of behaviors that occur over a period of time. This section discusses the foundations of defining each of these behaviors as criminal acts and discusses the legal and social implications of these crimes for society.

Defining and Identifying Stalking

Hollywood celebrities have long experienced acts that would be considered stalking according to today's definitions. Consider the actions of John Hinckley, Jr., who became infatuated with Jodi Foster when she first appeared as a child prostitute in the film *Taxi Driver*. Hinckley's obsession with Foster continued while she was a student at Yale, but he failed to gain her attention after numerous letters and phone calls. In 1981, Hinckley attempted to assassinate President Ronald Reagan in an effort to impress Foster. He was found not guilty by reason of insanity for his crimes and was committed to St. Elizabeth's Hospital for treatment. Today, he is allowed extended overnight visits outside of the hospital with his family, though he remains in the custody of the facility. Another example of celebrity stalking is Madonna's stalker Robert Dewey Hoskins. He was convicted in 1996 for making threats against the star—he told the star that he wanted to "slice her throat from ear to ear" and attempted to break into her house on two separate occasions. During one event, he successfully scaled the security wall of her home and was shot by one of her bodyguards. Other Hollywood victims of stalking include David Letterman, Sandra Bullock, Tyra Banks, and Lindsay Lohan, to name a few.

▲ **Photo 5.1** Many victims of stalking experience the constant fear of being followed and observed as they attempt to manage their daily lives. In this situation, the psychological terror that victims experience can be just as violent as any physical confrontation.

Indeed, it seems that a number of Hollywood personalities have been stalked by an obsessed fan at some point during their careers. While noteworthy events such as these brought significant attention to the crime of stalking, they did so in ways that reduced the social understanding of this crime as one that was limited to celebrities and the Hollywood circuit. Many of these cases involved perpetrators who suffered from mental disease or defect. This narrow definition had significant effects on the legitimization of this crime for "ordinary" victims of stalking.

According to the National Crime Victimization Survey, stalking is defined as "a course of conduct directed at a specific person that would cause a reasonable person to feel fear" (Baum, Catalano, Rand, & Rose, 2009, p. 1). Estimates by the Supplemental Victimization Survey (SVS) indicate that more than 5.9 million adults[1] experience behaviors defined as stalking[2] or harassment.[3] These data indicate that acts of stalking and harassment occur at a significantly higher rate than the majority of the public believes. In measuring the prevalence of stalking behaviors, the SVS collected data on seven different acts of stalking: (1) making unwanted phone calls; (2) sending unsolicited or unwanted letters or e-mails; (3) following or spying on the victim; (4) showing up at places without a legitimate reason; (5) waiting at places for the victim; (6) leaving unwanted items, presents, or flowers; and (7) posting information or spreading rumors about the victim on the Internet, in a public place, or by word of mouth. In most cases, the acts that constitute stalking, such as sending letters or gifts, making phone calls, and showing up

[1]The Supplemental Victimization Survey (SVS) only includes data on respondents ages 18 and older who participated in the National Crime Victimization Survey (NCVS) during January–June 2006. The data assess victimization incidents that occurred during the 12 months prior to the interview.

[2]According to these data, 3.4 million people are victims of stalking each year.

[3]Harassment is defined by the SVS as acts that are indicative of stalking behaviors, but do not incite feelings of fear in the victim.

Table 5.1 Prevalence of Stalking

Experienced at least one unwanted contact per week	46%
Victims were stalked for 5 years or more	11%
Experienced forms of cyberstalking	26.1%
Received unwanted phone calls or messages	66.2%
Received unwanted letters and e-mail	30.6%
Had rumors spread about them	35.7%
Were followed or spied on	34.3%
Experienced fear of bodily harm	30.4%
Believed that the behavior would never stop	29.1%

to visit, are not inherently criminal. These acts appear harmless to the ordinary citizen but can inspire significant fear and terror in victims of stalking. Table 5.1 illustrates the prevalence of these behaviors.

For many women, their relationships with their future stalker began in a very ordinary sense. They described these men as attentive, charming, and charismatic. But these endearing qualities soon disappeared and their interactions became controlling, threatening, and violent. Many women blamed themselves for not recognizing the "true colors" of their stalker earlier. This pattern of self-blaming affected their ability to trust their own judgment, and led these women to be hesitant about their decision-making abilities in future relationships as a result of their victimization.

Like many crimes, victims of stalking often do not report their victimization to police. According to SVS data, more than half of the individuals who were victims of stalking did not report their victimization. For many victims, their decision to not report these crimes stemmed from a fear of intensifying or escalating the stalking behaviors. Others dealt with their victimization in their own way, believing that their experience was a private and personal matter. Additionally, many believed that stalking was not a serious enough offense (or did not believe that a crime had occurred) to warrant intervention from the criminal justice system. Finally, some victims felt that nothing could be done to stop the behavior by their stalkers. For those individuals who did report their crimes, SVS data indicate that charges were filed in only 21% of these cases, further solidifying a belief for many victims that the criminal justice system was unable to effectively punish their stalkers in a court of law.

Victims engage in several different strategies in an effort to cope with their stalking victimization. Some victims attempted to solve the trauma through self-reflection and sought out therapeutic resources. Women also made significant changes to their behavior patterns. They might avoid community events out of a fear that their stalker would show up at the same function. Other women moved out of the area yet still expressed fear that their stalker would find them. Some victims tried to renegotiate the definitions of their relationship with their offender through bargaining, deception, or deterrence. Finally, some victims moved against their attackers by issuing warnings or pursuing a legal case against them (Cox & Speziale, 2009; Spitzberg & Cupach, 2003).

Laws on Stalking

Despite the number of acts against celebrities and ordinary citizens alike, stalking was not considered to be a crime for the majority of the 20th century. The first law criminalizing the act of stalking was created in 1990 by the State of California following the murder of actress Rebecca Schaeffer in 1989 by an obsessed fan. Schaeffer had risen to fame as an actress in the popular television show *My Sister Sam*. Robert Bardo had become obsessed with "Patti," the character played by Schaeffer on the show, and made several attempts to contact her on the set. He sent Schaeffer several letters and had built a shrine to her in his bedroom. However, she did not return his advances. Undeterred, he paid a private investigator $250 to obtain her home address. Bardo became fixated on Schaeffer and told his sister that "if he couldn't have Rebecca, no one else would." Upon making contact with Schaeffer at her residence, he shot her in the chest. Bardo was convicted of murder and sentenced to life in prison. Since the death of Rebecca Schaeffer and the creation of the first antistalking law in California, all 50 states, the District of Columbia, and the federal government have created criminal laws against stalking.

In addition to the general designation of stalking as a criminal offense, several jurisdictions have enacted additional laws that prohibit various forms of stalking behaviors. The U.S. Federal Interstate Stalking Law was passed in 1996 (and amended in 2000) and restricts the use of mail or electronic communications for the purposes of stalking and harassment. Other federal laws criminalize the use of wiretaps on telephones and prohibit the display or sale of personal information such as social security numbers to members of the public. Today, the majority of state laws on stalking include details on stalking via electronic methods. Some states include long lists of the types of behaviors covered under these statutes. However, the nature and forms of stalking change at a high rate of speed. While intended to provide legal protection to victims of these crimes, these laws are often unable to keep up with the fast-changing evolution of technology.

In order to prosecute someone for stalking, many state laws require victims to indicate that they experienced fear as a result of the offender's actions. While victims of other crimes may experience fear, stalking is unique in that a state of "fearfulness" is required in order to indicate that a crime occurred. A review of the literature indicates that women are more likely to experience fear as a result of being stalked, and comparisons of male and female victims of stalking find that women are 13 times more likely to indicate that they were "very afraid" of the person stalking them (Davis, Coker, & Sanderson, 2002). Using data from the National Violence Against Women Survey, Dietz and Martin (2007) found that nearly three fourths of women who were identified as victims of stalking behaviors indicated that they experienced fear as a result of the pursuit by their stalker. The levels of fear depended on the identity of the stalker (women indicated higher levels of fear when they were stalked by a current or former intimate or acquaintance) and how they stalked their victims (physical and communication stalking experiences generated higher levels of fear). But what about women who experienced behaviors consistent with the definition of stalking, but who did not feel fearful as a result of these interactions? Are these women not victims of stalking? In many states, they would not be considered victims, and the behaviors perpetrated against them would not be considered a crime.

Victims and Offenders of Stalking

Who are the victims of stalking? They are men and women, young and old, of every race, ethnicity, and socioeconomic status. Data indicate that there are certain groups who make up the majority of victims of stalking. A meta-analysis of 22 studies on stalking found that female victims make up 74.59% of stalking victims, while 82.15% of the perpetrators were male. In the majority of cases, the perpetrator was someone known to the victim, with 30.3% of all cases occurring as a result of a current or former intimate relationship. Only 9.7% of stalking cases involved someone who was a stranger to the victim (Spitzberg & Cupach, 2003).

Several researchers have attempted to identify a typology of stalkers. These studies distinguish between those stalkers who have a prior relationship with the victim and those without a previous relationship. Cases that involve a stranger relationship mirror many of the high-profile "Hollywood" cases of stalking, where the individual experiences delusions of love toward the victim. Another classification involves predatory stalkers who engage in behaviors with the intent to harm their victim. Others stalk in an effort to establish a romantic relationship with the victim. The perpetrators in these cases may be someone known or unknown to the victim. Finally, the majority of cases involve people who stalk following the rejection or severing of a relationship (Ravensberg & Miller, 2003).

Stalking is a common experience for victims of intimate partner violence. Indeed, the degree to which victims are stalked is directly related to the levels of physical, emotional, and sexual abuse that they experienced by their intimate partner—the greater the abuse in the relationship, the higher the levels of stalking can be. Several factors appear to influence whether a victim of domestic violence will be stalked. Women who are no longer in a relationship with their abuser are more likely to experience stalking compared to women currently involved in an intimate partner abusive relationship. Additionally, domestic violence abusers who are more controlling and physically violent toward their victims are more likely to stalk them. Finally, abusers who use drugs and alcohol are more likely to stalk their partners. The experience of stalking appears to intensify when women leave their partners. Research by Melton (2007) indicates that 92.1% of women who had been battered by their intimate partners experienced stalking following the closure of a domestic violence case in criminal court against their batterer. Over time, the prevalence of stalking decreased, with 56.3% of women reporting stalking behaviors after 6 months and 58.1% after 1 year. These data indicate that the prevalence of stalking remains quite high for victims of intimate partner violence, as more than half of the women continued to be stalked by their former partners a year after criminal justice system intervention. For most of these women, they indicated that their stalker was "checking up" on them. The behaviors included unwanted phone calls and being followed or watched by their stalker. For those women who had moved on to new relationships, almost three fourths of them indicated that their new partner was harassed, threatened, or injured by their stalker.

The experience of stalking has significant implications for the employment status of victims. The place of employment provides offenders with an easily accessible public venue to harass and intimidate their victim. Stalkers use a variety of tactics to prevent women from finding and keeping a job. Table 5.2 demonstrates some of the methods used by stalkers to affect the employment conditions for their victims. The experience of being stalked leads to on-the-job harassment and work disruption problems, as well as job performance problems for victims. Women who are stalked by current or former intimate partners often indicate that their partners use a variety of tactics to negatively affect their abilities to be successful in a work environment. They may not show up to watch the minor children, they may refuse to take victims to work or otherwise inhibit their transportation options, or they may harass them to the point where they call in sick due to high levels of stress. Other offenders engage in acts to sabotage their success at work by spreading rumors about them or by constant accusations that they are flirting or cheating with other men at their job. Some stalkers attempt to intimidate coworkers by posturing, or "staring down," the women's (mostly) male colleagues. In addition, the frequency and intensity of stalking affected the level of problems at work, as on-the-job stalking increased stress levels and distractions for the women. These actions also

Table 5.2 Employment Problems for Stalking Victims

	% of Stalked Victims
On-the-job harassment	87.9
Harass you on the phone at work	73.2
Harass you in person at work	51.0
Threaten you at work	50.8
Repeatedly follow and/or watch you while you are working	60.4
Work disruption	85.8
Undermine efforts to go to work or look for work	54.8
Physically restrain you from going to work	31.0
Steal the car keys or transportation money	50.2
Not show up to care for the children	27.2
Cause you to lose a job	27.6
Cause you to quit a job	38.5
Job performance	87.4
Unable to concentrate at work because of the abuse going on at home	81.4
Unable to perform your job to the best of your ability	71.5

created conflict or tension between stalking victims and their coworkers and supervisors. One woman described how her partner's harassment affected not only her current job, but also her future career opportunities in that field:

> I had to leave my last job because of [his harassment]. And they were glad to get [rid of me], and I don't think I could ever be rehired back in the clinic no matter what position I wanted because I think they think I'm a danger as long as I'm married to him. I'm a threat to them. So he caused me really to lose my last job. I'm surprised they didn't fire me, and I think they didn't because of my age. And they could never say anything bad about me. They couldn't even say anything about my work, what all I contributed or anything they needed, they couldn't say anything. And I already tried for another position and was very, very qualified, and when I called back to check on my application, it had disappeared. I knew then that [I would never be able to work in this kind of job again]. (Logan, Shannon, Cole, & Swanberg, 2007, p. 283)

Cyberstalking

The use of technology has changed the way in which many victims experience stalking. The use of devices such as e-mail, cell phones, and global-positioning systems (GPS) by offenders to track and monitor the lives of victims has had a significant effect on the experience of stalking. The term *cyberstalking* was created to address the use of technology as a tool in stalking. Of the 3.2 million identified victims of stalking by the SVS, one out of four individuals reported experiencing acts that are consistent with the definition of cyberstalking. Table 5.3 highlights examples of stalking aided by technology.

Like traditional methods of stalking, cyberstalking involves incidents that create fear in the lives of its victims. Just because cyberstalking does not involve physical contact does not mean that it is less damaging or harmful than physical stalking. Indeed, some might argue that the anonymity under which cyberstalkers can operate creates significant opportunities for offenders to control, dominate, and manipulate their victims, even from a distance, as there are no geographical limits for stalking within the domain of cyberspace. Indeed, someone can be stalked from just about anywhere in the world. For many victims of "traditional" stalking, cyberstalking presents a new avenue through which victims can be harassed, threatened, and intimidated.

While cyberstalking is a relatively new phenomenon, research indicates that the prevalence of these behaviors are expanding at an astronomical rate. Youth and young adults appear to be particularly at risk for these forms of victimization, given their connections to the electronic world through the use of the Internet, blogs, text messaging, and social networking sites such as Facebook. Research by Lee (1998) indicated that behaviors that can be identified as cyberstalking are rationalized among college-age students as a form of modern-day courtship and were not considered by the majority of the students to be of any particular significance, particularly in cases where the offender is known to the victim. Research by Alexy, Burgess, Baker, and Smoyak (2005) also utilized a sample of college students and found that only 29.9% of students labeled a simulated encounter as cyberstalking, even though 69% indicated that they felt threatened by the behavior. Given the limited understanding of these crimes by victims (and the larger society), it is important that advocates and justice professionals have an understanding about the realities of these crimes in order to provide adequate support for victims.

Table 5.3 Behaviors of Cyberstalking and Technology-Aided Stalking

- Monitoring e-mail communications
- Sending harassing, disruptive, or threatening e-mails or cell phone text messages
- Using computer viruses to disrupt e-mail and Internet communications
- Fraudulent use of victim's identity online
- Use of Internet services to gather and disseminate personal information with the intent of harassing the victim
- Use of global positioning devices to track movements of victim
- Use of caller identification services to locate whereabouts of victim
- Use of listening devices to intercept telephone conversations
- Use of spyware and keystroke logging hardware to monitor computer usage
- Use of hidden cameras to observe movement and activities of the victim
- Search of online databases and information brokers to obtain personal information about the victim

Policy Implications

Victims of stalking advocated the need for increased community awareness about stalking. Many women acknowledged that they didn't believe that what was happening to them was a criminal act since there was no experience of physical violence. One victim noted that in assessing whether a relationship is healthy, women should look at themselves and any changes in their personal behaviors rather than obsessing on the actions of their stalker. "Think about how you were before this happened and how happy you were, and I think once ladies reminisce on that, I think that's where strength comes from" (Cox & Speziale, 2009, p. 12). Others advised that women should not stay silent on the issue of stalking in order to protect their own safety, whether that meant filing a police report and obtaining a restraining order, or letting friends, family, and coworkers know of their victimization.

While protective orders are used to legally prohibit perpetrators from contacting or harassing their victim, it is unclear how effective such tools are for victims of stalking. Women who are victims of stalking were more likely to report that their perpetrator violated the protective order in their cases. Given the tactics that stalkers use to torment their victims, criminal justice agents may believe that these "minor" or "trival" acts are not substantive enough to warrant punishment for violating a restraining order. However, criminal justice officials need to understand the context of these incidents related to the larger picture of stalking in order to provide effective services for victims and enforcement of stalking laws against perpetrators (Logan, Shannon, & Cole, 2007).

The experience of stalking has a significant effect on a woman's mental health. Women who experience significant levels of stalking over time are more likely to be at risk for depression and posttraumatic stress disorder. These rates of depression and posttraumatic stress disorder are significantly higher for women who blame themselves for the behaviors of their perpetrator (Kraaij, Arensman, Garnefski, & Kremers, 2007). Victims indicate feelings of powerlessness, depression, sleep disturbances, and high levels of anxiety (Pathe & Mullen, 1997). They are also likely to develop a chronic disease or other injury since their experience with stalking began (Davis et al., 2002). It is clear that mental health services need to acknowledge how the experience of stalking affects the mental health status of victims and determine how to better provide services to this community.

Sexual Harassment

In July 1991, Clarence Thomas was nominated by President George H. W. Bush to fulfill a vacancy on the U.S. Supreme Court following the retirement of Thurgood Marshall. While many liberal groups opposed his nomination due to his conservative political views, the confirmation hearings exploded when Anita Hill, a law professor at University of Oklahoma, claimed that Thomas had sexually harassed her when he was head of the Equal Employment Opportunities Commission. Hill, who worked under Thomas at the EEOC, claimed that Thomas had made sexually inappropriate comments and gestures toward her following her rejection of Thomas as a romantic suitor. While Clarence Thomas was confirmed to the U.S. Supreme Court by the senate's vote of 52–48, the hearing significantly increased the public's awareness of sexual harassment.

The Hill-Thomas scandal is not the only sexual harassment case that has made national headlines in the political spectrum. In May 1994, Paula Jones filed a civil suit against President Bill Clinton for sexually harassing her in 1991 when he was governor of the State of Arkansas. While her case was eventually settled out of court for $850,000, her allegations of sexual impropriety were just the tip of the iceberg in the investigation of Clinton's sexual life. Cases of sexual harassment have also weaved their way into the storylines of

Hollywood, where both real-life and fictional accounts of sexual harassment have been portrayed in films such as *Disclosure,* starring Michael Douglas and Demi Moore, and *North County*, starring Charlize Theron.

While sexual harassment is not a new phenomenon, the legal enforcement against sexual harassment began forty years ago during the 1970s. One of the pioneers of this movement was Catharine MacKinnon, a legal scholar who pioneered the theory of sexual harassment as a form of sexual discrimination under Title VII of the Civil Rights Act of 1964. In her book *Sexual Harassment of Working Women: A Case of Sex Discrimination* (1979), she argues that acts of sexual harassment serve to perpetuate social inequalities against women. "Sexual harassment perpetuates the interlocked structure by which women have been kept sexually in thrall to men and at the bottom of the labor market. Two forces of American society converge: men's control over women's sexuality and capital's control over employees' work lives" (pp. 174–175). Her theory was first tested in the case of *Alexander v. Yale* (631 F. 2d 178, 1980). In this case, several female students joined together to protest against the lack of a grievance procedure for students who had experienced acts of sexual harassment on campus. Their efforts led to the creation of grievance procedures not only at Yale, but on college campuses nationwide. MacKinnon went on to represent a number of women in cases of sexual harassment, including the case of *Meritor Savings Bank v. Vinson* (1986), where the U.S. Supreme Court held that "sexual harassment is a form of sex discrimination that is actionable under Title VII" (477 U.S. 57).

Today, the sexual harassment of women remains a significant social problem. The experience of sexual harassment has significant direct and indirect effects on the lives of women. Sexual harassment can be found in a variety of avenues, including the workplace, education, housing, and public spaces. Offenders range from people known to the victim to strangers. While some of these variables have long been documented within the literature, new forums of sexual harassment are introduced every day. The age of the Internet has created a new forum through which people can be sexually harassed. Youths have also become a vulnerable target, as peer sexual harassment is a growing problem for adolescents.

Sexual harassment presents itself in a variety of ways. Most people tend to think of sexual harassment as verbal comments made of sexual reference or inappropriate physical touching. However, sexual harassment can also include derogatory looks or gestures; the sharing of sexually based jokes or stories; the display of sexually graphic pictures or writings; or commentary about a person's clothing, behaviors, or body. Like many other sexually based crimes, sexual harassment is about power, control, and domination.

One of the most difficult issues in identifying sexual harassment is that many victims do not acknowledge themselves as victims. Like many other crimes of victimization, acts of sexual harassment can be easily discarded by both offenders and victims as a minor issue. Billie Wright Dziech and Linda Weiner argue that many fail to acknowledge the presence of sexual harassment because it "trades on its victim's uncertainty about how to label their experiences . . . the very words 'sexual harassment' are ominous to some . . . they seem too legalistic, too political, too combative" (Dziech & Weiner, 1990). Victims may be confused or embarrassed by the events of harassment. They might blame themselves for causing the harassment. For others, they may be in denial and not want to acknowledge the behaviors of the perpetrator. Similar to myths about rape that suggest that women "ask for it" based on their dress and behaviors, or myths about intimate partner violence that suggest that women cause violence because they argued with their spouse or failed to cook a good meal, myths about sexual harassment suggest that women bring about their own victimization by choosing to work outside the home or in male-dominated occupations.

It is very difficult to understand the prevalence of sexual harassment. While the Equal Employment Opportunity Commission (EEOC) presents a definition of sexual harassment in the workplace, it may not represent all cases of sexual harassment. Additionally, many victims and perpetrators may not understand what it means to engage in acts that constitute sexual harassment. Finally, like many other sexually based crimes, victims are reluctant to report these incidents, often due to fear or shame about the harassment. Since 1997, the EEOC has received an average of 14,000 claims of sexual harassment each year. While women typically file the majority of claims (84–88% of claims were filed by women between 1997 and 2009), the number of cases filed by men has increased steadily over the past 12 years (with 16% of cases filed by men in 2009). In 2009, 11,948 cases were resolved and $51.5 million were awarded to the victims (not including any additional damages received due to civil court judgments; U.S. EEOC, n.d.).

In a report on sexual harassment in the federal government sector in 1995, 44% of women indicated that they had experienced sexual harassment in the workplace at some point during the previous 2 years. These rates are consistent with the levels of sexual harassment reported in 1987 and 1980 (42%). Even though new training programs have been implemented since this issue was first investigated, millions of dollars are spent due to job turnover, use of sick leave, and loss of employee productivity as a result of workplace harassment. While these rates may indicate that little has changed over time, it may also reference that more cases are coming forward due to increases in reporting rates (U.S. Merit Systems Protection Board, 1995). While these data are encouraging, they only scratch the surface in terms of the levels of sexual harassment experienced, both in and out of the workplace. Estimates indicate that at least one of every three women experiences some form of sexual harassment.

Perpetrators of Sexual Harassment

The literature divides perpetrators of sexual harassment into several different typologies. Research by Lucero, Middleton, Finch, and Valentine (2003) highlights four different offending typologies. The first type is referred to as the "persistent harassers." These harassers enjoy the rejection that they experience from their victim's denial to succumb to their demands. The second category is known as the "malicious" harasser. These individuals share many of the same characteristics with the "persistent harasser." They are quite aggressive in their harassment and experience pleasure by making their victims uncomfortable. Both persistent harassers and malicious harassers have multiple victims. However, malicious harassers transfer the blame for their behavior to the victim, arguing that the victim provoked the harassment. The third category refers to those whose position holds power and control over their target. Based on this power differential, these individuals are characterized as "exploitative harassers." The fourth category comprises individuals whose harassment stems from a desired romantic relationship with their target. These "vulnerable harassers" are significantly different from the other typologies, as they tend to suffer from lower self-esteem. They focus their attention on a single target and the harassment tends to persist over time.

Outside of these four dimensions, sexual harassment can be divided into additional examples of harasser identities. For example, the "power player" is a classic example of quid pro quo harassment whereby a superior threatens the victim's employment status or credibility. In this case, the victims can ensure their "safety" by succumbing to their perpetrator's sexual demands. A similar example to the "power player" is the "opportunist," who takes advantage of power differentials to exert the harassment. In contrast, "mother figures" seek to befriend a troubled colleague or student. Their portrayals of mentorship and guidance are a way to earn the trust of their victims and exploit them in times of crisis. The "mother figure" example is relatable to the "confidant," who seeks to gain the trust of the victim by sharing similar stories of victimization. Categories such as "one of the gang" or "group membership rituals" rely on the power of peer pressure and group acceptance to justify, maintain, and perpetuate sexual harassment. Victimization becomes a bonding experience between the perpetrators, who in most cases see no error in their ways. Being part of a group allows its members to diffuse cries from victims, who are deemed to be "overreacting" to simple gestures or jokes. Finally, "comedians" are those who engage in acts of sexual harassment because they believe that it is entertaining to do so (Sexual Harassment Support, n.d.).

Arenas of Sexual Harassment

Workplace Sexual Harassment

The U.S. Equal Employment Opportunity Commission (EEOC) defines sexual harassment as "unwelcome sexual advances, requests for sexual favors and other verbal or physical harassment of a sexual nature" (U.S. EEOC, n.d.). Harassment can also involve remarks that target a person's gender or sexual identity. Sexual harassment in the workplace has been linked with job-related stress and decreases in personal mental health, job performance, and general feelings of job satisfaction. Research indicates that employees who experience sexual harassment on the job report higher levels of illness, injury, and assault (Rospenda, Richman, Ehmke, & Zlatoper, 2005).

The U.S. Supreme Court has distinguished two forms of sexual harassment in the workplace. The first involves tangible examples of employment status. In these cases, a superior might threaten the employment status of an employee if he or she does not engage in sexual conduct. Known as quid pro quo ("this for that"), this form of harassment is only effective when a power differential exists between the two parties. The second form of sexual harassment is referred to as a hostile work environment. In these cases, the commentary or attitudes in the workplace by coworkers, customers, or supervisors creates an inappropriate, gender-biased, or sexually charged environment that causes discomfort for the victim. In both examples, the acts of the offender must be severe and pervasive in order to be identified as sexual harassment.

Recent research indicates that women in supervisory positions are most likely to experience workplace sexual harassment. Indeed, female supervisors were 137% more likely to be harassed sexually compared to women not in managerial positions (McLaughlin, Uggen, & Blackstone, 2009). Experiences of harassment on the job may also be linked to one's class status—working-class individuals may be less likely to report workplace harassment, particularly if they are faced with limited opportunities outside their current employment or have limited experience or training in their employment history. In these cases, people may be more likely to endure the harassment rather than deal with any perceived ramifications that may occur due to making the harassment public (McLaughlin, Uggen, & Blackstone, 2008).

How can employers protect women from sexual harassment? Research by DeCoster, Estes, and Mueller (1999) demonstrates that a positive, supportive relationship between supervisors and coworkers can serve as a protective factor against sexual harassment. Women are more likely to be victimized in male-dominated work arenas, compared to environments that are more gender balanced. "Training supervisors to be supportive, facilitating work-group solidarity, and creating a supportive work-group culture—in addition to implementing explicit sexual harassment policies—are all steps that organizations can take to deter sexual harassment of women in the workplace" (DeCoster et al., 1999, p. 43). However, some research indicates that the reporting of harassment may not appear to halt the behavior, as these individuals were 6.5 times more likely to experience subsequent incidents of harassment. These findings indicate that any censure or reprimanding of the harassment may do little to curb future behaviors of the offenders (McLaughlin et al., 2009).

Having a policy against sexual harassment alone is not enough for companies to exclude themselves from culpability in cases of workplace sexual harassment. The U.S. Supreme Court has held that corporations may not be liable for sexual harassment claims in the workplace in cases where one employee sexually harasses another as long as they provide harassment training prevention programming for their employees (*Burlington Industries, Inc. v. Ellerth* 524 U.S. 742, 1999). In addition, a 2002 ruling by New Jersey's state courts stated that providing training programming presents the image that companies are serious about combating sexual harassment in the workplace (*Gaines v. Bellino* 173 N J. 301, 2002). Several state legislatures have recognized the importance of sexual harassment training and have made laws requiring companies to provide training for their employees. California's law on sexual harassment requires companies to implement training programs on sexual harassment for all employees involved in supervisory positions. The basic provisions of California's sexual harassment law (California State Legislature Assembly Bill 1825, 2007) requires companies with 50 or more employees to provide 2 hours of training on sexual harassment every 2 years. The law also mandates the topics to be covered in training programs, including information on the federal and state laws on sexual harassment, remedies available to victims, and applied examples designed to educate supervisors on how to prevent harassment and discrimination in the workplace, as well as how to prevent retaliation against victims who file complaints.

Housing-Related Sexual Harassment

Housing-related sexual harassment is a significant issue, yet limited research exists on this topic. For example, the consequences of rebuffing sexual advances by a landlord have direct consequences on the housing status of the victims. Between 1988 and 2000, more than 8,500 acts of sexual discrimination were reported to the U.S. Department of Housing and Urban Development (U.S. Department of Housing [HUD], 2008) and HUD-affiliated agencies. Given the unequal distribution of power between a landlord and a tenant, women and the poor are at the greatest risk for harassment, particularly given the limited options of affordable housing. Women with children may be targeted in these situations, forced to deal with uncomfortable living situations and offensive landlord-tenant relationships in order to keep a roof over their head. A quid pro quo claim of sexual harassment in the housing arena can include cases whereby a tenant is evicted for refusing the sexual advances of a landlord or property manager. A claim of hostile housing environment involves significant sexual or derogatory comments by a landlord or property manager over time, which in turn creates a hostile residential setting for the tenant. Research by Reed, Collinsworth, and Fitzgerald (2005) suggests that sexual harassment in a residential environment may produce higher levels of fear than sexual harassment in the workplace, as the home is considered a sanctuary and safe environment. Given that landlords have access to a residence at any time, day or night, many women expressed significant concerns for their safety when they experienced unwanted sexual attention from their landlord. Many women endure the harassment because they believe they have no other options. In addition, landlords are aware of their position of power and use it to obtain sexual favors from their tenants (Tester, 2008). Like harassment in the workplace, harassment in housing has significant and detrimental consequences to a woman's well-being. While harassment in the workplace may be pervasive, it is limited to the workday and/or perhaps the occasional interaction outside of the office setting. "Women who experience sexual harassment in housing are unable to 'clock-out' and escape the harassment at the end of the day" (Tester, 2008, p. 362).

Policy Implications

In order to effectively combat the issue of sexual harassment both in and out of the workplace, we first need to understand the extent of the problem. Researchers indicate a need for nationally representative data and suggest that questions regarding sexual harassment should be a regular component of federally funded agencies that engage in research on the workplace, educational system, and social services. In addition, funding should be made available to access data on sexual harassment within communities of high risk, such as women who work in occupations that have been traditionally dominated by men (e.g., blue-collar industries such as the automobile industry and fire science, as well as white-collar positions such as upper-level administration within business, government, and educational settings).

While recent legislation has mandated sexual harassment training curricula for management personnel, there is a need for a greater understanding of the issues of sexual harassment for all members of the workforce. Research by Sipe, Johnson, and Fisher (2009) found that many college students do not grasp the significant nature of sexual harassment or believe that it can affect the workplace environment. When women did acknowledge the possibility of harassment occurring in the workplace, they did so believing that it couldn't happen to them personally. Given that many would-be members of the workforce do not see sexual harassment as a significant issue, training and education is imperative to effectively combat the risks and prevalence of sexual harassment in the workplace. Indeed, some research suggests that training on preventing sexual harassment in the workplace should begin during the college career in an effort to reinforce the tools for prevention.

In addition, protections need to be put into place to support victims of sexual harassment. Similar to the creation of rape-shield laws and whistleblower protections, victims of sexual harassment need to be protected from potential revictimizations that may occur as a result of pursuing claims of sexual harassment against employers or other institutional organizations. One of the best ways to protect victims is through the enforcement of policies on sexual harassment. Violations of workplace policies need to be enforced on a consistent basis. Such enforcement sends a clear message to perpetrators that acts of sexual harassment will not be tolerated. Ultimately, the best protection against harassment is to create an environment where women are viewed as equal to their male counterparts (Fitzgerald, 1993).

Summary

- Stalking is defined as a "course of conduct directed at a specific person that would cause a reasonable person to feel fear."
- Cyberstalking involves the use of technology to track and monitor the lives of victims of stalking.
- Many victims do not report their experiences of being stalked to law enforcement, as they fear that a report will escalate the behavior, or do not believe that it was a serious matter or that anything could be done to stop the behavior.
- The first criminal law on stalking was enacted in California in 1990 following the murder of Rebecca Schaeffer.
- Stalking is often related to incidents of intimate partner abuse.
- The experience of stalking generally intensifies when a victim severs her relationship with her stalker.
- Many victims of sexual harassment do not acknowledge themselves as victims, which affects the reporting rates for these offenses.

- The U.S. Supreme Court has acknowledged two forms of sexual harassment: quid pro quo and hostile work environment.
- Several typologies of sexual harassment perpetrators have been developed to explain differences between the types of offenders and the practices they engage in as part of their harassment.

KEY TERMS

Stalking

Harassment

Cyberstalking

Hollywood stalkers

Predatory stalkers

Romantic stalkers

Ex-intimate stalkers

Stalking and intimate partner abuse

Federal Interstate Stalking Law

Sexual harassment

U.S. Equal Employment Opportunity Commission (EEOC)

Workplace sexual harassment

Quid pro quo

Hostile work environment

Perpetrators of sexual harassment

QUESTIONS FOR DISCUSSION

1. How has the use of technology changed the way in which victims experience stalking? What challenges do these changes present for law enforcement and the criminal justice system in pursuing cases of cyberstalking?
2. How do victims cope with the experience of being stalked?
3. Discuss how the experience of stalking affects the employment status of victims.
4. How does sexual harassment affect the employment and residential status of victims?
5. What are the different categories of offenders of sexual harassment?
6. How can policies against sexual harassment improve the lives of women?

WEB RESOURCES

Stalking Resource Center: http://www.ncvc.org/src/Main.aspx

Stalking Victims Sanctuary: http://www.stalkingvictims.com

Sexual Harassment Support: http://www.sexualharassmentsupport.org/

Equal Rights Advocates: http://www.equalrights.org/publications/kyr/shwork.asp

U.S. Equal Opportunity Commission: http://www.eeoc.gov/facts/fs-sex.html

READING

This section uses the voices of women who experienced stalking within the context of an intimate partner violent relationship. The women talked about their partners' motivations for stalking, how stalking was conducted, and the effects of stalking. Not only is stalking common following the conclusion of a relationship, but these women highlight how stalking was an integral part of their experiences throughout their battering relationships. However, most women tend to minimize the extent of the violence that occurs within the context of stalking.

Stalking in the Context of Intimate Partner Abuse

In the Victims' Words

Heather C. Melton

Stalking is a relatively new concept. For example, the term *stalking* was first coined in 1989 (Coleman, 1997). In general, *stalking* refers to the willful, repeated, and malicious following, harassing, or threatening of another person (Coleman, 1997; Meloy, 1996; Meloy & Gothard, 1995; Tjaden & Thoennes, 1998); however, legal definitions differ from state to state, as do the estimates of stalking incidents. Conservative estimates suggest about 200,000 annual incidents occur nationally in the United States (Roberts & Dziegielewski, 1996; Wright et al., 1996). Other studies suggest that stalking may affect about 1.4 million victims annually in the United States: 1 million women and 400,000 men (Tjaden, 1997; Tjaden & Thoennes, 1998; U.S. Department of Justice, Violence Against Women Grants Office, 1998).

The relationships between stalkers and their victims may be characterized in one of three ways: intimates or former intimates, acquaintances, or strangers (National Institute of Justice [NIJ], 1996). Generally speaking, most research suggests that victims are most often the current or former spouses or intimate partners of their stalkers (Burgess et al., 1997; Coleman, 1997; Guy, 1993; NIJ, 1996; Roberts & Dziegielewski, 1996; Tjaden & Thoennes, 1998; U.S. Department of Justice, Violence Against Women Grants Office, 1998). Some researchers estimate that as many as 80% of all stalking cases involve a prior or current intimate relationship (Coleman, 1997; Roberts & Dziegielewski, 1996). Moreover, studies show a high correlation between stalking, verbal, and physical abuse in intimate relationships (Coleman, 1997; Davis, Ace, & Andra, 2000; Logan, Leukefeld, & Walker, 2000; McFarlane et al., 1999; Mechanic, Uhlmansiek, Weaver, & Resick, 2000; Mechanic, Weaver, & Resick, 2000; Tjaden, 1997; Tjaden & Thoennes, 2000; White, Kowalski, Lyndon, & Valentine, 2000). One study found that 80% of the victims of stalking reported having been physically assaulted and 31%

SOURCE: Melton, H. (2007). Stalking in the context of intimate partner abuse: In the victims' words. *Feminist Criminology, 2*(4), 347–363.

sexually assaulted by a partner who later stalked them (Tjaden, 1997; Tjaden & Thoennes, 1998). In addition, it is estimated that between 29% and 54% of all female murder victims are battered women, and it is believed that stalking preceded the murder in 90% of these cases (Guy, 1993; U.S. Department of Justice, Violence Against Women Grants Office, 1998). Thus, the developing research in the field suggests a strong association between stalking and intimate partner abuse (IPA) (Coleman, 1997; Davis et al., 2000; Logan et al., 2000; McFarlane et al., 1999; Mechanic, Uhlmansiek, et al., 2000; Mechanic, Weaver, et al., 2000; Tjaden, 1997; Tjaden & Thoennes, 2000; White et al., 2000). Less, however, is known about the context of stalking. In other words, there is much to be learned about why and how stalkers engage in this behavior and what effect it has on their victims.

Motivations for Stalking

Prior research has identified the various motivations for stalking. One study cited control, obsession, jealousy, revenge, and anger as possible motives for stalking (NIJ, 1996). Studies that have focused on stalking in the context of IPA have found that the need to control the victim or the desire to reestablish a relationship may motivate these stalkers (Roberts & Dziegielewski, 1996). Meloy and Gothard (1995) found that domestic partner stalkers are motivated by "abandonment rage arising out of a narcissistic sensitivity which appears to defend against the grief of object loss, which then drives the pursuit" (p. 262). Using data that specifically asked stalking victims their perceptions of their stalkers' motivations, the U.S. Department of Justice (1998) reported that 21% of victims said that their stalkers wanted to control them, 20% said that the stalker wanted to keep them in the relationship, 16% said that the stalker wanted to scare them, 12% were not sure, 7% said that the stalker was mentally ill or abusing drugs/alcohol, 5% said that the stalker liked or wanted attention, and 1% said that the stalker wanted to catch the victims doing something. Stalkers, then, appear to have many different reasons for engaging in stalking.

Stalking Behaviors

Stalking includes numerous different behaviors and has been defined in various ways. Using factor analysis, Coleman (1997) identified two categories of stalkers based on their types of stalking behaviors. Violent behavior stalkers are those stalkers who break or attempt to break into the victim's home or car, violate restraining orders, threaten or attempt to physically harm the victim, physically harm or threaten to harm themselves, steal/read mail, and damage property of the victim's new partner. Harassing behavior stalkers were typified by behaviors such as calling the victim at home, work, or school; following or watching the victim; making hang-up calls; arriving unwanted to the home, work, or school of the victim; sending unwanted gifts, letters, or photos; leaving unwanted messages on the answering machine; or making threats against or harming the victim's new partner (Coleman, 1997). Meloy (1996) identified the following stalking behaviors in his review of existing studies: the sending of aggressive letters, unwanted following, property damage, annoying phone calls, assaults, and unwanted gift giving. Finally, Burgess and her colleagues (1997) identified the following forms of stalking behavior: written and verbal communications; unsolicited and unrecognized claims of romantic involvement on the part of the victims; surveillance; harassment; loitering; and following that produces intense fear and psychological distress to the victim. Thus, researchers have identified a variety of behaviors that constitute stalking.

Stalking in the Context of IPA

Clearly, the majority of stalking occurs in an IPA context (Coleman, 1997; Davis et al., 2000; Logan et al., 2000; McFarlane et al., 1999; Mechanic, Uhlmansiek, et al., 2000; Mechanic, Weaver, et al., 2000; Tjaden, 1997; Tjaden & Thoennes, 2000; White et al., 2000). Research examining the link between IPA and stalking has found that stalking is highly correlated with psychological and physical abuse (Davis et al., 2000; Mechanic, Weaver, et al., 2000). The few studies that have examined this link focused on the relationship between abuse and stalking, when stalking begins in an IPA relationship, and the effects and consequences of stalking on the victim. In a study examining IPA victims' experiences with stalking, Mechanic, Weaver, et al. (2000) found that stalking appears to be more strongly correlated with psychological abuse than physical abuse; the emotional abuse variables in the study better predicted stalking than did the physical abuse variables. In contrast, in another study examining IPA victims' experiences with stalking, Mechanic, Uhlmansiek, and colleagues (2000) found that severe stalking was highly correlated with severe physical, sexual, and emotional abuse. The researchers (Mechanic, Weaver, et al., 2000) also found that the stalking in the relationship escalated among women who left their partners. In another study examining stalking victimization among a population of men and women who experienced IPA, Logan and colleagues (2000) concluded that stalking is a continuation of intimate partner violence (IPV) that occurs after the relationship has ended. Finally, regarding the effects of stalking, Mechanic, Uhlmansiek, and colleagues (2000) found that women who are severely stalked compared to infrequently stalked women are more likely to suffer from depression and posttraumatic stress disorder (PTSD). Similarly, Davis and his colleagues (2000) found that experiences with stalking in the context of IPA were related to increased levels of fear, anger, and distress. Therefore, the context of stalking is not entirely clear, and there are contradictory findings about the nature of the phenomenon.

The current study builds on this body of research and examines stalking through victims' experiences. To fully understand stalking and to be able to develop better stalking prevention and treatment, we must have an understanding of the problem from the victims' point of view. The effects of stalking are clearly linked to the motivations and tactics of stalkers, and additional information may provide important clues about how to address stalking more effectively. As such, the current study employed qualitative methodology to provide information about the context of stalking to help identify areas for future research that might be used to devise broader and more effective assistance for victims.

Method

Sample

The original sample consisted of female victims of IPA whose cases had entered the criminal justice system (i.e., the abuser was arrested) in one of three jurisdictions in the United States: a medium-sized Midwestern city, a large Western metropolitan area, and a Western, rural, college county. The respondents were recruited from the district attorney's offices in these three jurisdictions. Of the 178 women in the larger study, 21 women in two of the jurisdictions (the metropolitan area and the rural/college county) were randomly selected to participate in the current study and completed an in-depth interview pertaining to their experiences with stalking.

Of the 21 women, nine were White, seven were Latina, four were African American, and one identified as Middle Eastern. The average age was 36 years, with a range of 19 to 56 years. The majority of them (more than 75%) had a high school degree, and their income levels ranged from US$500 a month to $6800, with a mean of just under $2000 a month. Of the 21 women in the current sample, only one respondent was still with her abuser/stalker. Ten respondents (47.6%) experienced stalking over the two time periods, nine (42.8%) experienced stalking only in the first time period (prior to criminal justice intervention), and two respondents (9.5%) only experienced stalking during the second time period.

Findings

Motivations for Stalking in the Context of IPA

Women cited a variety of issues when asked about the motivations of their partner or ex-partners' stalking behaviors. These explanations included control, anger, and jealousy. Some women also stated their partners abused them in an effort to win them back, to scare them, and because their perpetrators were sick (mentally or physically). It is interesting to note that some women mentioned more positive reasons behind the stalking, namely that the stalking took place out of love or concern. These themes are illustrated below in the women's own words.

Six women (28.6%) cited control issues as motivating their partner or ex-partners' stalking behaviors. Jane, a 50-year-old White woman from the college/rural town, is an example of a woman who reported being controlled. She experienced stalking consistently before and after criminal justice intervention in her domestic violence case (i.e., Time 1 and Time 2) by her ex-partner. She said: "He's a controller. And that's a pattern. I read six books on domestic violence and that's the pattern. And I have heard that he could do this, have somebody else do this to me the rest of my life." Samantha, a 36-year-old White woman from the large metropolitan area, who experienced much violence and consistent stalking from her ex-husband, said,

> Because he still has not completely let go of the control. He has, he's such a controlling person you know, I mean. . . . But it's just, it's all control. All has to do with control. He has to know what I'm doing all the time. Why I don't know. To have, to be in control.

Kristen, a 49-year-old White woman also from the college/rural town, who experienced stalking only before criminal justice intervention (only Time 1), identified the motivation of her stalker as using fear to control her. She stated,

> You know, even though we separated and were going to get a divorce, it was still, he could never work it out in his head about a control issue. You can't control other people. And so this is their way of controlling you. By making you afraid. It's the only control they have over you.

Five respondents (23.8%), including some of the same women who cited control as a motivating factor, mentioned that their stalkers stalked them because they were angry. For example, Kristen also stated,

> It was just, I mean he's just a type of person, abusers are like this too. I mean they have a lot of anger in them that doesn't go away. I don't know, sometimes it stems from childhood, who knows where it all comes from but it's just something they don't get rid of.

It is interesting to note that Tina, a 41-year-old Latina woman from the large metropolitan area, who was consistently stalked and threatened by her ex-partner, cited her attempts to seek help from others for her abuse as the impetus for her partner's stalking:

> He's mad at me because I call the police on him. I wasn't even going to report it, but my friends talked me into it. I don't want him to come after me. So I says, if I don't report it, he'll continue on.

Some respondents (n = 3, 14.3%) mentioned jealousy as the primary motivation behind the stalking they experienced. For example, in terms of why he was stalking her, Joan, a 49-year-old White woman from the metropolitan area, whose ex-partner stopped stalking her after criminal justice intervention, mentioned that her partner stalked her to find out if she was with someone new: "Cause he would just ask me, 'Do you have somebody else or something.' He'd want to know if I was with anybody, if I had another man in the apartment or something."

Several respondents (n = 3, 14.3%) mentioned the stalker was trying to win her back by stalking. Crystal, a 34-year-old African American woman from the large metropolitan area, whose experience with stalking and violence from her ex-partner stopped after the criminal justice intervention, summarized this:

> You know that's what I used to tell him. You know I thought, if you wanted to prove to me, I was thinking that it was because he wanted to prove to me, to become a part of my life again. And I don't know why he thought that intimidation and scare tactics will make me change my mind. You know I thought, bring me flowers, bring me candy. Do something that is going

> make me think that you're a nice guy, but I truly think that maybe he thought that I, I don't know, maybe it was his last resort. He went from one thing of, you know, begging, and it wasn't going to work and then I thought, he went to, "you are going to go back to me." Which is crazy. What is he going to try to do? Is he going to try to force me? That's not going to work. If anything, it pushed me further from him. But in his mind he thought, "I'm going to force her to do this." And he's crazy. It wasn't going to happen.

Several respondents (n = 3, 14.3%) also reported that their stalkers were engaging in stalking behaviors to scare them. When asked why the stalker was doing the stalking, each of these three respondents responded with, "To scare me."

Two respondents (9.5%) mentioned that their stalker's motivation was, in part, due to either mental or physical illness. For example, Shari, a 52-year-old Latina woman from the metropolitan area, whose ex-partner was consistently stalking her and extremely violent (including almost daily threats to kill her), reported: "Well, he had a head injury. He was in an automobile accident.... He was violent anyway and I think it just added [to it]."

Last, several respondents (n = 3, 14.3%) mentioned that their stalkers were doing it for more positive reasons—out of love or concern. For example, when asked why her abuser stalked, Alice, an African American woman in her forties from the metropolitan city, who was stalked and abused consistently throughout both time periods, said: "You know, 'cause he loves me."

Clearly, victims perceived a variety of motivations for stalking. Although some of these motivations are consistent with prior research, the current study unearthed greater variety of motivations that have critical implications.

How Stalking Is Conducted

Several patterns emerged regarding the process of stalking. These patterns revolved around the stalking behaviors, the respondent's relationship with the abuser when the stalking started, how the stalking changed as the relationship changed, and whether the stalker had other "proxy stalkers" assisting him in the stalking process.

Respondents reported being stalked in several different ways and places. They were physically watched or followed in their homes, at their jobs, or in other public places. Respondents also reported receiving unwanted calls or messages or being sent unwanted letters by the stalker. All together, 16 of the 21 women interviewed (76.2%) reported being watched or followed in or en route to their homes, 10 (47.6%) reported being watched or followed at their jobs, and 9 (42.9%) reported being watched or followed in other places. Thus, stalking was conducted in a variety of spaces that illustrates that stalking is a method abusers use to assert themselves into their victims' lives, beyond the confines of their homes.

The following are examples of respondents' experiences with being followed or watched in or from their homes. Shelia, a 34-year-old African American from the metropolitan area, who experienced consistent stalking, said not only does her ex-partner come by the outside of her building but he also came into her building: "He comes around to check if the light is on in the bedroom at night. He comes in the building and listens at the door to see who is in there." Mary, a 43-year-old African American woman from the metropolitan city, who was the only respondent in the current sample of 21 who was still dating the person who was stalking her (not living together), said, "He seems to watch me. He knows everything. I didn't go to work. I was sick and he called later and says 'I saw your car there [her home].'" Shari, whose ex-partner had a head injury, said,

> He parked a mobile home two blocks from my house and watched me with binoculars. He stalked the neighborhood at 1:30 in the morning. And a neighbor of mine goes to work at 1:30 at the airport, so that's how I knew that. He would be directly across the street, sitting up in a tree, watching my house.

In addition to stalking at their homes, many women mentioned that they were also stalked where they worked. Lily, a 24-year-old Latina woman who, like many others, was stalked consistently before and after criminal justice intervention, said, "I work in day care and he would park out there and wait for me to get off my break." Crystal was also stalked at her place of work and said, "That was when he showed up at my job. And they'd say 'Some man's out there looking for you.' And I'd be like, 'Oh gosh.' So he would do that. Then he started breaking my car windows at work." Some women discussed being followed or watched in other places. These places included grocery stores, bars, bus stops, or just as they walked or drove around town. For example, Alice mentioned running into her stalker in the grocery store: "I went into the store, never knew he was there, and he came around the corner."

Another tactic used by men who stalked was to make unwanted calls or leave unwanted messages at the victim's home or work. Nine women (42.9%) reported experiencing this behavior. Denise, a 25-year-old Latina from the metropolitan area, who was stalked by her ex-partner over both time periods, said,

> Oh it was crazy. I mean in one instance he called me like thirteen times in a row, I remember. And finally I saw it there on my caller ID and I picked up the phone and said, "What!" You know and he was mad at me because I wasn't calling him. So it was all the time.

Samantha, who mentioned that her ex-husband stalked because of control, illustrated this by the behaviors he exhibits. About phone calls, she said,

> But, I am not giving him this house number; I am not putting in another line. I have done it before where I have a separate phone number and answering machine from the other line in the house, so that he can call and get a hold of [her daughter] and [her son] can get a hold of me and he abuses it. Totally abuses it. We would come home and the answering machine would be full of messages from him—"Where are you at? what are you doing? why can't you do this? you're never home."

It is important to examine when the stalking behaviors started during the course of the relationship to more fully understand the process of stalking. It is often assumed that stalking is a behavior that occurs primarily after an intimate couple has broken up (Logan et al., 2000; Mechanic, Uhlmansiek, et al., 2000). In particular, many researchers argued that stalking may be a continuation of domestic violence after the abuser no longer has physical access, and thus control, over their victim (Logan et al., 2000; Mechanic, Uhlmansiek, et al., 2000). It is surprising to note that a majority of the women ($n = 17$, 80.9%) stated that the stalking started in some form while they were still in a relationship. Ann said, "I think it [when the stalking started] was probably three months into [the relationship]." Kristen also mentioned that the stalking occurred while they were still involved: "No, he had done that before . . . He's a very controlling person and so yeah it had been going on for several years actually. We were married for six years and I would say a good four of those years." Finally, Nicky, a 28-year-old Latina from the larger metropolitan area, who was stalked consistently, mentioned that the stalking began very early in her relationship: "It started when in the relationship, within the first 3 months he started stalking me—always calling, wanting to be around."

Notably, 3 of the 17 women (17.6%) who stated that the stalking was occurring throughout the relationship mentioned that at the time they did not see it as stalking behavior. Ann continued,

> In the beginning, I think I sort of misread it as, "Oh that's really sweet that he's coming to visit me so much." But looking back now, I'm realizing that he had a job that he was afforded the luxury of never really needing to be in the office, so I was kind of thinking it was him coming to say hi. But I think it was more like him coming to check up on me, because it was never announced. I would never invite him. He just showed up.

Beth, a 21-year-old White woman from the metropolitan location, said,

> I never considered it that [stalking] when we were together, but I think it was constant. I just didn't see it as that at the time. When we were together, I thought it was cute that he wanted to know what I was up to. I blew it off. It was continuous though.

Barbara, a 56-year-old White woman from the metropolitan area, also said,

> [At the time, did you see it as stalking?] No, not right away. Because a couple of times it was really, really cold and he'd bring me some extra gloves or an umbrella. He was pretending like he was protecting me [the abuser was showing up, following her after she got out of computer classes].

Although most of the women indicated stalking started while the relationship was intact, 10 (47.6%) of the respondents discussed that as they broke up with the defendant, the stalking actually got more intense or increased. Samantha said,

> I think it actually did get worse when we separated . . . he would follow me and it was worse then because I would go with other men. And he, all he would do when I went back, was use this against me. So it did get worse and then after I finally did leave him, we were divorced, but I still lived with him. He divorced me and he wouldn't let me leave. And for the first year it was really bad with the following and watching me. I mean he would call and tell me exactly what I had done all day long. Whether I was at home or that I had left. I mean it was awful. I felt like I couldn't do anything. So then it was like, if I stay at home he's watching me and I mean I wasn't sleeping good or anything like that. It was bad for awhile.

And Lily, who was stalked consistently as previously stated, said,

> I would say in a way it got worse because it was to the point to where he was watching me from across the street. I work in day care and he would park out there and wait for me to get off from my break. And when I was at home he would sit out in the parking lot or on the street. Or he was having his friends come by. Or he would call and leave songs on my pager all the time. So, that kind of stuff. So in a way it did get worse. And then at one time, I had thought of going back to him just because the kids were always asking about their dad so much and he told me that he was going to change and do all this. I didn't and that triggered it again. So it just got worse. And one day, he slit my mom's tires.

Thus, abusers may resort to more abusive tactics when the relationship ends.

Another interesting pattern that emerged was the notion of "proxy stalking." More than one half of the interviewed women ($n = 11$, 52.4%) mentioned that not only had their current or former intimate partners stalked them but also that the stalker had others stalk them as well. Respondents most often mentioned friends and family members as the "proxy stalkers." Some examples of this follow. Ann said, "It's not just him [who stalks me], it's sort of his circle of friends. I don't want to run into them anywhere. Like I see them in the parking garage and stuff."

Lily stated,

> As far as following me, he used to have a couple of friends follow me. Like when I'd be out, over at my grandma's he'd tell me, "Yeah so and so told me they saw you at the Bingo." I was sometimes here and I could hear them go through the parking lot for an hour or something stupid. He would do it and he'd have a friend do it or with him.

Once again, Denise, who was stalked over both time periods by her ex-partner, said, "He'd have members of his family call for me. He had his brothers' cars, so they were helping him. I would see that as helping. And they had called for me before." Joan also mentioned, "He did tell me that he had people watching me and stuff. But who they was, I have no idea. I really don't know who they was." Finally, Shari, describing how her ex-partner used other people when he was incarcerated, said,

> He had a nephew. When he went to jail on this probation one, I was working at a new job. I thought I was safe. So then his nephew calls me . . . and told me, "My uncle told me to tell you that you can rest in peace now, 'cause he's going to jail for a while." I said, "Who is this?" He says, "[his nephew's name], his nephew." I says, "How in the heck did you get my phone number?" "Oh, my uncle know where you're at. He's been watching you."

If the stalker himself is incapacitated or under court order and cannot stalk, he may resolve this "set-back" and continue harassing his victim by stalking her through other people or "proxy stalkers."

Effects of Stalking

The effect stalking has on victims has important implications. To help the victims, it is important to know how stalking affects them. The women in the current sample were seriously affected by the stalking in many negative ways that can be categorized into tangible effects and emotional or mental health effects.

In terms of tangible effects, two women (9.5%) lost their jobs because of the stalking. Eight women (38.0%) were forced to move, in attempts to avoid the stalking, and an additional three women (14.3%) seriously considered moving because of the stalking. Three women (14.3%) discussed the expensive nature of the ordeal. For example, they felt compelled to get caller ID and change their phone numbers because of the threat and nuisance of the stalking, and incurred other costs from the damage the stalkers caused, such as the price to fix broken windows and cars.

The most common reported negative effect of stalking was related to the mental and emotional impact. Respondents reported feeling scared, depressed, humiliated, embarrassed, distrustful of others, and angry or hateful. Fifteen women (71.4%) reported that the stalking scared them. The following quotes illustrate just how scary this behavior was for victims. Sarah, a 49-year-old White woman who was continually threatened by her ex-partner, said,

> Scared to death. Because he doesn't give up. I mean my experience with him is with these other people it would be years, years that he would hold grudges. Um, and try to find out where they moved or he just doesn't let things go. So yeah, I'm scared. It's been a year, a year and a half almost and um that's nothing for him.

Lily stated,

> [And had you been scared when you were together?] Oh yeah. 'Cause he had threatened me, said that if I left him he would do something to my mom or my grandma, he'd do something so that I'd regret leaving him. So I didn't do anything and I didn't tell anybody nothing until after. And then as my relationship went on with him, like my cousin saw him pull my hat one time. My mom saw him pull me in the car. People started seeing things. So I just started saying stuff because my friend was telling me, "Well, what if he does something to you?" I didn't think the relationship was like that. So I told my mom everything that happened. My girlfriends, I started telling them. The only time I would talk to them was if I would like sneak and call from my mom's or something. But still I wasn't, I couldn't talk about it.

Kristen specifically mentioned that she is afraid of her stalker finding out that she has moved on in her life. She stated,

> Oh yeah, and I still am [afraid of him]. I'm afraid of what's going to happen when he

> finds out I've met someone else. That's what I'm afraid of. I'm worried about that. I don't know what to do. I'm thinking about moving for awhile. But that scares me.

Joan and Crystal discussed how their fear has affected their behavior and thoughts when outside. Joan stated, "It felt like when I was walking outside, I always had to be looking over my shoulder to make sure he wasn't behind me, or lurking behind a tree or something." Crystal reported,

> Sometimes I would be afraid to go outside of work. Especially with the change of season and 4:30, 5:00 and it is starting to get dark and the parking lot situation. You know, before I went to court and had the restraining order and everything I was always thinking, "Is he out there?"

Seven women (33.3%) reported their stalking led to distrust of others. Most women mentioned how difficult it is to contemplate getting involved with another person after the experience of stalking. Ann stated,

> Yeah, I'm extremely wary of getting involved with anyone. Probably since the initial separation. There's been two people who I've seen very casually and that was like for a two-week period tops. I think I'm just, I don't know, I think that having a child the dynamic is different. I don't want to put my child in a bad situation. And I think it's unfortunate that a lot of women would put themselves back in it again, except that they have a child so then they wouldn't. I think, women, I know it's hard because I've done it. It would be nice if we did it for our own well-being not just because we have a child.

Joan also mentioned how her experiences with stalking have affected her relationships with potential partners:

> I'm less trusting and I have a real hard time now trying to get into a relationship. There's times when I want to be in a relationship and then when I get involved with this person, I don't want to be around him. And the guy is completely different than [the abuser]. I don't know. I guess I'm real defensive. You know. It's hard to describe, how I feel. It's like I want a relationship, yet I don't want a relationship. I'm scared to commit.

Others talked about feeling humiliated (n = 5, 23.8%), depressed (n = 2, 9.5%), and angry or hateful (n = 2, 9.5%). For example, Crystal mentioned how it made her feel at her place of work. She stated,

> But it was embarrassing. You know, I was embarrassed because I had just started that job. To have this person come over that I didn't want a life with anymore. To be showing up, loaded. I mean I was just like, "Oh, god!" You know, and I was afraid that it was going to affect my job.

Mary, who is still with her abuser/stalker, is illustrative of how stalking can affect one's self-esteem. The stalking and the violence that she experienced put her in a deep depression. She had this to say:

> I'm awful. It's made me feel bad, real bad. I mean I uh, he makes me feel so horrible and makes me feel like I have to put up with him because I'm such an awful person that I don't have no business looking outside the house. And stuff like that you know.

Mary's body language was interesting as well. She was unable to look the interviewer in the eyes. Moreover, after talking about the above, she went on to say that she can barely leave her house now because she has a really hard time talking to anyone—from children on the street to cashiers at the grocery store.

Discussion and Conclusion

The current study examined victims' perceived motivations for perpetrators who stalked in the context of IPA. It asked victims about their experiences with the process and the effects of stalking. The current study found that control and anger were often perceived motivations for stalking. Victims also commonly felt that stalking was used to scare them and/or get them to reestablish the relationship. These findings are consistent with prior research (NIJ, 1996; Roberts & Dziegielewski, 1996; U.S. Department of Justice, 1998). It is important to note that like much of the research on IPA and, in particular the feminist research on IPA, control was the number-one cited motivation for the stalking (Dobash & Dobash, 1979; Dobash, Dobash, Cavanagh, & Lewis, 2000). It is important to note, however, that several respondents viewed the stalking as a result of more positive motivations (i.e., stalkers were doing it out of love or concern for the victim). This is in direct contrast with most of the research on stalking and needs to be explored in detail in future research. Although these victims clearly experienced stalking (i.e., they fit the criteria to be part of the current study) and acknowledged that the behavior was unwanted, they still tended to give perpetrators "the benefit of the doubt" in terms of motivations. This is consistent with research that discusses ways in which women who experience IPA tend to minimize their abuse (Smith, 1994). It also illustrates the problems that result when the social construction of love in a culture includes possessiveness and jealousy. As the women noted, many of them reported that they felt early stalking behaviors indicated that their partner was being considerate or expressing his love through "excessive" attention. This finding illustrates the need for more education and public awareness to heighten awareness and change the conceptualization of love so that jealousy and possessiveness are viewed as early warning signs of an abusive relationship.

Previous research identified how stalking affects women in IPA situations (Davis et al., 2000; Mechanic, Uhlmansiek, et al., 2000; White et al., 2000). The current study supports these findings. Stalking had a serious negative effect on women's lives. Women reported increases in levels of depression, fear, and anger. Women talked about having to change their lives, moving in an attempt to thwart the stalking. Women also discussed leaving jobs or having problems maintaining employment because of the stalking. The current study, however, highlighted ways that stalking affected other relationships in their lives. Some women discussed how stalking interfered with maintaining relationships with children, other family members, and friends. Respondents also discussed the ways stalking hampered their ability to trust other people, in general, and potential partners, in particular. Future research should examine this effect in more detail, especially because a failure to maintain other supportive relationships may enhance the effectiveness of the stalking. For example, women without support systems may be more apt to return to batterers who stalk them out of fear or because they simply lack the emotional resources to deal with the stalking any longer.

Previous literature also has identified behaviors used in the stalking process (Coleman, 1997; Mechanic, Uhlmansiek, et al., 2000; Mechanic, Weaver, et al., 2000; Tjaden & Thoennes, 2000). The current study confirmed these findings indicating that stalkers target victims at home and in public, and that women are often threatened with bodily harm. Victims also suffer unwanted phone calls and other types of persistent behaviors. The current study, however, identified the existence of "proxy stalkers." Although other studies (Mullen, Pathe, & Purcell, 2000; Sheridan & Davies, 2001) have reported this phenomenon, they have limited the discussion of proxy stalkers to relationships involving strangers.

Respondents in the current study discussed ways that other people "helped" the stalker by either standing in for the stalker (acting in place of him) or by supplementing the stalker's actions (by working with him to magnify the stalking experience). The use of proxy stalkers gives the stalker greater control over the victim than exists through his stalking alone. In many ways, this experience may be even more terrifying for victims as numerous people may be engaged in following them, watching them, and harassing them. They may feel even less safe than a victim who is stalked "only" by her abuser. Victims also may be in more danger when additional stalkers are added to the equation; perhaps the potential for violence and/or harm increases by the introduction of these "helpers." In addition, the proxy stalker may be engaging in stalking behavior when the primary stalker is unable to stalk (i.e., when the primary stalker is incarcerated or under a court order to stay away). For example, using proxy stalkers is a creative way to subvert the intention of protective orders, and many states may lack appropriate legislation to deal with proxy stalkers effectively. It is important that these proxy stalkers are held accountable by the criminal justice system if protective orders and stalking statutes are to be enforced with any efficacy. This issue needs additional research.

Another important finding from the current study relates to the function of stalking. The current study supports previous research that concludes that stalking behavior may intensify after the relationship is terminated (Logan et al., 2000; Mechanic, Weaver, et al., 2000). The current study, however, also found that the overwhelming proportion of victims reported experiencing stalking prior to the break-up period. This finding contradicts research that classifies stalking primarily as part of a continuation of IPA occurring after the relationship ends (see Logan et al., 2000; Mechanic, Weaver, et al., 2000). The current study indicates that stalking should also be seen as a variant of IPA during the relationship. Stalking may give the abuser greater control over his victim during the relationship and may be one of the many tactics batterers use to ensure that women stay in abusive relationships. Clearly much research indicates that batterers use a wide variety of tactics to control women in abusive relationships (Dobash & Dobash, 1979; Dobash et al., 2000; Tong, 1984); however, stalking is not usually discussed as one of these tactics. It more often is viewed as a way that abusers ensure that women return if they try to leave. Moreover, it was discovered that some women experienced stalking during the relationship without defining it as such. In fact, as previously noted, they defined the behavior as positive in the beginning (i.e., he is doing this because he loves me). Additional research is needed to better explore the ways that abusers may use stalking throughout the course of the relationship. Perhaps abusers who stalk during the relationship are more apt to stalk when the relationship has ended. Similarly, it is possible that abusers who employ stalking during a relationship are more dangerous stalkers when the relationship ends. These types of stalkers also may be more apt to use creative stalking techniques, such as employing proxy stalkers. These issues are important avenues for future research.

Because of the limitations in the current study (i.e., sample size of 21), it is unclear whether these findings may represent broader trends. Stalking, as a legal concept, is a relatively new phenomenon. There is a need to continue to explore the link between the motivations to stalk, the behaviors used to stalk, and the impact stalking has on victims. Qualitative research, such as the current study, can help identify new and important areas of research that require additional examination.

References

Burgess, A. W., Baker, T., Greening, D., Hartman, C. R., Burgess, A., Douglas, J. E., et al. (1997). Stalking behaviors within domestic violence. *Journal of Family Violence*, *12*(4), 389–403.

Coleman, F. L. (1997). Stalking behavior and the cycle of domestic violence. *Journal of Interpersonal Violence*, *12*(3), 420–432.

Davis, K. E., Ace, A., & Andra, M. (2000). Stalking perpetrators and psychological maltreatment of partners: Anger-jealousy, attachment insecurity, need for control, and break-up context. *Violence and Victims*, *15*(4), 407–425.

Dobash, R. E., & Dobash, R. (1979). *Violence against wives: A case against patriarchy.* New York: Free Press.

Dobash, R. P., Dobash, R. E., Cavanagh, K., & Lewis, R. (2000). Confronting violent men. In J. Hanmer, C. Itzin, with S. Quaid & D. Wigglesworth (Eds.), *Home truths about domestic violence: Feminist influences on policy and practice, a reader* (pp. 289–309). London: Routledge.

Guy, R. A. (1993). The nature and constitutionality of stalking laws. *Vanderbilt Law Review*, *46*(4), 991–1029.

Logan, T. K, Leukefeld, C., & Walker, B. (2000). Stalking as a variant of intimate violence: Implications from a young adult sample. *Violence and Victims*, *15*(1), 91–111.

McFarlane, J. M., Campbell, J. C., Wilt, S., Sachs, C. J., Ulrich, Y., & Xu, X. (1999). Stalking and intimate partner femicide. *Homicide Studies*, *3*(4), 300–316.

Mechanic, M. B., Uhlmansiek, M. H., Weaver, T. L., & Resick, P. A. (2000). The impact of severe stalking experienced by acutely battered women: An examination of violence, psychological symptoms and strategic responses. *Violence and Victims*, *15*(4), 443–458.

Mechanic, M. B., Weaver, T. L., & Resick, P. A. (2000). Intimate partner violence and stalking behaviors: Exploration of patterns and correlates in a sample of acutely battered women. *Violence and Victims, 15*(1), 55–72.
Meloy, J. R. (1996). Stalking (obsessional following): A review of some preliminary studies. *Aggression and Violent Behavior, 1*(2), 147–162.
Meloy, J. R., & Gothard, S. (1995). Demographic and clinical comparison of obsessional followers and offenders with mental disorders. *American Journal of Psychiatry, 152*(2), 258–263.
Mullen, P. E., Pathe, M., & Purcell, R. (2002). *Stalkers and their victims.* New York: Cambridge University Press.
National Institute of Justice. (1996). *Domestic violence, stalking, and antistalking legislation: An annual report to Congress under the Violence Against Women Act.* Washington, DC: U.S. Department of Justice, Office of Justice Programs. National Institute of Justice.
Roberts, A., & Dziegielewski, S. (1996). Assessment typology and intervention with the survivors of stalking. *Aggression and Violent Behavior, 1*(4), 359–368.
Sheridan, L., & Davies, G. (2001). Violence and the prior victim-stalker relationship. *Criminal Behaviour and Mental Health, 11*, 102–116.
Smith, M. D. (1994). Enhancing the quality of survey data on violence against women. *Gender & Society, 8*, 109–127.
Tjaden, P. (1997). *The crime of stalking: How big is the problem.* Washington, DC: U.S. Department of Justice, National Institute of Justice.
Tjaden, P., & Thoennes, N. (1998). *Stalking in America: Findings from the National Violence Against Women Survey.* Washington, DC: U.S. Department of Justice, National Institute of Justice.
Tjaden, P., & Thoennes, N. (2000). The role of stalking in domestic violence crime reports generated by the Colorado Springs Police Department. *Violence and Victims, 15*(4), 427–441.
Tong, R. (1984) *Women, sex, and the law.* Totowa, NJ: Rowman and Allanheld.
U.S. Department of Justice. (1998). *Bureau of Justice Statistics fact book: Violence by intimates* (NCJ167237). Washington, DC: Author.
U.S. Department of Justice, Violence Against Women Grants Office. (1998). *Stalking and domestic violence: The Third Annual Report to Congress Under the Violence Against Women Act.* Washington, DC: U.S. Department of Justice, Office of Justice Programs, Violence Against Women Grants Office.
White, J., Kowalski, R. M., Lyndon, A., & Valentine, S. (2000). An integrative contextual developmental model of male stalking. *Violence and Victims, 15*(4), 373–388.
Wright, J. A., Burgess, A. G., Burgess, A. W., Laszlo, A. T., McCrary, G. O., & Douglas, J. E. (1996). A typology of interpersonal stalking. *Journal of Interpersonal Violence, 11*(4), 487–502.

DISCUSSION QUESTIONS

1. What motivates people to engage in stalking behaviors against a current or ex-intimate partner?
2. How do the victims experience stalking in the context of intimate partner abuse?
3. What effects does the experience of being stalked have for victims?

READING

Over the past two decades, the Internet has expanded into a realm whereby everyday communications are commonplace. The electronic world has provided a new arena for individuals to engage in stalking behaviors in a new way, as it allows many offenders to instill fear in their victims without revealing their identity. This study explores the frequency of cyberstalking among college students and makes recommendations for future practices and policy decisions.

SOURCE: Paullet, K. L., Rota, D. R., & Swan, T. T. (2009). Cyberstalking: An exploratory study of students at a mid-Atlantic university. *Issues in Information Systems, 10*(2), 640–649.

Cyberstalking

An Exploratory Study of Students at a Mid-Atlantic University

Karen L. Paullet, Daniel R. Rota, and Thomas T. Swan

Introduction

The Internet has become a medium for people to communicate locally or globally in the course of business, education and their social lives. The Internet has made it easy for people to compete, meet a companion, or communicate with people on the other side of the world with the click of a mouse. In 2009, according to the Internet World Stats Report, 237,168,545 people used the Internet in the United States; as a result there is a concern for Internet safety [11]. The increased use of the Internet has created an impact on the number of online harassing, cyberstalking cases.

Since the 1990s, stalking and harassing have become more common via the Internet. Megan Meier was a teenage girl who regularly used social network sites. Meier, a 13-year-old, became friends with a boy named Josh on MySpace. For weeks, Meier was very happy with her new online romance when suddenly Josh became angry at Meier implying that she was not very nice to her friends. The last posting by Josh to Meier read, "The world would be a better place without you" [16]. On October 17, 2006, Meier hung herself in her bedroom 20 minutes alter receiving the message from Josh. Her reply to Josh read, "You're the kind of boy a girl would kill herself over" [16]. It was later found out that Meier was not communicating with a boy named Josh but with a 48-year-old woman named Lori Drew. Drew created a fake MySpace profile as Josh, to contact Meier in order to see what she was saying about her daughter. Drew's online stalking of Meier led to her death.

The United States Department of Justice defines cyberstalking as the "use of the Internet, e-mail, or other electronic communication devices to stalk another person" [20]. Offline stalking is a crime with which many people are familiar. Stalking is a "repetitive pattern of unwanted, harassing or threatening behavior committed by one person against another" [13]. Stalking that involves the use of multiple individuals to stalk, harass or threaten a victim is known as gang stalking [7]. Although offline stalking acts have been reported since the 19th century, cyberstalking is a crime that is just being examined and reported since the late 1990s. The U.S. Attorney General stated, "stalking is an existing problem aggravated by a new technology" [20]. Similarities have been noted between offline stalking and cyber stalking cases, including the fact that "the majority of cases involve stalking by former intimates, most victims are women, most stalkers are men and stalkers are generally motivated by the desire to control the victim" [20]. Using technology to stalk a victim can include, but is not limited to, the Internet, e-mail, text messaging, global positioning systems (GPS), digital cameras, video cameras and social network sites. One of the differences between cyberstalking and offline stalking is that cyberstalkers face no geographic boundaries. A person can live in Hawaii and be stalked by a person in Italy. The Internet makes it possible for a person to be stalked virtually anywhere in the world.

Purpose of Study

Citizens should be able to feel safe when using the Internet without being stalked or harassed. But, the increased use of the Internet has caused a national increase in the number of online cyberstalking/harassment cases. The purpose of this research study is to explore online harassing/cyberstalking experiences at a Mid-Atlantic university.

Stalking Defined

Offline stalking acts have been reported since the 19th century. Cyberstalking is a new crime that is just being examined and reported since the late 1990s. Many similarities exist between stalking and cyberstalking. In order to understand cyberstalking, it is necessary to define stalking. The U.S. Department of Justice defines stalking "as harassing and threatening behavior that an individual engages in repeatedly" [19]. These behaviors include, but are not limited to, following a person, repeated phone calls and phone messages, appearing outside a person's home or work, vandalism, taking an individual's mail or entering a person's home. The U.S. Department of Justice [19] reports most stalking laws require the perpetrator (the person committing the stalking) to make a credible threat of violence against the victim. Stalking, therefore, can be used to instill fear and/or intimidate the victim. This study argues that cyberstalking and harrassment will only decrease when the extent of the problem is fully understood and potential victims and law enforcement understand the protections necessary under the law.

A person commits stalking if they cause another person to fear for their safety. "Stalking is a crime of power and control" [14]. As defined by Tjaden and Thoennes[18], stalking is a course of conduct directed at a specific person that involves repeated (two or more occasions) visual or physical proximity, nonconsensual communication, or verbal, written or implied threats, or a combination thereof, that would cause a reasonable person fear.

Stalking has been addressed in books, movies, and publications. Stalking can even be recognized in music lyrics. The band The Police wrote a song called "Every Breath You Take." The lyrics of the song can be considered by some to be written about stalking.

> Every breath you take / Every move you make
>
> Every bond you break / Every step you take
>
> I'll be watching you.
>
> Oh can't you see / You belong to me [17]

Even if The Police were not talking about stalking, by reading the lyrics, the true meaning of stalking can be heard.

Cyberstalking Defined

The Internet and use of telecommunications technologies have become easily accessible and are used for almost every facet of daily living throughout the world. Cyberstalking is "the use of the Internet, e-mail and other electronic communication devices to stalk another person" [19]. For this study, cyberstalking will be referred to as online stalking and is similar to offline stalking, which is being aggravated by new technologies. Cyberstalking "entails the same general characteristics as traditional stalking, but in being transposed into the virtual environment as it is fundamentally transformed" [15]. Stalking itself is not a new crime, but cyberstalking is a new way to commit the crime of stalking while using the Internet or other forms of electronic communication devices.

Stalkers, both online and offline, "are motivated by the desire to exert control over their victims and engage in similar types of behavior to accomplish this end" [20]. The term cyberstalking can be used interchangeably with online harassment. "A cyberstalker does not present a direct threat to a victim, but follows the victim's online activity to gather information and make threats or other forms of verbal intimidation" [12]. A potential stalker may not want to confront and threaten a person offline, but may have no problem threatening or harassing a victim through the Internet or other forms of electronic communications. One can become a target for a cyberstalker through the use of the Internet in many forms. The victim can be contacted by email, instant messaging (IM) programs, via chat rooms, social network sites or the stalker attempting to take over the victim's computer by monitoring what they are doing while online. Bocij, Griffiths and McFarlane [3] conclude that there are no genuinely reliable statistics that can be used to determine how common cyberstalking incidents occur.

Cyberstalkers can choose someone they know or a complete stranger with the use of a personal computer and the Internet. Basu and Jones [1] remind us that growing up our parents told us not to talk to strangers, but one function of the Internet is to talk to strangers. The Internet, as a communication tool, has allowed people the freedom to search for information from anywhere and anyone in the world. Fullerton [6] states that Internet Service Providers (ISP's) e-mail, web pages, websites, search engines, images, listservs, instant chat relay (ICR's) are all cyberstalking tools. Other forms of communication used to stalk a victim include cell phones, text messaging, short message services (SMS), global positioning systems (GPS), digital cameras, spyware or fax machines. The information that is available about people on the Internet makes it easy for a cyberstalker to target a victim. With only a few keystrokes, a person can locate information on an individual via the Internet. The types of information that can be found include e-mail addresses, home telephone numbers, bank accounts, credit card information, and home addresses. Some services, such as Intelius and People Finders, charge to provide confidential information for any person that is willing to pay. Imagine a teacher posting a syllabus online to instruct students what date and time a particular class is in session. Someone that is a cyberstalker can use this small amount of information to follow the instructor to school or try to get inside the instructor's home since they know when she will be in class. Thanks to search engines such as Google, a cyberstalker can type in a person's home or work address and see where they live or work. Once the cyberstalker can physically see what the home or place of employment looks like the stalker can use the descriptions of the locations as a way to let the victim know they are being watched. "The fact that cyberstalking does not involve physical contact may create the misperception that it is more benign than physical stalking" [20]. It is not uncommon for cyberstalkers to progress into offline stalkers. "If not stopped early on, some cyberstalkers can become so obsessed with a victim that they escalate their activities to the level of physical stalking" [10]. Gregorie [9] indicates that people who do not have access to the Internet, or choose not to go online are not immune from cyber-based crimes. Databases of personal information available on the Internet can enable a person to find the necessary information to stalk or harass a victim.

Existing Laws

Stalking laws within the 50 states are relatively recent: the first traditional stalking law was enacted in 1990 in California. California's legal definition of stalking is "any person who willfully, maliciously, and repeatedly follows or harasses another person and who makes a credible threat with the intent to place that person in reasonable fear of their safety" (Cal. Penal Code § 646.9) [22]. Since California's enactment of the first stalking law in 1990, all 50 states and the federal government have anti-stalking laws. Most stalking cases are prosecuted at the state and local levels. Each state's stalking laws will vary in their legal definitions and the degree of penalty for the offense.

As of March 2009, 45 states have cyberstalking or related laws in place compared to 1998, in which only 16 states had cyberstalking and harassment laws. Two of the five states without cyberstalking laws have pending laws for the implementation of such acts. Cyberstalking is covered in some of the 45 states' existing stalking laws. Stalking laws that are written to include forms of stalking using electronic communication devices such as email, Internet, cell phone text messaging or similar transmissions cover the crime of cyberstalking. If a state's current stalking law covers forms of electronic communications that are punishable by law, a separate cyberstalking law is not required. If the stalking laws within the 50 states do not cover any forms of electronic communications such as the Internet, a separate law should be written. For example, the Pennsylvania stalking law states:

> (1) a person commits the crime of stalking when the person either engages in a course of conduct or repeatedly commits acts toward another person without proper authority, under circumstances which demonstrate either an intent to place such other person in

reasonable fear of bodily injury or to cause substantial emotional distress to such other person, or

(2) engages in a course of conduct or repeatedly communicates to another person under circumstances which demonstrate or communicate either an intent to place such other person in reasonable fear of bodily injury or to cause substantial emotional distress to such other person. 18 PA. Cons. Stat. Ann. § 2709.1 (a) (1) and (2) [22]

As used in the definition of stalking under Pennsylvania law, "communicates" is defined as:

> To convey a message without intent of legitimate communication or address by oral, nonverbal, written or electronic means, including telephone, electronic mail, Internet, facsimile, telex, wireless communication or similar transmission. 18 PA Cons. Stat. Ann. §2709.1 (f) [22]

Under Title 18 of the United States Code, federal law covers threatening messages transmitted electronically in interstate and foreign commerce 18 U.S.C §875 [22]. This means that a person who is being threatened in Pennsylvania via the Internet, from a person living in Florida, is protected by Federal law. Similarly, in Pennsylvania for example, local Pennsylvania law enforcement agencies may file stalking charges in Pennsylvania even if the electronically transmitted threat originated in another state, but only if the victim receives the threat in Pennsylvania.

Cyberbstalkers, if caught, can face criminal charges on three different levels based upon seriousness: a felony, misdemeanor, or summary offense. A felony is a serious crime, as defined under federal law, and in many states the offense can be punishable by death or imprisonment in excess of one year. A misdemeanor is a criminal net that carries a less severe punishment than a felony but more serious than a summary offense. "Misdemeanors in the U.S. generally have a maximum punishment of 12 months in jail" [5]. A summary is a minor violation of the law prosecutable without a full trial. An example of a common summary would be a traffic ticket.

Victims of cyberstalking need to obtain copies of all electronic forms of communication received from the stalker. The electronic evidence that is obtained can lead to a computer and not an individual. For example, if the stalker is using a computer in a library to send messages to a victim, the electronic trail will lead back to the computer in the library. Potentially, hundreds of people could have used that computer between when the stalking messages were sent and when the IP address was traced to the library.

Methodology

This study examined cyberstalking of undergraduate and graduate students at a mid-Atlantic university during September 2008. A quantitative methodology was selected for this research project as a means to examine students that have been stalked, harassed or threatened through the use of the Internet, email or other forms of electronic devices. In the student survey, participant responses were used to explore evidence of cyberstalking victims at the collegiate level. This data may be used to assist current and future victims of cyberstalking and assist law enforcement agencies in dealing with the problem.

The first page of the student questionnaire focused on participant demographics to include gender, age and education. The survey addressed the participant's use of the Internet to include the frequency of Internet use and what types of online activities are accessed. Online activities included email, bulletin boards, newsgroups, instant messaging, chat rooms, social network sites and dating sites. The study addressed the types of online activities in relation to becoming a victim of cyberstalking.

The survey asked participants if they could classify themselves as a victim of cyberstalking according to the definition supplied at the beginning of the survey. The study focused on how a victim was targeted and by whom. The methods that a victim could have been targeted by their stalker include email, bulletin boards, instant messaging, text messaging, chat rooms, social network sites, news groups, dating sites and eBay. Whether the victims personally knew their stalker was addressed, as well as the manner in which the stalker communicated to them. Communication used by the stalker included friendly, sexual, threatening, hateful, humorous and intimidating language. How long the communication lasted between the victim and the stalker was also addressed.

An important aspect that this study sought to determine was that at anytime during a victim's harassment did they fear for their safety. Stalked individuals that completed the survey were asked to identify the level of fear they possessed, if any. Victims were also asked if they reported the cyberstalking. Victims that answered yes were asked to whom they reported the incident to include law enforcement, Internet Service Provider, campus advisor, cell phone provider, web administer or online help organization, and if they received help.

Sample

The sample consisted of 302 undergraduate and graduate students at a mid-Atlantic university from a population of approximately 5,000 students.

Results

Cyberstalking was defined as threatening behavior or unwanted advances directed at another using the Internet and other forms of online and computer communications. Cyberstalkers can target their victims through threatening or harassing email, flaming (online verbal abuse), computer viruses, chat rooms, message boards, social network sites (such as MySpace), text messages, or tracing a person's Internet activities, among others. Approximately 13% of students identified themselves as victims of cyberstalking according to the definition provided at the beginning of the survey instrument.

All 302 students that completed the survey responded to yes to using the Internet.

The types of Internet use were analyzed to determine the respondent's online activities. All of the 302 participants (100%) indicated they use email. The majority of students, 211 (70%), access social network sites while 183 students (60%) use instant messaging.

The current survey of 39 victims reveals, 25 (64%) were female and 14 (36%) were male. These results are consistent with studies by Finn [4] and Bocij [2] that women are more likely to become a victim of cyberstalking than men. This current study revealed that women were almost twice as likely as men to become a victim of cyberstalking. As a point of relevance there were more males, 174, than females, 128, that completed the survey. To determine the relationship between victim and gender, chi-square was calculated (chi-square = 8.650, df = 1, $p < .003$) indicating that a relationship is found between gender and being a victim of cyberstalking. The observed significance level is .003, which is greater than the customary 0.05, indicating that the results did not happen by chance.

Research Question 1 queried the relationship between online activities and occurrences of cyberstalking. Students who were a victim of cyberstalking were more likely to receive harassment by email (chi-square = 5.769, df = 1, $p < .016$), text messaging, (chi-square = .026, df = 1, $p < .873$) and social network sites (chi-square = 4.333, df = 1, $p < .037$). Email and social network sites are statistically significant while no statistical significance was found when victims were contacted via text messaging. The observed significance level for email communication used by the stalker is .016, which is less than the customary 0.05 confidence level, signifying that the results did not happen by chance. This research study indicates that there is a relationship between email and social network sites as a communication device used by the stalker and being a victim of cyberstalking.

Of the 34 victims, 25 knew the identity of their stalker while 14 did not know the identity of their stalker. From the students that knew the identity of their stalker, 8 answered the person was a former boyfriend or girlfriend, 3 knew the person from work, 7 knew the person from school, 5 answered the stalker was a friend, and 2 met their stalker online. The result that

25 victims knew their stalker provides evidence that stalking is taking on a new form.

Research Question 2 sought to determine if women are more likely than men to report being a victim of cyberstalking. Three questions on the survey were used to answer the research question in which victims were asked if they reported the cyberstalking incident, to which they reported the incident to, and if they reported the incident, did they receive help. With the current study, the victims that reported the cyberstalking included 5 male (21%) and 19 females (79%). Of those that did not report the cyberstalking were 9 male (60%) and 6 female (40%). Women are more likely than men to report being cyberstalked. There is a statistically significant relationship between gender and reported cases of cyberstalking (chi-square = 6.154, df = 1, $p < .013$). Additionally, a total of 3 males (15%) and 17 females (85%) reported the cyberstalking to law enforcement (chi-square = 7.791, df = 1, $p < .005$). This concludes that there is a statistically significant relationship between gender and reporting the cyberstalking to law enforcement. A total of 34 students reported the cyberstalking either to law enforcement, their ISP, a cell phone provider or web administrator. Of the 34 students that reported the cyberstalking, only 11 received help while 23 victims did not receive any type of help. Surprisingly, 68% of victims that reported the incident did not receive any type of help.

Research Question 3 addressed the level of fear associated with being a victim of cyberstalking. Two questions on the survey were used to answer the research question in which students were asked if the harassment made them fear for their safety and what was their level of fear. Out of the 39 students that answered yes to being a victim of cyberstalking 2 males (9%) and 21 females (91%) feared for their safety while the stalking persisted. There is a statistically significant relationship between gender and fear (chi-square = 18.027, df = 1, $p < .000$). The remaining victims 12 men (75%) and 4 females (25%) did not fear for their safety at anytime during the stalking. The victims that had a low level of fear were comprised of 1 male (14%) and 6 females (86%). Those that reported a moderate level of fear were 1 male (8%) and 11 females (92%). The remaining victims, 0 males (0%) and 4 females (100%), reported a high level of fear. The victims were asked if they were currently being contacted by the cyberstalker. Out of the 39 victims, 28 students were no longer being contacted by the cyberstalker and 11 students were currently being stalked at the time the survey was administered.

Discussion

The Internet and use of telecommunications technologies have become easily accessible and are used for almost every facet of daily living throughout the world. This study found that approximately 13% of students surveyed were a victim of cyberstalking. The first research objective examined the relationship between online activities and occurrences of cyberstalking. This study discovered that there was not a correlation between how often a student accessed the Internet and becoming a victim of cyberstalking. University students that accessed the Internet one time a day or less were just as likely to become a victim of cyberstalking as someone that accessed the Internet more than once a day. Every participant that completed the survey, 302, indicated they use email.

Students who were a victim of cyberstalking were more likely to receive harassment by email and social network sites. Findings from the research support the first research objective in reference to email and social network sites. There was a direct relationship between email and social network sites as an online activity and being a victim of cyberstalking. Although not an online activity, but a form of electronic communication, almost half of the victims, 49%, were contacted by their stalker via text messaging. The use of text messaging was one of the most frequent ways that victims were contacted by their stalker.

The second research objective examined if women are more likely than men to report being a victim of cyberstalking. The research conducted by Finn [4] indicated that there were no demographic differences in regard to gender in relation to reporting incidents of cyberstalking. Gender and repotted cases of cyberstalking were not discussed in the Bocij [2] study. This current study revealed a significant relationship between gender and being a victim of cyberstalking. Women were four times more likely than men to report the cyberstalking incident. Goodson, McCormick and Evans [8] indicated in their study of online sexual harassment that female students were more likely than men to report being a victim. The current study revealed that 85% of victims that reported the cyberstalking to law enforcement were female while only 15% of males contacted law enforcement. There is a significant relationship between gender and reported cases of cyberstalking, as well as gender and reporting the incident to law enforcement.

The third research objective addressed the level of fear associated with being a victim of cyberstalking. Victims were asked if the stalking made them fear for their safety. Of the 39 victims, 21 females and 2 males feared for their safety while the stalking persisted. These numbers reveal a significant relationship between gender and fear. A total of 6 females and 1 male reported a low level of fear, 1 male and 11 females reported a moderate level of fear and 0 males and 4 females indicated a high level of fear.

Although not the intent of the original study, additional findings were discovered as a result of the statistical analysis. The student survey revealed that there are almost twice as many female victims (64%) than male (36%). Approximately 30% of the victims were between the ages of 18 and 25. Although there was not a relationship between age and level of education the study revealed that 7 victims, 14%, were doctoral students. The 14% of victims represented by doctoral students shows a slightly higher percentage of becoming a victim than the 13% revealed from the study. This number stands out because there were only 50 respondents from the doctoral program compared to 166 undergraduate student respondents and 86 master level respondents. There were three times as many undergraduate students and almost twice as many graduate masters' level students that completed the survey.

Recommended Approach to Preventing Cyberstalking

Based on my 11 years of experience working in the District Attorney's Office, specifically working with computer forensics along with prior research, the following list can help protect a person from being a victim of cyberstalking:

1. Never use your real name, nickname or any type of suggestive name while online.
2. When online, only type things you would actually say to someone face-to-face. Think about how what you say might be interpreted without eye contact, body language or voice.
3. THINK BEFORE YOU INK. Remember, once you send an electronic message it can remain in cyberspace indefinitely.
4. Log off immediately if you experience contact from someone that is hostile, rude or inappropriate.
5. Save all communications from the stalker as evidence.
6. Report the incident to your ISP, law enforcement agency, school administration or an online help agency such as www.haltabxise.org or www.cyberangels.org.

Conclusions

Studies are needed to improve our understanding of cyberstalking. The fast pace at which technology changes, as well as the inexpensive cost of technologies make it easier for a person to track and stalk a victim. Studies based on victim experiences need to be explored in depth so that the appropriate laws are written to protect victims of cyberstalking. A collaborative effort from victims, law enforcement, and private and public sectors is needed in order to combat cyberstalking and develop an effective response to the problem.

References

1. Basu, S., and Jones, R. (22, November 2007). Regulating Cyberstalking. *Journal of Information, Law and Technology*. Retrieved on January 22, 2009 from: http://go.warwick.ac.uk/jilt/2007_2/basu_jones/

2. Bocij, P. (2003). Victims of cyberstalking: An exploratory study of harassment perpetrated via the internet. *First Monday* Vol 8, No. 10. Retrieved February 29, 2008 from http://www.firstmonday.org/Issues/issue8_10/bocij/index.html

3. Bocij, P., Griffiths, M., and McFarlane, L. (2002). Cyberstalking a new challenge for criminal law. *Criminal Lawyer*, 122, 3–5.

4. Finn, J. (2004). A survey of online harassment at a university campus. *Journal of Interpersonal Violence*. Sage Publications. Retrieved January 20, 2008 from http://jiv.sagepub.com

5. Federal Defense Cases. (2007). Federal defense cases. Retrieved on October 28, 2008 from www.federaldefensecases.com/about/php

6. Fullerton, B. (2003, December 22). Features - cyberagestalking. *Law and technology for legal professionals*. Retrieved February 11, 2008 from http://www.llrx.com/node/1114/print

7. Gang Stalking: An overview. (2006, September 15). Retrieved on April 20, 2008 from http://educateyourself.org/cn/gangstalkingoverview 15 sep.06.html

8. Goodson, P., McCormick, D., and Evans, A. (2001). Searching for sexual explicit materials on the Internet: An exploratory study of college students' behavior and attitudes. *Archives of Sexual Behavior*, 30, 101–118.

9. Gregorie, T.M. (2001). Cyberstalking: Dangers on the information superhighway. *National Center for Victims of Crime*. Retrieved May 19, 2009 from http://www.ncvc.org/src/help/cyberstalking.html

10. Hitchcock, J.A. (2006). Net crimes and misdemeanors: Outmaneuvering web spammers, stalkers, and con artists. Medford, New Jersey: Information Today, Inc.

11. Internet World Stats (2007, November) Usage and population statistics. Retrieved January 19, 2008 from http://www.internetworldstats.com/stats14.htm

12. Jaishankar, K. and Sankary, U.V. (2006). Cyberstalking: A global menace in the information super highway. All India Criminology Conference. 16-18 2006. Madurai: India Madurai Kamaraj University.

13. Mechanic, M. (2000). Fact sheet on stalking. National Violence Against Women Prevention Research Center, University of Missouri at St. Louis. Retrieved January 19, 2007 from http://www.musc.edu/wawprevention.research/stalking/shtml

14. National Institute of Justice. (2002). Stalking. The Research and Evaluation Agency of the U.S. Department of Justice retrieved on September 16, 2007 from http://www.ojp.usdoj.gov/nij/topics/cri me/stalking/welcome.htm

15. Ogilvie, E. (2000). The internet and cyberstalking. Stalking: Criminal Justice Responses Conference, 7-8 December 2000. Sydney: Australian Institute of Criminology.

16. Steinhauer, J. (2008, November 26). Verdict in MySpace suicide. *New York Times*. Retrieved January 10, 2008 from www.nytimes.com/2008/11/27/us/27myspace.html

17. Sumner, Gordon. (1983). "Every Breath You Take." Stewart Copeland, Andy Sumers, Gordon Sumner (Sting), Hugh Padgham, Synchronicity A & M Records (1983).

18. Tjaden, P. and Theonnes, N. (1998). Stalking in America: Findings from the National Violence Against Women Survey. Washington, DC: US Department of Justice, Centers for Disease Control and Prevention.

19. U.S. Department of Justice. (2001). *Stalking and domestic violence*: NCJ 186157, Washington, DC: U.S. Government Printing Office.

20. U.S. Attorney General Report (1999). Cyberstalking. A new challenge for law enforcement and industry. (Electronic Version) Retrieved September 22,2007 from http://www.usdoj.gov.criminal/cybercrime/cyberstalking.htm

21. United States Census Bureau (2009). Retrieved February 12, 2009 from http://quickfacts.census.gov/qfd/states/4 21/42003.html

22. United States Code Annotated. (2009). Thomson/West.

DISCUSSION QUESTIONS

1. What is cyberstalking? Do college students identify cyberstalking as a criminal offense?

2. How does the experience of cyberstalking vary by gender?

3. How can cyberstalking be prevented?

READING

In the section introduction, you learned about the different forms of sexual harassment that women may experience. This reading expands on one particular form of sexual harassment: the unwanted attention from strangers in public. Not only did the women in this study indicate that these behaviors were commonplace, but that the experience of being sexually harassed by a stranger had significant negative consequences for their mental health.

Everyday Stranger Harassment and Women's Objectification

Kimberly Fairchild and Laurie A. Rudman

Introduction

The acknowledgment of stranger harassment, and the need to protect women from it, is virtually ignored in the social science and feminist literature. Stranger harassment is the "[sexual] harassment of women in public places by men who are strangers" (Bowman, 1993, p. 519). In other words, stranger harassment is perpetrated by men who are not known to the victim (i.e., not a co-worker, friend, family member, or acquaintance) in public domains such as on the street, in stores, at bars, or on public transportation. While the phenomenon has been defined, it is infrequently studied (cf. Gardner, 1995; MacMillan, Nierobisz, & Welsh, 2000). Why has stranger harassment been overlooked by social science researchers? As Bowman (1993) and Nielsen (2000) suggest in their analyses of stranger harassment from a legal point of view, the study of stranger harassment may be lacking because there is no legal recourse; it is nearly impossible to sue a stranger who disappears in a flash for sexual harassment, and it is likely that few would support laws limiting the freedom of speech in public places. Gardner (1995) goes even further to suggest that stranger harassment is so pervasive that it is a part of the social fabric of public life: "Women . . . currently experience shouted insults, determined trailing, and pinches and grabs by strange men and [are] fairly certain that no one—not the perpetrator and probably no official—will think anything of note has happened" (p. 4). Thus, stranger harassment may be perceived to be an innocuous part of daily life, and not an important topic for study (Gardner, 1995).

However, stranger harassment may not be so innocuous; in view of the multitude of negative effects that sexual harassment has on women (described below), it becomes clear that the gap in the literature

SOURCE: Fairchild, K., & Rudman, L. A. (2008). Everyday stranger harassment and women's objectification. *Social Justice Research, 21*, 338–357.

considering stranger harassment needs to be filled. In the current research, we take a first step toward a social psychological understanding of stranger harassment.

Sexual Harassment Versus Stranger Harassment

Over the past 25 years, sexual harassment research has boomed as researchers have sought to define the components of sexual harassment and elaborate its causes and consequences (Gutek & Done, 2001; Pryor & McKinney, 1995; Wiener & Gutek, 1999). To do so, sexual harassment has been commonly parsed into three main components: sexual coercion, gender harassment, and unwanted sexual attention (Gelfand, Fitzgerald, & Drasgow, 1995). Sexual coercion is the direct request or requirement of sexual acts for job or school related rewards (e.g., promotion or a better grade); this component aligns with the legal conceptualization of quid pro quo sexual harassment. Gender harassment involves degradation of women at the group level such as making jokes about women as sex objects or posting pictures of women as sex objects. Unwanted sexual attention involves degradation of women at the individual level, such as treating a woman as a sex object by sending her dirty e-mails, grabbing her inappropriately, or leering at her. Both gender harassment and unwanted sexual attention fall into the legal category of hostile environment sexual harassment. Gelfand et al. (1995) note that while women frequently label sexual coercion as sexual harassment, it is experienced by only 5–10% of samples, making it somewhat rare. Gender harassment is by far the most prevalent, experienced by approximately 50% or more of samples, followed by unwanted sexual attention, experienced by approximately 20–25% of samples.

Unfortunately, many sexual harassment researchers seem to assume that sexual harassment is a phenomenon experienced only in the workplace or at school. One of the most popular measures of sexual harassment is the Sexual Experiences Questionnaire (SEQ; Fitzgerald, Gelfand, & Drasgow, 1995a), which asks for respondents' experiences with a variety of behaviors (e.g., "unwanted sexual attention," "told suggestive stories," and "touching in a way that made you feel uncomfortable"). The bulk of behaviors listed in the SEQ can be applied to many situations, but the majority of researchers ask respondents to think about these experiences in the context of the workplace and school. As such, it is difficult to ascertain the prevalence of sexual harassment outside of these locales. Moreover, while researchers examining sex discrimination more broadly have recognized that harassment can occur in a variety of settings (i.e., beyond the workplace and school; e.g., Klonoff & Landrine, 1995; Klonoff, Landrine, & Campbell, 2000; Landrine et al., 1995; Landrine & Klonoff, 1997), they often fail to separate out the effects of being harassed by strangers (as opposed to known perpetrators). For example, Berdahl (2007a) assessed undergraduate students' experiences of sexual harassment using the SEQ. Since the students had little work experience, they were encouraged to consider their experiences in relation to school and time with friends and family. While this study expands the realms of where and with whom sexual harassment can occur, Berdahl does not parse the results based on location or source. However, in more theoretical work, Berdahl (2007b) argues that sexual harassment stems from a need to maintain social status and as such can occur in any situation in which a perpetrator's status is threatened.

In *Passing By: Gender and Public Harassment*, Gardner (1995) provides an empirical focus on stranger harassment as she details the contexts in which stranger harassment takes place, the participants in stranger harassment, the behaviors that are characteristic of stranger harassment, the interpretations people have of stranger harassment, and the strategies employed to avoid stranger harassment. Her evidence stems from information obtained from 506 interviews with 293 women and 213 men. From her qualitative analysis, it is clear that stranger harassment is highly akin to sexual harassment researchers' conceptualization of unwanted sexual attention. As Bowman (1993) describes it, stranger harassment "includes both verbal and nonverbal behavior, such as wolf-whistles, leers, winks, grabs, pinches, catcalls, and stranger remarks; the remarks are frequently sexual in nature and comment evaluatively on a woman's physical appearance or on her presence in public" (p. 523). The information provided by Gardner (1995) gives the reader a vivid sense of the experience of stranger harassment, but she overlooks the connection between stranger harassment and the established literature on unwanted sexual attention.

To date, MacMillan et al. (2000) provide the only known attempt to document differences between unwanted sexual attention from strangers and known perpetrators. Using data collected in 1993 from a national sample of Canadian women responding to the Violence Against Women Survey (VAWS; Johnson & Sacco, 1995), the authors focused on the data obtained from eight items measuring stranger and non-stranger sexual harassment. The stranger harassment items assessed "whether respondents had ever received an obscene phone call, received unwanted attention (i.e., anything that does not involve touching, such as catcalls, whistling, leering, or blowing kisses), been followed in a manner that frightened them, or experienced an indecent exposure" (p. 310). The items measuring non-stranger sexual harassment represented both quid pro quo and hostile environment sexual harassment. Their data show that 85% of the women reported experiencing stranger harassment, with the majority experiencing unwanted sexual attention (e.g., catcalls and leering). By contrast, 51% experienced non-stranger sexual harassment, with only 5% reporting having experienced quid pro quo sexual harassment. MacMillan et al.'s (2000) research indicates that stranger harassment may be a more pervasive problem than non-stranger harassment. Moreover, they found that stranger harassment has a more consistent and significant impact on women's fears than non-stranger harassment. Specifically, they noted that, "Stranger harassment reduces feelings of safety while walking alone at night, using public transportation, walking alone in a parking garage, and while home alone at night" (p. 319). MacMillan et al. (2000) were the first to show that stranger harassment is more prevalent than non-stranger sexual harassment, and that it has an impact on women's fears.

Consequences of Sexual and Stranger Harassment

Since sexual harassment and stranger harassment are conceptually related, they are likely to produce many of the same consequences. Since MacMillan et al. (2000) showed remarkably high rates of stranger harassment, it can be further inferred that stranger harassment may affect more women than sexual harassment. The work of Louise Fitzgerald and her colleagues (Fitzgerald, Drasgow, Hulin, Gelfand, Magley, 1997; Glomb et al., 1997; Magley, Hulin, Fitzgerald, & DeNardo, 1999; Schneider, Swan, & Fitzgerald, 1997) is among the most prominent for investigating the outcomes of sexual harassment. In their model of the antecedents and consequences of sexual harassment, Fitzgerald, Hulin, and Drasgow (1995b) propose that sexual harassment results in decreased job satisfaction and physical well-being. In addition, tests of their model suggest that sexual harassment has a negative impact on psychological outcomes; women who experienced low, moderate, and high levels of sexual harassment showed more negative psychological outcomes than women who experienced no sexual harassment (Schneider et al., 1997). Furthermore, Schneider et al. (1997) found that experiencing harassment has negative outcomes for women even if they do not label the events as sexual harassment. This finding was also supported by research that investigated the outcomes of self-labeling (Magley et al., 1999); specifically, the researchers found no differences in negative outcomes between women who labeled their experiences sexual harassment and women who did not label them as such. Thus, sexual harassment negatively impacts women's psychological well-being whether the harassment is mild or severe, labeled or not labeled. Unfortunately, while the sexual harassment research indicates negative psychological outcomes for women, it is unclear whether decreased psychological well-being refers to depression, anxiety, or some other mental health disorders. For example, Magley et al. (1999) used the Mental Health Index to assess psychological well-being. The Mental Health Index includes measures of depression, anxiety, and positive affect. However, the researchers used different variations of the index in their different samples, and did not separate depression and anxiety (combined as psychological distress). Moreover, no research on sexual harassment has examined self-objectification as a consequence, which has been linked to depression (e.g., Harrison & Fredrickson, 2003; Tiggemann & Kuring, 2004) and thus may account for some of the negative psychological outcomes. Additionally, sexual harassment research has not explored potentially significant consequences such as women's increased fear of rape or voluntarily restricting their movements. The present research on stranger harassment was designed to address these gaps in the harassment literature relating to self-objectification, fear of rape, and restriction of movement.

Objectification

Sexual objectification is a clear component of both sexual harassment and stranger harassment. In both cases, women are treated as objects to be looked at and touched, and not as intelligent human beings. The main tenet of self-objectification theory (Fredrickson & Roberts, 1997) is that the human body is not merely a biological system, but that "bodies exist within social and cultural contexts, and hence are also constructed through sociocultural practices and discourses" (p. 174). In American culture, women's bodies are constantly and consistently regarded as sexual objects through pornography, the mass media, and advertising. The unwanted sexual attention experienced in both sexual harassment and stranger harassment is another example of women being regarded as sexual objects. Despite the diversity of mechanisms through which sexual objectification can occur (e.g., pornography, advertising, and stranger harassment), "the common thread running through all forms of sexual objectification is the experience of being treated as a body (or collection of body parts) valued predominantly for its use to (or consumption by) others" (Fredrickson & Roberts, 1997, p. 174).

Self-objectification theory, as proposed by Fredrickson and Roberts (1997), provides a framework for understanding the psychological experience of sexual objectification. They argue that this experience is uniquely female and can lead to mental health problems. For Fredrickson and Roberts (1997), the consequences of objectification arise when the woman begins to objectify herself (i.e., self-objectify). Repeated exposure to sexual objectification increases the likelihood that women will objectify themselves. This leads women to regard themselves as mere sex objects, to experience body shame, and to chronically monitor their external appearance (Fredrickson & Roberts, 1997). Prior research shows that self-objectification is positively correlated with negative outcomes, including depression and disordered eating (e.g., Greenleaf, 2005; Harrison & Fredrickson, 2003; Muehlenkamp & Saris-Baglama, 2002; Muehlenkamp, Swanson, & Brausch, 2005; Slater & Tiggemann, 2002; Tiggemann & Kuring, 2004; Tiggemann & Slater, 2001). In the present research, we hypothesized that women who experience greater amounts of stranger harassment will be more likely to self-objectify. As such, it is a first attempt to test unwanted sexual attention (in the form of stranger harassment) as a predictor of self-objectification.

Fear of Rape and Restriction of Movement

The limited work on stranger harassment (MacMillan et al., 2000) suggests that it may increase women's fear of rape and therefore their willingness to limit their freedom of movement (e.g., Hickman & Muehlenhard, 1997; Swim et al., 1998).

In the present research, we hypothesized that women would fear sexual assault to the extent they reported being harassed in public by strangers. Research on the fear of rape among women suggests that women are more fearful of stranger rape than acquaintance rape, even though most women recognize that stranger rape is much less prevalent than acquaintance rape (Hickman & Muehlenhard, 1997). Research on sex differences in perception of danger and fear of victimization, such as murder or robbery, consistently illustrate that women are more fearful than men, although men are much more likely to be victims of crime than women (Ferraro, 1996; Harris & Miller, 2000). Ferraro's (1996) "shadow of sexual assault" hypothesis suggests that women are more fearful overall because the fear of rape permeates their fear of other victimizations. Since, for women, rape is a potential outcome of any face-to-face victimization, it may be a primary source of anxiety. In support of this hypothesis, Ferraro (1996) found that women's fear of rape predicted their fear of other personal crimes (e.g., murder, burglary). Fisher and Sloan (2003) replicated Ferraro's (1996) work finding that the fear of rape did indeed shadow other fears of victimization for women.

Similarly, Harris and Miller (2000) discovered that women, compared with men, are consistently more fearful of ambiguously dangerous situations involving men. They suggest that women's higher fear of victimization may stem from daily experiences of minor victimizations, which are likely to be ignored because of their non-criminal nature. Although they did not test this hypothesis, they specifically posited that the experience of "stares, whistles, condescending behavior, being interrupted when speaking, and harassment at work" socializes women to be more fearful and more perceptive of danger (Harris & Miller, 2000, p. 857). When taken together with Ferraro's (1996) and Fisher and Sloan's (2003) research, this suggests that stranger harassment may increase women's fear of rape, as well as their perceived risk of rape

Finally, the fear of rape literature suggests that women typically alter their behaviors by limiting how, when, and where they travel to protect themselves from rape (Hickman & Muehlenhard, 1997; Krahe, 2005; War, 1985). By avoiding walking alone at night or specific places (e.g., parking garages; Hickman & Muehlenhard, 1997), women voluntarily restrict their freedom to move about in the world. Similarly, Swim, Cohen, and Hyers (1998) note that women's tendency to avoid sites of sexual harassment restricts their freedom of movement. Thus, in addition to fear of rape, we predicted that women's voluntary restriction of movement would be a consequence of stranger harassment.

Coping With Stranger Harassment

Research on women's responses to sexual harassment suggests that the majority of women are likely to use passive, non-assertive coping strategies. Gruber's (1989) review of the literature found that less than 20% of women use assertive or active coping strategies. Women typically respond to harassment by ignoring it or attempting to avoid the harasser (see also Magley, 2002). Less frequently, women may cope with harassment by reporting or confronting the perpetrator, engaging in self-blame, or by perceiving the harassment to be a compliment or benign (Fitzgerald, 1990). While it is likely that many of the coping strategies used by women who are sexually harassed are similar to the strategies used by women who are stranger harassed (e.g., ignoring it), there may also be differences (e.g., there are no laws specifically against stranger harassment, so it is unclear to whom a stranger harasser would be reported).

For our purposes, we borrowed items from the Coping with Harassment Questionnaire (CHQ; Fitzgerald, 1990) that seemed most pertinent to stranger harassment and excluded items more descriptive of sexual harassment (e.g., "I filed a grievance," and "I told a supervisor or department head"). It was predicted that women who endorsed the active coping items (e.g., "I let him know I did not like what he was doing") would experience less objectification than women who endorsed the passive items (e.g., "I pretended nothing was happening") or who engaged in self-blame (e.g., "I realized I had probably brought it on myself"). In rejecting the harassment through active coping strategies, it is thought that these women will also be rejecting the objectified view of their bodies, thus limiting their self-objectification; on the other hand, women employing passive or self-blame strategies are not actively fighting the objectified view of their body and thus may be more likely to internalize the objectification. Finally, we had competing predictions about women who responded to stranger harassment as though it were benign (e.g., "I considered it flattering"). On the one hand, it was possible that these women would not be adversely affected by stranger harassment. On the other hand, women who perceived stranger harassment to be a compliment or innocuous might be already highly self-objectified. In essence, their response might reflect society's view of stranger harassment as something women should "expect" by virtue of their gender. If so, these women should show high levels of objectification depending on the frequency of stranger harassment.

Summary and Hypotheses

Women's experiences of sexual harassment in public places (i.e., stranger harassment) is an area of research that has been ignored by traditional sexual harassment research. Stranger harassment shares many common themes with sexual harassment, most specifically the component of unwanted sexual attention. However, stranger harassment is unique from sexual harassment in that it is perpetrated by strangers (as opposed to co-workers, teachers, or peers) and that it takes place in public domains such as on the street, in stores, and in bars (as opposed to the office or school).

The current research investigates the prevalence and hypothesized outcomes of stranger harassment, as well as potential moderators of stranger harassment's consequences. First, we sought to determine the frequency of stranger harassment experiences in a sample of female college students. Second, we predicted that frequent experiences with stranger harassment would lead to increased levels of self-objectification. Third, we expected that stranger harassment would positively predict women's fear of sexual assault and perceived risk of rape and, therefore, voluntary restriction of movement.

However, we also hypothesized that women's coping behaviors would moderate the relationship between stranger harassment and objectification. First, we expected that women who responded actively to stranger harassment (e.g., by confronting the harasser) would buffer themselves from self-objectification. Second, we predicted that women who responded passively (e.g., by ignoring the harassment) or who engaged in self-blame would be more likely to self-objectify with more experiences of stranger harassment. Finally, although women who viewed stranger harassment as benign might not be affected by their experiences, we suspected they might show high levels of objectification if their responses reflect being co-opted by society's view that women should expect to be sexually objectified.

Method

Participants

Female volunteers ($N = 228$) participated in exchange for partial credit toward their Introductory Psychology research participation requirement. About 44% (101) were White, 33% (75) were Asian, 8% (18) were Latina, 7% (16) were Black, and the remaining 8% reported another ethnicity. Participants ranged in age from 18 to 29, with a mean age of 19.3 years old. The majority (97%) reported being exclusively heterosexual.

Measures

Stranger Harassment

Experiences with stranger harassment were assessed using a modified version of the Sexual Experiences Questionnaire (SEQ; Fitzgerald et al., 1995a). Participants were first asked whether they had ever experienced nine different behaviors from strangers; these behaviors ranged in severity (e.g., "Have you ever experienced unwanted sexual attention or interaction from a stranger?"; "Have you ever experienced catcalls, whistles, or stares from a stranger?; "Have you ever experienced direct or explicit pressure to cooperate sexually from a stranger?"; and "Have you ever experienced direct or forceful fondling or grabbing from a stranger?"). Participants then responded to the same behaviors in terms of frequency (1 = once; 2 = once a month; 3 = 2–4 times per month; 4 = every few days; 5 = every day).

Following this, participants were instructed to think about how they typically respond to the experiences described above and to rate statements about potential reactions on scales ranging from 1 (not at all descriptive) to 7 (extremely descriptive). The reactions were selected from the Coping with Harassment Questionnaire (CHQ; Fitzgerald, 1990) to reflect active coping (e.g., "I talked to someone about what happened"), passive coping (e.g., "I just 'blew it off and acted like I did not care"), self-blame (e.g., "I realized he probably would not have done it if I had looked or dressed differently") or treating harassment as benign

Table 5.4 Reported Frequency (in Percent) of Women's Stranger Harassment Experiences

	Once a Month	Twice a Month	Every Few Days or More
Catcalls, whistles, or stares	32.0	33.3	30.9
Unwanted sexual attention	40.8	24.1	14.5
Crude or offensive sexual jokes	37.3	25.9	11.4
Sexist remarks or behaviors	40.8	22.4	11.4
Seductive remarks or "come ons"	30.0	24.6	15.8
Unwanted touching or stroking	36.0	11.4	2.7
Subtle pressure to cooperate sexually	30.3	6.1	8.1
Direct pressure to cooperate sexually	25.9	5.3	1.3
Forceful fondling or grabbing	26.3	4.8	1.3

or inconsequential (e.g., "I figured he must really like me," and "I treated it as a joke").

Objectification

Self-objectification was measured using McKinley and Hyde's (1996) Objectified Body Consciousness Scale (OBCS). The OBCS is comprised of three subscales (surveillance, body shame, and control beliefs) to which participants respond on scales ranging from 1 (strongly disagree) to 7 (strongly agree). For the current study, only the body surveillance and body shame scales were used, consistent with prior research (Muehlenkamp & Saris-Baglama, 2002; Tiggemann & Kuring, 2004; Tiggemann & Slater, 2001). The surveillance subscale assesses concern with body appearance over functioning (e.g., "I often worry about whether the clothes I am wearing make me look good," and "I am more concerned with how my body looks than with what it can do"). The body shame subscale assesses how respondents feel about their bodies' imperfections (e.g., "When I am not the size I think I should be, I feel ashamed"; "When I cannot control my weight, I feel like something must be wrong with me"). Both subscales showed adequate internal consistency (surveillance $\alpha = .86$; shame $\alpha = .87$). As in past research, the body surveillance and body shame scales were significantly correlated ($r = .50$, $p < .01$). Thus, they were averaged to form the Self-Objectification Index ($\alpha = .88$).

Fear and Risk of Rape

Women reported their fear of being raped by a stranger and an acquaintance on scales ranging from 1 (not at all afraid) to 10 (very afraid). Specifically, the items read, "How afraid are you of being raped by a stranger [acquaintance]?" They also responded to two items assessing perceived risk of being raped on scales ranging from 1 (not at all likely) to 10 (very likely).

Restriction of Movement

Women also responded to 10 items designed to assess restriction of movement, on a scale ranging from 1 (strongly disagree) to 7 (strongly agree). Sample items include "I feel safe walking around campus alone at night," "I would not feel comfortable walking alone in the city at night," and "If I need to go out of my house at night, I often try to have a male friend accompany me."

Procedure

Participants were escorted to private cubicles equipped with a desktop PC. The experimenter administered the instructions and informed consent and started a computer program for the participants. Participants completed the measures in the order described above. Items were presented randomly within each measure. Participants were then asked to report their age, ethnicity, and sexual orientation.

Discussion

The present findings represent a first step toward a social psychological analysis of stranger harassment. We found relatively high prevalence rates of stranger harassment for female college students. Approximately 11% reported experiencing unwanted sexual attention from strangers at least once a month, including sexist remarks or seductive "come ons," and nearly one-third reported harassment consisting of catcalls, whistles, or stares. In fact, 31% of our sample reported experiencing catcalls, whistles, and stares every few days or more. Moreover, over a quarter of our sample suffered experiences akin to sexual coercion or assault (e.g., forceful grabbing) at least once a month. These data support treating stranger harassment as a significant form of humiliation and indignity that targets women and is likely to undermine the quality of their lives. In essence, stranger harassment turns public spaces into an everyday hostile environment for women.

With respect to the consequences of stranger harassment, we predicted (and found) that it would positively predict women's self- objectification. Although our data cannot speak to causality, this finding suggests that one potential source of women's self-objectification may be their experiences with stranger harassment. Self-objectification reflects emphasizing the body's appearance over its function, and feeling ashamed of a less than ideal body. A large literature suggests that self-objectification predicts negative outcomes in women, including depression and disordered eating (e.g., Greenleaf, 2005; Harrison & Fredrickson, 2003; Muehlenkamp & Saris-Daglama, 2002; Muehlenkamp et al., 2005; Slater & Tiggemann, 2002; Tiggemann & Kuring, 2004; Tiggemann & Slater, 2001). As a result, it is conceivable that stranger harassment indirectly promotes psychological and behavioral problems in women, through its link to self-objectification.

Moreover, as expected, women's coping responses to stranger harassment were significantly related to self-objectification. First, active coping interacted with stranger harassment to predict less objectification. Thus, women who experience greater harassment and acknowledge the behavior as inappropriate by confronting or reporting the harasser, or talking the experience over with a friend, may be able to resist feeling sexually objectified. Second, women who responded passively (e.g., by ignoring or denying the harassment) reported feeling self-objectified. Since passive strategies were more prevalent than active (or any other type) of coping, the likelihood of women feeling objectified by stranger harassment is high. Third, self-blame responses were also positively related to self-objectification. As predicted, women who viewed the harassment as their own fault (i.e., as something they could have avoided) also reported feeling self-objectified. Finally, coping with harassment by viewing it as benign, innocuous, or complimentary was also positively related to self-objectification. By coping with the harassment as though it was a form of flattery (or "no big deal"), women may be capitulating to being sexually objectified. Even if they enjoy the attention from men, being objectified by others can lead to self-objectifying (Fredrickson & Roberts, 1997) which, as noted above, predicts serious outcomes in women such as depression and disordered eating.

We also predicted that stranger harassment would be positively related to women's fears of victimization and voluntary restriction of movement. However, with the exception of perceived risk of rape, our hypotheses were not supported. Nonetheless, the structural model suggested that stranger harassment may have indirect effects on fear of rape (through self-objectification) and restriction of movement (through fear of rape). Although past research has found that women who feared rape were more likely to curb their movements (e.g., to avoid going out alone at night; Hickman & Muehlenhard, 1997), we extended these findings to include stranger harassment and self-objectification as potential antecedents of victimization fears.

Limitations and Future Directions

One of the main limitations of the current research is that the sample consisted of college-aged women. It is quite possible that young women are more likely to experience stranger harassment than older women. However, MacMillan et al. (2000) found that 85% of Canadian women reported stranger harassment, suggesting that youth may not be a significant factor. Nonetheless, their research included behaviors that we did not assess (e.g., obscene phone calls), which may be experienced regardless of age. Thus, future research is necessary to lend confidence to the generalizability of our findings.

Another limitation is that the current research did not address the issue of where the harassment took place. While it is theorized that stranger harassment can occur in public places ranging from the street to stores to public transit, the unique characteristics of a college campus may present different "public" experiences than the average woman faces. For example, college women may be more likely to attend parties at fraternities or bars that allow for more harassment opportunities. A follow-up study is underway to ascertain some of the specifics about where stranger harassment is experienced on a college campus. In addition, evidence from Gardner (1995) suggests that women in metropolitan areas are more susceptible to harassment than women in suburban and rural areas. Future research needs to address the specifics of where stranger harassment is most frequent for a variety of settings.

The present research also suggests the need for further investigation of the link between objectification and sexual harassment. It seems likely that if women who are harassed by strangers experience self-objectification, women harassed by known perpetrators (e.g., in the workplace or school) may also suffer a similar outcome. Moreover, self-objectification and sexual harassment have been independently linked to negative psychological outcomes (e.g., depression and anxiety; Fitzgerald et al., 1997; Fitzgerald et al., 1995a; Fitzgerald et al., 1995b). Thus, future work should test the possibility that self-objectification may serve to mediate the relationship between sexual harassment and psychological dysfunction. Additionally, the current research assumes that the negative consequences of stranger harassment will be similar to the negative consequences of sexual harassment (i.e., decreased psychological well-being). Future research should directly assess the relationship of depression and anxiety to experiences of stranger harassment.

Further, women's strategies for coping with stranger harassment should be further investigated. For example, passive and self-blame responses may reflect women's gender role socialization (e.g., to avoid confrontation and blaming others), whereas active strategies may require more agency. Future research should explore a likely connection between women's acceptance of gender roles or stereotypes and their use of passive (versus active) strategies. Since passive and self-blame strategies were linked to self-objectification, future work may reveal a vicious cycle whereby women are taught to ignore or fault themselves for harassment, which then makes them more vulnerable to experiencing its negative effects. Results for self-blame were particularly poignant in this regard, as self-blame was related to perceived risk of rape. Although women who viewed stranger harassment as benign or complimentary were less likely to fear rape and restrict their movements, they also reported greater self-objectification. Feeling flattered by sexual attention from strangers may reflect women's acceptance of sexual objectification as normative—something women should expect from men as positive reinforcement (e.g., for being attractive). In this respect, stranger harassment may be similar in function to benevolent sexism (Glick & Fiske, 2001), in which women are praised for being a "good woman" but which actually has a pernicious influence by making them feel weak. Future research should examine whether women who respond to stranger harassment as though it were a compliment are also likely to be benevolent sexists.

Conclusion

Despite the wealth of sexual harassment research, women's analogous experience of public harassment by strangers has been largely ignored. The present findings suggest that stranger harassment is a remarkably common occurrence for many women, and that common means of coping with it may lead to increased self-objectification. Since self-objectification has negative consequences for women (e.g., depression and eating disorders), stranger harassment may be a serious form of discrimination. Moreover, through its link to objectification, stranger harassment may have indirect consequences that decrease the quality of women's lives, such as increased fear of rape and restriction of movement. Overall, stranger harassment appears to be a frequent and significant experience for women and therefore is deserving of future research designed to more fully elaborate the experience and its consequences.

References

Atwood, M. (1986). *The handmaid's tale*. New York, NY: Anchor Dooks.

Berdahl, J. (2007a). The sexual harassment of uppity women. *Journal of Applied Psychology*, 92, 425–437.

Berdahl, J. (2007b). Harassment based on sex: Protecting social status in the context of gender hierarchy. *Academy of Management Review*, 32, 641–658.

Bowman, C. G. (1993). Street harassment and the informal ghettoization of women. *Harvard Law Review*, 106, 517–580.

Ferraro, K. (1996). Women's fear of victimization: Shadow of sexual assault? *Social Forces*, 75, 667–690.

Fisher, D. S., & Sloan, J. J. (2003). Unraveling the fear of victimization among college women: Is the "shadow of sexual assault hypothesis" supported? *Justice Quarterly*, 20, 633–659.

Fitzgerald, L. F. (1990). *Assessing strategies for coping with sexual harassment: A theoretical/empirical approach.* Paper presented at the annual meeting of the Association for Women in Psychology, Tempe, AZ.

Fitzgerald, L. F., Drasgow, F., Hulin, C. L., Gelfand, M. J., & Magley, V. J. (1997). Antecedents and consequences of sexual harassment in organizations: A test of an integrated model. *Journal of Applied Psychology*, 82, 578–589.

Fitzgerald, L. F., Gelfand, M. J., & Drasgow, F. (1995a). Measuring sexual harassment: Theoretical and psychometric advances. *Basic and Applied Social Psychology*, 17, 425–445.

Fitzgerald, L. F., Hulin, C. L., & Drasgow, F. (1995b). The antecedents and consequences of sexual harassment in organizations: An integrated model. In G. P. Keita & J. J. Hurrell (Eds.), *Job stress in a changing workforce: Investigating gender diversity, and family issues* (pp. 55–73). Washington, DC: American Psychological Association.

Fredrickson, D. L., & Roberts, T. A. (1997). Objectification theory: Toward understanding women's lived experiences and mental health risks. *Psychology of Women Quarterly*, 21, 173–206.

Gardner, C. D. (1995). *Passing by: Gender and public harassment.* Berkley, CA: University of California Press.

Gelfand, M. J., Fitzgerald, L. F., & Drasgow, F. (1995). The structure of sexual harassment: A confirmatory analysis across cultures and settings. *Journal of Vocational Behavior*, 47, 164–177.

Glick, P., & Fiske, S. T. (2001). Ambivalent sexism. In M. P. Zanna (Ed.), *Advances in experimental social psychology*. Vol. 33 (pp. 115–188). Thousand Oaks, CA: Academic Press.

Glomb, T. M., Richman, W. L., Hulin, C. L., Drasgow, F., Schneider, K. T., & Fitzgerald, L. F. (1997). Ambient sexual harassment: An integrated model of antecedents and consequences. *Organizational Behavior and Human Decision Processes*, 71, 309–328.

Greenleaf, C. (2005). Self-objectification among physically active women. *Sex Roles*, 52, 51–62.

Gruber, J. E. (1989). How women handle sexual harassment: A literature review. *Sociology and Social Research*, 74, 3–9.

Gutek, D. A., & Done, R. S. (2001). Sexual harassment. In R. K. Unger (Ed.), *Handbook for the psychology of women and gender* (pp. 367–387). New York: Wiley.

Harris, M. D., & Miller, K. C. (2000). Gender and perceptions of danger. *Sex Roles*, 43, 843–863.

Harrison K., & Fredrickson, D. L. (2003). Women's sports media, self-objectification, and mental health in Black and White adolescent females. *Journal of Communication*, 53, 216–232.

Hickman, S. E., & Muehlenhard, C. L. (1997). College women's fears and precautionary behaviors relating to acquaintance rape and stranger rape. *Psychology of Women Quarterly*, 21, 527–547.

Japan tries women-only train cars to stop groping. (June 10, 2005). http://abcnews.go.com/GMA/International/story_id=803965&CMP=OTC-RSSFeeds0312. Accessed 14 June 2005.

Johnson, H., & Sacco, V. F. (1995). Researching violence against women: Statistics Canada's national survey. *Canadian Journal of Criminology*, 37, 281–304.

Klonoff, E. A., & Landrine, H. (1995). The schedule of sexist events: A measure of lifetime and recent sexist discrimination in women's lives. *Psychology of Women Quarterly*, 19, 439–472.

Klonoff, E. A., Landrine, H., & Campbell, R. (2000). Sexist discrimination may account for well-known gender differences in psychiatric symptoms. *Psychology of Women Quarterly*, 24, 93–99.

Krahe, D. (2005). Cognitive coping with the threat of rape: Vigilance and cognitive avoidance. *Journal of Personality*, 73, 609–643.

Landrine, H., & Klonoff, E. A. (1997). *Discriminating against women: Prevalence, consequences, and remedies*. Thousand Oaks, CA: Sage Publications.

Landrine, H., Klonoff, E. A., Gibbs, J., Manning, V., & Lund, M. (1995). Physical and psychiatric correlates of gender discrimination: An application of the schedule of sexist events. *Psychology of Women Quarterly*, 19, 473–492.

MacMillan, R., Nierobisz, A., & Welsh, S. (2000). Experiencing the streets: Harassment and perceptions of safety among women. *Journal of Research in Crime and Delinquency*, 37, 306–322.

McKinley, N. M., & Hyde, J. S. (1996). The objectified body consciousness scale: Development and validation. *Psychology of Women Quarterly, 20,* 181–215.

Magley, V. J. (2002). Coping with sexual harassment: Reconceptualizing women's resistance. *Journal of Personality and Social Psychology*, 83, 930–945.

Magley, V. J., Hulin, C. L., Fitzgerald, L. F., & DeNardo, M. (1999). Outcomes of self-labeling sexual harassment. *Journal of Applied Psychology*, 84, 390–402.

Muehlenkamp, J. J., & Saris-Daglama, R. N. (2002). Self-objectification and its psychological outcomes for college women. *Psychology of Women Quarterly*, 26, 371–379.

Muehlenkamp, J. J., Swanson, J. D., & Drausch, A. M. (2005). Self-objectification, risk-taking, and self-harm in college women. *Psychology of Women Quarterly, 29,* 24–32.

Nielsen, L. D. (2000). Situating legal consciousness: Experiences and attitudes of ordinary citizens about law and street harassment. *Law & Society Review, 34,* 1055–1090.

Pryor, J. D., & McKinney, K. (1995). Research advances in sexual harassment: Introduction and overview. *Basic and Applied Social Psychology, 17,* 421–424.

Schneider, K. T., Swan, S., & Fitzgerald, L. F. (1997). Job-related and psychological effects of sexual harassment in the workplace: Empirical evidence from two organizations. *Journal of Applied Psychology, 82,* 401–415.

Sheffield, C. J. (1989). The invisible intruder: Women's experiences of obscene phone calls. *Gender and Society, 3,* 483–488.

Slater, A., & Tiggemann, M. (2002). A test of objectification theory in adolescent girls. *Sex Roles, 46,* 343–349.

Sussman, A. (2006) In Rio rush hour, women relax in single sex trains. http://www.womensenews.org/article.cfm?aid=2750. Accessed 23 May 2006.

Swim, J. K., Cohen, L. L., & Hyers, L. L. (1998). Experiencing everyday prejudice and discrimination. In J. K. Swim & C. Stangor (Eds.), *Prejudice: The target's perspective* (pp. 37–60). San Diego: Academic Press.

Tiggemann, M., & Kuring, J. K. (2004). The role of body objectification in disordered eating and depressed mood. *British Journal of Clinical Psychology, 43,* 299–311.

Tiggemann, M., & Slater, A. (2001). A test of objectification theory in former dancers and non-dancers. *Psychology of Women Quarterly, 25,* 57–64.

Warr, M. (1985). Fear of rape among urban women. *Social Problems, 32,* 238–252.

Wiener, R. L., & Gutek, D. A. (1999). Advances in sexual harassment research, theory, and policy. *Psychology, Public Policy, and Law, 5,* 507–518.

DISCUSSION QUESTIONS

1. What effect does being stalked by a stranger have on perceptions of safety and fear of assault?
2. What coping strategies do women who are stalked by strangers use to deal with their victimization? How effective are these strategies?

International Issues for Women and Crime

Section Highlights

- International human rights for women
- International examination of the sexual assault and murder of women
- Human trafficking of women
- Honor killings

While there are a number of concerns involving the human rights of women around the world, this section highlights three examples of victimizations of women: the practice of honor-based violence against women, the issue of human trafficking and female slavery, and the femicides in Ciudad Juárez. Each of these crimes are related to the status of women within their communities, and suggestions for change are rooted within a shift of gendered normative values and the treatment of women in these societies. Each discussion focuses on the nature of the crime, the implications for women in these regions, and the role of these issues for discussions about criminal justice policies in an international context.

As you read through the experiences of the women and their victimizations, it is important to consider how the cultural context of their lives affects their victimization experience. The effects of culture are significant, as it can alter not only how these crimes are viewed by agents of social control (police, legal systems), but also how the community interprets these experiences. These definitions play a significant role in determining how these crimes are reported (or if reports are made), as well as any response that may arise from these offenses. It can be dangerous to apply a White, middle-class lens or "Americanized identity" to these issues—what we might do as individuals may not necessarily reflect the social norms and values of other cultures.

Honor-Based Violence

The category of "honor-based violence" (HBV) includes practices such as honor (honour) killings, bride burnings, customary killings, and dowry deaths. Each of these crimes involves the murder of a woman by a male family member, usually a father, brother, or male cousin. These women are killed in response to a belief that the women have offended a family's honor and have brought shame to the family unit. The notion of honor is one of the most important cultural values for members of these communities. "Honor is the reason for our living now . . . without honor life has no meaning. . . . It is okay if you don't have money, but you must have dignity (Kardam, 2005).

At the heart of the practice of honor-based violence is a double standard rooted in patriarchy, which dictates that women should be modest, meek, pure, and innocent. Women are expected to follow the rules of their fathers and, later, their husbands. In some cases, honor killings have been carried out in cases of adultery, or even perceived infidelity. Hina Jilani, a lawyer and human rights activist, suggests that the "right to life of women . . . is conditional on their obeying social norms and traditions" (Amnesty International, 1999). Women are viewed as a piece of property that holds value. Her value is based on her purity, which can be tainted by acts that many Western cultures would consider to be normal, everyday occurrences, such as requesting a love song on the radio or strolling through the park (Arin, 2001). For many women, their crime is that they wanted to become "Westernized" or participate in modern-day activities, such as wearing jeans, listening to music, and developing friendships. For other women, their shame is rooted in a sexual double standard where a woman is expected to maintain her purity for her husband. To taint the purity of a woman is to taint her honor. The concept of honor controls every part of a woman's identity. "When honor is constructed through a woman's body, it entails her daily life activities, education, work, marriage, the importance of virginity (and) faithfulness" (Kardam, 2005, p. 61).

Even women who have been victimized through rape and sexual assault are at risk of death via an honor killing, as their victimization is considered shameful for the family. In many cases, the simple perception of impropriety is enough to warrant an honor killing. Women who are accused of bringing negative attention and dishonor are rarely afforded the opportunity to defend their actions (Mayell, 2002).

> The distinction between a woman being guilty and a woman being alleged to be guilty of illicit sex is irrelevant. What impacts the man's honour is the public perception, the belief of her infidelity. It is this which blackens honour and for which she is killed. To talk of "alleged kari" or "alleged siah-kari" makes no sense in this system nor does your demand that a woman should be heard. It is not the truth that honour is about, but public perception of honour. (Amnesty International, 1999)

The practice of honor and customary killings are typically carried out with a high degree of violence. Women are subjected to acts of torture, and their deaths are often slow and violent. They may be shot, stabbed, strangled, electrocuted, set on fire, or run over by a vehicle. "A man's ability to protect his honour is judged by his family and neighbors. He must publically demonstrate his power to safeguard his honour by killing those who damaged it and thereby restore it" (Amnesty International, 1999). One would assume that the women in these countries would silently shame these acts of violence. Contrary to this belief, research indicates that the women in the family support these acts of violence against their sisters as part of the community mentality toward honor (Mayell, 2002).

While the United Nations estimates that there are 5,000 honor killings each year around the world, researchers and activists indicate that the true numbers of these crimes are significantly greater. Estimates indicate that tens of thousands of women are killed each year in the practice of honor-based violence. Yet, many of these crimes go unreported, making it difficult to develop an understanding of the true extent of the issue. According to research by Chesler (2010), which reviewed 230 cases of news-reported honor killings worldwide, the majority (95%) of the victims are young women (mean age = 23). In 42% of cases, there were multiple perpetrators involved in the killing, a characteristic that distinguishes these types of crimes from the types of femicide that are most commonly reported in Western countries. Over half of these women were tortured to death and were killed by methods such as stoning, burning, beheading, strangulation, or stabbing/bludgeoning. Nearly half (42%) of these cases involved acts of infidelity or alleged "sexual impropriety," while the remaining 58% of women were murdered for being "too Western" and defying the expectations that are set through cultural and religious normative values. Yet men are never criticized for their acceptance of Western culture. "Women are expected to bear the burden of upholding these ancient and allegedly religious customs of gender apartheid" (Chesler, 2010).

Even if justice officials do become involved in these cases, they are rarely punished to any extent. When human rights organizations and activists identify these incidents as honor-based violence, family members of the victim are quick to dismiss the deaths of their sisters and daughters as "accidents." In Turkish communities, if a woman has fractured the honor of her family, the male members of her family meet to decide her fate. In the case of "customary killings," the task of carrying out the murder is often given to the youngest male member of the family. In many cases, these boys are under the age of criminal responsibility, which further reduces the likelihood that any punishments will be handed down in the name of the victim (Arin, 2001).

One example of honor-based violence is the practice of karo-kari murder in Pakistan. Karo-kari is a form of premeditated killing and is part of the cultural traditions of the community. The terms "karo" and "kari" literally translate to "black male" and "black female" and are used in reference to someone who is an adulterer or adulteress; these terms do not refer to their racial or ethnic identity. In the majority of karo-kari cases, women are killed for engaging in acts of immoral behavior. These acts can include alleged marital infidelity, refusal to submit to an arranged marriage, or requesting a divorce from a husband (even in cases where abuse is present). In 2003, an estimated 1,261 women were killed in the name of karo-kari in Pakistan. Unofficial statistics may place this number as significantly higher given that many of these crimes occur within the family and may not be reported to the authorities. The practice of karo-kari in Pakistan is unique compared to honor killings in many other countries, as men can also be victims under karo-kari traditions. Generally speaking, cultural norms require the killing of both the man and woman involved in the infidelity in order to restore honor. This distinction is unique to karo-kari, whereas most other forms of customary killings involve only the woman. However, in some cases, the karo (man) is able to negotiate with the tribal counsel for the community to pay money or offer other forms of settlement (property of another woman) to compensate the "victim" (Amnesty International, 1999).

Honor-based violence is a violation of many international treaties and acts. Yet the laws in many countries contain provisions that permit acts of honor-based violence, including Syria, Morocco, Jordan, and Haiti. The practice of honor killings was condoned under the Taliban, and data indicate that since the fall of the Taliban regime, there has been an increase in reporting acts of violence against women (Esfandiari, 2006). Even in cases where there is no legal legitimization of the practice, cultural traditions and tribal justice sanction these crimes of violence and shield offenders from punishment (Patel & Gadit, 2008). Some countries may offer lesser penalties to men who murder female relatives. Other countries have laws on the books that identify customary killings as a criminal act, yet offer opportunities for the offender to escape punishment. One such example is Pakistan. While the Pakistani Parliament passed legislation in 2004 that punished honor killings with a 7-year sentence and would allow for the death penalty for the worst cases, many activists question whether such a punishment will ever be followed. In addition, the law contains a provision where the offender could "negotiate" a pardon with the victim's family members. Given that many of the offenders are indeed members of the victim's family, many believe that the retention of this provision will enable the offenders of these crimes to escape punishment (Felix, 2004; Masood, 2004).

There are few options for escape for women in these countries where honor-based violence prevails. International support for women at risk of becoming a victim of an honor killing is limited. There are few shelters for women seeking to escape their families. Any attempt to contact the police or other agents not only sends a woman back to the arms of her family where she is at the greatest risk, but her actions have created shame and dishonor for the family by exposing their life to the public community. The only customary option for these women is to escape to the tribal leader of the community (sadar), who can provide shelter while they negotiate a safe return to her home or to another community far from her family. However, this option is limited, as it still requires women to abide by traditional community standards and is not an option for women who are interested in asserting their rights or improving their status as women outside of their cultural norms (Amnesty International, 1999).

In their quest to improve the lives of women who may be victims of the practice of honor killings, Amnesty International (1999) outlines three general areas for reform:

1. Legal Measures. The current legal system in many of these countries does little to protect victims from potential violence under the normative structures that condone the practice of honor killings. Women have few, if any, legal rights that protect them from these harms. Legal reforms must address the status of women and provide them with opportunities for equal protection under the law. In cases where they survive an attempted honor killing, they need access to remedies that address the damages they experience. In addition, the perpetrators of these crimes are rarely subjected to punishment for their actions. Indeed, the first step toward reform includes recognizing that violence against women is a crime, and such abuses need to be enforced by the legal communities. International law also needs to recognize these crimes and enforce sanctions against governments that fail to act against these offenders. However, it is unclear how effective these legal measures will be for individual communities. In their discussions on what can be done to stop the practice of honor killings, Turkish activists did not feel that increasing the punishments for honor-based violence would serve as an effective deterrent, particularly in regions where the practice is more common and accepted within the community, as "punishments would not change the social necessity to kill and that to spend long years in jail can be seen as less important than lifelong loss of honor" (Kardam, 2005, p. 51).

2. Preventive Measures. Education and public awareness is the first step toward reducing honor-based violence toward women. These practices are rooted in culture and history. Attempts to change these deeply held attitudes will require time and resources aimed at opening communication on these beliefs. This is no easy task given the normative values that perpetuate these crimes. One of the first tasks may be to adopt sensitivity-training programming for judicial and legal personnel so that they may be able to respond to these acts of violence in an impartial manner. In addition, it is important to develop a sense of the extent of the problem in order to provide effective remedies. Here, an enhanced understanding of data on these crimes will help shed light on the pervasiveness of honor-based violence as a first step toward addressing this problem.

3. Protective Measures. Given the limited options for women seeking to escape honor-based violence, additional resources for victim services need to be made available. These include shelters, resources for women fleeing violence, legal aid to represent victims of crime, provisions for the protection of children, and training to increase the economic self-sustainability for women. In addition, the agencies that offer refuge for these women need to be protected from instances of backlash and harassment.

While these suggestions offer opportunities for change, many agents working in these regions indicate feelings of hopelessness that such changes are possible. Certainly the road toward reform is a long one, as it is rooted in cultural traditions that will present significant challenges for change. "When an honor killing . . . starts to disturb everybody . . . and when nobody wants to carry this shame anymore, then finding solutions will become easier" (Kardam, 2005, p. 66). Indeed, the first step in reform involves creating the belief that success is possible.

Human Trafficking

Rathana was born to a very poor family in Cambodia. When Rathana was 11 years old, her mother sold her to a woman in a neighboring province who sold ice in a small shop. Rathana worked for this woman and her husband for several months. She was beaten almost every day and the shop owner never gave her much to eat. One day a man came to the shop and bought Rathana from the ice seller. He then took her to a far-away province. When they arrived at his home he showed Rathana a pornographic movie and then forced her to act out the movie by raping her. The man kept Rathana for more than 8 months, raping her sometimes two or three times a day. One day the man got sick and went to a hospital. He brought Rathana with him and raped her in the hospital bathroom. Another patient reported what was happening to the police. Rathana was rescued from this man and sent to live in a shelter for trafficking survivors.

Salima was recruited in Kenya to work as a maid in Saudi Arabia. She was promised enough money to support herself and her two children. But when she arrived in Jeddah, she was forced to work 22 hours a day, cleaning 16 rooms daily for several months. She was never let out of the house and was given food only when her employers had leftovers. When there were no leftovers, Salima turned to dog food for sustenance. She suffered verbal and sexual abuse from her employers and their children. One day while Salima was hanging clothes on the line, her employer pushed her out the window, telling her, "You are better off dead." Salina plunged into a swimming pool three floors down and was rescued by police. After a week in the hospital, she returned to Kenya with broken legs and hands.

Katya, a student athlete in an Eastern European capital city, dreamed of learning English and visiting the United States. Her opportunity came in the form of a student visa program, through which international students can work temporarily in the United States. But when she got to America, rather than being taken to a job at a beach resort, the people who met her put her on a bus to Detroit, Michigan. They took her passport away and forced her and her friends to dance in strip clubs for the traffickers' profit. They controlled the girls' movement and travel, kept keys to the girls' apartment, and listened in on phone calls the girls made to their parents. After a year of enslavement, Katya and her friend were able to reach federal authorities with the help of a patron of the strip club in whom they had confided. Due to their bravery, six other victims were identified and rescued. Katya now has immigration status under the U.S. trafficking law. (U.S. Department of State, 2010)

Each of these scenarios represents a common story for many victims of human trafficking. These examples reflect a life experience where they have been manipulated, abused, and exploited. These are but a few examples of the crimes that make up the category of human trafficking.

The Trafficking Victims Protection Act (TVPA) defines severe forms of trafficking as

> sex trafficking in which a commercial act is induced by force, fraud or coercion, or in which the person induced to perform such an act has not attained 18 years of age; or the recruitment, harboring, transportation provision, or obtaining a person for labor or services through the use of force, fraud or coercion for the purpose of subjection to involuntary servitude, peonage, debt bondage, or slavery. (U.S. Department of State, 2009, p. 7)

Human trafficking is the second largest criminal activity and the fastest growing criminal enterprise in the world. Estimates by the United Nations suggest that approximately 2.5 million people from 127 countries are victims of trafficking. Due to the nature of these crimes, it is difficult to determine a precise number of human trafficking victims worldwide. According to data provided by the U.S. State Department, between 600,000 and 820,000 men, women, and children are trafficked across international borders every year. These numbers do not include the thousands, and potentially millions, of individuals who are trafficked within the boundaries of their homelands (U.S. Department of State, 2010).

Trafficking can involve cases within the borders of one's country as well as transport across international boundaries. Thailand is a well-known location for the sexual trafficking of women and girls who migrate from other southeast Asian countries such as Cambodia, Laos, Myanmar (Burma), and Vietnam, as well as other Asian countries such as China and Hong Kong. Others find their way to Thailand from the United Kingdom, South Africa, Czech Republic, Australia, and the United States (Rafferty, 2007). However, examples of trafficking are not limited to countries from the southeast Asian region. The sexual trafficking of women and children is an international phenomenon and can be found in many regions around the world, even in the United States. Between January 2007 and September 2008, there were 1,229 documented incidents[1] of human trafficking in the United States. An astounding 83% of these cases were defined as alleged incidents of sex trafficking, of which 32% involved child sex trafficking and 62% involved adults, such as forced prostitution and other sex crimes (Kyckelhahn, Beck, & Cohen, 2009).

Traffickers use several methods to manipulate women and girls into the sex trade and prey on their poor economic standing and desires for improving their financial status. These enticements include offers of employment, marriage, and travel. Each of these opportunities is a shield to trap women into sexual slavery. In some cases, women may be kidnapped or abducted, although these tactics are rare compared to the majority of cases, which involve lies, deceit, and trickery to collect its victims (Simkhada, 2008). In some cases, young children are recruited by "family friends" or community members or may even be intentionally sold into servitude by their own parents. According to Rafferty (2007),

> Traffickers use a number of coercive methods and psychological manipulations to maintain control over their victims and deprive them of their free will, to render them subservient and dependent by destroying their sense of self and connection to others, and to make their escape virtually impossible by destroying their physical and psychological defenses. The emotional and physical

[1]The Human Trafficking Reporting System (HTRS) is part of the Department of Justice and tracks incidents of suspected human trafficking for which an investigation, arrest, prosecution, or incarceration occurred as a result of a charge related to human trafficking.

> trauma, as well as the degradation associated with being subjected to humiliation and violence, treatment as a commodity, and unrelenting abuse and fear, presents a grave risk to the physical, psychological and social-emotional development of trafficking victims. (p. 410)

Victims are dependent on their traffickers for food, shelter, clothing, and safety. They may be trafficked to a region where they do not speak the language, which limits opportunities to seek assistance. They may be concerned for the safety of their family members, as many traffickers use threats to ensure cooperation (Rafferty, 2007). Girls who are imprisoned in a brothel are often beaten and threatened in order to obtain compliance. They are reminded of their "debts" that they are forced to work off through the sale of their bodies. Most girls have little contact with the world outside the brothel and are unable to see or communicate with the family members that are left behind.

While some girls are able to escape the brothel life on their own, most require the intervention of police or social workers. Girls receive services from "rehabilitation centers," which provide health and social welfare assistance to victims of trafficking. The intent of these agencies is to return girls to their homes; however, many of these girls indicate they experience significant challenges upon return to their communities. Many of these girls are not looked upon as victims, but rather as damaged goods who are shunned and stigmatized not only by society at large, but also by their family members (Simkhada, 2008).

Despite being aware of trafficking as a social issue, many jurisdictions have failed to effectively address the problem in their communities. Much of the intervention efforts against trafficking involve nongovernmental organizations (NGOs), national and international anti-trafficking agencies, and local grassroots organizations. While several countries have adopted legislation that criminalizes the sale and exploitation of human beings, many have yet to enact anti-trafficking laws. In some cases, countries may have laws on the books but have limited resources or priorities for enforcing such laws. Still other countries punish the victims of these crimes, often charging them with crimes such as prostitution when they seek out assistance from the police. While grassroots and anti-trafficking organizations have developed policies and practices designed to punish traffickers and provide assistance to the victims, few of these recommendations have been implemented effectively.

In the United States, legislation knows as the Trafficking Victims Protection Act (TVPA)[2] is designed to punish traffickers, protect victims, and facilitate prevention efforts in the community to fight against human trafficking. Enacted by Congress in 2000, the law provides that traffickers can be sent to prison for up to 20 years for each victim. In 2008, the Department of Justice obtained 77 convictions in 40 cases of human trafficking, with an average sentence of 112 months (9.3 years). Over two thirds of these cases involved acts of sex trafficking. At the state level, 42 states currently have anti-trafficking legislation in their jurisdictions and are active in identifying offenders and victims of these crimes (U.S. Department of State, 2008).

While the TVPA includes protection and assistance for victims, these provisions are limited. For example, victims of trafficking are eligible for a T-visa, which provides a temporary visa. However, there are only 5,000 T-visas available (regardless of the numbers of demand for these visas), and issuance of this type of visa is limited to "severe forms of trafficking (such as) involving force, fraud or coercion or any trafficking involving a minor" (Haynes, 2004, p. 241). In addition, applications for permanent residency are conditional on a victim's participation as a potential witness in a trafficking prosecution. In the 2 years following the implementation of the T-visa program, only 23 visas had been granted, a far cry from the demand given that over 50,000 people are trafficked into the United States alone each year (Oxman-Martinez, 2003). Similar to the TVPA, the European Union policies on trafficking prioritizes the

---

[2]Reauthorized by Congress in December 2008.

prosecution of offenders over the needs of victims, and visas are granted only for the purposes of pursuing charges against the traffickers. In addition, there is no encouragement or pressure by the EU for states to develop programs to address the needs of trafficked victims (Haynes, 2004). While the push to "jail the offender" of these crimes appears positive, the reality is that few prosecutions have succeeded in achieving this task. Even in cases where prosecutions are "successful" and traffickers are held accountable for their crimes, their convictions result in short sentences and small fines, the effect of which does little to deter individuals from participating in these offenses.

In contrast to the prosecution-oriented approach, several international organizations have developed models to fight trafficking that focus on the needs of the victim. These approaches focus on the security and safety of the victim, allow them to regain control over their lives, and empower them to make positive choices for their future while receiving housing and employment assistance. While this approach provides valuable resources for victims, it does little to control and stop the practice of trafficking from continuing.

Given the limitations of the "jail the offender" and "protect the victim" models, research by Haynes (2004) provides several policy recommendations that would combine the best aspects of these two approaches. These recommendations include

1. Protect, don't prosecute the victim—As indicated earlier, many victims find themselves charged with prostitution and other crimes in their attempts to seek help. Not only does this process punish the victim, but it serves to inhibit additional victims from coming forward out of fear that they too might be subjected to criminal punishments. Anti-trafficking legislation needs to ensure that victims will not be prosecuted for the actions in which they engaged as a part of their trafficked status. In addition, victims need to be provided with shelter and care to meet their immediate needs following an escape from their trafficker.

2. Develop community awareness and educational public service campaigns—Many victims of trafficking do not know where to turn for help. An effective media campaign could provide victims with information on how to recognize if they are in an exploitative situation, avenues for assistance such as shelters and safety options, and long-term planning support such as information on immigration. Media campaigns can also help educate the general public on the ways in which traffickers entice their victims and provide information on reporting potential victims to local agencies. Recent examples of prevention efforts in fighting trafficking have included raising public awareness through billboard campaigns; the development of a national hotline to report possible human trafficking cases; and public service announcements in several languages, including English, Spanish, Russian, Korean, and Arabic, to name a few (U.S. Department of State, 2009). These efforts help increase public knowledge about the realities of human trafficking within the community.

3. Address the social and economic reasons for vulnerability to trafficking—The road to trafficking begins with poverty. Economic instability creates vulnerability for women as they migrate from their communities in search of a better life. For many, the migration from their homes to the city places them at risk for traffickers, who seek out these women and promise them employment opportunities, only to hold them against their will for the purposes of forced labor and slavery. Certainly, the road to eradicating poverty around the world is an insurmountable task, but an increased understanding of how and why women leave could inform educational campaigns, which could relay information about the risks and dangers of trafficking and provide viable options for legitimate employment and immigration.

4. Prosecute traffickers and those who aid and abet traffickers—Unfortunately, in many of these jurisdictions, law enforcement and legal agents are subjected to bribery and corruption, which limits the assistance that victims of trafficking may receive. "Police are known to tip off club workers suspected of harboring trafficked women in order to give owners time to hide women or supply false working papers (and) are also known to accept bribes, supply false papers or to turn a blind eye to the presence of undocumented foreigners" (Haynes, 2004, p. 257). In order to effectively address this issue, police and courts need to eliminate corruption from their ranks. In addition, agents of justice need to pursue cases in earnest and address the flaws that exist within the system in order to effectively identify, pursue, and punish the offenders of these crimes.
5. Create immigration solutions for trafficked persons—An effective immigration policy for victims of trafficking serves two purposes: not only does it provide victims with legal residency rights and protections, but it also helps pursue criminal prosecutions against traffickers, especially since the few effective prosecutions have relied heavily on victim cooperation and testimony. At its most fundamental position, victims who are unable to obtain even temporary visas will be unable to legally remain in the country and assist the courts in bringing perpetrators to justice. In addition, victims who are offered immigration visas contingent upon their participation in a prosecution run the risk of jeopardizing potential convictions, as defense attorneys may argue that the promise of residency could encourage an "alleged" victim to perjure his or her testimony. Finally, the limited opportunities to obtain permanent visa status amount to winning the immigration lottery in many cases, as these opportunities are few and far between and often involve complex applications and long waiting periods.
6. Implement the laws—At the end of the day, policy recommendations and legislation do little good if such laws are not vigorously pursued and enforced against individuals and groups participating in the trafficking of humans. In addition, such convictions need to carry stern and significant financial and incarceration punishments if they hope to be an effective tool in solving the problem of trafficking.

While efforts to prioritize the implementation of anti-trafficking laws may slow the progress of these crimes against humanity, the best efforts toward prevention focus on eliminating the need for people to migrate in search of opportunities to improve their economic condition. An ecological perspective suggests that the cause of trafficking lies within issues such as poverty, economic inequality, dysfunction within the family, gender inequality, discrimination, and the demand for victims for prostitution and cheap labor. At its heart, human trafficking "is a crime that deprives people of their human rights and freedoms, increases global health risks, fuels growing networks of organized crime and can sustain levels of poverty and impede development in certain areas" (U.S. Department of State, 2009, p. 5). Until these large-scale systemic issues are addressed, the presence of trafficking will endure within our global society.

The Women of Juárez

The Mexican city of Ciudad Juárez sits across the Rio Grande from El Paso, Texas. A fast-growing industrial area, the region is known as a major manufacturing center for many American companies. With more than three hundred assembly plants (known as maquiladoras) in the region, Ciudad Juárez is a booming area for production. Since the 1994 passing of NAFTA (North American Free Trade Agreement), U.S. corporations such as Ford, General Electric, and DuPont (to name a few) have established manufacturing centers in a

▲ **Photo 6.1** A field of crosses stand today in the deserts of Ciudad Juarez, Mexico, where hundreds of bodies of women have been found. Local and international organizations estimate that thousands of other women have gone missing and have yet to be found.

region where labor costs are cheap and taxes are low, which result in high profits for companies. With four separate border access points, the region is a major center for exporting goods and transportation between Central Northern Mexico and the United States (Chamberlain, 2007). Awarded the "City of the Future" designation by *fDi* magazine and the Financial Times Group in 2008, Ciudad Juárez represents a region of opportunity and development. "It appears to be a win-win situation for the United States. Americans enjoy relatively inexpensive consumer goods, and American-owned corporations enjoy the free aspect of the free trade zone: it is free of unions, minimum wages and largely free of enforceable regulations" (Spencer, 2004, p. 505).

However, this "City of the Future" is also filled with extreme poverty. Drawn to the region with the promise of a better life, Mexican citizens arrive from rural towns only to discover a new form of economic disparity in border towns such as Juárez. While filled with factories, the city receives few benefits to stimulate its economy. The maquiladoras generated over 10 billion dollars of profit for these companies in 2000, yet the city of Juárez received less than $1.5 million in taxes to provide a sustainable community structure for residents. Shantytowns surround the maquiladoras, as there are few options for housing for the workers. There is no money to build schools or provide services to the residents of the city (Spencer, 2004). The high profit margins for companies come at a price for the maquiladores' workforce, where women make up more than 80% of the workforce and where cases of poor working conditions, low wages ($60 a week for their labor), and traumatic work environments have been documented (Althaus, 2010).

Mexican border towns are also known for their high levels of violence and narcotics trafficking. Today, Ciudad Juárez is considered one of the most dangerous cities in the world due to violent feuds between the drug cartels and police. Juárez is also dangerous for one particular population: young women. Since 1993, estimates suggest that over 400 women have been murdered in and around the city. While some of these girls are students who disappeared as they traveled to and from school, the majority of femicides in this region involve young women between the ages of 11 and 24 who traveled from their villages to Ciudad Juárez looking for work in the maquiladoras. Their bodies are discovered days, weeks, and months following their disappearance and are typically abandoned in vacant lots in Juárez and the surrounding areas. Many of these cases involve significant acts of sexual torture, including rape and the slashing of the breasts and genitals of the female victims (Newton, 2003). The women who are killed and tortured in this fashion become members of a club known as "las muertas de Juárez," or the Dead Women of Juárez.

In describing the murders of these women, several commentaries have pointed toward a clash between the traditional roles for women, a "machista" (chauvinistic) culture, and the rise of women's independence as an explanation for the violence. One author suggests that "these crimes are more murderous than murder, if such a thing is possible—they are crimes of such intense hatred that they seek to destroy the personhood of the women, negating their humanity and erasing their existence" (Revolutionary Worker, 2002). According to a 2003 report by the Inter-American Commission on Human Rights (IACHR), the crimes against women in Ciudad Juárez have received international attention due to the extreme levels of violence in the murders and the belief that these killings may have been the result of a serial killer. Their research indicates that these cases of femicide are part of a larger social issue related to a pattern of gender-based discrimination where the violence against women is not considered to be a serious issue. Given the relationship with gender in these cases, any official response to address these crimes must consider the larger social context of crimes against women and the accessibility of justice for women in these cases.

In searching to solve these crimes, police have jailed dozens of suspects for the murders throughout the years. Some of these presumed offenders were railroaded by a system desperate to quash an inquisition into police practices. Many of these alleged perpetrators had their confessions coerced from them. Some argue that the authorities have shown little concern for these crimes and its victims, sending a message that these women are unworthy victims. Indeed, victim-blaming tactics have often been used to explain the murders, suggesting that these women wore revealing clothing, frequented bars and dance clubs, and were prostitutes. The National Human Rights Commission has found that the "judicial, state and municipal authorities were guilty of negligence and dereliction of duty" (Agosin, p. 16).

The quest for justice by journalists, social activists, and the families of these young women has been a challenging road, as many of them have been threatened with violence if they continue their investigations. Others have been silenced due to the inaction by authorities. Given the poor treatment of victims' family members by the authorities, recent improvements have been made in the areas of legal, psychological, and social services. However, there is concern that there are limited funds allocated to meet the demands for these services (Inter-American Commission on Human Rights, 2003). The Mexican government has also created a victim services fund designed to provide monetary compensation to the families of the women and girls who have been murdered in Juárez. However, the program is poorly organized, and few families have been able to access the funds (Calderon Gamboa, 2007).

While the creation of a special prosecutor's office in 1998 did little to end the killings in Juárez, improvements have been made in recent times regarding the organization of evidence, the tracking of case details, and the streamlining of investigations and assignment of personnel. While Mexican authorities claim they have resolved the majority of the murders, their definition of "resolved" is based on a presumption of motive and the identification of a perpetrator and does not require that an offender be charged, tried, or convicted of the crime. Understandably, many families are dissatisfied by this definition of "resolved." As of 2003, only three convictions have been handed down, and the community has little faith in the validity of these convictions (Simmons, 2006). Many human rights and activist groups have blamed the Mexican government for the inadequate investigation of these crimes and the lack of accountability by police agents. Indeed, Mexico's failure to act in these cases constitutes a violation of international laws such as the American Convention of Human Rights and the Inter-American Convention on the Prevention, Punishment and Eradication of Violence Against Women (Calderon Gamboa, 2007).

While it is unclear whether the victims' families will ever receive closure in the deaths of their loved ones, human rights organizations have called for a systematic reform of conditions to ensure the future safety of women in Juárez. Their suggestions are presented within a framework designed to mend the cultural systems that historically have minimized the traumas of female victimizations. A key component of reform includes addressing the root causes of these murders by eliminating the machista culture that is prevalent in these communities. Suggestions include increasing employment opportunities for males in the maquiladores' labor force, gender-sensitivity training for the workplace, and the creation of safe public spaces for women to gather in and travel to, from, and within the city of Juárez (Calderon Gamboa, 2007).

Summary

- Honor-based violence involves the murder of women for violating gendered cultural norms. Death is required in order to restore the honor of a woman's family in the community.
- Women who have been killed in the name of honor are typically killed for acts of sexual impropriety, alleged infidelity, or for becoming too "Westernized."
- Most incidents of honor-based violence are committed by a male family member, such as a father, husband, brother, or cousin.
- Even though some jurisdictions have laws against the practice, offenders of honor-based violence are rarely punished, as the killings are an accepted practice within the communities.
- Efforts toward reducing or eliminating honor-based violence include legal reform, education and public awareness, and additional resources for victim services.
- Human trafficking involves the exploitation of individuals for the purposes of sexual slavery, domestic servitude, and forced labor. The majority of these cases involve sexual exploitation and abuse.
- Human trafficking is an international phenomenon, with hundreds of thousands of victims trafficked within and across borders every year.
- Traffickers prey on women from poor communities and appeal to their interests in improving their economic standing as a method of enticing them into exploitative and manipulative work environments.
- Many countries have failed to effectively address the problem of trafficking in their communities, where the majority of offenders go undetected. In the rare cases where a trafficker is brought to justice, their punishments rarely involve significant incarceration or financial penalties.
- Recommendations for best-practices against trafficking involve combining features from "jail the offender" and "protect the victim" models, seeking to improve victim services, increase public awareness about trafficking, and implement and enforce stricter laws against the practice.
- The pattern of femicides in Ciudad Juárez, Mexico, involves the violent rape and torture of hundreds of women and girls who travel from their homes in search of work in the factories where they endure poor working conditions and are paid little for their labor.
- There have been few convictions for these crimes, and most of the cases are unresolved. Victim blaming is a common practice, and many families and victim-rights groups have protested over the lack of attention paid to these incidents.
- Some scholars point toward the clash between the traditional roles of women, the machista culture, and the rise of women's independence as an explanation for the violence.

KEY TERMS

Maquiladores

Femicide

Machista

Las muertas de Juárez/Dead women of Juárez

Honor-based violence

"Westernized"

Karo-kari

Human trafficking

Sexual slavery

Trafficking Victims Protection Act of 2000

Jail the offender vs. protect the victim

QUESTIONS FOR DISCUSSION

1. How is the pattern of femicides in Ciudad Juárez linked to larger social issues such as patriarchy, masculine identity, and the entry of women into the workforce?
2. What steps have local and national agencies in Mexico taken to solve the cases of "las muertas de Juárez"? How have these efforts failed the families of the victims?
3. How is the concept of shame created in cultures where honor-based violence is prevalent?
4. To what extent are offenders in honor-based violence cases punished? What measures need to be implemented to protect women from these crimes?
5. How do women enter and exit the experience of sexual trafficking?
6. Compare and contrast the "jail the offender" and the "protect the victim" models of trafficking enforcement. What are the best practices that can be implemented from these two models to address the needs of trafficking victims?

WEB RESOURCES

Trafficking in Persons Report: http://www.state.gov/g/tip/rls/tiprpt/2010/

HumanTrafficking.org: http://www.humantrafficking.org

Polaris Project: http://www.polarisproject.org/

Not for Sale: http://www.notforsalecampaign.org/about/slavery/

Stop Honour Killings: http://www.stophonourkillings.com/

Women of Juárez: http://womenofjuarez.egenerica.com/

READING

This reading expands on what you learned in the section introduction on honor killings. Karo-kari is a particular form of honor killing that occurs in the rural and tribal areas of Sindh, Pakistan. This reading explains the practices of karo-kari and how issues of culture place women at risk for victimization.

Karo-Kari

A Form of Honour Killing in Pakistan

Sujay Patel and Amin Muhammad Gadit

Introduction

While generally categorized as unlawful, homicide has been justified under particular circumstances by some social and cultural groups. This includes the cultural sanctioning of premeditated killings of women perceived to have brought dishonour to their families, often by engaging in illicit relations with men. This violence exhibits a strong gender bias in that, in such settings, men who engage in similar behavior are typically subject to less severe punishments.

There is evidence of legal or cultural sanction for such practices in a number of ancient societies, including Babylon—where the predominant view was that a woman's virginity belonged to her family (Goldstein, 2002), ancient South- and Meso-American civilizations, and the Roman Empire. Incan laws permitted husbands to starve their wives to death as punishment for committing adultery and Aztec legal codes meted out death by stoning or strangulation for female adultery (Gardner, 1986). In Ancient Rome, the senior male within a household retained the right to kill a related woman who engaged in pre-marital or extra-marital relations (Goldstein, 2002).

Over the past decade, human rights groups have increasingly exposed various forms of gender-biased 'honour killing'. A number of countries have legislative positions that allow for partial or complete criminal defence against criminal charges on the basis of honour killing, including those of Argentina, Bangladesh, Ecuador, Guatemala, Turkey, Jordan, Syria, Egypt, Lebanon, Iran, Israel, Peru, Venezuela and the Palestinian National Authority (UNCHR, 2002). Honour killings have continued to occur in countries where they have been explicitly outlawed, such as Albania, Brazil, India, Iraq, Uganda and Morocco, as well as in immigrant communities in Europe and North America (UNCHR, 2002).

In Pakistan, honour killings were recently criminalized, but continue to occur frequently in many communities, particularly in four tribal regions of the country: Punjab, NWFP (North West Frontier Province), Baluchistan, and Sindh. The respective names given to the practice of honour killings in these regions are *kala-kali* (Punjab), *tor-tora* (NWFP), *siyahkari* (Baluchistan) and *karo-kari* (Sindh) (Malik, Saleem, & Hamdani, 2001). Although similar, each of these regional practices has a unique set of characteristics that sets them apart from one another. This review paper will focus on karo-kari, the most common type of honour killing in Pakistan.

'Karo-kari' is a compound word, which means 'black male' and 'black female', respectively, metaphoric

SOURCE: Patel, S., & Gadit, A. M. (2008). Karo-kari: A form of honour killing in Pakistan. *Transcultural Psychiatry, 45*(4), 683–694.

terms for those who commit illicit premarital or extramarital relations. A female is labelled a Kari because of the perceived dishonour that she had brought to her family through her illicit relationship with a man (other than her husband) who is subsequently labelled a *karo*. Once labelled a *kari*, male family members have the self-authorized justification to kill her and the co-accused *Karo* in order to restore family honour. Because men more commonly have access to economic resources, allowing them to either flee or buy a pardon from the dishonoured family, they are less often killed in such crimes of honour.

Given the paucity of information on this topic and the apparent spread of such violent acts, research in this area is crucial. By examining the motives underlying Karo-Kari and its epidemiological trends, we hope to gain insight into the relative roles of socio-cultural attitudes and psychopathology in shaping such homicidal practices.

Origin of Karo-Kari

The practice of karo-kari dates to the pre-Islamic period when Arab settlers occupied a region adjacent to Sindh, which was known as Baluchistan (Malik et al., 2001). These early settlers had strongly patriarchal traditions, with such practices as the live burial of unwanted newborn daughters. It is likely that karo-kari originated in, or was facilitated by, the subordination of women underlying these cultural practices. Such gender norms have become deeply entrenched in the social psyche of Sindh, leading to the preservation of karo-kari in the feudal social structure of local tribal communities. Although karo-kari, as a practice and term, is specific to Sindh, the general concept of honour has been independently described in many parts of the world.

Motives for Karo-Kari

The most commonly cited warrant for a karo-kari act is a woman's premarital or extra-marital relations with any man who is not her husband. This may include any form of relationship, regardless of whether it is coerced—as in the case of sexual assault—or consensual—as in the case of women who choose romantic partners without their family's approval. Perceptions of what constitutes dishonour to the family have broadened, with some honour killings linked to situations such as pregnancy out of wedlock, sexual misconduct, and marriage against family approval (Kulwicki, 2002).

In recent years, the label of karo-kari has been used to mask killings that likely occurred for reasons other than restoring family honour, in what have been called 'fake honour killings' (Amnesty International, 1999a). An example is the use of karo-kari as a camouflage for men who murder other men in personal disputes. The tradition has been manipulated by tribes to settle rivalries and vendettas; in Pakistan this was the case in the well-publicized story of Mukhtar Mai (Husain, 2006).

Women who demand divorces from their husbands may elicit a karo-kari attack under a false pretext. By categorizing these homicides as acts of karo-kari, men obtain the customary endorsement for their actions and avoid retribution. Another type of fictitious honour killing occurs in the poorer communities of Sindh, especially when a woman is felt to have become a financial burden on the household. These communities sometimes use karo-kari as a convenient way of acquiring wealth or land by declaring a woman of their household a kari. This allows the family to obtain the victim's share of inheritance, as well as appropriate compensation from the co-accused karo of their choice. Other financially motivated killings represented as karo-kari are often connected to marriages arranged within a family to retain property, or attempts to prevent a widow or divorced mother from remarrying in order to avoid the transfer of wealth to another family.

Epidemiology of Karo-Kari

The United Nations Population Fund estimates that at least 5,000 women worldwide are victims of honour killings each year (UNCHR, 2002). Many of these homicides occur in Pakistan. This figure is probably an underestimate since many cases of honour killing go unreported, especially in patriarchal societies that sanction this practice. Additionally, in countries where the practice of honour killing is outlawed but supported by portions of the population, communities attempt to cover up such acts to avoid legal authorities from disrupting a sacred cultural practice.

Although several agencies in Pakistan have attempted to quantify the national incidence rate of honour killings and karo-kari acts, these numbers are also likely to underestimate the actual incidence levels. The inaccuracy of these reported rates is reflected in the wide discrepancies between figures provided by various agencies. For instance, while the Human Rights Commission of Pakistan reports 1,464 honour killings occurred in Pakistan between 1998 and 2002 (HRCP, 2004), the Pakistan government reports 4,101 registered honour killings between 1998 and 2003 (HRCP, 2004). The Madadgaar helpline database gives a figure of 3,339 honour killings between 2000 and 2004 (Awan, 2004) and, finally a police report indicated that a total of 4,383 honour killings were reported in Pakistan between 2001 and 2004, with 2,228 of these occurring in Sindh (The Daily Jang, 2006).

It is unclear what proportion of these honour killings were of the karo-kari type. However, during the few years that Human Rights Commission of Pakistan has enumerated separate statistics for karo-kari, such deaths have accounted for the majority of honour killings in Pakistan. These data indicate that the total number of karo-kari deaths since 2004 is 416 (see Table 6.1).

Table 6.1 Characteristics of Homicides, Honour Killings, and Karo-Kari Deaths in Pakistan Between 2004 and Mid-2006 (Adapted from the Human Rights Commission of Pakistan)

	Year			
	2004	**2005**	**2006 (to June)**	**Total**
Homicides				
Total homicides	624	229	172	1025
Honour killings	302	152	114	568
Karo-kari deaths	277	113	26	416
Perpetrators of karo-kari				
Father	7	5	1	13
Brother	44	13	4	61
Husband	112	52	13	177
In-law	19	4	1	24
Relative	27	9	2	38
Son	5	4	0	9
Non-family	2	0	2	4
Action against perpetrator				
Accused of karo-kari	365	195	66	626
Arrested	68	36	6	110

Profile of Karo-Kari Victims, Perpetrators, and Accomplices

Victims of Karo-Kari are most often married adult females (HRCP, 2006). Nevertheless, those who are single or male may also be affected, and at any age. The same combination of economic vulnerability, limited social support and lack of knowledge regarding their legal rights, which prevent women from changing their subordinate status in society, put them at a higher risk for experiencing violence. Thus women who are unemployed, illiterate and live in impoverished conditions have a higher risk than others of becoming a karo-kari victim (Niaz, 2001).

The psychological burden placed on females in a patriarchal society is evident in the high prevalence of mental illness among women in Pakistan. Hence, it is possible that many victims of honour killings may have suffered from mental illness. Moreover, it is inevitable that as soon as a woman is labelled a kari she will endure significant psychological distress, which may even lead her to commit suicide prior to the inevitable homicide.

Almost all perpetrators of karo-kari are male family members, most commonly husbands, followed by brothers (Table 6.1). When possible, families choose males under the age of eighteen to carry out the murder, most likely because juvenile offenders serve the shortest prison terms. However, karo-kari may also be linked to the intense competition for honour, status and marriage ability among young men (Cohen, 1998). Other factors which may lead certain men to commit honour killings include socio-economic disadvantage and the pressure to fulfill the gender norms of their culture. Specifically, men who find themselves unable to meet their gender role expectations, may resort to violence against women in order to assert their masculinity or as an outlet for frustration (Krishnan, 2005).

In countries where honour killings are illegal, some lawyers have employed the 'honour defence' as an exculpatory strategy for perpetrators. Underlying this defence are the ideas that women are considered the property of men and that the protection of honour is a type of self-defence. Other perpetrators have attempted to use the 'temporary insanity' or 'crime of passion' defence, arguing that their homicidal acts were not premeditated but caused by extreme provocation which led to psychological distress, and subsequent loss of self-control and impaired judgment (Yale Law Journal, 1934; Can LII, 2006).

To date no studies have examined the presence of a psychopathological process leading perpetrators to commit karo-kari. However, many of the case reports of karo-kari reviewed indicated the presence of certain psychopathic traits in perpetrators, including a reckless disregard for the safety of women, a failure to conform to lawful behaviours and a lack of remorse. The fact that some have used the honour killing tradition to conceal other motives for their homicidal actions lends weight to the argument for an underlying psychopathic process. The majority of karo-kari killings are quite violent, most commonly committed using firearms. Other methods include stabbing, strangulation, hanging, electrocution or poisoning (HRCP, 2006). Finally, the fact that most perpetrators attempt to cover up their actions suggests that they have preserved insight into the criminal nature of karo-kari, although they believe that restoring family honour is more important.

The patriarchal cultural values prevailing in Pakistan have often led victims' relatives and community members, as well as legal and government authorities to act as explicit or implicit accomplices in karo-kari deaths. Because they view karo-kari as a culturally acceptable and even heroic act, many family and community members help to cover up such homicides by maintaining silence during investigations. Others keep silent because they fear retribution. Tribal courts have been known to impose death sentences on those who report honour killings to the police (*The Daily Times,* Pakistan, 2006).

Legal and government authorities often contribute to covering up karo-kari deaths simply by avoiding any involvement in such cases. Even when they do become involved, gender discrimination continues to create support for the perpetrator. Corruption among government authorities can result in a further disadvantage for females, who are less likely to have access to monetary resources or social capital. The small percentage of accused perpetrators who are arrested (Table 6.1) illustrates the lack of legal intervention in cases of karo-kari.

Socio-Cultural Influences and Karo-Kari

Some individuals and communities in Pakistan have maintained traditional patriarchal interpretations of Islam, which valorize female chastity and male superiority. The power dynamics of patriarchy tend to reduce women to their reproductive potential, in the process denying them agency as human beings.

In some communities women are considered to have monetary value and are the property of their male relatives. It is thought that the preservation of a woman's chastity and fidelity, through segregation and control, is the responsibility of the men to whom she belongs. By engaging in an illicit affair a woman comes into conflict with the socio-cultural framework of meanings prevalent in much of Pakistan, and is seen as having seriously violated the honour of her family. Some believe that in such cases, it is a man's duty to restore his family's honour by killing those who damaged it. Both the notion of honour and the concept of women as property are deeply woven into the socio-cultural fabric of Pakistan, leading many individuals, including women, to support the practice of honour killing, and others, such as legal authorities, to look the other way. Such widely-held values have also given perpetrators an excellent legal defence when they commit such acts.

Some national legal codes explicitly allow for 'honour killing'. For example, in Jordan, part of article 340 of the penal code states that 'he who discovers his wife or one of his female relatives committing adultery and kills, wounds, or injures one of them, is exempted from penalty'. Despite efforts to rewrite the law, the article was retained by parliament. Similar provisions appear in the penal codes of Syria (article 548), Morocco (article 418) and Haiti (article 269). On the other hand, persons found guilty of honour killings in Turkey are sentenced to life in prison (UNCHR, 2002).

In Sindh (Pakistan), many tribal communities employ an informal legal system based on feudal principles such as forced domestic labour and custodial violence. Despite being unrecognized by Pakistan's formal legal system, such tribal courts adjudicate the majority of karo-kari cases which obtain any legal hearing. Such courts generally sanction honour killings, and they are preferred by individuals in tribal communities because they provide inexpensive and expeditious access to justice.

Pakistan's formal justice system includes laws which seem to endorse the gender-biased rulings of tribal courts. For instance, the 1979 'Hudood Ordinance' criminalized extra-marital sex by females. The law also made it difficult for a woman to prove an allegation of rape by requiring that at least four adult male Muslim witnesses of good character attest to the act of sexual penetration. Furthermore, a male suspect had the opportunity to testify against the woman in court. A failure to prove the act of rape placed the woman at risk of prosecution for adultery, the punishment for which was death by stoning. Not only did this law play a role in giving legal sanction to women's subordinate social status, it gave implicit legal and cultural justification for the practice of karo-kari. Fundamentalist social forces in Pakistan strongly opposed any change in the law as they falsely believed it to be in accordance with the teachings of Islam.

In December 2004, under international and domestic pressure, the Pakistani government enacted a law that made the practice of karo-kari punishable by death, in the same manner as other homicides. On the other hand, in March 2005, the Pakistani government allied with opposition conservative Islamic parties to reject a bill, which would have strengthened the law against the practice of karo-kari. Finally, in November 2006 the 'Hudood Ordinance' was abolished and replaced by the new 'Women's Protection Bill' (Government of Pakistan, 2006). Despite these new laws, socio-cultural patterns and feudal attitudes remain largely unchanged: many Pakistanis feel that karo-kari homicides are justifiable, and therefore deserve legal pardon. As a result, perpetrators are rarely brought to justice. The few cases that go to court are usually plagued by gender discrimination resulting in lenient sentences or pardons for men. Also, because karo-kari is a crime of retaliation, judges have the option at their own discretion to allow victim's families to accept a simple apology, money, land or another female from the perpetrator as compensation for the crime. These factors make it almost inevitable that perpetrators will escape a severe punishment for such homicides.

Mental Health and Karo-Kari

While cultural assumptions about family relationships, the meaning of honour and appropriate gender roles clearly shape the accepting view many members of tribal communities take toward karo-kari, many other individuals with similar backgrounds do not take part in or endorse such acts. Thus, it is important to also consider the role of psychopathy in the perpetrators of karo-kari.

With respect to the victims, there is global consensus among researchers on the fact that oppression and violence not only violate women's basic rights but also threaten their health and the very state of their being. Patriarchal values, which greatly contribute to the persistence of karo-kari, often result in domestic disharmony and have an adverse psychological impact on many women in Pakistan (Niaz, 2004). The frequency and unexpectedness of karo-kari killings also contribute to the experience of uncertainty and fear among Pakistani women which, along with a lack of autonomy and equal opportunities, has the potential to erode their self-esteem, and thereby increase their risk for developing a variety of psychiatric disorders such as depression and anxiety.

Studies suggest that domestic strife is a primary cause of psychiatric illness in Pakistani women. A five-year study at the University Psychiatry Department in Karachi showed that out of 212 patients receiving psychotherapy, 65% were women who presented primarily due to conflicts with their spouse and in-laws (Zaman, in press). Interestingly, 50% of these women had no psychiatric diagnosis and were labelled 'distressed women' while 28% suffered from depression or anxiety. Additionally, a four-year study of psychiatric outpatients at a private clinic in Karachi found that 66% of the patients were females, of whom 60% had a mood disorder, 70% were victims of violence, and 80% struggled with domestic conflicts (Niaz, 1994).

Some women who are labelled 'kari' commit 'honour suicides' because of the shame they experience from committing a dishonourable act or because they fear being brutally attacked (Amnesty International, 1999b). Other karo-kari deaths categorized as suicides include cases in which a woman is forced to kill herself or is secretly poisoned. Such cases may partly account for the high rates of suicides among married Pakistani women reported in studies (Niaz, 1997).

In rural Sindh, there are rare instances when a woman manages to escape a karo-kari death. Although these women encounter a lack of social support, they do have the option to seek refuge in the home of a tribal holy man, known as a 'wadero'. The wadero will protect the woman for the rest of her lifetime as long as she obliges with his conditions, which may include acting as his unpaid servant. She may also be exposed to significant emotional and physical abuse by the wadero. As a result, even such 'safe havens' may have a profoundly negative psychological impact on women who escape from karo-kari.

An additional psychological burden is endured by children who witness domestic conflicts and karo-kari acts. These children face an increased risk for behavioural problems, substance abuse, anxiety and depression.

Conclusion

Mental health clinicians can play a number of important roles in dealing with psychiatric issues associated with karo-kari. At the clinical level it is important to take careful histories from women presenting to clinics in areas where 'honour killing' is common. Mental health practitioners can report concerns about a patient's safety to law enforcement agencies. While it may not be safe for medical professionals to publicly denounce karo-kari practices in these areas, they can become part of campaigns to raise awareness and educate the public on such issues. Unfortunately, concerns for personal security have generally kept many mental health professionals from becoming involved in such efforts.

Additional research on the relationship between culture and psychopathology in karo-kari will allow clinicians to identify persons at risk and to help potential victims and their families. Many studies have shown that the incidence rate of mental disorders among homicide offenders can be as high as 90% (Fazel & Grann, 2004). Better understanding of the sociocultural context that leads to karo-kari, would allow clinicians to intervene early when patients at risk for a homicide present with domestic disharmony. In the event of a killing, clinicians can also play a role in managing the psychological sequelae of karo-kari on the victim's families, as well as other community members. Finally, clinicians must consider the possible role of psychopathy, along with sociocultural influences, as factors that may predispose perpetrators to commit karo-kari acts.

References

Amnesty International. (1999a). *Pakistan: Honour killings of girls and women* (September). Report No.: ASA 33/18/99.

Amnesty International. (1999b). *Pakistan: Violence against women in the name of honour.* AI Index: ASA 33/17/99.

Awan, Z. (2004). *Violence against women and impediments in access to justice.* World Bank.

Canadian Legal Information Institute. Review of Judicial Processes. Availableat:http://www.canlii.org/on/cas/onca/2006/2006onca10275.html. Accessed: 10 November 2007.

Cohen, D. (1998). Culture, social organization, and patterns of violence. *Journal of Personality and Social Psychology, 75*, 408–419.

Fazel, S., & Grann, M. (2004). Psychiatric morbidity among homicide offenders: A Swedish population study. *American Journal of Psychiatry, 161*: 2129–2131.

Gardner, J. (1986). *Women in Roman law and society*. Bloomington, IN: Indiana University Press.

Goldstein, M. (2002). The biological roots of heat-of-passion crimes and honour killings. *Politics and the Life Sciences. 21*: 28–37.

Government of Pakistan (2006). Protection of Women (Criminal Amendment) Act, 2006. Available at: www.pakistani.org/pakistan/legislation/2006/wp6.html. Accessed: 18 July 2008.

Human Rights Commission of Pakistan (HRCP) (2004). Honour Killings. Available at: http://www.hrcp-web.org/Women.cfm. Accessed: 6 December 2007.

Husain, M. (2006). 'Take my riches, give me justice': A contextual analysis of Pakistan's honour crimes legislation. *Harvard Journal of Law & Gender, 29*, 221–246.

Krishnan, S. (2005). Do structural inequalities contribute to marital violence? Ethnographic evidence from rural South India. *Violence against Women, 11*, 759–775.

Kulwicki, A. D. (2002). The practice of honour crimes: A glimpse of domestic violence in the Arab world. *Issues in Mental Health Nursing, 23*, 77–87.

Malik, N., Saleem, I., & Hamdani, I. (2001). *Karo Kari, TorTora, Siyahkari, Kala Kali: There is no honour in killing.* National Seminar Report, findings from the Shirkat Gah 'Women, Law and Status programme' involving broad based and systematic research into honour crimes in Punjab, North Western Frontier Province and Sindh; November 25; Lahore, Pakistan. WLUML.

Niaz, U. (1994). Human rights abuse in family. *Journal of Pakistan Association of Women's Studies, 3*, 33–41.

Niaz, U. (1997). Contemporary issues of Pakistani women: A psychosocial perspective. *Journal of Pakistan Association of Women's Studies, 6*, 29–50.

Niaz, U. (2001). Overview of women's mental health in Pakistan. *Pakistan Journal of Medical Science, 17*, 203–209.

Niaz, U. (2004). Women's mental health in Pakistan. *World Psychiatry, 3*: 60–62.

Sev'er, A., & Yurdakul, G. (2001). Culture of honor, culture of change: A feminist analysis of honor killings in rural Turkey. *Violence Against Women, 7*, 964–998.

The Daily Jang, Pakistan (2006). 'Honour Killings'. Available at: http://www.jang.com.pk/thenews/nov2005-daily/27-11-2005/metro/k1.htm. Accessed: 2 February 2007.

The Daily Times, Pakistan (2006). 'Honour Killings'. Available at: http://www.dailytimes.com.pk/default.asp?page=2006%5C04%5C29%5Cstory_29-4-2006_pg7_1. Accessed: 5 May 2007.

The Yale Law Journal Company, Inc. (1934). Recognition of the honour defence under the insanity plea. *Yale Law Journal, 43*, 809–814.

United Nations Commission on Human Rights (2002). *Working towards the elimination of crimes against women committed in the name of honour.* 57th session of the United Nations Commission on Human Rights, United Nations, Report No.: 0246790.

United Nations Commission on Human Rights, (2002) 58th session 'Cultural practices in the family that are violent towards women' E/CN. 4/2002/83.

Zaman, R. (in press). Karachi University Psychology Department: Five-year survey (1992–1996).

DISCUSSION QUESTIONS

1. How is the tradition of karo-kari different than other forms of honor killing?
2. What role does patriarchy play in the continuing practice of karo-kari?
3. How can mental health practitioners help in communities where the practice of karo-kari is common?

READING

As you read in the section introduction, sex trafficking is a worldwide phenomenon. The following reading highlights the issues for women and girls in Nepal and their experiences in the sex trade. The article focuses on how women become trafficked (either voluntary or involuntary) as a result of the economic and social inequality that exists in their lives. The reading concludes with suggestions for prevention and intervention efforts to fight against trafficking and the treatment of women who become involved in the sex trade.

Life Histories and Survival Strategies Amongst Sexually Trafficked Girls in Nepal

Padam Simkhada

Introduction

The United Nations Protocol on Trafficking in Persons (UN, 2000) recognises human trafficking as a modern form of slavery and forced labour that relies on coercion, fraud or abduction. Trafficking in persons, especially women and children, is globally prevalent and a major international health and human rights concern. Globally, it is estimated that between 70 0000 (US Department of State, 2001) and four million (UNFPA, 2000) people are trafficked each year, the large differential in estimated numbers reflecting the difficulty in obtaining accurate data. Asia is seen as the most vulnerable region for human trafficking because of its huge population, growing urbanisation, lack of sustainable livelihoods and poverty (Asha-Nepal, 2006; Huda, 2006; Kamala Kampado and others, 2005).

India is a major destination country for sex-trafficked girls (Human Rights Watch, 1995; US Department of State, 2005) with large numbers of Nepalese, Bangladeshi and rural Indian females trafficked to Indian cities, particularly Mumbai (Bombay) (Nair, 2004). There is no accurate figure of the numbers trafficked; the International Labour Organization (ILO) estimates that 12 000 women and children are trafficked every year from Nepal (LLO/LPEC, 2002), whilst some non-governmental organisations (NGOs) give estimates as high as 30 000. Over 20 0000 Nepali girls are working in the sex industry in India (O'Dea, 1993)

Trafficking has been identified as a priority issue in Nepal since the early 1990s and many NGOs, community-based organisations and Government Ministries have developed social, cultural and economic programmes to address it. However, the lack of communication and coordination, duplication and competition amongst NGOs limits opportunities for good practice (Asha-Nepal 2006). Many preventative activities in Nepal are financed by donors willing only to support activities with specific objectives over a limited period of time. The international donor community has increased funding for related social issues, including women's and child welfare issues, bonded labour and human rights.

SOURCE: Simkhada, P. (2008). Life histories and survival strategies amongst sexually trafficked girls in Nepal. *Children and Society, 22*, 235–248.

There remains a need for conceptual clarity on the context and process of sex trafficking.

Many girls who become involved in sex work in Nepal do so because they are compelled by economic circumstances and social inequality. Some enter sex work voluntarily, others do so by force or deception, potentially involving migration across international borders. Nepalese girls who become involved in sex work via trafficking are the focus of this article. The overall aim of this study was to increase our understanding of the context of girl trafficking from Nepal to India for sex work. More specific objectives were to investigate: (i) the context of trafficking; (ii) the methods and means of trafficking; (iii) living conditions in brothels; and (iv) survival strategies amongst trafficked girls. In-depth qualitative interviews were used to identify the context and survival strategies of the study population. This was most appropriate given the exploratory nature of this research amongst an understudied population subgroup (Pope and Mays, 2006) and to provide rich, in-depth information about the experiences of individuals (DiCicco-Bloom and Crabtree, 2006). Young girls who have been trafficked for sex work are a hidden population, largely due to its illegal nature. Employers of trafficked girls may keep them hidden from public view and limit contacts with outsiders. Trafficked girls may not identify themselves as such through fear of reprisals from their employers, fear of social stigma from involvement in sex work or their HIV-positive status or from their activities being revealed to family members. Therefore, identifying trafficked girls and obtaining access to them for interviews is problematic. It is only once these trafficked girls have been identified through health workers, judicial institutions, NGOs and aid organisations that they can be identified. Any interview with trafficked girls is therefore likely to be 'retrospective', accessing formerly trafficked girls in transit homes, rehabilitation centres or in their communities of origin after return. The target population for this research was therefore girls trafficked to India and subsequently returned to Nepal.

Seven in-depth interviews with key informants were conducted to provide a broader understanding of the context surrounding trafficking in Nepal and to discuss access issues. These informants included directors of NGOs working on trafficking issues, co-ordinators of rehabilitation centres for trafficked girls and health workers whose clientele include former trafficked girls. The second stage of data collection involved in-depth interviews with 42 girls trafficked to India for sex work but who had since returned to Nepal. Respondents were identified through several methods of purposive (non-random) sampling. Researchers worked through relevant NGOs, women's organisations and health services in Nepal which also legitimised the research, helping to foster trust between researchers and respondents. In addition, respondents were recruited through 'snowballing'.

Research Findings

Characteristics of Trafficked Girls

Table 6.2 shows the socio-demographic characteristics of interviewed girls. Trafficked girls are thus typically unmarried, non-literate and very young, the majority being trafficked before the age of 18 years. The youngest was 12 years, none older than 25 years. More than one third of respondents were married at the time of trafficking. The predominant ethnic group of girls was Mongoloid or Dalit (untouchable) but other ethnic groups were represented.

Ways of Trafficking and Recruitment Tactics

Traffickers used a variety of means to draw girls into the sex trade. The four key tactics of sex trafficking identified included: (i) employment-induced migration via a broker; (ii) deception, through false marriage; (iii) visits offer; and (iv) force, through abduction (Table 6.3). The majority of respondents (55%) were trafficked through false job promises.

False Promises of Jobs

Jobs in carpet factories, providing Nepal's most important export, were the most common offer reported. Children from poor rural hill families are recruited

Table 6.2 Socio-Demographic Characteristics of Respondents ($N = 42$)

Grouping	%	*n*
Ethnicity		
Brahmin/Chhetri	21	9
Mongoloids (Gurung, Magar, Rai, Tamang)	36	15
Dalit (untouchable)	26	11
Others	17	7
Religion		
Hindu	74	31
Buddhist	21	9
Others	5	2
Marital status at the time of trafficking		
Unmarried	62	26
Married	36	15
Other (D/W/S)	2	1
Age at the time of trafficking/leaving home		
Below 15 years	31	13
16–18 years	55	23
Above 19 years	14	6
Education status at the time of trafficking		
Non-literate	86	36
Primary/non-formal education	12	5
Secondary education	2	1
Current education status		
Non-literate	33	14
Primary/non-formal education	50	21
Secondary education	17	7

D, divorced: W, widowed; S, separated.

Table 6.3 Route to Sex Trafficking, Traffickers, Destination and Mode of Exit From Indian Brothels ($N = 42$)

	%	*n*
Major motivating means		
False promises of jobs	54.8	23
Fraudulent marriage	19.0	8
Offer of visit/movie/holiday	14.3	6
Force and other	11.9	5
Trafficker		
Relatives	35.7	15
Known but not relatives	42.9	18
Unknown persons	21.4	9
Destination		
Mumbai	78.6	33
Delhi	11.9	5
Calcutta/other Indian city	9.5	4
Mode of exit		
Rescued	73.8	31
Escaped	16.7	7
Released by owner/self-return	9.5	4

from their villages and sold or apprenticed to factory owners. Brokers working within the carpet factories select likely girls, enticing them into leaving the factory with offers of better jobs elsewhere, a relatively easy task since many carpet workers are themselves caught in debt bondage where they receive no wages. The brokers arrange for their transport to India, frequently with friends' and family members' complicity.

When Dolma was 14, her stepfather took her from their village to Kathmandu, where his friend got her a job in a carpet factory. A few months later, a young male co-worker, introduced to Dolma as her 'nephew', suggested they leave the factory and go to Kakarbhitta, a town on the Indian border, where, he claimed, working conditions were better and they could earn more. Dolma agreed, and was taken out of the factory by her stepfather, her stepfather's friend and this young man. After 6 days' travelling by bus and train, they arrived in Mumbai and he sold her there.

In addition to factory recruitment, false offers of employment in other Indian and Nepali cities emerged as common forms of enticement. Sometimes older men promise girls employment in the city. Sabitri, another trafficked girl, reported that:

> one day I heard there was another factory nearby, paying higher wages than the factory I was currently working at. I went to the other factory to ask them if they had a job for me. 'You're in luck' said the manager, 'I need someone to accompany me and my wife to Hetauda, to collect wools for weaving. It will pay very well'. I immediately agreed and took this job. I did not think anything strange about it, especially since I would be travelling with his young wife. After a long journey I found myself in Mumbai where I was sold for Rs 40 000 by the manager.

In many cases the broker works from inside the factory, selects a girl, convinces her to go with him and then takes her to the border and sells her. When Tara was 12 years old she was taken to Kathmandu to weave carpets in a carpet factory. She worked in two carpet factories for 5 years.

> I met a boy while I was working in factory and we became very close. He told me that he would get me a good job. When he mentioned that I could earn a lot more money I instantly agreed to go with him. After 3 days we reached a big hotel in a new city. 'Why am I here?' I asked. 'You are going to do some cooking and cleaning work', he replied. A little while later they told me he had sold me.

Some girls, wanting to be independent, went to urban city for jobs and also ended up in Indian brothels.

Fraudulent Marriage

Fraudulent marriage offers are another common ruse employed by recruiters. In some cases, traffickers actually go through a marriage ceremony. In others, the marriage offer itself is enough to lure a woman away from home. The girl is either given a false promise of marrying the dalal (broker) who pretends to have settled down in India or she is told about a wealthy future husband, whom the dalal provides. Radha was one such victim. She readily agreed to marry an unknown person because of her family problem.

> When I was fifteen . . . I was married to a farmer. I lived with my husband for about one year and then returned to my mother's home. My husband came to collect me and my mother insisted I went with him. I had become pregnant. I could not work well, my husband did not treat me kindly, so again I returned back to my mother, where I gave birth to a son. My husband did not come to find me even though he knew that he had a son, so I stayed with my mother. When my son was four years old I heard that my husband had remarried. On the day of Shivaratri (Hindu festival) I went to the river to light a candle, where I met one of my relatives from my village who was with a few other men. My relatives introduced me to one and asked me if I would marry him. I didn't take the offer seriously. 'I can't get married', I replied, 'I have a son. Besides, I hardly know that man'. But my relative kept insisting. 'Come on' she replied. 'At least think about it. He lives in Hetauda and is a great person. You should not worry about your son'. I did think about it and the idea of remarrying gave me hope that perhaps happy days would come again. I agreed and I went with him. He took me to the restaurant and after this I cannot

> remember anything. When I awoke I found myself in the world of brothels. I had been sold for Rs 30 000 [about $450].

Visit Offer

In many cases a girl is lured by the trafficker or his agent, often a local young man who works in Kathmandu. After enough trust is established she is then offered a lucrative job in Mumbai as a maidservant, even as an actress, or she is told about an opportunity to set up a small business. In most people's minds, Mumbai stands for glamour, movies and prosperity, golden chances and escape from miserable lives. Priya's story is typical of this kind of trafficking. It indicates that not only poor girls, but also middle class girls are trafficked.

> My brothers used to worry about me and I used to quarrel with them. All my family would scold me, telling me to study harder but I did not listen. Even when they yelled at me I would just ignore them because I did not want to be a teacher, I was interested to be an actress. To be a good actress, you don't need to study and you don't need to go to school. One day, my friend Sita, her husband and I went to watch the movies together. After, they asked me if I would go to India with them. I couldn't refuse their request, as I wanted to be an actress. I also felt indebted to them for always welcoming me into their house. We spent three days travelling. Eventually we reached our destination—Bombay. At first we stayed at the Amar hotel but were soon taken to another place, where the women were decorated with expensive jewellery, clothes and scents. There we met a fat lady who Sita's husband introduced to us as a film director. She seemed very kind and generous.... Sita's husband told the 'film director' to let us rest and said that he would come back after he had been shopping. He never returned. Later we were told that it was a brothel and we had been sold for Rs 60 000.

Recruiters sometime seduce young girls by posing as potential boyfriends, pretending they are interested in the young girl, wanting to know her better. Recruiters ask the girls' names, addresses and people they know. When the girls become comfortable with the poseurs, the recruiters offer to treat them at restaurants nearby. As the recruiters gain the girls' trust, they ask the girls to accompany them on a visit to a relative in another town, or attend a party in towns nearby. In many cases the girl elopes with her new 'friend' without even telling her parents. Twenty-one-year old Ujeli told us that:

> my parents are agricultural labourers in the hills. I was able to attend school up to class 4, but then had to join my parents working on the fields, so I left school. At the age of 15, I went with a friend to watch a movie and met a young man called Kancha Lama with whom I became friendly. After some time he suggested going to a bigger southern town to buy cheap cloth with which to start my own small business. I went with him without asking my parents. Instead, he took me to India.

Abductions

Simple abductions also occur, although they are less common than cases of deceit. Some girls mentioned that they, or other girls in the brothels where they worked, had been drugged by their abductors. Girls who are abducted are often drugged before a journey during which they are sold to brothel owners in India.

> I was taken to India by neighbours, a mother and daughter, whom I knew quite well. They told me they had to go to a market far away to pick up something and asked me to come along. A taxi was waiting for them. They travelled a long way. It was very late when they finally arrived in Badi Bazaar. They got another taxi and arrived at a village house like my own. I was put in a room and the door was locked. A woman called Asa told me the woman she came with had gone out and

> would be back later, but she never came back. After three nights, I pleaded with her to let me go. I was told 'No, you have been sold and have to work. All Nepali girls have to work'.

Traffickers

In many cases family members, uncles, cousins and stepfathers also act as trafficking agents. Of the girls interviewed, 15 were trafficked to India with the help of family members or relatives. Likewise, 18 were trafficked by known persons but not relatives, and nine were trafficked by unknown persons. Traffickers are most typically men in their twenties or thirties or women in their thirties and forties who have travelled to the city several times, knowing hotels to stay in and brokers to contact. Traffickers frequently work in groups of two or more. Male and female traffickers are sometimes referred to as dalals and dalali (commission agents) who are either employed by a brothel owner directly, or operate more or less independently. In either case, to stay in business, they need the patronage of local bosses and the protection afforded by police bribes.

Women who are already in the sex trade and have graduated to the level of brothel-keepers, managers or even owners travel through their own and neighbouring districts in search of young girls. The following story encapsulates the essence of the dream of success and glamour that these women symbolise to the simple village girls. Female traffickers are referred to as didi or phupu didi (literally, paternal aunt) or sathi (best friend). Local women who have returned from India are also employed as recruiters. Usually these didis return to the villages to participate in local festivals and to recruit girls to take back to the cities. These women are well placed to identify potential trafficking victims because they know local girls and their families.

> After 2 years of my marriage, my husband brought a co-wife who gave birth to a son. I was then completely rejected from them. In the meantime, a woman who had came home for vacation promised me and my three other friends good jobs. We ran away with her and she took us to Calcutta. But instead of giving us good jobs she sold us to different brothels.

Not all dalals work independently. An unknown number are connected to different networks that operate on various levels and size of organisations. Some syndicates include government officials, border policemen and politicians.

Life in the Brothels

Nepali girls in India's red-light areas remain largely segregated in brothels located in what are known as Nepali kothas (compounds). The concentrations of Nepali vary between cities, but appear to be highest in the Mumbai neighbourhood of Kamathipura. Brothels vary by size, physical configuration, ethnicity of sex workers and price. Most Nepali girls are associated with gharwalies (brothel owner). Depending upon the gharwali, the number of girls and women per brothel ranges from 5–10 to 150–200, with an average of 90–100 girls and women per brothel. In all cases, movement outside the brothels is strictly controlled, inmates being subjected to both psychological and physical abuse. The cheapest brothels, no more than dark, claustrophobic rooms with cloth dividers hung between the beds, are known among Nepali as pillow houses. Certain lanes are known particularly as Nepali gallis [street]. The living conditions of Nepali girls in all brothels are very poor. A social worker familiar with the Indian brothel system told us:

> There are several grades of sex workers, based on beauty, hard work, 'talent'. The tops are call girls. Then comes 'bungalow' which is a higher grade, then 'pillow house' which is the lowest. Most girls start in pillow house and work up if they do well . . . some girls receive training, how to approach customers, languages. During training, girls are beaten and locked in a room like a jail . . . until they stop fighting. At first a girl gets two or three clients a day, then it escalates.

All interviewed girls had no previous experiences of sex work, and no intention of engaging in this trade. Jamuna recounted her early days in a brothel:

> When I entered the brothel I saw many girls who looked younger than 20 years of age. I did not know what they were supposed to do. They looked very strange to me. I had never seen girls wearing so much make-up and bright red lipstick. Their clothes were different too. They all had on very short skirts with lots of jewellery. They were not typical Nepali girls. The brothel-keeper told me to take a bath, get make-up and put on clean clothes. 'What is my job?' I asked. 'What's going on?' 'You will do what I tell you,' said the brothel-keeper, 'you will find out in a few hours.' 'I don't want to stay here' I replied. The brothel-keeper laughed and walked away. I looked at the others for help. 'There's no way out' they said 'you're going to be a prostitute'.
>
> Sarmila recounted her terrible experiences in the brothel,
>
> on my first day, a fat man came to my room. He had paid a large amount of money to rob me of my virginity. I locked myself in the bathroom but the brothel-keeper came and made me open the door. Again the fat man came into my room. I pleaded with him and eventually he left, giving me Rs 10. . . . The next day, however, a young boy came and I lost my virginity.

Every girl said that the brothel owner or manager forced her to work by invoking her supposed indebtedness. A girl's earnings depend on the type of brothel in which she is employed, her age, appearance and the nature of the sex acts she is compelled to perform. Although most business is conducted in the brothel, and is charged by the minute or hour, customers can pay extra to take women outside. A girl may be sent to a client's house or a hotel for the night. If a customer buys a woman's services for a longer period, her debt resumes upon her return. One customer paid a large amount of money and kept a woman in his home for 2 weeks. He returned her to the brothel, where she worked to repay the remaining debt.

None of the girls knew much about the monetary arrangements between the brothel owner, the agents and their families. But later on all were frequently reminded that they had to work to pay off their debts, and many were threatened for not earning enough. Some of the girls had a vague understanding that they would have to work for a specific length of time to pay off the debt, and that there was an agreed-upon amount of payment given them. Very occasionally, brothel owners might treat a girl more kindly, buying her clothes or giving her treats. This was rare, though; with few exceptions, girls were unable to communicate with anyone outside the brothel; some were even forbidden to take Nepali clients in case the latter helped girls escape. Even conversation with customers was sometimes forbidden:

> Only girls who pay off their 'loan', have gone on a holiday to their village and come back, are allowed to leave the brothel alone.

Very few of the interviewees were in occasional communication with their families. One girl found a customer who was willing to send word to her family.

> A Nepali man I met in the brothel wrote a letter to my family telling them what had happened to me. After few months my brother went to Bombay to see me there, but he was not allowed to do so. My family then brought charges against that trafficker and brothel. I was sent back to Nepal with the help of social workers.

Besides being compelled to serve customers, brothel owners sometimes forced sex workers to perform personal housework or childcare chores.

Return and Reintegration Into Community

Three major processes were involved in returning home. Girls are rescued, escape and released or self-returned (see Table 6.3). The majority were rescued by police and/or social workers. A few escaped on their own or with the help of other people. Only four girls were released by a brothel owner and/or self-returned (with the brothel owner's consent). Mainly, girls were being rescued and put into an Indian rehabilitation centre before returning to Nepal, or were then shifted to a Nepalese rehabilitation centre before returning to their family.

It is illegal for girls below 18 years to work in a brothel in India. Brothel-keepers always ask young girls to say that their age is more than 18 years if police raid the brothel. Some girls were able to escape from brothels with the help of others. Neela escaped with the help of her regular customer. When a girl is too old to attract customers she is released from the brothel. Some are thrown out when they are tested HIV positive, others only when they have full-blown AIDS. Sometimes girls were allowed to come back to Nepal for a short time. Some of those girls do go back to Indian brothels and some stay in Nepal. Kanchi was sold by a family member but after working for 5 years, she was able to return to Nepal. A small number of girls accepted their lives in the brothels and became brothel owners themselves.

Many trafficked girls spend some time in a rehabilitation centre in Nepal after exiting brothels in India. Rehabilitation centres are typically run by NGOs and provide health and social assistance to returned trafficked girls. In addition, girls are provided with literacy and skill-building classes to assist them to integrate back into their communities. However, these girls reported enormous problems in returning to community life, in particular reporting high levels of social stigma directed at trafficked girls. Frequently, not only society at large but also parents condemn their daughters morally, and repudiate them. They are fully aware that society looks down on them and therefore offers no hope for a dignified life. One girl mentioned:

> I do not want to go back to my home. I would rather prefer to stay at a rehabilitation shelter and continue my studies.

A common phrase cited by a number of respondents captured the social values surrounding girls involved in sex work:

> Kegarne chori cheli dimma jastai hunchha, ekchoti futepachhi, futyo, futyo. [What to do? Unmarried girls are like eggs, once broken you cannot join them.]

If girls who return home have managed to earn money, they are more easily accepted back into their communities, and may eventually marry. Those who escape the brothels before paying off their debts, who return without money, or who are sick and cannot work, are shunned by their families and communities. Many return to India.

Conclusions

The key routes to sex trafficking include employment-induced migration to urban areas, deception (through false marriage or visits) and abduction. Current findings underline the role of poverty in the sex trafficking of Nepali girls, with over half reporting being lured by traffickers through promises of economic opportunity. The predisposing factor of poverty has been previously highlighted regarding trafficking both within South Asia and in other regions worldwide (Huda, 2006; Okonofua et al., 2004; Woolman and Bishop, 2006). At the local level, trafficking stems from deep-rooted processes of gender discrimination, a lack of female education, the ignorance and naivete of rural populations, poverty and lack of economic opportunities in rural areas with the consequent marginalisation of particular social groups. Wider factors include the low social status of the girl child, corruption of officials, an open 1500 km border with India, lax law and weak enforcement machinery, and local political apathy (Acharya, 1998; Asha-Nepal, 2006; Asia Foundation and Population Council, 2001; Friedmann, 1996). These local level processes are in turn shaped by macro-level economic and social forces that are changing the way markets operate and the kind of labour that is required. None of this explains why some communities are more affected than others, however.

It is very hard to answer the question of how many girls were actually tricked or forced into the trade or how many went into the business of their own free will, because it is not clear where the dividing line is between choice and compulsion. As O'Dea (1993) noticed, the expression 'own free will' seems out of place in this context. The influence of poverty, family pressure, caste and gender discrimination has to be taken into account. Mere resignation due to lack of a viable alternative may seem a rational response. In the Nepali context, 'voluntary prostitution' is often considered a paradoxical term. However, it does not serve the reality of trafficked girls to fit their cases to a dichotomous system that only admits voluntary or forced prostitution. There are too many forces at work to decide.

Nepali girls are expected to work hard in the household. Studies indicate that to get rid of the poverty-stricken economy of the household, women and girls are always in search of economic opportunities within and outside the country (UNICEF 2006). The female crude economic activity rate in Nepal, reported over three censes, is far lower than in men (Shtrestha and Panta, 1995). Migration is playing an increasingly important part in Nepal's economy and social structure. As these factors lead to an increase in migration, more girls are found to be trafficked in the process, a finding consistent with earlier research (Asha-Nepal, 2006; Asia Foundation and Population Council, 2001; Rajbhandari, 1997). Many misconceptions or over-simplifications of the underlying causes of migration obscure the resources available to trafficked persons or their resiliency. For example, poverty is often cited as the reason for migration or accepting employment conditions of debt bondage, despite the common occurrence of migrants actually paying for transportation or transit services.

In Nepal, high-level decision makers, lawmakers and politicians at the local level are often accused of being the protector of the traffickers. Many commentators blame the lack of legal enforcement arguing that policies are sound in Nepal but not their implementation and that political commitment is required to implement public policies. Political leaders and higher authorities in bureaucracy are accused of releasing the arrested traffickers from custody and taking political and monetary benefits from them or having associations with brothel-keepers (Friedmann, 1996; Rajbhandari and Rajbhandari, 1997; Thapa, 1990). Malpractice in political and administrative levels in both places of origin (Nepal) and destination (India) of trafficking were reported by both victims and key informants in this study. Much Nepali and Indian literature has mentioned the hardships experienced and the poor economic structure of the household that leads girls to being vulnerable to trafficking and to their involvement in prostitution. Case studies (ABC Nepal, 1998; Rajbhandari and Rajbhandari, 1997) indicate that poor economic conditions are the most common factors identified by the girls. However, the possibility of their involvement in other sectors of economy is not detailed. Girls, once trafficked and forced to be in sex work, often accept their fate later, because there are no options or alternatives left.

The root causes of trafficking are thus multiple and complex. However, this study suggests that both trafficking and migration operate primarily through personal connections and social networks (such as an aunt who returns to the village and takes her niece back to the city), and through unregistered brokers who may or may not be strangers to the locality. Girls voice opinions like "my sister worked there before, so I went there", which further underline the significance of social networks in the sex trade.

It is still not known how many trafficked persons return without NGO assistance and what type of reintegration strategies they employ. There is some evidence in this study that some girls decide to settle in urban areas, setting up small businesses or, if they are sex workers, staying in the sex trade directly or indirectly as madams or brokers. At the same time, this study also noted that girls from communities where sex work is a common practice may find it easier to return home where they may marry and/or set up small businesses. Further research into coping and livelihood strategies employed by trafficked girls would assist in the development of more effective reintegration strategies. Society has traditional values that degrade brothel returnees, but brothel returnees also have a psychological stigma that makes them hesitate to face common people. It should not be implied that the brothel girls do not want to go back home; however, social norms and the possible reaction of the home community have become obstacles to restore them to normal life.

How then should we respond? The existence of specific and clearly defined networks of trafficking has implications both in terms of efficient use of resources and in terms of the effectiveness of the activities. In addition, the messages from NGOs in the form of leaflets are likely to be futile for the illiterate populations amongst whom they are distributed. Movement in and out of coercive and exploitative circumstances is a dynamic process that is well recognised in irregular migration, smuggling and trafficking. Interventions that intercept trafficking at its outcome point, rather than at the time or place when it first occurs, draw attention to the problem of identifying when movement within or between countries becomes exploitative and not voluntary, and could serve to protect an individual's right to migrate. A human rights analysis draws attention to the promotion of equality and non-discriminatory migration.

There is an overemphasis in the literature at present on legal responses. The legal response to trafficking, either through international conventions or state-sponsored regulations, can never be a complete response or a solution. Indeed, an over-reliance on legal mechanisms can produce results that are counterproductive. When laws are created to be as broadly encompassing as possible, an overgeneralisation occurs that actually restricts the application of the law, reducing its impact. Legal measures to restrict trafficking that lack specificity in terms of gender and age have been shown to mischaracterise the harm done by trafficking, and actually compound restrictions on the movement and employment of younger girls instead of protecting these rights (Huntington, 2002). Additionally, a rescued or escaped girl's rehabilitation efforts require a positive reflection of the society towards her for the rest of her life. In some cases, the hiding of brothel returnees would not be helpful, nor create general social acceptance. Traditional values and norms are hindrances to rehabilitation efforts. Social reintegration becomes much more painful for the person once involved in sex work and rehabilitated later, which may force girls to stay in the sex trade even if they return to Nepal; a significant proportion of girls have indeed reported their unwillingness to go back home after becoming sex workers. NGOs working against girl trafficking tend to focus only on the group of girls trafficked in the most exploitative way and publicise this picture. Existing interventions hardly cover the family-based trafficking which the present study has identified. Partial truths, from whatever side, do not help the issue and caution is required in assessing the situation; intervention strategies may otherwise be wasted. This study also highlights the role of violence in sex trafficking. Gender-based mistreatment in families appears often to contribute to girls' vulnerability to sex trafficking.

Significantly, the majority of trafficking victims reported being transported indirectly via carpet factories, representing a critical intervention opportunity. With appropriate training, carpet factory owners may be able to separate safely potential traffickers from victims, determine the true nature of the relationship, and secure victims' safety. Cultural factors may partially explain why such experiences place females at risk for trafficking. Nepali girls experiencing disruption via abuse or abandonment by their husbands often face extreme community ostracisation (UNICEF 2006). Families of such girls are also subject to stigmatisation and, therefore, may be reluctant to offer support or shelter based on fears of additional negative consequences for their status within the community, including marriageability of unmarried family members (Goel, 2005). Similarly, whilst traditional cultural norms associate sons with economic and social advantage, daughters are conversely constructed as burdens, particularly regarding dowry (Fikree and Pasha, 2004). Extended family members may be unwilling or unable to assume the costs of providing for unmarried/widowed females.

The study shows that stronger policy and strategy along with political commitment remain critical. There is a compelling need for interventions that actually empower women and girls in migration rather than seeking merely to protect them. The interaction of poverty and gender-based mistreatment of women and girls in families heightens the risk of sex trafficking. Prevention efforts should work to improve economic opportunities and security for impoverished women and girls, educate communities regarding the tactics and identities of traffickers, as well as promote structural interventions to reduce trafficking.

References

ABC Nepal. 1998. *Life in Hell: The True Stories of Girls Rescued from Indian Brothels.* ABC Nepal: Kathmandu.

Acharya U. 1998. *Trafficking in Children and the Exploitation in Prostitution and Other Intolerable Forms of Child Labour in Nepal: Nepal Country Report.* ILO-IPEC: Kathmandu.

Asha-Nepal. 2006. *A Sense of Direction: The Trafficking of Women and Children from Nepal.* Asha-Nepal: Kathmandu.

Asia Foundation and Population Council. 2001. *Prevention of Trafficking and the Care and Support of Trafficked Persons: In the Context of an Emerging HIV/AIDS Epidemic in Nepal.* The Asia Foundation and Horizons Project Population Council/Creative Press: Kathmandu.

DiCicco-Bloom B, Crabtree BF. 2006. Making sense of qualitative research. *Medical Education* 40: 314–321.

Fikree FF, Pasha O. 2004. Role of gender in health disparity: the South Asian context. *BMJ* 328: 823-826.

Friedmann J. 1996. Rethinking poverty: empowerment and citizen rights. *International Social Science Journal* 148: 161-172.

Ghimire D. 2001. *Prevention, Care, Rehabilitation and Reintegration of Rescued Girls* (ABC's Experience). Paper presented at the Technical Consultative Meeting on Anti-trafficking Programmes in South Asia, September.

Goel R. 2005. Sita's Trousseau: restorative justice, domestic violence, and South Asian culture. *Violence Against Women* 11: 639-665.

Hennink M, Simkhada P. 2004. Sex trafficking in Nepal: context and process. *Asian Pacific Migration Journal* 13: 305-338.

Huda S. 2006. 'Sex trafficking in South Asia'. International *Journal of Gynaecology and Obstetrics* 94: 374-381.

Human Rights Watch. 1995. *Rape for Profit: Trafficking of Nepali Girls and Women to Indian Brothels.* Human Rights Watch: New York.

Huntington D. 2002. *Anti-Trafficking Program in South Asia: Appropriate Activities, Indicators and Evaluation Methodologies.* Summary Report of a Technical Consultative Meeting. Population Council: New Delhi.

ILO/IPEC. 1998. *Trafficking in Children for Labour Exploitation, including Sexual Exploitation in South Asia: Synthesis Paper.* ILO/IPEC South Asian Sub-Regional Consultation: Kathmandu (Unpublished).

ILO/IPEC. 2002. *Internal Trafficking Among Children and Youth Engaged in Prostitution.* International Labour Organisation/International Programme on the Elimination of Child Labour: Kathmandu.

Kamala K, Sanghera J, Pattainhaik B. eds. 2005. *Trafficking and Prostitution Reconsidered: New Perspectives on Migration, Sex Work and Human Rights.* Boulder: Paradigm.

Khatri N. 2002. Nepal: *The Problems of Trafficking in Women and Children.* Paper Presented at the 7th Annual Meeting of the Asia Pacific Forum for National Human Rights Institutions, 11-13 November, New Delhi

Nair PM. 2004. *A Report on Trafficking of Women and Children in India: 2002-2003,* Vol. 1. UNIFEM, ISS, NHRC: New Delhi.

O'Dea P. 1993. *Gender Exploitation and Violence: The Market in Women, Girls and Sex in Nepal: An Overview of the Situation and a Review of the Literature.* UNICEF: Kathmandu.

Okonofua FE, Ogbomwan SM, Alutu AN, Kufre O, Eghosa A. 2004. Knowledge, attitudes and experiences of sex trafficking by young women in Benin City, South-South Nigeria. *Social Science and Medicine* 59: 1315-1327.

Pope C, Mays N. 2006. Qualitative methods in health research. *BMJ* 311: 182-184.

Rajbhandari R. 1997. *Present Status of Nepali Prostitutes in Bombay.* WOREC: Kathmandu.

Rajbhandari R, Rajbhandari B. 1997. *Girl Trafficking: Hidden Grief in the Himalayas.* WOREC: Kathmandu.

Shtrestha P, Panta P. 1995. *Economically Active Population. Population Monograph of Nepal.* Central Bureau of Statistics: Kathmandu; 205-238.

Thapa P. 1990. Keti bechbikhan: Lukeko Aparadh (Trade of Girls: A Hidden Crime) in Ghimire Durga. (ed). *Chelibetiko Abaidh Vyapar: Yasaka vivid Paksha (Illegal trade of girls: Its various aspects).* ABC Nepal, Kathmandu; 21-25.

UN. 2000. *Protocol to Prevent, Suppress and Punish Trafficking in Persons, Especially Women and Children.* Supplementing the UN Convention against Transnational Organized Crime, Annex II. United Nations Doc A/55/383, United Nations: New York.

UNFPA. 2000. *State of the World s Population.* UN Fund for Population Activities: New York.

UNICEF. 2006. *Situation of Women and Children in Nepal.* The United Nations Children's Fund: Kathmandu.

US Department of State. 2001. *Victims of Trafficking and Violence Protection Act 2000: Trafficking in Persons Report.* July. US Department of State: Washington DC.

US (2005) Trafficking in Persons Report. US Department of State: Washington DC.

Woolman S, Bishop M. 2006. State as pimp: sexual slavery in South Africa. *Development South Africa* 23: 385-400.

DISCUSSION QUESTIONS

1. What tactics do traffickers use to entice girls into the sex trade?
2. What are some of the challenges for girls and women when they return to their communities following working in a brothel?
3. What policy recommendations would you make to support victims of trafficking?

READING

In the section introduction, you learned about the numbers of women who have gone missing or who have been murdered without explanation in and around the region of Ciudad Juárez, Mexico. This article highlights the lack of response by the local government officials and law enforcement to resolve these cases. As a result, the family members of the victims and women's rights organizations have formed to fight not just for the rights of victims in these cases, but to advocate for women's and human rights in general.

Murder in Ciudad Juárez

A Parable of Women's Struggle for Human Rights

Mark Ensalaco

The abduction, sexual torture, murder, mutilation, and disappearance of hundreds of women in Ciudad Juárez and Chihuahua, Mexico, during the past decade and the failure of Mexican authorities to exercise due diligence in investigating the crimes have compelled women to mobilize in a campaign to stop the impunity for gender-based violence. The mobilization against the femicide in Ciudad Juárez is something of a parable of women's struggle for human rights because its significance extends beyond Mexico and even Latin America. This is so for three reasons: The economic, social, political, and cultural factors that cause gender-based violent crimes in Mexico are common in other regions of the world; the Mexican authorities' failure to investigate, prevent, and punish those crimes constitutes a form of gender discrimination that is prevalent in other regions of the world; and the imperative for Mexican women to mobilize in defense of their own human rights in the absence of government recognition that gender discrimination is the

SOURCE: Ensalaco, M. (2006). Murder in Ciudad Juárez: A parable of women's struggle for human rights. *Violence Against Women, 12*(5), 417–449.

cause of violence against women is an imperative shared by women in other regions of the world.

The mobilization of women's organizations in Mexico that began in the 1990s is reminiscent of the mobilization of victims' associations in Latin America in the 1970s and 1980s in response to the dirty wars of state terrorism against political dissidents, a response that ultimately led to the formation of the Latin American Federation of Associations of the Families of the Detained-Disappeared, FEDEFAM. Although what is at issue in Mexico is the state's failure to prevent violence by private individuals rather than the state's direct complicity in violent repression as in the 1970s and 1980s, the fact remains that the mobilization of civil society is critical to the protection of human rights. What is distinctive—and significant—about the women's mobilization in Ciudad Juárez is the integration of a gender perspective into this new struggle for human rights. The immediate objective of the mobilization is to end the violence and to achieve justice, but the larger objective is to achieve women's equality as a prerequisite of ending violence against women in its myriad of cultural manifestations.

This article examines the decade-long wave of gender-based violent crimes that began in 1993, the negligence of Mexican state and federal authorities to investigate and to prevent those crimes, and the emergence of a broad-based campaign involving women's organizations, human rights organizations, and regional and international human rights bodies with the common aim of ending the violence and promoting respect for women's rights as human rights. It concludes by noting that the struggle for human rights in Ciudad Juárez involves what may be called a double transformation. The first involves the transformation of women's organizations into human rights organizations that were compelled to engage in classic human rights advocacy by the state's failure to exercise due diligence. The second involves those organizations' deliberate effort to effect the transformation of human rights discourse and practice through the conscious integration of the gender perspective. The struggle for justice and equality in Ciudad Juárez reveals the benefit of these reinforcing transformations both to women's organizations and human rights organizations.

Murder in Ciudad Juárez

Ciudad Juárez, across the border from El Paso, Texas, in the Mexican state of Chihuahua, historically has experienced a high rate of violent crime. According to official Mexican data, Ciudad Juárez, with a population of less than 1.5 million, has the second highest homicide rate per capita after the capital city of 27 million residents. But in 1993, the city began to experience a phenomenon analogous to the political repression experienced in Central and South America in the 1970s and 1980s: the disappearance of citizens; the anguished search by family members; the discovery of corpses days, weeks, months, and sometimes years after the disappearance; and impunity for the perpetrators of the crimes. The critical difference is that those murdered in Ciudad Juárez were not political dissidents killed by state agents for reasons of the state. They were teenage women killed by private individuals for personal reasons, including sexual gratification. This is gender-based violence rather than political violence, but the murders in Ciudad Juárez have political significance and involve state responsibility.

The Statistics

Beginning in 1993, the rate of homicides of women in Ciudad Juárez accelerated, and the ratio of female-to-male homicides increased. This marked the beginning of a decade-long wave of gender-based violence that involved abduction, sexual torture and rape, murder, mutilation, and disappearance. The exact homicide figures are disputed.

The number of "disappeared" is more shocking than the number of dead, according to a source within the IACHR who spoke to us on background. That number approaches 300. This means there are as many women missing as there are known dead, with the implication that the true death toll could be near or above 600, an average of 60 murders each year. Moreover, beginning in 2000, the bodies of disappeared young women began to be discovered in the city of Chihuahua, some 230 miles from Ciudad Juárez. In an interview with the author, Esther Chavez commented, "The murders simply changed addresses."

The Categories

Analysis of the nearly 300 homicide case summaries provided by Casa Amiga and the 36 cases summarized by the Mexican National Commission on Human Rights suggests that the homicides fall into two categories. The first involves isolated or "situational" murders. Ciudad Juárez exhibits a high rate of murders of women associated especially with domestic violence but also ordinary crime and drug trafficking. The victims of these crimes were generally in their 20s or 30s; the oldest victim, a 50-year-old woman whose body was submerged in acid, appears to have been the victim of organized crime.

The second category involves pattern or "serial" killings. The victims range in age from 11 years to 20 years; most were between 13 and 19. Many had similar physical characteristics, skin complexion, and hair color and length. Many of the women were employed in the *maquilas*; most "disappeared"—that is, were abducted—late at night after leaving the maquilas or sometimes after leaving discos. The cases of the disappeared become especially important in connection with the attitude of authorities discussed later. According to Amnesty International (2003a, 2003b), many women were "abducted, held captive and sexually assaulted in the most ferocious manner before being murdered" (p. 4). The case of 17-year-old Lilia Alejandra Garcia Andrade, who was abducted in February 2001, is a tragic example. Lilia endured 5 days of sexual torture before strangulation. After the discovery of her daughter's body in a wasteland, Lilia's mother and the relatives of other victims formed Nuestras Hijas de Regreso a Casa (May Our Daughters Return Home), one of many organizations to form in the wake of the crimes (Casa Amiga, 2003; Nuestras Hijas de Regreso a Casa, 2004). Although exact figures are not available, a significant number of these victims suffered mutilation, possibly prior to death. To cite one gruesome example: The bodies of three of the four victims that residents discovered in September 1995 were mutilated; their left breasts were severed, their right nipples bitten off. The bodies of a number of other women were burned, in at least one case, prior to death (Amnesty International, 2003b).

The Causes

Several explanations have been offered as to the causes of the gender violence in Ciudad Juárez. These are general explanations, not theories of individual crimes. Observers point to a set of economic, social, political, and cultural factors that include (a) rapid population growth in a frontier city, (b) a transient population of economic immigrants and the breakdown of community ties, (c) low salaries and poor working conditions in the maquilas, and (d) weak or corrupt government, police, and judicial institutions. All of these are postulated to create an environment in which criminal violence is likely to become widespread. But most observers emphasize cultural factors for both the violence against women and the impunity that accompanies it. Lydia Alpízar (2003), a member of the Stop the Impunity campaign steering committee, argues that "women's entry into the labor market had an impact on gender relations and thus on increasing gender violence" (p. 27). That is, the economic empowerment of women produced a violent backlash by men in a classically *machista* society. Her conclusion is that "violence against women is legitimized in Mexican society because . . . it devalues women" (p. 29). The IACHR concurs that the crimes in Ciudad Juárez have the attributes of gender-based violence, and that "a significant portion of Juárez murders occur in the context of domestic or intra-family violence" (IACHR, 2002, para. 36). The problem is pervasive. As Esther Chavez commented, "I think there are 200 or 300 men out there that have killed women. There's impunity, it's attractive" (Stackhouse, 1999). Astrid Gonzalez, a psychiatrist who formed the Committee against Violence, said, "Juárez is the ideal place to kill women, because you are certain to get away with it" (Dillon, 1998).

The economic, social, political, and cultural factors that contribute to the gender-based violent crimes in Ciudad Juárez are not unique to that city or to Mexico, although a particular combination of those factors may account for the high homicide rate and the violence against women in Ciudad Juárez. The factors that cause and perpetuate violence against women are common in other regions of the world, including especially cultural factors. As Radhika Coomaraswamy (2002) noted in her penultimate report to the United Nations Commission on Human Rights in her capacity as UN special rapporteur on violence against women, its causes and consequences, "Gender-based violence" is often "related to the social construct of what it is to be either male or female," and "cultural constructions of masculinity have a direct impact on the lives of women." In Ciudad Juárez, the impact has been lethal.

The Failure to Exercise Due Diligence

From the perspective of international human rights law, the situation in Juárez has human rights implications because of the Mexican government's failure to exercise due diligence. Every impartial observer—the Mexican CNDH, Amnesty International, the UN special rapporteurs on extrajudicial executions and the independence of the judiciary, and the IACHR—has criticized Chihuahua state officials for their failure to investigate, prosecute, punish, and prevent these gender-based crimes. Moreover, the families of the murdered and disappeared have decried the government officials' attitude toward the victims and treatment of their families.

Failure to Respond to Missing Persons Reports

Among the most egregious failures is the failure to launch an immediate investigation of a reported disappearance. Even after the pattern of abductions and murders became apparent, local police continued to wait days before mounting a search for the disappeared. In one especially troubling case, in January 1999, a resident telephoned police to report the sound of a woman's screams coming from a drainage canal behind her home, only to be told that it was not that unit's responsibility to respond to emergency calls. A woman's body was found in a different location in Juárez the following day (Dillon, 1998). Two years later, police failed to respond to several residents' reports that two men were raping a young woman in the back of a car until after the men had driven off (Amnesty International, 2003a). In yet another case, police did not begin to search for a woman reported disappeared on February 18 until March 21—31 days after residents had already discovered her body (CNDH, 1998). The failure of authorities has had life and death consequences for some of the victims, as in the case of Lilia Garcia, who was held captive and subjected to brutal sexual violence before being killed. Families who reported a disappearance routinely reported that police instructed them to return in 48 hours to file an official missing person's report. Police sometimes provided legal reasons for the delay of an official investigation, but in very many cases police ignored the families' pleas for prompt action, telling the families that the missing girls were probably spending the night with their boyfriends. This reflects the dismissive attitude of authorities, discussed below.

Failure to Investigate

Observers also criticize local and state officials for their failure to conduct competent investigations. The recommendation forwarded by the CNDH to the governor of Chihuahua details multiple failures beginning with the discovery of a body; police and forensic experts failed to secure the crime scene and to gather all the evidence. The failure continued during the postmortem examination. Although in a substantial number of cases police discovered skeletal remains or bodies in an advanced state of decomposition, in the majority of cases authorities were able to perform postmortem examinations. In one case documented by the CNDH, medical examiners failed to perform an autopsy for a full month after the discovery of a victim. In other cases, medical examiners failed to search for evidence of traces of semen or pubic hair on victims who were obvious victims of sexual assault; in other cases, medical examiners failed to mention sexual assault in their reports, perhaps to conceal the sexual nature of these violent crimes. In fact, the CNDH observed that in the 36 cases it reviewed, authorities failed to perform many of the 37 separate procedures required in homicide cases involving sexual assault (CNDH, 1998). The state's failure extended to the identification of victims; a substantial number of victims have yet to be identified and, in several troubling cases, authorities incorrectly identified a victim and misinformed the family. Federal prosecutors who agreed to investigate the cases of 14 women that appeared to fall within the jurisdiction of federal authorities complained that Chihuahua state officials ignored their requests for case files (Castillo, 2003).

Failure to Prosecute

As of March 2003, when the IACHR published its report, Chihuahua state officials could claim to have "resolved" 27 of the 76 homicides they categorized as serial killings and 152 of the crimes they categorized as situational murders. But as the commission pointed out, that figure was not only low, it was misleading: Chihuahua state officials declared a case resolved when prosecutors merely presented evidence to a judge; resolution did not mean formal indictment, much less conviction. This was not just the result of insufficient resources or inadequate training. The IACHR (2002) criticized "the overall inefficiency of the administration of justice" (para. 34) as well as the "the negligence of authorities and the lack of political will" (para. 81). The commission concluded that "there had been no real commitment to an effective response" (para. 82).

There have been highly publicized arrests in the pattern or serial killings. In October 1995, police arrested Abdel Latif Sharif in connection with the rape and murder of 17-year-old Elizabeth Castro Garcia. Prosecutors attempted to connect him to many more homicides but managed to convict him in 1999 only for the Castro murder (Madigan, 1999; Newton, 1999). In April 1996, police arrested nine members of a gang, los Rebeldes (the Rebels) and attempted to implicate them in 17 murders and to connect them to Sharif in a conspiracy. In April 1999 police arrested a bus driver, Jesus Guardado Márquez, in connection with the rape of 14-year-old Nancy Villalba. Guardado confessed to the murder of four women and implicated four other bus drivers in connection with a total of 20 murders. In November 2001, after the discovery of eight bodies, police arrested two men: Gustavo Gonzalez Mesa and Javier Garcia Uribe. They were also bus drivers.

Police attempted to attribute as many murders to these defendants as possible to give the appearance of action, but human rights organizations and defense attorneys charge the police with extracting the suspects' confessions under torture. Moreover, in February 2002 police shot and killed Mario Escobedo, the attorney for one of the accused, as he drove along a rural road. Evidence, including the bullet holes in Escobedo's car, called into doubt the veracity of the police's explanation of the fatal shooting and called into question the possibility that police had murdered him. A year later, Gustavo Gonzalez Mesa was found dead in his cell.

Attitude Toward the Victims and Treatment of the Families

The public statements of high-placed government officials attest that they simply devalue the lives of the murdered women, and they therefore do not consider the homicides of women to be "worthy of serious prosecutorial efforts."

In 1999, the former Chihuahua state prosecutor commented, "Women with a night life who go out late at night and who come into contact with drinkers are at risk. It's hard to go out on the street when it's raining and not get wet" (Amnesty International, 2003b, p. 2). In a similar vein, Suly Ponce, who headed the FEIHM office, remarked, "Sometimes there are cases that a girl meets some person, he strikes up a relationship with her, they drink . . . and it ends up violently. It is difficult to know" (Moore, 2000). Federal authorities made similar comments. Jorge Lopez Molinar, the deputy federal prosecutor, commented, "Many of the women work in maquilas and because the pay is not enough to live on, from

Monday to Friday they work and during the weekends they engage in prostitution" (Moore, 2000). Lopez also implied that as many as 80% of the murdered women had fled dysfunctional families despite evidence to the contrary. When challenged by families to defend his remarks, he responded by saying, "The parents will always tell you the relations between them and their daughters were perfect. But we sometimes find out that the girls are not the saints the parents would have us believe" (Moore, 2000). He made the comment after the discovery of the body of 17-year-old Maria Sagrario Gonzales Flores, an employee of a General Electric maquila. Similarly, Felipe Tarrazas Morales, regional coordinator of the general prosecutor's office commented, "It cannot be affirmed whether or not they sold their bodies, what is certain is that they knew the nocturnal centers very well" (Herrick, 1998).

The UN special rapporteur on extrajudicial executions interpreted the officials' remarks to imply that the women "asked to be murdered," then criticized those remarks for contributing to impunity because they "emboldened the culprits" (Jahangir, 1999, para. 85). The attitude of the authorities was one reason the relatives of several victims, including relatives of Lilia Alejandra Garcia Andrade, formed May Our Daughters Return Home in 2001. The families charged that authorities had "defamed us and violated our most fundamental rights by transgressing the private and family life, which has been strongly questioned in order to stigmatize us as the culpable" ("Origen de Nuestra Organization," 2004). Casa Amiga's founder, Esther Chavez, added, "The moralists would have you believe women are bad if they dance. The problem of course goes much deeper than that" (Herrick, 1998). The deeper problem is gender discrimination. According to Astrid Gonzalez of the Committee against Violence, "There's a real insensitivity on the part of the government. The priority is to protect the image and the economic interests, not the people who are feeling the brunt of the violence" (LaFranchi, 1997).

The dismissive attitude of officials extends to grieving families. Police have failed to keep families informed of investigations; they have misidentified bodies and misinformed the families about the fate of their loved ones; they have neglected to make available psychological counseling or other services in a community that has experienced the trauma of a decade of abduction, sexual torture, murder, and disappearance. Just as Chihuahua state officials have not treated the gender-based crimes as serious crimes, they have not addressed the social psychological consequences of those crimes as serious matters (Jahangir, 1999).

The Mobilization of Women

The mobilization in reaction to the gender-based violence in Ciudad Juárez is reminiscent of the mobilization of human rights organizations and victims' associations in reaction to the political repression in Latin America in the 1970s and 1980s. As in the 1970s and 1980s, victims' associations gradually formed and joined forces with preexisting labor and human rights organizations. However, in Ciudad Juárez, women's organizations concerned with sexual and reproductive freedom were among the first to become concerned with the femicide in Chihuahua.

Esther Chavez's 8 of March Feminist Group was the first to actively document and denounce the violence against women in Ciudad Juárez. Formed originally to lobby against the proposed criminalization of reproductive choice in Chihuahua after the victory of the conservative National Action Party in gubernatorial elections, the 8 of March Feminist Group began to concentrate on the abduction and disappearances when the pattern became apparent. In 1999, Esther Chavez established the Casa Amiga Crisis Center (Madigan, 1999).

In the summer of 1997, Astrid Gonzalez's Citizens Committee against Violence also began denouncing the crimes. In November, the committee held a protest rally in Ciudad Juárez and delivered a letter to Governor Barrios with a set of demands, including the creation of a special prosecutor's office with adequate budgetary and technical resources and the initiation of community programs to prevent violence against women. A key demand was that the authorities keep the families and community informed of developments in the investigations of the proposed special prosecutor's office (LaFranchi, 1997; Ruiz-Brown, 1997). The state governor did not accede to any of the demands until the following year, after the discovery of more victims and the formation of the first victims' association.

In April 1998, residents discovered the body of 17-year-old Maria Sagrario Flores Gonzalez, the 16th victim discovered so far that year. In July 1998, Paula Flores, Maria Sagrario's mother, and the relatives of five other victims formed Voices sin Eco (Voices without Echo) for the purpose of pressuring the authorities to take action and to provide support for other families touched by the violence. Voices without Echo disbanded in July 2001, but by then another victims' association had emerged (Brandt, 1998; Castañon, 2001; Herrick, 1998). Castañon Norma Andrade and the mothers of six other victims formed Nuestras Hijas de Regreso a Casa (May Our Daughters Return Home) in February 2001, after the discovery of the remains of Lilia Alejandra Garcia Andrade, the young woman who was sexually assaulted during a 5-day period (Paterson, 2001). Lilia's teacher, Marisela Ortiz, serves as the organization's spokesperson. Finally, a third victims' association, Justicia para Nuestras Hijas (Justice for Our Daughters), formed in the city of Chihuahua in 2002, by the mother of Miriam Gallego Venegas, whose murder was the first of a series of murders there.

These were only the most prominent organizations whose representatives the Mexican and American press regularly quoted after each new grim discovery or development in the investigation. But there were many others. What was needed was unity of purpose, and in November 2001 another grim discovery prompted many of them to unify. On November 2, authorities discovered five bodies in a field on the outskirts of Juárez; they discovered three more the next day. The discovery of eight more victims was the catalyst of la Campaña Alto a la Impunidad: Ni una Muerte Más (the Stop the Impunity: Not One More Death Campaign; Amnesty International, 2003a; Knox, 2001; Pantin, 2001). Some 300 organizations participate in the campaign, spearheaded by Elige Red de Jovenes por los Derechos Sexuales y Reproductive (Choose Youth Network for Sexual and Reproductive Rights) and the nongovernmental Mexican Commission for the Promotion and Defense of Human Rights.

The emergence of these organizations and the launching of the Stop the Impunity campaign are reminiscent of the human rights movement that emerged in the 1970s and 1980s. In particular, the emergence of the victims' associations is reminiscent of the process that led to the creation of FEDEFAM. In fact, the FEDEFAM affiliate in Mexico was instrumental in arranging the visit of the UN special rapporteur on extrajudicial executions in 1999, discussed below. Like the human rights organizations that emerged in the 1970s and 1980s, these feminist organizations have engaged in the activities of classic human rights advocacy: providing direct assistance to the victims, documenting and denouncing the crimes, involving national human rights institutions, appealing to international and regional human rights bodies, and pursuing legal remedies under international human rights law—all for the purpose of shaming the Mexican state and federal governments into exercising due diligence.

Providing Direct Assistance

A critical failure of the Chihuahua authorities was their inattention to the families of the victims. The failure began with the first missing person report; it continued after the discovery of the bodies and throughout the grieving process. Because authorities failed to make available professional psychological counseling and related professional services, the women of Ciudad Juárez were compelled to provide those services themselves. In an interview with the author, Esther Chavez explained the magnitude of the problem this way: "The violence not only damages women, it damages their families, it damages society." The indicia of social trauma are present in Ciudad Juárez: clinical depression, alcoholism, self-destructive behavior, and suicide. The creation of the Casa Amiga Crisis Center, directed by Esther Chavez, was the most visible example of the families of Ciudad Juárez assuming the responsibility to address these problems. Casa Amiga's services are not only available to families of the city's murdered and disappeared women. Soon after Casa Amiga opened its door, its staff found themselves counseling victims of rape, domestic violence, sexual abuse, and incest, cases that were not directly related to the murders in Ciudad Juárez. The Casa Amiga Crisis Center has become a place where victims are safe to unburden themselves about suppressed secrets of interfamilial and sexual abuse—secrets that reveal that the murder of women in Ciudad Juárez is a symptom of a deep undercurrent of social pathology in the city.

Nuestras Hijas de Regreso a Casa (May Our Daughters Return Home) also provides mental health counseling for the families of the Ciudad Juárez victims and programs in nutrition, health, education, and housing—all within the framework of an integral human rights approach to the problems existing in the community. Finally, the role played by the Elige Red de Jovenes por los Derechos Sexuales y Reproductivos (Choose Youth Network for Sexual and Reproductive Rights) underscores an important aspect of women's struggle for equality: sexual and reproductive freedom. As the former UN special rapporteur for violence against women has explained,

> One of the greatest causes of violence against women is linked to the regulation of their sexuality . . . recognizing women's rights to sexual autonomy and sexual health will be a major step forward in eradicating violence against women. (Coomaraswamy, 2002)

Conclusion

One of the terrible ironies of the murders in Ciudad Juárez is that these gender-based crimes began the same year the United Nations World Conference on Human Rights declared women's rights to be human rights. As in all matters relating to human rights, declarations about human rights and state obligations are useful only insofar as they serve to mobilize action. For that reason, the mobilization of women in Ciudad Juárez is paradigmatic: Women's organizations have deliberately adopted a human rights approach to the struggle for women's equality in a cultural environment overtly hostile to women. They have applied the international law of human rights to the neglected realm where women are most vulnerable to mistreatment. They have employed domestic, regional, and international human rights institutions with demonstrable effect. And they seem poised to establish a precedent in international law based on the due diligence principle that will compel states to intrude in the private sphere.

In the process, those organizations have effected a double transformation. Women's organizations have effectively become human rights organizations, and they have altered human rights discourse and practice through the conscious incorporation of the gender perspective. The IACHR, in its report on the situation in Ciudad Juárez, observed, "Sexual violence is driven by gender inequality. While it is a serious human rights problem, it tends to be under-documented, under-reported and under-researched" (IACHR, 2002, para. 64). The integration of a gender perspective into human rights discourse and practice represents an advance toward rectifying this problem.

The women's organizations that mobilized to confront the gender-based crimes in Ciudad Juárez provide a paradigm in this regard. Those organizations resemble the victims' organizations that emerged in Latin America in the 1970s and 1980s to document and denounce the state practice of "disappearing" political enemies. Many of those organizations later became affiliated with FEDEFAM. Women played critical roles in the FEDEFAM-affiliated organizations, such as the famed Mothers of the Plaza of May in Argentina, and created a "new model human rights activity" based on "equality and ties of affection among members" (Guzman Bouvard, 1994, p. 219). However, in most respects, the FEDEFAM-affiliated organizations conducted traditional human rights activities. Missing was a distinctively feminist perspective that sought to incorporate women's issues into human rights discourse. By contrast, the women's organizations in Ciudad Juárez have effectively utilized traditional human rights strategies in pursuit of basic equality for women in recognition that discrimination against women is the underlying cause of the crimes occurring with impunity in their community.

By adopting the traditional strategies of human rights advocacy and by incorporating the gender perspective, the women's organizations in Ciudad

Juárez have completed a double transformation. Those organizations have transformed themselves into human rights organizations, and they also have transformed human rights discourse and practice. The gender perspective contributes to human rights in two ways. First, it infuses human rights discourse with the substantive concerns that are unique to women. Because the human rights of women are most often violated in the private sphere, the human rights violations of women do not fall easily into the traditional category of state action that international human rights treaties were codified to constrain. The principle of due diligence now offers the juridical means to address the substantive concerns of women and to engender the law.

Second, the gender perspective demands changes in the decision-making procedures of international human rights bodies. It is not enough that women's rights be mainstreamed into international human rights law; it is also crucial that women take leadership in setting the agenda of international and regional human rights bodies. There have been important developments. In 1994, the United Nations appointed a woman to serve as rapporteur on violence against women. That same year, the IACHR appointed a woman as rapporteur on the rights of women. The creation of these posts was the result of the declaration that women's rights are human rights made during the 1993 World Conference on Human Rights. The creation of the office of the UN High Commissioner for Human Rights was also a result of the world conference. In 1997, the United Nations appointed Mary Robinson to serve as High Commissioner for Human Rights. In 2004, the world body appointed Louise Arbour to that post.

Violence against women is rooted in gender discrimination and as such constitutes a violation of women's fundamental human right to equality and the right to physical integrity. The global struggle to eradicate violence against women in its myriad of cultural forms will benefit from the adoption of a human rights approach and the effective utilization of the strategies of human rights advocacy. But the transformation of human rights discourse and practice through the incorporation of the gender perspective is also crucial to the struggle against gender-based violence. In this regard, the mobilization of women in Ciudad Juárez is paradigmatic.

References

Alpízar, L. L. (2003, Fall). Impunity and women's rights in Ciudad Juárez. *Human Rights Dialogue*, pp. 27–29.

Amnesty International (2003a). *Intolerable killings: Summary and appeal of cases*. New York: Author.

Amnesty International (2003b). *Intolerable killings: 10 years of abductions and murders of women in Ciudad Juárez and Chihuahua*. New York: Author.

Brandt, M. (1998, October 19). A message in murder. *Newsweek*. Retrieved November 14, 2003, from http://takenbythesky.net/Juarez/oct19_1998.html

Casa Amiga. (2003). Grupo Ocho de Marzo and *Diario de Juárez* Estudio hemerográfio [Friends House] [8 of March Feminist Group and *Juárez Daily* Press Analysis]. Ciudad Juárez, Chihuahua: Author.

Castañon, A. (2001, July 9). No echo: Juárez murdered women's group disbands. *El Diario*. Retrieved March 2, 2004, from http://www.womenontheborder.org/Articles/no%20echo.htm

Castillo, E. E. (2003, May 6). Federal Mexican authorities accuse state prosecutors of blocking investigation in border slayings. *Atlanta Journal Constitution*. Retrieved on November 17, 2003, from http://takenbythesky.net/Juarez/may6_2003.html

Comisión Nacional de Derechos Humanos. (1998). *Recomendación 044/98*. Retrieved on February 5, 2004, from http://www.cndh.org.mx/Principal/document/recomen/1998/044_98.htm

Coomaraswamy, R. (2002, April 10). *Integration of the human rights of women and the gender perspective*. Presentation to the Commission on Human Rights, 58th session, Geneva, Switzerland.

Cumaraswamy, D. P. (2001). *Report of the special rapporteur on the independence of judges and lawyers*. Geneva, Switzerland: Commission on Human Rights.

Dillon, S. (1998). 70 deaths unsolved in Juarez. *New York Times*. Retrieved on November 15, 2003 from http://takenbythesky.net/Juarez/feb_1999.html

FBI helps Mexico investigate murders. (2002, July 19). *United Press International*. Retrieved on November 17, 2003, from http://takenbythesky.net/Juarez/july19_2002.html

Guzman Bouvard, M. (1994). *Revolutionizing motherhood: The mothers of the Plaza de Mayo*. Washington, DC: Scholarly Resources, Inc.

Herrick, T. (1998, May 10). *Feminists decry police handling of murders in Border City*. Retrieved from http://takenbythesky.net/Juarez/may10_1998.htm

Inter-American Court for Human Rights. Velazquez Rodriguez Case, Judgment of July 29, 1988, Inter-Am. Ct.H.R. (Ser. C) No. 4 1998: paragraphs 172–177.

Jahangir, A. Report of the Special Rapporteur on extrajudicial, summary or arbitrary executions, submitted pursuant to Commission on

Human Rights resolution 1999/35. Addendum. Visit to Mexico E/CN.4/2000/3/Add.3.

Knox, P. (2001). Maquila means murder zone. *Globe and Mail.* Retrieved on November 17, 2003 from http://takenbythesky.net/Juarez/nov30_2001.html

LaCrisis: Voz de la calle [The Crisis: Voice of the Street]. (2004, March 5). Muertes de Juárez: Vergüenza nacional [Dead women of Juarez: National shame]. Retrieved on March 5, 2004, from http:// www.lacrisis.com.mx/especial291203.htm

LaFranchi, H. (1997). Girls who find new roles in Mexico also face danger. *Las Vegas Sun.* Retrieved on November 15, 2003, from http://takenbythesky.net/Juarez/june3_1997.html

Lawyer to probe Mexican murders. (2004, January 31). *BBC News.* Retrieved August 26, 2004, from http://www.news.bbc.co.uk/1/hi/world/americas/3446697.stm

Madigan, T. (1999, March 18). Fighting to Change a Chauvinistic Society. *Fort Worth Star-Telegram.* Retrieved on November 15, 2003, from http://takenbythesky.net/Juarez/march18_1999.html

Moore, M. (2000, June 26). Justice elusive for slain women. *Washington Post.* Retrieved on November 17, 2003, from http://takenbythesky.net/Juarez/june26_2000.html

Newton, M. (1999). *Cuidad Juarez: The serial killer's playground.* Retrieved on November 14, 2003, from http:///www.crimelibrary.com

Nuestras Hijas de Regreso a Casa. (2004). *Origin de nuestra organización.* Retrieved on March 3, 2004, from http://www.geocities.com/pornuestrashijas/nuesorgig.html

Pantin, L. (2001). 250 murders prompt Mexico anti-violence campaign. *Women's e-news.* Retrieved on March 12, 2004, from http://takenbythesky.net/Juarez/dec21_2001.html

Paterson, K. (2001, March 13). *Reign of terror against women continues.* Retrieved on November 17, 2003, from http://www.takenbythesky.net/Juarez/mar13_2001.html

Reparations proposed in Mexican killings. (2003, November 13). *The New York Times,* p. 11.

Rodriguez, O. (2004, January 29). 13 Mexican cops linked to drug killings. *Associated Press.* Retrieved on January 29, 2004, from http://www.officer.com/article/article.jsp?id=9058&siteSection=1

Ruiz-Brown, A. M. (1997, December). Mothers of slain women demand justice. *Diario de Juarez.* Retrieved on November 15, 2003, from http://takenbythesky.net/Juarez/dec_1997.html.

Stackhouse, J. (1999). *Men killing women in Juarez, Mexico, with impunity.* Retrieved on November 17, 2003, from http://www.flipside.org

U.S. Department of State, Office of International Information Programs. (1999, December 3). *Drug czar on Mexico's suspected mass gravesites.* Washington, DC: Author.

Valdez, D. W. (2002, June 23). Families, officials claim cover-ups keeping killings from being solved. *El Paso Times.* Retrieved on November 17, 2003, from http://takenbythesky.net/Juarez/june23_2002 .html

Vazquez, B. (2004, March 11). Police investigate latest Juarez slaying. *Associated Press.* Retrieved on March 11, 2004, from www.yahoo.com/news?tmpl=story2&cid=589&u=/ap/20040310

DISCUSSION QUESTIONS

1. How have Mexican law enforcement officials failed to address the concerns of victims' family members in Juárez?
2. How have women's and victim rights groups transformed themselves in their attempts to address the pattern of femicides in Ciudad Juárez?
3. What can international organizations do to effect change in Juárez?

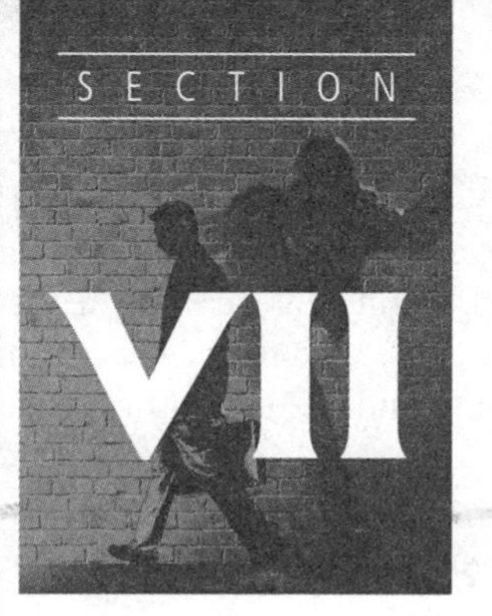

Girls and Juvenile Delinquency

Section Highlights

- The rise of the juvenile court
- The "double standard" for girls in the juvenile justice system
- Contemporary risk factors associated with girls and delinquency
- Gender-specific needs of young female offenders

While the majority of this book focuses on the needs of women and girls generally, this section highlights some of the specific issues facing girls within the juvenile justice system. Beginning with a discussion on the rise of the juvenile courts, this section highlights the historical and contemporary standards for young women in society and how the changing definitions of delinquency have disproportionately and negatively affected young girls. We then look at how these practices have manifested into today's standards of addressing cases of female delinquents. This section concludes with a discussion of reforms designed to respond to the unique needs of girls within the juvenile justice system.

The Rise of the Juvenile Court and the Sexual Double Standard

The understanding of adolescence within the justice system is a relatively new phenomenon, historically speaking. Originally, the development of the term "juvenile delinquent" reflected the idea that youths were "malleable" and could be shaped into law-abiding citizens (Bernard, 1992). A key factor in this process was the doctrine of *parens patriae*. *Parens patriae* began in the English Chancery Courts during the 15th century and evolved into the practice whereby the state could assume custody of children for cases where the child had no parents or the parents were deemed unfit care providers. As time passed, *parens patriae* became the government's justification for regulating adolescents and their behaviors under the mantra "in the best interests of the child" (Sutton, 1988).

Prior to the development of the juvenile court, the majority of cases of youthful offending were handled on an informal basis. However, the dramatic population growth, combined with the rise of industrialization, made it increasingly difficult for families and communities to control wayward youth. The doctrine of *parens patriae* led to the development of a separate system within the justice system designed to oversee the rehabilitation of youth who were deemed out of control.

Developed in 1825, the New York House of Refuge was one of the first reformatories for juvenile delinquents and was designed to keep youth offenders separate from the adult population. Unlike adults, youths were not sentenced to terms proportionate to their offenses in these early juvenile institutions. Instead, juveniles were committed to the institutions for long periods of time, often until their 21st birthday. Here, the doctrine of *parens patriae* was often used to discriminate against children of the poor, as the youth who were sent to the House of Refuge had not necessarily committed a criminal offense. Rather, youth were more likely to be described as "coming from an unfit home" or displaying "incorrigible behaviors" (Bernard, 1992). The practices at the House of Refuge during the 19th century were based less on controlling criminal behaviors and more on preventing future pauperism, which the reformers believed led to delinquency and crime (Sutton, 1988). Rather than address the conditions facing poor parents and children, reformers chose to respond to what they viewed as the "peculiar weaknesses of the children's moral natures" and "weak and criminal parents" (Bernard, 1992, p. 76).

The Progressive Era of the late 19th and early 20th century in the United States led to the child-saving movement, which comprised middle- and upper-class white citizens who "regarded their cause as a matter of conscience and morality [and] viewed themselves as altruists and humanitarians dedicated to rescuing those who were less fortunately placed in the social order" (Platt, 1969, p. 3). The efforts of the child-savers movement led to the creation of the first juvenile court in Chicago in 1899. The jurisdiction of the juvenile court presided over three youth populations: (1) children who committed adult criminal offenses, (2) children who committed status offenses, and (3) children who were abused or neglected by their parents (Chesney-Lind & Shelden, 2003).

Parens patriae significantly affected the treatment of girls who were identified as delinquent. During the late 19th and early 20th centuries, moral reformers embarked on an age-of-consent campaign, which was designed to protect young women from "vicious men" who preyed on the innocence of girls. Prior to the age-of-consent campaign, the legal age of sexual consent in 1885 ranged between 10 and 12 for most states. As a result of the efforts by moral reformers, all states raised the age of consent to 16 or 18 by 1920. While their attempt to guard the chastity of young women from exploitation was rooted in a desire to protect girls, these practices also denied young women an avenue for healthy sexual expression and identity. The laws that resulted from this movement were often used to punish young women's displays of sexuality by placing them in detention centers or reformatories for moral violations with the intent to incarcerate them throughout their adolescence. These actions held women to a high standard of sexual purity, while the sexual nature of men was dismissed by society as normal and pardonable behavior. In addition, the reformers developed their policies based on a White, middle-class ideal of purity and modesty—anyone who did not conform to these ideals was viewed as out of control and in need of intervention by the juvenile court. This exclusive focus by moral reformers on the sexual exploitation of White, working-class women led to the racist implication that only the virtues of White women needed to be saved. While reformers in the Black community were equally interested in the moral education of young women and men, they were unsupportive of the campaign to impose criminal sanctions on offenders for sexual crimes, as they were concerned that such laws would unfairly target men of color (Odem, 1995).

Age-of-consent campaigners viewed the delinquent acts of young women as inherently more dangerous than the acts of their male counterparts. Due to the emphasis on sexual purity as the pathway toward healthy adulthood and stability for the future, the juvenile reformatory became a place to shift the focus away from their sexual desire and train young girls for marriage. Unfortunately, this increased focus on the use of the reformatory for moral offenses allowed for the practice of net-widening to occur, and more offenders were placed under the supervision of the juvenile courts. Net-widening refers to the practice whereby programs such as diversion were developed to inhibit the introduction of youth into the juvenile justice system. However, these practices often expanded the reach to offenses and populations that previously were outside the reach of the juvenile justice system. The effects of this practice actually increased the number of offenders under the general reach of the system, whether informally or formally.

Beyond the age-of-consent campaign, the control of girls' sexuality extended to all girls involved in the juvenile court, regardless of offense. A review of juvenile court cases between 1929 and 1964 found that girls who were arrested for status offenses were forced to have gynecological exams to determine whether or not they had engaged in sexual intercourse and if they had contracted any sexually transmitted diseases. Not only were these girls more likely to be sent to juvenile detention than their male counterparts, they spent three times as long in detention for their "crimes" (Chesney-Lind, 1973). Indeed, throughout the early 20th century, the focus on female sexuality and sexually transmitted infections reached epic proportions, and any woman who was suspected to be infected with an STI was arrested, examined, and quarantined (Odem, 1995).

In addition to being placed in detention centers for engaging in consensual sex, young women were often blamed for "tempting defendants into immoral behavior" (Odem, 1995, p. 68) in cases where they were victims of forcible sexual assault. Other historical accounts confirm how sexual victimization cases were often treated by the juvenile court in the same manner as consensual sex cases—in both situations the girl was labeled as delinquent for having sex (Shelden, 1981). These girls were doubly victimized, first by the assault and second by the system. During these court hearings, a woman's sexual history was put on display in an attempt to discredit her in front of a jury, yet the courts did not request similar information about a man's sexual history as it would "unfairly prejudice the jury against him" (Odem, 1995, p. 70). These historical accounts emphasized that any nonmarital sexual experience, even forcible rape, typically resulted in girls treated as offenders.

The trend of using sexuality as a form of delinquent behavior for female offenders continued throughout the 20th and into the 21st century. The court system has become a mechanism through which control of female sexuality is enforced. Males enjoy a sense of sexual freedom that is denied to girls. In regards to male sexuality, the only concern generally raised by the court is centered on abusive and predatory behaviors toward others, particularly younger children. Here, probation officer narratives indicate that court officials think about sexuality in different ways for male and female juvenile offenders. For boys, no reference is made regarding noncriminalized sexual behaviors. Yet for girls, the risk of victimization becomes a way to deny female sexual agency. Here, probation officers would comment in official court reports about violations of moral rules regarding sexuality and displays of sexual behavior. In many cases, these officers expressed concern for the levels of sexual activity in which the girls were engaging (Mallicoat, 2007). The consequence of this desire to protect young girls results in their learning about sexuality in terms of disease and victimization, not in terms of pleasure and agency—"speak very often not of the power of desire but of how their desire may get them in trouble" (Tolman, 1994, p. 338). Regardless of the nature of offending for a woman, it was often interpreted as sexual in nature or was accompanied by a disturbance or unfavorable behavior that involved her sexuality (Triplett & Myers, 1995). Some researchers have argued that the continued control of female sexuality is related to the general control of women in society.

> Any discourse which legitimizes (a young woman's pleasure), acknowledges her sexual knowledge, values her performance and places it under her control is potentially threatening to his masculinity . . . Without a discourse of desire, but within discourses of victimization, we deny the female sexual subject, we deny girls sexual agency, they cannot speak about and we do not hear them speak about their sexuality or their sexual experience, both desired and imposed. (Alder, 1998, pp. 86–87)

The Nature and Extent of Female Delinquency

Girls are the fastest growing population within the juvenile justice system. Not only have the number of arrests involving girls increased, but also the volume of cases in the juvenile court involving girls has expanded at a dramatic rate. Despite the increased attention on females by the agents of the juvenile justice system and the public in general, it is important to remember that girls continue to represent a small proportion of all delinquency cases, as boys' offending continues to dominate the juvenile justice system (and the criminal justice system).

As discussed in Section I, the Uniform Crime Reports (UCR) reflects the arrest data from across the nation. This resource also includes information on juvenile offenders. Given that law enforcement officials represent the most common method through which juvenile offenders enter the system, arrest data provide a first look at the official processing of juvenile cases. Here, we can assess the number of crimes reported to law enforcement involving youth offenders, the most serious charge within these arrests, and the disposition by police in these cases. As a student of this text, you have already learned that the UCR data are not without their flaws. Given that juveniles are often involved in acts that are nonserious and nonviolent in nature, these practices of crime reporting and how the data are compiled can have a significant effect on the understanding of juvenile delinquency by society. Despite these flaws, the UCR remains the best resource for investigating arrest rates for crime (Snyder & Sickmund, 2006).

UCR data on juvenile offenders indicate that in 1980, girls represented 20% of juvenile arrests. By 2003, girls' participation in crimes increased to 29%. The majority of this increase occurred during the late 1980s to early 1990s when the rise of "tough on crime" philosophies spilled over into the juvenile arena. Table 7.1 illustrates data on the estimates of juvenile arrests and the percentages of females involved in these cases. A comparison of 1980 to 2003 data indicates that the female proportion of violent crime index offenses increased from 10% to 18%, while property offenses increased from 19% to 32%. These shifts in girls' arrests have certainly increased the attention of parents, juvenile court officials, and scholars, leading many to question the causes of such increases in girls' participation in delinquency. Even though boys are arrested more often than their female juvenile counterparts, the rate of arrests for girls has increased at a faster rate compared to boys over the past three decades (Knoll & Sickmund, 2010; Snyder & Sickmund, 2006). Clearly, the hype has become focused on the narrowing of the gender gap between males and females, despite the fact that females continue to represent a small proportion of the offending population.

Table 7.1 Uniform Crime Report Data on Female Juvenile Arrests

Most Serious Offense	Juvenile Arrest Estimates	Female Involvement (%)
Total	2,220,300	29
Violent Crime Index	92,300	18
Murder/non-negligent manslaughter	1,130	9
Forcible rape	4,240	2
Robbery	25,440	9
Aggravated assault	61,490	24
Property Crime Index	463,300	32
Burglary	85,100	12
Larceny-theft	325,600	39
Motor vehicle theft	44,500	17
Arson	8,200	12
Other (simple assault)	241,900	32
Forgery and counterfeiting	4,700	35
Fraud	8,100	33
Embezzlement	1,200	40
Vandalism	107,700	14
Weapons (possession, etc.)	39,200	11
Prostitution	1,400	69
Other sex offense	18,300	9
Drug abuse	197,100	16
Disorderly conduct	193,000	31
Curfew and loitering	136,500	30
Runaway	123,600	59

Much of the media attention throughout the past decade has fixated on the levels of violence and aggression among adolescent girls. The media present magazine cover stories and news headlines that raise the question, "Are girls becoming more violent?" This portrayal of "bad girls" by the media has been linked to data that reflected a significant increase in the number of arrests for crimes of violence involving girls. The increased attention on female delinquency by law enforcement has, in turn, affected the handling of these cases by the juvenile courts. In 2007, the U.S. juvenile courts were dealing with an estimated 1.7 million cases each year. Since the early 1990s, girls have represented a growing proportion of cases in the juvenile courts.

In 1991, girls made up 19% of delinquency cases, 26% in 2002, and 27% in 2007. By 2007, the juvenile courts were dealing with 448,900 cases with female defendants, which is more than two times the number of female juvenile court cases in 1985 (Knoll & Sickmund, 2010; Snyder & Sickmund, 2006). In addition to the increase in the number of cases referred to the juvenile court, cases today are more likely to be processed formally by the court instead of informally. Informal processing involves sanctions in which youth participate on a voluntary basis. Rather than file formal charges, informal processing allows youth to participate in community service, victim restitution, and mediation and voluntary supervision. If they complete these arrangements successfully, the case is closed without the formal processing, and youth are diverted from the system. In cases handled formally by the court, a petition is filed requesting a court hearing, which can result in the designation of being labeled a delinquent. In 1985, only 35% of girls' cases (and 48% of boys) were processed in a formal manner. In comparison, 50% of female cases (61% male) were handled formally by juvenile court authorities in 2007, signaling an end to informal sanctions by the court. While girls were slightly less likely to be adjudicated as delinquent (64% females compared to 67% males), this difference is negligible (Knoll & Sickmund, 2010; Snyder & Sickmund, 2006). Research by Steffensmeier, Schwartz, Zhong, and Ackerman (2005) found the increase in arrests and formal processing of juvenile cases has disproportionately impacted girls in three ways: (1) the practice of up-charging by prosecutors, whereby less serious forms of conduct were processed as assault cases; (2) an increase in cases of domestic disputes between intimates; and (3) a decrease in tolerance for girls who "act out." Brown, Chesney-Lind, and Stein (2007) add two additional examples of the disparate treatment of girls: (1) the shift from informal to formal processing for cases of family-based violence between parents and children; and (2) the rise of zero-tolerance policies for schools. Generally speaking, it appears that the juvenile justice system is less tolerant of girls who break the socially proscribed norms of gendered behavior and are punished by officials for "acting out," while boys benefit from a greater acceptance of these "unacceptable" behaviors (Carr, Hudson, Hanks, & Hunt, 2008).

While the data on increased arrests and formal processing of cases by the juvenile court lead many to assume that girls' delinquency is on the rise, self-report studies do not confirm this position, which indicates that the levels of violence have actually decreased for both boys and girls. In addition, data from the National Longitudinal Survey of Adolescent Health indicate that boys engage in significantly greater levels of delinquency overall compared to girls, and boys engage in higher rates of violent behaviors than girls (Daigle, Cullen, & Wright, 2007).

Given the contradiction between self-report and arrest data, scholars have concluded that the levels of violence amongst girls have not changed. Rather, it is the response by agents of social control that has shifted, leading to an increase in the arrest and formal processing of cases involving adolescent girls. A review of these cases indicates an overrepresentation of incidents of school- and family-based violence, cases that historically were handled on an informal basis (Brown et al., 2007). In an attempt to regain control over their daughters, many parents turn to the police and the juvenile court for help. For offenders who are not under the supervision of juvenile authorities, they may talk to or threaten the youth, using fear as a tool to gain compliance from the youth. Due to the high degree of discretion granted to police, officers have the ability to deal with these cases in an informal matter. However, this discretionary power can also be used to bring youth to the attention of juvenile court authorities in cases of minor disputes within the family. Typically, these cases are examples of the symbolic struggles of adolescence, whereby youth battle their parents for power, control, and freedom. Yet, recent data indicate that these family struggles have become the new road to the juvenile justice system, whereby youth are arrested in cases of domestic dispute and minor assault (Feld, 2009; Brown et al., 2007). Research with incarcerated girls at a juvenile facility supports this finding:

> Many of these family assault charges seem to have one theme in common. They appear to almost always involve a parent's attempting to physically block the daughter from taking some kind of action, such as leaving the house, and a daughter's subsequent push to resist the parent's restraining action results in the parent's alleging assault. (Davis, 2007, p. 422)

The formal processing of these cases becomes the basis to place a girl under probation supervision, and any subsequent power struggles between the parent and child become the grounds for a technical violation of probation. Over time, these power struggles become the basis to institutionalize youth or remand them to a group home environment. Ultimately, this attempt to reinstate and reinforce parental authority only increases the punitive consequences for youth who are labeled by the system as "out of control."

Given the increase in the number of female cases that are formally processed and adjudicated delinquent, it is no surprise that punishments have also increased for girls. While boys are more likely to be detained for their cases (22% of cases, compared to 17% of girls' cases), girls who are denied release generally spend significantly greater amounts of time in detention compared to boys (Belknap et al., 1997; Snyder & Sickmund, 2006). In addition, girls are subjected to longer periods of supervision, a practice that appears to increase the delinquency in girls due to excessive and aggressive monitoring techniques (Carr et al., 2008). Finally, the number of residential placements or sentences to formal probation terms has also increased for girls. For both boys and girls, the number of cases involving residential out-of-home placement between 1985 and 2002 increased 44%, though boys were more likely to be placed out of the home than girls (Snyder & Sickmund, 2006). Girls of color appear to be disproportionately affected by the shift to formal processing of delinquency cases, as Black and Hispanic girls are more likely to receive detention, whereas White girls are more likely to be referred to a residential treatment facility (Miller, 1994).

Technical Violations: The New Status Offense

Like the historical and contemporary control of female sexuality, status offenses are another realm where doctrines such as *parens patriae* allow for the juvenile court to intervene in the lives of adolescents. Status offenses are acts that are only illegal if committed by juveniles. Examples of status offenses include the underage consumption of alcohol, running away from home, truancy, and curfew violations. While the juvenile court was founded with the idea of dealing with both delinquency and status offenses, today's courts have attempted to differentiate between the two offense categories due to constitutional challenges on status offenses (Bernard, 1992). One of the elements of the Juvenile Justice and Delinquency Prevention Act of 1974 called for the decriminalization of status offenders in any state that received federal funds. Prior to its enactment, young women were much more likely to be incarcerated for status offenses compared to their male counterparts (Chesney-Lind & Shelden, 2004). While the institutionalization of sexually wayward girls officially ended with the JJDP act of 1974, several scholars suggest that the juvenile courts have negotiated ways around this legal roadblock. While youth can no longer be incarcerated specifically for status offenses, researchers contend that the practice of institutionalizing girls who are deemed out of control continues today (Acoca & Dedel, 1998; Chesney-Lind & Shelden, 2004). The modern-day practice of institutionalizing girls for status offenses is known as bootstrapping. The process of bootstrapping involves cases that are currently on probation or parole for a criminal offense who are prosecuted formally for a probation violation as a result of

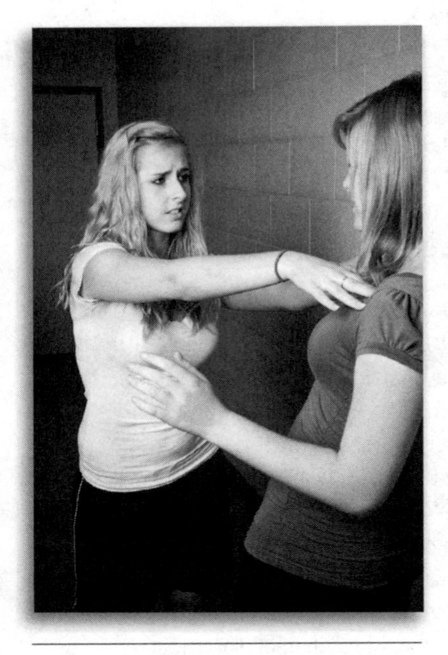

▲ **Photo 7.1** Recent years have seen an increase in the reporting of cases involving "violent girls." Are girls really becoming more violent, or have parents, schools, and police changed the way in which they respond to incidents of girls who are deemed to be "out of control"?

committing a status offense, such as running away from home or truancy (Owen & Bloom, 1998). While provisions of the revision of the Juvenile Justice and Delinquency Prevention Act in 1992 attempted to make the practice of bootstrapping more difficult for courts, evidence indicates that the practice continues in an inequitable fashion against girls. Research by Feld (2009) suggests that acts that were once treated as status offenses are now processed as minor acts of delinquency due to the expansion of the discretionary powers available to schools, police, and juvenile justice officials. The replacement of status offenses by probation violations has allowed justice officials to recommit girls to these residential facilities. While a commitment to a state institution or detention center for these type of status offenses is prohibited by the original authorization of the JJDP Act in 1974, it appears that these community-based residential facilities serve as a way to alternatively "incarcerate" young girls deemed "out of control" by the courts (Carr et al., 2008).

> The failure to provide alternatives to institutional confinement for "troublesome girls" creates substantial pressures within the juvenile justice system to circumvent DSO (deinstitutionalization of status offenders) restrictions by the simple expedient of relabeling them as delinquents by charging them with assault. (Feld, 2009, p. 261)

The increase of female delinquency and the changes in the processing of offenders have brought new concerns for the juvenile justice system. The majority of female offending is nonviolent in nature or results from technical violations of probation (Owen & Bloom, 1998). Caseloads for girls are more likely to reflect offenses like simple assault and larceny theft and are less likely to involve cases of robbery, burglary, vandalism, and drugs, compared to boys (Snyder & Sickmund, 2006). The increase in penalties for girls is also reflected in the limited availability for community-based and rehabilitative programming options. Some scholars suggest that it is the failure of other social systems such as schools, child welfare, and mental health programs, which indicates that these resources fail at the early stages of prevention and intervention and set the stage for future delinquency (Goodkind, 2005). The lack of resources both within and outside the juvenile justice system, combined with the "get tough on youth crime" agenda has led to the increased institutionalization of girls, many of whom are charged with less severe crimes and who are not a risk to public safety (Beger & Hoffman, 1998). This practice of net-widening ensures that once in the system, girls find it difficult to escape the watchful eye of the juvenile justice system (Chesney-Lind, 1997).

Risk Factors for Female Delinquency

Earlier sections of this text have highlighted the historical failures of criminology to address the unique causes of women's and girls' offending. The theoretical inattention to these issues has significantly affected the identification and delivery of services for women and girls. As Chesney-Lind and Shelden (2004) suggest, it is a failure for policy makers and practitioners to assume that, just because girls typically engage in nonviolent or nonserious acts of crime and delinquency, their needs are insignificant. Indeed, a historical review of the juvenile justice system finds that programs and facilities are ill equipped to deal with the needs of girls. Indeed, many of the official actions of the juvenile system have placed them at a greater risk for victimization. While boys and girls can exhibit many of the same risk factors for delinquency, such as family dysfunction, school failures, peer relationships, and substance abuse, the effects of these risk factors may resonate stronger for girls than they do for boys (Moffitt, Caspi, Rutter, & Silva, 2001). In addition, research indicates that girls possess significantly higher risk factors toward delinquency than boys. It is interesting to note that while White girls tend to exhibit significantly higher levels of risk for certain categories (such as substance abuse), youth of color, particularly African American youth, are significantly overrepresented in the juvenile court (Gavazzi, Yarcheck, & Lim, 2005). Given these failures of the juvenile courts, it is important to understand the risk factors for female delinquency in an effort to develop recommendations for best practices for adolescent delinquent and at-risk girls. For juvenile girls, the most significant risk factors for delinquency include a poor family relationship, a history of abuse, poor school performance, negative peer relationships, and issues with substance abuse. In addition, these risk factors are significantly interrelated.

The influence of the family unit is one of the most commonly cited references in the study of delinquency. Social control theorists have illustrated that a positive attachment to the family acts as a key tool in the prevention of delinquency. The family represents the primary mechanism for the internalization and socialization of social norms and values (Hirschi, 1969). Yet, research indicates that girls may have stronger attachments to the family compared to boys, which can serve as a protective factor against delinquency. However, families can only serve as a protective factor when they exist in a positive, prosocial environment. Research indicates that girls benefit from positive communication, structure, and support in the family environment (Bloom, Owen, Deschenes, & Rosenbaum, 2002b). Just as the family unit can protect girls from delinquency, it can also serve as a pathway to crime and lead girls into delinquency at a young age. Youth may turn to delinquency to enhance their self-esteem or to overcome feelings of rejection by their families (Matsueda, 1992). Research has indicated that negative family issues constitute a greater problem for girls than boys. Family fragmentation due to divorce, family criminality, and foster care placements, in addition to family violence and negative family attachment, have been identified as family risk factors for female delinquents. Families with high levels of conflict and poor communication skills, combined with parents who struggle with their own personal issues, place girls at risk for delinquency (Bloom, Owen, Deschenes, & Rosenbaum, 2002b). Research involving incarcerated girls in California indicated that several of the girls had experienced or witnessed the death of one or both of their parents, and half of their mothers and/or fathers had been incarcerated at some point in their lives (Acoca & Dedel, 1998). Regardless of the reason behind the loss of a parent in their life, girls who are raised in a single-parent home have a greater risk for delinquency involvement compared to girls raised in a two-parent household (McKnight & Loper, 2002). In addition, once a girl becomes immersed in the juvenile justice system, her delinquency can serve to increase the detachment between her and her family (Girls Incorporated, 1996).

Sexual, physical, and emotional abuse have long been documented as significant risk factors for female offenders. The impact of abuse is intensified when it occurs within the family. Such abuse can be detrimental to the positive development of the adolescent female and can result in behaviors such as running away, trust issues, emotional maladjustment, and future sexual risk behaviors. Girls are 3–4 times more likely to be abused than their male counterparts, and 92% of incarcerated girls in a California study reported having been subject to at least one form of abuse (emotional 88%, physical 81%, sexual 56%; Acoca & Dedel, 1998). While sexual abuse is the most studied form of abuse for girls, other forms of maltreatment can have a significant effect on the development of girls. Longitudinal research studies conducted by Cathy Spatz Widom (1989, 1991) indicate that youth (boys and girls) who experience abuse and neglect throughout childhood are at significant risk for delinquent behaviors and juvenile court intervention. In many cases, girls who experience violence as victims later engage in acts of violence toward others. While girls tend to engage in minor acts of delinquency, these violations are often the tip of the iceberg for the issues that affect preteen and adolescent females. In many cases, status offenses such as running away from home reflect an attempt to escape from a violent or abusive home environment. Unfortunately, in their attempt to escape from an abusive situation, girls often fall into criminal behaviors as a mechanism of survival. Widom (1989) found that childhood victimization increases the risk that a youth will run away from home, and that childhood victimization and running away increase the likelihood of engaging in delinquent behaviors. A history of sexual abuse also affects the future risk for victimization, as girls who are sexually abused during their childhood are significantly more likely to find themselves in a domestically violent relationship in the future (McCartan & Gunnison, 2010).

Research has also focused extensively on the role of peer relationships in adolescent offending. The presence of delinquent peers presents the greatest risk for youth to engage in their own acts of delinquency. While much of the research suggests that girls are more likely to associate with other girls (and girls are less likely to be delinquent than boys), research by Miller, Loeber, and Hipwell (2009) indicates that girls generally have at least one friend involved in delinquent behaviors. While girls in this study indicated that they associated with peers of both genders, it is not the gender of the peers that can predict delinquency. Rather, it is the number of delinquent peers that determines whether a youth engages in problem behaviors. Here, the effects of peer pressure and the desire for acceptance often lead youth into delinquency, particularly if the majority of the group is involved in law-violating behaviors.

Several factors can affect one's association with delinquent peers. First, scholars indicate the shift toward unsupervised "free time" among youth as a potential gateway to delinquency, as youth who are involved in after-school structured activities are less likely to engage in delinquency (Mahoney, Cairns, & Farmer, 2003). Given the slashing of school-based and community programs due to budgetary funds, there are fewer opportunities to provide a safe and positive outlet for youth in the hours between the end of the school day for youth and the end of the work day for parents. Second, age can also affect the delinquent peer relationship. For girls, peer associations with older adolescents of the opposite sex have an impact on their likelihood to engage in delinquent acts if the older male is involved in crime-related behaviors (Stattin & Magnusson, 1990). Finally, negative family attachment also affects the presence of delinquent peers, as girls whose parents are less involved in their daily lives and activities are more likely to engage in problem behaviors (such as substance abuse) with delinquent peers (Svensson, 2003).

School failures have also been identified as an indicator of concern for youth at risk. Truancy is a concern for educational professionals, as it can be an indication of school failures such as suspension, expulsion, or being held back. In research by Acoca and Dedel (1998), 85% of incarcerated girls in their study compared their experience in school to a war zone, where issues such as racism, sexual harassment, peer violence, and disinterested school personnel increased the likelihood of dropping out. For girls, success at school is tied to feelings of self-worth—the more students feel attached to the school environment and the learning process and develop a connection to their teachers, the less likely they are to be at risk for delinquency (Crosnoe, Erickson, & Dornbusch, 2002). Additionally, the slashing of prosocial extracurricular activities has also negatively affected girls. Here, activities that involve creativity, build relationships, and enhance personal safety help to build resiliency in young women and guard against delinquent behaviors (Acoca & Dedel, 1998). Finally, the involvement of a parent in his or her daughter's school progress can help build resiliency for girls (Bloom et al., 2002b).

Several risks have been identified for adolescent females' involvement in alcohol and drug use: early experimentation and use, parental use of drugs and alcohol, histories of victimization, poor school and family attachments, numerous social opportunities for use, poor self-concept, difficulties in coping with life events, and involvement with other problem behaviors (Bodinger-deUriarte & Austin, 1991). Often, substance abuse highlights the presence of other risk factors for delinquency (Bloom et al., 2002b). Substance abuse affects female delinquency in two ways. First, girls who experience substance abuse in their families may turn to behaviors such as running away to escape the violence that occurs in the home as a result of parental drug and alcohol use. Second, girls themselves may engage in substance abuse as a mechanism of self-medication to escape from abuse histories (Chesney-Lind & Shelden 2003). In addition, research indicates that the use of substances can be a gendered experience. While boys tend to limit their drug use to marijuana, girls experiment with and abuse a variety of substances, including methamphetamines, cocaine, acid, crack, and huffing chemicals. Not only did their poly-drug use indicate significant addiction issues, their substance abuse altered their decision-making abilities, influenced their criminal behaviors, and placed them at risk for danger (Mallicoat, 2007). While substance abuse increases the risk for delinquency for girls, the absence of substance abuse serves as a protective factor against delinquency (McKnight & Loper, 2002).

Meeting the Unique Needs of Delinquent Girls

While girls may make up a minority of offenders in the juvenile justice system, their needs should not be absent from juvenile justice policies. As indicated earlier, girls have a number of different and interrelated issues that historically have been ignored by the system. The 1992 reauthorization of the Juvenile Justice and Delinquency Prevention Act acknowledged the need to provide gender-specific services to address the unique needs of female offenders. The allocation of funds by Congress to investigate gender-specific programming provides that each state should

> 1) conduct an analysis of the need for an assessment of existing treatment and services for delinquent girls; 2) develop a plan to provide needed gender-specific services for the prevention and treatment of juvenile delinquency; and 3) provide assurance that youth in the juvenile system are treated fairly regarding their mental, physical and emotional capabilities, as well as on the basis of their gender, race and family income. (Belknap & Holsinger, 1998, p. 55)

Over the past two decades, research has highlighted the factors that may affect a young woman's road to delinquency, and the reauthorization of the JJDP mandates that states incorporate this understanding into the assessment tools and programming options for girls.

In an effort to respond to the declaration for gender-specific services, many states have embarked on systematic evaluations designed to investigate the needs of girls in their facilities and develop recommendations of best practices for girls. Here, the goal is that information will be used by program providers in developing and implementing prevention and intervention strategies for girls. In a report on gender-specific services for adolescent females, Belknap et al. (1997) write,

> When examining gender-specific programming, it is important to recognize equality does not mean sameness. Equality is not about providing the same program, treatment and opportunities for girls and boys. . . . Equality is about providing opportunities that mean the same to each gender. The new definition legitimizes the differences between boys and girls. Programs for boys are

> more successful when they focus on rules and offer ways to advance within a structured environment, while programs for girls are more successful when they focus on relationships with other people and offer ways to master their lives while keeping these relationships intact. (p. 23)

What should gender-specific programming look like? Programs must be able to address the wide variety of needs of the delinquent girl—given that many of the risk factors for delinquency involve a number of interrelated issues, programs need to be able to address this tangled web of needs rather than attempt to deal with issues on an individual and isolated basis. Research identifies that a history of victimization and a lack of attachment to their family (or a negative family attachment) are perhaps the most significant foundational issues facing at-risk and delinquent girls. The greatest long-term success comes from those programs that provide support not just for the individual girl, but for her support system as well. Unfortunately, many family members resist being involved in programming, as they fail to accept responsibility for the role that they may have played in the development of their daughter's delinquency (Bloom et al., 2002a). This lack of involvement raises significant concerns for the family environment of these girls. Although more than one half of girls reported that they do not get along with the parents (51%) and reported that their relationship with their parents was involved in getting them in trouble (59%), 66% of the girls stated that they would return to live with their parents after their release. The fact that two thirds of the girls will return to their homes upon release is problematic given that 58% of girls surveyed reported experiencing some form of violence in their home. It is impossible to develop programs for incarcerated females without reevaluating policies that contribute to the destruction of the family. Gender-specific programming for adolescent females needs to focus on rebuilding the family unit and developing resiliency factors such as positive role modeling. In addition, programs such as family counseling and drug and alcohol treatment can positively affect troubled families.

The prevalent nature of a victimization history in adolescent females raises this issue to one of central importance in gender-specific programming. Not only do programs need to provide counseling services for these girls, but placement services for females need to be expanded, as punishment in detention is not an appropriate place for girls who run away to escape their abusers. Because early childhood victimization often leads to risky sexual behaviors with the conclusion of teenage pregnancy and parenthood, education should be offered to these girls as a preventive measure for pregnancy and sexually transmitted diseases. Similar to the needs of high-risk pregnancies, juvenile justice facilities are often ill equipped to deal with the physical and mental health needs of incarcerated females. The emotional needs of developing teenagers combined with the increase in the prevalence of mental health disorders of incarcerated females makes this an important component for gender-specific programming for female populations. Physical and mental health complaints by youth need to be interpreted by staff and facilities as a need, not as a complaining or manipulating behavior. Additionally, such interventions must be established on an ongoing basis for continual care, versus limited to an episodic basis (Acoca & Dedel, 1998).

Although traditional research on female offenders has targeted risk factors that lead to negative behaviors, recent research has shifted to include resiliency or protective factors to fight against the risks of delinquent behavior. These factors include intelligence; brilliance; courage; creativity; tenacity; compassion; humor; insightfulness; social competence; problem-solving abilities; autonomy; potential with leadership; engagement in family, community, and religious activities; and a sense of purpose and belief in the future. While these resiliency factors typically develop within the context of the family, the support for such a curriculum needs to come from somewhere else, since many delinquent girls often come from families in crisis (Acoca & Dedel, 1998).

While the intent to provide gender-specific services indicates a potential to address the unique needs of girls, not all scholars are convinced that girls will be able to receive the treatment and programs that are so desperately needed. While many states embarked on data-heavy assessments reflecting the needs of girls, few of these adventures have translated into effective programmatic changes (Chesney-Lind & Shelden, 2003). Funding remains the most significant barrier in providing effective services for girls. Even when gender-specific programming options exist, the need for these services can outweigh the available options. The limited number of placements, combined with long waiting lists for such services, often make treatment options unavailable for most girls (Bloom et al., 2002a). However, several individual and community factors also affect program delivery, including lack of information or difficulties in accessing services, resistance toward programming by girls and their families, and distrust of service providers. In addition, racial, economic, and cultural issues can affect whether communities will seek out assistance and the degree to which these services will reflect culturally relevant issues (Bloom et al., 2002b). In order to develop effective and available programming, the system needs to place the allocation of resources as a priority in identifying and addressing the needs of girls in the juvenile justice system.

Summary

- The age-of-consent campaign raised the age of sexual consent to 16 or 18 in all states by 1920.
- Historically, women were denied an avenue for healthy expression of sexuality.
- Female victims were blamed for tempting men into immoral acts in cases of forcible sexual assault.
- Arrest data and self-report data present contradictory images on the nature and prevalence of female violence.
- Scholars suggest that agents of social control have altered the way in which they respond to cases of female delinquency, particularly in cases of family or school violence.
- Many incidents of family violence stem from symbolic struggles for adolescent freedom between girls and their parents.
- For juvenile girls, the most significant risk factors for delinquency include a poor family relationship, a history of abuse, poor school performance, negative peer relationships, and issues with substance abuse.
- Effective gender-specific programming needs to provide long-term programming for girls and their social support network that addresses the causes of delinquency in girls' lives.
- Programming that includes resiliency or protective factors plays a significant role in gender-specific programming.
- Programs face significant barriers in implementing services for girls.

KEY TERMS

Juvenile delinquency

Parens patriae

Age-of-consent campaign

Net-widening

Bootstrapping

Informal processing

Formal processing

Status offenses

Juvenile Justice and Delinquency Prevention (JJDP) Act of 1974

1992 Reauthorization of the Juvenile Justice and Delinquency Prevention (JJDP) Act

Gender-specific programming

Risk factors for female delinquency

Resiliency

DISCUSSION QUESTIONS

1. How did the age-of-consent campaign punish girls and deny healthy expressions of sexuality? What effects of this movement remain today?
2. How have girls continued to be punished for status offenses, despite the enactment of the JJDP Act of 1974?
3. What risk factors for delinquency exist for girls?
4. What should gender-specific programming look like? What challenges do states face in implementing these programs?

WEB RESOURCES

Girls Study Group: http://girlsstudygroup.rti.org/

Office of Juvenile Justice and Delinquency Prevention: http://www.ojjdp.gov

National Center for Juvenile Justice: http://www.ncjj.org

READING

In the section introduction, you learned about the decriminalization of status offenders. This article explores how families have found new ways to deal with their troubled daughters. Here, families seek out police as a way to manage family conflicts, which ultimately places young girls under the control of the juvenile court whereby they are bootstrapped for "crimes" that were once considered status offenses.

At-Risk Girls and Delinquency

Career Pathways

Carla P. Davis

Before the 1974 Juvenile Justice and Delinquency Prevention (JJDP) Act, girls were arrested or detained primarily for status offenses (offenses that would not be regarded as wrongdoing if committed by an adult). Although decriminalization of status offenses should have resulted in a diminished presence of girls in the justice system, there has instead been a recent increase. More than a quarter of the youths arrested every year are girls. Delinquency cases involving girls increased by 83% between 1988 and 1997, with data indicating an increase of 106% for African American girls, 74% for Anglo girls, and 102% for girls of other races (American Bar Association and National Bar Association, 2001; Federal Bureau of Investigation [FBI], 1998). The use of detention for adolescent girls increased 65% between 1988 and 1997, with African American girls, as well as African American women, comprising nearly 50% of those in secure detention (American Bar Association and National Bar Association, 2001; Bureau of Justice Statistics, 2001, 2003).

Through ethnographic fieldwork and interviews, this article reveals current negotiations and practices of parents and the juvenile justice system after the deinstitutionalization of status offenses. Through examining the contexts of the girls' offenses, family conflict over issues of parental authority emerged as a salient factor underlying the girls' initial contact with the system and the construction of their recorded offenses. This article illustrates the ways that parental responses to challenges to their authority influence entry and movement through the system. These illustrations suggest the possibility that although the 1974 act decriminalized status offenses, challenges to parental authority are now being constructed and processed under other "official" categories of crimes or delinquencies.

Getting Into the System: Status Offenses and the Sexual Double Standard

Previous literature documents how girls in the juvenile justice system have historically been disproportionately sanctioned for status offenses (Belknap & Holsinger, 1998; Chesney-Lind & Shelden, 2004; Cohn, 1970; Datesman & Scarpitti, 1977; Gibbons & Griswold, 1957;

SOURCE: Davis, C. P. (2007). At-risk girls and delinquency: Career pathways. *Crime and Delinquency, 53*(3), 408–435.

Odem, 1995; Odem & Schlossman, 1991; Schlossman & Wallach, 1978). The phenomenon of girls entering the system through status offenses is part of the broader historical development of the creation of organizations to monitor the social and moral behavior of troubled youths who do not commit serious offenses. This evolution led to the establishment of the first juvenile court in 1899. In essence, the purpose was to distinguish delinquent from criminal behavior (Platt, 1969).

In the 19th century, with the growth of urbanization, industrialization, immigration, and increased geographic mobility, communal mechanisms of social control collapsed. An underclass emerged, labeled as poor, who were perceived as living in slums regarded as unregulated and lacking in social rules (Platt, 1969). Children of the poor were primary among those whom it was believed could benefit from rehabilitation. Most were recent immigrants, with ethnic backgrounds different from that of established American residents. In general, these children were not considered hardened offenders but instead were considered to be vagrant or wayward youths whose noncriminal behavior could be rehabilitated. They were often thought of as "incorrigible" or "beyond control" and considered to be living in environments likely to foster delinquency and criminality (Chesney-Lind & Shelden, 2004). Social reformers emphasized the temporary and reversible nature of adolescent crime and believed that delinquent children must be saved by preventing them from pursuing criminal careers. Institutions proliferated to reform the behaviors of these youths. This child-saving movement crafted a system of government that had unprecedented authority to intervene in the lives of families, and particularly in the lives of youths (Empey, 1982; Platt, 1969; Zatz, 1982).

The moral behavior of girls was of specific concern to the child savers (Belknap & Holsinger, 1998; Chesney-Lind & Shelden, 2004). As part of this reform movement, White middle-class women reformers sought to protect White, working-class girls from straying from moral paths. These middle-class women's activities revolved around monitoring the moral and sexual behaviors of working-class, particularly immigrant, girls (Chesney-Lind & Shelden, 2004; Gordon, 1988; Odem, 1995). On the basis of middle-class ideals of female sexual propriety, reformers assumed they had the authority to define appropriate behavior for working-class women and girls. Girls who did not conform to these ideals were labeled as *wayward* and deemed to be in need of control by the state in the form of juvenile courts, reformatories, and training schools (Chesney-Lind & Shelden, 2004; Odem, 1995). Black female delinquents were placed in adult institutions or sent out of state until it became practically or fiscally unfeasible to do so (Young, 1994). Research examining court practices after the court's initial inception reveals a preoccupation with girls' sexuality, as revealed in charges relating to some form of waywardness or immorality, and views contemporary status-type offense charges as buffer charges for suspected sexual activity (Chesney-Lind & Shelden, 2004; Odem, 1995; Odem & Schlossman, 1991; Schlossman & Wallach, 1978; Shelden, 1981).

Studies also document the historical pattern of parental use of the status offense category in referring their daughters to authorities for a variety of activities (Andrews & Cohn, 1974; Belknap & Holsinger, 1998; Chesney-Lind & Shelden, 2004; Ketchum, 1978; Odem & Schlossman, 1991; Teitelbaum & Gough, 1977). This, coupled with the vagueness of status offense statutes and the precedence of authorities and courts to uphold parental authority, makes the misuse of the status offense category particularly likely (Sussman, 1977). The recognition that some parents turn to the courts to enforce their authority is thought to be a primary reason for many girls' presence in the juvenile justice system (Chesney-Lind & Shelden, 2004).

After establishment of the juvenile court, the next major attempt to address noncriminal, troubled youths was in the 1960s when another reform movement attempted to redefine categorization of wayward and/or noncriminal offenders (Empey, 1973; Zatz, 1982). The result of these reform efforts was the 1974 JJDP Act. The federal government recognized a specific category of offenders—"status offenders"—and ordered measures that would remove or divert this group of juveniles away from the juvenile justice system and away from incarceration. Because historically, girls have been disproportionately sanctioned for status offenses, the immediate impact of the 1974 JJDP Act was greater for girls. Girls' institutionalization rates for status offenses fell by 44% (Krisberg & Schwartz, 1983). However, this decline in the institutionalization of status offenders leveled off between 1979 and 1982, and there is continuing concern that status offenders are not being sufficiently differentiated from delinquents and that they are still largely represented in the justice system (Arthur D. Little, Inc., 1977; Chesney-Lind & Shelden, 1998, 2004; Federle & Chesney-Lind, 1992; Schwartz, Jackson-Beeck, & Anderson, 1984; Zatz, 1982).

Previous literature suggests two opposing views for the impact of deinstitutionalization on the discretionary powers of the police, with some asserting that it has increased discretionary powers, allowing circumvention of the 1974 JJDP Act (Austin & Krisberg, 1981; Klein, 1979; Lemert, 1981), and others arguing that deinstitutionalization has weakened the discretionary powers of the police (Schwartz, 1989).

Previous literature has suggested that status-type offenses are being relabeled as *criminal offenses* (Klein, 1979; Mahoney & Fenster, 1982) and that some actions constituting nonserious family conflicts have been relabeled upward as violent or assault charges (Acoca, 1999; Acoca & Dedel, 1998; Chesney-Lind & Shelden, 1998, 2004; Mayer, 1994). This is considered one reason underlying the increase in assault charges for girls, particularly a rise in the category of "other" assaults. This is particularly thought to be a significant factor in African American girls' prevalence in the system (Bartollas, 1993; Robinson, 1990). Although authorities may have historically incarcerated White immigrant working-class girls for protectionist reasons during the child-saving movement, Black and Brown girls are also likely to be detained or incarcerated because, like Black and Brown boys, they are seen as dangerous.

The findings in this article begin to provide evidence of the processes of relabeling status-type offenses as criminal offenses. The data illustrate how these families in their interactions with authorities have negotiated alternative methods for dealing with troublesome teenage girls since the 1974 Deinstitutionalization Act has restrained courts from responding vigorously to status-type offenses.

Theoretical Framework: The Insanity of Place

Conflicts over parental authority form the basis on which parents eventually turn to the justice system to seek help in restoring their authority with their daughters. Although enforcing the sexual double standard is an underlying source of tension between parents and their daughters, this in itself does not explain how the girls come into contact with the system. Goffman's (1971) "The Insanity of Place" provides a useful framework for understanding the processes by which girls come in contact with the juvenile justice system. In "The Insanity of Place," Goffman examined the processes in families by which a person comes to be classified as mentally ill and committed to hospitalization. According to Goffman, the productive functioning or welfare of a family depends on family members' supporting the expected internal order of authority relationships. In supporting the internal order of a family, members know and keep their places in the family structure. When a family member fails to support the internal order by not keeping his or her expected social position in relation to others within the family, this threatens one of the fundamental elements on which family unity is based.

A breakdown in family solidarity begins when a family member, for whatever reason, feels the life that others in the family have been according him or her is no longer sufficient, thus the member makes demands for change. In response, other family members may accept these demands for change as valid and modify the structure of the family in accordance, or they may refuse to recognize the demands and attempt to maintain the existing social structure. It is the latter—when family members decide not to honor or recognize the family member's demands for change and the demanding family member's refusal to fall back to the status quo—that results in increasing tension. In this standoff, the family member may either voluntarily withdraw from the family organization or remain a part of the family. If he or she chooses to remain with the family but refuses to fall back to the status quo in family relationships, the member in effect promotes himself or herself in the family hierarchy, thus beginning his or her "manic" activity. According to Goffman (1971) the demands of the "maniac" are not necessarily bizarre in themselves but are bizarre coming from the person with respect to his or her social location in the family. At this breaking point, the stronger of the two participants may form a collaborative arrangement with a third party to control the weaker party's environment and definition of the situation.

The parents (primarily mothers) act as informal agents of control by attempting to assert authority over their daughters' actions. Rather than accepting their parent(s)' authority, these girls challenge parental assertions of authority and demand the same autonomy over their actions as have the parents. The girls' promotion of themselves in the family hierarchy to that of an equal plane as their parents is disruptive to the internal organization of the family. In these instances, the parents neither physically withdraw, nor do they reconstitute the family organization to accommodate the self-assumptions of autonomy of their daughters. Instead, in attempting to restore their authority with their daughters, these parents form collaborative arrangements with the justice system to either threaten the girls into obeying parental authority or having the girls removed by detention if they do not act accordingly.

Frequently, it is the collusion of the families and the justice system that places the girls in the category of delinquent. This collusion is accentuated by racial and ethnic stereotypes of formal social control agents that contribute to perceptions of Black and Brown girls as dangerous. Although Goffman's (1971) framework provides a base for understanding these family dynamics, an understanding of this collusion is deepened when Goffman's framework is coupled with literature (Bishop & Frazier, 1992; Bridges & Steen, 1998; Gaarder, Rodriguez, & Zatz, 2004; Leiber & Stairs, 1996; Miller, 1999; Wordes, Bynum, & Corley, 1994) exploring how racial and ethnic stereotypes affect perceptions and subsequent processing by formal social control agents.

As this article will illustrate, tensions escalate as the girls persist in their autonomous conceptions and accompanying demands and/or actions, while their parents exercise great efforts in attempts to bring the girls back into appropriate relationship to them: a relationship in which their daughters keep a "child's place." It is not so much that their daughters' actions represent delinquent acts as much as that the actions are out of line with daughters' expected social position in the family. The remainder of this article will use Goffman's (1971) framework in examining how issues of authority are negotiated between the girls and their parent(s) or guardians and how a breakdown in these negotiations may facilitate contact, entry, and movement through the juvenile justice system. Since the 1974 Deinstitutionalization Act restrained courts from responding vigorously to status-type offenses, families in their interactions with authorities have negotiated alternative methods for dealing with troublesome teenage girls.

Data and Methods

Sample Population

The data on which this article is based consist of field notes from 2 years of participant observations of 50 girls, periodic taped interviews with 30 of those girls during incarceration, and field notes and interviews with 7 of the 30 girls after they were released from incarceration. The data in this article are part of a larger study in which I examined not only the girls' pathways but also their institutional adaptation patterns while they were incarcerated. At the time I began the research, all of the girls were incarcerated in a coed public detention facility for minor offenders between the ages of 13 and 18. I gained access to this facility through a juvenile court order granting me permission to do a participant observation and interview study of all activities of the girls at the facility, after meeting with the facility's clinical director and clinical staff. The ethnic composition of the girls is predominately African American and Latina (Salvadoran and Mexican), and they are from predominately underprivileged to lower- and/or working-class neighborhoods. The most common charges included assault, assault and battery, assault with a deadly weapon, and prostitution.

The interviews occurred after months of having established rapport, by "hanging out" with the girls, during which time I recorded extensive field notes. The interviews were continuous and unstructured (more in the form of conversations) and primarily consisted of the girls telling me about their lives before and after they entered the system. Although I did not formally interview their parents, I was able to substantiate the girls' accounts through informal conversations with parents and probation officers and by sitting in on family group therapy sessions.

Of my sample interview population of 30 girls, 57% had family or guardian conflicts that facilitated either contact with the system (13%), movement toward detention and incarceration (34%), or both contact and movement (10%). Of the remaining 43%, 17% were girls who did not have families and came into the system from the Department of Child and Family Services (DCFS). Girls from DCFS will often move into the juvenile justice system as a result of the same types of control struggles that are faced by girls coming from families. Just as the standoff between girls and their parents often results in the parents' reliance on the justice system to help them establish authority over the actions of their daughters, the standoff between girls and foster parents or group home staff often results in foster parents' or staff reliance on the system to help them establish authority.

Of those girls who did not come into the system from DCFS, the majority lived with their mothers and stepfathers. The remaining were almost evenly divided between girls living with grandparents, girls living in two-parent homes, and girls living in single-mother homes. For simplification, I will use the term *parent* in the sense of adult guardian. The actual person may be a grandmother or other relative. For the most part, even in families that included either male parents or guardians, maintaining authority with the girls was primarily left to the female parent or guardian.

Parental Authority Versus Girls' Autonomy: "Going Out"

At the root of these family conflicts are struggles over control, in which the parent attempts to establish authority over the girl's actions, whereas the girl resists this authority in an attempt to gain autonomy. These struggles revolve around such issues as doing chores, going to school, seeing boys, hanging out with friends, observing curfews, and spending time talking on the phone. However, the most contentious issue in these control struggles is the girls' desire to have the freedom to "go out"—to spend time away from home. Parents (or grandparents) attempt to establish authority over their daughter's freedom to go out or spend time away from home by using varying strategies to control or restrict it, whereas the girls use various strategies to counter these restrictive measures. The following excerpts illustrate how the girls and parents negotiate going out even before they enter the juvenile justice system. Subsequent sections will illustrate how these struggles may facilitate contact and entry into the system.

One way that parents may assert their authority is by placing restrictions on their daughters' time away from home. Their daughters may respond in a variety of ways. If the internal order of the family (family solidarity) is intact, the daughters may accept this authority restricting time away from home and simply abide by these restrictions. Alternatively, children may attempt to alter the prohibition by pleading and agreeing to be back by a certain time:

> MARA (Age 16): Going out. Wanting to go out is a big thing in my house. Cause they'll be like, "No, you don't wanna come back at this time and da-dada-da-da." I'll be like, "Well, why don't you let me come back at ten o'clock? I'ma be here." "No, cause it's a school night, and it's this and it's that."

At least at this point, although Mara is challenging authority by pleading to go out, she appears to be sticking to her social place in the family. She makes demands for change, which are denied by her grandparents, and she appears to acquiesce to their position, thus internal order or parental authority appears to be intact.

Alternatively, a girl may not agree to her parent's restrictions on going out but instead may choose to ignore the restrictions:

> CATHY (Age 17): She didn't like me going out partying . . . and she tried—convincing me

> not to go—. . . she would say, "Please be home by two." And I would never be home by two . . . it's not like you're gonna be home by two, you know, it's like—*two?* I leave at *twelve*, how am I gonna be back by two—you know?

In the earlier excerpt, Mara's grandmother appears to have some semblance of authority in that she denies Mara's pleas to go out, and Mara appears to accept these restrictions. However, in Cathy's scenario, a breakdown in family solidarity is evident by the mother's, rather than the daughter's, resort to pleading. With restrictive command statements not being enough, Cathy's mother resorts to pleading with her daughter not to go out. When it becomes apparent that these appeals will not be heeded, Cathy's mother attempts to set curfew limits by pleading for her to be back by 2:00 a.m. Cathy denies both the legitimacy and the feasibility of her mother's pleas and ignores both the restrictions on leaving and the 2:00 a.m. curfew. In so doing, Cathy has promoted herself in the family hierarchy by refusing to keep a child's place in that she refuses to comply with her mother's restrictions over her time away from home.

The previous scenario illustrated the parent's resort to pleading in the case of a breakdown in the internal order of the family. When the breakdown of parental authority reaches a point where resorts to pleading are to no avail, family members may then resort to the use of physical coercion. When a parent uses physical coercion in attempting to restrict the daughter from leaving the house, the daughter may counter with physical force:

> MONICA (Age 17): I remember one night, they [friends] came to pick me up and—I was—running down the stairs to go out the house—my aunt was blocking one door, my dad was blocking the other, and my grandma was blocking one—I said okay, you know, who do I go for—so, I went—I went to try to go through the door my grandma was at—and it ended up—I didn't hit her, but it ended up, I pushed her—out of the way—to go out there, because they're [friends] honking, I'm like, "Shit, I wanna get high, oh my god,"—next thing I know, my dad has me by my hair on the floor, "You fucking little bitch—don't you *ever* hit my mother," you know—so it was really bad—so I ended up not going out that night.

A breakdown in the functioning of the internal order of the family is signaled by the fact that Monica's parents are apparently unable to gain compliance with prohibitions on going out and thus resort to physical coercion by blocking the doors. The daughter attempts to counter with physical force, but the parent's physical force prevails. Because of physical strength, using physical force to challenge girls' attempts to go out generally only works when the challenger is a male.

When there is a breakdown in authority, physical force may prevent overt attempts of a girl to go out, but it does not address covert attempts. To counter covert attempts to go out, parents may resort to more manipulative measures in attempting to restrict their daughters from going out, such as maintaining locked doors and keeping close reign on the keys. However, the girls may counter with various strategies, as illustrated in the following scenario in which Mara's grandmother keeps all of the doors locked and carries the keys with her on a chain around her waist. Mara finds a way to remove the keys without being detected:

> MARA (Age 16): I would sneak out the house . . . she [grandmother] watches movies on the floor on the mat, and she falls asleep . . . she likes holding me when she sleeps—like I'm her teddy bear . . . when she fell asleep . . . I acted like I had fallen asleep . . . I pushed the hook, and I pulled it off . . . I held the keys with me . . . I wiggled away from her . . . and she just turned back over. I had the keys and I ran to the back door, and I unlocked the door, and I came running back and I, um, I slipped them on her and . . . I'll just act like I'll turn on my T.V. in my room . . . and I'll go to the back door, and I'll just leave for like 2 or 3 hours and she won't even know I'm gone, she'll be asleep, and I'll

> come back—and I'll—get the keys again and lock the door, and she'll *still* be asleep.

The grandmother's practice of keeping all of the doors locked and the keys around her waist symbolizes a breakdown in the internal order and functioning of parental authority. A simple prohibitive command and honoring of that command is apparently not enough to gain the granddaughter's compliance to prohibitions on going out. The fact that the grandmother has resorted to such extreme measures indicates that she is aware that her authority is not intact and must resort to more coercive measures. However, whereas an institution is equipped to effectively exercise such surveillance measures, the family is not set up to effectively exercise coercion. Pure coercion (with no trust and legitimacy) is generally an unreliable method of maintaining control in a family and home (Goffman, 1971). In addition, covert challenges to parental authority generally do not lead to overt friction between daughters and parents as long as the measures remain undetected by the parent.

Coming Into Contact With the System: Calling Police to Help Restore Authority

Previous literature has suggested the phenomenon of parents calling on police to act as family disciplinarians (Joe, 1995). This section illustrates what happens when parents call police to help them restore authority. Similar to Andrews and Cohn's (1974) findings in New York, parents were likely to refer their children to the court for a variety of reasons. One way that these family conflicts over parental authority may contribute to placing girls at risk for initial contact with the juvenile justice system is when parents solicit assistance from police. When their authority is challenged and attempts to try to restore authority internally by resorting to mechanisms such as pleading, bargaining, and physical coercion have failed, these parents frequently appeal to the police as a means to restore or reestablish parental control. At least initially, the police reaction is primarily symbolic—they are constrained from and/or hesitant to react punitively by making arrests; rather, they "talk to," lecture, cajole the girls into behaving. The following scenario provides an example of how this unfolds. Mara's grandmother began calling the police to her home when Mara was 11:

CARLA:	What did she used to call the cops for?
MARA (Age 15):	When we wouldn't listen to her.
CARLA:	And she—what would she say when she called them?
MARA:	My—my grandchildren aren't listening to me—they don't wanna do what I tell them—I don't want them here.
CARLA:	And they would come out to the house?
MARA:	And they would pull us outside and just talk to us and that was it. . . . Umm—maybe about 12 times . . . The *same* ones would come . . . They said, "We're always coming over here, your grandma's always telling us something new, or it's always the *same* thing. And you know what?—It's always something stupid, and we're tired of you guys not listening to your grandmother—they're old—your grandparents are old—they—they—they should be retired, they should be—just kicking back in rocking chairs, and you guys shouldn't be—they shouldn't be handling you, and we think we should take you away—" and I'd be like, "No—no—no—no—no—no—I'm gonna be good, I'ma listen to my grandmother, don't take me away, I don't wanna go anywhere—"

A breakdown in family solidarity is evident by Mara and her brother not supporting the expected internal order of authority relationships. With her authority not intact, Mara's grandmother resorted to appealing to the police for assistance. The police responded by lecturing and threatening Mara and her brother with "We think we should take you away." Just as a breakdown in family solidarity may result in parents' mechanisms for asserting authority becoming more progressive, the same may be true for police techniques attempting to restore authority. When lecturing and threatening seem not to be effective, police may resort to scare tactics in attempting to restore parental authority:

> MARA (Age 15): They [police] were forever like, "You ain't got no fucking respect for your grandmother, and you're gonna end up in the Halls, and you don't know what it's like—people—be um—people be *raping* people, and—" . . . they were just like constantly in my face . . . cussing at me—they were *basically* trying to scare me . . . but it never worked.

Similarly, Cathy estimates that before she entered the system, her mother called the cops to the house approximately 20 times. Although the specific points of contention ranged from not washing the dishes to staying away from home for extended periods of time, her mother's overall complaint is that Cathy does not listen to her and does not follow her rules:

> CATHY (Age 17): She's called them before, but for stupid stuff like I didn't wash the dishes . . . I got home late . . . I didn't come home . . . total times, she's called the cops I'd say around . . . maybe 20 . . . just for incidents like I didn't come home, and—stuff like that or—you're not listening to your mom—you're not following the rules—stuff like that. . . . They would just tell me, "Well, listen to your mom, just begin—" just like whatever—and they just wanted to leave—they were like, "Well, ma'am . . . we have things to handle—we have things to do—you just can't be calling us for *this*."

These excerpts illustrate how the police, at least initially, treat these calls as interpersonal problems rather than as criminal matters. However, when the police run out of patience, they may consider arresting and processing the matter as a criminal case, bringing the girl officially into the juvenile justice system. The following is an example of what happens when the police run out of patience in these situations. In this example, Mara's family calls the cops for a third time to report her as a runaway. Whereas the first couple of times, the cops simply retrieved her and returned her home, their patience is tried the third time they receive this request, and they threaten to take her to juvenile hall:

> MARA (Age 15): So I was just sitting in the cop car, and I was crying—I was like, "No, I don't wanna go to the—um—Juvenile Hall—I don't wanna go, I don't wanna go," they were like, "Well, that's it—that's your last chance, this is your third time running away, and—um, we're sick of being called, and we're tired of all this bullshit and da-da-da-da-da," . . . my aunt comes over, and it's—you know—my aunt's sitting there *crying* to the cops, "*Please* don't take her, she'll be good, she'll be good, she'll be good." . . . Just let her come home with me, and she'll be—you know—she'll start acting right" . . . she [aunt] convinced him, and then. . . . They let me go under the condition that I stayed at my aunt's house, and I said, "Okay."

This time when the family called to report Mara's running away, rather than doing the usual and returning her home, the police prepare (or at least feign preparation) to take her to the police station for processing. Mara's aunt's intervention once the police arrive on the scene illustrates how family may influence whether a girl enters the system in the first place by influencing police decisions about taking her in for processing. In this instance, although the aunt was the one who initially called in the police, she diverted her niece from entering the system at this time by meeting the cops on the scene and persuading them to let her niece go home with her. It also illustrates the amount of discretion that police have in determining when to cease treating a family conflict as a personal matter and to instead begin treating it as a criminal matter.

Struggles Culminating in Assault Charges: Relabeling Domestic Disputes

The previous sections of this article explored the types of family conflicts over parental authority and how a breakdown in family solidarity or internal order of family may lead to parents' soliciting help from outside authorities. This is often many of the girls' initial contact with the system before they actually enter it, and this section will explore how these family conflicts over authority result in the girls actually entering the system. Although Joe (1995) found that with respect to runaways, the police's net of social control does not appear to be widening, others suggest that police discretionary powers allow them to circumvent the restraints of deinstitutionalization in other circumstances (Acoca, 1999; Acoca & Dedel, 1998; Austin & Krisberg, 1981; Chesney-Lind & Shelden, 1998, 2004; Klein, 1979; Lemert, 1981; Mahoney & Fenster, 1982; Mayer, 1994). This happens when the police relabel family conflicts or domestic disputes as assaults. This section explores the process by which relabeling of family conflicts into assaults occurs.

Girls' arrests for assaults have dramatically increased since 1970 (Chesney-Lind & Shelden, 2004; FBI, 1971, 1981, 1995, 2001). The increases are greatest for two categories of assaults: aggravated assaults and "other" assaults. Arrest rates for girls for aggravated assault increased by 364% between 1970 and 1995 (Chesney-Lind & Shelden, 2004; FBI, 1971, 1981, 1995). Arrest rates for girls for "other" assaults increased by 343% between 1970 and 1995 (Chesney-Lind & Shelden 2004; FBI, 1971, 1981, 1995). One explanation offered for this increase is the possibility of "greater attention to normal adolescent fighting and/or girls fighting with parents" (Chesney-Lind & Shelden 2004, p. 11). In the past, these conflicts may have been more likely dealt with informally. Previous literature has suggested that greater attention to conflicts between girls and parents may at least partly be attributed to increased attention to domestic violence and/or changes in domestic violence laws, which encourages more active involvement of police in family conflicts (Chesney-Lind & Shelden, 1998, 2004; Gaarder et al., 2004).

Family-related assault charges often represent the highest degree of escalation in family control struggles. By the time assault charges occur, struggles over parental authority have been ongoing for some time and have reached a breaking point. By the time families reach this point, their attempts to establish authority through conventional mechanisms (routine commands and/or requests) or through less conventional methods (pleading, bargaining, or manipulation) have failed. Parents then resort to attempting to exert parental authority with coercive physical control, which is likely to be countered with similar physical resistance from the daughters. Earlier examples in this article described instances where parents called police to their homes to restore order and police responded by going to the homes and talking, cajoling, or threatening the girls. In situations where the girls may use physical force to counter parental attempts of controlling with physical force, both parents and police seek a restoration of order by removing and detaining the girl at least temporarily. Many girls either enter the system with assault charges or are subsequently incarcerated for these charges after having already entered the system.

There may be evidence of assault in some cases. However, in many cases, there often seems to be a lack of evidence supporting these assault charges. These instances often seem to be opportunities for parents to appeal to the juvenile justice system for assistance in their overall control struggles with their daughters. When police arrive on the scene, they are much more likely to believe the parent or guardian rather than the girl's version of what transpired. Countering coercive physical control with physical resistance presents the opportunity for parents to begin the process of having disobedience classified as "delinquent." Changing the plea from "she won't obey me" to "she assaulted me" gives police something on which to act. Many of these family assault charges seem to have one theme in common. They appear to almost always involve a parent's attempting to physically block the daughter from taking some kind of action, such as leaving the house, and a daughter's subsequent push to resist the parent's restraining action results in the parent's alleging assault. As the following passages will illustrate, it is not uncommon for family assault charges to emerge from the most contentious point underlying control struggles—a daughter's freedom to leave the house or "go out."

In the following excerpt, Renee describes what happened when she returned home to pick up some items after one of her extended periods away from home:

> RENEE (Age 16): So we went back to my house, and I was picking up my makeup . . . shampoo . . . lotion. . . . And—my mom's like, "No, this isn't a motel, you're not just gonna come in and out whenever you please, you're only *fourteen*," . . . she was really upset . . . she was trying to keep me in the house . . . and—basically I—I pushed her. I had a whole bunch of stuff in my hands, like all these bottles and stuff—and I pushed her out of my way when I was coming out of the bathroom . . . then . . . they [parents] called the police . . . "she needs to stay here," . . . she's out of control," and he [stepfather] would not let me go anywhere . . . the police came, and I was cussing them out . . . they arrested me . . . they're like, "Well, I'm getting you for battery on your mom,"—"Battery on my mom, I didn't do anything to her"—"You pushed her, didn't you?" I'm like, "Yeah." Shit—okay, they got me for battery.

Being able to come and go as one pleases is distinctly an adult prerogative. Renee's mother suggests her frustration lies in part by Renee's not keeping a child's place in the family order when she says, "You're not just gonna come in and out whenever you please, you're only *fourteen*." In this case, it is not just that Renee is not keeping her place by coming or going as she pleases; she has returned home with concrete justification for parental attempts to police her sexual activities—a boyfriend. In attempting to prevent Renee from leaving the house again, her mother first attempts to block her, and Renee responds by pushing her out of the way. Her stepfather then intervenes to block and restrain Renee while they call the police and continues to block Renee from leaving until the police arrive. In this situation, the parents attempted to restore authority or order in the family by calling in outside authorities (police) to remove and detain Renee. This illustrates the police role or influence in defining the situation upon arriving on the scene and shows the importance of social control agents in defining what constitutes a criminal act—in this case, battery. This also challenges Joe's (1995) conclusions that the consequence of deinstitutionalization was a narrowing and restricting of police discretion and social control.

Similarly, the following excerpt also illustrates how conflict over leaving the house or going out resulted in assault charges. In this instance, a breakdown in authority relationships is symbolized by Teresa's not adhering to her mother's restrictions on going out. Consequently, Teresa's mother resorted to attempts to manipulate the situation by making it more difficult for Teresa to leave the house. Because getting dressed is a precondition of leaving the house, by placing Teresa's clothes in her (mother's) closet, Teresa's mother attempted to keep Teresa from leaving. This made it more difficult for Teresa to leave the house because Teresa would have to go through her mother to retrieve her clothes. The following illustrates what happened when Teresa attempted to retrieve some items of clothing before she left on one of her extended stays away from home:

> TERESA (Age 16): So, she took my clothes—she put it in . . . the closet in her room—so that I couldn't get to it. And I needed something to wear that day, so I went to—the closet—in *her* room—and I was going to get my clothes—so she walked to—over from

> her bed—and she tried to block me—she tried to hold the closet door closed. And then ... she tripped over my little brother's toy that was on the floor ... that's when she called the police ... she said that I hit her and almost made her fall.

Teresa's mother's attempts to gain authority over Teresa's leaving the house by manipulating placement of her clothes failed and resulted in Teresa's mother resorting to coercive, physical control. However, in attempting to block Teresa from retrieving her clothes from the closet, either her mother tripped or perhaps some sort of altercation occurred prompting her mother to trip. Either way, the event provided the opportunity for Teresa's mother to gain assistance in her overall authority struggles with Teresa by constructing Teresa's actions as being serious enough to warrant removal from the home and detention. The police had come to the house twice before in response to Teresa's mother's call and, as with this time, indicated that they found no evidence to apprehend the daughter. It seems that the mother's persistence and suggestions that her daughter was out of control may have influenced the police to take her in for processing and incarceration in Juvenile Hall. In addition to describing the possible context for assault charges, this case also illustrates how the police serve as mediators in family conflicts well before the girls actually enter the system. In this case, not only was there no evidence of assault when the police appeared on the scene for the third time, but the charges were subsequently dropped because Teresa's mother did not show up in court to testify.

Although family control struggles seem to most commonly occur between parents (primarily mothers) and daughters, these struggles may also occur between the girls and older siblings:

> BRENDA (Age 15): I had got into it—the one—the brother (age 23) I said I don't like—one day—um—my mama had found some Weed in my room—and I had got mad and then—I started cussing and stuff—and he don't like me cussing—especially with my mom right there—so like I cussed, and I ran outside—and [he] came behind me—and he was like holdin me—he threw me on the ground, and he started holdin me—like—and I started kickin him and stuff, and he was still holdin me ... I was just kickin him so hard, and he slipped ... he slapped me ... he didn't mean to do it—I was just kickin him so hard, and he slipped ... then I went to go get a knife, and then I called the police—then they end up takin *me* ... he always big and bad like when I ain't got nothing—but if I like pick up a stick, then he wanna move back—so I picked up a knife ... and then the police came ... I threw the knife before they came, and my sister—let me know—she like—"The police out there," ... I just threw 'em [knife] down—and then they came and—and they—took me.

One of the interesting dimensions of this scenario is that when the police arrived on the scene, the struggle had subsided, and there was no knife in sight:

> CARLA: They didn't even *see* you with the knife.
>
> BRENDA: They she didn't—that's what—like the police—he came to court too—he said he didn't see me and um—my attorney [public defender] was telling 'em—he was like, "If she wasn't swingin' the knife—it can't be assault with a deadly weapon."

That the police were ready to remove Brenda with no apparent evidence of struggle raises the question of the basis for removing Brenda from her home:

> BRENDA: My mom told 'em like—she need a break ... she told 'em that um—like, "Take her cause she goin crazy." ... At trial, my mom was telling him [judge] like—she just wanted me

> gone for like 2 weeks just to—give me a lesson or whatever—and she was just sayin stuff like—she wanted me home . . . she just wanted to give me a lesson—but my judge—he was like—he ain't goin for none of that—he deal with all my mama' kids—like most of 'em—they done been in here . . . he was like—"What makes you think if I release you—like you ain't gon be like them and get in trouble again?" So, he was like, "Nah—I'ma keep you."

As illustrated in previous excerpts, this particular control struggle provided an opportunity for Brenda's mother to appeal to authorities for "a break" by pleading with them to remove Brenda. Age or minor status of the girls leaves them powerless in defining what actually occurred when the police arrive on the scene. The tradition of upholding parental authority over the rights of children results in authorities nearly always accepting the parent's or guardian's definition of the situation. Although filing assault charges may be one strategy that parents or guardians use as a last resort to gain help in their overall control struggles with their daughter, in most instances, they only intend for their daughters to be gone for short periods of time, such as a short-term stay in Juvenile Hall to "teach her a lesson." This is particularly the case for those parents who are familiar with the way the system works. However, the above passage illustrates how sometimes this strategy may backfire. In this situation, the judge did not see authority in the home as viable and sentenced Brenda to incarceration.

Brenda's situation breaks with the general pattern of girls not being incarcerated after their first contact with the system and/or first arrest if the parent is willing to take them home. This illustrates how a judge's sentencing decision may be based not on the girl's offense but on the judge's perception of whether the internal order of the family's authority relationships is intact. In this situation, the judge's decision to incarcerate Brenda seems to evolve from a sense that Brenda's family lacks a viable authority or control structure, based on the fact that several of Brenda's siblings had previously been in this judge's court.

How Control Struggles May Shape Differential Juvenile Justice Outcomes

Not only are many girls in the system for status offenses, but also a substantial number of girls are in the system for violating court orders. Many of these acts of violations of court orders are in essence status offenses, such as running away. Previous literature suggests judges' use of violations of court orders as a technique of "bootstrapping" girls into detention (Costello & Worthington, 1981). Once a girl enters the system, control struggles are reproduced as violations of probation, and with each of these violations, punitive measures are likely to increase. Whereas going out and/or running away were not crimes before the girl entered the system, once in the system (on probation), going out, especially for extended periods of time, now becomes a violation of probation, or in other words, going out against parental prohibitions now becomes a crime.

Just as family control struggles may contribute to placing a girl at risk for entering the system, these struggles may also influence differential outcomes after she enters the system. Once the juvenile justice system becomes involved, it is not uncommon for the judge to consult with the parent before rendering a decision. A parent's input may influence whether a girl is able to come home or whether she is institutionalized after arrest. For example, if a parent stipulates to the judge that her daughter may come home if she abides by her rules, then the judge may send the girl home on house arrest. House arrest is a more severe measure than simply being placed on probation because it prohibits the girls from leaving the house for any purposes other than school. For this reason, house arrest appears to be a favorable measure for parents in their struggles to control their daughter's time spent away from home, or going out. However, this is rarely the case. Because many of the girls' entries into the justice system are predicated on family authority struggles, house arrest often fails because nothing is done to address the underlying control struggles.

> MIRANDA (Age 14): House arrest—that couldn't even *fade* me . . . House arrest cannot fade me, cause I would be at home, and I'm

> supposed to be in the house—I would *leave*—
> I'd just get up and *leave*.

Sentencing a girl to house arrest represents the court's attempt to reinforce parental authority by supplementing parental prohibitions with court-sanctioned prohibitions, thus the penalties for disobeying restrictions to go out are much harsher. However, what house arrest does is add another layer of authority, assuring more punitive consequences for the girls' strategies to gain more freedom away from home. Specifically, house arrest makes the girls' extended periods of time away from home crimes for which they can be incarcerated, which usually means recurrent stays of approximately 30 days in Juvenile Hall.

Alternatively, if a parent or guardian refuses to accept the girl back home, the judge may institutionalize her or send her to a group home:

> MARA (Age 15): They said, "Okay, then we're gonna let you go on probation," and my grandmother ended up saying, "No, I don't want her home. She can't listen to me . . . She doesn't obey me, she's always answering back . . . I don't want nothing to do with her." . . . So . . . they arrested me. They put me in a soda pad [temporary home] and gave me a couple more court dates. And my grandmother still said she didn't feel I was changing, and she didn't want me home. So, they put me in placement.

Mara's grandmother confirmed that at the court hearing, the judge was ready to release Mara home with probation, but the grandmother refused to accept her. Not yet willing to sentence her to incarceration or placement outside of the home, the judge placed her in a temporary shelter, and she was given a couple more court dates, giving the grandmother a longer time to reconsider. Only after the grandmother seemed steadfast in her decision not to accept her granddaughter home did the judge sentence her to an open placement (group home).

Although sentencing to a foster home or open placement (group home) is considered to be a viable alternative to incarceration, for many girls in the juvenile justice system, it is often the first step to eventual incarceration. Girls who are sent to foster or group homes rather than returning to their parent's home face the same dilemmas. Control struggles are often reconstructed in foster and group homes, which are deemed community alternatives to incarceration. Not surprisingly, the girls want the same freedom of action that they wanted at home, whereas the foster parents or group home staff wish to establish authority and curtail girls' freedom:

> SABRINA (Age 18): Yeah—it was two of 'em—parents [foster parents]—they were like pastors at a church. And it was just like something I wasn't used to, and it was like I couldn't be myself—you know?—I'm the type of person where I like to blast my music, I like to—smoke cigarettes, I like to—go out partying, and I like to talk on the phone—and they wasn't having it, and I didn't—like it—you know? So, my social worker made an agreement with me—"Well, this is gonna be your curfew—you do whatever you have to do on that when you're out,"—you know?—and it worked for a little bit, and then I got tired of it, and—I left [after about 6 months] . . . I went to a group home I—awoled and I never went back . . . finally turned myself in to my social worker, and she took me to *another* foster home.

Control struggles between Sabrina and her foster parents are evidenced by Sabrina's assertion that she is "the type of person where I like to blast my music, I like to—smoke cigarettes, I like to—go out partying, and I like to talk on the phone—and they wasn't having it." Sabrina's social worker attempts to ease the standoff between Sabrina and her foster parents by suggesting, if not implying, to Sabrina that in exchange for abiding by her curfew, she may do whatever she wishes during her time out or away from the home. This practice of "don't ask, don't tell" about activities while away as long as home by curfew was also a tactic some mothers used after release, which resulted in diminishing, if not eliminating, control struggles. Because of these control struggles, Sabrina went from a foster home, to a group home, to a foster home, then back to a group home.

Just as the standoff between girls and their parents often results in the parents' reliance on the justice system to help them establish authority over the actions of their daughters, the standoff between girls and group home staff often results in staff reliance on the system when there is a breakdown in authority:

> MARA (Age 15): And I went to S. for my first [group home] ... all girls.... And so—they were trying to get us to go to a boy's [group home] and have this dance ... I didn't want to go.... And you know, I just said, "Well, if you take me there, I swear, I'm just going to go off on everybody there, you know, and they ain't gonna like it." So they [group home] sent me to a mental hospital. And the mental hospital—I didn't want to go there ... and I ended up staying there for 2 weeks ... I got sent back to [group home]. And I didn't want to go back to [group home], and they [group home] ended up keeping me again, and when I came back, they were trying to get me to go to school. I was like, "I'm not going to school," ... they took me to [Juvenile Hall] trying to—you know—admit me in, and [juvenile hall] said they couldn't take me.... So they took me back to the [mental] hospital. In the hospital, I stayed there a week ... the one I liked—the exact same one. And—I stayed there for a week, and—the [group home] came to pick me up again. I told them, "No—I'm not going back. I am *not* going back." So I refused [group home], and [juvenile hall] just came, and they picked me up right away.

As illustrated in the above scenario, group home control struggles result in girls' moving from institution to institution (often from group home to group home), with incarceration in juvenile halls in between each group home. This occurs not because the girls commit new offenses but because the same control struggles that took place between parents and daughters take place between group home staff and the girls.

After release from institutionalization, girls in the juvenile justice system are likely to return to the same family control struggles that contributed to their entering the system or being incarcerated. Their desires for autonomy are often not likely to be dampened after time spent incarcerated:

> MONICA (Age 17): I had just got out, so my mom was still kinda tripping, like—you know—"Oh, you're already trying to go out, and—blah-blah—whatever." ... So, I started getting really like—I don't know—like—just anxious—I wanted to leave the house—I started going out a lot ... she tried to put twelve o'clock on me, but—that never happened—we would—me and my sister would pour into the house like two or three in the morning ... my mom's like, "Well—no, I don't want you going, you have school tomorrow," you know,—this and this and that—"No, you're not gonna go,"—and I was like, "No, I *am* gonna go." I was like, "I'll be right back."—"Oh Monica, don't go,"—"No, I *am* gonna go."—"Okay, okay, that's how you wanna play it—go ahead—go Monica," and I said, "I'm going,"—so I took off.

This illustrates how struggles over parental authority have not been modified from their pre-incarceration form. Monica's mother first attempts to exercise her authority with a direct command, "No, you're not gonna go." Monica dismisses her mother's command with "No, I *am* gonna go," to which her mother resorts to some semblance of a pleading. After realizing that this battle was lost, Monica's mother resigns—"Okay, okay, that's how you wanna play it—go ahead."

In addition, if family control struggles resume after a girl is released, she may face the constant threat that her parent may report her misbehaviors to the court or other authorities. The girls often fear that their freedom is dependent on their parents. In the following passage, Mara expresses concern that at her upcoming court date, her grandmother will tell the judge that she should not be home and they will reincarcerate her:

> MARA (Age 15): My grandma—since we've been fighting, she's been threatening me, that's she gonna tell the—um—the court that I shouldn't be home . . . so I hope she doesn't do that. Because they'll put me right back in the system . . . little things we'll argue over . . . it starts off with something little, and then it works—very big. I don't know. But every *day*, there's a fight . . . and then she ends up picking up the phone, and I end up leaving the house. Cause I don't wanna stick around . . . to see the cops if she does call 'em.

In this instance, the grandmother's leverage in their arguments consists of constant threats to report Mara to either the police or to court officials at her upcoming hearing. It is debatable whether this is an effective tool to bring her granddaughter in line, but it is enough to keep Mara in a heightened sense of anxiety, which serves to exacerbate tensions.

The level of family discordance is likely to have significant consequences for the quality of life of girls after they are released, thus having consequences for whether they are successfully able to get off of probation. This is significant because most girls are reincarcerated not on the basis of new offenses but on probation violations. Because periods of incarceration are unlikely to contribute to improving the family control struggles that propelled the girls into the system, after the girls are released, these control struggles are one of the most difficult challenges with which they must contend.

Discussion and Conclusion

In finding that family dynamics often generate what is officially considered an offense, this research supports previous literature suggesting that some deviance is in large part the product of the response of the group, either family or community (Becker, 1963; Goffman, 1971; Lemert, 1951; Perrucci, 1974). These girls are largely classified as delinquent through their families' appeals to the justice system for help. The parents (primarily mothers or female guardians) act as informal agents of control until a breakdown in family solidarity prompts appeal to more formal measures of control. In many instances in attempting to restore their authority with their daughters, these parents form collaborative arrangements with the justice system to either threaten the girls into obeying parental authority or having the girls removed by detention if they do not act accordingly. Although nearly all families have these struggles, the mechanisms and/or options used by families depend on their social locations along hierarchies of race, ethnicity, and class. The most marginalized along these hierarchies are less likely to have resources other than appealing to police and the justice system. Furthermore, unlike those at other social locations, these families are in environments that more readily bring them in contact with police.

The system's response to these parents' pleas for help shows continuation of historical practices of state intervention into the families of children perceived to come from environments regarded as unregulated and lacking appropriate values and structure thought necessary to foster obedience, self-discipline, and hard work. Whereas in the 19th century, children of European immigrant families who had ethnic backgrounds different from those of established American residents were primarily the target population of excessive state intervention, in the 20th and 21st centuries, these demographics have shifted to children of primarily African American and Latino families. Whereas intervention in the 19th century was primarily imposed from the outside, these contemporary parents (lacking alternative familial or community resources) are themselves calling the police, thus initiating state intervention.

Although there are previous studies on how macro structural factors may converge to shape fractured emotional attachments between parents and children in impoverished African American families (Duncan, Brooks-Gunn, & Klabanov, 1994; Henriques & Manatu-Rupert, 2001; Leadbeater & Bishop, 1994; Sampson & Laub, 1994), there are no studies of how these factors may shape problematic authority relationships in these families. Simultaneous intersecting structures of race/ethnicity, class, and gender converge to place African American women and Latinas in positions of extreme powerlessness. Although the powerlessness of African

American women and Latinas has been acknowledged relative to the larger society, the powerlessness of these women in their own communities and in their families has been less explored. These women have no power and are often struggling economically and emotionally just to survive.

To the extent that challenges to parental authority play an instrumental role in the girls' coming in contact and moving through the justice system, this suggests the continuing significance of status offenses (offenses that would not be considered wrongdoing if committed by an adult) in the arrest and incarceration of girls, even though the 1974 JJDP Act officially decriminalized status offenses. The findings in this article suggest that not only has the problem of differentiating youths whose actions are more affronts to parental and local authority than violations of law not been resolved but challenges to parental authority are now being constructed and processed under other "official" categories of crimes or delinquencies.

The family dynamics revealed in this article play a significant role in the girls' contact, entry, and movement through the justice system and are important not only for implications for the continuing significance of status offenses but also for subsequent program and policy planning for girls in the system. The data in this article suggest the extent to which families lack resources to navigate these conflicts. Certainly, the institution provides a remedy to family control struggles over "going out," by locking the girls in; however, the institution's resources and measures for ensuring discipline and obedience are not something that can be readily transferred to the family. As reflected by one mother's comments to her daughter, "Well, of course you get up and go to school here [institution], you have a whole team of people to help with that."

The girls return to the same family control struggles that existed before they entered placement. Policy and program designers should take into consideration the significance and nature of these family control struggles in facilitating girls' contact and entry into the system, and programming should be aimed at developing community resources to address these issues. Any program that does not also address the needs of the families, particularly the mothers or other female guardians, will fall woefully short because this is to whom the girls return. Although directing resources toward building programs within institutions may be expedient, the girls and their families would be better served by emphasizing and developing community, rather than institutional, programs.

References

Acoca, L. (1999). Investing in girls: A 21st century challenge. *Juvenile Justice*, *6*, 3–13.

Acoca, L., & Dedel, K. (1998). *No place to hide: Understanding and meeting the needs of girls in the California juvenile justice system*. San Francisco: National Council on Crime and Delinquency.

American Bar Association and National Bar Association Joint Report. (2001). *Justice by gender: The lack of appropriate prevention, diversion and treatment alternatives for girls in the justice system*. Washington, DC: Author.

Andrews, R., & Cohn, A. (1974). Ungovernability: The unjustifiable jurisdiction. *Yale Law Journal*, *83*, 1383–1409.

Arthur D. Little, Inc. (1977). *Responses to angry youth: Cost and service impacts of the deinstitutionalization of status offenders in ten states*. Washington, DC: Author.

Austin, J., & Krisberg, B. (1981). Wider, stronger, and different nets: The dialectics of criminal justice reform. *Journal of Research in Crime and Delinquency*, *18*, 165–196.

Bartollas, C. (1993). Little girls grown up: The perils of institutionalization. In C. Culliver (Ed.), *Female criminality: The state of the art* (pp. 30–47). New York: Garland.

Becker, H. (1963). *Outsiders*. New York: Free Press.

Belknap, J., & Holsinger, K. (1998). An overview of delinquent girls: How theory and practice have failed and the need for innovative changes. In *Female offenders: Critical perspectives and effective interventions* (pp. 31–64). Gaithersburg, MD: Aspen.

Bishop, D. M., & Frazier, C. (1992). Gender bias in juvenile justice processing: Implications of the JJDP Act. *Journal of Criminal Law and Criminology*, *82*, 1162–1186.

Bridges, G., & Steen, S. (1998, August). Racial disparities in official assessments of juvenile offenders: Attributional stereotypes as mediating mechanisms. *American Sociological Review*, *63*, 554–570.

Bureau of Justice Statistics. (2001). *Prisoners in 2000*. Washington, DC: U.S. Department of Justice.

Bureau of Justice Statistics. (2003). *Prisoners in 2000*. Washington, DC: U.S. Department of Justice.

Chesney-Lind, M., & Shelden, R. (1998). *Girls, delinquency, and juvenile justice*. Belmont, CA: Wadsworth.

Chesney-Lind, M., & Shelden, R. (2004). *Girls, delinquency, and juvenile justice* (3rd ed.). Belmont, CA: Wadsworth.

Cohn, Y. (1970). Criteria for the probation officer's recommendations to the juvenile court. In P. G. Garbedian & D. C. Gibbons (Eds.), *Becoming delinquent* (pp. 66–80). Chicago: Aldine.

Costello, J., & Worthington, N. L. (1981). Incarcerating status offenders: Attempts to circumvent the Juvenile Justice and Delinquency Prevention Act. *Harvard Civil Rights-Civil Liberties Law Review, 16*, 41–81.

Datesman, S., & Scarpitti, F. (1977). Unequal protection for males and females in the juvenile court. In T. N. Ferdinand (Ed.), *Juvenile delinquency: Little brother grows up* (pp. 45–62). Beverly Hills, CA: Sage.

Duncan, G. J., Brooks-Gunn, J., & Klabanov, P. K. (1994). Economic deprivation and early childhood development. *Childhood Development, 65*, 296–318.

Empey, L. T. (1973). Juvenile justice reform: Diversion, due process and deinstitutionalization. In L. E. Ohlin (Ed.), *Prisoners in America* (pp. 13–48). Englewood Cliffs, NJ: Prentice Hall.

Empey, L. T. (1982). *American delinquency.* Homewood, IL: Dorsey.

Federal Bureau of Investigation. (1971). *Uniform crime reports: Crime in the United States.* Washington, DC: U.S. Department of Justice.

Federal Bureau of Investigation. (1981). *Uniform crime reports: Crime in the United States.* Washington, DC: U.S. Department of Justice.

Federal Bureau of Investigation. (1995). *Uniform crime reports: Crime in the United States.* Washington, DC: U.S. Department of Justice.

Federal Bureau of Investigation. (1998). *Uniform crime reports: Crime in the United States.* Washington, DC: U.S. Department of Justice.

Federal Bureau of Investigation. (2001). *Uniform crime reports: Crime in the United States.* Washington, DC: U.S. Department of Justice.

Federle, K. H., & Chesney-Lind, M. (1992). Special issues in juvenile justice: Gender, race, and ethnicity. In I. M. Schwartz (Ed.), *Juvenile justice and public policy: Toward a national agenda* (pp. 165–195). Indianapolis, IN: Macmillian.

Gaarder, E., Rodriguez, N., & Zatz, M. (2004). Criers, liars, and manipulators: Probation officers' views of girls. *Justice Quarterly, 21*(3), 547–578.

Gibbons, D., & Griswold, M. J. (1957). Sex differences among juvenile court referrals. *Sociology and Social Research, 42*, 106–110.

Goffman, E. (1971). The insanity of place. In *Relations in public* (pp. 335–390). New York: Harper & Row.

Gordon, L. (1988). *Heroes in their own lives.* New York: Viking.

Henriques, Z., & Manatu-Rupert, N. (2001). Living on the outside: Women before, during, and after imprisonment. *The Prison Journal, 81*(1), 6–19.

Joe, K. A. (1995). The dynamics of running away, deinstitutionalization policies and the police. *Juvenile and Family Court Journal, 46*(3), 43–55.

Ketchum, O. (1978). Why jurisdiction over status offenders should be eliminated from juvenile courts. In R. Allinson (Ed.), *Status offenders and the juvenile justice system* (pp. 33–56). Hackensack, NJ: National Council on Crime and Delinquency.

Klein, M. (1979). Deinstitutionalization and diversion of juvenile offenders: A litany of impediments. In N. Morris & M. Torry (Eds.),*Crime and justice* (pp. 145–201). Chicago: University of Chicago Press.

Krisberg, B., & Schwartz, I. (1983). Re-thinking juvenile justice. *Crime & Delinquency, 29*, 381–397.

Leadbeater, B. J., & Bishop, S. J. (1994). Predictors of behavioral problems in preschool children of inner-city Afro-American and Puerto Rican adolescent mothers. *Child Development, 65*, 638–648.

Leiber, M. J., & Stairs, J. (1996). Race, contexts, and the use of intake diversion. *Journal of Research in Crime and Delinquency, 36*, 76–78.

Lemert, E. (1981). Diversion in juvenile justice: What hath been wrought. *Journal of Research in Crime and Delinquency, 18*, 34–46.

Lemert, E. M. (1951). *Social pathology: A systematic approach to the theory of sociopathic behavior.* New York: McGraw-Hill.

Mahoney, A., & Fenster, C. (1982). Female delinquents in a suburban court. In N. Hahn & E. Stanko (Eds.), *Judge, lawyer, victim, thief: Woman, gender roles and criminal justice* (pp. 26–42). Boston: Northeastern University Press.

Mayer, J. (1994, August). *Girls in the Maryland juvenile justice system: Findings of the female population taskforce.* Paper presented at the Gender Specifics Services Training, Minneapolis, MN.

Miller, J. (1999). An examination of disposition decision-making for delinquent girls. In M. D. Schwartz & D. Milovanovic (Eds.), *Race, gender and class in criminology: The intersections* (pp. 219–246). New York: Garland.

Odem, M. (1995). *Delinquent daughters: Protecting and policing adolescent female sexuality in the United States, 1885–1920.* Chapel Hill: University of North Carolina Press.

Odem, M. E., & Schlossman, S. (1991). Guardians of virtue: The juvenile court and female delinquency in early 20th century Los Angeles. *Crime & Delinquency, 37*, 186–203.

Perrucci, R. (1974). *Circle of madness.* Englewood Cliffs, NJ: Prentice Hall.

Platt, A. (1969). *The child savers.* Chicago: University of Chicago Press.

Robinson, R. (1990). *Violations of girlhood: A qualitative study of female delinquents and children in need of services in Massachusetts.* Unpublished doctoral dissertation, Brandeis University, Waltham, MA.

Sampson, R. J., & Laub, J. H. (1994). Urban poverty and the family context of delinquency: A new look at structure and process in a classic study. *Child Development, 65*, 538.

Schlossman, S., & Wallach, S. (1978). The crime of precocious sexuality: Female delinquency in the Progressive Era. *Harvard Educational Review, 48*, 65–94.

Schwartz, I. (1989). *(In) justice for juveniles: Rethinking the best interests of the child.* Lexington, MA: D. C. Heath.

Schwartz, I., Jackson-Beeck, M., & Anderson, R. (1984). The hidden system of juvenile control. *Crime & Delinquency, 30*, 371–385.

Shelden, R. G. (1981). Sex discrimination in the juvenile justice system: Memphis, Tennessee, 1900-1917. In M. Q. Warren (Ed.), *Comparing male and female offenders* (pp. 88–102). Beverly Hills, CA: Sage.

Sussman, A. (1977). Sex-based discrimination and the PINS jurisdiction. In L. E. Teitelbaum & A. R. Gough (Eds.), *Beyond control: Status offenders in the juvenile court*. Cambridge, MA: Ballinger.

Teitelbaum, L., & Gough, A. (1977). *Beyond control: Status offenders in the juvenile court*. Cambridge, MA: Ballinger.

Wordes, M., Bynum, T., & Corley, C. (1994, May). Locking up youth: The impact of race on detention decisions. *Journal of Research in Crime and Delinquency, 31*, 149–165.

Young, V. D. (1994). Race and gender in the establishment of juvenile institutions: The case of the South. *The Prison Journal, 732*, 244–265.

Zatz, J. (1982). Problems and issues in deinstitutionalization: Historical overview and current attitudes. In *Neither angels nor thieves: Studies in deinstitutionalization of status offenders* (pp. 55–72). Washington, DC: National Academy Press.

DISCUSSION QUESTIONS

1. What effect does a breakdown in parental authority have on delinquent behaviors?
2. How are girls "bootstrapped" into the system as a result of family conflicts?
3. How does family conflict shape court decision-making processes and case outcomes for delinquent girls?

READING

As you've learned, girls were often institutionalized during the early history of juvenile delinquency for being incorrigible or out of control. With the deinstitutionalization of status offenders, policy makers and practitioners have found new ways to relabel female status offenders. There has been a significant amount of hype in the media about the violent girl. This reading explores how the changing definitions of violence and delinquency have resulted in the increases in identifying cases of "crime" for female offenders.

Violent Girls or Relabeled Status Offenders?

An Alternative Interpretation of the Data

Barry C. Feld

Over the past decade, policy makers and juvenile justice officials have expressed alarm over a perceived increase in girls' violence. Official statistics report that police arrests of female juveniles for violent offenses such as simple and aggravated assault either have increased more or decreased less than those of their male counterparts and thereby augured a gender convergence in youth violence (Federal Bureau of Investigation, 2006; Steffensmeier, Schwartz, Zhong, & Ackerman, 2005). Reflecting the

SOURCE: Feld, B. C. (2009). Violent girls or relabeled status offenders? An alternative interpretation of the data. *Crime and Delinquency, 55*(2), 241–265.

official statistics, popular media amplify public perceptions of an increase in "girl-on-girl" violence, "bad girls gone wild," "feral and savage" girls, and girl-gang violence (Scelfo, 2004; Sanders, 2005; Kluger, 2006; Williams, 2004). One possible explanation for the perceived narrowing of the gender gap in violence is that gender-specific social structural or cultural changes actually have changed girls' behaviors in ways that differ from boys.

On the other hand, the supposed increase in girls' violence may be an artifact of decreased public tolerance for violence, changes in parental attitudes or law enforcement policies, or heightened surveillance of several types of behaviors, such as domestic violence and simple assaults, which disproportionately affect girls (Garland, 2001; Kempf-Leonard & Johansson, 2007; Steffensmeier et al., 2005). Steffensmeier et al. (2005) compared boys' and girls' official arrest rates with other data sources that do not depend on criminal justice system information (e.g., longitudinal self-report and victimization data) and concluded that "the rise in girls' violence . . . is more a social construction than an empirical reality" (p. 397). They attributed the changes in female arrests for violent crimes to three gender-specific policy changes: a greater propensity to charge less serious forms of conduct as assaults, which disproportionately affects girls; a criminalizing of violence between intimates, such as domestic disputes; and a diminished social and family tolerance of female juveniles' "acting out" behaviors.

Their data and analyses support a social constructionist argument that the recent rise in girls' arrests for violence is an artifact of changes in law enforcement policies and the emerging "culture of control" rather than a reflection of real changes in girls' behavior (Garland, 2001; Steffensmeier et al., 2005). Although cultural and police policy changes likely contribute to a greater tendency to arrest girls for minor violence, the social construction of girls' violence also may reflect policy changes that occurred within the juvenile justice system itself, especially the deinstitutionalization of status offenders (DSO). After federal mandates in the mid-1970s to deinstitutionalize status offenders, analysts described juvenile justice system strategies to "bootstrap" and/or "relabel" female status offenders as delinquents to retain access to secure facilities in which to confine "incorrigible" girls (Bishop & Frazier, 1992; Feld, 1999).

In this article, I focus on patterns of arrests and confinement of boys and girls for simple and aggravated assaults over the past quarter century. The analysis bolsters Steffensmeier et al.'s (2005) contention that much of the seeming increase in girls' violence is an artifact of changes in law enforcement activities. However, I attribute some of the increase in girls' arrests for violence to federal and state policies to remove status offenders from delinquency institutions. Initially, laws that prohibited confining status offenders with delinquent youth disproportionately benefited girls, whom states most often confined under that jurisdiction. But they provided an impetus to relabel status offenders as delinquents to continue to place them in secure institutions. Within the past two decades, deinstitutionalization policies have coincided with the generic "crackdown" on youth violence in general and heightened concerns about domestic violence in particular, further facilitating the relabeling of status offenders by lowering the threshold of what behavior constitutes an assault, especially in the context of domestic conflict.

I first examine the historical differences in juvenile justice system responses to male and female delinquents and status offenders. The next section focuses on the 1974 federal Juvenile Justice and Delinquency Prevention (JJDP) Act, which mandated DSO. In the following section, I analyze arrest data on boys and girls for certain violent crimes—simple and aggravated assault—to highlight differences in the seriousness of the crimes for which police arrest them. The analyses suggest that some girls' arrests for simple assault may be a relabeling of incorrigible girls as delinquents. I then focus on the offender–victim relationship of boys' and girls' assaults, which differentially affects the likelihood of girls' arrests for family conflicts in domestic disputes. Then I examine differences between patterns of incarceration for boys and girls sentenced for simple and aggravated assault. A discussion of the findings and conclusions follows.

Historical Differences in Juvenile Justice System Responses to Boys and Girls and DSO

The progressive reformers who created juvenile courts combined two visions, one interventionist and the other divisionary (Zimring, 2002). They envisioned a specialized court to separate children from adult offenders—diversion—and to treat them rather than to punish them for their crimes—intervention (Platt, 1977; Rothman, 1980; Ryerson, 1978; Tanenhaus, 2004). The juvenile court's delinquency jurisdiction initially encompassed only youths charged with criminal misconduct. However, reformers quickly added status offenses—noncriminal misbehaviors such as "incorrigibility," running away, "immorality," and "indecent and lascivious conduct" (Feld, 2004)—to the definition of delinquency. Historically, juvenile courts responded to boys primarily for criminal misconduct and to girls mainly for noncriminal status offenses (Schlossman, 1977; Sutton, 1988). The status jurisdiction reflected progressives' cultural construction of childhood dependency as well as their sexual sensibilities (Kempf-Leonard & Johansson, 2007; Schlossman & Wallach, 1978). From the juvenile courts' inception, controlling adolescent female sexuality was a central focus of judicial attention and intervention (Sutton, 1988; Tanenhaus, 2004). Historians consistently report that judges detained and incarcerated girls primarily for minor and status offenses and at higher rates than they did boys (Platt, 1977; Schlossman, 1977; Tanenhaus, 2004).

Although juvenile courts' status jurisdiction potentially encompassed nearly all juvenile misbehavior, by the early 1970s, critics argued that juvenile courts incarcerated noncriminal offenders with delinquents in secure detention facilities and institutions, stigmatized them with delinquency labels, discriminated against girls, and provided few beneficial services (Feld, 1999; Schwartz, Steketee, & Schneider, 1990). Judicial intervention at parents' behest to control their children also exacerbated intra family conflicts and enabled some caretakers to avoid their responsibilities (Sussman, 1977). In the early 1970s, states charged about three quarters of the girls whom juvenile courts handled as status offenders rather than as criminal delinquents (National Council on Crime and Delinquency, 1975; Schwartz et al., 1990).

The 1974 federal JJDP Act (42 U.S.C. § 223[a][12]) prohibited states from confining status offenders with delinquents in secure detention facilities and institutions and withheld formula grant money from states that failed to develop plans to remove them (Schwartz, 1989). The increased procedural formality and administrative costs of adjudicating delinquent offenders after *In re Gault* (1967) and the JJDP Act's deinstitutionalization goals provided impetus to divert status offenders to services and programs in the community. A 1980 amendment to the JJDP Act, adopted at the behest of the National Council of Juvenile and Family Court Judges, allowed states to continue to receive federal funds and to confine status offenders if juvenile court judges committed them to institutions for violating "valid court orders" (Schwartz, 1989). This exception allowed judges to bootstrap status offenders, disproportionately girls, into delinquents and to incarcerate them for contempt of court for violating court-ordered conditions of probation (Bishop & Frazier, 1992; Hoyt & Scherer, 1998). The 1992 reauthorization of the JJDP Act required states to analyze and provide "gender-specific services" to prevent and treat female delinquency, but most states used the funds to collect data about girls in the juvenile systems rather than to develop new programs (e.g. Bloom, Owne, Deschenes, & Rosenbaum, 2002; Community Research Associates, 1998; Kempf-Leonard & Sample, 2000; MacDonald & Chesney-Lind, 2001).

As a result of the 1974 DSO initiatives, the number of status offenders in secure detention facilities and institutions declined dramatically by the early 1980s. Because states disproportionately confined girls for noncriminal misconduct, they were the primary beneficiaries (Chesney-Lind, 1988; Handler & Zatz, 1982; Krisberg, Schwartz, Lisky, & Austin, 1986; Maxson & Klein, 1997). An early evaluation of the JJDP Act's DSO mandate by the National Academy of Sciences reported a substantial reduction in the detention and confinement of status offenders (Handler & Zatz, 1982). By 1988, the number of status offenders held in secure facilities had declined by 95% from those detained prior to adoption of the JJDP Act (U.S. General Accounting Office, 1991).

Although the JJDP Act prohibited states from incarcerating status offenders, it did not require states to appropriate adequate funds or to develop community-based programs to meet girls' needs. Even as policy makers and lawmakers struggled to find other options to respond to these "troublesome" youths, early analysts warned that states could evade deinstitutionalization requirements by relabeling status offenders as delinquents, for example, by charging them with simple assault rather than incorrigibility (Handler & Zatz, 1982).

Three decades after passage of the JJDP Act, states' failure adequately to fund or inability to offer appropriate community services provides a continuing impetus to use the juvenile delinquency system to circumvent DSO (Hoyt & Scherer, 1998; Maxson & Klein, 1997). "Status offenders" are not a unique or discrete category of juveniles, and they share many of the same characteristics and behavioral versatility as other delinquent offenders. As a result, the juvenile justice system simply could charge a female status offender with a minor crime, adjudicate her as a delinquent, and thereby evade deinstitutionalization strictures (Costello & Worthington, 1981; Federle & Chesney-Lind, 1992; Kempf-Leonard & Sample, 2000).

Macro structural economic and racial demographic changes during the 1970s and 1980s led to the emergence of an urban Black underclass and increased the punitiveness of juvenile justice policies, and these changes indirectly affected girls' susceptibility to arrest for violence. In the late 1980s and early 1990s, the epidemic of crack cocaine spurred increases in gun violence and Black male homicide, and states adopted punitive laws to "get tough" and "crack down" on youth crime (Blumstein, 1996; Feld, 1999; Zimring, 1998). States changed their laws to transfer more juveniles to criminal courts for prosecution as adults, and these amendments reflect a broader cultural and jurisprudential shift from rehabilitative to retributive and managerial penal policies (Feld, 2003; Garland, 2001; Tonry, 2004). Most of the punitive legislative agenda affected boys, particularly urban Black boys, charged with serious, violent crimes (Feld, 1999). Even though girls were not originally the intended subjects of the changes, the shift in juvenile justice responses to youth violence adversely affected girls, whom states could charge with assault (Chesney-Lind & Belknap, 2004; Poulin, 1996). Because the crackdown on youth violence and the rise in girls' arrests for assault coincided with DSO, focusing on the juvenile system's responses to girls provides an indicator of its changing mission and adaptive strategy.

Arrests of Boys and Girls for Violence: Simple and Aggravated Assaults

Police arrest and juvenile courts handle fewer girls than their proportional makeup of the juvenile population. As Table 7.2 reports, in 2003, police arrested an estimated 2.2 million juveniles. Girls constituted fewer than one third (29%) of all juveniles arrested and fewer than one fifth (18%) of those arrested for Violent Crime Index offenses. Girls constituted about one quarter (24%) of all the juveniles arrested for aggravated assaults and about one third (32%) of juvenile arrests for simple assault. Girls' arrests for simple assault constitute the largest proportion of their arrests for any violent crime. Arrests for Violent Crime Index offenses—murder, forcible rape, robbery, and aggravated assault—account for a very small proportion (4.2%) of all juvenile arrests, and aggravated assaults constitute two thirds (66.6%) of the Violent Crime Index offenses (Snyder & Sickmund, 2006). Significantly, however, police arrested about 85% of all girls arrested for Violent Crime Index offenses for aggravated assault (Federal Bureau of Investigation, 2006). By contrast, police arrested fewer than two thirds (62%) of boys for aggravated assaults and a much larger proportion for the most serious Violent Crime Index crimes of murder, rape, and robbery.

Changes in gender patterns of juveniles' arrests may reflect real differences in rates of offending by boys

Table 7.2 Juvenile and Female Arrest Estimates for Violence, 2003

Crime	Total Juvenile Arrest Estimates for All Offenses	Percentage Female Share of Arrests
Total	2,220,300	29
Violent Crime Index[a]	92,300	18
Aggravated assault	61,490	24
Simple assault	241,900	32

SOURCE: Snyder and Sickmund (2006).

a. Violent Crime Index includes murder, forcible rape, robbery, and aggravated assault.

and girls over time, or they may be justice system artifacts reflecting differences in the ways police and courts choose to respond to boys and girls (Girls Inc., 1996). Although girls constitute a smaller portion of juvenile arrestees than boys, the two groups' arrest patterns have diverged somewhat over the past decade. This divergence distinguishes more recent female delinquency from earlier decades, when male and female offending followed roughly similar patterns and when modest female increases were concentrated primarily in minor property crimes rather than violent crime (Steffensmeier, 1993).

As Table 7.3 indicates, arrests of female juveniles for various violent offenses have either increased more or decreased less than those of their male counterparts. From 1996 to 2005, the total number of juveniles arrested dropped by about 25%, primarily because arrests of boys decreased by 28.8%, whereas those of girls decreased only less than half as much (14.3%). Arrests of boys for Violent Crime Index offenses decreased substantially more than those of female offenders. Over the past decade, arrests of boys for Violent Crime Index offenses declined by 27.9%, whereas those of girls decreased by only 10.2%. Aggravated assaults constitute two thirds of all juvenile arrests for offenses included in the Violent Crime Index. Boys' arrests for aggravated assaults decreased by nearly one quarter (23.4%), whereas girls' arrests declined much more modestly (5.4%). By contrast, girls' arrests for simple assaults increased by one quarter (24%), whereas boys' arrests declined somewhat (4.1%). Thus, the major changes in arrest patterns for juvenile violence over the past decades are the sharp decrease in boys' arrests for aggravated assaults and the parallel increase in girls' arrests for less serious assaults.

Although the percentages reported in Table 7.3 reflect changes in the numbers of arrests, Figure 7.1 shows changes in the arrest rates per 100,000 male and female juveniles aged 10 to 17 years for Violent Crime Index offenses between 1980 and 2005. Overall, police arrested male juveniles at much higher rates than they did female juveniles. Consistent with Table 7.3, arrest rates for both groups peaked in the mid-1990s, and then the male rates exhibited a much sharper decline

Table 7.3 Percentage Changes in Male and Female Juvenile Arrests, 1996 to 2005

Crime	Girls	Boys
Total crime	−14.3	−28.8
Violent Crime Index	−10.2	−27.9
Aggravated assault	−5.4	−23.4
Simple assault	24.0	−4.1

SOURCE: Federal Bureau of Investigation (2006).

Figure 7.1 Male and Female Juvenile Arrest Rates, 1980 to 2005, Violent Crime Index Offenses

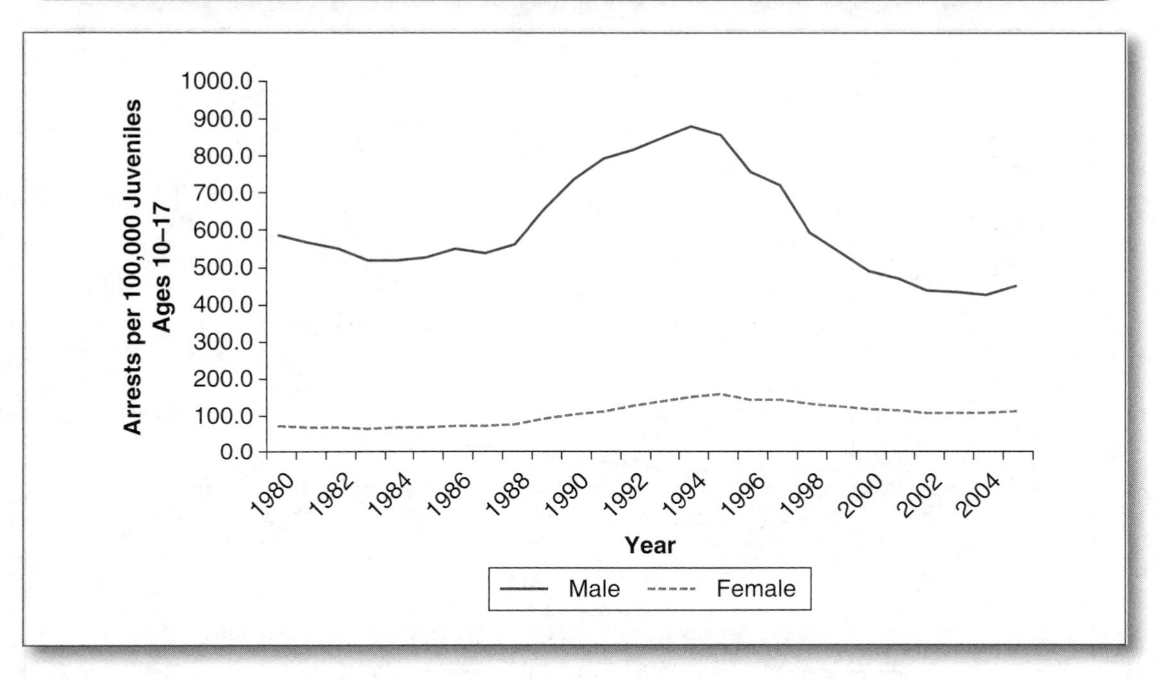

SOURCE: National Center for Juvenile Justice (2008).

than the female rates. Indeed, the male juvenile arrest rate for Violent Crime Index offenses in 2005 was nearly one quarter (23.3%) lower than in 1980. By contrast, girls' arrest rate for Violent Crime Index offenses rose from 70.4 to 106.9 per 100,000 over the same period, a 51.8% increase. In 1980, Violent Crime Index arrest rates for male juveniles were about 8 times higher than those of female juveniles, whereas by 2005, they were only 4 times higher. Thus, the juvenile "crime drop" of the past decade reflects primarily a decline in boys' arrests.

Arrests for aggravated assault constituted the largest component of the Violent Crime Index, and arrests for simple assault constituted the largest component of non–Violent Crime Index arrests. Over the past quarter century, clear changes have occurred between boys' and girls' patterns of arrests for these offenses. As Figure 7.2 indicates, boys' and girls' arrests for aggravated assault diverged conspicuously. The female arrest rate in 2005 was nearly double (97%) the arrest rate in 1980 (88.8 vs. 45 arrests for girls per 100,000). Although police arrested male juveniles for aggravated assault about 3 times more frequently than they did female juveniles, the boys' proportional increase (11.8%) was much more modest than that exhibited by the girls over the same period (267.8 vs. 239.4 arrests for boys per 100,000).

Police arrest juveniles for simple assaults much more frequently than they do for aggravated assaults. Again, changes in the arrest rates of female juveniles for simple assaults over the past quarter century greatly outstripped those of their male counterparts. The rate at which police arrested girls for simple assault in 2005 was nearly quadruple (3.9) the rate at which they arrested them in 1980 (499.8 vs. 129.7 female arrests per 100,000). Although the male arrest rate for simple assaults started from a higher base than the female rate, it only doubled (2.1) over the same period (948.9 vs. 462.7 arrests per 100,000).

Figure 7.2 Male and Female Juvenile Arrest Rates, 1980 to 2005, Simple and Aggravated Assaults

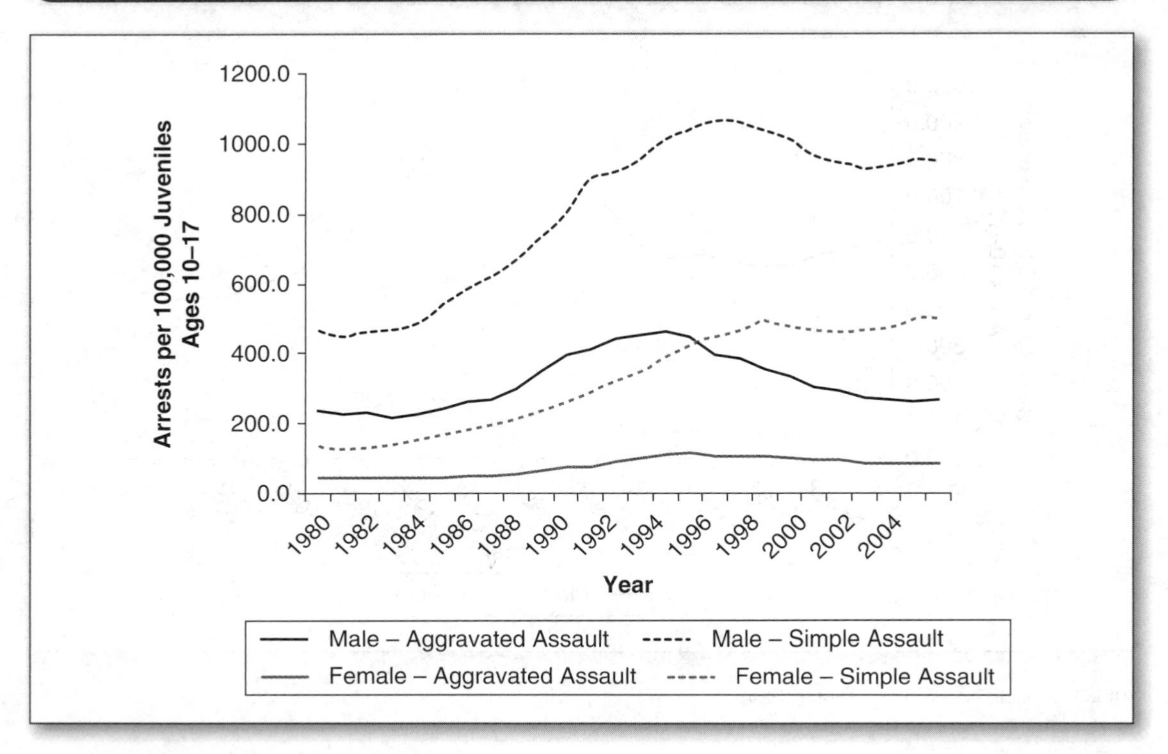

SOURCE: National Center for Juvenile Justice (2008).

To gauge the relative seriousness of most juveniles' arrests for violence, Figure 7.3 depicts the ratios of arrest rates for simple assaults and aggravated assaults for boys and for girls. In 1980, police arrested girls for simple assaults about 3 times (2.9) as often as they did for aggravated assaults. They arrested boys for simple assaults about twice (1.9) as often as they arrested them for aggravated assaults. Thus, police arrested girls more frequently than they did boys for less serious types of violence. In part, boys more often use weapons and inflict physical injuries on their victims than do girls, thereby aggravating many of their assaults. By 2005, police arrested girls more than 5 times (5.6) as often for simple assaults as they did for aggravated assaults. By contrast, the ratio of boys' arrests for simple to aggravated assaults only trebled (3.5). Thus, police are arresting even more girls for the least serious forms of violence than they did previously, and that ratio increased more so than for boys. These changes in ratios of arrest rates reflect the two different patterns reported in Table 2. The nearly one quarter (23.4%) decline in boys' arrests for aggravated assaults over the past decade increased their ratio of simple to aggravated assaults. By contrast, the nearly one quarter (24%) increase in girls' arrests for simple assaults over the same period substantially increased their ratio of simple to aggravated assaults. Thus, by all the measures—arrests, arrest rates, and ratios of simple to aggravated assaults—the increase in girls' arrests for simple assaults and boys' decrease in arrests for aggravated assaults constitute the most significant change in youth violence over the decades.

Despite these dramatic and gender-linked changes, it remains unclear whether the increase in

Figure 7.3 Ratios of Simple to Aggravated Assault Arrests for Boys and Girls, 1980 to 2005

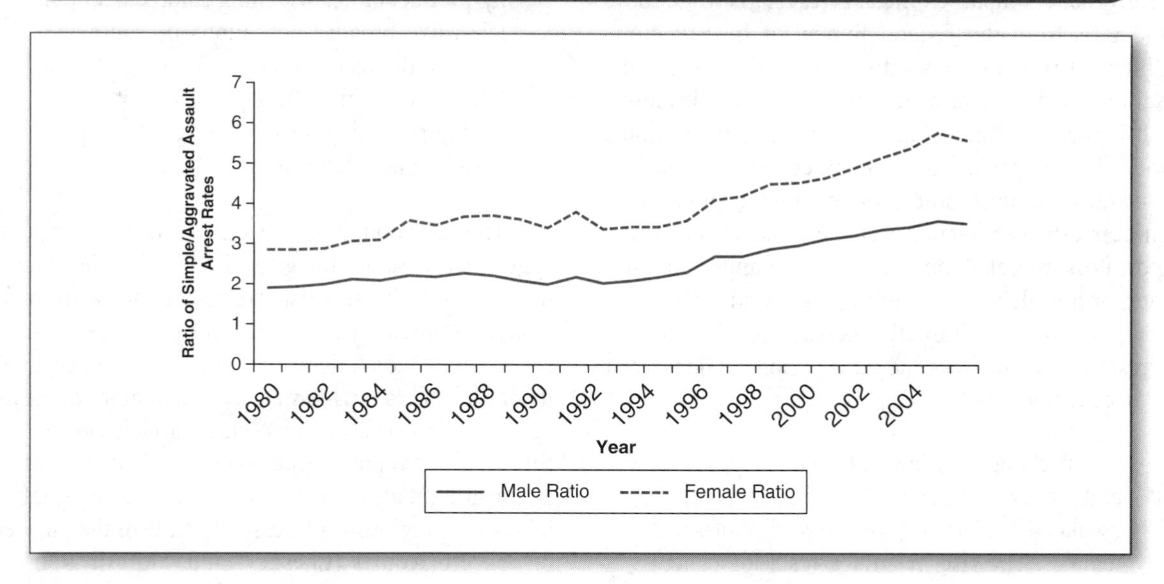

SOURCE: National Center for Juvenile Justice (2008).

girls' arrests signifies a real change in girls' underlying violent behavior or reflects police reclassification of assault offenses with a gender-specific component (Steffensmeier et al., 2005). Unlike crimes such as murder and robbery, which have relatively well defined elements and clearer indicators, police exercise considerably more discretion when they characterize behavior as an assault at all and whether they classify it as a simple or an aggravated assault, and these meanings have changed over time (Blumstein, 2000). An increase in proactive policing of disorder and minor crimes; a lower threshold to arrest or charge those types of offenses, especially among youth; and more aggressive policing in private settings may create the appearance of a juvenile "crime wave" when none actually exists. Zimring (1998) emphasized the role of police discretion and policy changes in the increase in arrests of youths for assaults. He argued that

> since 1980, there is significant circumstantial evidence from many sources that changing police thresholds for when assault should be recorded and when the report should be for aggravated assault are the reason for most of the growth in arrest rates.... Any reduction in the threshold between simple and aggravated assault and any shift in the minimum standard for recording an offense would have the kind of statistical impact on assault arrests that has occurred since the late 1980s. (pp. 39–40, 46)

Analysts of the changing characteristics of assaults over the past decades have compared ratios of aggravated assaults to homicides (e.g., Zimring, 1998) or of assaults to robberies (e.g., Snyder & Sickmund, 2006; Zimring & Hawkins, 1997) to demonstrate the malleable and changing definitions of assaults. Because arrests for aggravated assaults increased without any corresponding rise in arrests for homicides or for robberies, they have attributed the escalation in assault arrests to changes in law enforcement policies, such as changing offense seriousness thresholds or responses to domestic violence, rather than to real increases in assaults per se. Similarly, Steffensmeier et al. (2005) compared official arrest statistics for boys and girls from the Federal Bureau of Investigation's Uniform Crime Reports with victims' responses to the National Crime Victimization Survey and juveniles' self-reports in Monitoring the Future and the National Youth Risk Behavior Survey to assess whether the victim and self-report indicators mirrored the increase in girls' arrests for violence over the same period. These indicators revealed no systematic changes in girls' rates or prevalence of offending compared with that of boys, despite the dramatic increase in girls' official arrests for violence over the same period. Steffensmeier et al. concluded that

> recent changes in law enforcement practices and the juvenile justice system have apparently escalated the arrest proneness of adolescent females. The rise in girls' arrests for violent crime and the narrowing of the gender gap have less to do with underlying behavior and more to do, first, with net-widening changes in law and policing toward prosecuting less serious forms of violence, especially those occurring in private settings and where there is less culpability, and, second, with less biased or more efficient responses to girls' physical or verbal aggression on the part of law enforcement, parents, teachers, and social workers. (pp. 387–90)

The demarcation between status offenses and delinquency is as imprecise, malleable, and manipulable as the definition of assaults. "Because many status offenders are not simply runaways or truants but also engage in delinquent activities, it is possible for many such youths to be 'relabeled' delinquents rather than remain classified as status offenders" (Castallano, 1986, p. 496). The ambiguous difference between incorrigible or "unruly" behavior (status offenses) and the heterogeneous and elastic nature of violent behavior, particularly in the context of domestic discord, likely contributes to girls' increased arrests for simple assault. Steffensmeier et al. (2005) argued that

> female arrest gains for violence are largely a by-product of net-widening enforcement policies, like broader definitions of youth violence and greater surveillance of girls that have escalated the arrest-proneness of adolescent girls today relative to girls in prior decades and relative to boys. (p. 357)

The near doubling (1.9) in the ratio of simple to aggravated assaults for girls (2.9 vs. 5.6; Figure 7.3) indicates that most girls' arrests are increasingly for violent offenses at the lowest end of the seriousness scale. School "zero tolerance" policies and police "quality of life," "broken windows," and mandatory domestic violence arrest strategies cumulatively lower the threshold for reporting behavior as an assault or for aggravating it and lead to the arrests of more girls for behaviors previously addressed outside of the purview of police or courts (Chesney-Lind, Morash, & Irwin, 2007). Steffensmeier et al.'s analyses demonstrated that such policies can create an artificial appearance of a girls' violent crime wave when the underlying behavior remains much more stable. Indeed, such policies "tend to blur distinctions between delinquency and antisocial behavior more generally, lump together differing forms of physical aggression and verbal intimidation as manifesting interpersonal violence, and elevate interpersonal violence (defined broadly) as a high-profile social problem (particularly among youth)" (p. 363).

Victims of Boys' and Girls' Violence: Gender-Specific Domestic Disputes

Changing public attitudes and police practices toward domestic assaults have contributed to a growth in reports and arrests for simple assaults that victims and officers previously ignored (Blumstein, 2000; Miller, 2005). Mandatory arrest policies for domestic violence may have increased girls' risk for arrest by reducing social tolerance for girls' delinquency (Chesney-Lind, 2002; Miller, 2005). The heightened sensitivity to domestic violence combined with the prohibitions on incarcerating status offenders may encourage police to arrest girls more frequently for assault. Charging girls with simple assault rather than with a status offense, such as incorrigibility or unruly conduct, enables families, police, and juvenile courts to relabel the same behaviors as delinquency and thereby evade the prohibitions of the JJDP Act (Chesney-Lind & Belknap, 2004; Girls Inc., 1996; Mahoney & Fenster, 1982; Schneider, 1984).

> Family problems, even some that in past years may have been classified as status offenses (e.g., incorrigibility), can now result in an assault arrest. This logic also explains why violent crime arrests over the past decade have increased proportionately more for juvenile females than males. (Snyder, 2000, p. 4)

Parents' expectations for their sons' and daughters' behavior and obedience to parental authority differ (Chesney-Lind, 1988), and these differing cultural expectations affect how the justice system responds to girls' behavior when they "act out" within the home (Krause & McShane, 1994; Sussman, 1977). Girls who deviate from traditional gender norms such as passivity or femininity may be at greater risk for arrest for domestic violence (Miller, 2005). Girls fight with family members or siblings more frequently than do boys, whereas boys fight more often with acquaintances or strangers (Bloom et al., 2002; Hoyt & Scherer, 1998). Some studies report that girls are 3 times as likely to assault family members as are boys (Franke, Huynh-Hohnbaum, & Chung, 2002). Parents who in the past could have charged their daughters with being unruly or incorrigible now may request that police arrest them for "domestic violence" arising out of the same family scuffle (Russ, 2004). A study in California found that the female share of domestic violence arrests increased from 6% in 1988 to 17% in 1998 (Bureau of Criminal Information and Analysis, 1999).

> Some experts have found that this growth [in girls' assault arrests] is due in part not to a significant increase in violent behavior but to the re-labeling of girls' family conflicts as violent offenses, the changes in police practices regarding domestic violence and aggressive behavior, [and] the gender bias in the processing of misdemeanor cases. (American Bar Association & National Bar Association, 2001, p. 3)

Policies of mandatory arrest for domestic violence, initially adopted to restrain abusive men from attacking their partners (Miller, 2005), provide parents with another tool with which to control their unruly daughters. Regardless of who initiates a "violent" domestic incident, it is more practical and efficient for police to identify the youth as the offender when a parent is the caretaker for other children in the home (Gaarder, Rodriguez, & Zatz, 2004). As one probation officer observed,

> if you arrest the parents, then you have to shelter the kids. . . . So if the police just make the kids go away and the number of kids being referred to the juvenile court for assaulting their parents or for disorderly conduct or punching walls or doors . . . the numbers have just been increasingly tremendously because of that political change. (Gaarder et al., 2004, p. 565)

Analyses of girls' assault cases referred to juvenile court report that about half were "family centered" and involved conduct that parents and courts previously addressed as incorrigibility cases (Chesney-Lind & Pasko, 2004).

Many cases of girls charged with assault involved nonserious altercations with parents, who often may have been the initial aggressors (Acoca, 1999; Acoca & Dedel, 1998). Probation officers describe most girls' assault cases as fights with parents at home or between girls at school or elsewhere over boys (Artz, 1998; Bond-Maupin, Maupin, & Leisenring, 2002; Gaarder et al., 2004). School officials' adoption of zero-tolerance policies toward youth violence increases the number of youths referred for schoolyard tussles that they previously handled internally (Steffensmeier et al., 2005).

Girls typically perpetrate violence at home or at school and against family members or acquaintances, whereas boys are more likely to commit violent acts against acquaintances or strangers (Steffensmeier et al., 2005). Two pieces of evidence provide indicators of differences between boys and girls in offender–victim relationships and support the inference that more girls' violence arises in the context of domestic conflicts. Obviously, homicide is not an instance of the relabeling of status offenses, but the offender-victim relationship in homicides provides one indicator of gender-specific differences in violent offending. Table 7.4 reports the victim-offender relationships for boys and girls who committed homicides between 1993 and 2002. In more than one third (36%) of cases in which girls killed, their victims were family members, contrasted with only 7% of boys' homicides. By contrast, boys murdered strangers more than twice as frequently as did girls (38% vs. 18%). Thus, the most lethal forms of violence committed by girls were far more likely than for boys to occur in a domestic context.

Figure 7.4 examines the offender–victim relationships of youths involved in aggravated and simple assaults and provides another instance of gender-specific differences in violent offending in domestic disputes. The Federal Bureau of Investigation's National Incident-Based Reporting System is an incident-based crime reporting program that collects, among other data, information about offenders, victims, and their relationships (Snyder & Sickmund, 2006). More than one quarter of girls (28%), compared with fewer than one fifth (16%) of boys, committed aggravated assaults against family members. By contrast, boys assaulted acquaintances more frequently than did girls, and they assaulted strangers twice as often as girls. A similar pattern occurred for boys and girls involved in simple assaults. Girls' assaults occurred more frequently within the family than did boys' assaults, whereas boys more often assaulted acquaintances or strangers. Some of the increase in girls' arrests for simple assaults can be attributed to their greater likelihood than boys to "victimize" family members, the decrease in public and police tolerance for all forms of domestic violence, and the ease with which police may reclassify incorrigible behavior as assault.

Table 7.4 Victims of Murders Committed by Juveniles, 1993 to 2002

Victim-Offender Relationship	Boys	Girls
Family	7	36
Acquaintance	55	46
Stranger	38	18

SOURCE: Snyder and Sickmund (2006, p. 69).

> The rise in girls' arrests for violent crime and the narrowing gender gap have less to do with underlying behavior and more to do, first, with net-widening changes in law and policing toward prosecuting less serious forms of violence, especially those occurring in private settings and where there is less culpability, and, second, with less biased or more efficient responses to girls' physical or verbal aggression on the part of law enforcement, parents, teachers, and social workers. (Steffensmeier et al., 2005, p. 387)

Several studies provide evidence of the juvenile justice system's relabeling status offenders as delinquents to incarcerate them. A comparison of juvenile court petitions filed against girls before and after Pennsylvania repealed its status jurisdiction in the mid-1970s found that the proportion of girls charged with assaults more than doubled (from 14% to 29%) following the change (Curran, 1984). In response to the JJDP Act's DSO mandate, the proportion of girls confined in training schools for status offenses declined from 71% in 1971 to 11% in 1987, while there was a commensurate increase in the proportion of girls confined for minor delinquencies during the same period (Schwartz et al., 1990). Moreover, states appear to confine girls for less serious offenses than they do boys. In 1987, juvenile courts confined over half (56%) of girls for misdemeanor offenses, compared with only 43% of boys (Schwartz et al., 1990).

Figure 7.4 Male and Female Offender-Victim Relationships

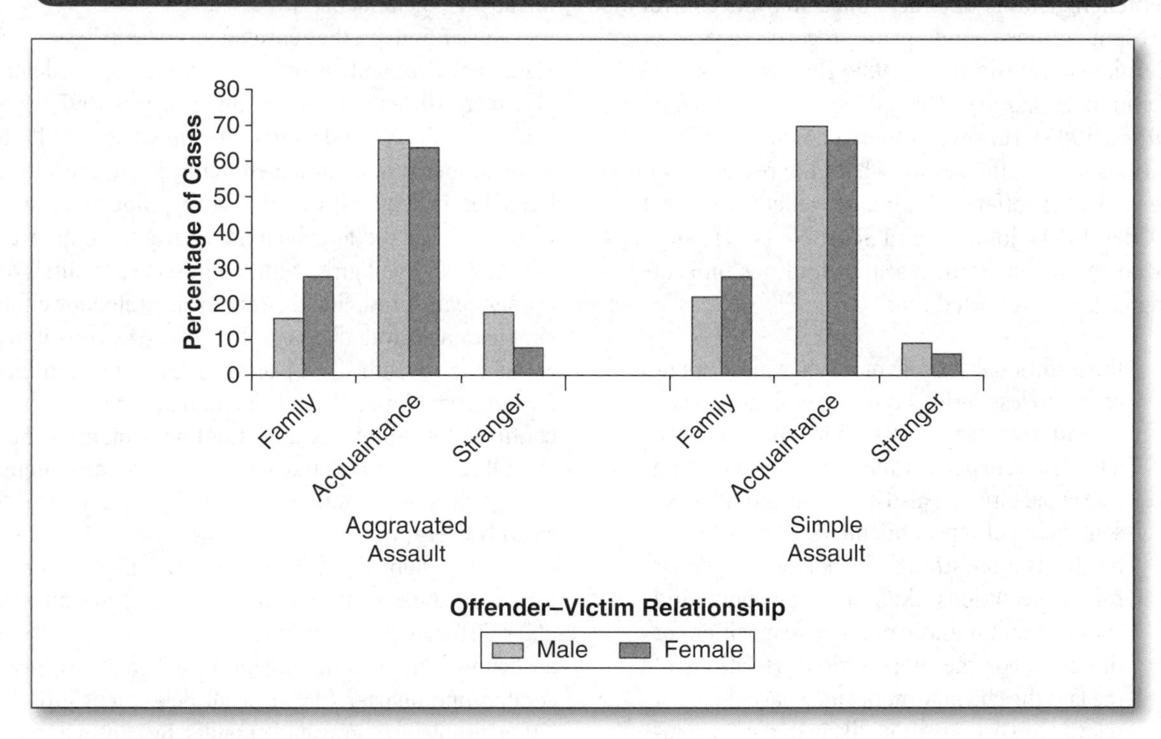

SOURCE: Snyder and Sickmund (2006, p. 145).

Offense Characteristics of Delinquent Boys and Girls in Confinement

Juvenile court judges possess a wide range of options to sentence delinquents: dismissal, continuance without a finding, restitution or fine, probation with or without conditions, out-of-home placement in a public or private facility or group home, confinement in a county institution or state training school, or placement in another secure public or private setting. Because male juveniles commit most of the serious crimes, evaluations of juvenile court sentencing practices typically focus on boys and examine racial rather than gender disparities (e.g., Feld, 1999; McCord, Widom, & Crowell, 2001).

Some sentencing research on gender bias focuses on "chivalrous" or lenient treatment of delinquent girls to explain why girls receive less severe sanctions than do similarly charged boys (Hoyt & Scherer, 1998). Other analysts invoke "protectionist" or "paternalistic" explanations to account for why juvenile courts intervene more actively in the lives of sexually active females and status offenders than they do boys charged with minor offenses (e.g., Chesney-Lind, 1977, 1988; Johnson & Scheuble, 1991; Schlossman, 1977; Schlossman & Wallach, 1978). Earlier research consistently reported a gender double standard in the sentencing of girls and boys. Juvenile courts incarcerated proportionally more girls than boys charged with status offenses and sentenced boys charged with delinquency more severely than they did girls (e.g., Bishop & Frazier, 1992). More recent studies have reported fewer gender differences in sentencing status offenders once analysts control for present offense and prior record (e.g., Corley, Cernkovich, & Giordano, 1989; Hoyt & Scherer, 1998; Teilman & Landry, 1981; U.S. General Accounting Office, 1995). However, others contend that the definitions of the offenses for which the research control (e.g., status offenses) already reflect gender bias (Alder, 1984). Johnson and Scheuble (1991) summarized the inconsistent research findings on sentencing girls and reported that

> the traditional sex role model has more application to less serious types of violations, such as status offenses, for which females are given a more severe penalty than males for violating role expectation. It also has application for the sentencing of repeat offenders. Such behavior by girls is more strongly in violation of gender role expectations than it is for boys and should result in more punitive disposition for the girls. For the more serious violations of the law, the chivalry model may have the most relevance. Girls are more likely to receive leniency and protection from the consequences of the more serious crimes. (p. 680)

Bishop and Frazier (1992) analyzed juvenile courts' use of contempt power to sanction male and female status offenders who violated valid court orders and reported differential treatment and bootstrapping of girls that covertly perpetuated gender bias.

The next analyses look at characteristics of youths in juvenile residential facilities. Police arrest and juvenile courts file petitions, detain, adjudicate, and place boys in institutions at higher rates and for more serious offenses than they do girls. However, the juvenile justice system processes girls for aggravated and simple assaults at higher rates than it does girls charged with other types of offenses, such as property, drugs, and public order crimes (Feld, in press; Snyder & Sickmund, 2006). Rather than examining the cumulative process by which judges place youths in correctional facilities, the Census of Juveniles in Residential Placement provides a 1-day count of youths in residential placements on a biennial basis.

Table 7.5 adapts the Census of Juveniles in Residential Placement data and reports on juveniles in residential placement (detention and confinement) in 1997, 1999, 2001, and 2003. In 2003, girls constituted about 14% of all delinquents in confinement and 13% of those confined for violent crimes, and both proportions have increased over the four biennia. Nearly three quarters (about 72%) of all girls confined in secure facilities for crimes against individuals were incarcerated for either simple or aggravated assault. In 2003, girls constituted about one seventh (13%) of all delinquents confined for aggravated assault and one fourth (25%) of those confined for simple assault. Confinement for simple assault represents the largest proportion for any offense for which states confine girls, and it has increased steadily over the census years.

To highlight the differences between the offenses for which states confine male and female juveniles, in 2003, girls constituted only about one in seven (14%) of all delinquents in confinement. However, states incarcerated one quarter (25%) of all delinquent girls for either simple or aggravated assaults. By contrast, states confined boys for a more heterogeneous mix of offenses, of which simple and aggravated assaults accounted for only about one seventh (15%).

When changes in confinement for assault are examined, it is found that in each succeeding biennial census, the proportion of girls confined for aggravated and simple assaults increased. Even though boys constituted 92% of all delinquents confined for Violent Crime Index offenses (Sickmund, Sladky, & Kang, 2005), the proportion of girls confined for aggravated assaults, as a percentage of all delinquents confined for aggravated assaults, increased from 12% to 16%. In all four biennia, states confined a majority of all boys for aggravated assaults (62%, 60%, 54%, and 51%) rather than simple assaults. By contrast, the majority of girls whom states confined for assaults were incarcerated for simple assaults rather than aggravated assaults (45%, 45%, 40%, and 38%). Although violent girls may violate gender norms and thereby appear more serious (Schaffner, 1998), by

Table 7.5 Confinement of Boys and Girls for Simple and Aggravated Assaults, 1997 to 2003

Variable	1997	1999	2001	2003
Total delinquents confined	98,222	102,958	99,297	91,831
Female proportion of all delinquents in confinement	11%	12%	13%	4%
Number of girls confined for all person offenses	3612	4365	4443	4401
Proportion of delinquent offenders confined for all person offenses who are female	10%	12%	13%	13%
Number of girls confined for simple and aggravated assault	2,535	3,147	3,211	3,198
% of total delinquents confined for aggravated assault who are female	12	14	15	16
% of total delinquents confined for simple assault who are female	22	23	24	25
% of girls confined for simple and aggravated assaults as a proportion of all girls' delinquency confinements	23	25	25	25
% of boys confined for simple and aggravated assaults as a proportion of all boys' delinquency confinements	16	16	15	16
Girls'% aggravated assaults to all assaults	45	45	40	38
Boys'% aggravated assaults to all assaults	62	60	54	51

SOURCE: Adapted from Sickmund, Sladky, and Kang (2005).

contrast with the boys, larger proportions of girls are confined for less violent and injurious crimes than their male counterparts. The incarceration of larger numbers and proportions of girls for simple assaults suggests a process of relabeling other status like conduct, such as incorrigibility, to obtain access to secure placement facilities.

Conclusion and Policy Implications

Juvenile courts adapt to changes in their organizational environment, and institutional maintenance may explain juvenile courts' continued endurance at least as well as their professed ability to achieve their rehabilitative goals (Schwartz, Weiner, & Enosh, 1998; Sutton, 1988). The breadth and mutability of the juvenile court's mission enable it to redefine the boundaries of social control it administers (Sutton, 1988) and allow court personnel to maintain operational stability in the face of the delinquent male "crime drop," with an offsetting increase in female cases (Federle, 2000). DSO coincided with the emergence of a "culture of control," greater emphases on proactive policing, and aggressively addressing minor disorder and law violations (Garland, 2001). "The trend has been to lower the threshold of law enforcement, in effect to arrest or charge up and be less tolerant of low-level crime and misdemeanors, and to be more inclined to respond to them with maximum penalties" (Steffensmeier et al., 2005, p. 363).

The broad discretion available to parents, police, prosecutors, and juvenile court personnel allows them to charge many status offenders as minor delinquents and to "bring status offenders under the jurisdiction of the court at a rate almost as great as had existed prior to the [decriminalization] reform" (Schneider, 1984, p. 367). Courtroom observers report that following DSO, prosecutors charged many girls with criminal offenses for behavior that they previously charged as status offenses (Mahoney & Fenster, 1982). After Washington State temporarily decriminalized status offenders, some police and courts "redefined" them as minor criminal offenders so that juvenile courts could retain jurisdiction and authority over them (Castallano, 1986; Schneider, 1984). Analyses of the changing handling of girls' simple and aggravated assaults strongly suggest that the perceived growth in girls' "violence" may reflect a "criminalization of intra-familial conflicts and aggressive behavior," rather than an actual change in girls' behavior (American Bar Association & National Bar Association, 2001, p. 14).

After three decades of DSO, the juvenile justice system remains committed to protecting and controlling girls, but without responding to their real needs. When Congress passed the JJDP Act in 1974, neither the federal nor state governments made substantial or systematic efforts to provide girls with adequate programs or services in the community (Chesney-Lind & Shelden, 2005; Maxson & Klein, 1997). Although the 1992 reauthorization of the JJDP Act included provision for "gender-specific services," the implementation of that mandate has languished. The failure to provide alternatives to institutional confinement for "troublesome girls" creates substantial pressures within the juvenile justice system to circumvent DSO restrictions by the simple expedient of relabeling them as delinquents by charging them with assault.

References

Acoca, L. (1999). Investing in girls: A 21st century strategy. *Juvenile Justice, 6*, 3–13.

Acoca, L., & Dedel, K. (1998). *No place to hide: Understanding and meeting the needs of girls in the California juvenile justice system.* San Francisco, CA: National Council on Crime and Delinquency.

Alder, Christine. (1984). Gender bias in juvenile diversion. *Crime & Delinquency, 30*, 400–414.

American Bar Association & National Bar Association. (2001). *Justice by gender: The lack of appropriate prevention, diversion and treatment alternatives for girls in the justice system.* Washington, DC: American Bar Association.

Artz, S. (1998). *Sex, power, and the violent school girl.* Toronto, Canada: Trifolium.

Bishop, D. M., & Frazier, C. (1992). Gender bias in juvenile justice processing: Implications of the JJDP Act. *Journal of Criminal Law and Criminology, 82*, 1162–1186.

Bloom, B., Owne, B., Deschenes, E. P., & Rosenbaum, J. (2002). Improving juvenile justice for females: A statewide assessment in California. *Crime & Delinquency, 4*, 526–552.

Blumstein, A. (1996). Youth violence, guns, and the illicit-drug industry. *Journal of Criminal Law and Criminology, 86*, 10–36.

Blumstein, A. (2000). Disaggregating the violence trends. In A. Blumstein & J. Wallman (Eds.), *The crime drop in America* (pp. 13–44). New York: Cambridge University Press.

Bond-Maupin, L., Maupin, J. R., & Leisenring, A. (2002). Girls' delinquency and the justice implications of intake workers' perspectives. *Women & Criminal Justice, 13*, 51–77.

Bureau of Criminal Information and Analysis. (1999). Report on arrests for domestic violence in California, 1998. *Criminal Justice Statistics Center Report Series, 1*(2), 5–6.

Castallano, Thomas C. (1986). The justice model in the juvenile justice system: Washington state's experience. *Law and Policy, 8*, 397–418.

Chesney-Lind, M. (1977). Paternalism and the female status offender. *Crime & Delinquency, 23*, 121–130.

Chesney-Lind, M. (1988). Girls and status offenses: Is juvenile justice still sexist? *Criminal Justice Abstracts, 20*, 144–165.

Chesney-Lind, M. (2002). Criminalizing victimization: The unintended consequences of pro-arrest policies for girls and women. *Criminology & Public Policy, 1*, 81–90.

Chesney-Lind, M., & Belknap, M. (2004). Trends in delinquent girls' aggression and violent behavior: A review of the evidence. In M. Puytallaz & P. Bierman (Eds.), *Aggression, antisocial behavior and violence among girls: A developmental perspective* (pp. 203–222). New York: Guilford.

Chesney-Lind, M., & Pasko, L. (2004). *The female offender: Girls, women, and crime* (2nd ed.). Thousand Oaks, CA: Sage.

Chesney-Lind, M., & Shelden, R. (1997). *Girls, delinquency, and juvenile justice* (2nd ed.). Pacific Grove, CA: Brooks/Cole.

Chesney-Lind, M., Morash, M., & Irwin, K. (2007). Policing girlhood? Relational aggression and violence prevention. *Youth Violence and Juvenile Justice, 5*, 328–345.

Community Research Associates. (1998). *Juvenile female offenders: A status of the states report.* Washington, DC: Office of Juvenile Justice and Delinquency Prevention.

Corley, C. J., Cernkovich, S., & Giordano, P. (1989). Sex and the likelihood of sanction. *Journal of Criminal Law and Criminology, 80*, 540–556.

Costello, J. C., & Worthington, N. L. (1981). Incarcerating status offenders: Attempts to circumvent the Juvenile Justice and Delinquency Prevention Act. *Harvard Civil Rights—Civil Liberties Law Review, 16*, 41–81.

Curran, D. J. (1984). The myth of the "new" female delinquent. *Crime & Delinquency, 30*, 386–399.

Federal Bureau of Investigation. (2006). *Uniform crime reports: Crime in the United States 2005*. Washington, DC: U.S. Department of Justice.

Federle, K. H. (2000). The institutionalization of female delinquency. *Buffalo Law Review, 48*, 881–908.

Federle, K. H., & Chesney-Lind, M. (1992). Special issues in juvenile justice: Gender, race, and ethnicity. In I. Schwartz (Ed.), *Juvenile justice and public policy: Toward a national agenda* (pp. 165–195). New York: Lexington.

Feld, B. C. (1999). *Bad kids: Race and the transformation of the juvenile court*. New York: Oxford University Press.

Feld, B. C. (2003). Race, politics, and juvenile justice: The Warren court and the conservative "backlash." *Minnesota Law Review, 87*, 1447–1577.

Feld, B. C. (2004). *Cases and materials on juvenile justice administration* (2nd ed.). St. Paul, MN: West.

Feld, B. C. (2009). Girls in the juvenile justice system. In M. Zahn (Ed.), *The delinquent girl*. Philadelphia: Temple University Press.

Franke, T. M., Huynh-Hohnbaum, A.-L. T., & Chung, Y. (2002). Adolescent violence: With whom they fight and where. *Journal of Ethnic & Cultural Diversity in Social Work, 11*(3–4), 133–158.

Gaarder, E., Rodriguez, N., & Zatz, M. S. (2004). Criers, liars, and manipulators: Probation officers' views of girls. *Justice Quarterly, 21*, 547–578.

Garland, D. (2001). *The culture of control: Crime and social order in contemporary society*. Chicago: University of Chicago Press.

Girls Inc. (1996). *Prevention and parity: Girls in juvenile justice*. Indianapolis, IN: Author.

Handler, J. F., & Zatz, J. (Eds.). (1982). *Neither angels nor thieves: Studies in deinstitutionalization of status offenders*. Washington, DC: National Academy Press.

Hoyt, S., & Scherer, D. G. (1998). Female juvenile delinquency: Misunderstood by the juvenile justice system, neglected by social science. *Law and Human Behavior, 22*, 81–107.

In re Gault, 387 U.S. 1 (1967).

Johnson, D. R., & Scheuble, L. K. (1991). Gender bias in the disposition of juvenile court referrals: The effects of time and location. *Criminology, 29*, 677–699.

Kempf-Leonard, K., & Johansson, P. (2007). Gender and runaways: Risk factors, delinquency, and juvenile justice experiences. *Youth Violence and Juvenile Justice, 5*, 308–327.

Kempf-Leonard, K., & Sample, L. L. (2000). Disparity based on sex: Is gender-specific treatment warranted? *Justice Quarterly, 17*, 89–128.

Kluger, J. (2006). Taming wild girls. *Time, 167*(18), 54–55.

Krause, W., & McShane, M. D. (1994). A deinstitutionalization retrospective: Relabeling the status offender. *Journal of Crime and Justice, 17*, 45–67.

Krisberg, B., Schwartz, I., Lisky, P., & Austin, J. (1986). The watershed of juvenile justice reform. *Crime & Delinquency, 32*, 5–38.

MacDonald, J. M., & Chesney-Lind, M. (2001). Gender bias and juvenile justice revisited: A multiyear analysis. *Crime & Delinquency, 47*, 173–195.

Mahoney, A. R., & Fenster, C. (1982). Female delinquents in a suburban court. In N. H. Rafter & E. A. Stanko (Eds.), *Judge, lawyer, victim, thief: Women, gender roles and criminal justice* (pp. 221–236). Boston: Northeastern University Press.

Maxson, C. L., & Klein, M. W. (1997). *Responding to troubled youth*. New York: Oxford University Press.

McCord, J., Widom, C. S., & Crowell, N. A. (2001). *Juvenile crime, juvenile justice*. Washington, DC: National Academy Press.

Miller, S. L. (2005). *Victims as offenders: The paradox of women's violence in relationships*. New Brunswick, NJ: Rutgers University Press.

National Center for Juvenile Justice. (2008, October 24). Juvenile arrest rates by offense, sex, and race. Available at http://ojjdp.ncjrs.org/ojstatbb/crime/excel/jar_2007.xls

National Council on Crime and Delinquency. (1975). Jurisdiction over status offenders should be removed from the juvenile court: A policy statement. *Crime & Delinquency, 21*, 97–99.

Platt, A. M. (1977). *The child-savers: The invention of delinquency*. Chicago: University of Chicago Press.

Poulin, A. B. (1996). Female delinquents: Defining their place in the justice system. *Wisconsin Law Review*, 1996, 541–575.

Rothman, D. (1980). *Conscience and convenience: The asylum and its alternative in progressive America*. Boston: Little, Brown.

Russ, H. 2004. The war on catfights. *City Limits, February,* 19–22.

Ryerson, E. (1978). *The best-laid plans: America's juvenile court experiment*. New York: Hill & Wang.

Sanders, J. (2005, June 23). How to defuse "girl on girl" violence. *Christian Science Monitor*. Retrieved from http://www.csmonitor.com/2005/0623/p09s01-coop.html

Scelfo, J. (2004). Bad girls go wild: A rise in girl-on-girl violence is making headlines nationwide and prompting scientists to ask why. *Newsweek*. Available at http://www .newsweek.com/id/50082

Schaffner, L. (1998). Female juvenile delinquency: Sexual solutions, gender bias, and juvenile justice. *Hastings Women's Law Journal, 9*, 1–25.

Schlossman, S. L. (1977). *Love and the American delinquent: The theory and practice of "progressive" juvenile justice 1825-1920*. Chicago: University of Chicago Press.

Schlossman, S. L., & Wallach, S. (1978). The crime of precocious sexuality: Female juvenile delinquency in the progressive era. *Harvard Educational Review, 48*, 655–694.

Schneider, A. L. (1984). Divesting status offenses from juvenile court jurisdiction. *Crime & Delinquency, 30*, 347–370.

Schwartz, I. M. (1989). *(In)justice for juveniles: Rethinking the best interests of the child*. Lexington, MA: Lexington Books.

Schwartz, I. M., Steketee, M. W., & Schneider, V. W. (1990). Federal juvenile justice policy and the incarceration of girls. *Crime & Delinquency, 36*, 511–520.

Schwartz, I. M., Weiner, N. A., & Enosh, G. (1998). Nine lives and then some: Why the juvenile court does not roll over and die. *Wake Forest Law Review, 33*, 533–552.

Sickmund, M., Sladky, T. J., & Kang, W. (2005) *Census of Juveniles in Residential Placement databook*. Available at http://www.ojjdp.ncjrs.org/ojstatbb/cjrp/

Snyder, H. (2000). *Challenging the myths*. Washington, DC: U.S. Department of Justice, Office of Juvenile Justice and Delinquency Prevention.

Snyder, H. N., & Sickmund, M. (2006). *Juvenile offenders and victims: 2006 national report*. Washington, DC: U.S. Department of Justice, Office of Justice Programs, Office of Juvenile Justice and Delinquency Prevention.

Steffensmeier, D. (1993). National trends in female arrests, 1960-1990: Assessment and recommendations for research. *Journal of Quantitative Criminology, 9*, 411–441.

Steffensmeier, D., Schwartz, J., Zhong, S. H., & Ackerman, J. (2005). An assessment of recent trends in girls' violence using diverse longitudinal sources: Is the gender gap closing? *Criminology, 43*, 355–405.

Sussman, A. (1977). Sex-based discrimination and PINS jurisdiction. In L. E. Teitelbaum & A. R. Gough (Eds.), *Beyond control: Status offenders in the juvenile court* (pp. 179–199). Cambridge, MA: Ballinger.

Sutton, J. (1988). *Stubborn children: Controlling delinquency in the United Sates, 1640–1981*. Berkeley: University of California Press.

Tanenhaus, D. S. (2004). *Juvenile justice in the making*. New York: Oxford University Press.

Teilman, K. S., & Landry, P. H., Jr. (1981). Gender bias in juvenile justice. *Journal of Research in Crime and Delinquency, 18*, 47–80.

Tonry, M. (2004). *Thinking about crime: Sense and sensibility in American penal culture*. New York: Oxford University Press.

U.S. General Accounting Office. (1991). *Noncriminal juveniles: Detentions have been reduced but better monitoring is needed*. Washington, DC: Author.

U.S. General Accounting Office. (1995). *Minimal gender bias occurred in processing noncriminal juveniles*. Washington, DC: Author.

Williams, C. (2004, December 28). Where sugar and spice meet bricks and bats. *The Washington Post*, p. B01.

Zimring, F. E. (1998). *American youth violence*. New York: Oxford University Press.

Zimring, F. E. (2002). The common thread: Diversion in juvenile justice. *California Law Review, 88*, 2477–2495.

Zimring, F. E., & Hawkins, G. (1997). *Crime is not the problem: Lethal violence in America*. New York: Oxford University Press.

DISCUSSION QUESTIONS

1. What role do family conflicts play in the rise of girls' rates of delinquency?
2. How have status offenses "reemerged" in the juvenile court and affected the processing of youth offenders?
3. What recommendations would you make for the juvenile justice system in dealing with girls?

READING

Throughout this book, you have learned about how theories and programming for offenders is based on a male standard and does little to represent the needs and lives of girls and women. In this reading, the authors talk with young girls incarcerated in a juvenile justice facility to listen to what girls think they want and need in order to rehabilitate their lives and remain crime-free. These perspectives are then compared to what the literature says about the needs of girls, and recommendations are made to improve the lives of girls.

SOURCE: Garcia, C. A., & Lane, J. (2009). What a girl wants, what a girl needs: Findings from a gender-specific focus group study. *Crime and Delinquency*. Published online April 3, 2009, as doi:10.1177/0011128709331790.

What a Girl Wants, What a Girl Needs

Findings From a Gender-Specific Focus Group Study

Crystal A. Garcia and Jodi Lane

Until the mid-1980s few studies focused on juvenile female offenders (Belknap, 2001; Chesney-Lind, 1997; Chesney-Lind & Pasko, 2004; Leonard, 1982; Morris, 1987). Despite the important work of Belknap, Bloom, Chesney-Lind, Covington, Deschenes, Holsinger, Leonard, Owen, and others regarding the life experiences of female offenders and their pathways to crime, few justice jurisdictions have attempted to implement gender-specific programming. It is understandable that male offenders are the primary focus, because they commit more crimes and more violence (Snyder & Sickmund, 2006), but arrest patterns are changing. Female delinquency and arrest rates in the United States recently have outpaced those of males for many crimes (Federal Bureau of Investigation, 2003; Poe-Yamagata & Butts, 1996; Snyder & Sickmund, 2006). Although most girls still enter the juvenile justice system for status offenses and property crimes, more girls are being arrested for violence (American Bar Association and National Bar Association, 2001; Chesney-Lind, 2004; Chesney-Lind & Okamoto, 2001; *Juvenile Female Offenders*, 1998; Kakar, Friedmann, & Peck, 2002).

Girls' arrest patterns may be changing, but the way the system responds to them has changed little. Most programs were developed to treat young male offenders (Chesney-Lind & Shelden, 2004; Shearer, 2003) and ignored the specific developmental, social, and psychological needs of girls (American Bar Association and National Bar Association, 2001; Bloom, Owen, Deschenes, & Rosenbaum, 2002b). Academics, practitioners, and policy makers finally agreed that increases in juvenile female offending could no longer be ignored (Poe-Yamagata & Butts, 1996). Consequently, federal funds were made available to the states through Challenge Grants if the states survey their juvenile justice systems and identify and address gaps in gender-specific services (Juvenile Justice and Delinquency Prevention Act of 2002).

Indiana's approach to assess gender-specific services was Indiana's Gender-Relevant Programming Initiative (IGRPI). The initiative's purpose was to determine what was known about the gender-specific needs of Indiana youth and to identify the availability of such programs. IGRPI researchers used a three-pronged approach: an in-depth literature review, focus groups with various constituencies, and a statewide survey of juvenile justice and youth service providers (see Ziemba-Davis, Garcia, Kincaid, Gullans, & Myers, 2004). This article reports findings from the focus groups with at-risk and delinquent girls.

What Girls Need: Gender-Specific Programming

Some have argued that girls who come to the attention of the authorities differ from their "brothers" in terms of their path to and their needs from the juvenile justice system (American Bar Association and National Bar Association, 2001; Greene, Peters, & Associates, 1998).

Others claim that those who work with juveniles generally have ignored the important influence of a gendered society and therefore the importance of gender on offending (Bloom, Owen, & Covington, 2004; Loeber, Farrington, & Petechuck, 2003). Consequently, justice practitioners place girls into programs that appear gender neutral but were actually developed for males (Kivel, 1992). Perhaps more troubling is what Bloom, Owens, Deschenes, and Rosenbaum (2002a) found in their study—both staff and girls believed that girls were "invisible" in the juvenile justice system and that few programs were developed with girls' needs in mind (p. 535).

Because the pathways and triggers that lead girls to delinquency can differ (Chesney-Lind & Bloom, 1997), some scholars argue that girls need gender-specific intervention and correctional programs (Bloom, Owens, & Covington, 2003). Gender-specific programming uses "a comprehensive approach to female delinquency rooted in the experience of girls.... It bridges theory into practice by combining female adolescent theory with juvenile justice practices" (Greene et al., 1998). Specifically, gender-specific programming focuses on girls' particular psychological, social, and developmental needs.

Although the various criminological theories that have examined the impact of gender on offending (general strain, learning, control, and feminist theories) differ in their explanations for why girls participate in delinquency, all theories point to the very important role that family and relationships play in girls' lives (Agnew, 1992; Burton, Cullen, Evans, Alarid, & Dunaway, 1998; Chesney-Lind, 1989, 1997; Gottfredson & Hirschi, 1990; Heimer & DeCoster, 1999; Mears, Ploeger, & Warr, 1998). If girls are more influenced by poor family dynamics, parental loss, physical and/or sexual abuse, and romantic relationships with delinquent partners than boys are, then the types of treatment programs provided to girls must address these specific psychological needs (Bloom et al., 2003). Such programs should be offered in safe, nurturing environments that use culturally competent treatment models, offer dignity and respect to clients, and allow for bonds to be established between treatment staff and the girls involved (Bloom et al., 2003). Moreover, such programs should address abandonment, neglect, and abuse issues and revolve around building trusting, healthy relationships with the service providers and others, particularly females in the girls' families and social circles (Belknap, Dunn, & Holsinger, 1997b; Bloom et al., 2003; Valentine Foundation, 1990). These programs should also address the myriad of resulting psychological problems many adolescent girls battle (Belknap & Holsinger, 1998), such as eating disorders, depression, substance abuse, self-mutilation, and personality disorders (Bloom et al., 2003).

Programs that are gender-specific must also recognize the critical role of romantic relationships and sexuality in girls' views of themselves and their responses to the world. Programs need to not only teach girls how to build "psychologically healthy" relationships but also include components that focus on girls' unique developmental needs (Bloom et al., 2003). For instance, girls need to learn about birth control, safe sex practices, parenting skills, hygiene, and body image (Beckman, 1994; Bloom & Covington, 1998; Greene et al., 1998).

Moreover, gender-specific programs should consider the particular social needs of girls. Beyond just dating, girls need to learn to navigate occurrences of relational aggression—for example, other girls using harassment, bullying, and ruining reputations for the purpose of socially isolating someone (Piper, 1994). Programs should also teach girls how to identify prosocial activities, foster positive peer networks, and choose nondelinquent romantic partners (Beckman, 1994).

Girls' programming must also address their future-oriented needs, both developmental and social. Specifically, programming should use techniques that allow girls to identify and correct any educational deficits as well as build on their current assets (Greene et al.,1998). Programs should assist girls in identifying career aspirations and attaining the knowledge, skills, and abilities needed to achieve these goals. Many occupational and vocational training programs administered by justice agencies provide training in traditionally gendered vocations (e.g., cosmetology and janitorial services for girls and welding, carpentry, and machine shop work for boys). Although these are respectable vocations, such training should introduce young women to "realistic employment opportunities that allow for self support" (Bloom et al., 2005, p. 4) and to career paths once thought to be outside of traditional women's work (Greene et al., 1998).

Respondents in Bloom and colleagues' (2002a) study also suggested the importance of providing comprehensive services, offering aftercare and transitional assistance, and teaching girls skills to overcome their problematic histories. These respondents also indicated the need for specialized training regarding how to treat girls (Belknap, Dunn, & Holsinger, 1997a; Belknap

& Holsinger, 1998; Bloom et al., 2002b). Greene et al. (1998) noted that gender-specific programming should include information about the consequences of life choices, teaching girls skills such as decision making, problem solving, and anger management; providing positive alternatives to previously problematic behavior; and garnering community resources to help girls. Moreover, programs should use girls' input in decisions about service delivery, use female staff, provide mentors for girls, and train staff about gender-specific programming (Bloom et al., 2003; Greene et al., 1998; Juvenile Justice Evaluation Center, 2006).

Although informative, much of this previous work has ignored the girls' opinions. Consequently, the purpose of the current article is to (a) report what at-risk and delinquent girls claim they want and need from the system, (b) determine whether what the girls say they want is similar to what the literature says they need, and (c) provide practical recommendations that practitioners can use to improve the status of girls in their care.

Results

Demographics

The research team spoke with 112 girls: 10 in shelter care, 51 in detention, and 51 in state correctional facilities. As Table 7.6 indicates, participants ranged in age from 11 to 18. Not unexpectedly, girls in state correctional facilities were slightly older. The majority of shelter care girls identified their racial/ethnic group as

Table 7.6 Focus Group Demographics

	Shelter Care ($n = 10$)	Detention Center ($n = 51$)	State Facility ($n = 51$)
Age range, years	12–17	11–18	13–18
Average age, years	15.2	15.6	16.5
Race: White, %	90	53	49
Education level, %			
<8th grade	50	51	30
9th–10th grade	50	39	48
11th–12th grade	0	10	22
Average age at first arrest, years	13.9	13.4	13.7
Parent's marital status, %			
Never married	30	33	28
Married	20	29	37
Divorced	50	38	35
% with at least one child	0	13	13
% having a family member incarcerated	90	96	79

White, whereas only one half of the other girls did. The majority of non-White girls were African American, although 10 self-identified as Hispanic, 2 as biracial, and 2 as Native American.

Average age at first arrest was consistent across the three groups, ranging from 13.4 to 13.9 years. There were some differences among the groups in terms of who came from intact families—20% of shelter care, 29% of detention, and 37% of state facility girls. Overall, 32% of the girls' parents were married. This is slightly lower than the percentage of women prisoners (42%) that the Bureau of Justice Statistics reported as living with intact families as girls (1994). Additionally, 13% of detention and state facility girls claimed that they had at least one child.

A majority of the girls in the study (between 79% and 96%) responded that they had at least one family member who had been incarcerated. Of the 90% of shelter care girls who reported familial incarceration, many reported it was their father (67%), mother (44%), and/or sibling (44%). More detention girls (96%) reported familial incarceration. Fifty percent of their fathers, 23% of their mothers, and 50% of their siblings had been locked up. Although familial incarceration was high among state facility girls (79%), fewer of them had fathers (27%), mothers (9%), and siblings (39%) who had been incarcerated. Overall, approximately 30% had parents who had been incarcerated. This is fewer than the proportion reported by Acoca (1999) (between 46% and 54%) but similar to that reported by Dannerbeck (2005).

Paths to Delinquency

The first substantive question asked was "What kinds of things get girls into trouble?" Many responses offered by the groups were similar (see Table 7.7). Drugs; sex, sexually transmitted diseases (STDs), and pregnancy; boyfriends; relational aggression; running away; emotional, physical, and sexual abuse; fighting with parents; drinking alcohol; friends and peer pressure; and smoking were discussed in all sessions. Shelter care girls also identified familial instigation (e.g., their parents or other family members introduced them to drugs, alcohol, or other crime), although the detention and state girls did not. This is not surprising, because most of these girls explained that they were in shelter care because of problems that their parents created and/or their parents' inability to care for them attributable, in part, to drugs and alcohol problems.

Detention and state facility girls also mentioned dating older men, fighting and battery, and truancy. Only state facility girls noted gang involvement, school problems, and parents' divorce. It is not surprising that only state facility girls mentioned gang involvement, because they had traveled longer paths in the juvenile justice system and were more likely to be introduced to gangs. State girls also said cutting oneself got girls into trouble. Cutting was discussed in most sessions, but the other girls did not suggest that it could get girls into trouble.

Drugs. As seen in Table 7.7, drug use was identified as the number one thing (across all groups) that gets girls into trouble. In all of the sessions, marijuana, alcohol, methamphetamine (e.g., crank and speed), cocaine, and ecstasy were cited as commonly used by friends. In addition, state facility girls said that inhalants, prescription pain killers (e.g., Vicodin, OxyContin, and Hydrocodone), hash, tranquilizers (e.g., Xanax, Valium, and Klonopin), and over-the-counter medications (specifically massive doses of Robitussin, Coricidin, and Benadryl) were regularly used by themselves or friends. The research team noted during debriefings that prescription pain killers and tranquilizers were mentioned most often by White girls, whereas over-the-counter medications were discussed more often by non-White girls. When follow-up questions were asked about why girls start using drugs, many indicated that drug use was learned in the home (see Kakar et al., 2002). One state facility resident offered,

> Some girls do drugs with their parents. Like someone I knew, uh, their father taught them how to roll blunts and how to smoke them and everything.

Table 7.7 Things That Get Girls Into Trouble: Percentage of Discrete Responses Within Groups

Responses	Shelter Care ($n = 10$)	Detention Center ($n = 41$[a])	State Facility ($n = 51$)	Total ($N = 102$)
Friends/peer pressure	30	12	12	14
Boyfriends	50	46	45	46
Dating older men	0	12	33	22
Sex, STDs, and pregnancy	60	49	43	47
Smoking	20	5	24	16
Alcohol	10	22	16	18
Drugs	90	63	67	68
Fighting (not with parents)	0	24	53	36
Relational aggression	20	29	25	26
Familial instigation	40	2	0	5
Fighting with parents, incorrigibility	20	17	18	18
Parents' divorce	0	0	22	11
Truancy	0	27	37	29
Running away	30	29	16	23
Sexual abuse	10	7	14	11
Physical abuse	70	17	16	22
Emotional abuse	30	7	18	15
Gangs	0	0	12	6
School problems	0	0	18	9
Cutting	0	0	8	4

NOTE: STD = sexually transmitted diseases.

a. Fifty-one detention girls participated in the study; however, the information provided in this table includes responses from 41. In one detention center group, the tape recorder was accidentally not turned on for this question.

Others explained that their drug use (and their friends') was a way to cope with the pressures of a difficult home life. One detention center girl remarked,

Some of the effects on the females come from their family background about how they was raised in the homes, what went on, what drugs

> was used or if they were abused. If somethin' went on at home and they didn't know how to handle it they turn to drugs for help.

Motherhood as an escape. Getting pregnant is seen by some girls as a way to gain legal independence and to escape from undesirable family situations. One shelter care girl explained, "Maybe their home life isn't the best and so that's [pregnancy] a way out." Another remarked,

> It is really sad that girls want to get pregnant that are our age just for a way out. But, it's true because there are so many more rights for a pregnant teenager than just a regular teenager.

Boyfriends. The notion that boyfriends play a role in leading girls "astray" was a common theme in all focus groups. One detention center girl explained,

> I think sometimes like the girls want to revolve their life around their boyfriend and they feel since they "been with them" or whatever, they want. Well, they feel that they're always right so like if he says to go do this or go do that, they'll probably jump up and do it.

Older men. Girls in detention and the state facilities mentioned that dating older men could bring trouble. Although these relationships often were problematic, the girls explained that they were sometimes logical from an economic and utilitarian perspective:

> They got jobs, they can buy you things, give you a place to stay, they got cars . . . they got money, you can get your hair done.

> Like 'cause this day and time most people's father wasn't around much in their life and so they never had a male role model in their life, and so they're turning to [older] guys to get comfort and support that their father never gave them.

> Like 'cause this day and time most people's father wasn't around much in their life and so they never had a male role model in their life, and so they're turning to [older] guys to get comfort and support that their father never gave them.

Problems with other girls. Physical fighting and relational aggression also were commonly cited. Although physical fighting was not noted among shelter care girls, it was a major topic of discussion among state facility girls.

Trouble at home. Many girls explained that trouble with their parents (either fighting with them or not following their rules) also caused trouble.

One resident at a state correctional facility explained that her home life contributed to her troubles:

> I think for me, its family problems because it's like when you are home you get hit on a lot or something, it causes you to run away, to skip school, and you get into a lot of trouble with, you know, your friends, and start doing drugs just so you don't feel pain.

Cutting. Another theme mentioned in all but one group (but more common among the state facility girls) was "cutting" (self-mutilation). Many girls intimated that they or other girls they knew got into trouble for cutting their limbs. When asked why girls cut themselves, several explained that it provided them a way to deal with emotional pain and gave them a sense of control over their situations (Belknap, 2001; see Kakar et al., 2002). One girl at a state facility described her experience with cutting:

> It's a relief . . . yeah like when a whole bunch of feelings just bunch up inside and then I don't know, it's weird. When you cut yourself, it's just sort of like all of the emotions are going away.

Getting Arrested

When girls were asked what most often gets girls arrested, the top three reasons given by all groups were drugs/alcohol (use and possession), battery (often on parents), and running away (from both parents and placements). Because the answers to this question were similar to the answers to what things get girls into trouble, only a few issues are highlighted here.

Clearly, poor family relations played a large role in the risky behaviors these girls demonstrated. Several participants offered what they considered to be reasonable explanations for battering parents and running away. One detention center resident remarked, "Some's point of view is when their mom and dad hit on them, so you hit back."

Why they run. Two other detention center girls illustrated how running away from home might appear to be a rational choice for some girls:

> You want to get away, even though you already know the consequences already. But, you don't care because you know that sometimes being locked up will be better than being at home.
>
> There should be further investigation into why we are running away from home . . . why we don't want to be there . . . because obviously something isn't right that we keep runnin' from.

Other commonly cited reasons for arrest were truancy, theft, auto theft, incorrigibility, and driving without a license. The only difference in responses between the groups came from the state facility girls—some claimed that gang activity precipitated arrest among their friends.

Where They've Been

The girls were asked to discuss their experiences with out-of-home placements and to explain whether they found these placements helpful. Many had been in more than one placement. Even shelter care girls had been in placements beyond shelter care. Not many girls found their placements helpful. The overriding theme in all groups was that placement staff do not care what happens to girls. A total of 13 girls (12%) expressed the futility of some programs, and the system in general, in that they rarely address existing problems in girls' environments. For example, one shelter care girl answered,

> To give you the truth, I really didn't think they help. I mean as soon as you get out of the place, you are going to be put right back in the negativity and you are going to be put back in the place you came from and it's just the same.

Staff. Other participants found some placements useful, but argued that to be most effective they needed to have staff who cared. One detention center girl said,

> Some people, some staff have a lot of influence on us, like talkin' to us or sharing their experiences or sharing what might be helpful for us. And then some are just like, "I don't care, I still leave at 2:30 so it don't matter to me."

One detention center girl echoed the sentiment of others when she noted that staff should not live by the "do as I say, not as I do" mantra:

> The places that have helped me the most have staff that listen, but also hold me accountable for everything that I did. Other places the staff was smokin' and tellin' me not to.

However, there were several instances when girls said that there was an adult who they believed helped them or had a positive impact—and these adults were almost always front-line staff. As one detention girl noted:

> I've had some staff that could pour out their heart. They would do everything but commit a serious crime to help you out.

Another detention girl told the group about how supportive her counselor was:

> I can call my counselor any time of the day and she can come talk to me. So, it's really relieving if you can talk to somebody that you trust and just get your anger out right there.

Programming

The types of programming that the girls participated in varied across and within groups. The most common types of programming mentioned by the girls were counseling (63%), anger management (34%), drug treatment (24%), Thinking for a Change (20%), sexual abuse treatment (15%), parenting and life skills classes (10% and 13%, respectively).

Individual and group counseling. Participants were asked about types of counseling they had experienced and whether it was useful. Nearly two thirds of the girls said they had had "traditional" counseling (individual or group), both before and after formally entering the system. Unfortunately, most girls were unable to provide more detail—meaning they could not state whether the counseling was based on a cognitive–behavioral approach, was psychodynamic, and so on. The girls were split about how helpful counseling was. The girls who thought that counseling was really helpful preferred one-on-one counseling. However, others did not care for it, citing trust issues. They did not believe that their conversations with therapists were confidential because their probation officers, parents, and judges always seemed to know about conversation content. Many girls did not appreciate counseling at all if they were forced to participate. When they voluntarily chose to participate, they preferred specific programs that used group counseling such as Thinking for a Change (T4C), "a cognitive–behavioral change program for offenders that includes cognitive restructuring, social skills development, and development of problem solving skills" (*Thinking for a Change,* 2006). State facility girls explained that T4C forces participants to understand their thought processes and to choose how they want to respond in given situations. The girls in one state facility group agreed when a girl explained why she liked T4C:

> It teaches you about giving feedback, understanding your anger, understanding other people's anger. . . . It helps a lot.

State facility girls also identified two other types of counseling that they found to be particularly helpful—grief and loss and sexual abuse counseling. One state facility girl stated,

> It [grief and loss counseling] helped a lot. It got me, uhm, how to deal with my feelings and let go of the stuff that happened in the past.

Girls in all five state facility focus groups believed counseling for sexual abuse was useful. One girl explained how counseling for sexual abuse helped her:

> Cuz, it made me, uh, think about like, it brought up bad memories, but it helped me with my problems so I can get beyond that.

Participants also had experience with other types of programming, including teen court, peer mediation, and classes like anger management.

What they liked. Beyond T4C, grief and loss, and sexual abuse counseling, the girls appreciated prenatal/parenting classes and believed that their own parents would also benefit from participation in these classes. However, participants noted that the information provided should be more specific. They also liked independent living, peer mediation/teen court, and 12-step programs.

Problems with probation. The girls who mentioned that they had been on probation were not impressed with their experiences. The biggest complaint was their probation officers' perceived lack of interest. The girls thought that they should meet with their probation officers far more often. For example, one detention center girl argued, "I know I was on probation, but I ain't never really seen her 'til I come to court." Another claimed,

> We should get to see our POs and social workers when we need to, especially when we are locked up. I've written my PO twice since I been here, staff even let me call and she hasn't come to see me. My dad tried, too.

Fairness and Equity

The girls were also asked whether they were treated fairly by the system. Not unexpectedly, some girls thought that they should never be detained or incarcerated. Others complained that there was no equity—meaning that other girls with the same or similar charges were dealt with more leniently and that punishment depended on what probation officer or judge a girl "got stuck with." As one state facility girl noted:

> I got sent to Girls' School after like three chances and I have never been on house arrest. And, everybody, I've seen people do worse than me and they've been with gun charges and not gone there.

Surprisingly, many girls believed they were treated fairly. These girls claimed that they had been given several chances, but that it was the girls that screwed up.

The only group that was adamant about the "unfairness" of their plight was the shelter care girls, who felt doubly victimized. They argued that first they were victims at the hands of parents and guardians. Then, they felt victimized again when they were removed from their homes and kept in shelter care. One shelter care girl said it was not fair: "I'm only here because of running away from an abusive mom." Another shelter care girl commented on the precariousness of her status as a victim:

> In like the last year my role switched. And I'm actually what they would consider a victim. And, I've noticed that you have more rights when you are the criminal than when you're the victim. I had so many rights, and the court stood by me 100% when I was doing something wrong. But as soon as I needed their help, they went against me. And they haven't helped me at all.

Gender inequities. Another interesting line of discussion dealt with whether girls and boys received differential treatment. Most girls believed that boys and girls received similar punishments for serious offenses, although some believed that boys received harsher sentences for delinquency. However, the girls felt that boys received preferential treatment once locked up. Two detention girls noted:

> The detention officers treat them [the boys] better than us. It's like ok, you know when you gotta ask for a pass to the cafeteria? The boys will get up and get what they want, the girls can't.
>
> I know boys who get worse sentences but they'll get out way quicker than girls get out. Like the programs, like, it's easier for them to get out of the programs than it is for females.

The girls also believed differential treatment existed regarding less serious charges. Several girls argued (rather passionately) that they faced harsher treatment for status offenses, particularly running away. They claimed that most boys were not locked up for running away and that for status offenses the system did not do much to boys. One detention center girl (claiming to be a status offender) explained that judges are tough on girls for minor things:

> Boys are gonna be boys and they are gonna do, get into trouble and be all the bad ones, but they want us girls to be perfect angels. . . . The judge told me himself, that he told me, "I am harder on girls." He told me that himself.

What Girls Want and Need

There was some consensus among the groups regarding what girls claimed they wanted and needed from the system (see Table 7.8). The most common responses are illustrated below.

Table 7.8 What Girls Want and Need From the System: Percentage of Discrete Responses Within Groups

Responses	Shelter Care (*n* = 10)	Detention Center (*n* =51)	State Facility (*n* = 51)	Total (*N* = 112)
Have a voice and to be understood	60	47	13	40
Fair and caring staff	50	31	13	31
Independent living skills	50	27	0	17
To understand the process/system	0	16	16	12
Respect	0	12	12	11
More counseling (particularly for sexual abuse)	20	12	8	11
More than incarceration	0	18	4	10
Earlier consequences	0	14	8	10
Help for their parents	0	6	4	5
Parenting classes	30	6	0	5
Positive female role models/mentors	0	12	0	5
Help with their education	0	6	4	4
More emotional support	0	0	8	4
More involvement by probation officers	0	2	4	3
Help with developing healthy relationships	0	6	0	3
Trust	0	0	2	1
Drug treatment	0	0	2	1
Help with transitioning to the community	0	0	2	1

A voice. One theme that emerged in the detention and state facility focus groups was that the girls wanted to play an active role in their cases. As the next few quotes illustrate, the girls wanted to have a "voice" in their proceedings and to be heard. As one shelter girl said,

> The kid has a story and the parent has a story, and a person or a social worker is looking through 'em. The parent is always right. I mean always!

A detention center girl indicated,

> This should be court, everybody gets a chance to say what they want to say and they get their chance in court. 'Cause they only hearing one side, the bad side. They should hear both sides and come to a conclusion.

The girls also wanted their probation officers, counselors, and case workers to listen to and not judge them. A detention center girl commented that she learned early on not to tell staff when she was having problems because that information would eventually be used against her. She commented, "You learn early, don't talk, don't tell, don't trust."

Caring program staff. The girls also wanted custody staff who cared about them and were not there just for the money. Two shelter care girls echoed the feelings of several girls about seemingly non caring staff:

> There is more people, more staff that care, but there is a great percentage that don't . . . it puts me in a really bad mood when I feel or find out that staff doesn't care.

> The ones that care, like, they can talk to you and you can tell them anything. But, there are some staff that I feel I can't tell stuff to and really don't care because they're here for a paycheck.

Practical life skills. There was some agreement among the groups regarding what they wanted in programs and services. The girls said they wanted programs that teach practical skills and career development—like how to make a living or go to college. The girls acknowledged that they could do more with their lives, but they did not know what to do or how. For example, many girls discussed independent living classes. A shelter care girl explained,

> We need more independent living. Independent living helped me prepare for what is ahead. It helped me find a job and it will pay for tuition.

Girls in another group claimed that they would like to take these classes. One participant exclaimed rather enthusiastically,

> I wanna take these classes! Like, probably taking these classes gives you an understanding of what's really important when you're on your own!

A few girls who had been involved in independent living discussed the possibility of going to college, but this was not a common topic. One detention center girl (who had not had independent living classes) explained that she did not think of college because "some people don't have parents or counselors that tell them about that."

Understanding the process. The last major thing that girls identified as a "want" was for the people who work in these programs to explain the process (i.e., what is happening to them and why it is happening). Even though the girls acknowledged that they understood that their actions had consequences, these girls felt that they never really understood what could happen to them in the system. Two state facility girls noted,

> A lot of kids that go into court aren't, um, aware of all the rights that they have and how to, um, express those rights and you just take advantage of them. And, um, my judge personally used that against me to put me here.

> What my probation officer did, she never saw me, she never was like trying to talk to me. The only time I saw her was either when I was in court or like a probation meeting that was mandatory. That was the only time I saw her, and then she's trying to make decisions for me that affects me for my life. And she doesn't even know, what like I need . . . or doesn't even care to ask.

Respect. Some girls were adamant about needing respect; one state facility girl explained,

> I would change basically disrespectful staff that don't respect us, that cuss at us, but expect us to not cuss back. And, I

> would change the way that people talk to us, talk less, pay more attention to us, understand our problems.

Facing abuse. The girls said that they wanted more programs that deal with physical and sexual abuse and their long-term consequences. Although not directly asked about personal victimization, all of the girls said that they knew victims of both types of abuse, and many volunteered that they personally had been victimized (see Bloom et al., 2002b; Kakar et al., 2002). These girls seemed to understand that such victimization could affect them throughout their lives but could not articulate how or why.

Mentors and role models. Most of the girls were familiar with mentoring programs and liked the concept but had not had a mentor. A girl in detention explained that she would like to participate in a big sister program. "A big sister program would help us, encourage us, and give us insight." However, most of these girls will not likely have a big sister from Big Brothers/Big Sisters of America because delinquent children are not eligible for their general programs.

More than incarceration. Finally, the girls explained that they needed more than just punishment. The participants who discussed this issue did not indicate that they should not be held accountable; rather, they wanted more. They explained that they needed "help" but did not feel help was a priority. One detention center girl noted,

> They say they are there to help you, you know, and get you help, but it seems like all they ever want to do is lock you up for as long as they can.

Surprisingly, there was little denial of personal responsibility. As one detention center girl commented, "I'm not gonna blame it on the state when it's my fault that I'm in here." Blaming others for their situation only occurred systematically in the shelter care groups. This may be understandable, given that some were removed from their homes because of the actions of their parents/guardians. For example, 3 participants in the shelter care groups claimed or alluded to the fact that they were in shelter care because of parental abuse or their parent/guardian's refusal to take custody of them. As one girl explained,

> I'm here because Mom wants nothing to do with me; Dad is in Florida. I am here because there was no place to stick me.

The girls also felt that they were given mixed messages at every step in the process by different people who they believed had differing agendas. The girls just wanted to know what was going to happen to them. Moreover, they wished they had more structure and more consequences along the way, rather than having the court unexpectedly impose stiff penalties. One shelter care girl noted that simply threatening punishment is not effective:

> I got threatened a lot with Girls' School and getting sent to away to a boot camp type thing, all kinds of things. And every time I messed up I never got sent. It was almost like, yeah, ok. Sure you'll send me.

Whereas 2 detention center girls wished they had experienced "real" consequences earlier:

> If I would a gotten disciplined, I probably wouldn't be sittin' at this table.

> Well he [the judge], gave me a bunch of chances, like, fifty chances and I kept on doing the same thing . . . then he sent me to the Girls' School, which ain't gonna get me no help or anything.

Summary, Conclusions, and Implications

Summary—What Gets Girls Into Trouble

Drugs; sex, STDs, and pregnancy; boyfriends; troubled home lives; truancy; running away; relational aggression; delinquent peers; physical, emotional, and sexual abuse; drinking alcohol; and smoking were referred to (in varying magnitudes) in all groups as key issues that lead girls into delinquency. Dating older men was predominant in the discussions among state girls but only in a marginal sense among detention girls; gang involvement, school problems, and issues of self-mutilation were mentioned only by state girls.

Summary—What Girls Say They Need

Girls claimed they wanted to have a voice in their proceedings; participate in programming headed by caring, respectful staff; and be taught practical life skills that translate into their daily lives and help them support themselves. They wanted more programming to help them cope with abuse histories. Finally, participants argued that they needed to understand the justice process and experience incremental consequences rather than face incarceration after receiving many chances.

Conclusion—Consensus

There was consistency regarding what gets girls into trouble and what girls need from the system. Furthermore, the findings of this study are remarkably consistent with those of previous research that highlight (a) girls' opinions of their experiences in the system and what they believe their programming needs are and (b) what the gender-specific literature says they need. Girls want a voice, respect, and understanding of the process (Belknap et al., 1997a). They also want to learn practical skills such as parenting and conflict resolution (Bloom & Covington, 1998) and to receive vocational training that provides realistic employment opportunities (Bloom et al., 2003; Greene et al., 1998). Moreover, they want programming that addresses sexual and physical abuse, poor family relations, and building healthy relationships (Acoca, 1999; Bloom & Covington, 1998; Bloom et al., 2002a; Valentine Foundation, 1990). Some girls also highlighted their need for positive role models, mentors, and female staff who are adequately trained and want to work with them (Belknap et al., 1997b; Bloom et al., 2003; Greene et al., 1998; Juvenile Justice Evaluation Center, 2006). Such strong consensus among the girls in this study, girls in other studies, and the extant literature might lead some to conclude that *what girls want is what they need.*

Implications for Practice

First, the uniformity in responses (and their similarities with previous research) should not be ignored. With so much agreement among the girls (and prior research) about the things that lead girls into delinquency and what the system can do, the issues highlighted by the girls should be considered when revising current practices and developing new programming (Holsinger, 2000; Juvenile Justice Evaluation Center, 2006).

Second, although practitioners conduct risk/needs assessments of clients during incarceration and preparation for reentry, we rarely consider the family and environment that produced these youth. Thirteen girls acknowledged this when discussing why they believe interventions do not work. It may be nearly impossible for justice agencies to improve conditions of poverty, crime, and community disorganization, but they may be able to instigate positive family changes. Until parents and other key family members are required to participate in their daughters' treatment, we cannot expect to have widespread, lasting effects on the girls' behavior.

Third, the system should employ staff who want to work with girls. Several girls remarked that they wanted help from people they could trust, who would listen to and try to understand them, and who cared about more than a paycheck. During this study, we met caring staff who were committed to the girls they served; however, we also encountered many who truly disliked working with girls and did not care who knew it. Previous research (see Belknap et al., 1997a, 1997b) has also noted that girls do not feel heard or respected by the people who have power over them. Some of the dissatisfaction of girls may be chalked up to teenage angst. But, when so many speak so loudly, the system needs to hear them and respond.

What Can Be Done?

Justice resources are overtaxed; however, a few simple things can be done to improve the plight of girls: (a) carefully screen potential employees for gender biases; (b) recruit more women to work in juvenile facilities; (c) implement gender-specific program models suggested by the literature that follow the principles of effective correctional intervention (Andrews et al., 1990); and (d) evaluate programs.

Implications for Future Research

Those interested in seeing whether these findings apply to girls generally could conduct similar studies with a more rigorous methodology (e.g., using more quantitative measures or a random sample). Also, the current study did not ask participants about their individual criminal or abuse histories. To develop a more precise measure of the prevalence of these problems among girls, researchers could conduct life history interviews and match those data with official records and welfare case histories. Researchers should also consider including boys in focus groups to ascertain whether what they want and believe they need from the system is substantively different than what girls report.

References

Acoca, L. (1999). Investing in girls: A 21st century strategy. *Juvenile Justice, 6*, 3–13.

American Bar Association and the National Bar Association. (2001). *Justice by gender: The lack of appropriate prevention, diversion and treatment alternatives for girls in the justice system.* Retrieved September 15, 2004, from http://www.abanet.org/crimjust/juvjus/girls.html

Agnew, R. (1992). Foundation for a general strain theory of crime and delinquency. *Criminology, 30*, 47–87.

Andrews, D. A., Zinger, I., Hoge, R. D., Bonta, J., Gendreau, P., & Cullen, F. T. (1990). Does correctional treatment work? A clinically relevant and psychologically informed metaanalysis. *Criminology, 28*, 369–404.

Beckman, L. (1994). Treatment needs of women with alcohol problems. *Alcohol, Health & Research World, 18*, 206–211.

Belknap, J. (2001). *The invisible woman: Gender, crime, and justice* (2nd ed.). Belmont, CA: Wadsworth.

Belknap, J., Dunn, M., & Holsinger, K. (1997a). *Gender specific services work group: A report to the Governor.* Columbus, OH: Office of Criminal Justice Services.

Belknap, J., Dunn, M., & Holsinger, K. (1997b). *Moving toward juvenile justice and youth serving systems that address the distinct experience of the adolescent female.* Ohio: Gender Specific Services Work Group Report. Columbus.

Belknap, J., & Holsinger, K. (1998). An overview of delinquent girls: How theory and practice have failed and the need for innovative changes. In R. T. Sapling (Ed.), *Female offenders: Critical perspectives and effective interventions* (pp. 31–64). Gaithersburg, MD: Aspen.

Bloom, B., & Covington, S. (1998). *Gender-specific programming for female offenders: What is it and why is it important?* Paper presented at the 50th Meeting of the American Society of Criminology, Washington, DC. November 24.

Bloom, B., Owens, B., & Covington, S. (2003). *Gender-responsive strategies: Research, practice, and guiding principles for women offenders.* Washington, DC: National Institute of Corrections, U.S. Department of Justice. Retrieved August 6, 2006, from http://www.nicic.org/pubs/2003/018017.pdf

Bloom, B., Owens, B., & Covington, S. (2004). Women offenders and the gendered effects of public policy. *Review of Policy Research, 21*, 31–48.

Bloom, B., Owens, B., Deschenes, E., & Rosenbaum, J. (2002a). Improving juvenile justice for females: A statewide assessment in California. *Crime & Delinquency, 48*, 526–552.

Bloom, B., Owens, B., Deschenes, E., & Rosenbaum, J. (2002b). Moving toward justice for female juvenile offenders in the new millennium: Modeling gender-specific policies and programs. *Journal of Contemporary Criminal Justice, 18*, 37–56.

Bureau of Justice Statistics. (1994). *Women in prison.* Washington, DC: U.S. Department of Justice.

Burton, V. S., Cullen, F. T., Evans, T. D., Alarid, L. F., & Dunaway, R. G. (1998). Gender, self-control, and delinquency. *Journal of Research in Crime and Delinquency, 35*, 123–147.

Chesney-Lind, M. (2004, August). Girls and violence: Is the gender gap closing? *National Electronic Network on Violence Against Women.* Retrieved November 1, 2006, from http://www.vawnet.org/DomesticViolence/Research/VAWnetDocs/ARGirlsViolence.php

Chesney-Lind, M. (1997). *The female offender: Girls, women, and crime.* Thousand Oaks, CA: Sage.

Chesney-Lind, M. (1989). Girls' crime and woman's place: Toward a feminist model of female delinquency. *Crime and Delinquency, 35*, 5–29.

Chesney-Lind, M., & Bloom, B. (1997). Feminist criminology: Thinking about women and crime. In B. MacLean & D. Milovanovic (Eds.), *Thinking critically about crime* (pp. 54–65). Vancouver, BC: Collective Press.

Chesney-Lind, M., & Okamoto, S. K. (2001). Gender matters: Patterns in girls' delinquency and gender responsive programming. *Journal of Forensic Psychology Practice, 1*, 1–28.

Chesney-Lind, M., & Pasko, L. (2004). *The female offender: Girls, women and crime* (2nd ed.). Thousand Oaks, CA: Sage.

Chesney-Lind, M., & Shelden, R. G. (2004). *Girls, delinquency, and juvenile justice* (3rd ed.). Belmont, CA: Wadsworth.

Dannerbeck, A. M. (2005). Differences in parenting attributes, experiences, and behaviors of delinquent youth with and without a parental history of incarceration. *Youth Violence and Juvenile Justice, 3*, 199–213.

Federal Bureau of Investigation. (2003). *Crime in the United States, 2002.* Washington, DC: U.S. Department of Justice.

Gottfredson, M. R., & Hirschi, T. (1990). *A general theory of crime.* Stanford, CA: Stanford University Press.

Greene, Peters, & Associates. (1998). *Guiding principles for promising female programming: An inventory of best practices.* Washington, DC: Office of Juvenile Justice and Delinquency Prevention. Retrieved February 1, 2004, from www.ojjdp.ncjrs.org/pubs/principles/contents.html

Heimer, K., & De Coster, S. (1999). The gendering of violent delinquency. *Criminology, 37*, 277–317.

Holsinger, K. (2000). Feminist perspectives on female offending: Examining real girls' lives. *Women & Criminal Justice, 12*, 23–51.

Juvenile female offenders: A status of the states report. (1998). Washington, DC: Office of Juvenile Justice and Delinquency Prevention, U.S. Department of Justice. Retrieved March 15, 2003, from http://ojjdp.ncjrs.org/pubs/gender

Juvenile Justice and Delinquency Prevention Act (2002, Sec. 223(a)(7)(A)-(B)).

Juvenile Justice Evaluation Center. (2006). *Gender-specific programming.* Retrieved March 5, 2006, from www.jrsa.org/jjec/programs/gender/state-of-evaluation.html

Kakar, S., Friedmann, M., & Peck, L. (2002). Girls in detention: The results of focus group discussion interviews and official records review. *Journal of Contemporary Criminal Justice, 18*, 57–73.

Kivel, P. (1992). *Men's work: Stopping the violence that tears our lives apart.* Center City, MN: Hazelden.

Leonard, E. (1982). *Women, crime, and society.* New York: Longman.

Loeber, R., Farrington, D. P., & Petechuk, D. (2003). *Child delinquency: Early intervention and prevention.* Washington, DC: Office of Juvenile Justice and Delinquency Prevention, U.S. Department of Justice. Retrieved November 1, 2004, from http://www.ncjrs.org/pdffiles1/ojjdp/186162.pdf

Mears, D. P., Ploeger, M., & Warr. M. (1998). Explaining the gender gap in delinquency: Peer influence and moral evaluations of behavior. *Journal of Research in Crime and Delinquency, 35*, 251–266.

Morgan, D. L. (1998a). *Planning focus groups.* Thousand Oaks, CA: Sage.

Morgan, D. L. (1998b). *The focus group guidebook.* Thousand Oaks, CA: Sage.

Morris, A. (1987). *Women, crime and criminal justice.* New York: Blackwell.

Piper, M. (1994). *Reviving Ophelia: Saving the selves of adolescent girls.* New York: Ballantine.

Poe-Yamagata, E., & Butts, J. A. (1996). *Female offenders in the juvenile justice system: Statistic summary.* Washington, DC: Office of Juvenile Justice and Delinquency Prevention.

Shearer, R. A. (2003). Identifying the special needs of female offenders. *Federal Probation*, 67, 46–51.

Snyder, H. N., & Sickmund, M. (2006). *Juvenile offenders and victims: 2006 national report.* Washington, DC: Office of Juvenile Justice and Delinquency Prevention.

Thinking for a change: Cognitive behavioral program for offenders. (2006). Washington, DC: National Institute of Corrections. Retrieved August 7, 2006, from http://nicic.org/WebPage_220.htm

Valentine Foundation. (1990). *A conversation about girls.* Bryn Mawr, PA: Author.

Ziemba-Davis, M., Garcia, C. A., Kincaid, N. L., Gullans, K., & Myers, B. L. (2004). *What about girls in Indiana's juvenile justice system.* Indianapolis: Indiana Criminal Justice Institute.

DISCUSSION QUESTIONS

1. What role does gender play in the risk factors for girls and delinquency?
2. What types of programming did the girls find most useful? What aspects were least effective for their lives and why?
3. What do girls indicate that they need from a treatment or program option in order to be successful in their exit from delinquency and criminal behavior?

Female Offenders and Their Crimes

Section Highlights

- The role of female offenders in the "war on drugs"
- Women who kill their children
- Prostitution and sex work
- Female gang offending

Women engage in every type of criminal activity. Much like their male counterparts, females are involved in property offenses, simple assault, robbery, and even murder. While males have always engaged in greater numbers of criminal acts, women's participation in crime is increasing. Research over the past several decades has focused on the narrowing of the gender gap, which refers to the differences in male and female offending for different types of offenses. But what does this really mean? Are women becoming more violent, as media reports have suggested? Is the rise in women's incarceration a result of more women engaging in serious criminal acts? How do we investigate these questions?

The image of the female offender has become a sensationalized topic as a result of the increased media attention on female offending. True crime documentaries and fictionalized television dramas that highlight women's participation in criminal activity give the perception that the rates of female offending, particularly in cases of violence, have increased dramatically. The promotional webpage for the cable television show *Snapped* (Oxygen network) focuses on true crime cases of women who kill and their motivations for crime. Yet rates of crime for women in these types of cases have actually decreased since 2000, findings that are contrary to the increased attention by the media on female offenders.

Table 8.1 2008 UCR Arrest Data: Males Versus Females

	Percentage of Offense Type Within Gender (%)	
	Males	Females
Violent crime	4.39	3.12
Homicide	.10	.04
Forcible rape	.21	.01
Robbery	1.03	.44
Aggravated assault	3.06	2.63
Property crime	10.10	17.26
Burglary	2.41	1.32
Larceny-theft	6.86	15.39
Motor vehicle theft	.71	.47
Arson	.11	.07

NOTE: Total male arrests = 7,872,871; total female arrests = 2,489,838.

Table 8.2 2008 UCR Arrest Data: Males Versus Females

	Percentage of Offense Type by Gender (%)	
	Males	Females
Violent crime	3.34	.75
Homicide	.08	.01
Forcible rape	.16	.00
Robbery	.78	.10
Aggravated assault	2.32	.63
Property crime	7.67	4.15
Burglary	1.83	.32
Larceny-theft	5.21	3.70
Motor vehicle theft	.54	.11
Arson	.08	.02

NOTE: Total number of arrests = 10,362,709.

In Section I, you learned about the changes in male and female crime participation over a 1-year and 10-year period using arrest data from the Uniform Crime Reports. Using these same data, we can investigate the gender gap in offending. Table 8.1 compares the percentage of males and females in different offense types. These data illustrate that the proportion of violent crime cases is greater for males than females. In contrast, the proportion of property crimes is greater for females than males, as property crimes comprise 17.26% of all female arrests, compared to 10.10% of male arrests. While these data illustrate that the gender gap may be narrowing in terms of gender proportions of crime, it is important to note that the number of male arrests is more than three times greater than the number of arrests of women for all crimes. Taking these differences into consideration, Table 8.2 highlights the percentages of crimes in selected offenses for both male and female offenders and compares the percentage of male and female involvement for all arrests (male and female). With more than 10 million arrests in 2008, women represent a smaller percentage in offending for all categories compared to men. For example, women's participation in crimes of homicide is less than 1% of all arrests, while men's participation in this crime equals approximately 3.5%. Despite the increased media attention on the growth of violent female offending, women's participation in crimes of aggravated assault represent less than 1% of all arrests (.63%) while male arrests for aggravated assault are more than three times greater (2.32%). The narrowest gap between male and female offending can be found in property crimes, particularly for cases of larceny-theft. While the participation of women in offending behaviors has increased throughout the 20th and 21st century, they remain a small proportion of offending.

While Tables 8.1 and 8.2 illustrate the small proportion of offending for women, these data are based on a single snapshot in time. How might these findings change if they were compared across different time periods? Research by Steffensmeier and Allan (1996) compares the proportion of male and female arrests for three separate years: 1960, 1975, and 1990. Their findings indicate that females make up 15% (or less) of arrestees for most types of major crimes (such as crimes against persons and major property crimes) across all time periods. For minor property offenses, the greatest increases are noted between 1960 and 1975 arrest data. Here, the female percentage of arrests increased from 17% in 1960, to 30% in both 1975 and 1990. The only exception where women make up the majority of arrests is for the crime of prostitution (where women make up between two thirds and three fourths of all arrests across all three time periods).

While Steffensmeier and Allan (1996) compared data across three time periods in the late 20th century, what has happened since then? Research by Rennison (2009) compared offending data from the National Crime Victimization Survey for the nine years between 1992 and 2001 and indicated that there had been negligible differences in the gender gap between male and female offending behaviors. Indeed, any differences in the gender gap result not from the increases of female offending, but, rather, the decreases in male offending rates for particular offenses, which fell at a greater rate than the decrease in female offending rates.

While women participate in many different types of crimes, the remaining focus of this section highlights four different categories of crime. The first category focuses on crimes committed by both men and women. Here, the section focuses on a topic that is at the heart of the dramatic rise of female participation in the criminal justice system: the war on drugs. The second category deals with crimes that are unique to women due to the gendered expectations of women. Here, the section turns to a crime that garners significant media attention: mothers who kill their children. The third category deals with crimes committed by women that are typically female dominated. Here, the section highlights the role of women in prostitution and the debate about the decriminalization of sex work. The fourth category deals with crimes that are committed by women that are typically dominated by men. Here, the section investigates the role of gender within the gang culture.

Women and Drugs

Throughout the majority of the 20th century, women were not identified as the typical addict or drug abuser. In many cases, the use of prescription and illegal substances was normalized by society, often in response to the pressures of gender-role expectations. For example, cocaine and opiates were legally sold in pharmacies and were frequently prescribed by doctors for a variety of ailments. Historically speaking, "women's addiction (was) constructed as the product of individual women's inability to cope with changing versions of normative femininity" (Campbell, 2000, p. 30). Examples of this can be found in advertisements depicting women and anti-anxiety medications in an effort to calm the frenzied housewife who is overwhelmed with her duties as a wife and mother. Drug use has even been promoted as desirable with the image of the heroin chic fashionista of the 1990s, exemplified in images of the supermodel Kate Moss.

Beginning in the late 20th century, the literature illustrates a similar pathway for drug use for women, regardless of race, ethnicity, or drug of choice. Whether the discussion focuses on women addicted to crack cocaine in lower-income communities, or middle-class women who abuse alcohol or prescription drugs, women have similar pathways of depression, abuse, and social and economic pressures that lead them toward substance use and abuse as a method of coping with their lives (Inciardi, Lockwood, & Pottiger, 1993). Many women were introduced to illegal substances at an early age, and in many cases this introduction came courtesy of family members or friends. Others began their drug use in an effort to cope with the traumas of early childhood abuse and violence. For some, using drugs was a way to bring excitement to a life that many described as depressed or boring. Yet the excitement seemed to fade as the women expressed a desire to leave the drug lifestyle but felt they had few options available to improve their life. Despite the struggles within the lifestyle, some women believed that the realities of sobriety would be too painful and challenging to deal with. Here, addiction remained a way to escape life, if only on a temporary basis (Roberts, 1999).

As the behaviors of addicted women began to shift toward criminal activity in an effort to support their drug habit, the perception that drug addiction is "dangerous" spread. For many substance-addicted women, their need for drugs led them to sex work and property crimes in order to support their habit. Drug use became something to fear by members of society. Indeed, the images of the pregnant addicted mother and crack babies of the '80s and '90s represent the greatest form of evil found in the drug-abusing women.

The War on Drugs

The heightened frenzy about the "dangerousness" of drugs has fueled the war on drugs into an epidemic. The war on drugs first appeared as an issue of public policy in 1971, when President Richard Nixon called for a national drug policy in response to the rise of drug-related juvenile violence. Over the next decade, controlled substances such as cocaine were illegally smuggled into the United States by drug kingpins and cartels throughout Mexico and South America (National Public Radio, n.d.).

Since the 1980s and the passage of the anti–drug abuse act, the incarceration rates for both men and women have skyrocketed. Yet the majority of persons imprisoned on these charges are not the dangerous traffickers who bring drugs into neighborhoods and place families and children at risk. Rather, it is the drug user who is at the greatest risk for arrest and imprisonment. In response to the social fears about crack cocaine in the inner city, lawmakers developed tough-on-crime sentencing structures designed to increase the punishments for crack cocaine. Sentencing disparities between powder and crack cocaine created a system whereby drug users were treated the same as mid-level dealers. In 1995, the U.S. Sentencing Commission released a report highlighting the racial effects of the crack and powder cocaine sentencing practices and advised Congress to make changes to the mandatory sentencing practices to reduce the discrepancies. Their suggestions fell on deaf ears among congressional members who did nothing to change these laws. For the next 15 years, cases of crack and powder cocaine perpetuated a 100 to 1 sentencing ratio, whereby offenders in possession of 5 grams of crack were treated the same as dealers in possession of 500 grams of powder cocaine. In 2010, President Obama signed the Fair Sentencing Act, which reduced the disparity between crack and powder cocaine sentences to a ratio of 18 to 1. Under the new law, offenders will receive a 5-year mandatory minimum sentence for possessing 28 grams of crack (compared to 5 grams under the old law) and a 10-year sentence for possessing more than 280 grams of crack cocaine.

Prior to the war on drugs and mandatory sentencing structures, most nonviolent drug conviction sentences were handled within community correction divisions. Offenders typically received community service, drug treatment, and probation supervision. The introduction of mandatory minimum sentencing represented a major change in the processing of drug offenders. While these sentencing structures are applied equally to male and female defendants, the role of women's participation often differs substantially from male involvement in drug-related crimes. With the elimination of judicial discretion, judges were unable to assess the role that women played in these offenses. The result was a shift from community supervision to sentences of incarceration, regardless of the extent of women's participation in criminal drug-related activities (Merolla, 2008).

The shift to incarceration from community supervision had a detrimental effect on women. Between 1986 and 1991, the incarceration rates of women for drug-related offenses increased 433%, compared to a 283% increase for males (Bush-Baskette, 2000). Drug-convicted women make up 72% of the incarcerated population at the federal level (Greenfeld & Snell, 2000). Most of these cases involve women as users of illegal substances. Even in the small proportion of cases where women are involved in the sale of drugs, they rarely participate in mid- or high-level management in the illegal drug market, often due to sexism within the drug economy (Maher, 2004). In addition, the presence of crack shifted the culture of the street economy, particularly for women involved in acts of prostitution. The highly addictive nature of crack led more women to the streets in an effort to find a way to get their next high. At the same time, the flood of women in search of sex work created an economy whereby the value of sexual services significantly decreased.

While recent changes in federal drug sentencing laws have reduced the disparities in sentencing, the damage has already been done. The effects of these laws created a new system of criminal justice where the courts are overloaded with drug possession and distribution cases, and the growth of the prison economy has reached epic proportions. Yet these efforts appear to have done little to stem the use and sale of such controlled substances. Indeed, the overall rates of crimes other than drug-related cases have changed little during the last 40 years. The effects of these policies have produced significant consequences for families and communities, particularly given the increase in the incarceration rates of women. Section IX explores in depth the consequences in the incarceration of women, both for herself and her family, as well as her community. It is these consequences that have led some scholars to suggest that the war on drugs has in effect become a war on women (Chesney-Lind, 1997).

Mothers Who Kill their Children

While the crime of filicide is a rare occurrence, it raises significant attention in the media. The case of Andrea Yates is one of the most identifiable cases of filicide in the 21st century. After her husband left for work on June 20, 2001, Yates proceeded to drown each of her five children one at a time in the bathtub of the family home. Her case illustrates several factors that are common to incidents of maternal filicide. Yates had a history of mental health issues, including bipolar disorder, and she had been hospitalized in the past for major depression. She was the primary caretaker for her children and was responsible for homeschooling the older children. She and her husband were devout evangelical Methodists. Yates indicated that she felt inadequate as a mother and wife, believed that her children were spiritually damaged, and stated that she was directed by the voice of Satan to kill her children (Spinelli, 2001).

▲ **Photo 8.1** Andrea Yates appears before a Texas courtroom in the case of the drowning of her five children, Yates was ultimately found not guilty by reason of insanity.

The case involving the children of Andrea Yates is just one tragic example of a mother engaging in filicide, or the killing of her children. There are several different categories of filicide. Neonaticide refers to an act of homicide during the first 24 hours after birth, compared to cases of infanticide, which includes acts whereby a parent kills his or her child within the first year of life. Here, the age of the child distinguishes these cases from general acts of filicide, which include the homicide of children older than 1 year of age by their parent. While the practice of filicide does not exclude the murder of a child by its father, mothers make up the majority of offenders in cases of infanticide and neonaticide.

What leads a woman to kill her child? There are several different explanations for this behavior. Research by Resnick (1970) distinguishes five different categories of infanticide. The first category represents cases where the infant was killed for altruistic reasons. In these incidents, the mother believes that it is in the best interests of the child to be dead and that the mother is doing a good thing by killing the child. Here, the mother believes (whether real or imagined) that the child is suffering in some way and that the child's pain should end. Based on Resnick's (1970) typology, Yates would be identified as a mother who kills her children out of altruistic reasons. A review of Yates' case indicates two themes common to altruistic filicide. The first theme reflects the pressure that exists in society for women to be good mothers. For Yates, this pressure was influenced by her religious fundamentalism, which placed the importance of the spiritual life of her children under her responsibility. The pressure to be a perfect mother was exacerbated by her history of mental illness. The second theme reflected the pressure of bearing the sole responsibility to care for the children. Here, Yates expressed feeling overwhelmed by the demands of their children's personal, academic, and spiritual needs, in addition to the responsibilities of caring for the family home. She also lacked any support from outside of the family, which further contributed to her feelings of being overburdened (West & Lichtenstein, 2006).

The second category in Resnick's typology refers to the killing of a child by an acutely psychotic woman. These cases are closely linked with explanations of postpartum psychosis where the mother suffers from a severe case of mental illness and may be unaware of her action or be unable to appreciate the wrongfulness of her behaviors. Examples of this type of filicide may involve a woman who hears voices that tell her that she needs to harm her child. The third category represents the killing of an unwanted infant. In many cases, these are cases of neonaticide. Research indicates that there are similar characteristics within the cases of mothers who kill their children within their first day of life. These women tend to be unmarried, under the age of 25, and generally concealed their pregnancy from friends and family. Some women may acknowledge that they are pregnant, but their lack of actions toward preparing for the birth of the child indicate that they may be in denial that they may soon give birth. Others fail to acknowledge that they are pregnant and explain away the symptoms of pregnancy (Miller, 2003). They typically give birth without medical intervention and generally do not receive any form of prenatal care. The majority of these women do not suffer from any form of mental illness, which would help to explain the death of their children. Instead, most of the cases of homicide of an infant are simply a result of an unwanted pregnancy. In these instances, the children are typically killed by strangulation, drowning, or suffocation (Meyer & Oberman, 2001). The fourth category involves the "accidental" death of a child following incidents of significant child abuse and maltreatment. Often, the death of a child occurs after a long period of abuse. The fifth category represents cases where the death of a child is used as an act of ultimate revenge against another. In many cases, these vengeful acts are against the mother's spouse, significant other, or ex-intimate: the father of the child (Resnick, 1970).

Postpartum Syndrome and Filicide

Mothers who kill their children present a significant challenge to the cultural ideals of femininity and motherhood. Society dictates that mothers should love and care for their children, behave in a loving and nurturing manner, and not cause them harm or place their lives in danger. In many cases, the presence of a psychological disorder makes it easier for society to understand that a mother could hurt her child.

Information on postpartum syndromes is used at a variety of different stages of the criminal justice process. Evidence of psychosis may be used to determine whether a defendant is legally competent to participate in the criminal proceedings against her. However, this stage is temporary, as the woman would be placed in a treatment facility until such a time that she is competent to stand trial. Given that postpartum syndromes are generally limited to a short period of time (compared to other forms of psychiatric diagnoses), these court proceedings would only be delayed temporarily.

More often, information about postpartum syndromes is used as evidence to exclude the culpability of the woman during a trial proceeding. In some states, this evidence forms the basis of a verdict of "not guilty by reason of insanity." Here, the courts assess whether the defendant knew that what she was doing at the time of the crime was wrong. "The insanity defense enables female violence to coexist comfortably with traditional notions of femininity. It also promotes empathy toward violent women, whose aberrance becomes a result of external factors rather than conscious choice" (Stangle, 2008, p. 709). In cases where an insanity defense is either not available or is unsuccessful, evidence of postpartum syndromes can be used to argue for the diminished capacity of the offender.

A third option allows for courts to find someone guilty but mentally ill (GBMI). Here, the defendant is found guilty of the crime, but the court may mitigate the criminal sentence to acknowledge the woman's mental health status. For many offenders, this distinction can allow them to serve a portion of their sentence in a treatment hospital or related facility (Proano-Raps & Meyer, 2003). While Andrea Yates was convicted of murder and sentenced to 40 years to life by the State of Texas in 2002, her conviction was later overturned. In her second trial, she was found not guilty by reason of insanity and was committed to a state mental health facility for treatment.

Prostitution

Hollywood images of prostitution depict a lifestyle that is rarely found in the real world. Movies such as *Pretty Woman*, *Leaving Las Vegas,* and *Taxi Driver* paint a picture of the young, beautiful prostitute who is saved from her life on the streets. In reality, there are few Prince Charmings available to rescue these women. The reality that awaits most of these women is one filled with violence, abuse, and addiction—deep scars that are challenging to overcome.

Prostitution involves the act of selling or trading sex for money. Prostitution can take a variety of forms, including escort services or work in brothels, bars, and truck stops. Approximately 15% of women in prostitution find themselves working in street-level sex work. For these women, money may not be the only commodity available in exchange for their bodies, as they also trade sex for drugs or other tangibles such as food, clothing, and shelter. While these women do not make up the majority of women in prostitution, they do experience the greatest levels of risk for violence and victimization.

The journey into prostitution is not a solitary road. Rather, it involves a variety of individual, contextual, and environmental factors. Trends in the literature have acknowledged a variety of risk factors for women in prostitution, including abandonment, abuse, addiction, and poverty. A history of abuse is one of the most commonly referenced risk factors for prostitution, and research by Dalla (2000) indicates that drug addiction almost always paves the way for work in prostitution. Another common pathway for women in prostitution is the experience of early childhood sexual victimization. While there is no direct link that indicates that the experience of incest is predictive of selling one's body, research indicates that there is a strong correlation between the two (Nokomis Foundation,

2002), and one prostitution recovery program indicates that 87% of their participants experienced abuse throughout their early childhood, often at the hands of a family member. For these women, incest became the way in which they learned about their sexuality as a commodity that could be sold and traded. This process of bargaining became a way in which these victims could once again feel powerful about their lives (Mallicoat, 2006). A history of child abuse is also highly correlated with running away from home during the teen years. Once on the streets, girls turn to prostitution and pimps in an effort to survive on the streets (Chesney-Lind & Shelden, 1998).

Women in prostitution experience high levels of violence during their careers. On the streets, they witness and experience violence on a daily basis. More than 90% of women on the streets are brutally victimized (Romero-Daza, Weeks, & Singer, 2003). They are robbed, raped, and assaulted by their customers and pimps (Raphael & Shapiro, 2004). Many do not report these incidents out of fear that they will be arrested for engaging in prostitution, coupled with a belief that the police will do little to respond to these crimes. Indeed, women often return to the streets immediately following their victimization. This temporary intervention is viewed as a delay in work, rather than an opportunity to search for an exit strategy. One woman characterized her experience as normal—"society and law enforcement consider a prostitute getting raped or beat as something she deserves. It goes along with your lifestyle. There's nothing that you can do" (Dalla, 2000, p. 381). They also witness significant acts of violence perpetrated against their peers, an experience than leads to significant mental health issues. Drug use becomes a way to cope with the violence in their daily lives. As the pressure to make money increases in order to sustain their substance abuse addiction or to provide a roof over their head at night, women may place themselves in risky situations with their customers (Norton-Hawk, 2004). In an effort to protect against potential harms, women rely on their intuition to avoid potentially violent situations. Many girls indicate that they won't leave a designated area with a client and generally refuse to get into a car with a client. Others carry a weapon such as a knife. Despite the risks, some women reference the thrill and power they experience when they are able to survive a violent incident (Dalla, Xia, & Kennedy, 2003). Many women are surprised when they reflect back on the levels of violence that they experienced on the streets. Some may dissociate themselves from the realities of this journey and believe that the experience was not as traumatic as they originally believed. However, the battle scars from their time on the streets provide the evidence for the trauma they endured, both physically and mentally.

The role of substance abuse is central to the discussion of risk for prostituting women. About 70% of women in prostitution have issues with drug addiction. Some women begin their substance use prior to their entry in prostitution to cope with the pain associated with past or current sexual violence in their lives. They then resort to prostitution to fund their drug habits (Raphael, 2004). For others, the entry into substance abuse comes later in an effort to self-medicate against the fear, stress, and low self-esteem resulting from the selling of sex (Nixon, Tutty, Downe, Gorkoff, & Ursel, 2002). As their time on the streets increases, so does their substance addiction. Indeed, the relationship between drug use and prostitution may be a self-perpetuating circle in which they feed off one another. A sample of women in jail for prostitution had significantly higher rates of drug use compared to women arrested for non-prostitution-related offenses (Yacoubian, Urbach, Larsen, Johnson, & Peters, 2000).

In recent years, media accounts have focused significant attention on the use of crack cocaine by street prostitutes. Research has linked the presence of crack to an increase of individuals working on the street, which in turn decreases the price that women receive for their services. Addiction to drugs like crack has created an economy where money is no longer traded for sex. Rather, sexual acts become a commodity to be exchanged for the drug. The levels of violence associated with the practice of selling sex increases in their drug-fueled economy (Maher, 1996).

While drug addiction presents a significant health concern for women in prostitution, additional issues exist for women in terms of long-term physical health effects. Women are at risk for issues related to HIV, hepatitis, and other chronic health concerns, including dental, vision, neurological, respiratory, and gynecological problems (Farley & Barkin, 1998). Finally, the death rate of women in prostitution is 40 times higher than the death rate of the overall population (Nokomis Foundation, 2002).

Mental health concerns are also a significant issue for women in the sex trade. Cases of posttraumatic stress disorder (PTSD) are directly related to the levels of violence that women experienced on the streets. Two thirds of prostituted women experience symptoms of PTSD (Schoot & Goswami, 2001). Prostitutes suffering from PTSD may be unable to accurately assess the levels of threat and violence that surround their lives, which in turn places them in increased risks of danger (Valera, Sawyer, & Schiraldi, 2000).

The Legalization Debate

The question of whether prostitution should be considered a criminal activity is one of considerable debate. In Nevada, legal prostitution is limited to counties with a population under 400,000, excluding high-traffic areas such as Reno and Las Vegas from offering legalized brothels.[1] The laws within Nevada focus almost exclusively on the minimization of risk and reduction of violence for women in prostitution. Since 1986, Nevada has required that prostitutes who work in a brothel must submit to weekly exams to assess for any sexually transmitted infections or the presence of HIV. Brothels also implement a variety of regulations such as audio monitoring and call buttons in the rooms, a limit on services outside of the brothel environment, and control on any potentially negative behaviors of the clients in an effort to maintain the safety and security of the brothel environment and the women who work there. Research indicates that women who work within the brothel setting express that they feel safe and rarely experienced acts of violence while working as a prostitute. Indeed, it is these safety mechanisms that led women to believe that brothel sex work was by far the safest environment in which to engage in prostitution, compared to the violence and danger that street prostitutes experience (Brents & Hausbeck, 2005).

In the Netherlands, the legalization of brothels in 2000 created a new way to govern the sex trade. While the act of prostitution has been legalized since the early 20th century, it was the brothel environment (popularized by the red light district and "window" shopping in the city of Amsterdam and other cities) that was illegal. At the time of brothel legalization, the practice of prostitution in the Netherlands was not an uncommon phenomenon, and estimates indicate that over 6,000 women were working per day in prostitution-related activities (Wagenaar, 2006). The effects of the legislation lifted the formal prohibition of the brothel, even though many municipalities tolerated their presence and agents of social control such as law enforcement and the courts largely refrained from prosecuting cases. By creating a system whereby brothels had to be licensed, authorities were able to gain control over the industry by mandating public health and safety screenings for sex workers. As part of the process of decriminalization of prostitution, the state created the opportunity for brothel owners to have a legal site of business. Labor laws regarding the working conditions for prostitutes were put into effect. In addition, it created a tax base in which revenue could be generated (Pakes, 2005). The goals of decriminalization allowed for the Dutch government to improve the lives of women in prostitution by creating safe working conditions; to create a system of monitoring of the sex trade; and to regulate illegal activities that might be

[1]However, evidence exists that street prostitution and escort services are still prevalent within these regions.

associated with the selling of sexuality, such as streets crimes associated with prostitution, the exploitation of juveniles, or the trafficking of women into the sex industry (Wagenaar, 2006).

By creating a sustainable economy of prostitution, some critics suggest that not only are the needs of the customer met, but these regions create an economic strategy for women, particularly women within challenged economic situations. However, creating a system of legislation is no guarantee that laws will be followed; even with the legalization of prostitution in New South Wales, Australia, the majority of brothels fail to register their businesses and pay little attention to the regulatory rules for operation. In addition, illegal sexual practices have continued to flourish—the Netherlands is identified as a leading destination for pedophiles and child pornographers under the belief that the promotion of legalized prostitution has created opportunities for illegal prostitution in these regions, as well (Raymond, 2004).

Other legislation focuses on the criminalization of the demand for sexual services. In addressing the issue of prostitution in Sweden, legislatures have focused on making the purchasing of sex from women a criminal act. The belief here is that by criminalizing the male demand for sex, it may significantly decrease the supply of women who engage in these acts. By criminalizing the "johns," Sweden has taken a stand against a practice that they feel constitutes an act of violence against women (Raymond, 2004). In the passing of these laws, the parliament indicated, "it is not reasonable to punish the person who sells a sexual service. In the majority of cases . . . this person is a weaker partner who is exploited" (Ministry of Labour, Sweden, 1998, p. 4). In the United States, even in an environment where both the purchaser and seller of sex can be subjected to criminal prosecution, the data indicate that women are significantly more likely to face sanctions for selling sex, compared to men who seek to purchase sex (Farley & Kelly, 2000). While the focus on demand is an important characteristic in the selling of sex, it is not the only variable. Indeed, larger issues such as economics, globalization, poverty, and inequality all contribute to a system where women fall victim to the practices of sexual exploitation.

Farley and Kelly (2000) suggest that even with the legalization of the brothel environment, prostitution remains a significant way in which women are brutalized and harmed. The social stigma of women who engage in the selling of sex does not decrease simply because the act of prostitution becomes legal. Indeed, the restriction of brothels to specific regions only further isolates women from mainstream society and magnifies the stigma that they experience (Farley, 2004). In cases of victimization, women continue to experience significant levels of victim blaming when they are victimized, even if prostitution is decriminalized. The system of public health, which is promoted as a way to keep both the prostitute and her client safe, fails to meet some of the most critical needs of women in this arena, as these efforts toward promoting safety are limited exclusively to physical health, and little to no attention is paid to the mental health needs of women in prostitution (Farley, 2004).

Research indicates that women involved in street prostitution want to leave the lifestyle, but they express concern over how their multiple needs (including housing, employment, and drug treatment) may limit their abilities to do so. There are few programs that possess adequate levels of services to address the multiple needs of women during this transition. A review of one prostitution recovery program found that affordable safe housing is the greatest immediate need for women in their transition from the streets (Mallicoat, 2011). Homelessness puts women at risk for relapse; "without reliable housing, it is challenging to escape the cycle of prostituting" (Yahne, Miller, Irvin-Vitela, & Tonigan, 2002, p. 52). In addition, women must possess the skills and have access to support in order to facilitate this process. Women exiting the streets indicate a variety of therapeutic needs, including life skills, addiction recovery programming, and mental health services designed to address the traumas they experienced. An exit strategy needs to acknowledge the barriers to success and continuing struggles that women will experience as a result of these traumas.

Girls and Gangs

While girls make up a small proportion of gang members, there has been significant media attention on the rise of violent girls and gang girls throughout the late 20th century.

It is difficult to determine the extent of female gang membership. Surveys conducted by law enforcement agencies in the 1990s estimated that between 8 and 11% of gang members were female (Moore & Terrett, 1998). These rates have remained consistent throughout 2007 and reflect that there has been no significant increase in female gang participation, contrary to the image that is perpetuated by the media (National Youth Gang Center, 2009). However, not all law enforcement jurisdictions include girls in their counts of gang members, a practice that can skew data on the number of girls involved in gangs (Curry, Ball, & Fox, 1994). Self-report studies during this same time frame reflect a higher percentage of female gang participation compared to law enforcement data and indicated that 38% of the self-identified gang members between the ages of 13 and 15 were female (Esbensen, Deschenes, & Winfree, 1999).

Who are female gang members? Much of the early literature on girls and gangs looked at female gang members as secondary to male gangs. Classic studies by Campbell (1984) and Moore (1991) illustrated that girls entered the gang lifestyle as a result of a brother or boyfriend's affiliation. Here, the men in their lives paved the way for girls to engage in gang-affiliated crimes and activities. Girls in the gang were often distinguished from their male counterparts by their sexuality. This sexualization manifested in several ways: (1) as a girlfriend to the male gang member, (2) as one who engages in sex with the male gang members, and (3) as one who uses her sexuality in order to avoid detection by rival gang members and law enforcement (Campbell, 1995). Modern research indicates that female gangs are not only increasing their membership ranks, but also expanding their function and role as an independent entity separate from the male gang. Girls in the gang are no longer the sexual toy of the male gang, but have become active participants in crimes of drugs and violence.

Recent literature draws upon two theoretical perspectives to understand the nature and roles of girls within the gang. The liberation hypothesis is influenced by the emancipation/liberation theories of the 1970s (Adler, 1975; Simon, 1975) and suggests that the increase of female participation in gangs is related to an increase in opportunities to participate in other traditionally male domains of crime. Another interpretation suggests that girls may experience autonomy and feelings of empowerment as a result of their gang activities (Nurge, 2003).

Research by Joe and Chesney-Lind (1995) disagrees with the position that gang life is liberating for girls. Their work suggests that gang participation is not about adolescent rebellion, but rather provides a peer-support network to help both girls and boys cope with the violence and disorder that runs throughout their families and community. The second theme is known as the social injury hypothesis, which posits that girls in gangs experience higher levels of risk, danger, and injury compared to their male counterparts (Moore, 1991). Here, girls in the gang experience greater levels of risk compared to any benefits that gang membership may provide. While these two perspectives offer value on their own terms, they can also work together. While girls involved in the gang lifestyle may experience a new sense of freedom and thrill from their lives, their membership also places them in danger for violence and victimization (Nurge, 2003).

The lives of girls in gangs tell a story filled with violence, poverty, racism, and limited resources. They come from families who struggle to make ends meet in economically depressed areas. In these communities, opportunities for positive prosocial activities are significantly limited, and the pressure to join a gang runs rampant. Many of the girls have limited achievements in the classroom, and their educational experience had little to do with books or teachers. Instead, they shared stories of disorder, threats, and crime (Molidor, 1996). The majority of their parents had never married, and the presence of domestic abuse within the home was not uncommon. Many of the girls had a parent or other family members who were involved in the criminal justice system and were either currently incarcerated, or had been during some part of their lives.

For some girls, membership in a gang is a family affair, with parents, siblings, and extended family members involved in the gang lifestyle. Research by Miller (2000) indicated that 79% of the girls who were gang involved had a family member who was a gang member, and 60% of the girls had multiple family members in gangs. For these girls, gang affiliation comes at an early age. During the childhood and preteen years, their gang activities may consist of limited acts of delinquency and drug experimentation. During junior high, girls exhibit several risk factors for delinquency, including risky sexual behavior, school failures, and truancy. By the time these girls become teenagers, they are committed to the gang and criminal activity and participate in a range of delinquent acts, including property crimes, weapons offenses, and violent crimes against persons. The later adolescent years (ages 15–18) represent the most intense years of gang activity (Eghigian & Kirby, 2006).

While the gang is a way of life for some girls, many others find their way to the gang in search of a new family. Many girls involved in gangs have histories of extensive physical and sexual abuse by family members during early childhood. Many of the girls run away from the family residence in an attempt to escape the violence and abuse in their lives. In an attempt to survive on the streets, the gang becomes an attractive option for meeting one's immediate and long-term needs such as shelter, food, and protection. Not only does the gang provide refuge from these abusive home environments, but it provides a sense of family that was lacking in their families of origin (Joe & Chesney-Lind, 1995). Research by Miller (2000) indicates that it is not so much a specific risk factor that propels girls into the gang, but rather the relationship among several factors, such as a neighborhood exposure to gangs, a family involvement in the lifestyle, and the presence of problems in the family, that illustrates the trajectory of girls into the gang lifestyle.

The literature on female gangs indicates that the lifestyle, structure, and characteristics of female gangs and their members are as diverse as male gangs. Some girls hang out with gangs in search of a social life and peer relationships, but typically do not consider themselves as members of the gang. The structure of the girl gang ranges from being a mixed-gender gang to functioning as an independent unit. For girls involved in mixed-gender gangs, their role ranged from being an affiliate of the male gang unit, to even, in some cases, having a "separate but equal" relationship to their male counterparts (Schalet, Hunt, & Joe-Laidler, 2003).

One historical example of a female auxiliary gang is the Vice Queens, which was an affiliate of the Vice Lords, a Black conflict gang in Chicago during the 1960s. Unlike the Lords, the Queens lacked cohesion within the group and had no formalized leadership structure. Much of their connection to the Kings was through their relationships, and significant portions of their time were spent "hanging out" with the boys and encouraging the boys to engage in sexual acts with them. While there were few examples of formal "dating" between the members, there was a sense of temporary monogamy. In most of the "relationships," sex was a common tool used by the women. Rarely did the girls view these relationships as something that had promise for long-term romance and stability. However, the Queens also participated in activities independent of their male counterparts. Their involvement in delinquency ranged from running away, truancy, and shoplifting, to the occasional foray into more masculine crimes such as auto theft and grand larceny. Most of their delinquent activity with the Lords involved instigating fights with rival gang members. Their experience with fighting and hustling provided the girls with a unique skill set, which allowed them to survive the violence and trauma that existed in their neighborhood (Fishman, 1995).

The initiation process for girls ranged from being "jumped" in or "walking the line," whereby the girls were subjected to assault by their fellow gang members, to being "sexed" in or "pulling a train," an experience that involved having sex with multiple individuals, often the male gang members. However, not all of these initiation rites came with a high degree of status within the gang, as those girls who were sexed in generally experienced lower levels of respect by fellow gang members (Miller, 2000). Girls who had been "sexed into the gang" were subjected to continued victimization from within the gang. While not all girls were admitted to the gang in this manner, this image negatively affected all of the girls. "The fact that there was such an option as 'sexing in' served to keep girls disempowered, because they always faced the question of how they got in and of whether they were 'true' members. In addition, it contributed to a milieu in which young women's sexuality was seen as exploitable" (Miller, 1998, p. 444).

Recent media attention has targeted the gang girl and the perception that violence by these girls is increasing. Research by Fleisher and Krienert (2004) indicated that among girls who described themselves as active members of a gang, almost all (94%) had engaged in a violent crime during the previous 6 months, and two thirds (67%) had sold drugs during the past 2 months. More than half (55%) had participated in property crimes such as graffiti or destruction to property, while two thirds (67%) engaged in economic crimes such as prostitution, burglary, robbery, or theft in the previous 6 months. However, research indicates that this increased attention is less about the crimes that girls commit and more about their violations of gender-role expectations. The negotiation of the "bad girl" identity allows for girls to appear tough, command respect, and legitimize their independence and power within the gang. However, the terms "tough," "respect," and "power" take on a very different meaning for girls compared to boys. Traditional notions of femininity such as appearance and restrained sexuality are inherent in commanding respect within the gang (Joe-Laidler & Hunt, 2001). At the same time, respect is about making sure that no one takes advantage of you and that you stand up for yourself. For example, much of the assaultive behavior in which gang girls participate involves fights with members of rival female gangs.

For girls in the gang lifestyle, violence is more than just engaging in criminal offenses. Indeed, the participation in a delinquent lifestyle that is associated with gang membership places girls at risk for significant victimization. Girls who are "independent" of a male gang hierarchy tend to experience high levels of violence as a result of selling drugs and their interactions on the streets with other girls. These "independent" girls are aware of the potential risk they face and take a number of precautionary measures to enhance their safety, such as possessing a weapon, staying off the streets at night, and traveling in groups. While the close relationship with the male gang can often serve as a protective factor, it can also place the girls at risk of rape and sexual assault by their "homeboys" (Hunt & Joe-Laidler, 2001). In addition, girls whose gang membership is connected to a male gang unit tend to experience higher levels of violence on the streets compared to girls who operate in independent cliques. These girls are at a higher risk of victimization due to the levels of violence that they are exposed to from assaults and drive-by shootings involving the male gang members. Indeed, many of these crimes (and potential risks of victimization) would not be present if they were not involved in the gang lifestyle (Miller, 1998).

The exit from the gang lifestyle for girls can occur in several ways. For most girls, this exit coincides with the end of adolescence. They may withdraw from the lifestyle, often as a result of pregnancy and the need to care for their young children. For others, their exit is facilitated by an entry into legitimate employment or advanced education. Others will be removed from their gangs as a result of incarceration in a juvenile or adult correctional facility. While some may choose to be "jumped out," most will simply diminish their involvement over time rather than be perceived as betraying or deliberately going against their gang peers (Campbell, 1995). The few women who choose to remain in the gang have several pathways from which to choose. They may continue their gang participation as active members and expand their criminal resume. Their relationships with male gang members may continue with their choice of marriage partners, which allows them to continue their affiliation in either a direct or indirect role (Eghigian & Kirby, 2006).

Summary

- Women engage in every category of crime, yet their participation is significantly less than male offending practices.
- Regardless of race, ethnicity, class, or drug, women have similar pathways to addiction: depression, abuse, and social and economic pressures.
- The war on drugs has led to increased incarceration rates for both men and women, but has had particularly damaging effects for women.
- Evidence of postpartum syndrome is often used in criminal cases of women who kill their children.
- There are several different reasons why mothers may kill their children, but not all involve issues of mental illness.
- Approximately 15% of women in prostitution work on the streets.
- Street prostitutes face the highest levels of risk for violence and victimization.
- Women in prostitution face significant mental and physical health issues as a result of their time on the streets. These issues lead to significant challenges as they try to exit prostitution and make a new life off the streets.
- There are several different types of girl gangs, including mixed-gender gangs, affiliate or auxiliary gangs, and independent girl gangs.
- Sexuality can be a component of the gang life for some girls, but it is not necessarily the experience for all girls involved in gangs.

KEY TERMS

Gender gap

Zero-tolerance policies

Felony Drug Provision of the Welfare Reform Act of 1996

Neonaticide

Infanticide

Filicide

Altruism

Posttraumatic stress disorder (PTSD)

Legalization

Decriminalization

Brothel

Street prostitution

Postpartum syndrome

Liberation hypothesis

Social injury hypothesis

Vice Queens

Walking the line

Pulling a train

Independent female gangs

Mixed-gender gangs

Affiliate

QUESTIONS FOR DISCUSSION

1. Why is the media obsessed with the images of the female offender? What implications does this have on understanding the realities of female offending?
2. What does research say about the gender gap in offending?
3. How have drug addiction and the war on drugs become a gendered experience?
4. What role does mental illness play in cases of women who kill their children?
5. What are the risk factors for prostitution? How do these issues affect a woman's ability to exit the streets?
6. Why are jurisdictions reluctant to legalize or decriminalize prostitution?
7. How do girls use their gender within the gang context?
8. Discuss the exit strategies for girls in gangs.

WEB RESOURCES

The Sentencing Project: http://www.sentencingproject.org

Women and the War on Drugs: http://www.drugpolicy.org/communities/women

Prostitution Research and Education: http://www.prostitutionresearch.com

Children of the Night: http://www.childrenofthenight.org

Prostitutes Education Network: http://www.bayswan.org

National Gang Center: http://www.nationalgangcenter.gov/

READING

This reading provides a case study of a woman struggling with addiction within the criminal justice system. Her story highlights how the drug laws have negatively affected women, in terms of both the rising incarceration rates and the social implications of being labeled a convicted drug offender.

Locked Up Means Locked Out

Women, Addiction, and Incarceration

Vanessa Alleyne

Vivian

Vivian[1] is a 31-year-old African American woman who used marijuana and alcohol in slowly increasing amounts over the past several years as she became more heavily involved with a man who was an illicit drug trafficker. She lived in public housing and graduated from high school with dreams of pursuing a career in art design before becoming pregnant in her senior year of high school. She soon had two children and worked part-time as a cashier in a local store.

Vivian was arrested during a raid of the bodega in which she worked and was found with an unprescribed pill, Vicodin, in her pocket. Ironically, Vivian had occasionally used Vicodin for legitimate pain purposes, with a long-term herniated disc in her back, but would usually get them "locally" rather than through a prescription. However, this was Vivian's second arrest for drugs. The first was for being in a car with a man who had drugs. They were stopped by the police one night and both were arrested. She was given probation, which she did not complete to the probation officer's satisfaction (she continued to associate with the drug involved man). Thus, she had a violation of probation on her record at the time of her second arrest.

Vivian was arrested and sent to the county jail. Her children, a boy and a girl, were put into foster care. She had no viable extended family available to take the children. She remained in jail awaiting arraignment for one week and met her public defender for the first time during that brief court session, where she pleaded not guilty to charges of illegal possession of a controlled substance. Vivian could not post the $5000 bail (10% cash) set by the judge, so she was returned to jail to await trial or a plea offer.

As with so many others, self-advocacy from inside proved to be extremely difficult for Vivian. She was horrified to learn that most of her female colleagues in jail waited anywhere from three to six months before hearing back from their attorneys for bail reduction requests, plea offers, or trials. Frustrated and angry. Vivian spent hours each week trying to establish contact with her children through Child Protective Services. She would write to the case worker and to her children weekly, and occasionally had access to a telephone via rare trips to the Law Library to contact the CPS worker about them.

SOURCE: Alleyne, V. (2006). Locked up means locked out: Women, addiction and incarceration. *Women and Therapy, 29*(3/4), 181–194.

She was told eventually that the children were well, had been sent to two separate foster homes, and that she could write letters to them through the agency. The foster families had not agreed to bring her children to the jail for a visit. Thus, she wrote letters to her children and sent them to the case worker, but couldn't really tell whether or not her children were getting them.

Three months later, Vivian's public defender arrived with a plea offer from the prosecutor: three years in state prison, out in nine months with good behavior. Vivian felt that getting substance abuse treatment would be a better option for her though, because without it she believed she might not qualify to get her children back. She refused the offer and asked the attorney again for treatment. The public defender agreed to go back to the judge and make a case for treatment instead of incarceration.

The public defender put Vivian's name on the list for his office's investigator to seek treatment options for her. There was no counselor, case manager, or social worker at the jail to provide this critically needed function. The public defender's investigator, with hundreds of similar and worse cases, had to arrange for the jail to transport women to and from the treatment facilities for treatment interviews. The interviewing and assessment process took two and a half months for Vivian, which she later learned was actually less than usual.

During this time, Vivian was evicted from her apartment because she was not there to pay rent. All her possessions, including her children's clothing and furniture, were lost.

After several months, the judge in Vivian's case approved treatment in lieu of incarceration. However, he approved long-term intensive treatment of 12 to 18 months in an inpatient program for women.

Approximately one month later Vivian was notified that a bed had opened up for her at one of the two long-term intensive treatment programs for women in her part of the state. Now nine months into this jail experience, Vivian faced the decision of whether to spend another twelve to eighteen months in a severely restrictive "therapeutic community" for substance-abusing women or nine months in state prison. By this time, Vivian had heard about drug court as well, but shied away from this when she learned that she would be required to participate in its monitoring program for the next four years. Given her earlier probation violation for merely associating with her former boyfriend, Vivian knew this was an impossible task for her.

Missing her children, worried about them, without further word from her boyfriend, profoundly isolated and depressed in nearly a year already spent in jail, Vivian opted for the shortest route to freedom: via state prison. She decided that she would rather attend outpatient day treatment after prison and have her children with her at home while doing so. Thus, she pleaded guilty to a felony conviction, was sentenced to three years, and returned to jail to await the bus to state prison.

The women's state prison, operating at 138% of capacity for many years, took another six weeks to transport her from the jail. Vivian knew she would get credit for some of her time in jail, but didn't know how much or when it would be applied to reduce her sentence.

When she arrived at the state prison, it look another month for Vivian to move through the classification process, be placed and audited for time served. She learned that she would have to serve another two months in order to become eligible for release. Vivian was exhausted and frightened for much of the first two weeks at the prison. While in jail she had heard many stories about women and guards in prison, and didn't want trouble from anyone. She was concerned about being approached by gangs, about potential violence, and about how to stay out of the way of the corrections officers, who had a reputation for being particularly rigid and unforgiving.

Two months later, Vivian was given a release date and two referral slips: one to an outpatient drug treatment program in her area, the other to a shelter.

On the day of her release, Vivian was given a bus ticket to a major city near her neighborhood and five dollars.

As Vivian boarded the bus for her home area, she felt that she'd aged 20 years in a little more than one. The stresses of the past 18 months washed over her, and she sobbed out loud for the first time in years. In fact, she looked much older than her 32 years, and her hair had thinned significantly during her incarceration. She returned to her neighborhood, placed her name on the shelter waiting list for that evening, and went to the social service center to apply for reinstatement for public assistance.

Two nights later, a very shaky Vivian lay in her shelter bed at 3:00 a.m., beside herself with anxiety, frustration, and anger at the downward spiral from which she felt unable to emerge. Aside from the inhumane and unsafe conditions she was forced to endure in the shelter, she'd just learned that day that she was now ineligible for welfare benefits because of her felony conviction. (Section 115 of the welfare reform act of 1996 provides that persons convicted of a state or federal felony drug offense for using or selling drugs are subject to a lifetime ban on receiving cash assistance and food stamps. While 34 states have since modified or eliminated this draconian legislation in recent years, she unfortunately lived in one of the 16 slates that has not. Vivian was one of 35,000 African American women who were directly affected by this ban.)

Without assistance, Vivian knew that she had no visible means of support, and as such could not demonstrate to Child Protective Services that she could become self sufficient and provide adequately for her children. Her children were two of over 135,000 who were also negatively and directly impacted by the ban. Thus, her greatest desire, to be reunited with her two children, now separated from her for well over a year, seemed further away than ever.

With a felony conviction on her record, Vivian was soon to learn that she was locked out of receiving public housing and food stamps as well. She was ineligible for a wide range of jobs that screen out those with criminal records or felony convictions. Even her dream to pursue art design in college was to elude her, as she was now ineligible for federal student loans or grants. She could not even vote during election time to support candidates who could overturn these so-called "acts of reform."

Gradually Vivian began to realize that her prison sentence had continued on the outside. Being locked up had in effect locked Vivian out of all legitimate routes away from a drug involved lifestyle. She felt like a caged bird. Consequently, she began a slow drift back toward her former boyfriend, not because she saw a real future with him, but rather out of fiscal necessity and emotional neediness. He rescued her, but with a high price tag—involvement in exactly the lifestyle that Vivian wanted and needed so much to avoid.

Nonetheless, with no other useful options in sight, Vivian decided to move in with him. She attended a day drug treatment program in order to qualify to get her children back as soon as possible. She got her old job back at the bodega, paid mostly under the table, and began to work on creating a reasonably believable, yet clearly fictitious, presentation to the system which now threatened to permanently remove from her the only family she had left.

The criminal justice system in which Vivian became engulfed was itself immersed in failure to keep pace with its own voracious appetite for incarceration. Its inability to handle her relatively minor substance abuse transgression in a timely and effective manner was but a symptom of a system run amok. Societal costs of such a system are enormous: $20,000 to $30,000 per year for Vivian to be incarcerated; $3,600 to $14,000 (depending on the state) a year per child for placement in the foster care system; elimination of Vivian's current income and reduction of future work opportunities; the cost of her homelessness and further descent into poverty; and the eradication of Vivian's and her family's upward mobility by defunding educational access (Freudenberg, 2002; Kassebaum, 1999).

The psychosocial costs of Vivian's incarceration may be even greater. How does one begin to quantify the short- and long-term effects of trauma experienced by Vivian's small children as they were suddenly removed from their homes and placed with strangers? Of their vulnerability to a myriad of future psychological and educational difficulties related to this sudden devastation of home, family, and all things familiar? How can we measure the psychological devastation that Vivian experienced in a harsh, overcrowded, isolated prison system designed to create and enforce conditions of extreme deprivation? What value should be placed on the lost opportunity for comprehensive substance abuse treatment that could have been offered while Vivian was made to languish in jail for nine months before ever reaching state prison? (Freudenberg, 2002; Kassebaum, 1999; CASA, 2001).

Recent experience with trauma associated with terrorism, war, and natural disasters (e.g., 9/11 attacks, Hurricane Katrina and its aftermath in Louisiana) has shown how devastating the impact of trauma can be for individuals, families, and society as a whole. The traumatic events experienced by this family, sadly enough, must be understood to reverberate throughout society, as more and more women with substance abuse difficulties are drawn into a grossly underprepared, overburdened criminal justice system.

The United States of America leads all nations on earth in the rate of incarceration of its citizens (Karberg & Beck, April 16, 2004). Last year, 726 of 100,000 persons in the United States were under correctional supervision, including parole, probation, or in pre-trial detention. This rate is significantly ahead of other countries next on the list—Belarus and Russia, each at 523 per 100,000.

Within this population of more than 2.1 million persons, more than 185,000 are women (8.7%). The United States ranks 15th in the world in the percentage of women in its population who are incarcerated. In this era of newer, higher, and greater levels of achievement and progress for women, another milestone has quietly been reached: never before have more women been confined in correctional facilities.

The growth of women's incarceration is nearly double the rate for men over the past two decades. These rates are disproportionately due to the American political war on drugs, given that women in prison are more likely than men (30% vs. 20%) to be serving a sentence for a drug charge (Matter, Potler, & Wolf, 1999).

A confluence of national policies and actions have created the current crisis which disproportionately impacts poor women and women of color. Three related national phenomena occurred during the 1980s which shifted and sharply increased rates of incarceration for women, disrupting millions of families and wreaking havoc on overburdened drug treatment and criminal justice systems in the process. Those phenomena are discussed in detail here.

The Democratization of Drug Use

Cocaine, once a major drug of abuse for middle- and upper-middle-class whites in the 1970s, began its advance across class and racial lines when new forms of use were discovered. The introduction of cocaine in freebase and crystallized forms ("crack") lowered economic barriers posed by the powdered version of the drug, thereby democratizing its use. Cocaine became available to anyone with access to three dollars and the simple chemical recipe to convert the powder to a purer, more potent smokable form. Thus, larger numbers of individuals who had heretofore been shut out of using the trendy drug were quickly drawn into its vortex. Unaware that the pace of the addictive process was highly accelerated with the use of this more potent version of the drug, thousands of women in urban and rural settings began to experience the relentless intensity of the brain-produced cravings for more of the drug. The resulting increase in associated nonviolent criminality, e.g., theft, prostitution, to satisfy drug cravings led to a sharp spike in the numbers of women in detention (Alleyne, 2004; Pallone & Hennessy, 2003).

"Just Deserts"

On the federal level, a shift occurred partly in response to the rise of cocaine related crimes which continues to reverberate to this day. Pallone and Hennessy (2003) describe a "just deserts" model of correctional practice which began to take hold in the 1970s, partly in response to cocaine's spread, but also in reaction to reports which emerged casting doubt on the effectiveness of rehabilitation for prisoners. The "just deserts" model (Allen, 1981; Morris, 1974; von Hirsch, 1984) argued for prisons to be places of precise and inflexible punishment and incapacitation. Congress then supported this ideology through the passage of two major pieces of legislation which became the cornerstone of the Reagan Administration's War on Drugs. The first, the Criminal Sentencing Reform Act of 1981, focused on mandatory sentencing for many drug crimes, thereby removing judicial discretion for many types of offenses. The second, the Omnibus Crime Reduction Act of 1986, mandated incarceration for numerous drug use, possession, and sale offenses which heretofore had ended in probation, fines, or significantly lower sentences. A particularly onerous example of the draconian nature of these changes was seen in the "three strikes you're out" federal law created to send three time convicted felons to prison for life *without the possibility of parole.*

Zero Tolerance Policing

Simultaneously, a new pattern of police practice began to emerge in major U.S. cities. In New York City, for example, a new philosophy of police practice known as "zero tolerance" used computer technology and directed daily law enforcement efforts toward eliminating "quality of life" crimes (Bratton, 1998; Greene, 1999), as opposed to an earlier focus on large scale drug operation organizers. These hyperaggressive strategies were implemented by NYC Police Commissioner William Bratton, who posited that cities could once again be made more livable if smaller crimes could be prosecuted, e.g.. panhandling, turnstile jumping, public vagrancy, prostitution. In the process of pursuing these daily "nuisance" crimes, thousands of new individuals (mostly poor, mostly African American and Latino) would be drawn into the police fingerprint database. Operating on the hypothesis that larger crimes are often committed by individuals with prior records, Bratton saw this as a two-pronged victory in that petty misdeeds would decrease while the police fingerprint database was built up for easier apprehension of future criminals.

These practices had a devastating impact on poor women and people of color. Violations which earlier could have been satisfied through fines, community service, or continued without a finding were now offenses which landed women in jail. The inability to meet bail guaranteed that these women would serve time in city or county jails awaiting disposition of their cases. An overburdened criminal justice system with too few public defenders then virtually assured that poor women would languish behind bars, often for longer than what the eventual sentence would require.

Falling Crime, Rising Incarceration?

The Sentencing Project (2004) reported that crime rates have fallen for the last 14 years, yet prison incarceration rates have risen 52% during the same period. Karberg and Beck's (2004) astute analysis of Bureau of Justice statistics indicates that the *entire increase in prison incarceration rates* is due to the changes in sentencing policy and practice. While government and political figures are often quick to publicize and align themselves with the decline in crime rates in the popular media, little discussion of the continuing rise in incarceration rates typically follows (Jackson & Naureckas, 1994).

Double Disproportion: Women of Color

It must be noted that quantum increases in women in prison have disproportionately affected Black and Latina women. They are sentenced to prison for drug offenses at rates that far outpace their numbers in the population, and their rates of arrest. Thirty-two percent of New Yorkers are African American or Hispanic women, for example, yet they constitute 91% of the people who are sentenced to prison for drugs in the state (Mauer et al., 1999). Most women receive prison sentences for non-violent drug related crimes.

Thus, disparities in arrests, hyperaggressive policing, and sentencing policy have been disproportionately felt by poor women of color and their families. Vivian's story highlights the subtle shift in resources which has taken place in many states, away from public policy which provides a modicum of education and health care support for low-income women of color and their families to the subsistence standards found in correctional facilities and foster care systems.

Substance Abuse, Motherhood, and Incarceration

For a mother, to be faced with the option of long-term treatment isolated from loved ones, versus being reconnected with children through prison bars, is tantamount to having no choice at all. The plea bargain system and judges' failure to consider fully the range of substance abusers that are now caught in the criminal justice web creates a mockery of the ideals of both justice (let the punishment fit the crime) and drug treatment (let the treatment fit the client). Contemporary substance abuse treatment has made great progress in differentiating among substance abusers. Earlier notions of "one size fits all" treatment, where early stage abusers and chronic relapsing long-term users were sent to 28-day inpatient rehabilitation, have given way to evidence based decisions supported by patient placement criteria (Mee-Lee, 2001). Yet many judges have failed to note the distinctions. Additionally, judicial decision making appears to be further constrained by what may be a politically driven need to preserve incarceration-like conditions by referring many more women to long-term intensive treatment than may actually qualify for it based on clinical evidence. This, coupled with the restricted number of slots available for mental health and substance abuse treatment, results in outcomes that are all too often detrimental to women, children, and families.

Mental Health and Drug Treatment

The need for substance abuse treatment in the United States greatly outstrips resources directed to this serious public health concern. On any given day in this country, an estimated 22.2 million individuals would diagnostically qualify for substance abuse treatment in this country, yet fewer than 2 million receive it (SAMHSA, 2005). While acknowledging that many of the millions who need treatment may not be actively seeking it, there still remains a disparity in service provision that is unmatched in other areas of health.

In recent years, social science literature has begun to report on the shift away from deinstitutionalization to criminalization of mental health illnesses (Freudenberg, 2002; Navasky, 2005). Far more persons with mental health concerns are put behind bars than given treatment in hospitals. Large numbers of mentally ill, chemically addicted individuals are sent to, and remain, in jails and prisons for much longer periods of time for relatively minor offenses, which are usually directly traceable to their mental illness (Butterfield, 1998). Perhaps the most stark indicator of this public policy shift can be seen in state and federal government budgets, where in recent years spending for prison expansion in some states has exceeded that for spending on police and other local budgetary items (Butterfield, 2002).

The move to "criminalize rather than medicalize" has been particularly damaging to women, who are seven times more likely to enter prisons with histories of untreated post traumatic stress, sexual abuse or assault, and depression. Often women report using drugs or alcohol as forms of self-medication, in order to offset the impact of these difficult histories (CASA, 1998).

Mental illness and drug addiction are psychological and physiological phenomena, not character flaws. Yet women are caught in the triple stigmatization that comes from being female, in prison, and addicted. Women who may be psychologically fragile are often further damaged by the punitive, harsh environments found in correctional facilities.

Strategies for Change

Change for incarcerated women must emerge first from public awareness. As is the case for life behind bars, all too often those not directly impacted would rather look the other way and ignore the systematic troubles that are now visited upon hundreds of thousands of women and their families. Social and economic analysts and policy makers may be more directly aware of the injustices visited upon incarcerated addicted women, but are loathe to act, fearing political backlash. Yet with the numbers of women behind bars rising yearly, along with larger portions of state and federal budgets, in the face of falling crime rates, that posture is more and more difficult to maintain.

Ideas for change must begin with an appreciation that the perspective and experience of women is not the same as for men. "Few research studies on female inmates have been conducted, but most of those conclude that women exhibit differences in the severity and uniqueness of certain needs compared with male inmates" (Brennan & Austin, 1997). In 2003 the National Institute for Corrections issued a groundbreaking report on gender-responsive strategies for women offenders (Bloom, Covington, & Raeder, 2003). The report called for acceptance of a guiding principle that gender makes a difference as a *starting point* for change. If that perspective is brought to bear in correctional facilities, it can result in structural changes that appreciate that women's pathways into prison are drug related more often than not; that relationships emerging from untreated trauma and addiction are all too often at the root of incarceration; and that family connections are often a significant motivating factor throughout the criminal injustice process.

Even a small reallocation of existing prison dollars would begin to address the huge gap between women who need treatment and those who receive it behind bars. Estimates of treatment need indicate that 25% of those in prison who need treatment, receive it (CASA, 1998). Given the overrepresentation of incarcerated women with drug-related problems, it is a fair assumption that this statistic would be much higher if limited to women. From a clinical change perspective, it is clear that drug use is reduced and eliminated more effectively when treatment is obtained (Baekeland & Lundwall, 1975; Simpson, Joe, Rowan-Szal, & Greener, 1997). Thus, prisons and society as a whole would be well served to consider these issues and move to implement changes based on these ideas. Until these strategies and others are brought to the fore, we are all caged birds, trapped in a system of our own creation.

Note

1. Vivian's story is a composite of several clients with whom the author worked in a New Jersey jail. Her name and other identifying information have been changed.

References

Angelou. M (1970). *I know why the caged bird sings.* New York, NY: Random House.

Allen, F. A. (1981). *The decline of the rehabilitative ideal: Penal policy and social purpose.* New Haven: Yale University Press.

Alleyne, V. L. (2004). *The relationship between Black racial identity, motivation, and retention in substance abuse treatment.* Unpublished doctoral dissertation. Columbia University, New York, NY.

Baekeland, F., & Lundwall, L. (1975). Dropping out of treatment: A critical review. *Psychological Bulletin, 82,* 738–783.

Bloom, B., Covington, S., & Raeder, M. (2003). *Gender-responsive strategies: Research, practice, and guiding principles for women offenders.* Washington. DC: National Institute of Corrections.

Bratton, W. (1998) *Turnaround: How American's top cop reversed the crime epidemic.* New York, NY: Random House.

Brennan, T., & Austin, J. (1997). *Women in jail: Classification issues.* Washington, DC: National Institute of Corrections. http://niclc.org/pubs/1997/013768.pdf.

Butterfield, F. (1998, March 5). Prisons replace hospitals for the nation's mentally ill. *New York Times,* p. A1.

Butterfield, F. (2002). *Study finds increase at all levels of government in cost of criminal justice.* New York Times, p. A14.

CASA. (1998). *Behind bars: Substance abuse and America's prison population.* New York. NY: National Center on Addiction and Substance Abuse at Columbia University.

CASA. (2001). *Shoveling up: The impact of substance abuse on state budgets.* New York, NY; National Center on Addiction Studies at Columbia University.

Freudenberg, N. (2002). Adverse effects of US jail and prison policies on the health and well-being of women. *American Journal of Public Health, 92*(12), 1895–1899.

Greene, J. A. (1999). Zero tolerance: A case study of police policies and practices in New York City. *Crime & Delinquency, 45*(2), 171–187.

Jackson, J. A., & Naureckas, J. (1994). *Crime contradictions: U.S. News illustrates flows in crime coverage: Fairness & Accuracy in Reporting* (FAIR).

Karberg, J. C., & Beck, A. J. (April 16, 2004). *Trends in US correctional populations: Findings from the Bureau of Justice Statistics.* Paper presented to the National Committee on Community Corrections, Washington. D.C.

Kassebaum, P. (1999). *Substance abuse treatment for women offenders: Guide to promising practices. Technical Assistance Publication Series* 23 [DHHS Publication No. (SMA) 00-3454], Rockville, MD: U.S. Department of Health and Human Services.

Mauer, M., Potler, C., & Wolf, R. (1999). *Gender and justice: Women, drugs and sentencing policy.* Washington, DC: The Sentencing Project.

Mee-Lee, D. (Ed.). (2001). *ASAM patient placement criteria for the treatment of substance-related disorders* (2nd-Revised ed.). Chevy Chase, MD: American Society of Addiction Medicine.

Morris, N. (1974). *The future of imprisonment.* Chicago: University of Chicago Press

Navasky, M., O'Connor, K. (2005). *Frontline: The new asylums.* Boston. MA: WGBH Educational Foundation.

Pallone, N. J., & Hennessy, J. J. (2003). To punish or to treat: Substance abuse within the context of oscillating attitudes toward correctional rehabilitation. In N. J. Pallone (Ed.), *Treating substance abusers in correctional contexts: New understandings, new modalities* Binghamton, NY: The Haworth Press, Inc.

SAMHSA (2005). *2003 National survey on drug use and health: Findings.* Retrieved October 30, 2005, from http://oas.samhsa.gov/nhsda/2k3nsduh/2k3Results.htm

Simpson, D., Joe. G. W., Rowan-Szal, G. A., & Greener, J. M. (1997). Drug abuse treatment process components that improve retention. *Journal of Substance Abuse Treatment,* 14(6, 1997 Nov-Dec.), 565–572.

von Hirsch, A. (1984). The ethics of selective incapacitation: Observations on the contemporary debate. *Crime & Delinquency, 30(2),* 175–194.

DISCUSSION QUESTIONS

1. What affect did Vivian's incarceration have on her children?
2. What negative long-term consequences did Vivian experience as a result of her guilty plea for a drug-related charge?
3. How can policies such as the War on Drugs be reformed to help women who deal with addiction issues?

READING

While the media paints women who kill their children as monsters, the reality is that there are a number of variables that distinguish the different categories of women who engage in acts of filicide. This article adds to what you have already learned in the text section by expanding on the different classifications of women who kill their children and what distinguishes these women from each other.

Infanticide and Neonaticide

A Review of 40 Years of Research Literature on Incidence and Causes

Theresa Porter and Helen Gavin

Introduction

Becky Sue Marrow concealed her pregnancy for 9 months then dismembered and burned her newborn son. According to her lawyers, she was in a "dissociated" mental state at the time of the crime. However, as Marrow had tried to hide a pregnancy in the past, had attempted to divert witnesses from the fire pit where she burned the infant's corpse, and did not suffer from any symptoms of amnesia, it is likely that the psychiatrist

SOURCE: Porter, T., & Gavin, H. (2010). Infanticide and neonaticide: A review of 40 years of research literature on incidence and causes. *Trauma, Violence and Abuse, 11*(3), 99–112.

was correct when he told the court, "She had clear insight. She knew exactly what she was doing" (The Canadian Press, 2008). However, she was only found guilty of "offering an indignity to a dead human body and disposing of the dead body of a child with the intent to conceal its birth" and sentenced to 2 months house arrest (CBC News, 2008).

This case highlights many of the issues involved in infanticide and neonaticide, alleged mental illness, culpability, recidivism, future risk, pregnancy concealment, and appropriate responses by the courts. This article will review the state of the research on these issues and provide an analysis for future directions.

Infanticide and Neonaticide Incidence

Infanticide is the killing of young children, whereas neonaticide is the killing of the infant within the first 24 hours after birth. Both should be distinguished from the more general term of filicide, which is the killing by a parent of any child of their own.

The killing of infants and newborns is one of the most common forms of murder by women. Unwanted babies, particularly the female or handicapped ones, were left to the elements in ancient times, possibly with the hope that they would be taken in by passersby to be raised as slaves, perhaps with the awareness that they would die from exposure or the attentions of wild animals. In the 20th and 21st centuries, the murder of infants and children remains a significant problem. In the last 30 years, while child deaths due to diseases, accidents, and congenital defects have decreased (Finkelhor & Ormrod, 2001), the incidence of homicide for children younger than the age of 1 year has increased in some areas (Finkelhor, 1997) and is currently estimated to be 8.0 per 100,000 in the United States. In Canada, however, the incidence is estimated to be less than 3.0 per 100,000 (Hatters-Friedman, Horwitz, & Resnick, 2005).

The majority of the murders of infants and newborns are by the biological mother. The U.S. Department of Health and Human Services estimates that, of the 2,000 children killed annually in the United States, 1,100 are killed by the biological mother (Kohm & Liverman, 2002). In 1999, in a single U.S. state, more than 50 infants were abandoned in dumpsters (McKee, 2006). Furthermore, approximately 31,000 newborns are abandoned in U.S. hospitals annually (McKee, 2006). Two separate studies of the infanticide rate in England, Scotland, and Wales found that infants less than 1 year old are at 4 times greater risk of being murdered than any other age group, with the 1st day of life being the highest risk. For Scotland, the rate of infant murder was 43 per million, compared to the rate of 29 per million for young adults during the same study period (Marks & Kumar, 1993, 1996). In Germany, there is a three times higher homicide risk for infants born to eastern German women than those born to western German women (Spiegel, 2008).

Examining the incidence statistics is difficult, as this is a crime that it is either unrecorded or recorded with other offences. Most countries do not have a government agency mandated to track infant deaths and when tracking does occur, it is often as part of an aggregate with other types of death (Kohm & Liverman, 2002). However, seven incidence studies between 1994 and 2006 suggest that the rate of infanticide/neonaticide in industrialized countries (England, Scotland, Wales, United States, Canada, New Zealand) ranges from 2.4 per 100,000 to 7.0 per 100,000 (see Table 8.3).

Government statistics on child homicide can be confusing. First, they are only an accounting of known homicides and it is likely that infants are the class of homicide victims least identified due to the ease of hiding the corpse. For example, the National Center for Health Statistics (United States) discovered that birth certificates could not be found for 2.8% of all deaths between 1983 and 1991, with infants less than 30 days old being the group most commonly missing birth certificates (Overpeck, Brenner, Trumble, Trifiletti, & Berendes, 1998). The implication is that these cases represent infants whose existence was hidden until the corpse was found. Furthermore, the cause of death of an infant may be difficult to establish and may be falsely attributed to sudden infant death syndrome (SIDS). It has been estimated that up

Table 8.3 Incidence of Homicide per 100,000

Author	Nation	Incidence per 100,000
Siegel et al., 1996	United States	3.1
Cummings, Mueller, Theis, & Rivara, 1994	United States	6.9
Herman-Giddens, Smith, Mittal, Carlson, & Butts, 2003	United States	2.1
Bennet et al., 2006	United States	2.5
Marks & Kumar, 1996	United Kingdom	4.3
Bropokman & Nolan, 2006	United Kingdom	6.3
P. Dean, 2004	New Zealand	4.5

to 10% of SIDS cases are actually undetected homicides (Leven & Bacon, 2004). Furthermore, due to the 1969 addition of an "Undetermined" category to the International Classification of Diseases (ICD), as well as many related changes in the standard certificate of death used in the United States, infant homicides in the United States may have been disproportionately underrecorded since 1968 (Jason, Carpenter, & Tyler, 1983). This study also reports a sudden drop in infant homicide rates the year these classifications were changed, strongly suggesting that, when given the opportunity, people would rather avoid designating an infant's death as a homicide. Furthermore, several studies have indicated that, despite the medical examiner reporting grounds for homicide in an infant's death, the police may not make a report and the local courts may not prosecute the case (McKee, 2006).

The way sources aggregate information may also be problematic. The Bureau of Justice Statistics website has a page listing Homicide Trends in the United States, but it combines deaths by parents with deaths by stepparents, thereby distorting the information. In the United States, stepfathers do have a high murder rate, but when one compares biological parents, mothers kill their children at a higher rate than fathers (Kohm & Liverman, 2002).

Characteristics of Neonaticidal Women Versus Infanticidal Women

Romina Tejerina and Stephanie Collins never met and yet they share a kind of frightening sisterhood; they both murdered their newborn babies immediately after giving birth. Tejerina, a 23-year-old Argentine, became pregnant as the result of a sexual assault and was afraid to tell anyone. Within minutes of giving birth to a daughter at home, Tejerina stabbed the baby repeatedly, who died 2 days later (La Nacion, 2005). Collins, a 27-year-old American, had previously had an abortion and lived in a state with a newborn Safe Haven[1] program. Collins concealed her pregnancy, gave birth out of the hospital, then immediately killed the newborn boy, and put his body in the garbage (Angier, 2005). Despite their apparent differences, these cases represent typical neonaticide; a woman without mental illness hides an unwanted pregnancy, and when presented with a newborn infant whose existence she finds undesirable, kills the infant, thereby eliminating her problem.

In 1970, Resnick published his groundbreaking study of neonaticide. He performed a literature review of documented newborn murders from the mid-18th century until 1968 in 13 different languages and found that women who kill their newborns were substantively different from women who kill their infants or toddlers. His work has been supported in numerous other international studies that show that women who murder newborns tend to be younger than 25 years old, emotionally immature, unmarried, often living with their parents, unemployed, or attending school. They do not seek prenatal care and are often no longer involved with the baby's father. For example, Emerick, Foster, and Campbell (1986) in their study of infant homicides in Oregon found that the infant deaths were associated with lack of prenatal care and non-hospital birthing.

Resnick (1970) and several other studies have confirmed that the majority of neonaticidal women are not mentally ill at the time of the murder and maternal suicide after neonaticide is rare. Hatters-Friedman, Heneghan, and Rosenthal (2007), in their review of 81 women who either denied or concealed their pregnancies, found that none had psychotic denial and a psychiatry consult was only requested on four of the women. In 2001, Meyer and Oberman reviewed 37 cases of neonaticide and found that most of the perpetrators did not have a major mental illness. In D'Orban's British study, she found that the majority of the neonaticidal women were not suffering from psychosis or depression. Haapasalo and Petäjä (1999), in their Finnish study of 15 neonaticides, proposed that mental illness was not a relevant variable, with less than 30% of the women claiming any psychological issues. Similarly, the Finnish sample of Putkonen, Collander, Weizmann-Henelius, and Eronen (2007) of 14 psychiatrically evaluated cases found only four cases with psychotic symptoms.

The majority of infants killed in the 1st day are born out of a hospital, usually at the woman's home (Paulozzi & Sells, 2002), although there are recorded cases of neonaticides in birthing units (Mendlowicz, da Silva, Gekker, de Moreas, Rapaport, & Jean-Louis, 2000, cited in Hatters-Friedman & Resnick, 2009). Newborns who are the second child of a woman under age 19 are at an increased risk of homicide (Overpeck cited in Spinelli, 2003). The hallmark example of neonaticide is that the newborn is unwanted and so the woman, after concealing the pregnancy for 9 months, gives birth alone, and then kills the newborn via non-weapon methods such as suffocation, strangulation, or drowning (Meyer & Oberman, 2001).

Women who murder infants who are older than 1 day are significantly different to women who murder newborns. They tend to be older than 25, use weapon as well as non-weapon methods of murder, are often married, and well educated (Resnick, 1970).

These women tend to premeditate their murders (Logan, 1995, in Dalley, 1997) and may murder the infant as retaliation against another person, during an episode of abuse, or to remove an unwanted child (D'Orban, 1979).

Mental Illness and Infanticide

In 2004, 35-year-old Texan Dena Schlosser heard voices commanding her to remove the arms of her 11-month-old daughter as a sacrifice to god, and she obeyed the voices. This was not Schlosser's first psychotic episode. She had another psychotic episode after the birth of her middle child, which resulted in both a hospitalization and an investigation by child protective services (CPS). However, the case was closed after 7 months and CPS concluded that Schlosser was not a threat to her children. After the death of her daughter, Schlosser was diagnosed with bipolar disorder, found not guilty by reason of insanity, and sent to the North Texas State Hospital. She was later released by court order to outpatient care. The judge, perhaps mindful of the infamous Yates[2] case, mandated that Schlosser not only attend weekly psychiatric treatment but also comply with compulsory medication and birth control (Hundley, 2008).

A subset of women who murder their infants do have a definitive mental illness that can be shown to have strongly influenced their behaviors. For example, in the study by Kauppi, Kumpulainen, Vanamo, Merikanto, and Karkola (2008) of 10 Scandinavian women who murdered infants, 6 had psychotic symptoms, and in the study by Lewis and Bunce (2003) of 55 filicidal women, 52.7% were psychotic. In the study by Krisher, Stone, Sevecke, and Steinmeyer (2007) of 57 infanticidal women, 24% were initially found incompetent to stand trial due to mental illness.

However, as with neonaticides, a large number of infanticide cases do not involve a severe mental illness that precluded the woman from being aware of the wrongfulness of her actions (Hatters-Friedman & Resnick, 2009). For example, Beyer, Mcauliffe-Mack, and Shelton (2008) suggested that all of the 40 infanticidal women in their sample had personal gain for their actions, that is, living their lives unencumbered by an infant. Bourget, Grace, and Whitehurst (2007) noted that suicide is much lower among infanticidal women than among women who murder older children. In the British study by D'Orban (1979), only 26% of the 89 cases involved mental illness; 60% of the infanticides were for other reasons, such as revenge against the infant's father or to remove an unwanted infant.

Other factors besides severe mental illness are involved in infanticide. For example, Spinelli (2003) reports that women who drop out of school are eight times more likely to murder their infant than women who had a college education, even when age is controlled. Furthermore, research has found that infanticide is associated with the woman's anger. Krischer, Stone, Sevecke, and Steinmeyer (2007) in a cluster analysis of 57 infanticidal and neonaticidal women found that infanticide was associated with the woman's anger as well as her youth. Two separate studies reported women in their samples, who were not designated as mentally ill admitted to having fantasies of harming their infants (Jennings, Ross, Popper, & Elmore, 1999; Levitzky & Cooper, 2000). Jennings et al. (1999) compared clinically depressed and non-depressed mothers and found that 7% of the non-depressed women admitted to having thoughts of harming their offspring. Levitzky and Cooper (2000) questioned 23 women with infants experiencing colic syndrome and found that 70% reported explicit fantasies and thoughts of aggression toward the infant, but only 26% said the thoughts solely occurred during a colic episode. There are also no major distinctions in other factors, for example, personality test results for women who murdered their children do not significantly differ from those of women who murder adults (McKee, Shea, Mogy, & Holden, 2001).

Often, infanticide is attributed to a woman being in a postpartum state of extreme hormonal fluctuation. However, there have been surprisingly few studies in this area and those that exist largely show that hormone changes do not have a significant impact on a woman's mental health, despite long-standing assumptions to the contrary (Wisner & Stowe, 1997). For example, Harris' (1994) review reported that no association had been found between progesterone, estrogen, or cortisol and postpartum mood or psychosis. In 1997, Wisner and Stowe reported that studies to date had shown a lack of evidence that serum levels of gonadal hormones account for mood disturbance in women. C. Kumar et al. (2003) studied 29 women with clinical histories of mania, hypomania, or schizoaffective episodes during pregnancy. The women were given transdermal regimens of estrogen in either 200, 400, or 800 μg/day within 48 hrs of delivery, tapered by one half every 4 days, for total of 12 days. A neuroendocrine challenge test was used on the fourth day. Not one of the dosages was found to reduce the rate of relapse during a postpartum interval in these women. To date, there has not been a definitive study that showed conclusively that the hormonal changes associated with childbirth cause clinical-level mood changes.

The terms "baby blues," "postpartum depression," and "postpartum psychosis" are often used as if synonymous, despite their significant differences. The "blues" are the most common, occurring in between 25% and 85% of women. These mild symptoms of crying and irritability begin within a few days of childbirth and end by the 2nd week (Dobson & Sales, 2000).

Depression, a clinical diagnosis that meets the Diagnostic and Statistical Manual of Mental Disorders (Fourth Edition, Text Revision; DSM-IV-TR) standards, affects between 7% and 19% of women and generally lasts up to a few months (Campbell & Cohn, 1991). Despite being temporally associated with childbirth, multiple studies have shown that this is not a form of mental illness specific to postpartum women. For example, Cooper, Campbell, Day, Kennerly, and Bond (1988), in their examination of 483 postnatal women, found that the prevalence and incidence of non-psychotic psychiatric symptoms in the year after giving birth was no greater than that of non-puerperal women. Troutman and Cutrona (1990) compared 128 primiparous adolescents with 114 matched non-childbearing adolescents and found that there was no significant difference in the rate of depression in the two groups. This was further supported by Cox, Murray, and Chapman (1993), who compared 232 six-month postnatal women with a matched control group, who had not given birth within the previous year and found no significant difference in the prevalence of depression. Rather than being the result of childbearing, it appears to be an episode of clinical depression in women predisposed to this form of mental illness and who are likely to go on to experience further episodes separate from their puerperal status (R. Kumar & Robson, 1984). Factors such as personality, negative feelings toward the infant, and ambivalence over parenthood appear to be relevant factors to the onset of symptoms in these predisposed women. For example, in 1991, O'Hara, Schlechte, Lewis, and Varner compared 361 childbearing and non-childbearing women and found the predictors for depression in the childbearing women were previous episodes of depression and a vulnerability/stress interaction, while for non-childbearing women, the only predictor was the vulnerability/stress interaction. R. Kumar and Robson (1984) found that depressive symptoms during the first 3 months postnatally were associated with marital conflict and maternal ambivalence regarding the infant. Verkerk, Denollet, Van Heck, Van Son, and Pop (2005) in their study of 277 women both at 32 weeks gestation and 3, 6, and 12 months postpartum found that the combination of high neuroticism and high introversion was the predictor of clinical depression in the 1st year postpartum, even when they controlled for clinical levels of depression during the pregnancy. A history of depression was the only other predictor.

Psychosis occurs in postpartum women at a rate of less than 1 case per 1,000 births (Terp & Mortensen, 1998) and generally requires an inpatient hospitalization to stabilize the symptoms of hallucinations and delusions. Its onset appears to be within 2 weeks to 2 months of childbirth and to be related to an underlying predisposition to mania and bipolar disorder rather than being caused by childbirth or hormone changes (Hay, 2009). For example, Kendell, Chalmers, and Platz (1987) examined records in more than 54,000 births over a 12-year period, looking for psychiatric admissions within 90 days postpartum. They found 120 cases, most within the initial 30 days and with those who had a history of bipolar disorder being at highest risk. Similarly, Sit, Rothschild, and Wisner (2006), in their review of research from 1966 to 2005, found that the start of puerperal psychosis began within the first 4 weeks postpartum and was associated with bipolar disorder. C. Dean and Kendell (1981) compared the hospitalization lengths for postpartum and non-postpartum women with mania and found no difference in type of treatment or hospitalization length. Whalley, Roberts, Wentzel, and Wright (1982) compared the morbidity risks for postpartum and non-postpartum psychoses in relatives of women with histories of psychosis (either postpartum or not). They found that the risk for psychosis was the same in both groups of relatives. Furthermore, they also assessed the frequencies of human leukocyte antigen (HLA)-A, -B, and -C locus antigens, nine blood group antigens, and 10 red blood cell isoenzymes in those relatives.[3] In this study, those genetic markers did not distinguish puerperal from non-puerperal affective psychoses.

Long-term follow-up studies indicate that most women who have an episode of psychosis in the 2 to 3 months after giving birth will go on to have more episodes, regardless of future childbearing (Lewis & Bunce, 2003; Valdimarsdottir, Hultman, Harlow, Cnattingius, & Sparen, 2009). For example, Reich and Winokur (1970), in their study of 20 parous women with bipolar disorder and 29 of their female relatives who also had both children and an episodic affective disorder, found that 50% of the patients and 25% of their relatives went on to have future episodes. Robling, Paykel, Dunn, Abbott, and Katona (2000) followed 64 women who had been psychiatrically hospitalized within 6 months of giving birth. The women were followed for a mean of 23 years and 75% of them had at least one additional psychiatric episode, generally unrelated to childbirth; 37% of the women had more than three future episodes; and 29% had episodes during subsequent postpartum events. Van Gent and Verhoeven (1992) studied the use of lithium prophylaxis in pregnant women with histories of bipolar disorder and found that the majority of those who did not use medication had a relapse in symptoms and a majority of those who did use the medication did not have a relapse. Videbech and Gouliaev (1995) in their long-term follow-up study of 50 women with postpartum episode psychosis found that the relapse rate was 60%, 40% of the women remained at least partially disabled and unable to work at full capacity due to mental illness, and cases of exclusive puerperal relapse were rare (4%). As with depression, psychosis during a postpartum period is not a unique form of the illness (Tschinkel, Harris, Lenoury, & Healy, 2007; Whalley et al., 1982). Wisner et al. (in Spinelli, 2003) recommended that acute onset psychosis in postpartum should be viewed as bipolar episode until proven otherwise. The above studies strongly point to a genetic tendency in women who experience acute onset psychosis in postpartum periods.

Psychosis-related infanticide is extremely rare as the underlying illness itself is rare (Hatters-Friedman et al., 2005). Furthermore, infanticides related to psychosis tend to involve desired children rather than unwanted ones and immediate confessions rather than attempts to hide culpability (Lewis & Bunce, 2003). Psychotic infanticidal women are less likely to have prior involvement in CPS than non-psychotic infanticidal women.

Malingering of mental illness can arise in infanticide cases, as with other types of homicide. A review of Brazilian neonaticide cases over 95 years found an increase in the number of claimed amnesia cases after the enactment of a 1940 infanticide statute that emphasized the role of mental illness as a mitigating factor (Mendlowicz, Rapaport, Mecler, Golshan, & Moraes, 2002). This strongly suggests that many of these women were malingering an untestable symptom to avoid being held accountable for their actions.

In summary, while some women experience minor emotional disruptions following childbirth, psychosis and clinical levels of depression related to giving birth are rare. In both cases, however, the research to date indicates that these symptoms of mental illness are not unique forms caused by hormonal changes but rather are manifestations of preexisting mental illnesses. The majority of infanticides and neonaticides are not related to the woman's mental illness.

Language About the Victims

The perpetrators of infanticide often use language in a manner that deflects their responsibility and distances them from the event. Stanton and Simpson (2006) noted that, during interviews with women who had murdered their children, the women made statements such as "when my baby died," rather than "when I killed my baby." A similar manner of using language to diffuse the woman's responsibility is seen in many articles on infanticide. While studies may use the term "mother" to discuss the murderer of the infant, the infant is often termed "victim." In English, language is generally used in a dichotomous way, with either opposites or matched pairs, black/white, salt/pepper. It is usual to hear or read pairings of "mother/child" or "mother/infant," "victim/perpetrator" or "victim/murderer." The pair "mother" with "victim" suggests a reluctance to view women as murderers.

Several commentators have suggested that infanticide is caused by the woman's feeling "trapped" or "stressed" due to child care. However, parenting is not generally seen as such a stressful burden that homicide is a foreseeable outgrowth. Rather, in neonaticide and infanticide, the women have negative attitudes toward their infants and "do not wish to spend their physical, emotional and social energy raising them" (Palermo, 2002, p. 141). One of the most obvious factors involved in infanticide is the unwillingness of the perpetrator to view the victim as an autonomous person, with her own rights and identity. This can be seen in suicide notes in cases of infanticide followed by suicide by the perpetrator, where the perpetrator requests that they be buried in the same coffin or makes statements that she and the victim will be "together forever." Crimmins, Langley, Brownstein, and Spunt (1997) suggest that, to infanticidal women, the infant is symbolic of some other object, rather than an autonomous living person.

Infanticidal and Neonaticidal Means

Due to their small size and inability to defend themselves, the murder of an infant does not require either strength or skill. Therefore, smothering, strangling, suffocating, and drowning are all common methods of infanticide, although many other means are used including starving, burning, stabbing or cutting, shooting, exposure, gross assault, gassing, scalding, poisoning, and defenestration. Finkelhor and Ormrod (2001) suggest that women are more likely to use their hands as a weapon and less likely to use firearms, compared to men, but Lewis, Baranoski, Buchanan, and Benedek (1998; in Palermo, 2002) indicated that up to 25% of women who murdered their children use weapons.

Gender of Murdered Infants/Neonates

While infanticide of females appears to be higher than infanticide of males in India and China (Sahni et al., 2008), many studies report a higher rate of infanticide of male than female infants in industrialized western nations. Lester (1991) proposes that this discrepancy is related to the higher murder rate of males in general. The problem with this argument is that the majority of murders of adult men are perpetrated by other men (Fox & Zawitz, 2007; U.S. Dept. of Justice, 2006). It does not necessarily follow that, while most adult male murders are perpetrated by other males, infant male murders would be perpetrated by females. Women commit the majority of all infant murders, so why are they killing more male infants?

Marleau, Dube, and LeVeille (2004) suggest that more males than females are born and therefore a higher infanticide rate is simply an artefact of the higher availability of male infants. While the ratio of male to female live births is slowly moving toward 1:1 in industrialized nations, currently approximately 1–2% more boys are born than girls. Marleau et al. reported that, in their sample of 420 infanticide cases, 58.3% of the murdered infants were male. Beyer et al. (2008), Bropokman and Nolan (2006), Crimmins et al. (1997), and Hodgins and Dube (1995) all report a similar 5–6% higher rate of infanticide of males. Even accounting for the differences in live births, about 6% more male infants were murdered than female infants in industrialized nations. One possible explanation is that a male infant is more "other" to a woman than a female infant. Is the male infant somehow symbolic of the woman's male sex partner? At this point, the issue is unclear and will require more research.

Denial/Concealment

"Infant's body discovered in trash" (*LA Times*, February 9, 1996).

"Infant's body discovered near the 2-9 Dumpsters" (*Daily Trojan*, 10/11/05).

"The case of the frozen babies" (Schpoliansky & Childs, 2009).

"Three babies found in deep freeze in Germany's infanticide epidemic" (Boyes, 2008).

"Infant's body discovered at plant" (*NY Times*, May 4, 2006).

"Infant's body found inside plastic bag" (CBC News, April 2, 2009).

"Infant's body found in Onslow County garbage truck" (Capitol Broadcasting Company, October 27, 2008).

"Infant's body found in Erfurt freezer" (*The Local*, May 27, 2009).

"Dead newborn found at abandoned apartment in northern Japan" (*Bay Ledger News*, May 31, 2007).

As these news headlines show, the discovery of a newborn's corpse is an unfortunately frequent occurrence. In a study of neonaticides published in 1990, 64% of the newborn's corpses were discovered by accident in garbage cans or other refuse sites. None of the newborns could be matched to a missing person report, which indicated that the woman had intentionally concealed her pregnancy and abandoned the infant upon birth (Crittenden & Craig, 1990).

One of the hallmark factors in neonaticide compared to infanticide is the secrecy of the pregnancy and subsequent birth. As noted above, women who commit neonaticide are markedly different from women who commit infanticide. Neonaticidal women tend to be younger, emotionally immature, and do not desire to become a parent at the time of the homicide. Many of these women manage to conceal their pregnancies from their parents and others for the entire 9 months, although, according to Beyer et al. (2008), at least one other person was aware of the pregnancy in 83% of the cases.

The issue of whether the woman was aware of her pregnancy would be a relevant factor in a neonaticide trial. Given the multiple signs and symptoms of pregnancy, especially by the end of the 9th month, including very specific location of weight gain, amenorrhea, and fetal movement, it seems unlikely that most women could remain passively oblivious to pregnancy, without other factors contributing to the lack of awareness (Vallone & Hoffman, 2003). According to Beyer et al. (2008), in their sample of 40 neonaticidal women, over half had a history of previous pregnancies, giving them a clear frame of reference for their physical signs and precluding any logical claims that they were unaware of their pregnancies. Beyer et al. (2008) further notes that in the majority of cases the women went through labor and then murdered the infant within close proximity to others, without disturbing anyone, let alone calling for help. This behavior suggests intentional concealment. If a woman found herself experiencing labor and giving birth, somehow without ever having known she was pregnant, an expected reaction would be to call for help. Even if she mistakenly believed the baby was stillborn, calling for help would still be the expected behavior, not placing the infant in the trash as is done in these cases.

Is it ever possible to be totally unaware of a pregnancy until the point of labor? The research by Wessel, Endrika, and Buscher (2002), based on a Berlin population study, found that 1 in 2,455 cases reach labor without the woman being aware that she was pregnant. While it does occur, it is clearly rare. Furthermore, the women in this study were not accused of neonaticides and so their claims of obliviousness lack secondary gain.

Several studies use the term "denial" rather than concealment to discuss the unwanted pregnancy that results in neonaticides. They suggest that the woman denies her pregnant state to herself. Dulit (2000) posits that there are three types of "denial" seen in neonaticides cases. The first seems to be less a form of denial and more a simple desire: I hope I'm not pregnant. The second involves deliberate deception of others and seems best described by concealment. The third, "true" denial, is a function of actively pushing the facts away: "I can't think about this now." To deny something first requires an acknowledgment that the reality exists and then requires active inattention (Kohm & Liverman, 2002). Therefore, these are not cases of chronic mentally ill women who deny their pregnancies due to mental disturbance. Rather, these women are cognitively aware of their pregnancies but their behaviors and emotions do not match that awareness (Lee, Li, Kwong, & So, 2006). The woman makes a decision not to alter her behaviors (get a gynecological exam, pregnancy test, abortion, prenatal care), not to form any maternal prenatal attachments, and, ultimately, to murder the newborn infant.

Beier et al. (2006), after reviewing both forensic and obstetric cases of denial and concealment of pregnancy, found that these groups were identical and devised the term "pregnancy negation." They suggest that pregnancy negation might be viewed as a type of conversion or adjustment disorder; however, pregnancy negation lacks the egodystonic factor necessary to be a type of conversion disorder and does not fit the time frame necessary for an adjustment disorder.

Recidivism

In 1996, American Waneta Hoyte was convicted of the deaths of five of her infant children after she confessed that their deaths were not due to SIDS as previously thought (Busch, 1997). In 1999, American Marie Noe was convicted of the deaths of her eight infants, who had all previously been thought to have died due to SIDS (Begley & Underwood, 1998). In May 2003, Australian Kathleen Folbigg was found guilty of the death of her four infants. Their deaths had been believed to be due to SIDS but Folbigg detailed the murders in a diary that was later discovered (Glendinning, 2003). In 2005, German courts convicted Sabine Hilschenz of the death of eight infants, whose bodies she hid in flower containers at her home; she is suspected of the death of a ninth infant but the statute of limitations had expired for that case. In 2007, in the northern German town of Kiel, the corpse of an infant was found in the garbage; the DNA matched that of another infant corpse found in the trash in the area in 2006 (Fox News, 2007). In June, 2009, Frenchwoman Veronique Courjault was found guilty of the death of the two infants whose bodies were found in her freezer. She also confessed to a third infanticide of which the police had no knowledge (Schipoliansky & Childs, 2009). These women are all examples of repeated neonaticide.

What is the likelihood that a woman who murders her infant will go on to commit another act of violence toward an infant in the future? At this point, there are very limited data available on that topic. A group of 47 Argentine infanticidal women were followed for 237 days following discharge from prison and 11% engaged in maladaptive behavior during the follow-up period. However, as the follow-up period was less than 1 year and recidivism rates are usually reported in 5-year, 10-year, and 15-year periods, it remains unclear whether infanticidal women go on to commit more acts of violence.

Stanton and Simpson (2006), in their interviews of mentally incompetent infanticidal women, noted that these women did not express concerns about living with their illness upon release and did not reference valuing professional psychological help during their hospital stays. This suggests that these women underestimated their risk for relapse and therefore risk to others in the future.

For infanticides and neonaticides unrelated to maternal psychosis, the above cases indicate that the risk for recidivism clearly exists and should not be underestimated or based on the woman's lack of prior criminal arrests. The majority of women who kill their children had no prior arrest history (McKee, 2006), but as the cases of Hoyte, Noe, Folbigg, and Courjault show, a lack of previous arrests does not mean a lack of criminal or violent behavior.

Typologies

Resnick (1970) was the first to suggest that typologies could be developed to describe and categorize infanticidal events and introduced the concept of neonaticide as a separate category. Although the expression "typology" may not be completely relevant here, as such terminology and classification had not been established at this point, Resnick has still described classifications that could illustrate distinctions in the motivations for killing. These include

- killings for "altruistic" reasons;
- killing by an acutely psychotic woman;
- killing of an unwanted infant;
- accidental killing via severe child abuse;
- killing for revenge against another; and
- neonaticides.

For Resnick, altruistic reasons for killing a child could include ending the child's real or imagined suffering. For example, a mother who plans to kill herself may feel that her child would suffer if left behind in the world and so kills the child too.

From Resnick's initial list, multiple other categorization forms have been designed, of varying length, comprehensiveness, and utility (see Table 8.4).

Both Resnick's and D'Oban's typologies have been validated internationally and remain the best known, although McKee's typology appears to be highly useful for forensic purposes.

Table 8.4 Categorizations of Infanticidal/Neonaticidal Perpetrators

Author	Typology
Scott, 1973	• Child was unwanted • Mercy killing to relieve real suffering • Aggression due to gross mental pathology • Murder as a result of stimulus arising outside of victim • Murder as a result of stimulus arising from victim
D'Orban, 1979	• Battering women • Mentally ill women (neonaticide) • Retaliating women • Women who killed unwanted children • Mercy killing
Guileyardo, Prahlow, & Bernard, 1999	• Altruistic • Euthanasia • Acute psychosis • Postpartum mental disorder • Unwanted child • Unwanted pregnancy (infanticide) • Angry impulse • Spousal revenge • Sexual abuse • Munchausen's syndrome by proxy • Violent child • Negligence/neglect • Sadistic punishment • Drug/other abuse • Seizure • Innocent bystander
Bourget & Gagne, 2005	• Pathological filicide • Accidental filicide • Retaliating filicide (neonaticide) • Paternal filicide
McKee, 2006	• Detached mothers • Abusive/neglectful mothers • Psychotic/depressed mothers • Retaliatory mothers • Psychopathic mothers

There are problems with the current typologies, stemming from the subjectivity of the terms. As Mugavin (2005) points out, most typologies include a discussion of "violent" and "non-violent" methods to murder the infant. For example, he cites Resnick who compared infanticidal methods used by men to those used by women. The methods men used are described as "more active" while the methods women used are described as "passive." Is it possible to classify murder as non-violent or passive? It is unlikely that the experience of being slowly drowned or smothered seems non-violent to the victim.

"Altruism" as a descriptor of motivation for infanticide is also problematic. As Harder (1967; in Stanton, 2002) suggests, the idea that the murder of an infant was somehow altruistic probably comes from the continual myth that women are always loving mothers, even during murder. This view represents a social perspective rather than scientific objectivity and could, according to Lewis and Bunce (2003), have ramifications in court proceedings. Scott's terminology, "result of stimulus arising from victim" is also problematic, as it seems to imply that the infant was somehow culpable of his or her own murder.

Furthermore, as with any typology system, there is the problem of overlap; cases that are complex and do not fit neatly into any single category. For example, the American press recently published a story of a chronic schizophrenic woman who murdered her infant, following a fight with the baby's father. Should this be classified as due to mental illness or for revenge? For forensic purposes, a more functional classification system might be to divide between infanticides and neonaticides and between killings by psychotic and by not-psychotic women.

Legal Issues

In 2008, the Canadian courts tried a 27-year-old woman for the murder of her infant sons in 1998 and in 2002. Because the original crime occurred when she was a minor, she remained unnamed. Both deaths had previously been ruled as due to SIDS, but she later admitted to suffocating both infants intentionally. A psychiatrist testified that the woman "clearly overstated" psychotic symptoms and said "false reporting of serious symptoms raised pretty significant red flags." Some of the alleged symptoms, such as olfactory hallucinations and reading words backward, are not symptoms commonly attributable to psychosis but are the type of symptoms a layperson might imagine was a sign of severe mental illness. It was also revealed during the trial that she had read a book about postpartum depression, which could have been the source of the falsely reported symptoms (Tracey, 2008). Despite this information, she was found guilty on two counts of infanticide. This is markedly different from being found guilty of first-degree murder, as infanticide laws are predicated on the concept that childbirth causes a woman's mind to be "disturbed" and thereby diminishes her capacity to control her behavior.

While a minor subset of women who kill infants are psychotic at the time, the majority appear to have willfully murdered their unwanted offspring. However, courts seem reluctant to hold women accountable for their actions in these cases. Courts consistently give lighter sentences to women convicted of infanticide than men convicted of the same crime (D'Orban, 1979, in Drescher-Burke, Krall, & Penick, 2004; Wilczynski, 1997). It does not appear that anyone asks about the mental state of a man who murders an infant nor is there a question as to his disposition; infanticidal males are remanded to prison. Infanticidal women, however, are given a more complex and gender-focused disposition.

The United Kingdom currently uses the Infanticide Act of 1938, which reduced the death of a child up to the age of 1 year by his or her mother to manslaughter. Canada, Ireland, Australia, New Zealand, Brazil, Denmark, and Sweden use similar laws. These laws implicate childbirth and lactation as a ground for mental disturbance despite the lack of any scientific basis for this belief. The accused woman does not have to prove she was actually incapacitated by psychosis at the time of the infanticide, as her mental state is presumed to be disrupted by virtue of having given birth. Furthermore, the time covered is set arbitrarily at 1 year, despite the World Health Organization (WHO) defining the postpartum period as only 6 weeks long. Because the laws are erroneously based on antiquated suppositions about biology, they cannot extend to men, making the laws inherently biased. These laws largely disallow the idea that a woman may kill her infant for reasons other than mental illness. The result is that the majority of women in these nations who murder their infants are remanded to counseling and probation rather than jail (Stangle, 2008).

Are special laws necessary for infanticide cases? As Dobson and Sales (2000) state,

> The basic issue is whether or not the year following childbirth represents a special time, when psychological and biological forces interact to cause mental illness so severe that there should be an assumption that a woman should not be held fully responsible for an act of murder committed against her newly born child. (p. 1099)

As discussed above, the evidence is clear that, for the majority of women, the postnatal year is not biologically or psychologically dangerous; it is no different for postnatal women than for women who adopt infants. A very small subset of women who have underlying predisposition to psychosis or mania may have an episode of symptoms within this time frame, but the evidence strongly suggests that these women would have manifested their symptoms regardless of their postnatal state (Dobson & Sales, 2000; Sit, Rothschild, & Wisner, 2006). Childbirth and lactation, and the hormones involved in these events, do not cause women to become mentally ill. Therefore, in most cases, the preexisting statutes for dealing with mentally ill defendants should suffice rather than requiring a gender-specific law. For example, in the United States, the accused may use an insanity defense. This defense is estimated to be used successfully in less than 1 of 1,000 cases. However, in maternal filicide cases that use the insanity defense, the women are acquitted in up to 65% of the cases (Stangle, 2008). This strongly suggests that a separate infanticide law is unnecessary.

Are those women acquitted on an insanity plea actually insane? Alternatively, is the high acquittal rate more indicative of society's inability to conceptualize "mothers" as "killers"? D'Orban (in Lambie, 2001) reported that, despite being categorized as "mentally ill," only 26% of the infanticidal women in her study met the criteria for psychosis. Wilczynski and Morris (1993; in Rapaport, 2006) report that about half of the women who are convicted of infanticide are not suffering from any identifiable mental disorder. This strongly suggests that there is something else going on in the legal proceedings other than an assessment of the women's culpability.

Not all cases of infanticide even make it fully to a court hearing, again suggesting a gender issue. Putkonen et al. (2007) found that only 27% of all neonaticides in Finland between 1980 and 2000 were ever prosecuted and convicted. Wilczynski (1997), comparing British court dispositions of infanticide by men and women, found that less than 50% of the cases involving women were prosecuted compared to 90% of the men's cases. Women were more likely to be granted bail than men (50% vs. 0%), and, while the majority of men received incarceration dispositions (84.2%), the majority of women received treatment dispositions (87.2%). In the late 20th century, a British hospital covertly videotaped women attempting to smother their infants with either their hands or with pillows. Despite the video evidence, every woman denied attempting to smother the infant and only one received a custodial sentence (Adshead, Brooke, Samuels, Jenner, & Southall, 2000). Even when women were incarcerated, their sentence lengths averaged less than half the sentence lengths of men convicted of similar offenses. She cites similar results in Australia, Canada, America Denmark, and Sweden.

This aversion to holding women accountable for their actions, often termed chivalric justice, has significant negative implications for society. It suggests that our society values the lives of children far less than we value the lives of adults. In addition, it reinforces the stereotype that women are irrational beings under the control of their biology, unlike men.

Recently, there has been mention in the U.S. courts of a "neonaticide syndrome." In the U.S. court system, it is necessary to pass a Frye test to establish the admissibility of supposedly scientific evidence. To claim the existence of a "syndrome" mitigating an individual's responsibility in a homicide case, there would need to be evidence that the existence of this "syndrome" is generally accepted by a meaningful percentage of the scientific community. In this case, the supposed "neonaticide syndrome" is not likely to pass a Frye test for several reasons (Bourget et al., 2007). According to the DSM-IV-TR

(American Psychiatric Association, 2000), a syndrome is a "grouping of signs and symptoms based on their frequent co-occurrence that may suggest a common underlying pathogenesis, course, familial pattern or treatment selection." There is no research to demonstrate that neonaticidal women all experience the same symptoms or any symptoms at all. International research indicates that neonaticide is carried out by immature women who view their newborns as a threat to their lifestyles (Hatters-Friedman & Resnick, 2009; Resnick & Hatters-Friedman, 2003). These women often claim to feel passive or powerless, yet their actions of concealing the pregnancy, labor, and corpse strongly suggest they are able to function in their own best interest. In addition, a single behavior, especially murder, does not, by itself, define a mental illness. It has been suggested that a subset of these women "dissociate" during the murder of the infant, thereby claiming a lack of responsibility (Spinelli, 2001). However, as dissociation cannot be objectively measured in a manner that precludes malingering, such claims are best met with skepticism (Resnick & Hatters-Friedman, 2003).

Conclusion

At this point, research has given a fairly accurate understanding of who commits infanticide, why and by what means:

> Neonaticide is generally committed by women who often conceal the pregnancy, give birth away from a hospital and then suffocate, strangle or drown the unwanted newborn before hiding the corpse. Neonaticidal women generally do not have an incapacitating mental illness.

Infanticides are generally committed by more mature women who use a variety of violent methods, may premeditate the crime, and engage in infanticide for reasons ranging from retaliation against another adult to child abuse or neglect to removal of an unwanted child.

A subset of infanticides are committed by women during a psychotic episode. These women are likely to have significant psychiatric problems for the remainder of their life, regardless of their childbearing, as psychosis is not the result of maternal hormone fluctuations.

While this produces a greater understanding of the causes of neonaticide and infanticide, there is still a lack a strong conceptualization as to how to prevent these crimes from occurring. For those infanticide cases related to the woman's psychosis, it may be useful to educate gynecologists, obstetricians, and birthing unit staff regarding this issue and for them to educate partners and parents of women giving birth. Open conversations with women regarding their and their family's history of mental illness would assist in identifying some women with predispositions to psychosis. Public service messages by the media on this topic would also be useful to educate the public of warning signs and symptoms. For neonaticide, because the perpetrators largely avoid any obstetric contact, another avenue of education would be needed. Social networking sites are becoming a common communication venue for women younger than 30 years and could be useful in increasing awareness about concealed pregnancies and neonaticide risks among young people. School personnel could be educated regarding the signs of concealed pregnancy. It is also important that the courts and policy makers focus on the problems of chivalric justice as well as the possibility of serial or repeat neonaticide/infanticide.

Notes

1. Safe Haven laws in most U.S. states allow the mother of a newborn to leave the infant at a designated site (e.g., hospital) anonymously without risk of prosecution. These laws were designed to discourage infanticide but lack evidence of efficacy.

2. Andrea Yates is an American woman who drowned five young children and is currently committed to a psychiatric facility in Texas. The case was notorious due to Yates' history of multiple psychotic episodes following childbirth and continuing procreation despite repeated warnings from mental health professionals.

3. HLA-A, -B, and -C locus antigens are part of the major histocompatibility gene complex on chromosome 6 and are relevant to human immunity; antigens are molecular antibody generators; isoenzymes are catalysts.

References

Adshead, G., Brooke, D., Samuels, M., Jenner, S., & Southall, D. (2000). Maternal behaviors associated with smothering: A preliminary descriptive study. *Child Abuse & Neglect, 24,* 1175–1183.

American Psychiatric Association. (2000). *Diagnostic and Statistical Manual of Mental Disorders, Fourth Edition, Text Revision.* Washington, DC: American Psychiatric Association.

Angier, D. (2005). *Fountain mom gets 25 years for killing newborn.* 7/23/05 News Herald. Retrieved September 25, 2009, from http://www.newsherald.com/articles/class-76048-killing-bodycopy justified.html

Begley, S., & Underwood, A. (1998, August 17). Death of the innocents. *Newsweek.* Retrieved September 3, 2009, from http:// www./archive/5473-newsweek/august-1998.html

Beier, K., Wille, R., & Wessel, J. (2006). Denial of pregnancy as a reproductive dysfunction: a proposal for international classification systems. *Journal of Psychosomatic Research, 61,* 723–730.

Bennet, M., Jr., Hall, J., Frazier, L., Jr., Patel, N., Barker, L., & Shaw, K. (2006). Homicide of children aged 0–4 years, 2003–04: Results from the National Violent Death Reporting System. *Injury Prevention, 12,* ii39–ii43.

Beyer, K., McAuliffe-Mack, S., & Shelton, J. (2008). Investigative analysis of neonaticide: An exploratory study. *Criminal Justice and Behavior, 35,* 525–535.

Bourget, D., & Gagne, P. (2005). Paternal filicide in Québec. *Journal American Academy Psychiatry Law, 33,* 354–360.

Bourget, D., Grace, J., & Whitehurst, L. (2007). A review of maternal and paternal filicide. *Journal American Academy Psychiatry Law, 35,* 74–82.

Bropokman, F., & Nolan, J. (2006, July). The dark figure of infanticide in England and Wales: Complexities of diagnosis. *Journal of Interpersonal Violence, 21,* 869–888.

Busch, F. (1997). *A mother on trial.* Retrieved September 25, 2009, from www.nytimes.com/books/97/09/14/reviews/970914.14buscht.html

Campbell, S., & Cohn, J. (1991). Prevalence and correlates of postpartum depression in first time mothers. *Journal of Abnormal Psychology, 100,* 594–599.

The Canadian Press. (2008). *The Canadian press: Woman who put newborn in firepit gets conditional sentence 2/8/08.* CBC News 2008. Retrieved September 25, 2009, from http://www.cbc.ca/canada/new-brunswick/story/2008/02/08/morrow-trial.html

Cooper, P., Campbell, E., Day, A., Kennerly, H., & Bond, A. (1988). Nonpsychotic psychiatric disorder after childbirth; a prospective study of prevalence, incidence, course and nature. *British Journal of Psychiatry, 152,* 799–806.

Cox, J. L., Murray, D., & Chapman, G. (1993). A controlled study of the onset, duration and prevalence of postnatal depression. *British Journal of Psychiatry, 163,* 27–31.

Crimmins, S., Langley, S., Brownstein, H., & Spunt, B. (1997, February). Convicted women who have killed children: A self psychology perspective. *Journal of Interpersonal Violence, 12,* 49–69.

Crittenden, P., & Craig, S. (1990). Developmental trends in the nature of child homicide. *Journal Interpersonal Violence, 5,* 202–216.

Cummings, P., Theis, M., Mueller, B., & Rivara, F. (1994). Infant injury death in Washington State, 1981 through 1990. *Archives of Pediatrics & Adolescent Medicine, 148,* 1021–1026.

D'Orban, P. T. (1979). Women who kill their children. *The British Journal of Psychiatry, 134,* 560–571.

Dalley, M. (1997/2000). *The killing of Canadian children by a parent(s) or guardian(s): Characteristics and trends 1990–1993.* Royal Canadian Mounted Police January 1997 & 2000 Missing Children's Registry & National Police Services. Retrieved September 25, 2009, from http://www.rcmp-grc.gc.ca/omc-ned/resear-recher/kill-tuer-eng.pdf

Dean, C., & Kendell, R. E. (1981, August). The symptomatology of puerperal illnesses. *British Journal of Psychiatry, 139,* 28–33.

Dean, P. (2004). Child homicide and infanticide in New Zealand. *International Journal of Law and Psychiatry, 27,* 339–348.

Dobson, V., & Sales, B. (2000). The science of infanticide and mental illness. *Psychology, Public Policy and Law, 6,* 1098–1112.

Drescher-Burke, K., Krall, J., & Penick, A. (2004). *Discarded infants and neonaticides: A review of the literature.* Berkeley, CA: National Abandoned Infants Assistance Resource Center, University of California at Berkeley.

Dulit, E. (2000). Girls who deny a pregnancy. Girls who kill the neonate. *Adolescent Psychiatry, 25,* 219–235.

Emerick, S. J., Foster, L. R., & Campbell, D. T. (1986, April). Risk factors for traumatic infant death in Oregon 1973–1982. *Pediatrics, 77,* 518–522.

Finkelhor, D. (1997). The homicides of children and youth: a developmental perspective. In G. K. Kantor & J. L. Jasinski (Eds.), *Out of the darkness: Contemporary perspectives on family violence,* pp. 17–34. Thousand Oaks, CA: SAGE.

Finkelhor, D., & Ormrod, R. (2001). *Homicides of children and youth.* U.S. Department of Justice Office of Justice Programs Office of Juvenile Justice and Delinquency Prevention 10/2001. Retrieved September 25, 2009, from http://www.nexthorizon.unh.edu/ccrc/pdf/jvq/CV34.pdf

Fox, J., & Zawitz, M. (2007). US Department of Justice, Bureau of Justice Statistics Web Site. Retrieved July 28, 2009, from http://www.ojp.usdoj.gov/bjs/homicide/homtrnd.htm

Fox News. (2007). *Baby-drops introduced in Germany as infanticide cases spike.* 3/27/07. Retrieved September 25, 2009, from http://www.foxnews.com/story/0,2933,261588,00.html

Glendinning, L. *Inside the mind of a killer mother.* The Age. October 25, 2003. Retrieved September 3, 2009, from http://www.theage.com.au/articles/2003/10/24/1066974316572.html

Guileyardo, J., Prahlow, J., & Bernard, J. (1999). Filicide and filicide classification. *The American Journal of Forensic Medicine and Pathology, 20,* 286–292.

Haapasalo, J., & & Petäjä, S (1999). Mothers who killed or attempted to kill their child: Life circumstances, childhood abuse, and types of killing. *Violence and Victims, 14,* 219–239.

Harder, T. (1967). The psychopathology of infanticide. *Acta Psychiatr Scand, 43,* 196–245.

Harris, B. (1994). Biological and hormonal aspects of postpartum depressed mood. *British Journal of Psychiatry, 164,* 288–292.

Hatters-Friedman, S., Heneghan, A., & Rosenthal, M. (2007, March-April). Characteristics of women who deny or conceal pregnancy. *Psychosomatics, 48,* 117–122.

Hatters-Friedman, S., Horwitz, S., & Resnick, P. (2005). Child murder by mothers: A critical analysis of the current state of knowledge and a research agenda. *Am J Psychiatry, 162,* 1578–1587.

Hatters-Friedman, S., & Resnick, P. (2009). Neonaticide: Phenomenology and considerations for prevention. *International Journal of Law and Psychiatry, 32,* 43–47.

Hay, P. (2009). Post-partum psychosis: Which women are at highest risk. *PLoS Medicine, 6,* e1000027.

Herman-Giddens, M. E., Smith, M. J., Mittal, M. Carlson, J., & Butts, J. (2003). Newborns killed or left to die by a parent. *JAMA, 289,* 1425–1429.

Hodgins, S., & Dube, M. (1995). *Parents who kill their children: A cohort study.* Retrieved September 3, 2009, from http://homicideworking group.cos.ucf.edu/include/documents/hrwg95.pdf#page=151

Hundley, W. (2008). The Dallas morning news, 11/8/08 *Deana Schlosser, plano mom who cut off baby's arms, moving to output care.* Retrieved September 3, 2009, from http://www.dallasnews.com/sharedcontent/dws/dn/latestnews/stories/110808dnmetschlosser.18ca635cc.html

Jason, J., Carpenter, M., & Tyler, C. (1983). Underrecording of infant homicide in the United States. *Am J Public Health, 73,* 195–197.

Jennings, K., Ross, S., Popper, S., & Elmore, M. (1999). Thoughts of harming infants in depressed and nondepressed mothers. *Journal of Affective Disorders, 54,* 21–28.

Kauppi, A., Kumpulainen, K., Vanamo, T., Merikanto, J., & Karkola, K. (2008). Maternal depression and filicide: Case study of ten mothers. *Archives of Women's Mental Health, 11,* 201–206.

Kendell, R., Chalmers, J., & Platz, C. (1987). Epidemiology of puerperal psychosis. *British Journal of Psychiatry, 150,* 662–673.

Kohm, L., & Liverman, T. (2002). Prom mom killers: the impact of blame shift and distorted statistics on punishment for neonaticides. *William and Mary Journal of Women and the Law, 9,* 43–71.

Krischer, M., Stone, M., Sevecke, K., & Steinmeyer, E. (2007). Motives for maternal filicide: results from a study with female forensic patients. *International Journal of Law and Psychiatry, 30,* 191–200.

Kumar, C., McIvor, R., Davies, T., Brown, N., Papadopoulos, A., Wieck, A., & Marks, M. (2003). Estrogen administration does not reduce the rate of recurrence of affective psychosis after childbirth. *The Journal of Clinical Psychiatry, 64,* 112–118.

Kumar, R., & Robson, K. (1984). A prospective study of emotional disorders in childbearing women. *The British Journal of Psychiatry, 144,* 35–47.

La Nacion, 2005. *Condenaron a Tejerina a 14 anos de prision.* June 10, 2005. Retrieved September 24, 2009, from http://www.lanacion.com.ar/nota.asp?nota_id¼711765

Lambie, I. (2001). Mothers who kill: The crime of infanticide. *International Journal of Law and Psychiatry, 24,* 71–80.

Lee, A. C. W., Li, C. H., Kwong, N. S., & So, K. T. (2006). Neonaticide, newborn abandonment, and denial of pregnancy—Newborn victimisation associated with unwanted motherhood. *Hong Kong Medical Journal, 12,* 61–64.

Lester, D. (1991). Murdering babies. A cross-national study. *Social Psychiatry and Psychiatric Epidemiology, 26,* 83–85.

Levene, S., & Bacon, C. J. (2004). Sudden unexpected death and covert homicide in infancy. *Archives of Disease in Childhood, 89,* 443–447.

Levitzky, S., & Cooper, R. (2000). Infant colic syndrome: Maternal fantasies of aggression and infanticide. *Clinical Pediatrics, 39,* 395–399.

Lewis, C. F., Baranoski, M., Buchanan, J., & Benedek, E. (1998). Factors associated with weapon use in maternal filicide. *Journal of Forensic Sciences, 43,* 613–618.

Lewis, C. F., & Bunce, S. C. (2003). Filicidal mothers and the impact of psychosis on maternal filicide. *J Am Acad Psychiatry Law, 31,* 459–470.

Logan, M. (1995). Mothers who murder: A comparative study of filicide and neonaticide. *RCMP Gazette, 57,* 2–10.

Marks, M., & Kumar, R. (1993). Infanticide in England and Wales. *Medical Science and the Law, 33,* 320–339.

Marks, M., & Kumar, R. (1996). Infanticide in Scotland. *Medical Science and the Law, 36,* 299–305.

Marleau, J., Dube, M., & LeVeille, S. (2004). Neonaticidal mothers: Are more boys killed? *Medicine, Science and the Law, 44,* 311–316.

McKee, G. (2006). *Why mothers kill: A forensic psychologist's casebook.* UK: Oxford University.

McKee, G., Shea, S., Mogy, R., & Holden, C. (2001). MMPI 2 profiles of filicide, mariticidal and homicidal women. *Journal of Clinical Psychology, 57,* 367–374.

Mendlowicz, M., Rapaport, M., Mecler, K., Golshan, S., & Moraes, T. (1998). Case controls study on the socio-demographic characteristics of 53 neonaticidal mothers. *International Journal of Law and Psychiatry, 21,* 209–219.

Mendlowicz MV, da Silva, Filho JF, Gekker M, de Moraes TM, Rapaport MH, Jean-Louis F. (2000). Mothers murdering their newborns in the hospital. *Gen Hosp Psychiatry.* Jan-Feb;22(1):53–5.

Mendlowicz, M., Rappaport, M., Fontenelle, L., Jean-Louis, G., & De Moraes, T. (2002, March). Amnesia and neonaticides. *American Journal of Psychiatry, 159,* 111–113.

Meyer, C., & Oberman, M. (2001). *Mothers who kill their children: understanding the acts of moms from Susan Smith to the "prom mom."* New York: NYU Press.

Mugavin, M. (2005). A metasynthesis of filicide classification systems: psychosocial and psychodynamic issues in women who kill their children. *Journal of Forensic Nursing, 1,* 65–72.

Mulryan, N., Gibbons, P., & O'Connors, I. (2002). Infanticide and child murder—Admissions to the Central Mental Hospital 18502000. *Ir J Psych Med, 19,* 8–12.

O'Hara, M., Schlechte, J., Lewis, D., & Varner, M. (1991, February). Controlled prospective study of postpartum mood disorders:

Psychological, environmental and hormonal variables. *Journal of Abnormal Psychology, 100,* 63–73.

Overpeck, M., Brenner, R., Trumble, A., Trifiletti, L., & Berendes, H. (1998, October). Risk factors for infant homicide in the United States. *New England Journal of Medicine, 339,* 1211–1216.

Palermo, G. (2002). Murderous parents. *International Journal of Offender Therapy and Comparative Criminology, 46,* 123–143.

Paulozzi, M., & Sells, M. (2002, March 8). Variations in homicide risk during Infancy: US 1989–1998. *CDC Morbidity and Mortality Weekly Report, 51,* 187–189.

Putkonen, H., Collander, J., Weizmann-Henelius, G., & Eronen, M. (2007). Legal outcomes of all suspected neonaticides in Finland 1980-2000. *International Journal of Law and Psychiatry, 30,* 248–254.

Rapaport, E. (2006). Mad women and desperate girls: Infanticide and child murder in law and myth. *33 Fordham Urb. L.J. 527* (2005–2006).

Reich, T., & Winokur, G. (1970, July). Postpartum psychosis in patients with manic depressive disease. *The Journal of Nervous and Mental Disease, 151,* 60–68.

Resnick, P. (1970, April). Murder of the newborn: A psychiatric review of neonaticide. *Am J Psychiatry, 126,* 1414–1420.

Resnick, P., & Hatters-Friedman, S. (2003, August). Book review: Spinelli's infanticide. *Psychiatric Services, 54,* 1172.

Robling, S., Paykel, E., Dunn, V., Abbott, R., & Katona, C. (2000). Long-term outcome of severe puerperal psychiatric illness: A 23 year follow-up study. *Psychological Medicine, 30,* 1263–1271.

Rouge-Maillart, C., Jousset, N., Gaudin, A., Bouju, B., & Penneau, M. (2005). Women who kill their children. *The American Journal of Forensic Medicine and Pathology, 26,* 320–326.

Sahni, M., Verma, N., Narula, R., Varghese, R., Sreenivas, V., & Puliyel, J. M. (2008). Missing girls in India: Infanticide, feticide and made-to-order pregnancies? Insights from hospital-based sex-ratio-at-birth over the last century. *PLoS ONE, 3:e2224.* Published online May 21, 2008.

Saunders, E. (1989). Neonaticides following "secret" pregnancies: seven case reports. *Public Health Reports, 104,* 368–372.

Schipoliansky, C., & Childs, D. (2009). *When parents kill their kids (the case of the frozen babies).* ABC News. June 18, 2009. Retrieved September 3, 2009, http://abcnews.go.com/Health/ MindMood News/story?id=7864712&page=1

Schmidt, P., Grab, H., & Madea, B. (1996). Child homicides in Cologne (1985-1994). *Forensic Science International, 79,* 131–144.

Scott, P. D. (1973). Parents who kill their children. *Medicine, Science and the Law, 13,* 120–126.

Siegel, C., Graves, P., Maloney, K., Norris, J., Calogne, B., & Lezotte, D. (1996). Mortality from intentional and unintentional injury among infants of young mothers in Colorado 1986–1992. *Archives of Pediatrics & Adolescent Medicine, 150,* 1077–1083.

Silverman, R., & Kennedy, L. (1988, Summer). Women who kill their children. *Violence & Victims, 3,* 113–127.

Sit, D., Rothschild, A. J., & Wisner, K. L. (2006). A review of postpartum psychosis. *Journal of Women's Health, 15,* 352–368.

Spiegel. (2008). *German politician blames communism for child killings.* February 25, 2008. Retrieved September 3, 2009, from http://www.spiegel.de/international/germany/0,1518,537577,00.html

Spinelli, M. (2001). A systematic investigation of 16 cases of neonaticide. *Am J Psychiatry, 158,* 811–813.

Spinelli, M. (2003). *Infanticide: Psychosocial and legal perspectives on mothers who kill.* Washington, DC: American Psychiatric Publishing Inc.

Stangle, H. (2008). Murderous Madonna: Femininity, violence, and the myth of postpartum mental disorder in cases of maternal infanticide and filicide. *William & Mary Law Review, 50,* 699.

Stanton, J., & Simpson, A. (2002). Filicide: a review. *International Journal of Law and Psychiatry, 25,* 1–14.

Stanton, J., & Simpson, A. (2006). The aftermath: Aspects of recovery described by perpetrators of maternal filicide committed in the context of severe mental illness. *Behavioral Sciences and the Law, 24,* 103–112.

Taguchi, H. (2007). Maternal filicide in Japan: Analyses of 96 cases and future directions for prevention. *Psychiatria et Neurology Japonica, 109,* 110–127.

Terp, I., & Mortensen, P. (1998). Postpartum psychosis: clinical diagnoses and relative risk of admission after parturition. *The British Journal of Psychiatry, 172,* 521–526.

Tracey, S. (2008). *Tears, fury follow infanticide verdict Guelph Mercury.* September 12, 2008. Retrieved September 3, 2009, from http://www.canadiancrc.com/Newspaper_Articles/Guelph_Mercury_Tears_fury_follow_infanticide_verdict_12SEP08.aspx

Troutman, B., & Cutrona, C. (1990). Nonpsychotic postpartum depression among adolescent mothers. *Journal of Abnormal Psychology, 99,* 467–474.

Tschinkel, S., Harris, M., Lenoury, J., & &Healy, D. (2007). Postpartum psychosis: two cohorts compared 1875–1924 and 1994–2005. *Psychological Medicine, 37,* 529–536.

U.S. Dept of Justice, Bureau of Justice Statistics. (2006). *Homicide trends in the U.S.* Retrieved July 27, 2009, from http://www.ojp. usdoj.gov/bjs/homicide/gender.htm

Valdimarsdottir, U., Hultman, C., Harlow, B., Cnattingius, S., & Sparen, P. (2009). *Psychotic illness in first time mothers with no previous psychiatric hospitalizations: A population based study. PLoS Medicine.* February 10, 2009. Retrieved May 28, 2009, from http://www.plosmedicine.org/article/info%3Adoi%2F10.1371%2Fjournal.pmed.1000013

Vallone, D. C., & Hoffman, L. M. (2003) Preventing the Tragedy of Neonaticide. *Holistic Nursing Practice:* September/October, 17 Issue 5, 223–230.

Van Gent, E., & Verhoeven, W. (1992). Bipolar illness, lithium prophylaxis and pregnancy. *Pharmacopsychiatry, 25,* 187–191.

Vanamo, T., Kauppi, A., Karkola, K., Merikanto, J., & Räsänen, E. (2001). Intra-familial child homicide in Finland 1970–1994: incidence, causes of death and demographic characteristics. *Forensic Science International, 117,* 199–204.

Verkerk, G., Denollet, J., Van Heck, G., Van Son, M., & Pop, V. (2005). Personality factors as determinants of depression in postpartum women: A prospective 1 year follow-up study. *Psychosomatic Medicine, 67,* 632–637.

Videbech, P., & Gouliaev, G. (1995). First admission with puerperal psychosis: 7–14 years of follow-up. *Acta Psychiatrica Scandinavica, 91,* 167–173.

Wessel, J., Endrika, J., & Buscher, U. (2002). Frequency of denial of pregnancy: results and epidemiological significance of a 1-year prospective study in Berlin. *Acta Obstetricia et Gynecologica Scandinavica, 81,* 1021–1027.

Whalley, L., Roberts, D., Wentzel, J., & Wright, A. (1982). Genetic factors in puerperal affective psychoses. *Acta Psychiatrica Scandinavica, 65,* 180–188.

Wilczynski, A. (1997). Mad or bad? Child killers, gender and the courts. *British Journal of Criminology, 37,* 419–436.

Wilczynski, A., & Morris, A. (1993). Parents who kill their children. *Criminal Law Review, 31,* in Rapaport, E. (2006). Mad women and desperate girls: Infanticide and child murder in law and myth. *Fordham Urban Law Journal, 33,* 527.

Wisner, K., & Stowe, Z. (1997). Psychobiology of postpartum mood disorders. *Seminars in Reproductive Medicine, 15,* 77–89.

Newspaper Quotes

Boyes, R. (2008). Three babies found in deep freeze in Germany's infanticide epidemic. Retrieved September 3, 2009, from http://www.timesonline.co.uk/tol/news/world/europe/article3874969.ece

Dead newborn found at abandoned apartment in northern Japan. (May 31, 2007). Bay Ledger News. Retrieved September 3, 2009, from http://www.blnz.com/news/2007/05/31/Dead_newborn_found_abandoned_apartment_apan.htm

Holl, J. (2006, May 4). Infant's body discovered at plant. NY Times. Retrieved September 3, 2009, from http://query.nytimes.com/gst/fullpage.html?res=9801E4D91E3FF937A35756C0A9609C8B63

Infant's body discovered in trash. (1996, February 9). LA Times. Retrieved September 3, 2009, from articles.latimes.com/1996-0209/local/me-34143_1_trash-truck.

Infant's body found in Erfurt freezer. (May 27, 2009). The Local. Retrieved September 3, 2009, from http://www.thelocal.de/ national/20090527-19545.html

Infant's body found in Onslow County garbage truck. (2008, October 27). Capitol Broadcasting Company. Retrieved September 3, 2009, from http://www.wral.com/news/news_briefs/story/3826815/

Infant's body found inside plastic bag. (2009, April 2). CBC News. Retrieved September 3, 2009, from http://www.cbc.ca/canada/british-columbia/story/2009/04/02/bc-vancouver-charles-dead-babyfound.html

Schpoliansky, C., & Childs, D. (2009). The case of the frozen babies. Retrieved September 3, 2009, from http://abcnews.go.com/Health/MindMoodNews/story?id=7864712&page=1

DISCUSSION QUESTIONS

1. What are the differences between neonaticidal and infanticidal women?
2. How can "safe haven" laws prevent acts of neonaticide? What do these policies lack in their efforts to protect children from harm?
3. How are cases of neonaticide and infanticide handled by the legal system? How should the system respond in these cases?

READING

This section uses the voices of women engaged in street-level prostitution to illuminate some of the concepts that you learned about in the section introduction. Here, you'll read details about the motivations, risks, protective strategies, and experiences of street prostitution.

SOURCE: Williamson, C., and Folaron, G. (2003). Understanding the experiences of street level prostitutes. *Qualitative Social Work, 2*(3), 271–287.

Understanding the Experiences of Street Level Prostitutes

Celia Williamson and Gail Folaron

The Career of Prostitution

Four studies provide an emic view of prostitution and represent the best information on understanding street prostitution over time. First, Davis's Drift theory (1971) outlines a process by which women essentially drift into prostitution and over time professionalize into the role. Determinants of entrance involve poor interactions with conventional society, family instability, a desire for new experiences, and having idle time. Silbert and Pines (1982) add financial reasons to the list of primary motivators for entrance into street work. Other studies concur (Benjamin and Masters, 1964; James, 1978).

After entrance, women enter a transitional phase of moving in and out of conventional and deviant society. Over time, conventional ties become weaker and superficial while deviant ties become increasingly stronger. Once the street prostitute has internalized the identity as prostitute, she cannot see herself living the conventional life (Davis, 1971).

Second, a report released by the National Center for Missing and Exploited Children (1992) identifies common elements in the decision to exit street work. According to their findings, coercion and brutality from pimps often occurred and young women became trapped in a life of street prostitution as juveniles. Over time, a growing concern about incarceration, becoming less marketable, health issues, the realization of little financial security, and an increased dislike of themselves and their customers cause some to exit prostitution.

Information on life after prostitution is scant. Mansson and Hedlin (1998) describe life after prostitution for ex-prostitutes in Sweden. They identify four primary challenges women face when leaving prostitution, including emotional work to understand past experiences, struggles with shame, difficulty maintaining intimate and close relationships, and coping with a financially marginal living situation.

Little information is available on the full continuum of street prostitution as a lifestyle. Mathews (1986) provides us with the most complete analysis of street prostitution from entrance to exit using his 'social effects model,' for understanding needs, skills, and values of the women involved. He concluded that economic determinants as well as psychological and status needs were important in the decision to enter street prostitution. These same key concepts are used to explain maintenance in the prostitution life from 'entrance' to 'entrenchment.' Common factors of young women who exited prostitution included disillusionment with the lifestyle, receiving support from outside prostitution, escaping the life and acquiring legitimate work or training, and unlearning the values and patterns of the lifestyle.

These studies provide the conceptual framework of what constitutes the continuum of prostitution from entrance to exit. The congruence of many of the earlier process studies with one another and to the findings of this study lends credence to the belief that a basic social process of street prostitution exists and, over time remains the same. In building on this knowledge, emphasis is placed on the effects of three common risk factors, including violence (Miller, 1993; Williamson and Folaron, 2001), drugs (Ratner, 1993), and the deterioration of emotional and physical health (Farley and Hotaling, 1995; Vanwesenbeeck, 1994). These risks must be considered by helping professionals in their response to prostitutes.

Data Collection and Analysis

Through a process of purposive sampling, 21 women, aged 18–35 years old, participated in in-depth, face to face interviews lasting approximately two hours. The researcher had previously worked as a social worker in a community neighborhood center in the center of the city where she built relationships with area street informants and social service outreach workers who introduced her to the study participants. The interviews were conducted at places identified by each respondent. Each interview was audio recorded and transcribed verbatim. Informed consent was obtained prior to interview. Follow-up interviews were necessary in some cases to gather additional data and clarify responses. Study participants were not compensated for their interviews.

The sample included 13 white Appalachian women, 7 African-Americans, and 1 Hispanic woman. The time spent in prostitution ranged from 3 months to 13 years. None of the participants were currently involved in street prostitution activities. The exits from prostitution ranged from six days to five years. In all, 6 of the women previously had pimps and 15 worked without a pimp. Thirteen women entered prostitution before any drug addiction occurred, and eight entered already addicted to crack cocaine. All of the women lived on low-income budgets prior to entrance into prostitution.

The overriding study question was 'What has been your experience in prostitution? Tell me stories about what you remember happening. Include your thoughts, feelings, and perceptions as you remember them. Continue to describe each experience until you feel it is complete.' Open-ended, guided questions were also used representing broad categories such as participation in drug use, customer-related experiences, HIV risks and protection, emotional health, violence, roles and rules surrounding prostitution, and the process around how each occurred.

Data analysis included a process of substantive and theoretical coding. Substantive coding consists of 'open coding' or 'coding the data in every way possible' (Glaser, 1978: 56). This involved coding and categorizing the data line by line. Interviews and data analysis occurred simultaneously in a 'constant comparative analysis.' Constant comparative analysis is a 'process of taking information from data collection and comparing it to emerging categories' (Creswell, 1994: 57).

Once the basic social process was uncovered, subsequent interviews consisted of selective coding for the basic social process of street prostitution. The coding was then 'limited to those variables that relate[d] to the core variable in sufficiently significant ways' (Glaser, 1978: 61). The core variable or basic social process became a guide to further data collection. The focus was then placed on discovering the causes, consequences, contingencies, contexts, covariances, and conditions of the basic social process. This approach was derived from Glaser's grounded theory method (Glaser, 1978).

Findings

Socialization into the prostitution lifestyle requires commitment as women move through a series of phases. Each phase includes benefits, risks and losses. As women get caught up in the lifestyle, the benefits and risks often become blurred. Exiting the lifestyle requires personal determination and access to social and economic supports from outside the lifestyle. The following information lays out the phases of the lifestyle along with the motivations, benefits, risks, and losses inherent in each phase.

Enticement Into the Lifestyle

As women make the decision to enter street prostitution, two preconditions are necessary. First, they must allow themselves to be enticed by the prospects of substantial financial gain. Second, they must learn to shed any moral objections to prostitution work. The motivation to enter street level sex work is rooted in the conscious desire to want something better for themselves and/or their families. While the actual event of sex-for-money marks formal 'entrance' into prostitution, the social process of entrance into street work begins with a period of enticement. The most pervasive form of enticement is financial gain. Women describe having few alternatives as financially promising as prostitution. For Cara, selling sex seemed to be her only alternative: 'I was broke, living on the streets wondering how I was gonna eat.' After seeing other young women out on the streets Tina, who was 14 years old and living in poverty, decided to give prostitution a try.

> I seen how fast their money was and I seen how they were spending it. And at first I'm saying, 'damn, I want some money, man' I want to look good like all these other kids around here.' . . . I just figured I'm gonna go out here and just try this because it's one of the things I just seem to be able to do. (Tina)

Once women come to the conclusion that only sex work will provide enough needed money, they rationalize any conflicting moral issues with sex work. Helping to solidify these rationalizations are friends who introduce them into the life. They report meeting and socializing with those involved in the prostitution lifestyle. Cara was enticed by the amount of money that could be made working the streets:

> This girl asked me if I would walk with her because she was scared to walk by herself. And . . . I said, 'well what do you mean walk with you? If somebody tries to hurt you, they'll hurt me too.' She said, 'no, you just watch the type of car I get in and try to get a look at the license plate number in case I don't come back, you know who I went with. You know, to where I can identify a car. . . . So then she was gone like 5 minutes and back and had 50 dollars and I said, 'damn!' (Cara)

Overcoming Barriers to Prostitution: Meeting Financial Needs

Upon entrance, prostitution is not viewed as a negative, but is instead perceived as a survival strategy to help alleviate financial burdens. Women perceived obstacles in the formal economy that presented barriers to financially sustaining themselves and a family. Women who entered under legal age reported being too young to legally work in conventional employment settings and sought out alternative avenues for making money. Tina explains, 'Me being a kid and everything, I couldn't go out and get a job because of my age.'

Emotional burdens from a stressful home life, including abuse, neglect, and poverty motivated some women to enter prostitution. Eight women in the study came from homes where they were sexually abused, seven left alcoholic homes, two had a family history of prostitution with older sisters serving as role models, two were physically abused and one came from a home where a parent suffered from a severe mental illness and was both neglectful and physically abusive. All of the women lived in poverty. Those who had come from abusive homes or were homeless as a result of prior abuse saw prostitution as an improvement over their present circumstances. When Elsie turned 15 years old, prostitution seemed a viable option:

> My step-dad was raping me and my mom was [neglecting] us . . . passing us off to her friends so she could go out. [They] never spent no time with us. I've been in and out of foster care my whole life. My real dad's in prison, so I had a lot of problems going on. . . . I was curious. I seen other girls doing it . . . and they had some things I didn't have and I asked them how they got it and they would tell me, 'you know, just go out and do this . . ., and you can make this much money'. . . . And you know, they worked for pimps and stuff. . . . And as long as they're making money, the pimp takes care of them. So I was like, 'oh that sounds fun.' At first I thought it was fun. (Elsie)

Learning the Lifestyle: Feeling Powerful

Prostitution is often referred to as 'the game' or 'the lifestyle'. In street level prostitution the term 'game' is often used by pimps to refer to the prostitution business. The term 'lifestyle,' is often used by the prostitutes themselves who must learn to both survive and thrive within the world of street level prostitution.

Learning the lifestyle requires adapting to a new environment and learning the rules of behavior for success and survival. In this learning phase, the rewarding aspects of prostitution are prominent and women focus on the benefits of the lifestyle. Prostitution life is fast paced. There is always something to do, whether it is working or spending the money just made. Life is no longer mere survival; it becomes fascinating.

> I was hanging out at bars and getting quick and easy money and spending it quick as easy. I was staying at hotels, like when you got credit cards, taking cabs every damn where. There would be times when I would go get me a whole bunch of money and I would go shopping. There would be times when I would make three, four hundred dollars and go shopping, get my hair done. I liked taking my little sister shopping. (Chris)

Women reported a new sense of financial security. Even those involved with a pimp believe their contributions make a significant difference between success and failure in the lives of the pimp and themselves. With financial means to support themselves, women believe in their ability to control and influence activities in their lives in a way they previously had not been able. As a result, they reported a sense of accomplishment and mastery over their world. It is empowering.

In this learning phase of the prostitution lifestyle, the women begin to incorporate new habits and new ways of thinking. They find a way to silence moral restraints and begin to value exploitation. The rule on the street is, 'don't hate the player, hate the game.' With that sentiment, women are taught to take it in stride and learn from their mistakes.

> Oh, she tricked me. The game out here is like, 'you give me a girl and I'll give you this,' you know what I'm saying? So she said, 'hey Nina, I want you to meet this guy, you know, he's really crazy and cool and everything. He'll take you to these real slick places' and all this.... I'm like 'well cool,' you know. He pulls up in this gold Cadillac, you know, and we're getting in. I'm sittin in the back, you know, they up front talking all under their breath and stuff you know, talking to each other. And the next thing you know he's reaching and handing her something and she jumps out. And he says, 'jump up front sweety.' You know, so I jump up front. I'm talkin to him and the next thing you know, I'm out here . . . she took and actually sold me to some one and I didn't know it. You know, I'm young in the game, and she sells me to this man for an 8-ball of crack cocaine. And you know . . . the next thing I know, I'm down on Broadway clockin' $400 a day for him. (Nina)

Protective Strategies and Meeting the Demands of Daily Hassles

At this phase of the prostitute lifestyle, daily hassles are the most common source of stress. These include interactions with the police, efforts to anticipate and prevent customer-related violence, and the stress associated with the stigma associated with prostitution. It becomes imperative that the prostitute learn strategies to protect herself to remain productive in the game. Protective strategies are targeted responses to potentially negative situations.

Beyond settling on a price, getting the money, and providing the agreed upon services, women who use strategies to assess for the police believe they reduce their chances for arrest.

> I ask them 'are you police?' I ask them to pull it out and I ask them to touch me. If you have a suspicion that it's a police officer you say, 'well, that depends'. You don't never say yes or no answers. (Wiley)

Protective strategies used to screen out potentially dangerous dates consist of ritualistic behavior and intuitive assessment of customers.

> You don't want to get hit on, you don't want to get beat up, you don't want them to take the

> money back. . . . But you gotta learn to be smart about the shit. You stay within the vicinity of where you're picked up. If you go further . . . that's a chance you take. (Cara)

> If you don't feel right, don't get in the car. . . . Listen to their conversation. Does it seem right? Watch their hands, they might try to go for a weapon . . . you can even check under the seats. I've done it a million times. (Carol)

> At first I'd let them do most the talking. Feel them out. Find out where their head was, what they were interested in. If I felt that things weren't right, I wouldn't do nothing. I'd get out of the car. (Monica)

These strategies may also include carrying a small weapon while working. Nina described carrying a razor blade for protection:

> A girl I knew taught me how to flip a blade in my mouth. . . . She showed me how to give a blow job without cutting the man, without cutting myself, and when I need it, just flip my tongue and it was there to do my thing. (Nina)

Women typically fared worse than their assailants, however, even with the possession and use of weapons.

> Me and my friend went to this motel room to date these two guys. My friend brought a gun because she was scared that something might happen. . . . My friend ended up getting killed with her own gun by one of the dates. . . . I was the only one that walked out of that hotel room that night that wasn't shot. She shot both the customers before one of them got a hold to the gun and shot her. (Brenda)

Condoms are commonly used as protection from pregnancy, HIV, and other sexually transmitted diseases. Money to purchase condoms and gaining male consent to wear them are disease protection barriers women face. Women were creative in overcoming male resistance regarding condom compliance. For example, some women hold condoms in the side of their cheek while negotiating for the date, slip it on during oral sex and slide it off with her hand, the male never being the wiser. Intercourse requires even more skill and women meet the challenge.

Condoms are also used to inhibit emotional intimacy. Emotionally, the condom means the difference between dissociation and being emotionally present. With lovers, women report being emotionally present.

> when you're doing it like 3,4,5,7, times a night, you kinda just go somewhere else . . . you know, you're mind ain't there. I mean it's happening to you physically . . . and I'm not saying I have two personalities or something like that, but you kinda just like . . . your body's there, but . . . your mind is at Disney World buying T-shirts. You know . . . but with someone you love and care about you're there physically . . . emotionally. . . . So it's different. (Debbie)

Living the Lifestyle: Trusting in the Game

The amount of money that can be acquired is limited only by time and effort. Money is tangible proof of accomplishments. It can be personally satisfying and a source of positive reinforcement to continue. During this phase, women report being addicted to the lifestyle.

> A lot of the lifestyle is money. The money is real hard to give up. You can make a lot of money, but you don't keep it, but it's the idea you're making all this money . . . and you always have access to money. (Debbie)

> And you can spend it on what you want and get more. (Nina)

> Once you know that, it's addicting and hard to stop. (Debbie)
>
> . . . But it's also about being out late . . . traveling and having fun . . . and being with who you wanna be with . . . and living like you rich . . . like you got credit cards. (Carol)

The most prominent feature of this phase is the increased time spent in prostitution. Prostitutes in this phase distance themselves from conventional connections such as school, church, and other political and social institutions. Time is spent working and socializing with others in the lifestyle. Three distinct activities result from immersion into the lifestyle. First, increased socialization with underground social networks encourages increased drug use. Second, women experience a broader range of customer encounters. Third, because of increased customer contact, women experience more customer-related violence.

Facing Sterner Realities: Lessons About Chance and Skill

Although the prostitution lifestyle is valued and enjoyed, women in this phase are spending more time working and therefore place themselves at higher risk for violence. As women are repeatedly victimized, the work is no longer seen as 'easy money.'

> There's no way. There is no safe way. No matter what car you get in or what truck, you don't know if that guys gonna beat you up and leave you for dead. There is no life in it. (Monica)

Police protection is viewed as the luck of the draw. Because women have little means of protection outside of their own protective strategies, an element of 'chance' is embedded in any customer prostitute exchange. 'Chance' is the uncontrollable and unpredictable probability that an encounter with a customer may turn violent and even deadly. 'Skill' is the knowledge and use of protective strategies. In street level prostitution, there is a substantial amount of uncertainty regarding customers that represents skill and chance conditions. During the learning period, these 'chance conditions' seemed minimal, something that could be minimized with more skill training, experience and effort. With extended time in the field and repeated assaults, the element of 'chance' has to be acknowledged as prostitutes live the lifestyle.

> I was walking outside of a bar and I was drunk and high. . . . I told [this guy] I had a place where we could go date. And he took me over to a truck terminal. . . . He ripped my clothes off. He beat me. He threw me out for dead and drove off. A truck came in and that's what made him stop. . . . I think he thought he killed me. I ended up in intensive care. (Monica)

Skill and chance conditions are strategies that may be effective in one instance but not in another. The extent to which a strategy is effective depends on a number of factors, including the customer's motivation and intent for an encounter and the worker's judgments about the customer's intent. The customer's true motivation is unknown until after an exchange encounter has occurred. Therefore, each exchange encounter involves the skill of the sex worker both to 'read' her date and make critical judgments regarding the customer's intent, and to accept the element of chance from not knowing the customer's true intent. A mismatch between a worker's appraisal and the customer's true intent may result in threats, harm, assault, or murder. Because a worker can never, with all certainty, know a customer's intentions, a level of vulnerability is always present.

> It's real dangerous no matter how you look at it. You don't know if you're gonna make it home alive or not. I seen a girl die in this life. Thrown out a truck and ran over. I watched it happen. (Monica)

Up to this point, women appeared to focus on active, problem-focused protective strategies. When women have not fared well in violent encounters with men, their responses have been to shift from using problem-focused protective strategies to using predominately emotion-focused strategies. Depression was a universal occurrence among participants by this phase. In an attempt to respond to the emotional pain associated with depression, drug use also shifted from recreational use to functional use. Women used drugs to counter recurring depression and continue working. Mary explains, 'I stay depressed.... The drugs take away the feelings.' Cindy describes the desire to be high, 'cause when you on drugs you forget everything until that high go down and then you start thinking about stuff' (Cindy).

Caught Up in the Lifestyle: Accumulating Burdens

Women eventually find themselves 'caught up' in a wave of chronic depression, drug abuse, and learned helplessness. The most common reaction to depression is drug abuse. Cara, for example, began spending up to US$500 per day on drugs in an attempt to ward off her depression. 'It seemed like the higher I got, the more money I spent, the more depressed I got.' Michelle described her life during this phase as 'a big cycle: drugs, work, get high, drugs, work, get high.'

In this phase, life consists of two activities, drug taking and drug seeking. Social networks shift to drug users and drug dealers. Friends, family, or customers are often the targets of exploitation, manipulation, and deceit as the drug-addicted prostitute steals from her family to feed her addiction.

> I have ran off with people's money.... I have took from my family a couple times. I have took from friends that trusted me. One time... I cleaned this guy's whole apartment out. I took everything, and then told him I didn't do it and he believed me. (Kay)

Immediate gain becomes the goal of almost every opportunity, later provoking deep remorse and shame. 'When you start remembering everything, you know, people that you hurt, that comes along with prostitution, because you tend to rip people off a lot. It makes you feel bad' (Maureen).

Safety becomes a secondary concern as women place themselves at increased risk to obtain money to buy drugs. During this time women accept the element of chance and the little control they have over customer–prostitute encounters. At this time the women are willing to get into the car of almost any customer regardless of chance.

> What would make me not get into the car with somebody? Really nothing. I would just get in there. I wouldn't even care if they were the police or if they were a mass murderer or something ... I'm gonna get in that car and I'm gonna try it because if it's some money that I need then I'm gonna do anything for it. (Nina)

Women addicted to crack cocaine, prior to entrance into street prostitution, view sex work as a means to finance their habit. Drug-addicted women enter already 'caught up' in a wave of drug addicting behaviors and drug addiction needs. In order to serve the addiction, addicts new to prostitution engage in a substantial amount of customer contact. The probability that they would employ protective strategies is slim, having never taken the initial time to enter, learn, and adjust to the mores of the lifestyle. Without sufficient 'skill' training, women leave the encounters largely to chance and report many incidents of violence, rape, and occurrences of torture.

Not all women turn to drugs. In total, 5 of the 21 women interviewed were not involved in crack cocaine. Three were restricted from usage by pimps and two were interviewed prior to becoming caught-up. The three women not involved in drug abuse nonetheless reported the same paralyzing depression from being 'caught up' in something bigger than they were.

> I ended up in the psychiatric hospital in this crazy ward.... I was diagnosed with major

> depression. I could only stay there a little while. When they released me I didn't have nowhere to go so I went back to my pimp and started working again. (Tonya)

For the majority of the respondents, physical deterioration accompanies depression and drug abuse. Because of excessive drug use, many respondents suffer substantial weight loss. Most were not prone to seek professional help for illnesses until they became chronic. When physical conditions became chronic, women were more likely to visit emergency rooms than doctors' offices. Prescribed care was not typically followed and follow-up visits were rare. As a result, medical conditions took longer to heal and persistent medical conditions, without proper care, worsened.

> Sometimes I'm in so much pain, I can't hardly stand it. My hip surgery was scheduled three times, but I'd get high or just be out there, cracked out, or not have a ride or something. But it's scheduled again. (Cara)

In addition to a gradual decline in one's general physical condition, over time the devaluation of a woman's physical assets can be emotionally costly. In a culture that values youth over experience, life becomes more difficult for the street women. They can take no solace in personal accomplishments, societal contributions, or in nurturing relationships they failed to build throughout the years. After many years, what is left is a weakened emotional state and deteriorated physical body. Wanting something better in life is all many of them have and even that is damaged by low self-esteem and persistent depression. By this phase, it is often a struggle to seek help.

> The day I decided I wanted to get out of prostitution, I got on the bus.... I knew if I could get to Genesis House that I would be ok. I just turned a trick and I had money. It was so hard. I had money to buy more drugs. I felt so bad inside. And when I feel bad, I get high. It took everything I had to stay on that bus and take myself to [that program]. So many times I started to get off and go buy some dope. I kept telling myself, 'if I get to Genesis House, I'll be ok.' (Patrice)

Women immersed in 'living the lifestyle' work, get high, and work again in an incessant attempt to alleviate the physical and emotional pain. Life becomes a desperate plea to turn more dates, get more drugs, and feed the addiction. Women at this phase report losing social support from informal social networks that included meaningful connections with family.

> When I was out there on the streets, I would send my kids money and they would tear it up and send it back. I would make promise after promise, promise them I wasn't smoking [crack]. I kept lying. My kids was ashamed of me. (Cara)

Leaving the Lifestyle: Taking Stock and Getting Out

Exiting the lifestyle requires a time of reflection. Nearing the exit, the women contemplate what will happen to them if they continue to live as a drug addicted prostitute or a prostitute with chronic depression. They recognize that what they have accumulated as a result of their financial dream, amounts to little but a collection of arrest records, a blur of experiences, and a path of abandonment by those whom they cared about. They realize that the skills they have learned while in prostitution are not marketable.

When women finally make the decision that they can no longer hurt themselves and their families, they experience intense remorse for the prostitution activities that have hurt their children and other loved ones. They set out to repair broken and abandoned relationships. They want something better. Working the streets has become too dangerous and degrading. They can no longer tolerate their lives or themselves and develop a disdain for what they have become. Six days after exit, Sonya reflected:

> My daughter . . . they took her from me. I started into drugs and now I'm tired . . . I'm tired of being a piece of meat and men slobberin' all over me. It's nice money, but it's not worth losing my life over. And you know I mainly gotta do it for myself, because I'm my worse nightmare. I turned into someone I always told myself I would not be. . . . I mean I was strictly against drugs you know. . . . I told myself I would never turn into [that]. (Sonya)

It is the sum total of daily hassles, acute traumas, and chronic conditions that precipitate a woman's decision to exit prostitution. There is some evidence that institutional pressures also have an effect on a woman's decision to leave, particularly pressures from law enforcement and child protective services. A combination of arrests, time in jail, and probation mandates put pressure on women to consider exiting the life. Outside pressure is successful when it is accompanied by a strong, personal desire to exit the life. Without a personal desire to quit, outside restrictive measures are only temporary. When the restrictions lift, the individual returns to business as usual.

Of the 14 women involved in the study who had children, 7 had been involved with child protective services. Six of the seven women had been unable to regain custody of their children at the time of the interview. Reclaiming their families was often a long-term goal that motivated some women to want to make a change.

Both Tracey and Chris were among the many women in this study who were both emotionally distraught and physically distressed when they decided to stop prostituting. 'I ended up losing . . . everything. . . . I mean I lost my kids' (Tracey). 'I still cry at night asking God to forgive me. . . . [I believe] God has forgiven me, it's just me forgiving me.' (Chris).

Over time, one's own body begins to undermine intention and provide its own form of restrictive pressure. Physical deterioration and degeneration as a result of the lifestyle led some women to decide they could no longer physically continue to meet the demands of the profession. It became harder for wounds to heal on a body that lacked proper nutrition and suffered months or years of neglect and abuse.

Relationships are impacted by a woman's decision to enter and to exit prostitution. Close relationships are severed through death, court orders (such as in the case of children), and the dissolution of family ties. Exit may include the end of valued sexual companions including, but not limited to, one's pimp.

Re-entrance Into the Lifestyle: The Loss of Options

Whatever means prompt women to exit, without help and support, the most gallant tries often result in re-entrance into the prostitution lifestyle following a stressful event. After exiting and moving in with a boy friend, Cara found herself beat up and eventually homeless, sleeping on a porch with no family support:

> My eye was swollen up and all cut open. . . . The police and ambulance came. . . . The next morning after attending my mom's funeral, I showed up at [my sister's] house with a patch on my eye and stitches in my face. She wouldn't let me stay. I got all beat up and didn't have nowhere to go. [My sister] didn't want me to stay with her. . . . Mom was dead and [my older brother] was dead. . . . So I just went from there. I walked to the bar and ran into a couple friends. They seen I was down and out and let me sleep on their porch. And my family knew it. And do you think that one of them would say, 'Come on Cara.' Nope, so I said, 'fuck it, my life is this damn bad?' . . . And I started again. (Cara)

As women reflect back on their life of prostitution, most report that poverty was also a leading factor in their re-entrance back into the life. Tracey left prostitution the first time because she could no longer withstand the violence. Shortly after leaving, however, Tracey found that she was once again living in poverty. She re-considered prostitution.

> I always told myself I'd never smoke crack, you know, because I'd seen too many girls out here hurt, too many girls lost everything to crack. But I had lost everything, too. If I'd been smoking crack, it just would've happened quicker.... I mean with 2½ years in recovery, not drinking, I went back to the lifestyle.... I never really gave it up. I mean, I gave it up cause I wasn't actually doing it. But there was still doubt in my mind that there's another way I can do it that will be successful. I mean I had to go through it. For me to actually let it go, I had to get back in and realize nothing's gonna change. There's no way to do it the 'right way'. It's crazy, I mean the men are still beating you and it's just never gonna change. It's always gonna be the same. (Tracey)

Discussion

The passage through prostitution is not a linear, one-way process. Life circumstances may help move women in and out of prostitution many times before they permanently leave. Whether a woman re-enters prostitution or not will largely depend on the motivating factors for entrance in the first place, namely needing financial support and having the desire to alleviate emotional burdens.

In our attempts to work with street prostitutes, social workers should remember that 'prostitution for women is not considered merely a temporal activity as it is for men who are clients, but rather a heavily stigmatized social status which in most societies remains fixed regardless of a change in behavior' (Pheterson, 1990: 399).Therefore, the need to disguise one's lifestyle in the presence of health and social service professionals is often the practice of these women. It will be necessary for the social worker to create an atmosphere of trust to encourage client and patient disclosure.

The costs to women involved in prostitution are high for both her and her family. Once involved in the prostitution lifestyle, women get caught up in situations beyond their control. Helping women find gainful employment and resolve emotional conflicts resulting from childhood abuse are often primary needs of entering street prostitutes. Once involved, the women greatly increase their risk for physical abuse, disease, drug addiction and malnutrition.

Outside restraints alone, such as arrests or Child Protective Services involvement, are not enough to permanently impact a women's progression through the lifestyle phases. The desire to change must come from within. It is the responsibility of the worker to identify the stage of involvement, risks, needs, and motivations of female prostitutes to effectively intervene and support a woman's exit from the street life.

Social workers working with street prostitutes must identify their clients' lifestyle phase in order to develop targeted and effective interventions.Assessments should call attention to both physical complaints and emotional traumas. Interventions should address the risks faced by their clients and provide services that directly target the reasons women initially entered prostitution.

Conclusion

This study adds to the body of knowledge devoted to identifying the phases of street prostitution. The study acknowledges the multiple realities of prostitution as both empowering and dehumanizing. In considering the risk factors and emotional burdens, we find overwhelming negative consequences of the lifestyle over time. The findings from this research strongly suggest that the longer women are involved in street level prostitution the greater their risk for negative consequences and fewer their opportunities for positive life changes. Finally, the study uncovers and connects those daily hassles, acute traumas, and chronic conditions associated with prostitution over time.

In the final analysis, one constant remained throughout all phases of the prostitute lifestyle. From entrance to exit, all 21 women desired 'something better' for themselves.

References

Alegria, Margarita, Vera, Mildred, Freeman, Daniel. H., and Robles, Rafaela (1994) 'HIV Infection, Risk Behaviors, and Depressive Symptoms among Puerto Rican Sex Workers', Journal of Public Health 84(12): 2000–2.

Benjamin, Harry and Masters, Robert E. L. (1964) Prostitution and Morality. New York: Julian Press.

Cohen, Judith B. and Alexander, Priscilla (1995) 'Female Sex Workers: Scapegoats in the AIDS Epidemic', in Ann O'Leary and Loretta S. Jemmott (eds) Women at Risk: Issues in the Primary Prevention of AIDS, pp. 195–218.New York: Plenum Press.

Creswell, John W. (1994) Research Design: Qualitative & Quantitative Approaches. Thousand Oaks, CA: Sage.

Davis, Nanette J. (1971) 'The Prostitute: Developing a Deviant Identity', in James M. Henslin (ed.) Studies on the Sociology of Sex, pp. 297–322. New York: Appleton-Century-Crofts.

Farley, Melissa and Hotaling, Norma (1995) 'Research Study of Prostitutes', San Francisco Examiner, 16 April.

Glaser, Barney (1978) Theoretical Sensitivity. Mill Valley, CA: Sociology Press.

Inciardi, James A., Lockwood, Dorothy and Pottieger, Anne E. (1993) Women and Crack Cocaine. New York: Macmillan.

James, Jennifer (1978) 'The Prostitute as Victim', in Jane R. Chapman and Margaret Gates (eds) The Victimization of Women, pp. 175–201. Beverly Hills, CA: Sage. http://www.webgrrls.com/eva/feminism.html (consulted 3 May 1998).

Mansson, Sven-Axel and Hedin, Ulla-Carin (1998) 'Breaking the Matthew Effect on Women Leaving Prostitution', paper presented at 93rd Annual American Sociological Association Meeting, San Diego, CA. Available from: Sociology Express 1-800-752-3945.

Miller, Jody (1993) 'Your Life is on the Line Every Night You're on the Streets: Victimization and the Resistance among Street Prostitutes', Humanity & Society 17(4): 422–6.

National Center for Missing and Exploited Children (1992) Female Juvenile Prostitution: Problem and Response. Arlington, VA: National Center for Missing and Exploited Children.

Pheterson, Gail (1990) 'The Category Prostitute in Scientific Inquiry', Journal of Sex Research 27(3): 397–407.

Silbert, Mimi and Pines, Ayala (1982) 'Victimization of Street Prostitutes', Victimology 7: 122–133.

Vanwesenbeeck, Ine (1994) Prostitutes' Well-being and Risk. Amsterdam: VU University Press.

Williamson, Celia and Folaron, Gail (2001) 'Violence, Risk, and Survival Strategies of Street Prostitution', Western Journal of Nursing Research 23(5): 463–75.

DISCUSSION QUESTIONS

1. Discuss the motivations, benefits, risks, and losses that exist in the prostitution lifestyle.
2. What collateral damages do women experience as a result of their prostitution career?
3. What protective strategies do women engage in to keep themselves safe on the streets?

READING

In the section introduction, you learned about the role of gender in gang membership. This reading explores the issues of gender and gangs in the United Kingdom. Here, the author exposes a number of issues with current research on gender and gangs, including issues in defining gang membership and the limited availability of research using female samples. This reading seeks to add to the literature on gender and gangs by interviewing girls who are involved in gangs to inform how gang membership has both positive and negative influences in the lives of girls.

SOURCE: Batchelor, S. (2009). Girls, gangs, and violence: Assessing the evidence. *Probation Journal, 56*(4), 399-414.

Girls, Gangs, and Violence

Assessing the Evidence

Susan Batchelor

Introduction

In recent years, growing attention has been paid—both in the academic literature and criminal justice policy and practice—to the phenomenon of the 'gang' (Hallsworth and Young, 2008). With some notable US exceptions (Campbell, 1984, 1990; Chesney-Lind and Hagedorn, 1999; Joe and Chesney-Lind, 1995; Joe Laidler and Hunt, 1997; Miller, 2001, 2008), there has been a tendency in such discussions to ignore the experiences of girls and young women, or to write about them solely from the perspective of young men.[1] This lack of empirical evidence makes it difficult to keep anecdotal evidence about violent 'girl gangs' in perspective and balanced by facts. In dominant discourses, 'girl gangstas' are either depicted as sexually liberated 'post-feminist' criminals or sexually exploited victims. The current article takes issue with these dichotomous portrayals, drawing on emerging UK research on girls, gangs and violence to demonstrate that whilst girls and young women are more likely to be the victims of peer-based violence than the perpetrators, they can and do engage in violent offending for many of the same non-pecuniary reasons as boys and young men. Most of this violence, however, is not gang-related.

Nuts, Sluts and the 'Post-Feminist Criminal': Stereotypes of Gang-Involved Girls

Stories about the growing 'problem' of girl gangs, roaming the streets and randomly attacking innocent victims, have been a recurring feature of the British media since the mid-1990s (see, for example, Bale, 2009; Bracchi, 2008; Carroll, 1998; Knowsley, 1994,1996; Laville, 2005; Leadbetter, 2006; Oakeshott, 2002; O'Hara, 2007; Stephen, 1999; Thompson, 2001, 2004). Typical accounts suggest that 'girl violence is on the increase in an alarming way' (Lambert, 2001), fuelled by a 'ladette' binge-drinking culture (Clout, 2008) in which 'young women are aping and mimicking the traditional behaviour [of] young men' (Geoghegan, 2008). This so-called 'masculinization' is often portrayed as 'the dark side of girl power' (Prentice, 2000), an unfortunate by-product of young women seeking equality with young men (Batchelor, 2007b; Chesney-Lind and Irwin, 2008). More recently, attention has focused on the sexual exploitation of girls by gangs, with journalists drawing on anecdotal evidence from police and youth justice professionals to distinguish between 'two types of girls who become involved [in gangs]: those who are "as tough as the boys" and fight to defend themselves, and those who become involved with, and can be sexually exploited by, gangs of boys, sometimes under the auspices of being "initiated" or accepted into the group' (O'Hara, 2007).

Anecdotal evidence also features prominently in policy documents and statements made by high profile public servants. In 2002, Lord Warner, the then Chairman of the Youth Justice Board in England and Wales, expressed concern that girls and young women were being used by male gang members for sexual services, or to conceal weapons and drugs: 'We have heard anecdotal stories of young women being coerced into sexual activity as part of gang culture', he said. 'Sex is the chosen form of physical intimidation of girls' (Warner, quoted in Burrell, 2002). This concern, again not discernibly informed by any research evidence, was reiterated in the recent Gangs and Group Offending Guidance for Schools, issued by the Department for Children, Schools and Families (2008). Girls and young women, the report stated, 'are subservient in male gangs and even submissive, sometimes being used to carry weapons or drugs, sometimes using their sexuality as a passport or being sexually exploited, e.g., in initiation rituals, in revenge by rival gangs or where a younger group of girls sexually service older male gang members' (DCSF, 2008: 7). In April 2008, the Lord Advocate Elish Angiolini made headlines in Scotland when she appeared before the Parliament's Equal Opportunities Committee and stated that she and others in the Crown Office and Procurator Fiscal Service were worried about the increasing number of girls in groups using knives: 'I've seen anecdotal evidence,' she said, 'that many women are not just simply the collaborators—going along with a dominant male partner, being an accessory, carrying knives for boyfriends, assisting in cleaning up after a murder, hiding weapons, etc.' (Angiolini, quoted in BBC Online, 2008). Some are, she claimed, often the 'prime movers,' attacking others within their own group: 'This can be gang-related or it can just be that there is someone in a group who is quite persecuted by the gang leader or their cohorts. That is the kind of machismo behaviour that hitherto we would only see from a male offender' (Angiolini, quoted in Naysmith, 2008).

One reason that such dichotomized representations hold sway is the dearth of empirical research in this area, particularly in the UK. Patterns of female invisibility in thinking about gangs have been largely set by male-centred research investigations (Campbell, 1984; Miller, 2001). Historically, the field of criminology has been a masculinist enterprise, primarily interested in understanding the more dramatic or exciting offending of (predominantly lower-class) boys and young men (Millman, 1975). As a result, most empirical research and theoretical explanations of gangs have tended to focus on gangs as a male phenomenon, discussing girls and young women 'solely in terms of their . . . relations to male gang members' (Campbell, 1 990: 1 66). As one recent review of the literature concluded:

> 'Sex objects or tomboys'—these are the images that, until recently, dominated the literature on female gang members. Individual females were portrayed in terms of their sexual activity, with an occasional mention of their functions as weapon carriers for male gang members. . . . Even when describing female gang members as tomboys, researchers emphasized that the females' motivations were focused on males.[2] (Moore and Hagedorn, 2001: 2)

In the US, such images have been challenged by feminist researchers, who have attempted to provide a more 'nuanced portrayal of the complex gender experiences of girls in gangs' (Miller, 2001: 16). This research has demonstrated that female gang members not only adhere to rigid gender expectations and experience heightened risks for physical and sexual victimization, but also claim that gang membership fosters a sense of belonging and empowerment, offering them a refuge from abusive families and the means by which to resist dominant gender stereotypes (Campbell, 1990; Joe and Chesney-Lind, 1995; Joe Laidler and Hunt, 2001; Miller, 2001, 2008; Moore, 1991; Nurge, 2003).

In the UK, the literature on gangs generally, and on girls and gangs specifically, is less well developed. This is largely due to the rejection of the gang paradigm by British researchers (Campbell and Muncer, 1989; Sanders, 2002). Early attempts to apply American gang theory to the UK failed to find evidence of structured, street gangs (Downes, 1966; Parker, 1974; Scott, 1956), leading to a shift in focus, from a concern with delinquent gangs towards the study of leisure-based youth subcultures (where offending is one of a number of areas of investigation) (Hall and Jefferson, 1976; Muncie, 2009). As Hallsworth and Young (2008) have observed, British resistance towards the gang paradigm has meant 'data on gangs have not been routinely collected or disseminated as they are in the USA. . . . In other words, in the UK there is no sound evidential base to prove the case [for the proliferation of violent street gangs] one way or the other' (Hallsworth and Young, 2008: 177). Recent research is expanding our knowledge about UK gangs (cf. Aldridge and Medina, 2008; Bannister and Fraser, 2008; Bennett and Holloway, 2004; Bradshaw, 2005; Communities that Care, 2005; Kintrea et al., 2008; Pitts, 2007; Sharp et al., 2006;

Young et al., 2007), but the majority of these studies emphasize gangs as a male phenomenon with little or no attention paid to girls or young women.

In Search of a Ghost: Defining and Measuring 'Gangs'

Gang researchers have perennially struggled with the task of defining what constitutes a 'gang' and, as a result, there is little consensus within the literature as to the meaning of the term (Aldridge et al., 2008; Ball and Curry, 1995; Esbensen et al., 2001). Katz and Jackson-Jacobs characterize this debate as being 'essentially an argument over the correct description of a ghost' (Katz and Jackson-Jacobs, 2004: 106); that is, an argument over an invisible symbol, the projection of our worst fears (Fraser, 2008). A variety of different definitions exist, ranging from a (relatively benign) group of young people who spend time hanging out together in public space (e.g., Thrasher, 1 927) through to a strictly hierarchical, malevolent and organized criminal network (e.g., Jankowski, 1991). While realist accounts view gangs as objective entities with fixed characteristics—e.g., group identity, permanence, involvement in illegal activity (Esbensen, 2000)—social constructionist perspectives acknowledge the fluidity of gangs, which are seen to evolve and adapt over time, often in response to hostile social reaction (e.g., Klein, 1971). These different definitions have important implications for estimates of gang prevalence (Hallsworth and Young, 2008; Young et al., 2007). As demonstrated below, some studies rely on respondents' self-definition, simply asking survey participants to report if they belong to a gang (e.g., Communities that Care, 2005; Smith and Bradshaw, 2005). Others utilize practitioners' perceptions and definitions (e.g., Bullock and Tilley, 2002; Stelfox, 1996). Both such approaches can be problematic as they rely on respondents' understandings and perceptions of 'gangs', which may differ dramatically (Sharp et al., 2006). For example, some respondents may refer to their group of friends as a 'gang' even if they do not engage in any delinquent or criminal activity; others may reserve the use of the term to refer to more serious and organized crime groups. Alternative approaches include providing respondents with a precise definition of what the researchers perceive to be a gang, perhaps in conjunction with a series of 'filters' or 'funnels' specifying distinctive behavioural, cultural or structural features (e.g., Bennett and Holloway, 2004; Sharp et al., 2006; Young et al., 2007). Of course the problem here is that different researchers can employ different definitions, making comparison difficult. This has led some to utilize the protocols developed by the Eurogang Network,[3] which view gangs as 'any durable, street-oriented youth group whose involvement in illegal activity is part of their group identity' (cited in Sharp et al., 2006 and Young et al., 2007).

Regardless of the definition used, research suggests that gang membership in the UK is relatively rare, with most self-report surveys reporting prevalence rates of between four and six per cent (cf. Bennett and Holloway, 2004; Communities that Care, 2005; Sharp et al., 2006). Quantitative datasets also demonstrate the age-related nature of gang membership. The Edinburgh Study of Youth Transitions and Crime,[4] a longitudinal study of around 4300 young people starting secondary school in Edinburgh in 1998, found that the overall proportion of respondents who self-defined as gang members dropped from around 18 per cent at age 13, to 12 per cent at age 16, and five per cent at age 17 (Smith and Bradshaw, 2005). These findings were consistent with those of Sharp et al. (2006), who—based on data from the Offending, Crime and Justice Survey[5] in England and Wales (and utilizing a series of filtering questions developed by the Eurogang Network)—discovered that involvement in 'delinquent youth groups' (DYG) peaked at age 14-15 (12%), falling to nine per cent at 16-17 years and two per cent at 18-19 years. Both studies found comparable rates of gang membership amongst girls and boys in the younger age groups, but reported that involvement peaked earlier for girls, at around 13-15 years, before tailing off in their late teens. By age 16, prevalence was considerably higher amongst boys than girls. According to Smith and Bradshaw (2005), this is because girls reach maturity earlier than boys.

Because most of the gang activity identified in self-report studies starts young and then rapidly declines, it is fair to assume that most 'gangs' or 'delinquent youth groups' are not engaged in serious criminal activity. Indeed, some such groups may not be involved in offending at all, particularly in those studies relying on self-definition. The questions used to define gang membership in both the Safer London Youth Survey[6] (Communities that Care, 2005) and the Edinburgh Study, for example, made no reference to delinquent or criminal behaviour.[7] That said, research does demonstrate that individuals identified as gang members report higher levels of offending than non-gang members, across all offence categories (Bennett and Holloway, 2004; Sharp et al., 2006; Smith and Bradshaw, 2005).[8] This pattern holds for girls as well as boys, although the level of offending amongst girls is much lower than amongst boys. Compared to boys, girls are less likely to be members of single-sex DYGs (Sharp et al., 2006) or 'hard core' gangs that have 'both a name and sign or saying' (Smith and Bradshaw, 2005: 10).[9] The limited UK data on antecedent factors associated with DYGs suggest that girls and boys share a number of common risk factors for gang involvement (i.e., having friends in trouble with the police, having run away from home, and having been expelled or suspended from school), but gender specific factors include having been drunk in the past year and attitude to certain criminal acts (amongst boys) and disorder problems in the local area, little or nothing to do in the local area, and poor perception of school (amongst girls) (Sharp et al., 2006).[10]

Nuts and Sluts Revisited: Girls' Role in UK Gang Research

Whereas self-report surveys suggest an almost equal level of female gang membership in the UK, qualitative studies suggest that gangs are much more male dominated and, as a result, tend to pay lesser attention to the views and experiences of girls (Esbensen et al., 1999). The two most recent studies in this vein are Aldridge and Medina's (2008) account of youth gangs in an English city[11] and John Pitts' (2007) research with armed gangs in Waltham Forest. Other qualitative studies include Kintrea et al.'s (2008) exploratory study of young people and territoriality (in Bradford, Bristol, Glasgow, Peterborough, Sunderland and Tower Hamlets) and Young et al.'s (2007) interviews with young people involved in group offending (in five towns and cities in England and Wales). As will become apparent, only the latter account includes any detailed consideration of the views and experiences of girls and young women.

Aldridge and Medina devote only one paragraph of their end of award report to a discussion of gender. In it, they highlight the difficulty of identifying and accessing gang-involved girls: 'Our ethnographic data . . . indicate that females are seen as playing a secondary role in most of the gangs we had access to. In one of the gangs . . . we gathered reports of a greater involvement but were unsuccessful in talking to female members' (2007: 7). Kintrea and colleagues devote two paragraphs to girls and young women in their study, but all of the quotations used to back up the claims made are drawn from interviews with (male) adult practitioners, thus perpetuating the stereotype of girls as auxiliary members:

> Most participants who were reported to be involved in territoriality were boys or young men. . . . Girls and young women took a more minor part; they were less often involved in gang conflict and they were less constrained by territoriality in their personal dealings, and it was believed that the impact on their life chances was much less than for boys. . . . A typical comment about girls' role was: 'The girls play a background role to the gang. They are there, and they are there for their boys but they are not as territorial as the boys are. They are proud of their areas and they are proud of where they come from and they stick by their lads, but they are not as visible . . . for the girls

> it's part of hanging out with the lads'. (Kintrea et al., 2008: 25)

In line with mainstream research conducted in the US, the report also portrays girls as relegated to gender-specific crimes, claiming that girls 'play an important role in encouraging gang activity,' through 'wanting to have a boyfriend who is the biggest, baddest guy in their scheme' and '[making] boys in their area jealous by deliberately cultivating friendships with guys from other areas' (derogatively referred to as 'scheme-hopping') (Male practitioner, quoted in Kintrea et al., 2008: 25). This is a view reiterated by the male gang members and adult practitioners interviewed by Pitts (2007), who devotes one paragraph to a discussion of gender. 'Girls,' his interviewees report, 'play an ancillary role, sometimes carrying or hiding guns or drugs for the boys' (Pitts, 2007: 40). They are apparently 'attracted to the "glamour" and "celebrity" of gang members' (Pitts, 2007: 40) but often find themselves being sexually exploited, sometimes in exchange for drugs:

> The relationship [between gang members and their girlfriends] tends to be abusive; one of dominance and submission. Some senior gang members pass their girlfriends around to lower ranking members and sometimes to the whole group at the same time. Unreported rape by gang members, as a form of reprisal or just because they can, is said to occur fairly frequently and reports to the police are rare. (Pitts, 2007: 40)

'There are other girls,' Pitts claims, who 'do not perform the same sexual role as the "girlfriends" of gang members' (Pitts, 2007: 40). In London, these girls are said to 'regard themselves as "soldiers" and concentrate on violent street crime' (Pitts, 2007: 40). In Glasgow, they form adopt imitative 'she' gangs (i.e. auxiliaries to male gangs) (Kintrea et al., 2008: 26). The only example found of an all-female gang was discovered by Aldridge and Medina (2008), but this (small) group was said to engage in acquisitive as opposed to violent offending.

A View From the Girls: Friendship Groups as a Source of Fulfilment and Fun

The only UK gangs research to include any detailed consideration of the views and experiences of girls and young women is Young et al.'s (2007) study of groups, gangs and weapons for the Youth Justice Board in England and Wales.[12] Like previous studies, this research had some difficulty in identifying female gang members (see also Batchelor, 2001; Batchelor et al., 2001), resulting in their having to extend the definition of gang membership to include young women who were 'known to have offended with other people.' This wider definition produced a sample of 25 young women aged between 14 and 20 years. Seven had been involved in street robbery, five had committed an assault and seven had been arrested for shoplifting. A smaller number had been involved in sex work ($n = 3$), or been arrested for possession of drugs and/or small time drug dealing ($n = 3$). All of the young women interviewed described growing up in 'bad areas', characterized by poverty and deprivation. Many had difficult family backgrounds and often related experiences of bereavement and loss, as well as bullying and neglect by parents and carers. This frequently resulted in experience of the care system and an inability to meet the demands of mainstream education.

Young et al.'s (2007) research is important because it demonstrates that when researchers engage directly with girls and young women, a different picture emerges of their 'gang' involvement. Unlike the UK studies reported above, which tend to portray girls and young women in terms of their status as 'girlfriends,' Young et al.'s interviewees said that the mixed-sex groups they belonged to were composed of peers whose principal relation to each other was friendship. All denied that their group was a gang. Sometimes the young women went out with older group members, some of whom were abusive, but this was said to be uncommon. Unlike in the US literature (e.g., Miller, 2001), the young women did not 'join' the group as matter of ritual (the group emerged from friendships forged at schools, in the care system, or in the estates where they lived) and there was no evidence to suggest that they were subject to initiation rites (such as being 'jumped in' or 'sexed in'). What's more, Young et al. (2007) uncovered evidence of all-female groups, whose principal points of reference were each other and not their male associates:

> Seven young women belonged to all-female groups and although they would periodically hang about with the local young men, this was not because these relationships with males were considered to be important or necessary. Indeed, from their testimonies it was evident that these women did not consider the males around them as friends or even friendly. Nor did this group enter into intimate relationships with the young men they associated with. These young women determined when they associated with the males in their social circle and were not significantly influenced by the actions of males or male-dominated groups. (Young et al., 2007: 143)

Whilst the main activity that the young women engaged in with their friends was 'hangin' out' and ''aving fun,' some also participated in interpersonal physical violence and street robbery (or 'jacking'). Most violence occurred within the peer group, often as a result of rumours or excessive teasing, as retribution, perceived rule infraction or injustice, or jealousy. Fighting was also associated with 'being pissed', although most young women did not drink with the intention of causing trouble, but rather to combat boredom and 'for the pleasures that came with intoxication' (Young et al., 2007: 148). Street robbery was similarly pursued as a source of excitement (and power), but sometimes 'took on a more instrumental sheen because the victim was both robbed of her possessions as well as being physically humiliated for something she had done, or for some slight she was believed to have occasioned' (Young et al., 2007: 151). Far from playing a minor role in group violence, these young women claimed that 'females were more likely to pursue thrills, engage in fights and cause more trouble than their male counterparts' (Young et al., 2007).

These findings clearly challenge the claims made by Pitts (2007) and Kintrea et al. (2008), not least because they demonstrate the active and assertive role that young women can play within their peer networks. However, given the interviewees' reluctance to define these friendship groups as a 'gang,' they also cast doubt on the levels of female gang membership reported by Sharp et al. (2006) and Smith and Bradshaw (2005). In short, Young et al.'s research suggests that whilst some young women engage in violent crime for much the same reasons as young men, this violence is not gang-related.

Ambivalence and Agency: Girls and Violence

In addition to Young et al.'s study of 'girl gangsters', there are a small number of (mainly qualitative) studies that have looked at girls and violence in an attempt to 'bring the voices of young women to the centre of theoretical and methodological debates' (Batchelor, 2005: 361). These studies report strikingly similar findings to those discussed above, in regard to girls' attitudes and experiences of violence. However they also paint a more complex picture of the role of victimization and agency in the lives of young women who offend.

In an exploratory study of teenage girls' views and experiences of violence carried out in Scotland, Burman and colleagues found little evidence of a huge rise in physical violence by girls, nor of girl gangs (Batchelor, 2001; Batchelor et al., 2001; Burman, 2004; Burman et al., 2001, Burman and Batchelor, 2009).[13] Although exposure to and fear of violence were fairly common, only a small number of girls reported using physical violence frequently. This group of girls had disproportionate experience of violence in their own lives, at the hands of both their families and their peers. However, such violence tended to be normalized and the girls showed a high tolerance for physical violence, particularly in self-defence. A fairly high level of verbal abuse was uncovered across the sample, and gossip, teasing and name-calling were reported as common precursors to fights between girls:

> Contrary to its literal meaning, 'talking behind someone's back' could be construed as an overt and challenging expression of aggression, generating intense anger, annoyance, and the need to act in 'self-defence'.... When the effects of 'gossip' and 'bad-mouthing' were considered within the context of girls' friendships, insights emerged as to why they were considered to be a powerful catalyst for physical violence. The premise of 'close' friendships between teenage girls is sharing, trust, loyalty and the keeping of secrets. Girls in the study commonly described their friendships with other girls as 'the most important thing' in their lives, and spending time and hanging out with friends was their main social activity. This means that girls can react powerfully to fall-outs with friends and breaches of confidence. (Batchelor et al., 2001: 129)

Exposure to routine violence was also common amongst the young women convicted for violent offences interviewed by Batchelor (2005, 2007a, 2007b).[14] Like the young women in Young et al.'s study, Batchelor's participants reported significant histories of family disruption, bereavement and neglect, as well as experiences of physical and sexual abuse, sometimes at the hands of male partners. Despite these circumstances they often demonstrated great loyalty to their families, friends and boyfriends, alongside unresolved feelings of disappointment, anger and grief (which they dealt with by using drugs and/or alcohol as a form of self medication). Violence was perceived as a form of self-defence, 'an attempt to pre-empt [further] bullying or victimization through the display of an aggressive or violent disposition' (Batchelor, 2005: 369). Fights usually arose over issues of personal integrity, including instances of false accusation, gossiping behind backs, and pejorative remarks about sexual morality and/or the young woman's abilities as a mother (see also Campbell, 1981). One of the unwritten rules of violence, then, was that 'You need tae stick up fer yourself to get respect' (Batchelor, 2007b).

Of course girls and young women do not solely experience their everyday lives in relation to the perceived threat of physical (or sexual) danger; rather, there is a strong sense that they also engage in risk-seeking behaviour where the pursuit of excitement, thrills, and pleasure take precedence (Batchelor, 2007a). In line with the findings of Matza and Sykes (1961), along with work carried out under the rubric of 'cultural criminology' (for an overview, see Ferrell, 1999), girls and young women in both Burman et al. (2003) and Batchelor's (2007a) research cited the adrenaline 'rushes' involved in offending, stating that violence could be 'fun.' For young women involved in street robbery, for example, the value of the goods stolen was often said to be of less importance than the sense of euphoria and exhilaration associated with 'putting one over' on someone. Thus violence presented some young women with a measure of self-esteem and self-efficacy; a sense that they had crossed the boundaries into someone else's world and 'gotten away with it'. This sense of status and esteem was sometimes linked to the supposedly 'masculine' nature of the offences they committed, since confronting expectations that women should not engage in violence provided an additional source of excitement, pleasure, self-respect and status.

Taken together, these findings provide an important challenge to essentialist arguments about the emergence of a new breed of 'girl gangsters' who simply seek to emulate the violent behaviour of young men. Criminally violent young women are not liberated young women, but young women who are severely constrained by both their material circumstances and attendant ideologies of working-class femininity. They are not determined by these circumstances, however. By pointing to the risk-seeking nature of much of violence perpetrated by girls and young women, qualitative research with young women demonstrates the positive contribution violent behaviour can have in terms of their sense of self and self-efficacy. In short, such research acknowledges that subordination and agency are simultaneously realized in young women's lives, and thereby demonstrates that there is no such thing as the essential 'gang girl'.

Conclusion

Stories about 'girl gangs' and 'violent young women' appear regularly in the UK media, where violence by girls is presented as a new and growing social problem. Yet, despite increasing concern, little is actually known about girls' attitudes towards or experiences of gang involvement. As with other areas of criminological enquiry, UK gang research has involved two different methodological approaches: (i) quantitative analyses of risk factors identified by self-report studies; and (ii) qualitative (observational/interview-based) accounts of social, situational and experiential factors. These different approaches have resulted in a lack of consensus concerning not only the extent of girls' gang involvement, but also the nature of that involvement.

Depending on what definition of a group or gang is adopted, and the age range of the sample, self-report surveys indicate a level of membership amongst youth of between 2 and 20 per cent. Female involvement appears comparable with male involvement, particularly in the younger age categories, although girls report engaging in offending at a much lower rate than their male peers. Qualitative research, in contrast, suggests that male gang members consistently outnumber female members, or indeed they fail to prove the existence of gang girls at all. Interviews with adult practitioners and boys who are involved in group offending suggest that girls play a minor role in most gangs and are subjected to high levels of sexual and physical victimization. Interviews with young women, however, point to the positive features of group involvement for girls, as well as highlighting girls' varied motivations for (predominantly low level) violence. Perhaps most notably, this latter research demonstrates that, contrary to media reports and statements made by prominent public servants, there is little to suggest that recent rises in individual violent crime among girls and young women are at all gang-related.

Regardless of whether they are defined as 'gang' members or not, some young women clearly spend much of their time hanging around on the streets with delinquent peer groups, and this has important implications for their lives. Spending time with friends is a prime social activity for most young people, but girls in particular commonly describe their friendships as 'the most important thing' (Burman et al., 2003; Griffiths, 1995; Hey, 1997). For young women coming from backgrounds characterized by disruption, abuse and neglect, the peer group takes on heightened significance as a source of identity, approval, support and protection (Joe and Chesney-Lind, 1995; Miller, 2001). For these young women, participating in home-centred 'bedroom cultures' (McRobbie and Garber, 1976) is unlikely to be an option and, where affordable and accessible leisure facilities are not available locally, they are more likely to use the streets as places of leisure (Skelton, 2000). Participation in unstructured leisure activities has been shown to be highly correlated with delinquent and violent behaviours amongst youth (Agnew and Peterson, 1989; McNeill and Batchelor, 2002). As the findings discussed above demonstrate, most of the violence that girls and young women experience, as both perpetrators and victims, takes place within either the family or the friendship group. This implies that social work and probation practitioners need to heed the familial and peer contexts of young women's offending, recognizing that both groups can be simultaneously harmful and protective.

In their recent article 'Gang Talk and Gang Talkers: A Critique,' Hallsworth and Young exhort officials, academics, and practitioners to look 'beyond and behind mystifications like gang culture' and be 'wary about imposing misleading labels' (Hallsworth and Young, 2008: 192). To this I would add the need to resist simplistic accounts of girls' involvement that rely on dichotomous portrayals of male and female behaviour and thereby reinforce limiting gender stereotypes. There has been a tendency, in both academic and policy discourse, to identify the 'typical' female offender, thereby homogenizing what is in fact a diverse group

(Burman and Batchelor, 2009; Carlen, 1985). Statistical representations do not break down data by gender and ethnicity, for example, nor do existing qualitative accounts address the complexity of raced, classed and gendered subjectivities. Given the difficult family backgrounds and levels of physical and sexual abuse experienced by many women who offend, it is unsurprising that responses to female offending have tended to focus on women's status as 'victim,' depicting their actions as symptomatic of individual pathology or, alternatively, the result of circumstances beyond their control. However, such women are not merely victims, they are also agentic social actors who have the ability to make choices and impose them on the world, albeit it in circumstances not of their own choosing. As I have argued previously (Batchelor, 2005), if we are to effect change in the lives of young women who offend, we need to respect this agency by maximizing involvement and participation.

In short, effective interventions should provide opportunities for girls and young women to participate in positive relationships, not just with probation or social work staff, but with their families and friends as well. This implies a need for affordable and accessible leisure activities, some of which are geared to the specific needs of girls and young women. Such activities need to be staffed by specialist workers, who are attuned to, and equipped to deal with, girls' bullying and victimisation.

Notes

1. I am using the term 'girls' to refer to females aged under 16 years and 'young women' to refer to those aged 16-20 years.

2. Miller, for instance, explained that 'the behaviour of the [girls] . . . appeared to be predicated on the assumption that the way to get boys to like you was to be like them rather than [sexually] accessible to them' (Miller, 1973: 34, cited in Moore and Hagedorn, 2001: 2).

3. The Eurogang Network was formed with the remit of agreeing a consistent definition, questions and methodologies allow comparative international 'gang' research. URL (accessed 27 March 2009): http://www.umsl.edu/~ccj/eurogang/euroganghome.htm

4. The Edinburgh Study of Youth Transitions (ESYT) focused on one whole year cohort of young people in the City of Edinburgh. Rather than selecting a sample of young people every school in Edinburgh was asked to participate. The City of Edinburgh comprises enormous diversity, including all the extremes of poverty, wealth, high and low crime areas and high and low areas of drug abuse. URL (accessed 27 March 2009): http://www.law.ed.ac.uk/cls/esytc/

5. The Offending, Crime and Justice Survey (OCJS) is a nationally representative, longitudinal, self-report offending survey for England and Wales, commissioned by the Home Office (Sharp et al., 2006). The sample of respondents is drawn from persons aged 10-25 years, resident in private households in England and Wales. URL (accessed 27 March 2009): http://www.homeoffice.gov.uk/rds/offending_survey.html

6. The Safer London Youth Survey questioned 11,400 young people, aged 11–15 years, in six inner London boroughs (all areas with high levels of crime—notably gun crime—and deprivation).

7. The only quantitative surveys that do use involvement in illegal activities as an additional filter are Sharp et al. (2006) and Young et al. (2007).

8. For example, Sharp et al. (2006) reported that 63 per cent of gang-involved youth had committed a 'core' offence (robbery, assault, burglary, criminal damage, thefts from or of cars and drugs sales) in the last year compared to 26 per cent of non-members.

9. In the Edinburgh study, the proportion of boys defined as being members of 'hard core' gangs was three times that of girls at age 16/17.

10. US research gives greater emphasis to the role played by family problems in preceding girls' gang involvement. Factors cited by Moore (1991) include: childhood abuse and neglect, domestic abuse between parents, family drug and alcohol addiction, witnessing the arrest of a family member, having a family member who is chronically ill, and experiencing death in the family during childhood.

11. In order to protect the identity of the research site, Aldridge and Medina removed all references to the city in which they conducted their research from the final report. Within 'research city,' fieldwork was concentrated in two sites: 'Inner West, a corridor of historically marginalized areas with a substantial black population and a recognized gang problem that has resulted in a sequence of lethal shootings' and 'Far West, a large, predominantly white council estate with a gang problem that is not officially recognized' (Aldridge and Medina, 2008: 16).

12. Interviews were conducted in five case study areas, which varied in their geographical spread, crime rates and levels of deprivation, and the age structure and ethnic mix of their local populations.

13. The girls and violence study employed a range of methods including self-report questionnaires (n = 671), small group discussions (n = 18) and in-depth life-history interviews

(n = 12) (Burman et al., 2000). Whilst not representative, the sample included a cross-section of girls drawn from inner city, town and rural areas and included girls from ethnic minority backgrounds, girls who had a disability, girls living in isolated locations, and those accommodated by the local authority.

14. Batchelor's research was conducted with young women in prison in Scotland. All of her participants were single and all were white; ages ranged from 16 to 24 years. As a group, their lives were characterized by poverty, addiction, abuse and/or psychological harm.

References

Agnew, R. and D. Petersen (1989) 'Leisure and Delinquency', Social Problems 36: 332–49.

Aldridge, J. and J. Medina (2008) Youth Gangs in an English City: Social Exclusion, Drugs and Violence (ESRC End of Award Report). Swindon: Economic and Social Research Council.

Aldridge, J., J. Medina and R. Ralphs (2008) 'The Problems and Dangers of doing Gang Research', in F. van Gemert, D. Peterson and H. Lien (eds) Youth Gangs, Ethnicity and Migration, pp. 31–46. Cullompton: Willan Publishing.

Bale, D. (2009) 'Teen was Attacked by Girl Gang', Norwich Evening News, 16 March 2009.

Ball, R.A. and D.G. Curry (1995) 'The Logic of Definition in Criminology: Purposes and Methods for Defining "Gangs"', Criminology 33: 225–45.

Bannister, J. and A. Fraser (2008) 'Youth Gang Identification: Learning and Social Development', Scottish Journal of Criminal Justice Studies 14: 96–114.

Batchelor, S. (2001) 'The myth of girl gangs', Criminal Justice Matters 43: 26-27. [Reprinted in Y. Jewkes and G. Letherby (eds) (2002) Criminology: A Reader. London: SAGE]

Batchelor, S. (2005) '"Prove me the Bam!" Victimization and Agency in the Lives of Young Women who Commit Violent Offences', Probation Journal 52(4): 358–75.

Batchelor, S. (2007a) '"Getting Mad Wi' it": Risk-seeking by Young Women', in K. Hannah-Moffat and P. O'Malley (eds) Gendered Risks, pp. 205–28. London: Glasshouse Press.

Batchelor, S. (2007b) '"Prove me the Bam!" Victimization and Agency in the Lives of Young Women who Commit Violent Offences', unpublished doctoral thesis, University of Glasgow.

Batchelor, S., M. Burman and J. Brown (2001) 'Discussing Violence: Let's Hear it from the Girls', Probation Journal 48(2): 125–34.

BBC Online (2008) 'Drink Linked to Female Offenders, BBC Online, 22 April 2008.

Bennett, T. and K. Holloway (2004) 'Gang Membership, Drugs and Crime in the UK', The British Journal of Criminology 44: 305–23.

Bracchi, P. (2008) '"The Feral Sex": The Terrifying Rise of Violent Girl Gangs', The Mail, 16 May 2008.

Bradshaw, P. (2005) 'Terrors and Young Teams: Youth Gangs and Delinquency in Edinburgh', in S.H. Decker and F.M. Weerman (eds) European Street Gangs and Troublesome Youth Groups: Findings from the Eurogang Research Program, pp. 241–57. Walnut Creek, CA: AltaMira Press.

Bullock, K. and N. Tilley (2002) Shootings, Gangs and Violent Incidents in Manchester: Developing a Crime Reduction Strategy. London: Home Office.

Burman, M. (2004) 'Turbulent Talk: Girls Making Sense of Violence', in C. Alder and Worrall (eds) Girls' Violence: Myths and Realities. Albany, NY: SUNY Press.

Burman, M. and S. Batchelor (2009) 'Between Two Stools: Responding to Young Women who Offend', Youth Justice 9(3).

Burman, M., S. Batchelor and J. Brown (2001) 'Researching Girls and Violence: Facing the Dilemmas of Fieldwork', British Journal of Criminology 41 (3): 453–9.

Burman, M., Brown., J. and Batchelor, S. (2003) '"Taking it to Heart": Girls and the Meaning of Violence', in E. Stamko (ed.) The Meanings of Violence. London: Routledge.

Burman, M., K. Tisdall, J. Brown and S. Batchelor (2000) A View from the Girls: Exploring Violence and Violent Behaviour (ESRC End of Award Report). Swindon: Economic and Social Research Council.

Burrell, I. (2002) 'Children Drawn into Crime as a "Lifestyle Choice"', The Independent, 18 March 2002.

Campbell, A. (1981) Girl Delinquents. Oxford: Basil Blackwell.

Campbell, A. (1984) The Girls in the Gang. Oxford: Basil Blackwell.

Campbell, A. (1990) 'Female Participation in Gangs', in C.R. Huff (ed.) Gangs in America, pp. 163–82. Newbury Park, CA: SAGE.

Campbell, A., and S. Muncer (1989) 'Them and Us: A Comparison of the Cultural Context of American Gangs and British Subcultures', Deviant Behavior 10: 271–88.

Carlen, P. (1985) 'Introduction', in P. Carlen, D. Christina, J. Hicks, J. O'Dwyerand Tchaikovsky (1985) Criminal Women. Cambridge: Polity Press.

Carroll, R. (1998) 'Gangs Put Boot into Old Ideas of Femininity', The Guardian, 22 July 1998.

Chesney-Lind, M. and J.M. Hagedorn (eds) (1999) Female Gangs in America: Essays on Girls, Gangs and Gender. Chicago: Lake View Press.

Chesney-Lind, M. and K. Irwin, (2008) Beyond Bad Girls: Gender, Violence and Hype. New York: Routledge.

Clout, L. (2008) 'Violent Women: Binge Drinking Culture Fuels Rise in Attacks by Women', The Telegraph, 31 July 2008.

Cohen, A. K. (1 955) Delinquent Boys: The Culture of the Gang. New York: The Free Press.

Communities that Care (2005) Findings from the Safer London Youth Survey 2004. London: Communities that Care.

Department for Children, Schools and Families (2008) Gangs and Group Offending Guidance for Schools. London: Department for Children, Schools and Families.

Downes, D. (1966) The Delinquent Solution: A Study in Subcultural Theory. London: Routledge and Kegan Paul.

Esbensen, F.A. (2000) Preventing Adolescent Gang Involvement. Washington: US Department of Justice.

Esbensen, F.A., E.P. Deschenes and T.L. Winfree (1999) 'Differences between Gang Girls and Gang Boys: Results from a Multisite Survey, Youth & Society 31 (1): 27–53.

Esbensen, F.A., T.L. Winfree, N. He and T.J. Taylor (2001) Youth Gangs and Definitional Issues: When is a Gang a Gang, and Why Does It Matter? Crime & Delinquency 47'(1): 105–30.

Ferrell, J. (1999) 'Cultural Criminology', Annual Review of Sociology 25: 395–418.

Fraser, A. (2008) 'The Meanings of Crime, Community and Territoriality for Young People in Glasgow.' Paper presented at the British Society of Criminology Conference, Huddersfield, July 2008.

Geoghegan, T. (2008) 'Why are Girls Fighting like Boys?', BBC Online, 5 May 2008.

Griffiths, V. (1995) Adolescent Girls and their Friends: A Feminist Ethnography. Aldershot: Avebury.

Hall, S. and T. Jefferson (eds) (1976) Resistance through Rituals: Youth Subcultures in Post-war Britain. London: Hutchinson.

Hallsworth, S. and T. Young (2008) 'Gang Talk and Gang Talkers: A Critique', Crime, Media, Culture 4(2): 175–95.

Hey, V. (1997) The Company She Keeps: An Ethnography of Girls' Friendships. Buckingham: Open University Press.

Jankowski, M.S. (1991) Islands in the Street: Gangs and American Urban Society. California: University of California Press.

Joe, K. and M. Chesney-Lind (1995) '"Just Every Mother's Angel": An Analysis of Gender and Ethnic Variations in Youth Gang Membership', Gender and Society 9(4): 408–30.

Joe Laidler, K. and G. Hunt (1997) 'Violence and Social Organization in Female Gangs', Social Justice 24: 148–69.

Joe Laidler, K. and G. Hunt (2001) 'Accomplishing Femininity among the Girls in the Gang', British Journal of Criminology 41: 656–78.

Katz, J. and C. Jackson-Jacobs (2004) 'The criminologists' gang' in C. Sumner (ed.) Blackwell Companion to Criminology. London: Blackwell Publishers.

Kintrea, K., J. Bannister, J. Pickering, M. Reid and N. Suzuki (2008) Young People and Territoriality in British Cities. York: Joseph Rowntree Foundation.

Klein, M.W. (1971) Street Gangs and Street Workers. Englewood Cliffs, NJ: Prentice-Hall.

Knowsley, J. (1994) 'Earrings, Bracelets and Baseball Bats: Girl Gangs in Spotlight after Attack on Hurley', The Sunday Telegraph, 17 November 1994.

Knowsley, J. (1996) 'Girl Gangs Rival Boys in Battle to Rule the Streets', The Sunday Telegraph, 5 May 1996.

Lambert, A. (2001) 'Is This What We Meant by Feminism?', The Independent, 1 March 2001.

Laville, S. (2005) 'Girl Gang Leaders Get Life for Murder', The Guardian, 9 February 2005.

Leadbetter, R. (2006) 'City Chief Promises to Take Action on Girl Gangs', The Evening Times, 29 August 2006.

McNeill, F. and S. Batchelor (2002) 'Chaos, Containment and Change: Undertaking a Local Analysis of the Problems of Persistent Offending by Young People', Youth Justice 2(1): 27–43.

McRobbie, A. and J. Garber (1976) 'Girls and Subcultures: An Exploration', in S. Hall and T. Jefferson (eds) Resistance through Rituals. London: Hutchison.

Matza, D. and G. Sykes (1961) Juvenile delinquency and subterranean values. American Sociological Review 26(5): 712–19.

Miller, J. (2001) One of the Guys: Girls, Gangs and Gender. New York: Oxford University Press.

Miller, J. (2008) Getting Played: African American Girls, Urban Inequality and Gendered Violence. New York: New York University Press.

Millman, M. (1975) A Feminist View of the Sociology of Deviance', Sociological Inquiry 45(3): 251–77.

Moore, J. (1991) Going Down to the Barrio: Homeboys and Homegirls in Change. Philadelphia: Temple University Press.

Moore, J. and J. Hagedorn (2001) Female Gangs: A Focus on Research. Washington: US Department of Justice.

Muncie, J. (2009) Youth and Crime: A Critical Introduction, 3rd Edition. London: SAGE.

Naysmith, S. (2008) 'Crime Crosses the Gender Gap', The Herald, 10 June 2008.

Nurge, D. (2003) 'Liberating yet Limiting: The Paradox of Female Gang Membership', in D. Brotherton and L. Barrios (eds) The Almighty Latin King and Queen Nation: Street Politics and the Transformation of a New York City Gang, pp. 161–82. New York: Columbia University Press.

Oakeshott, I. (2002) 'Teenage Girls in Shocking Epidemic of Crime: Terrifying Rise of the Girl Gangs who Plague our Streets', Daily Mail, 9 May 2002.

O'Hara, M. (2007) 'Hidden Menace', The Guardian, 6 June 2007.

Parker, H. (1974) View from the Boys: A Sociology of Down-town Adolescents. Newton Abbot, Devon: David and Charles.

Pitts, J. (2007) Reluctant Gangsters: Youth Gangs in Waltham Forest. Luton: University of Bedfordshire. URL (accessible 7 August 2008): http://www.walthamforest.gov. uk/ reluctant-gangsters.pdf.

Prentice, E. (2000) 'Dark Side of Girl Power', The Times, 22 November 2000.

Sanders, B. (2002) Youth Crime and Youth Culture in the Inner City. London: Routledge.

Scott, P. (1956) 'Gangs and Delinquent Groups in London', British Journal of Delinquency 7: 4–24.

Sharp, C., J. Aldridge and J. Medina (2006) Delinquent Youth Groups and Offending. London: Home Office.

Skelton, T. (2000) 'Nothing to Do, Nowhere to Go? Teenage Girls and "Public" Space in the Rhondda Valleys, South Wales', in S. Holloway and G. Valentine (eds.) Children's Geographies: Playing, Living, Learning, pp. 80–99. London: Routledge.

Smith, D.J. and P. Bradshaw (2005) Gang Membership and Teenage Offending. Edinburgh: University of Edinburgh.

Stelfox, P. (1996) Gang Violence: Strategic and Tactical Options. London: Home Office.

Stephen, C. (1999) 'Girl Gangs Roam in Portlethen', The Northern Leader, 21 May 1999.

Thompson, T. (2001) 'Girls Lead the Pack in New Gangland Violence', The Observer, 15 April 2001.

Thompson, T. (2004) 'Teen Love, Teen Hate: Girl Gang Takes Savage Revenge', The Observer, 18 April 2004.

Thrasher, F.M. (1 927) The Gang: A Study of 1313 gangs. Chicago, IL: University of Chicago Press.

Young, T., M. Fitzgerald, S. Hallsworth and I. Joseph (2007) Guns, Gangs and Weapons. London: Youth Justice Board.

DISCUSSION QUESTIONS

1. What are some of the stereotypes about girls in gangs?
2. What challenges are researchers faced with in defining and measuring girl gangs and gang membership?
3. How can communities respond to the needs of girls who are exploring gangs and delinquent behavior?

Processing and Sentencing of Female Offenders

Section Highlights

- Processing and sentencing of female offenders
- Treatment of female offenders
- The role of patriarchy, chivalry, and paternalism in processing and sentencing female offenders

As you learned in Section VII, the gender gap in crime has remained consistent since 1990, contrary to the perceptions that dominate media representation of crime. For most crime types, the increase in female arrests reflects not an increase in offending rates of women, but rather a shift in policies to arrest and process cases within the criminal justice system that historically had been treated on an informal basis (Rennison, 2009; Steffensmeier & Allan, 1996; Steffensmeier, Zhong, Ackerman, Schwartz, & Agha, 2006). This section highlights the different ways in which gender bias occurs in the processing and sentencing of female offenders.

How might we explain the presence of gender bias in the processing of female offenders? Research highlights that women and girls can be treated differently than their male counterparts by agents of social control, such as police, prosecutors, and judges, as a result of their gender. Gender bias can occur in two different ways: (1) women can receive lenient treatment as a result of their gender, or (2) women may be treated harsher as a result of their gender. These two competing perspectives are known as the chivalry hypothesis and the evil woman hypothesis. The chivalry hypothesis suggests that women receive preferential treatment by the justice system. As one of the first scholars on this issue, Otto Pollak (1950) noted that agents of the criminal justice system are reluctant to criminalize women, even though their behaviors may be just as criminal as their male counterparts. However, this leniency can be costly, as it reinforces a system whereby women are denied an equal status with men in society (Belknap, 2007). While most research indicates the presence of chivalrous practices toward women, the potential for sex discrimination against women exists whereby they are treated more harshly than their male counterparts, even when charged with the same offense. Here, the evil woman hypothesis suggests that women are punished not only for violating the law, but for breaking the socialized norms of gender-role expectations (Nagel & Hagan, 1983).

Throughout the past forty years, research has been inconclusive about whether or not girls receive chivalrous treatment. While the majority of studies indicate that girls do receive leniency in the criminal justice system, the presence of chivalry is dependent on several factors. This section focuses on four general themes in assessing the effects of chivalry on the processing and treatment of female offenders: (1) the stage of the criminal system, (2) the race of the offender, (3) the effect of legal and extralegal characteristics, and (4) the impact of sentencing guidelines on the processing of female offenders.

Stage of the Criminal Justice System

Chivalry can occur at different stages of the criminal justice system. Much of the research on whether women benefit from chivalrous treatment looks only at one stage of the criminal justice process. At the pretrial stage, research indicates that women are more likely to be treated leniently than men. Here, the power of discretion is held by the prosecutor, who determines the charges that will be filed against an offender and whether charge-reduction strategies will be employed in order to secure a guilty plea. Charge-reduction strategies involve instances where an offender agrees to plead guilty in exchange for a lesser charge and a reduction in sentence. Some research indicates that women are less likely to have charges filed against them, or are more likely to receive charge reductions, compared to their male counterparts (Albonetti, 1986; Saulters-Tubbs, 1993). Research by Spohn, Gruhl, and Welch (1987) found that women of all ethnic groups were more likely, compared with men of all ethnic groups, to benefit from a charge reduction. Given the shift toward determinant sentencing structures and the reduction of judicial discretion, the power of the prosecutor in this practice increases. While research by Wooldridge and Griffin (2005) indicated an increase in the practice of charge reductions under state sentencing guidelines in Ohio, their results indicated that women did not benefit from this practice any more or less than male offenders. While seriousness of crime and criminal history remain the best predictors of receiving a charge reduction, research is inconclusive on the issue of the effect of gender on this process.

Of all the stages of the criminal justice system, the likelihood of pretrial release is the least studied; given that this stage has one of the highest potentials for discretion by prosecutors and the judiciary, it is important to assess whether gender plays a role in the decision to detain someone prior to trial. Research by Demuth and Steffensmeier (2004) found that females are less likely to be detained at the pretrial stage than men, controlling for factors such as offense severity and criminal history. Several factors can influence the presence of chivalrous treatment for women at this stage. Offense type affects this process, as female offenders who were charged with property-based offenses were less likely to receive pretrial detention compared to males with similar offenses (Ball & Bostaph, 2009). Generally speaking, females are typically viewed as less dangerous than their male counterparts, which results in the lenient treatment of females, making them less likely to be detained during the pretrial process (Leiber, Brubacker, & Fox, 2009). Women are also more likely to have significant ties to the community, such as family and childrearing duties, which make it less likely that they will fail to appear for future court proceedings (Steffensmeier, Kramer, & Streifel, 1993). Women who are charged with drug or property crimes are significantly less likely to be detained prior to trial compared to women who engage in crimes against persons (Freiburger & Hilinski, 2010). Indeed, this bias appears throughout the pretrial process, as women are 30% less likely than men to be detained prior to trial and also receive lower bail amounts than men (and therefore run less risk of being forced to remain in custody due to an inability to make bail). While this preferential treatment exists for women compared to men regardless of race/ethnicity, White women do receive the greatest leniency compared to Hispanic and Black women. Here, research indicates that women of color are less likely to be able to post bond, resulting in their detention at the pretrial stage, compared to White women.

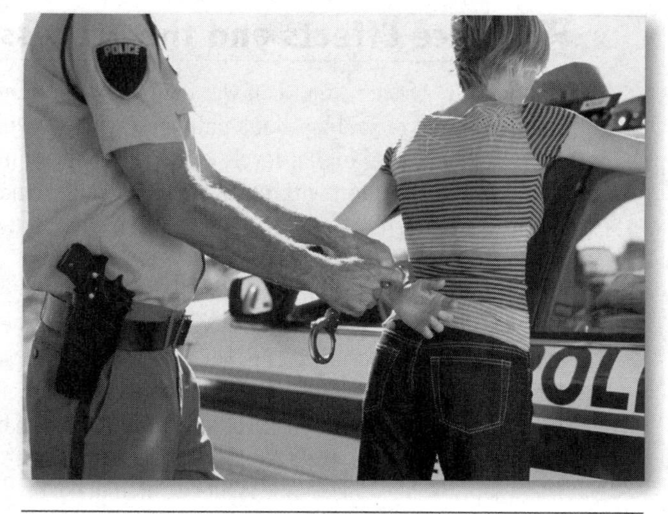

▲ **Photo 9.1** Throughout the 20th century, the number of arrests of females increased dramatically, though much of these data reflect changes in policy regarding the processing of women and girls by the justice system versus a direct increase in female offending rates.

The appearance of preferential or chivalrous treatment in the early stages of criminal justice processing also affects how women and girls will be treated in later stages, as females who already receive more favorable treatment by prosecutors continue to receive such chivalrous treatment as their cases progress. The majority of research indicates that women are more likely to receive chivalrous treatment at sentencing. At this stage of the criminal justice process, women are less likely to be viewed as dangerous (Freiburger & Hilinski, 2009) and are less likely to recidivate (Daly, 1994). Women without a prior criminal record, and who receive a reduction in charges, are more likely to receive leniency at sentencing, as they were less likely to receive a jail sentence and more likely to be sentenced to probation supervision (Farnworth & Teske, 1995). Indeed, women are viewed as better candidates for probation supervision compared to male offenders (Freiburger & Hilinski, 2009). Research on the decision to incarcerate reflects that women are less likely to be sent to jail or prison for their crimes, compared to men (Spohn & Beichner, 2000). However, research on the length of sentences for women who are incarcerated is a bit mixed. While most research indicates that women receive shorter sentences of incarceration, a deeper look reveals that women are more likely to receive shorter sentences to jail, but no gender differences exist for length of prison sentences (Freiburger & Hilinski, 2009). Offense type also affects the relationship between gender and sentencing, as women are less likely to receive prison sentences for property and drug cases than their male counterparts. In those cases where women are incarcerated for these crimes, their sentence length is significantly shorter compared to the sentence length for male property and drug offenders. Here, the disparity in sentencing can be attributed to the levels of discretion exercised by judges in making sentencing decisions (Rodriguez, Curry, & Lee, 2006).

Race Effects and the Processing of Female Offenders

Historically, African American women have been punished harsher than White women. This punishment reflected not a racial bias, but a pattern consistent with their levels of offending, as women of color traditionally engaged in higher levels of crimes than White women and the types of offenses they engaged in had more in common with male offenders than female offenders. Over time, research indicated that the offending patterns of White women shifted such that women, regardless of race or ethnic status, engaged in similar levels of offending.

Significant bodies of research address concerns over the differential processing of male offenders on the basis of race and ethnicity. Here, research consistently agrees that men of color are overrepresented at every stage of the criminal justice system. Given these findings, what effect does discrimination have for female offenders? Several scholars have suggested that chivalry is selective and is more likely to advantage White females over women of color. "As a rule, women of color, poor women, younger women, women immigrants, and lesbians are afforded less leniency or are processed more prejudicially than other females" (Belknap, 2007, p. 154). While White women make up the majority of the prison population in 2008 (47.9%), African American women constitute 27.8% of incarcerated women and Hispanic/Latina women make up 16.7% of the population. Given that U.S. Census data for 2009 (http://www.census.gov) indicate that African Americans make up 12.9% of the population and persons of Hispanic origin make up 15.8%, these statistics reflect a practice of overrepresentation of ethnic minorities among incarcerated women. In addition, women incarcerated at the federal level for drug offenses are predominantly African American. Given these findings, some researchers have questions on whether discriminatory views about women offenders, and particularly women of color, may negatively influence prosecutorial and judicial decision-making processes (Gilbert, 2001). Given the significant powers of prosecutors in making charge decisions, offering plea agreements and charge reductions, and making sentence recommendations, the potential effects of racial and ethnic bias can be significant.

In the early stages of crime processing, the interactions between gender and race can give the impression that women of color are treated more harshly by the criminal justice system. However, research findings indicate that the bias may be one of economics, rather than race. Katz and Spohn (1995) found that White women are more likely to be released from custody during the pretrial stages, compared to Black women, as a result of the ability to fulfill demands for bail. When defendants cannot make bail, there may be incentives to accept a plea deal that would limit the time spent in custody. Yet this "freedom" comes at a cost, as the label of an "ex-felon" can affect them and limit the available options for the future for the rest of their lives.

Despite the evidence that women of color do not receive chivalrous treatment by the criminal justice system, some research indicates the presence of leniency for girls and women of color. Here, scholars suggest that the preferential treatment of African American girls by judges is seen as an attempt to remedy the biased decision making of criminal justice actors during earlier stages of the criminal justice process that may have led to harsher attention (lack of pretrial release and bail options, less likely to receive charge reductions, etc.; Leiber et al., 2009). Research indicates that race can have an effect on the sentencing practices for both adult and juvenile offenders. In one study on sentencing outcomes for juveniles, Guevara, Herz, and Spohn (2008) indicated that race effects did not always mean that girls of color were treated more harshly than White girls. Their results indicate that White females were more likely to receive an out-of-home placement. While many would suggest that an out-of-home placement is a more significant sanction, their research indicates that juvenile court officials may be engaging in "child saving" tactics in an effort to rehabilitate young offenders. Because girls of color were less likely to receive this sentencing option, juvenile justice officials might be suggesting that White girls were more likely to receive help from an out-of-home placement. This conclusion relies heavily on racialized stereotypes about the likelihood to rehabilitate and creates a bias in the treatment of offenders.

The Effects of Legal and Extralegal Factors on Sentencing Women

The assessment of whether women benefit from chivalrous treatment by the criminal justice system is not as simple as comparing the sentences granted to men and women in general. Many factors must be considered, including the severity of the offense, the criminal record of the offender, the levels of injury experienced by the victim, and the culpability or blameworthiness of the offender. For example, women generally have a smaller criminal history than males and are less likely to engage in violent offenses or play a major role in criminal offenses.

In assessing whether women receive chivalrous treatment, it is important to control for these legal and extralegal variables. Research indicates that legal variables do affect the decision-making process for both males and females, albeit in different ways. In addition, the effects of gender vary with each stage of the criminal justice system. For example, offense type influences whether a defendant will be detained during the pretrial stages. Here, women who were charged with a property crime were less likely to be detained or denied bail compared to women charged with a drug offense. In comparison, males charged with a property crime were more likely to be detained compared to those charged with a drug offense (Ball & Bostaph, 2009). Second, the presence of a prior record or being under the supervision of the criminal justice system significantly increased the likelihood of being detained at the pretrial stage for both males and females; however, the effect of these variables had a stronger effect for female offenders than males (Freiburger & Hilinski, 2010). Finally, women who were convicted of a lesser offense had significantly lower odds of receiving an incarceration sentence compared to men who were convicted of the same charge (Spohn & Beichner, 2000).

Research by Spohn and Beichner (2000) indicated that not only do legal variables affect sentencing decisions, but the gender effect of these indicators can vary by jurisdiction. For example, female offenders from Kansas City were more likely than men to be incarcerated if they were currently on probation at the time of the current offense. In comparison, the presence of a prior record of incarceration had a greater effect for men than women in Chicago.

Not only do legal factors such as criminal history and offense severity appear to affect the pretrial decision process for women, extralegal factors such as the type of attorney also affect the likelihood of pretrial release for women. Women who were able to hire a private attorney were 2½ times more likely to make bail, compared to those women who were reliant on the services of a public defender. Clearly, the ability to hire (and financially afford) a private attorney is linked to the ability to satisfy the financial demands of bail as set by the court. In comparison, women who were represented by a public defender were twice as likely to be detained at the pretrial stage (Ball & Bostaph, 2009).

Ties to the community (such as family life) can also serve as an extralegal factor that can mediate sentencing practices. For example, motherhood appears to mitigate a sentence of incarceration, as women with dependent children are less likely to be incarcerated compared to women who do not have children. In these cases, judges appear to consider the social costs of imprisoning mothers and the effects of incarceration on children, particularly in cases of nonviolent or drug offenses (Spohn & Beichner, 2000). In a review of departures from strict sentencing guidelines, research by Raeder (1995) indicates that single parenthood and pregnancy may be used as rationale for granting women a reduced sentence. Here, it is not gender specifically that accounts for mitigation, but rather concern for the family (as non-familied women do not receive similar instances of leniency in sentencing). These departures have been confirmed by the courts in cases such as U.S. v. Johnson (964 F.2d 124 [2d Cir. 1992]). Indeed, such departures are not reserved exclusively for women but can also benefit male defendants who are the primary caregiver for minor children (see U.S. v. Cabell, 890 F. Supp. 13, 19 [D.D.C. 1995], which granted a departure from the sentencing guidelines for a male offender who was the primary caregiver for the children of his deceased sister's children).

Gender-Neutral Laws and Sentencing Guidelines

Throughout most of history, judges have had high levels of discretion in handing out sentences to offenders. In most cases, judges were free to impose just about any terms and conditions at sentencing, from probation to incarceration. Essentially, the only guidance for decision making came from the judge's own value system and belief in justice. This created a process whereby there was no consistency in sentencing, and offenders received dramatically different sentences for the same offenses, whereby the outcome depended on which judge heard their case. While this practice allowed for individualized justice based on the needs of offenders and their potential for rehabilitation, it also left the door open for the potential of bias based on the age, race, ethnicity, and gender of the offender.

During the 1970s, the faith in rehabilitation for corrections began to wane and was replaced with the theory of "just deserts," a retributive philosophy that aimed to increase the punishment of offenders for their crimes against society. In an effort to reform sentencing practices and reduce the levels of discretion within the judiciary, many jurisdictions developed sentencing guidelines to create systems by which offenders would receive similar sentences for similar crimes. At the heart of this campaign was an attempt to regulate sentencing practices and eliminate racial, gender, and class-based discrimination in courts. As part of the Sentencing Reform Act of 1984, the U.S. Sentencing Commission was tasked with crafting sentencing guidelines at the federal level. Since their implementation in November 1987, these guidelines have been criticized for being too rigid and unnecessarily harsh. In many cases, these criticisms reflect a growing concern that judges are now unable to consider the unique circumstances of the crime or characteristics of the offender. Indeed, the only standardized factors that are to be considered under the federal sentencing guidelines are the offense committed, the presence of aggravating or mitigating circumstances, and the criminal history of the offender.

Prior to sentencing reform at the federal level, the majority of female offenders were sentenced to community-based programs, such as probation. Under federal sentencing guidelines, not only are the numbers of incarcerated women expanding, but the length of time that they will spend in custody has increased as well. In addition, these changes in sentencing practices also led to the abolition of parole at the federal level and in many states—meaning that women have few options to reduce their sentences due to their rehabilitative efforts.

Research by Koons-Witt (2006) investigates the effects of gender in sentencing in Minnesota. Minnesota first implemented sentencing guidelines in 1980. Like the federal sentencing guidelines, Minnesota founded their guidelines on a retributive philosophy focused on punishment for the offender.

The guidelines were designed to be neutral on race, gender, class, and social factors. However, the courts can consider aggravating and mitigating factors such as the offender's role in the crime if they make a decision outside of the sentencing guidelines. Koons-Witt investigated the influence of gender at three distinct points in time: prior to the adoption of sentencing guidelines in Minnesota, following their introduction (early implementation 1981–1984), and in 1994, 14 years after the sentencing guidelines were implemented (late implementation). Her research indicated that female offenders were more likely to be older than their male counterparts and have a greater number of dependent children. In contrast, men were faced with more serious crimes, were more likely to be under community supervision at the time of the current offense, and had greater criminal histories. Prior to the implementation of sentencing guidelines, gender did not appear to have an effect on sentencing guidelines. This finding contradicted the findings of other research, which illustrated that judges did treat female offenders in a chivalrous fashion. The one exception for Koons-Witt's findings was that sentences were reduced for women who had dependent children. Following the early implementation of sentencing guidelines (1981–1984), several legal outcomes increased the potential for incarceration regardless of gender. These legal factors include prior criminal history and pretrial detention. This pattern was repeated during the late implementation time period (1994). However, the influence of extralegal factors reappeared during this time period, whereby women with dependent children were more likely to receive community correctional sentences compared to women who did not have children. In these cases, Koons-Witt suggests that the courts may be using the presence of dependent children as a mitigating factor in their decision to depart from the sentencing guidelines, producing an indirect effect for the preferential treatment of women.

Some critics of gender-neutral sentencing argue that directed sentencing structures like the federal sentencing guidelines have affected women in a negative fashion. These sentence structures assume that men and women have an equal position in society and, therefore, the unique needs of women do not need to be considered when making sentencing decisions. While the intent behind the creation of sentencing guidelines and mandatory minimums was to standardize sentencing practices so that offenders with similar legal variables received similar sentences, the effect has been an increased length of incarceration sentences for both men and women. Given the inability of judicial officials to consider extralegal factors in making sentencing decisions, these efforts to equalize sentencing practices have significantly affected women.

Conclusion

This section reviewed how and when preferential treatment is extended to female offenders. Whether chivalry exists within the criminal justice system is not an easy question to answer, as it is dependent on the stage of the criminal justice system, the intersectionality of race and ethnicity, legal and extralegal factors, and the implementation of determinate sentencing structures. Even in cases where research suggests that chivalrous treatment serves women through shorter sentences and an increased likelihood to sentence offenders to community-based sanctions over incarceration, not all scholars see this preferential treatment as a positive asset for women. For many, the presence of chivalry is also linked to these gender-role expectations whereby "preferential or punitive treatment is meted out based on the degree to which court actors perceive a female defendant as fitting the stereotype of either a good or bad woman" (Griffin & Wooldredge, 2006, p. 896). Not only does the potential for women to be punished for breaking gender role expectations exist (i.e., the evil woman hypothesis), but there exists a double-edged sword in fighting for special treatment models. Gender equality does not necessarily mean "sameness." Rather, this perspective suggests that women possess cultural and biological differences that should be considered in determining the effects of "justice." However, there is a potential danger in treating women differently as a result of these cultural and biological indicators. Given that the law affords reductions in sentencing based on mental capacity and age (juvenile offenders), to extend this treatment toward women can suggest that women "cannot be expected to conform their behavior to the norms of the law . . . thus when women are granted special treatment, they are reduced to the moral status of infants" (Nagel & Johnson, 2004, p. 208).

Summary

- Chivalry can occur at different stages of the system. Women who receive preferential treatment in the early stages of processing may continue to benefit from chivalry in the later stages.
- Generally speaking, women are more likely to be released during pretrial stages, receive charge reductions, and receive a jail or probation supervision sentence.
- When women are incarcerated, they typically receive shorter sentences compared to men.
- While research generally finds that men of color are treated more harshly than White men, research on race and gender is mixed, with some studies indicating that women of color are treated more harshly than Whites, and other researchers finding that women of color are treated in a more lenient fashion compared to White women.
- Legal factors such as criminal history and offense type affect the processing of women.
- Extralegal factors such as the type of attorney can affect the likelihood of pretrial release for women.
- Women who are responsible for the care of young children are more likely to be afforded chivalrous treatment by the courts at the time of sentencing.
- Gender-neutral laws, such as sentencing guidelines, have significantly increased the number of women serving time in U.S. prisons.

KEY TERMS

Chivalry

Evil woman hypothesis

Legal factors

Extralegal factors

Gender-neutral laws

Sentencing guidelines

DISCUSSION QUESTIONS

1. Why is it important to study the processing of female offenders at each stage of the criminal justice system, versus just during the final disposition?
2. How do prosecutors and judges use their discretion to give preferential or chivalrous treatment toward women?
3. Which legal and extralegal factors appear to have the greatest impact on the processing of females? Which variables indicate preferential treatment of women in unexpected ways?

READING

In this article, you'll learn how race and ethnicity affect the sentences for female juvenile offenders. The authors investigate two hypotheses: (1) race and ethnicity will affect sentencing outcomes, regardless of offense severity and prior history; and (2) race and ethnicity effects on sentencing vary based on offense severity and criminal history. Their findings indicate that race and ethnicity serve both as an advantage and disadvantage in sentencing dispositions.

Racial and Ethnic Disparities in Girls' Sentencing in the Juvenile Justice System

Lori D. Moore and Irene Padavic

Recent trends point to the need for renewed theoretical and research attention to sentencing decisions for juvenile girls of different races and ethnicities. Girls are the fastest growing group of offenders in the juvenile justice system (Office of Juvenile Justice and Delinquency Prevention, 2009; Stahl et al., 2007), but important variations exist among them. Racial/ethnic minority girls are overrepresented in the juvenile system (Schaffner, 2006), and they tend to receive harsher punishment than White girls (e.g., Guevara, Herz, & Spohn, 2006). Questions about why court decision makers might treat similarly offending girls differently falls in the domain of feminist theory, and sociologists and criminologists have begun drawing on such theory to unpack the sentencing process.

This article draws on the population of Black, Hispanic, and White girls who were judicially disposed in the Florida Department of Juvenile Justice (FL DJJ) to examine hypotheses deriving from theory about the power of racialized gender expectations. In particular, we ask two questions informed by theory about how social control operates through gender-role expectations: Do racial and ethnic disparities exist between Black and White and between Hispanic and White girls that legally relevant factors fail to explain? And how do factors associated with gender role ideology affect the relationship between race/ethnicity and disposition severity?

These questions arise from feminist scholarship that attempts to understand how gender bias appears in the criminal justice and juvenile justice systems. Women and girls (compared to men and boys) in these institutions are devalued, excluded, typecast, or neglected, primarily because the institutions' goals are not geared toward them (Carr et al., 2008, p. 26).

The problem of women's poor fit with the goals of these justice systems appears in all areas that include women, from guards to judges to prisoners. Female prison guards, for example, are stereotyped as having better interpersonal skills and being more adept at defusing conflicts than male guards, but these qualities have little cachet in the gendered organizational logic of

SOURCE: Moore, L. D., & Padavic, I. (2010). Racial and ethnic disparities in girls' sentencing in the juvenile justice system. *Feminist Criminology, 5*(3), 263–285.

the prison system (Britton, 1997). Women lawyers and judges in the Florida legal system found sex typing and harassment on the part of men judges to be marked (Padavic & Orcutt, 1997). In the juvenile justice system, girls compared to boys face harsher punishment for lesser offenses, are monitored and reprimanded more frequently, are sentenced to treatment facilities that fail to take into account the gendered pathways to delinquency, and are less likely to receive substance abuse treatment (Belknap & Holsinger, 2006; Carr et al., 2008; MacDonald & Chesney-Lind, 2001; Tracy et al., 2009). In sum, women and girls involved in the criminal and juvenile justice systems experience them differently from men and, in the case of girl offenders, they travel gendered pathways to offending and find themselves in gendered organizations that reinforce gender inequality.

Gender-Role Expectations and Punishment

Feminist scholarship has challenged the notion that the primary function of the legal system is to protect and rehabilitate women and girls (the chivalry hypothesis). Research on gender and punishment in the juvenile justice system, for example, tends to find that girls are more likely than boys to receive harsher punishment for similar offenses and for status offenses (e.g., running away or truancy), and they are more likely to receive harsher punishment at younger ages (Belknap & Holsinger, 2006; Carr et al., 2008; Chesney-Lind, 2006; Chesney-Lind & MacDonald, 2001; Sherman, 2005; Tracey et al., 2009; Zimring, 2005). Researchers tend to attribute these findings to a patriarchal justice system that devalues women and girls (e.g., Carr et al., 2008; Chesney-Lind, 2006).

Research has found that girls tend to be punished for sex-inappropriate crimes and for expressions of anger typically associated with boys; in other words, anger and sex-role inappropriate behavior in girls evoke sanctions. An experimental study, for example, showed that college students sought to mete out harsher punishment to girls who engaged in gender-atypical crimes (e.g., selling pornography) than to girls involved in gender-typical crimes (e.g., check kiting; Sinden, 1981). A study of juvenile offenders in a Midwestern county (Mallicoat, 2007) found that probation officers characterized the sexually inappropriate behavior of girls (such as having unprotected sex) as showing a lack of moral character, while tending to ignore similar behavior by boys (unless such behavior was criminally punishable). As for emotional expressions, girls behaving in stereotypically male ways (e.g., acting aggressively, using weapons, being violent, and expressing anger rather than sadness) can elicit exceptionally harsh reactions from officials. In a study of girls held at a juvenile detention facility, a probation officer said "Claudia thinks she's one of the guys. She's not very ladylike, and that gets her into a lot of trouble. Her first response is to start jumpin' bad [being physically aggressive]" (Schaffner, 2006, p. 129). The same study found that girls who behaved aggressively were sanctioned more harshly than other girls, both formally (with legal sanctions) and informally (with verbal reprimands). In a study of police encounters with juvenile suspects, police officers more often arrested girls who behaved in stereotypically masculine ways (e.g., aggressive or hostile) than girls who behaved in stereotypically feminine ways (e.g., deferent or apologetic; Visher, 1983). Finally, based on a study of the case files of girls in an Arizona juvenile justice court, Gaarder and colleagues (2004, p. 576) concluded that, "When girls did not adhere to 'feminine' behaviors or attitudes, there was often an assumption that they were 'becoming more like boys,' and should be treated as boys would be." In sum, the "simplistic notions of 'good' and 'bad' femininity . . . permit the demonization of some girls and women if they stray from the path of 'true' (passive, controlled and constrained) womanhood" (Chesney-Lind, 2006, p. 11).

Racialized Gender-Role Expectations and Punishment

Intersectionality theory (Collins, 2000; Crenshaw, 1991) posits that a "matrix of domination" creates a hierarchy of privilege and oppression that can simultaneously locate an individual in a position of advantage and disadvantage, depending on the reference group being used for comparisons. Intersectionality theory identifies gender, race, class and sexuality as interlocking points of disadvantage or advantage (Bettie, 2003; Collins, 2000, 2004; Crenshaw, 1994; McCall, 2005). According to Collins (2000, p. 278), "Oppression is filled with . . . contradictions . . . because each individual derives varying amounts of penalty and privilege from the multiple systems of oppression which frame everyone's lives." Intersectionality theory implies that researchers seeking to understand how patriarchy operates in the criminal and juvenile justices system must center their analyses on the "race/gender/crime nexus" (Chesney-Lind, 2006, p. 10).

The economic, social, and spatial marginalization of women and girls of color tends to locate them in positions of disadvantage in terms of offending and official reactions to their offending. In a study of Hawaiian parolees, for example, Native Hawaiian women were more likely than women of other ancestries to offend at earlier ages, hold juvenile records, recidivate, and be incarcerated (Brown, 2006). The researcher linked these outcomes to the marginalization of Native Hawaiian women (Brown, 2006). Marginalization also affects how Black women and girls are policed in urban neighborhoods. A study of Black youth in St. Louis showed that police presence had a negative effect on girls, because police were on hand to harass them for minor status offenses, such as curfew violations, but were slow to respond (or failed to respond) when girls reported an emergency or were in danger (Brunson & Miller, 2006).

Race and ethnicity shape how the criminal and juvenile justice systems punish women and girls. Although some research shows that girls tend to receive harsher punishment than boys, the findings become more complex when researchers consider the intersection of race/ethnicity and gender. In a study of disparities in the Florida juvenile justice system, for example, White and minority girls were less likely than their same-race male counterparts to receive detention rather than be released, but minority girls were more likely to receive detention than Whites of either sex (Bishop & Frazier, 1996). In a study of gender and race in one Midwestern county's juvenile justice system, White girls received harsher punishment than Black boys, and for Black girls, leniency was concentrated only in the early stages (e.g., intake and petition; Leiber et al., 2009). Other research on sentencing in the criminal justice system has found that Black women (e.g., Kramer, 2002; Kramer & Ulmer, 2009) and Black and Hispanic women (e.g., Steffensmeier & Demuth, 2006) are not disadvantaged vis-à-vis White women. Thus the intersection of gender with race/ethnicity will change any sentencing prediction based solely on one of the statuses. This variability in outcomes arises because an individual might be nested in a privileged position on the matrix of domination by virtue of gender but disadvantaged by virtue of race, or vice versa.

Stereotypes that court officials hold about the gender-role expectations for women and girls of different races and ethnicities have been thought to underlie disparities in the criminal and juvenile justice systems (e.g., Bickle & Peterson, 1991; Guevera, Herz & Spohn, 2006; Leiber et al., 2009; Mallicoat, 2007). White girls are stereotyped as passive, in need of protection, nonthreatening, and amenable to rehabilitation (Bickle & Peterson, 1991; Franklin & Fearn, 2008; Gaarder et al., 2004). Black girls and women, in contrast, are stereotyped as independent, aggressive, loud, pushy, rude, sexual, unfeminine, deserving of violence, and crime-prone (Collins, 2004, pp. 122–125 & 131; Miller, 2008; Moore & Hagedorn, 1996, p. 216; Sinden, 1981). Hispanic women (and girls by logical extension) are seen as dependent and submissive, family oriented, domestic, and highly sexual (or "sexual firebrands"; Cofer, 1993, pp. 143 & 145; Espin, 1984, p. 425; Segura & Pierce, 1993, pp. 296 & 300). Some of these stereotypes have been shown to be active in the criminal and juvenile justice systems. A study of female death row inmates accused of murdering their husbands, for example, found that the media and the court portrayed Black and Hispanic defendants as "explosive or hotheaded" and White defendants as "coldhearted or cunning" (Farr, 1997, p. 267). In the juvenile justice system, one study (Bridges & Steen, 1998), based on 233 narratives from probation officers in the western United States, found that they held negative attributions about Black youthful offenders, whom they stereotyped as lacking respect for the law and holding bad attitudes that they described as stemming from character flaws. In contrast, they attributed White juvenile offending to family structure or delinquent peers, which are elements outside of individual character. The researchers concluded that probation officers' stereotypes about the causes of offending influenced their recommendations, making them more likely to propose harsher sentences for Black than for White offenders. Finally, in a study of girls on probation in an Arizona county, researchers found that probation officers held views about appropriate feminine behavior that failed to match up with girls' social realities or experiences (e.g., race, culture, class, gender, mental health, and substance abuse histories; Gaarder, Rodriguez & Zatz, 2004). Thus court officials' stereotypes about race/ethnicity and gender also may play a role in how girls are sentenced.

Black and Hispanic girls, because of their lower order position on the matrix of domination and their deviation from appropriate femininity in the eyes of court officials, are likely to face harsher sentencing than White girls, although refinements of racialized gender-role expectations can complicate this general pattern. A study of eight criminal court districts found that "gender-based role expectations" created sentencing disparities between White and Black women (Bickle & Peterson, 1991). In particular, being a caretaker (i.e., having a husband or children) afforded White women greater leniency than similar Black women. Yet the results were more complex; performing this role well afforded Black women greater leniency than it afforded White women. The implication is that court officials stereotype White women as naturally good at caretaking and thus do not grant them credit for performing an expected role, whereas they expect Black women to be poor wives and mothers and, therefore, reward them with shorter sentences when they outperform relative to expectations. Another example of the complexity in the link between gender, race expectations, and sentencing comes from a Los Angeles-based study that found that juvenile justice probation officers stereotyped Black girls' offending behavior as the product of poor lifestyle choices or criminality, whereas they regarded White and Hispanic girls' offending behavior as the product of low self-esteem and the influence of delinquent peers (Miller, 1994, pp. 232–239). In the case of stereotypically masculine offenses, officers' stereotypes led to harsher sentences for Black and Hispanic than for White girls. In another study of race, gender, and punishment in two Midwestern juvenile courts, White girls were more likely than non-White girls to be charged and receive an out-of-home placement versus a less harsh sentence (probation; Guevara, Herz, & Spohn, 2006). In sum, interactions between race/ethnicity and gender yield complex patterns of punishment outcomes in the criminal and juvenile justice systems.

Hypotheses and Analytic Strategy

The sections above inform two hypotheses. The first pertains to our overall predictions about race/ethnicity and disposition severity after controlling for other factors (such as age, current offense severity, and prior record). The second pertains to how the courts might punish girls according to a racialized gender-role ideology where disposition decisions are made based on a combination of the offender's race/ethnicity and her offending behavior. We note that the data do not allow for a direct test of the impact of gender ideology in sentencing decisions; rather we can only indicate whether our findings are consistent or inconsistent with such an interpretation.

Our first hypothesis predicts racial disparities in sentencing, net of girls' offense severity and prior records. Offense severity and prior record may matter in court officials' sentencing decisions for several reasons. These factors bear on court officials' goals, which are to hand down decisions that are appropriate to the seriousness of the crime (fair retribution), that reduce the likelihood of recidivism, maximize the likelihood of rehabilitation, and decrease risks to the community (Steffensmeier, Ulmer, & Kramer, 1998). In general, minority girls compared to White girls tend to have more prior contacts with the juvenile justice system and are often charged with more serious offenses. Yet, if the system turns a blind eye toward a girl's race/ethnicity, we should expect to see similar sentences for White and for minority girls, once offense severity and prior record are controlled. If we do not, then it lends support to the notion that judges are bringing to bear attributes of the girls that are not part of their record but nonetheless differentially affect sentencing decisions—a finding consistent with the sentencing disparities literature.

Our second hypothesis predicts that the effects of race/ethnicity on disposition severity are contingent on the offense and prior record. We predict harsh sentences for minority compared to White girls at the lowest levels of offense severity and prior record and harsh sentences for White compared to minority girls at the highest levels. Such findings would provide indirect support for the notion of a racialized gender-role ideology, where girls are punished according to their location in the race/ delinquency hierarchy. These findings would also imply that additive models (e.g., race/ethnicity, plus current offense severity, plus prior record) are not sufficient to explain disposition disparities. Our hypotheses are the following:

> *Hypothesis 1:* Black and Hispanic girls will receive harsher dispositions than White girls net of age, current offense severity, and prior record.
>
> *Hypothesis 2:* White girls whose current offenses are minor (below the index mean) will receive more lenient dispositions than similar minority girls, but White girls whose current offenses are severe (above the index mean) will receive harsher dispositions than similar minority girls.
>
> *Hypothesis 3:* White girls whose prior records are minor (below the index mean) will receive more lenient dispositions than similar minority girls, but White girls whose prior records are severe (above the index mean) will receive harsher dispositions than similar minority girls.

Our analytic strategy is as follows. To test our first hypothesis, we regress disposition severity on race/ ethnicity, age, current offense severity, and prior record to establish if there are racial/ethnic disparities in dispositions. We test the second and third hypotheses to determine if and how the effects of race on disposition depend on the current offense severity and prior record.

Data and Method

Data come from the Florida Department of Juvenile Justice Offense File for the 2006 fiscal year, which contains the entire population of juveniles referred to the FL DJJ. A major strength of the dataset is its comprehensiveness, since the system tracks juveniles from referral through disposition and maintains records of all encounters with the FL DJJ from age 10 through 18. One other element of the dataset stands out as a strength: It contains information on a large number of Hispanic youth. A weakness, however, is that because the dataset lacks an indicator of jurisdiction, we cannot control for between-jurisdiction variation, which has been shown in prior research to affect juvenile justice outcomes (e.g., Burruss & Kempf-Leonard, 2002; DeJong & Jackson, 1998; Feld, 1991).

The data were originally organized around referrals rather than individuals, so we used each offender's case identity number to reorganize the unit of analysis as individuals. We restricted our analyses to girls because our main theoretical focus is on disposition disparities among girls. We selected girls whose last offense occurred between July 1, 2005 and June 30, 2006 ($n = 18{,}472$) and captured offending history for 3 years beginning on July 1, 2003 and ending on the date of their current offense.

Dependent Variable

We measured disposition using four dummy-coded categories of severity: (a) nonjudicial, (b) probation, (c) commitment to a facility, and (d) transfer to adult court. Florida's juvenile justice system is made up of a series of decision points that include arrest, intake, nonjudicial disposition, and judicial dispositions (probation, commitment, or transfer to adult court; FL DJJ, 2009). We focused our analysis on nonjudicial and judicial dispositions because our theory concerns decision making at the end of this sequence. Nonjudicial dispositions are the least severe category, as they do not require an appearance before a judge and stipulate punishments such as apologizing to the victim, attending workshops, or performing community service. The second category, probation, is more severe because it requires an appearance before a judge. The third category, commitment to a facility, is more severe than nonjudicial and probation dispositions not only because it requires an appearance before a judge but also because the DJJ holds the youth in custody. The fourth category, transfer to adult court, is the most severe because the DJJ relinquishes its jurisdiction to the adult court, where the record becomes nonconfidential and the youth becomes established as an adult offender (Office of Juvenile Justice and Delinquency Prevention [OJJDP], 1999). Studies tend to find that youthful offenders transferred to adult court are sentenced more harshly than in the juvenile court; however, researchers note that adult courts treat juvenile offenders more leniently than adult offenders (e.g., Kupchik, 2006). Because our analyses focus on the juvenile court only, we conceived of transfer to the adult court as the most severe outcome. Compared to other juvenile justice systems, Florida's is considerably more punitive and may be increasingly so, especially in terms of commitments and transfers to adult court (OJJDP, 1999). In Florida, the "get tough" sentiment is captured by former Governor Jeb Bush in a 2002 statement: "Florida's youth custody officers have apprehended thousands of juvenile offenders who think our 'Tough Love' message is just words. The juveniles are finding out the consequences are very real" (FL DJJ, 2002).

Independent Variables

The primary independent variable is race/ethnicity. Blacks, Hispanics, and Whites were measured as dummy variables, with Whites used as the reference group. Other racial/ethnic groups were excluded because of their small numbers.

Research has identified age as affecting juvenile sentencing, and we measured it as a continuous variable ranging from 10 to 18 years, with an average of almost 16. Court officials often take age into consideration when deciding how to mete out punishment, and age has been found to afford younger juvenile offenders more leniency than older offenders because youthfulness is assumed to be associated with diminished culpability (e.g., Leiber & Johnson, 2008). The Florida juvenile justice system has no lower bound for juvenile court jurisdiction and an upper bound of 21 (although in some circumstances, young adults aged 21 and older can be handled in the juvenile court; Snyder & Sickmund, 2006). We restricted the sample to youth age 10 to 18 because of small sample size at the lower and upper limits and because similar age ranges are commonly used in the literature (e.g., Bishop & Frazier, 1996; Guevara, Herz, & Spohn, 2006; MacDonald, 2003).

Legal variables—current offense severity and prior record—have also been identified in the literature. The current offense is the last, most serious offense committed during the study period, but because offenders can be charged with multiple offenses on one day, the most serious offense committed on the last date of offense was selected. Current offense severity is a discrete measure that ranges from 1 to 3, with 1 being the least and 3 being the most. Offense categories were coded as follows: 1 = misdemeanor, 2 = felony, 3 = felony against a person. We conceived of felony against a person as being the most severe category and separate it from other felonies because girls (and disproportionately girls of color) are increasingly coming into contact with the juvenile justice system with person or violent offenses (Chesney-Lind & Irwin, 2008; Snyder & Sickmund, 2006). Assaultive offenses, which school officials may be more likely to accuse Black than White girls of committing (American Civil Liberties Union, 2007) also were coded as felonies against persons. We note that our data are limited in that they only contain delinquency records and not status offenses, which have been shown to differ in punishment outcomes by gender and race (e.g., Chesney-Lind, 1997; Chesney-Lind & Sheldon, 2004; Snyder & Sickmund, 2006). Other studies that have focused on delinquency case processing have found significant gender and racial differences in outcomes (e.g., Leiber et al, 2009). Prior record also has been identified as important in explaining racial disparities (see Engen, Steen, & Bridges, 2002). To create the prior record index variable, we divided the sum of each juvenile's prior offense severity by the number of prior offenses (Bishop & Frazier, 1996).

To test the prediction that the effects of race/ethnicity on disposition severity depend on the offense severity and the prior record, we constructed interaction terms. One set is Black × Current offense severity and Hispanic × Current offense severity, where the reference group is White × Current offense severity. Interactions with prior record were measured similarly. These measures allowed us to determine whether patterns are consistent or inconsistent with claims about judges acting on a racialized gender-role ideology.

Tables 9.1 and 9.2 show descriptive statistics. Average age is almost 16, and approximately 40% of the girls are Black, 49% White, and 11% Hispanic. A majority (66%) received a nonjudicial disposition, 26% received probation, 6% received a commitment, and approximately 1% received transfer dispositions. Table 9.2 shows the number of Black, White, and Hispanic girls by offense severity and disposition.

Table 9.1 Descriptive Statistics for Girls

Variable	Coded Value	Mean/Prop	Range
Race			
Black	1 = Black, 0 = other	0.40	0–1
White	1 = White, 0 = other	0.49	0–1
Hispanic	1 = Hispanic, 0 = other	0.11	0–1
Age	Continuous	15.70	10–18
Offense severity	1 = Misdemeanor, 2 = Felony, 3 = Felony Person	1.25	1–3
Prior record	Index—Sum of prior offense severity/No. of prior offenses	2.02	0–8
Disposition			
Nonjudicial	1 = Nonjudicial, 0 = other	0.66	0–1
Probation	1 = Probation, 0 = other	0.26	0–1
Committed	1 = Committed, 0 = other	0.06	0–1
Transfer	1 = Transfer, 0 = other	.009	0–1

NOTE: $N = 18{,}472$.

Table 9.2 Number of Black, White, and Hispanic Girls by Offense Severity and Disposition

Disposition	Misdemeanor			Felony			Felony Person		
	Black	White	Hispanic	Black	White	Hispanic	Black	White	Hispanic
Nonjudicial	3,831	5,509	1,164	284	505	132	421	263	101
Probation	1,785	1,728	397	163	243	51	297	206	64
Commitment	411	434	87	46	50	10	61	43	10
Transfer	30	14	3	25	38	7	36	16	7

Results

Table 9.3 shows cross-tabulations of mean offense severity by race and ethnicity by disposition severity and reveals that within each disposition, Black and Hispanic girls compared to White girls have worse current offenses.

Table 9.4 shows cross-tabulations of mean prior record by race and ethnicity by disposition severity and reveals that within each disposition, Black and Hispanic girls have worse prior records than White girls.

In Table 9.5, Model 1 regressed disposition on race/ethnicity, age, current offense, and prior record. Contrary to our expectation that Hispanic girls would receive harsher dispositions than White girls, Hispanic ethnicity was not significant in all models, lending no support to Hypothesis 1 for Hispanics. Our expectations for Black girls were confirmed: Holding age, current offense severity, and prior record constant, their odds of receiving a probation versus a nonjudicial disposition were 1.27 times greater than White girls', 1.19 times greater for commitment rather than a nonjudicial disposition, and 1½ times greater for a transfer rather than a nonjudicial disposition.

Model 2 includes interaction terms for race/ethnicity and current offense severity. Black girls' odds of receiving a probation versus a nonjudicial disposition are about 28% higher than White girls' (odds ratio = 1.28) when current offense severity is at its centered mean, which means that Black girls are more likely than White girls to receive the harsher disposition for an average offense.

Figure 9.1 shows how race is contextualized by current offense severity for the probation versus a nonjudicial disposition decision. For all levels of current offense severity, Black girls were more likely than White girls to receive a probation disposition versus a nonjudicial

Table 9.3 Offense Severity Means for Black, White, and Hispanic Girls by Disposition

Disposition	Black	White	Hispanic	Total
Nonjudicial	1.25	1.16	1.24	1.20
Probation	1.34	1.30	1.35	1.32
Commitment	1.32	1.26	1.28	1.29
Transfer	2.07	2.03	2.24	2.07
Total	1.29	1.21	1.28	1.25

Table 9.4 Prior Record Means for Black, White, and Hispanic Girls by Disposition

Disposition	Black	White	Hispanic	Total
Nonjudicial	1.62	1.16	1.44	1.36
Probation	3.25	3.15	3.24	3.21
Commitment	3.70	3.53	4.00	3.65
Transfer	4.00	3.91	4.44	4.01
Total	2.29	1.80	2.05	2.02

Table 9.5 Odds Ratios for Multinomial Logistic Regression of Disposition on Extralegal and Legal Variables for Girls

	Model 1			Model 2			Model 3		
	Probation	Commitment	Transfer	Probation	Commitment	Transfer	Probation	Commitment	Transfer
Extralegal variables									
Race/ethnicity									
Black	1.27***	1.19*	1.52*	1.28***	1.18*	1.88**	1.31***	1.22**	1.58*
Hispanic	0.96	0.80	0.96	0.96	0.80	0.94	1.00	0.78	0.96
Age	1.14***	1.13***	4.10***	1.14***	1.13***	4.11***	1.14***	1.13***	4.10***
Legal variables									
Current offense severity	1.20***	1.06	3.55***	1.31***	1.10	4.37***	1.19***	1.06	3.54***
Prior record	1.49***	1.62***	1.64***	1.49***	1.62***	1.64***	1.57***	1.69***	1.72***
Interaction terms									
Black × Current offense				0.85*	0.96	0.68*			
Hispanic × Current offense				0.92	0.89	0.95			
Black × Prior record							0.91***	0.92**	0.91
Hispanic × Prior record							0.90***	0.95	0.94
Pseudo R^2	0.13			0.13			0.13		
−2 log likelihood	13,622.19			13,617.38			13,605.90		
N	18,472			18,472			18,472		

NOTE: Coefficients are multinomial odds ratios. Base outcome is nonjudicial.

*$p < .05$; **$p < .01$; ***$p < .001$.

Figure 9.1 Probability of Probation Versus a Nonjudicial Disposition for Black Girls Compared to White Girls, by Current Offense Severity

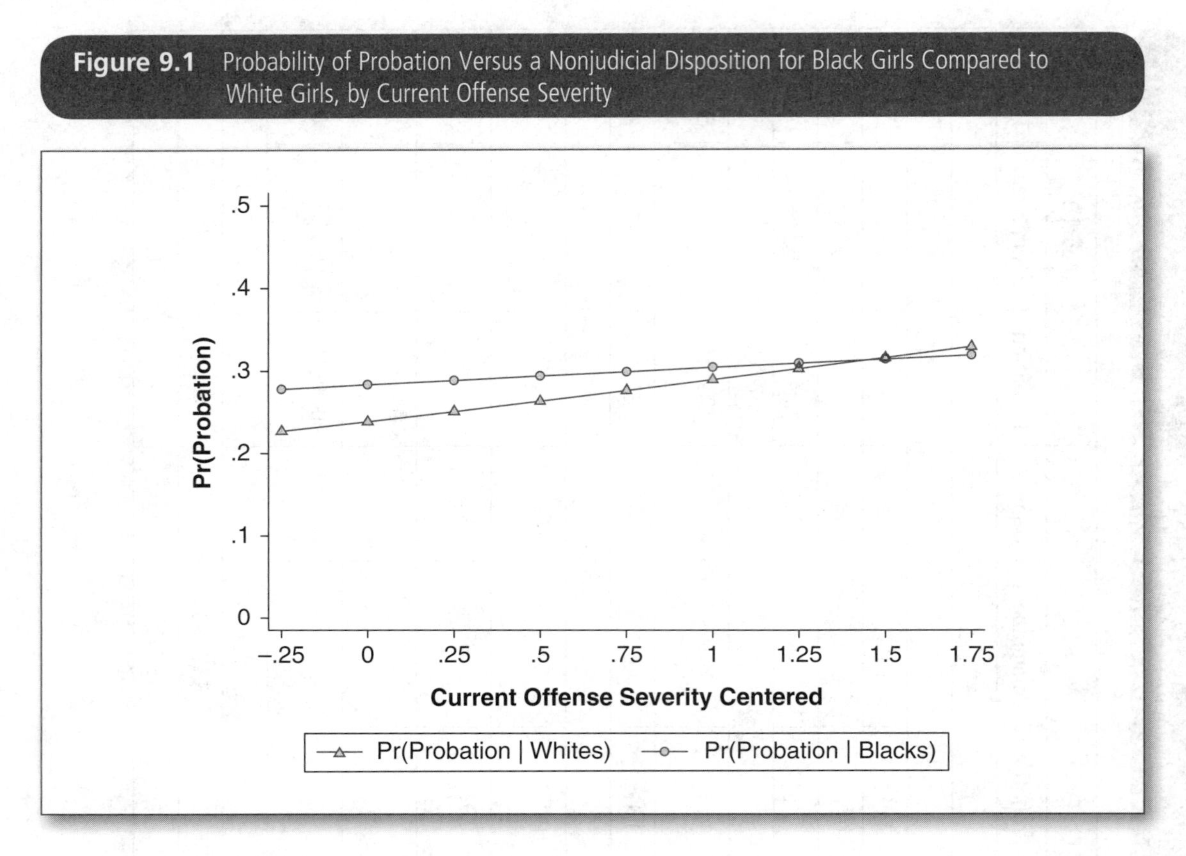

disposition, except at the highest levels of offense severity, where the probability converges. Figure 9.2 shows a similar pattern for the transfer versus the nonjudicial disposition. These findings are consistent with the notion of a racialized gender-role ideology (and with Hypothesis 2). It appears that although initially the juvenile justice system granted White girls leniency, it became increasingly intolerant as their delinquency increased in severity.

Model 3 includes interaction terms for race/ethnicity and prior record to test Hypothesis 3, about the effects of prior record and race on sentencing outcomes. Black girls were more likely than White girls to receive the harsher disposition for an average prior record.

Figure 9.3 shows that at the least extensive levels of prior record, Black girls were more likely than their White counterparts to receive a probation disposition relative to a nonjudicial disposition. As prior record becomes more extensive, however, Black girls were less likely than their White counterparts to receive probation relative to a nonjudicial disposition. This finding is consistent with racialized gender-role ideology. The juvenile justice system punishes Black girls more harshly than White girls but only for the group of girls with average or below-average prior records. For girls with above-average prior records, however, the juvenile justice system metes out harsher punishment to White girls than Black girls. Figure 9.4 shows that for lower levels of prior record, Black girls were more likely than White girls to receive a commitment disposition relative to a nonjudicial disposition, whereas at the higher levels, White girls were more likely to receive the harsher disposition, which demonstrates that the juvenile justice system grants White compared to Black girls leniency only up to a certain point.

As Figure 9.5 illustrates, Hispanic ethnicity was insignificant for all models, with one exception.

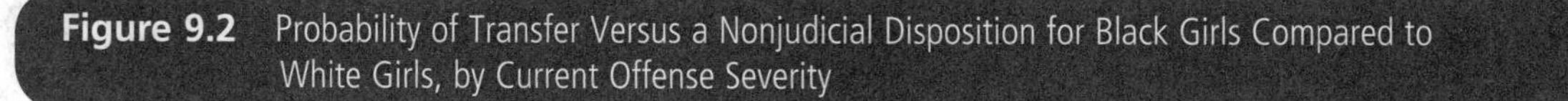

Figure 9.2 Probability of Transfer Versus a Nonjudicial Disposition for Black Girls Compared to White Girls, by Current Offense Severity

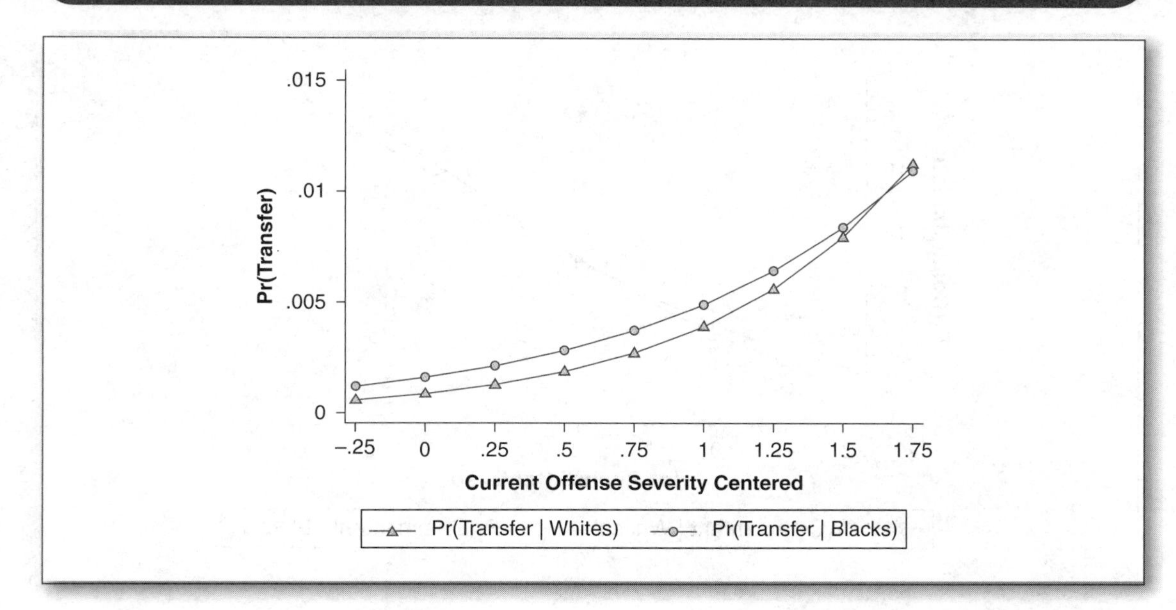

Figure 9.3 Probability of Probation Versus a Nonjudicial Disposition for Black Girls Compared to White Girls, by Prior Record

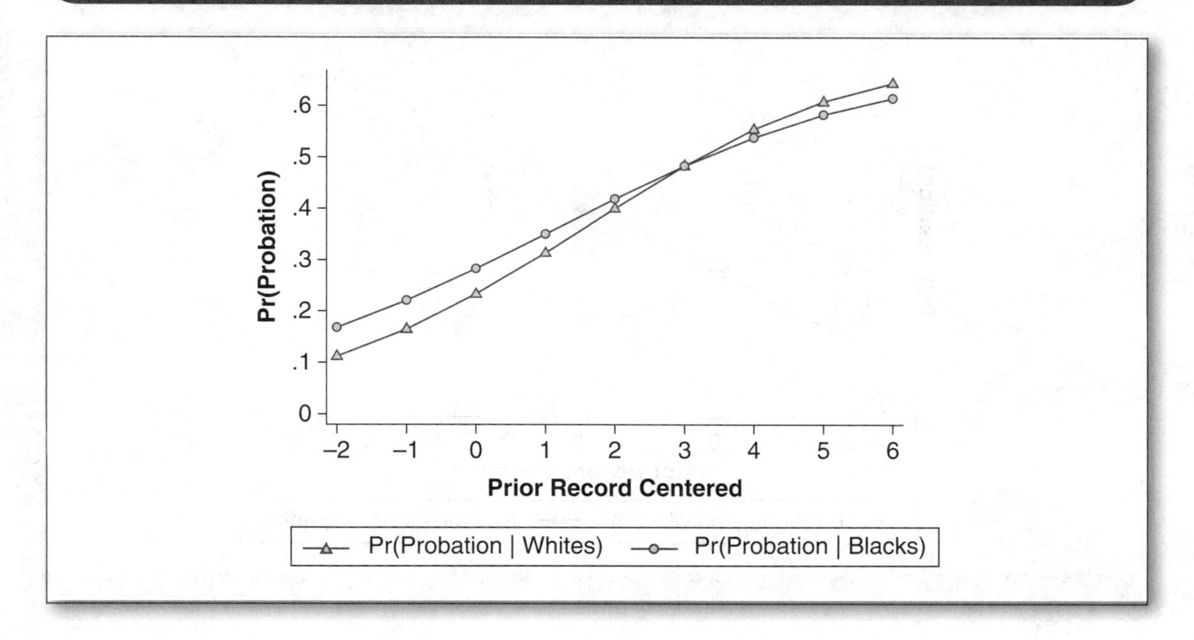

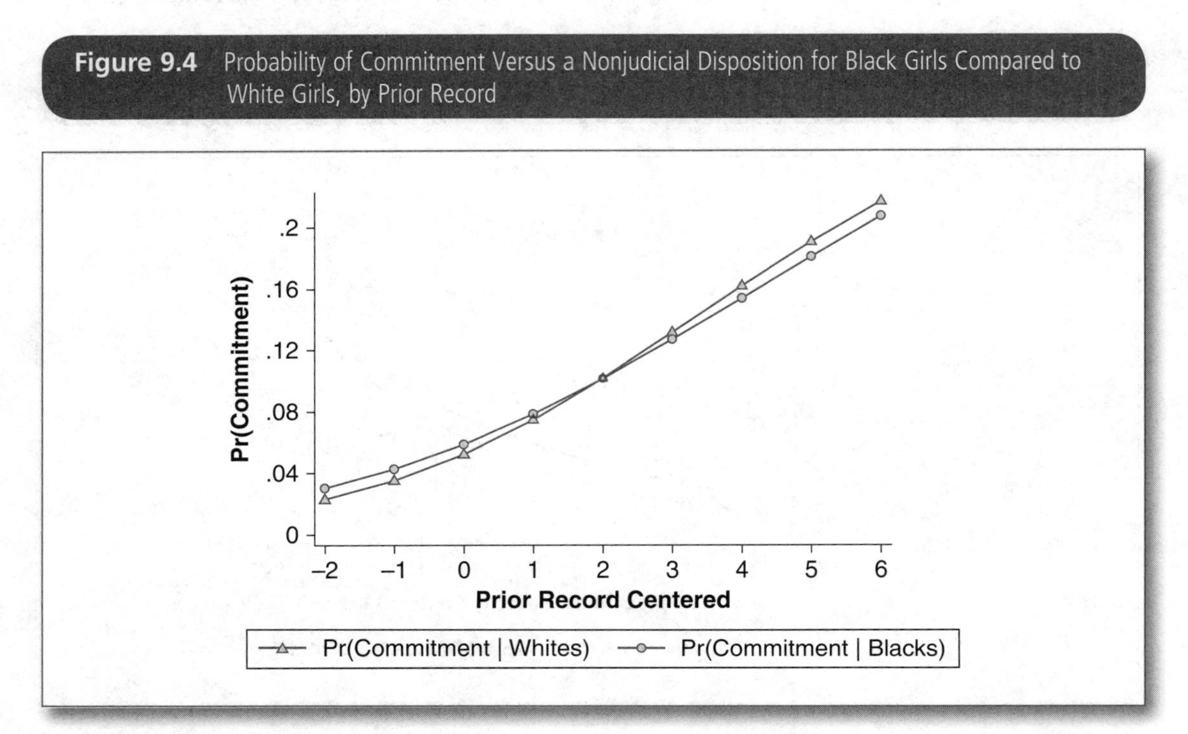

Figure 9.4 Probability of Commitment Versus a Nonjudicial Disposition for Black Girls Compared to White Girls, by Prior Record

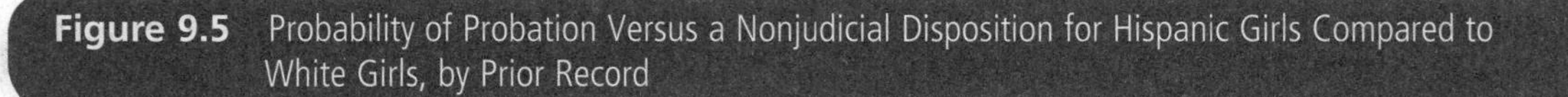

Figure 9.5 Probability of Probation Versus a Nonjudicial Disposition for Hispanic Girls Compared to White Girls, by Prior Record

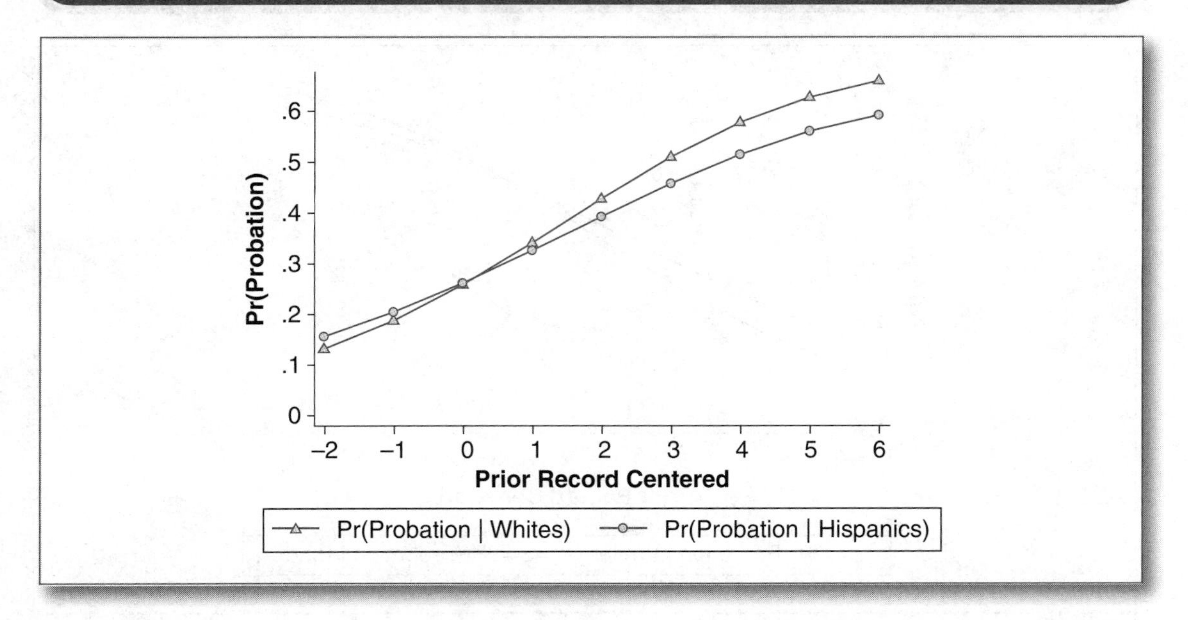

Hispanic girls with below-the-mean prior records were more likely than their White counterparts to receive probation relative to a nonjudicial disposition; in other words, they received harsher sentences. At the mean value for prior record, the predicted probability for Hispanic and White girls converged. For every subsequent value for prior record, Hispanic girls were less likely than their White counterparts to receive probation relative to a nonjudicial disposition, meaning that the worst-offending Hispanic girls were granted more leniency than the worst-offending White girls. This finding is consistent with racialized gender-role ideology, which predicts that White girls compared to racial and ethnic minority girls receive leniency in sentencing, but that White girls lose their advantage when they deviate too far beyond average offending behavior.

Discussion and Conclusion

Based on data for the population of Black, White, and Hispanic female youthful offenders in Florida, we have demonstrated how the interaction of race/ethnicity and legal variables (current offense severity and prior record) situates girls in positions of relative advantage and disadvantage vis-à-vis sentencing decisions. We argue that a racialized gender-role ideology underlies this disparity. Below, we summarize our findings and discuss their implications.

Consistent with past research concerning the main effects of race/ethnicity on disposition, we found that racial and ethnic minority girls received harsher punishment than White girls, with one important exception: the case of Hispanic girls in some circumstances. As expected, compared to White girls, Black girls received more severe dispositions even after taking into account the seriousness of the offense, prior record, and age. This finding provides evidence of Black-White racial bias in the juvenile justice system. Hispanic girls, in contrast, were not disadvantaged vis-à-vis White girls. Our main effects model and all but one of the interaction models showed no statistically significant differences in disposition severity. This finding is surprising in light of previous research that found a Hispanic girl disadvantage (e.g., Miller, 1994), and could be because of stereotypes about the reasons for Hispanic girls' delinquency. It is possible that court officials hold a gender-role ideology that locates Hispanic and White girls similarly in the matrix of domination because stereotypes associated with White women and girls (e.g., passive and nonthreatening; Bickle & Peterson, 1991) overlap with stereotypes associated with Hispanic women and girls (e.g., submissive and dependent; Cofer, 1993). Juvenile court officials have been found to be lenient toward Hispanic girls because of their attributions about Hispanic girls' motivations for offending and about the girls' family lives. A study of 174 girls' case files in Arizona, for example, showed court officials directing sympathy toward Hispanics whom they believed had acted out as a result of struggles between their ethnic and American identities and toward those whose families spoke English (because they viewed these families as more cooperative and functional; Gaarder, Rodriguez, & Zatz, 2004). Furthermore, although the Miller (1994) study found a Hispanic girl disadvantage, the attributions court officials made about their pathways to delinquency were similar to those they made for White girls—low self-esteem and peer pressure. Future research on how court officials assign attributes to Hispanic girls would further researchers' understanding of how ethnicity and gender play out in sentencing. Black girls, however, failed to enjoy the leniency granted to White and Hispanic girls.

How does a racialized gender-role ideology create disposition disparities? Our analyses revealed that the effects of race/ethnicity on disposition severity were conditioned by girls' current and prior offending behavior. In four of the six tests, White girls compared to Black girls were granted leniency in disposition decisions, but only up to a threshold, at which point their probabilities of receiving a harsher disposition either converged or surpassed their racial/ethnic minority counterparts. These findings suggest that the juvenile justice system is tolerant of White girls with minor-to-average offense severity levels and low-to-average prior records but relatively intolerant of their Black counterparts. As White girls surpass what the juvenile justice system considers acceptable offending behavior for their racial group, it reacts in an increasingly punitive manner. The juvenile justice system appears to be unmoved by above-average levels of Black girls' offending behavior, perhaps because judges expect high levels of deviance from this group. If the juvenile justice system holds behavioral expectations for girls based on their membership in different racial/ethnic groups, then a failure on the part of the girl to conform to that expected behavior could result in negative sanctions (i.e., harsher dispositions), an interpretation consistent with our findings.

We caution that unmeasured factors (e.g., family status, school status, gang membership, parental status, jurisdictional variation, and so on) might be correlated with race/ethnicity in ways that could render our race/ethnic differences spurious. We also were unable to test for interaction effects found significant in previous research, such as the interaction between race and court location (e.g., Eisenstein, Flemming, & Nardulli, 1988; Feld, 1991; Ulmer & Johnson, 2004) and between race and SES (Zatz, 2000). We also note that because we lack data on decision-makers' gender-role ideology, our findings allow us only to point to patterns consistent with such an ideology, not to claim that sentencing derives from it. Given the trends of increasing rates of girls' incarceration and the racialized dimensions of the problem, it is crucial to gain a better understanding of the links between sentencing and decision-makers' ideologies about gender and race. Future research that incorporates mixed methods (interviews with court actors and offenders, courtroom observations, along with quantitative data on offender behaviors) similar to Daly's (1994) examination of gender, race, and the criminal justice system, is likely to yield the type of data that allow a thorough examination of these dimensions.

What are the implications of our findings for research on racial/ethnic and gender disparities in the juvenile justice system and for juvenile justice policy? The primary implication is the utility of considering an intersectional approach to determine if and how the intersection of race/ethnicity, gender, and other variables creates justice disparities. By using such an approach, which leads to testing for interaction effects, we discovered that the effects of race/ethnicity on disposition severity are contextualized by the level of girls' offending behavior. In particular, we found several threshold effects for current offense severity and prior record for White compared to Black girls and one for White compared to Hispanic girls. Researchers examining the possibility that disparities in dispositions are complicated by the intersections of race/ethnicity, gender, and other factors will be poised to make important contributions to understanding these disparities and how court officials react to girls based on a combination of their race and behavior.

What are the implications of our findings for juvenile justice policy? Our findings speak to the need for gender-specific programming that takes into account girls' needs and particular pathways into delinquency. Researchers have called attention to the lack of programming that considers girls' varied cultural backgrounds and addresses substance abuse, motherhood, employment, and sexual abuse (Bloom, Owen, & Covington, 2005; Chesney-Lind, 2001). Policies and programs aimed at taking seriously the girls' lives, delinquency pathways, and social and cultural backgrounds are likely to help, rather than harm, girls who come into contact with the juvenile justice system.

References

American Civil Liberties Union. (2007). *Locating the school-to-prison pipeline.* Retrieved July 16, 2010 from http://www.aclu.org/crimjustice/juv/35553res20080606.html

Belknap, J., & Holsinger, K. (2006). The gendered nature of risk factors for delinquency. *Feminist Criminology, 1,* 48–71.

Bettie, J. (2003). *Women without class: Girls, race and identity.* Berkeley: University of California Press.

Bickle, G., & Peterson, R. (1991). The impact of gender-based family roles on criminal sentencing. *Social Problems, 38,* 373–394.

Bishop, D. M., & Frazier, C. E. (1996). Race effects in juvenile justice decision-making: Findings of a statewide analysis. *Journal of Criminal Law and Criminology, 86,* 392–414.

Bloom, B., Owen, B., & Covington, S. (2005). *Gender-responsive strategies for women offenders: A summary of research, practice and guiding principles for women offenders.* Washington, DC: U.S. Department of Justice, National Institute of Corrections.

Bridges, G., & Steen, S. (1998). Racial disparities in official assessments of juvenile offenders: Attributional stereotypes as mediating mechanisms. *American Sociological Review, 63,* 554–570.

Britton, D. (1997). Gendered organizational logic: Policy and practice in men's and women's prisons. *Gender & Society, 11,* 796–818.

Brown, M. (2006). Gender, ethnicity, and offending over the life course: Women's pathways to offending in the Aloha state. *Critical Criminology, 14,* 137–158.

Brunson, R., & Miller, J. (2006). Gender, race and urban policing: The experience of African American youths. *Gender & Society, 20,* 531–552.

Burrass, G., Jr., & Kempf-Leonard, K. (2002). The questionable advantage of defense counsel in juvenile court. *Justice Quarterly, 19,* 37–68.

Carr, N., Hudson, K., Hanks, R., & Hunt, A. (2008). Gender effects along the juvenile justice system: Evidence of a gendered organization. *Feminist Criminology, 3,* 25–43.

Chesney-Lind, M. (1997). *The female offender.* Thousand Oaks, CA: Sage.

Chesney-Lind, M. (2001, Spring). Are girls closing the gender gap in violence? *Criminal Justice Magazine, 16*(1).

Chesney-Lind, M. (2006). Patriarchy, crime, and justice: Feminist criminology in an era of backlash. *Feminist Criminology, 1,* 6–26.

Chesney-Lind, M., & Irwin, K. (2008). *Beyond bad girls: Gender, violence and hype.* New York: Routledge.

Chesney-Lind, M., & Shelden, R. (2004). *Girls, delinquency and juvenile justice.* Los Angeles: Wadsworth.

Cofer, J. O. (1993). The myth of the Latin woman: I just met a girl named Maria. In E. Disch (Ed.), *Reconstructing gender: A multicultural anthology* (4th ed.). (pp. 142–149). Boston: McGraw Hill.

Collins, P. H. (2000). *Black feminist thought: Knowledge, consciousness, and the politics of empowerment.* New York: Routledge.

Collins, P. H. (2004). *Black sexual politics: African Americans, gender, and the new racism.* New York: Routledge.

Crenshaw, K. (1991). De-marginalizing the intersection of race and sex: A Black feminist critique of antidiscrimination doctrine, feminist theory, and antiracist politics. In K. Bartlett & R. Kenney (Eds.), *Feminist legal theory* (pp. 57–80). Boulder, CO: Westview.

Crenshaw, H. (1994). Mapping the margins: Intersectionality, identity politics, and violence against women of color. In M. A. Fineman & R. Mykitiuk (Eds.), *The public nature of private violence* (pp. 93–118). New York: Routledge.

Daly, K. (1994). *Gender, crime and punishment.* New Haven, CT: Yale University Press.

DeJong, C., & Jackson, K. (1998). Putting race into context: Race, juvenile justice processing and urbanization. *Justice Quarterly, 15,* 487–504.

Eisenstein, J., Flemming, R., & Nardulli, P. (1988). *The contours of justice: Communities and their courts.* Boston: Little, Brown.

Engen, R. L., Steen, S., & Bridges, G. S. (2002). Racial disparities in the punishment of youth: A theoretical and empirical assessment of the literature. *Social Problems, 49,* 194–220.

Espin, O. M. (1984). Cultural and historical influences on sexuality in Hispanic/Latin women: Implications for psychotherapy. In M. L. Andersen & P. H. Collins (Eds.), *Race, class and gender: An anthology* (pp. 423–428). Belmont, CA: Wadsworth.

Fair, K. (1997). Aggravating and differentiating factors in the cases of White and minority women on death row. *Crime and Delinquency, 19,* 260–270.

Feld, B. (1991). Justice by geography: Urban, suburban, and rural variations in juvenile justice administration. *Journal of Law & Criminology, 82,* 156–210.

Florida Department of Juvenile Justice. (2009). *An overview of the delinquency disposition process.* Retrieved July 21, 2010, from http://www.djj.state.fl.us/Research/CAR/CAR_2009/%282008-09CAR%29-History-of-DJJ.pdf

Florida Department of Juvenile Justice. (2002, October 16). *Governor's tough love reform averts 1,000 juvefelonies, youth custody officers apprehend 4,000 juvenile offenders.* Florida Department of Juvenile Justice, p. 1.

Florida Department of Juvenile Justice Offense File. (2005-2006). [Computer File]. Tallahassee, FL: Florida Department of Juvenile Justice [distributor].

Franklin, C, & Fearn, N. (2008). Gender, race, and formal court decision-making outcomes: Chivalry/paternalism, conflict theory or gender conflict. *Journal of Criminal Justice, 36,* 279–290.

Gaarder, E., Rodriguez, N., & Zatz, M. (2004). Criers, liars, and manipulators: Probation officers' views of girls. *Justice Quarterly, 21,* 547–578.

Guevara, L., Herz, D., & Spohn, C. (2006). Gender and juvenile justice decision making: What role does race play? *Feminist Criminology, 1,* 258–282.

Kramer, J., & Ulmer, J. T. (2002). Downward departures for serious violent offenders: Local court "corrections" to Pennsylvania's sentencing guidelines. *Criminology, 40,* 601–636.

Kramer, J., & Ulmer, J. T. (2009). *Sentencing guidelines: Lessons from Pennsylvania.* Boulder, CO: Lynne Rienner.

Kupchik, A. (2006). *Judging juveniles: Prosecuting adolescents in adult and juvenile courts.* New York: New York University Press.

Leiber, M., Brubaker, S., & Fox, K. (2009). A closer look at the individual and joint effects of gender and race in juvenile justice decision making. *Feminist Criminology, 4,* 333–358.

Leiber, M., & Johnson, J. (2008). Being young and Black: What are their effects on juvenile justice decision making? *Crime & Delinquency, 54,* 560–581.

MacDonald, J. M. (2003). The effect of ethnicity on juvenile court decision making in Hawaii. *Youth and Society, 35,* 243–263.

MacDonald, J., & Chesney-Lind, M. (2001). Gender bias and juvenile justice revisited: A multiyear analysis. *Crime & Delinquency, 47,* 5–29.

Mallicoat, S. (2007). Gendered justice: Attributional differences between males and females in the juvenile courts. *Feminist Criminology,* 2, 4–30.

McCall, L. (2005). The complexity of intersectionality. *Signs, 30,* 1771–1800.

Miller, J. (1994). Race, gender and juvenile justice: An examination of disposition decisionmaking for delinquent girls. In M. D. Schwartz & D. Milovanovic (Eds.), *The intersection of race, gender and class in criminology* (pp. 219–246). New York: Garland.

Miller, J. (2008). Violence against urban African American girls. *Journal of Contemporary Criminal Justice, 24,* 148–162.

Moore, J. W., & Hagedorn, J. M. (1996). What happens to girls in the gang? In C. R. Huff (Ed.), *Gangs in America* (2nd ed., pp. 205–220). Thousand Oaks, CA: Sage.

Office of Juvenile Justice and Delinquency Prevention. (1999). *A study of juvenile transfers to criminal court in Florida.* Retrieved July 16, 2010 from http://www.ncjrs.gov/pdffilesl/ fs99113.pdf

Office of Juvenile Justice and Delinquency Prevention. (2009). *Juvenile arrests 2008.* Retrieved July 16, 2010 from http://www.ncjrs.gov/pdffilesl/ojjdp/228479.pdf

Padavic, I., & Orcutt, J. (1997). Perceptions of sexual harassment in the Florida legal system: A comparison of dominance and spillover explanations. *Gender & Society, 11,* 682–698.

Schaffher, L. (2006). *Girls in trouble with the law.* New Brunswick, NJ: Rutgers University Press.

Segura, D. A., Pierce, J. L. (1993). Chicana/o family structure and gender personality: Chodorow, familism, and psychoanalytic sociology revisited. In K. V. Hansen & A. I. Garey (Eds.), *Families in the U.S.: Kinship and domestic politics* (pp. 295–314). Philadelphia: Temple University Press.

Sherman, F. (2005). *Detention reform and girls: Challenges and solutions.* Baltimore: The Annie E. Casey Foundation. Retrieved July 16, 2010 from http://www.aecf.org/upload/ publicationfiles/jdai_pathways_girls.pdf

Sinden, P. (1981). Offender gender and perceptions of crime seriousness. *Sociological Spectrum, 1,* 39–52.

Snyder, H. N., Sickmund, M. (2006). *Juvenile offenders and victims: 2006 national report.* Washington, DC: National Center for Juvenile Justice.

Stahl, A., Puzzanchera, C, Livsey, S., Sladky, A., Finnegan, T. A., Tierney, N., et al. (2007). *Juvenile court statistics: 2003–2004.* U.S. Department of Justice. Retrieved July 16, 2010, from http://www.ncjrs.gov/htmFojjdp/218587/index.html

Steffensmeier, D., & Demuth, S. (2006). Does gender modify the effects of race-ethnicity on criminal sanctioning? *Journal of Quantitative Criminology, 22,* 241–261.

Steffensmeier, D., Ulmer, J. & Kramer, J. (1998). The interaction of race, gender, and age in criminal sentencing: The punishment cost of being young, Black, and male. *Criminology, 36,* 763–798.

Tracy, P., Kempf-Leonard, K., & Abramoske-James, S. (2009). Gender differences in delinquency and juvenile justice processing: Evidence from national data. *Crime & Delinquency, 55,* 171–215.

Ulmer, J., & Johnson, B. (2004). Sentencing in context: A multilevel analysis. *Criminology, 42,* 137–177.

Visher, C. A. (1983). Gender, police arrest decisions, and notions of chivalry. *Criminology, 21,* 5–28.

Zatz, M. (2000). The convergence of race, ethnicity, gender, and class on court decision making: A look toward the 21st century. *Policies, Processes and Decisions of the Criminal Justice System, 3,* 503–552.

Zimring, F. E. (2005). *American juvenile justice.* New York: Oxford University Press.

DISCUSSION QUESTIONS

1. How do gender-role expectations vary by race and ethnicity?
2. How do race and ethnicity affect sentencing outcomes in the juvenile court?
3. At what point of the system do White girls no longer benefit from chivalrous treatment? How does one's current offense and prior history affect the benefits of chivalrous treatment for White girls?

READING

In the section introduction, you learned about the practice of chivalry and how it can alter the processing and sentencing of female offenders with punishments that are more lenient than those of male offenders. This reading examines whether chivalry exists in the sentencing practices of female offenders in Finland. In addition to looking at offense severity and criminal history, the authors assess whether gender norms, such as motherhood (e.g., women staying home to raise their children) and employment status (with men being the providers for their families) affect sentencing practices.

Ages of Chivalry, Places of Paternalism

Gender and Criminal Sentencing in Finland

Candace Kruttschnitt and Jukka Savolainen

Over two decades ago, in what was the first major review of the relationship between gender and criminal court processing decisions, Nagel and Hagan (1983: 134) concluded that (1) female offenders receive preferential treatment over male offenders; (2) the preferential treatment accorded to females is more pronounced in the decisions pertaining to the granting of probation, suspended sentences or fines; and (3) gender has little effect on 'the most severe outcomes, such as imprisonment'. Two decades later, with considerably more rigorous research, the findings in the USA remain largely unchanged. Most studies analysing sex-based disparities in sentencing find that women are favoured over men in the 'in/out' decision (or the decision to imprison as opposed to granting probation) but not in length of prison sentence (see e.g. Daly and Bordt 1995; Kruttschnitt 1996; Daly and Tonry 1997).

Reviews of the relationship between sex and criminal court sentences offer essentially three explanations for the relatively persistent findings of discrimination in favour of female offenders. Some scholars argue that gender interacts with particular statuses and expectations that are associated with more or less severe dispositions. So, for example, the relatively lenient treatment accorded female offenders is explained by their greater childcare responsibilities and perceptions that they are less culpable and less of a threat to society than male offenders (Steffensmeier et al. 1993; Daly and Bordt 1995). Others voice concerns over the methodological quality of this research (see e.g. Wooldredge 1998), leading to the conclusion that any evidence of sex disparities is an artefact of inadequate statistical controls and, building on the prior rationale, the way in which gender can be 'embedded' in offence severity and prior record

SOURCE: Kruttschnitt, C., & Savolainen, J. (2009). Ages of chivalry, places of paternalism: Gender and criminal sentencing in Finland. *European Journal of Criminology, 6*(3), 225–247.

(Daly and Tonry 1997: 231). And still others argue that, despite the invocation of sentencing reforms, designed to limit judicial discretion based on extra-legal criteria such as sex and race, these reforms have had only a limited effect (Koons-Witt 2002; Griffin and Wooldredge 2006).

Although each of these factors may be important for understanding the extant research on sex and sentencing, we aim to advance this field by taking a different approach. We examine the effect of sex on sentencing outcomes with data from Finland. The use of Finnish data allows us (1) to draw on some of the innovations in feminist theory that have implications for criminologists and (2) to shed light on whether the prior findings are culturally, and in some respects economically and socially, unique to the United States (see also Boritch 1992). We begin with a review of the theoretical paradigm that has underpinned virtually all of the sex-sentencing research.

Sex Role Theory

To explain sex-based disparities in sentencing decisions, scholars have drawn primarily on sex role theory. The notions of chivalry and paternalism, although etymologically distinct, are frequently invoked to justify the leniency accorded to women (Moulds 1980). Chivalry connotes male protection of females, while paternalism—a slightly more pejorative concept—implies status and power differences between men and women; the status of women is likened to the status of children, who are essentially defenceless and propertyless (Nagel and Hagan 1983: 114). The opposing hypothesis is the 'evil woman thesis', which is based on the notion that women who commit crimes, particularly male-dominated offences, will be treated more severely than men because their crimes violate normative sex roles. Nagel and Hagan (1983: 135–6) correctly noted that these hypothetically opposing theses—chivalry and evil woman—are, in fact, more complementary than contradictory simply because both explain gender-based disparities in sentencing with the sex role attitudes and stereotypes of deviance processing agents.

Subsequent research shifted attention to the levels of informal control and family responsibilities in the lives of men and women and the ways in which the legal system responded to these controls (Kruttschnitt 1982; Kruttschnitt and Green 1984; Daly 1987, 1989a). By drawing attention to judges' and probation officers' reactions to men and women who adopt or abdicate normative gender scripts, whether in the form of women's economic dependency and its attendant social control or the gender division of labour that gives women more childcare responsibility than men, these paradigms have much in common with the paternalism and evil woman theses. Essentially, 'preferential or punitive treatment is meted out based on the degree to which court actors perceive a female defendant as fitting the stereotype of either a good or bad woman' (Griffin and Wooldredge 2006: 896).

Although sex role theory implies the potential for change (e.g. by altering our expectations about, and opportunities for, women), it cannot explain why individuals are rewarded for conforming to stereotypes or sanctioned for deviating from them without reverting to the rather circular notion that people prefer what is customary. More generally, Connell (1987: 47–54) provides a particularly cogent discussion of why sex role theory has lost its utility as a framework for the social analysis of gender. For our purposes, the previously noted point is the most important: that is, the theory creates a 'normative case' against which all individuals are compared and it fails to acknowledge how situational contexts and structural constraints make gender identities fluid, sometimes reinforcing normative patterns and at other times reflecting more androgynous patterns of action. Research on street crime in the USA, for example, reveals that, although many criminal activities are quite constrained by gender stereotypes (e.g. Steffensmeier 1983; Maher 1997; Mullins and Wright 2003), not all are; and the overlay of economic distress and racial inequalities can blur commonsense notions about women's involvement in, and rationales for, violent crime (see e.g. Miller 1998; Miller and White 2003; Kruttschnitt and Carbone-Lopez 2006). These concrete accounts of men's and women's involvement in violent crimes suggest that in some circumstances expectations based on sex roles are less relevant to their specific behaviours than are concerns over their livelihood or reputation. Perhaps failing to acknowledge the contingent and fluid nature of sex roles explains why scholars in the USA have not been very successful at predicting and explaining the sentences accorded non-white male and female offenders (see e.g. Daly 1989b; Bickle and Peterson 1991; Crew 1991; Griffin and Wooldredge 2006). What are thought to be normative gender roles are generally prototypes of white middle-class persons (West and Fenstermaker 1995: 18).

Advances: Feminist Criminology

There has been increasing interest in the situational and contingent nature of gender relations from both quantitative and qualitative feminist criminologists. Miller and Mullins (2006) provide an excellent review of this research and, in so doing, drew our attention to the potential relevance of Connell's (1987, 2002) gender theory for understanding sex-based disparities in criminal sanctions. Connell (2002: 10) begins her thesis by arguing that:

> Gender patterns may differ strikingly from one cultural context to another, but are still 'gender.' Gender arrangements are reproduced socially (not biologically) by the power of structures to constrain individual action, so often they appear unchanging. Yet gender arrangements are in fact always changing, as human practice creates new situations and as structures develop crisis tendencies.

Each of these points is explored in her work as she examines gender across multiple levels—the gender order (patterns of power relations), the gender regime (patterns of gender arrangements within and between institutions) and gender relations (the interactions between individuals)—to try to help to make sense of different patterns of gender politics. In pursuing these themes, Connell builds on ethnographies that show how, on the micro level, gender is accomplished in social action (e.g. Thorne 1993) and how, on the macro level, gendered regimes can preserve or disrupt taken for granted differences between men and women, including differences in power and equality (West and Fenstermaker 1995). Finally, Connell notes that gender reforms over the last two decades of the 20th century were uneven, correcting imbalances in some societies (for example, with the arrival of women *en masse* to party politics in Scandinavian countries) and maintaining them in others, with a gender backlash (Connell 2002: 128–35).

How is this relevant to sex-based disparities in sentencing? It suggests that gender orders and regimes should be fundamental to our understanding of sex disparities in criminal sentencing. Yet relatively little research of this type exists.[1] Bridges' and Beretta's (1994) examination of how the characteristics of states (their criminal laws, their mental health services and the economic standing of women) influence the likelihood of imprisonment for men and women is a notable exception. Controlling for arrest rates, they found that states with higher rates of female labour force participation have higher rates of female imprisonment. From the perspective of gender theory this may seem counterintuitive but, lacking other indicators of the gender order in these states (e.g. women's political representation, family-friendly employment laws, etc.), it may indicate what Connell (2002: 71) refers to as a crisis tendency, or the 'internal contradictions or tendencies that undermine current patterns and force change'. A simple example of this would be the contractions between the growth in women's labour force participation and the gender gap in

[1]Research in criminology that directs attention to the relationship between the gender order and gender relations can be found in studies of domestic violence (Dugan 2003; Dugan et al. 2003) and homicide victimization. The work on homicide victimization is more developed but the results vary considerably depending on whether the absolute status or the relative status of women is examined. When women's absolute status is examined, the evidence suggests that women's higher status (measured by employment, education and/or income) provides them with some protection against victimization (e.g. Gartner 1990; Bailey and Peterson 1995; Vieraitis et al. 2007). However, when women's status is examined relative to men's, their equality increases their chances of being victimized (e.g. Gartner et al. 1990; Whaley and Messner 2002; Vieraitis and Williams 2002; Pridemore and Frellich 2005). Although it is difficult to derive implications from these findings for criminal court sanctions, they do suggest that gender orders may be an important determinant of gender relations.

pay or, for working women, household responsibilities. Work that has focused on the historically specific nature of gender roles and criminal sanctions also suggests that such crisis points produce more severe sanctions for female offenders relative to male offenders.

In one of the few longitudinal analyses of sex disparities in sentencing, Boritch (1992) found that women were treated more harshly than men both in the decision to imprison and in sentence length in the late 19th century in Canada. She explains this seeming anomaly by noting how the public mores of the 'church-going' middle class were challenged in the Urban Reform Era by the growing number of working-class women who were migrating to the cities and challenging conventional economic and social roles. This suggests not only that gender orders evolve over time in an uneven fashion but also that the ways in which they unfold have important implications for our understanding of sex-based disparities in punishment (see also Boritch and Hagan 1990).

In what follows we present an overview of some of the most notable contrasts in the gender orders and gender regimes of Finland and the USA. Although our data and analyses focus only on the impact of offenders' legal and social characteristics on sentencing decisions in Finland, we make these comparisons to help us develop our hypotheses, highlighting how the structural relations between men and women in Finland might produce very different sex-based sentencing patterns relative to what scholars have reported for the USA. We then present our data, analytic methods and findings.

The Institutional Contexts of Women's Lives in Finland

Finland is a small homogeneous nation relative to the USA and its move to social equality between women and men has proceeded much more quickly, albeit unevenly, than it has in the United States. For most of the 20th century, Finland's economic needs ensured that women would be an important part of the industrial and agricultural labour force (Haavio-Mannila and Kauppinen 1992: 230). Nevertheless, some of the most substantial changes in Finnish women's lives occurred over the last several decades of the 20th century. The proportion of women from age 20 to retirement who reported being gainfully employed increased dramatically over this time period, easily surpassing the USA by the end of the last century. In 1960, women as a percentage of the total workforce were relatively comparable in these two nations (39 percent in Finland and 32 percent in the USA) but by 1997 women represented 59 percent of the workforce in Finland but only 43 percent of the workforce in the USA (Niemi 1995; Messner and Savolainen 2001). The vast majority of Finnish women (80 percent) also remain employed even if they have children below school age. Despite these relative advances in employment, the labour market in Finland, as is true of many countries with well-developed family policies, remains somewhat segregated by sex; although mothers with young children often work, they tend to hold low-paying 'female' jobs, not managerial positions (Manninen 1999; Mandel and Semyonov 2005).

The domestic position of women in Finland and the USA also differs substantially. Time-use studies indicate that Finnish women's proportional efforts in the home are less than their female counterparts' in the USA (Haavio-Mannila and Kauppinen 1992) and women in Finland are far less likely than their American counterparts to be legally married. In fact, cohabitation is almost as common as marriage in Finland but not in the United States, where couples are four times more likely to be married than to be cohabiting (Messner and Savolainen 2001: 41). Couples who elect to have children are also treated with relative equality in the employment sector in Finland: either parent can stay at home until the child is 3 years of age, with financial compensation and a job guarantee, or communities must arrange for child care for the parent who returns to work (Haavio-Mannila and Kauppinen 1992). Work–family policies in Finland (like those in Sweden and Norway) have been described as embodying 'a new gendered vision of society in which both women and men balance informal care work and employment' (Misra et al. 2007: 809). They are able to do this because employers and the government are expected to acknowledge the demands and responsibilities of men and women who are both working and raising children by allowing them to reduce their work week and by providing them with services that will assist them with their childcare responsibilities.

Finnish women, then, are less likely than American women to be dependent on a male partner and/or to assume full-time childcare responsibilities. These factors may be a product of, or contribute to, their more dominant presence in political life compared with American women. Women in Finland gained the right to vote and hold office somewhat earlier (1906) than American women and by the 1990s they represented 38 percent of members of parliament (Manninen 1999).

Hypotheses

The arguments to this point suggest two different, and conflicting, hypotheses. On the one hand, the relatively high political and labour force representation of women in Finland points to substantial changes in the gender order of this society. These shifts would lead us to expect that, controlling for salient legal, case and social characteristics, sex will have no impact on the sentences handed down in Finnish criminal courts. On the other hand, scholars studying the impact of work–family policies on status attainment and income have found that, although women's economic dependency has been substantially reduced in Scandinavian countries by their gains in labour force participation (which is undergirded by liberal maternal leave), a risk remains of ghettoizing women into low-paying jobs that offer little chance for advancement (Mandel and Semyonov 2005). Thus, to the degree that 'family-friendly' policies have resulted in gender-segregated occupations and care-giving patterns that reinforce a traditional gender regime, punishments should be relatively more lenient for women than for men.

To shed further light on these conflicting hypotheses, we also consider whether gender-specific normative statuses (e.g. parenting for women and employment for men) have additive or interactive effects on the sentences of male and female offenders. If the gender order and gender regime in Finland have shifted to ensure that parental needs with respect to 'caretaking labour' and employment are gender neutral, punishments for men and women should not be conditioned by these social statuses.

Data, Variables and Method of Analysis

Our research uses a data set created at Statistics Finland, the central statistical agency of the Finnish government. As a matter of law, Statistics Finland has access to most official population registers in Finland and the record-keeping practices used by various government agencies make it possible to link individual-level information about, for example, criminal records, educational attainment, family situation and employment status. Given the sensitive nature of these data, Statistics Finland does not release total population data sets, only samples. As such, our data set is a random sample of sentencing outcomes from 1996 and 2000. It features 40 percent of all the prison sentences and 20 percent of all remaining sentencing outcomes for these two years.[2]

These data also have some formatting restrictions, which were designed to ensure that the privacy of individual subjects would be protected. Specifically, some of the information on individual characteristics is available only in ordinal (rather than interval) units. For example, information about the age of the offender is not available in one-year intervals and income data are aggregated into six categories. (The data on dates of criminal justice processing are available in one-day units but as a cumulative number with an arbitrary starting point unknown to us.)

[2]Owing to the low annual number of sexual felony offenders, Statistics Finland refused to release any data pertaining to this class of convictions. This data set was originally created, in 2004, to study criminal recidivism across different sentencing outcomes. The selection of the two time points, 1996 and 2000, stems from considerations serving those purposes of research.

Dependent Variable

The present study is focused on three basic sentencing outcomes: prison, community service and probation. These are the most severe forms of penal sanction available in the Finnish penal code. Given our interest in the sex difference in the likelihood of incarceration, we limit our attention to cases that reach the judicial neighbourhood of a prison sentence. In practice, this decision means that we have excluded cases where the most serious sanction was a fine.

In Finland, *community service* was developed as an alternative sanction to incarceration. A prison sentence of eight months or less can be commuted to community service depending on a number of screening criteria. Usually individuals who are deemed capable of functioning in the community, and who are deemed to be low risks for recidivism, are eligible for community service (Lappi-Seppälä 2004: 8). Despite the growth in the use of this sanction during the 1990s, community service remains underutilized in the Finnish courts. In 1999, fewer than 40 percent of eligible cases were actually sentenced to community service (Lappi-Seppälä 2000: 208).

'Conditional imprisonment' is a literal translation of the sentence that corresponds to probation or a suspended prison sentence in the Finnish penal code. An offender facing a prison sentence of up to two years is eligible for a conditional (suspended) sentence provided that 'the maintenance for the general respect of the law' does not demand incarceration (Lappi-Seppälä 1998: 6). In practice, probation is used widely when dealing with young and/or first-time felons. It is not uncommon for a Finnish offender to accumulate multiple suspended sentences prior to his or her first unconditional prison sentence (Lappi-Seppälä 2000: 192).

The Finnish penal code does not make a distinction between prison and jail sentences. All sentences leading to any period of incarceration, long or short, are referred to as (unconditional) prison sentences. Our data on prison sentences are limited to outcomes that do not exceed 3.5 years in length. Owing to the relatively low annual incidence of long prison sentences, Statistics Finland refused to release the data on such cases. This restriction means that our sample features considerable homogeneity across the three sentencing categories. As explained above, individuals facing incarceration can have their sentence commuted to either community service (if the prison sentence is for eight months or less) or probation (if the sentence corresponds to two years or less in prison). Therefore, what might be considered, by American standards, limited variation in the length of incarceration in fact gives us maximum ability to determine how sex, net of other relevant predictors, influences case outcome (i.e. whether the offender receives a prison sentence or is allowed to remain in the community). Thus, the dependent variable in these analyses is a dichotomy: prison vs. no prison, and the latter category includes community service and probation.[3]

Independent Variables

In order to isolate the sex effect, it is critical to control for a number of case characteristics that influence the sentencing decision. Our research attends to the characteristics of the instant offence as well as the offender's prior criminal history. In addition to the nature (type and severity) of the instant offence, we have data on any mitigating circumstances that may have influenced the sentencing outcome. Our measures of mitigating factors include whether the offence was committed in self-defence, whether the offender was deemed to have played a subordinate (accessory) role in the commission of the crime, whether (s)he was pressured or coerced to offend, and whether the offender qualifies for Youthful Offender status (younger than 18 years of age). To address differences in criminal history, we hold constant the number of prior convictions, prior prison sentence and if any of the prior (or instant) convictions involved a violent offence.

[3]Initially, we considered the option of separating the two community-based sanctions. However, since this alternative had no impact on the findings relative to the sex of the offender, we adopted the more parsimonious measure.

The sex of the defendant is obviously our key predictor variable. Other demographic characteristics in our analysis include three indicators of socioeconomic status (income, employment status and housing situation) and two indicators of family status (family situation and the number of children). Family situation captures cohabitation status and the presence or absence of children in the household in a single variable. Note that this variable treats cohabiting and marital unions equally. This decision reflects the fact that, in Finland, these two styles of union formation are equal in terms of both legal status and social norms (Cherlin 2004: 849; Kiernan 2004).

Analysis and Findings

Descriptive Statistics

Table 9.6 displays the distribution of the variables used in our research within categories of sex. These statistics show that 16.5 percent of the men but only 6.3 percent of the women in our sample received a prison sentence. However, as can also be seen in Table 9.6, male and female offenders are dissimilar across several characteristics that are related to this sentencing outcome. For example, nearly one-half of the women but only one-fifth of the men are first-time offenders. Females are also much less likely than males to have been convicted of a violent felony. The analysis of the demographic variables reveals that, irrespective of sex, the majority of the individuals in this sample are socioeconomically marginalized. Only about one-in-three defendants were employed at the time of the conviction and nearly two-thirds score below the bottom quartile of the income distribution. Male offenders are also somewhat younger and less likely to have children than female offenders.

The relatively high number of 'pensioners' in these data may seem curious for those unfamiliar with the Finnish system of welfare benefits. This category is not limited to people who have retired from the labour force after decades of gainful employment. Indeed, the majority (65 percent) of the 'pensioners' in our sample are below the age of 50. This group includes all the individuals who qualify for permanent public assistance. In the case of working-age people, this benefit is typically based on serious disability. Problems related to 'mental health and behavioral disorders' constitute the most common diagnosis leading to disability pension in contemporary Finland (Social Insurance Institution of Finland 2007). This general category includes problems related to excessive use of alcohol and drugs.

Multivariate Models

As previously noted, the bivariate relationship between sex and imprisonment showed that, relative to females, males are 2½ times more likely to receive a prison sentence. However, this statistic does not address characteristics that differentiate male and female offenders and that are likely to influence the decision to incarcerate. To address this omission, Table 9.7 describes findings from a logistic regression equation that controls for other relevant demographic characteristics and a wide range of legal factors.

First, as can be seen in Table 9.7, the coefficient for sex (male) is positive but non-significant. Unlike what is traditionally found in disposition studies in the USA, we see no evidence of preferential treatment for female offenders relative to male offenders in the decision to incarcerate. The large difference observed in the bivariate analysis (Table 9.6) is a function of pre-existing differences in other characteristics.

Second, we also estimated a multivariate model based only on the legally relevant variables (i.e. the social and demographic variables were removed from the equation). This model (not shown here) allows us to determine whether the sex differences in case characteristics can account for the difference we observed in the bivariate analysis. In a model featuring legal case characteristics as the only control variables, the odds ratio associated with sex is reduced further, relative to the odds ratio for sex in the full model presented in Table 9.7 (1.03 vs. 1.11). This finding suggests that, although family status and related extra-legal factors

Table 9.6 Descriptive Statistics of Dependent and Control Variables by Gender

Variables	Men		Women	
	N	%	*N*	%
Prison sentence	1546	16.5	60	6.3
Instant offence				
Theft	1237	13.2	72	7.5
Fraud/counterfeiting	502	5.3	171	17.9
Robbery/extortion	135	1.4	10	1.0
Other property	346	3.7	30	3.1
Aggravated assault	176	1.9	32	3.4
Simple assault	879	9.4	69	7.2
Manslaughter	24	0.3	0	0.0
Aggravated DUI	5938	63.3	562	58.8
Aggravated traffic safety	148	1.6	9	0.9
Year of sentence				
1996	4697	50.0	439	46.0
2000	4688	50.0	515	54.0
Region				
South	3675	39.2	393	41.2
West	3128	33.3	333	34.9
East or North	2582	27.5	228	23.9
Youthful offender	385	4.1	24	2.5
Mitigating circumstances	320	3.4	32	3.4
Prior convictions				
No priors	1942	20.7	455	47.7
1	1448	15.4	161	16.9
2	995	10.6	92	9.6
3	719	7.7	63	6.6
4–9	2279	24.3	115	12.1
10+	2002	21.3	68	7.1
Prior prison sentence	2656	28.3	83	8.7
Violent offender	3964	42.2	238	24.9
Age				
Under 25	2182	23.2	153	16.0

Variables	Men		Women	
	N	%	*N*	%
25–29	1211	12.9	122	12.8
30–39	2272	24.2	291	30.5
40–49	2181	23.2	250	26.2
50 or older	1539	16.4	138	14.5
Income				
No income	4414	47.0	447	46.9
Lowest 20%	1675	17.9	154	16.1
20–40%	1580	16.8	176	18.4
40–60%	687	7.3	97	10.2
60–80%	566	6.0	65	6.8
Highest 20%	461	4.9	15	1.6
Employment status				
Unemployed	3704	39.9	356	37.7
Pension	1642	17.7	137	14.5
Other	1118	12.1	131	13.9
Employed	2809	30.3	320	33.9
Housing situation				
Other/unknown	1227	13.2	82	8.6
Ownership	3557	38.2	294	31.0
Rental	4522	48.6	572	60.3
Family status				
Not in family population	4231	48.1	310	33.8
Union, no children	1296	14.7	194	21.2
Union with children	2388	27.1	192	20.9
Single parent	887	10.1	221	24.1
Number of children				
0	6157	65.6	546	57.2
1	1632	17.4	206	21.6
2	1043	11.1	118	12.4
3+	553	5.9	84	8.8
Total	9385	100.0	954	100.0

Table 9.7 Predictors of Prison Sentence in Finland: Logistic Regression Equation (N = 10,339)

Legal/Case Characteristics	b	SE	OR
Age at conviction	–0.024**	0.004	0.977
Year of sentencing (2000)	–0.022	0.081	0.978
Region of sentencing			
South (reference)			
West	0.206*	0.091	1.229
East/North	–0.153	0.100	0.858
Current offence			
Theft (reference)			
Fraud	–0.366*	0.171	0.694
Robbery/extortion	0.575*	0.290	1.777
Other property	0.151	0.197	1.163
Aggravated assault	–0.078	0.245	0.925
Simple assault	–0.587**	0.146	0.556
Negligent manslaughter	0.955	0.932	2.600
Aggravated DUI	–0.660**	0.112	0.517
Aggravated traffic safety	0.003	0.329	1.003
Youthful offender	–0.954*	0.394	0.383
Mitigating circumstances	–0.051	0.202	0.951
Prior convictions			
No priors (reference)			
1	1.844*	0.784	6.323
2	3.380**	0.732	29.385
3	3.393**	0.736	29.747
4–9	4.223**	0.715	68.256
10+	5.150**	0.719	172.422
Prior prison sentence	1.769**	0.109	5.867
Sociodemographic variables			
Sex (male)	0.103	0.199	1.108
Income			
No taxable income			
Lowest 20%	0.333**	0.099	1.395
20–40%	–0.066	0.127	0.936
40–60%	–0.439	0.253	0.644

Legal/Case Characteristics	b	SE	OR
60–80%	−1.293**	0.450	0.275
Highest 20%	−2.211*	1.024	0.110
Employment status			
Employed (reference)			
Unemployed	−0.333*	0.152	0.717
Pensioner	0.232*	0.114	1.261
Other/unknown	0.796**	0.109	2.216
Housing situation			
Rental (reference)			
Own	0.035	0.103	1.036
Other/unknown	0.224	0.123	1.251
Family situation			
Not in family population (reference)			
Union, no children	−0.127	0.119	0.881
Union with children	−1.209*	0.480	0.298
Single parent	−1.002*	0.491	0.367
Number of children			
0 (reference)			
1	0.708	0.483	2.030
2	0.799	0.493	2.224
3 or more	0.395	0.517	1.485
Pseudo-R^2 (Nagelkerke)		54.8%	

$^*p < .05$; $^{**}p < .01$.

affect the sentences accorded women and men in criminal court, their impact on the outcomes these offenders receive is trivial by comparison to criminal case characteristics.

Controlling for the offender's sex, several sociodemographic factors also have a significant effect on the decision to incarcerate. In terms of odds ratios, the risk of a prison sentence associated with individuals in the lowest income category (excluding those with no taxable income) is 12 times higher than for those in the highest income bracket. Family situation is also an important predictor of sentencing outcome. The Finnish courts are considerably more reluctant to sentence a parent to prison than an individual without children—independent of the parent's sex. Clearly, then, sentencing decisions in Finland are not immune to extra-legal considerations. But the fact that a defendant's sex is not among this group of influential extra-legal factors provides support for the hypothesized relationship between shifts in the gender order and regime and the treatment of women and men before the law.

To further explore these hypotheses, we turn now to the interaction effects. As previously noted, the central theoretical paradigm in the gender-disposition literature has been sex role theory. From this perspective, adherence to (or deviation from) normative gender scripts should influence sentencing outcomes. So, for example, in a culture characterized by sharp gender divisions of labour, we would expect mothers will be treated more leniently than fathers. Likewise, to the extent that labour market employment is conceived of as a predominantly masculine activity, unemployed men will be penalized more severely than similarly situated women. However, this perspective offers little room for understanding how situational contexts and structural shifts can alter seemingly normative gender scripts. Drawing on Connell's gender theory, we have identified Finland as a national context with seemingly greater levels of gender neutrality in both family policy and labour market participation. And, to this point, the evidence suggests that changes in the gender regime and gender order have neutralized the effects of an offender's sex on criminal court dispositions. But we have also noted, consistent with the analyses of Mandel and Semyonov (2005) and others (Gornick and Meyers 2003; Jacobs and Gerson 2004), that the outcome of 'family-friendly' policies may not be all that easy to detect. If such welfare state interventions created a two-tiered labour market—where women occupy primarily lower-paying 'female jobs' and men are found in a wider range of jobs including those that lead to managerial positions—the effects of parenthood and employment status on courtroom decisions should continue to be conditioned by the defendant's sex. Therefore, to provide a more stringent test of these hypotheses, we consider the interactive effects of sex and parental and employment responsibilities on case outcomes. The findings are presented in Table 9.8.

As the first step in the analysis of the interaction effects, we estimated the full regression equation (described in Table 9.7) separately for women and men. The relevant findings are presented in Table 9.8. To conserve space, we report findings pertaining only to the two socioeconomic variables of interest: employment status and family situation.

In the female sub-sample, we find that neither employment status nor family situation has a significant effect on the imprisonment decision. For the male sample, the opposite is true: individuals with weak attachments to the labour market are penalized more heavily than their employed counterparts and men who are fathers are treated more leniently than those who have no children. However, this sex difference is likely to be an artefact of the difference in sample sizes: the male sample is nearly 10 times larger than the female sample. Examining the magnitude of the relevant coefficients in the two samples, we see no substantial variation in the odds ratios associated with 'union with children' for females and males. Yet this household arrangement is statistically significant only in the male sample.

There is, however, a notable gender difference in the *direction* of the unemployment effect in Table 9.8. Unemployment increases men's risks of incarceration; the corresponding effect for women is negative albeit statistically insignificant.[4]

Finally, we should note that, in the course of the analysis, we also estimated multinomial models featuring three outcome categories: prison, community service and probation. This approach was motivated by concern that our dichotomous outcome may be masking sex differences in the decisions related to the two community-based sanctions. We found no support for this proposition. Female offenders were no more or less likely to be sentenced to community service as opposed to probation than their male counterparts. We also examined sex differences in the length of the sentence among those who received a prison

[4]We also find that the effect of having an 'other/unknown' employment status on criminal court sanctions differs for women and men. It is, however, difficult to interpret this outcome in light of the uncertain nature of this category. We could speculate that, in the Finnish welfare state, the inability to determine employment status is a sign of extreme marginalization and/or membership in the illegal economy. Perhaps males in this category consist mostly of homeless/institutionalized alcoholics, whereas prostitutes dominate the female category. Whatever the explanation, it seems likely that this outcome reflects some gender differences in the social characteristics of the individuals inhabiting this residual category.

Table 9.8 The Influence of Employment Status and Family Situation on the Sentence of Incarceration: Gender-Specific Logistic Regression Models

	Women ($N = 954$)			Men ($N = 9385$)		
Predictors	b	SE	OR	b	SE	OR
Employment status						
Employed (reference)						
Unemployed	−1.028	0.813	0.358	0.253*	0.116	1.288
Pensioner	0.984	0.559	2.674	0.793**	0.112	2.209
Other/unknown	1.076	1.131	2.932	−0.350*	0.154	0.705
Family situation						
Not in family population (reference)						
Union, no children	−0.392	0.626	0.676	−0.119	0.122	0.888
Union with children	−1.277	2.172	0.279	−1.292**	0.515	0.275
Single parent	−0.665	1.964	0.514	−1.068*	0.525	0.344

NOTE: The estimates reported in Table 9.8 are based on models featuring the full set of control variables (see Table 9.7).

*$p < .05$; **$p < .01$.

sentence. There was no statistically significant sex difference in this outcome even at the bivariate level. However, as noted previously, our data were restricted to relatively short prison sentences.

Discussion

The aim of our analysis was to provide a new paradigm for examining the relationship between an offender's sex and criminal court outcomes. To this end, we drew on Connell's gender theory and we moved outside of the US context to examine the decision to incarcerate in Finland—a country with an ostensibly very different gender order and gender regime than that of the USA. Work–family policies in Finland are designed to make it easy for both men and women to have young families while maintaining their paid jobs, providing extended leave for either parent when a child is born and the guarantee of child care when they return to work. Although these 'family-friendly' policies have been designed to level the playing field for males and females, their effects, as others have noted, are not entirely clear. Certainly they have moved more women into the labour market and in so doing reduced some aspects of gender inequality, particularly women's degree of economic dependency on men. But they have also created a two-tiered labour market that offers women fewer opportunities for advancing out of low-paying jobs. Does women's increased visibility outside of the home create enough of an impetus in the gender order to mitigate the traditional leniency judges extended to them in criminal case outcomes? Or does their marginal position in the labour market guarantee continued paternalism?

The findings suggest that shifts in the gender order may well reduce sex disparities in sentencing. In Finland, gender has no effect on the decision to incarcerate, net of other social and legal case characteristics. In fact, as we showed, relative to case characteristics, a defendant's sex is minimal in determining criminal court sanctions. Importantly, this finding resonates with others who have found that, when 'blameworthiness' is taken into account, the effects of sex on the decision to incarcerate are substantially reduced (Steffensmeier et al. 1993). However, our findings speak to more than just the importance of legal factors in understanding why male and female offenders traditionally have been treated so differently by the judiciary. To understand the unique effects of the different social positions of men and women and the important implications these may or may not have for sanctioning, we considered whether gender continues to moderate the effects of both parental and employment statuses on the sanctions accorded to male and female offenders. We found no evidence of such interactive effects.

The gendered nature of family responsibilities may not have entirely disappeared in Finland. As others have shown, the implementation of policies that are designed to reduce the conflict between paid work and child care are directed primarily at women, which 'lowers women's work effort and encourages employers' discrimination against women' in the form of limiting women's access to high rungs of the labour market (Mandel and Semyonov 2005: 965). Despite these presumably unintended consequences of 'family-friendly' policies, it appears from our analysis that the upshot has served to shift the gender order sufficiently to produce changes in gender relations in the criminal court. Offenders who are parents or who are legally employed, regardless of their sex, are treated equally in the eyes of the court, at least with respect to the decision to imprison.

Of course, there are alternative explanations for our findings. Finland is a country with a very different criminal justice system than the one found in the USA. It is a more 'rational' legal system that is not influenced by juries or judicial elections. Although this could contribute to more impartial sentencing practices, another relevant, and related, explanation for the differences we observe in the sex–sentencing paradigm lies in the institution most directly linked to sentencing outcomes—that is, the judiciary. In 2005, 40 percent of the judges in the Finnish court system were women (Liljeroos 2008). In the USA, although women have made significant inroads into the legal community, their presence on the bench remains substantially lower, at 25 percent (Center for Women in Government & Civil Society 2006). This difference in the gender composition of the Finnish and US courts resonates with Connell's (2002) arguments pertaining generally to how gender differences in power and equality can be disrupted and specifically to how gender is accomplished and shaped by everyday social actions. In the absence of such shifts in the judicial makeup of the court, it may be impossible to move beyond stereotypes that reify paternalistic notions of women as powerless and in need of protection. It may also be impossible to determine the validity of perspectives that challenge the sex role paradigm that has so dominated and perhaps even stifled disposition research.

Certainly, we will need to see more studies, and more fine-grained studies, of the relationship between an offender's sex and criminal court sanctions. Cross-national research provides one avenue but another welcome venue is state comparisons (e.g. Bridges and Beretta 1994). The benefits of such work are potentially far-reaching, applying not only to theory development and testing but also to providing us with a greater understanding of how we can influence the gender dynamics and disparities in criminal courts and other social contexts.

References

Bailey, W. C. and Peterson, R. D. (1995). Gender inequality and violence against women: The case of murder. In J. Hagan and R. D. Peterson (eds) *Crime and inequality*, 174–205. Stanford, CA: Stanford University Press.

Bickle, G. and Peterson, R. D. (1991). The impact of gender based family roles on criminal sentencing. *Social Problems 38*, 372–94.

Boritch, H. (1992). Gender and criminal court outcomes—an historical analysis. *Criminology 30*, 293–326.

Boritch, H. and Hagan, J. (1990). A century of crime in Toronto: Gender, class and patterns of social control. *Criminology 28*, 567–95.

Bridges, G. S. and Beretta, G. (1994). Gender, race and social control: Toward an understanding of sex disparities in imprisonment. In G. S. Bridges and M. A. Myers (eds) *Inequality, crime and social control*, 158–75. Boulder, CO: Westview Press.

Center for Women in Government & Civil Society (2006). Women in state policy leadership, 1998–2005: An analysis of slow and uneven progress. URL (accessed 31 December 2008): http://www.cwig.albany.edu/APMSG2006.htm.

Cherlin, A. (2004). The deinstitutionalization of American marriage. *Journal of Marriage and Family 66*, 848–61.

Connell, R. W. (1987). *Gender and power. Society, the person and sexual politics*. Stanford CA: Stanford University Press.

Connell, R. W. (2002). *Gender*. Cambridge: Polity Press.

Crew, B. K. (1991). Sex differences in criminal sentencing: Chivalry or patriarchy? *Justice Quarterly 8*, 59–83.

Daly, K. (1987). Discrimination in the criminal courts: Family, gender and the problem of equal treatment. *Social Forces 66*, 152–75.

Daly, K. (1989a). Rethinking judicial paternalism: Gender, work–family relations and sentencing. *Gender and Society 3*, 9–36.

Daly, K. (1989b). Neither conflict nor labeling nor paternalism will suffice: Intersections of race, ethnicity, gender, and family in criminal court decisions. *Crime and Delinquency 35*, 136–68.

Daly, K. and Chesney-Lind, M. (1988). Feminism and criminology. *Justice Quarterly 5*, 497–538.

Daly, K. and Bordt, R. L. (1995). Sex effects and sentencing: An analysis of the statistical literature. *Justice Quarterly 12*, 141–75.

Daly, K. and Tonry, M. (1997). Gender, race and sentencing. In M. Tonry (ed.) *Crime and justice. Volume 22*, 210–52. Chicago, IL: University of Chicago Press.

Dugan, L. (2003). Domestic violence legislation: Exploring its impact on the likelihood of domestic violence, police involvement, and arrest. *Criminology and Public Policy 2*, 283–312.

Dugan, L., Nagin, D. and Rosenfeld, R. (2003). Exposure reduction or retaliation? The effects of domestic violence resources on intimate partner homicide. *Law & Society Review 27*, 169–98.

Gartner, R. (1990). The victims of homicide: A temporal and cross-national comparison. *American Sociological Review 55*, 92–106.

Gartner, R., Baker K. and Pampel, F. C. (1990). Gender stratification and the gender gap in homicide victimization. *Social Problems 37*, 593–612.

Gornick, J. C. and Meyers, M. K. (2003). *Families that work: Policies for reconciling parenthood and employment*. New York: Russell Sage Foundation.

Griffin, T. and Wooldredge, J. (2006). Sex-based disparities in felony dispositions before versus after sentencing reform in Ohio. *Criminology 44*, 893–923.

Haavio-Mannila, E. and Kauppinen, K. (1992). Women and the welfare state in the Nordic countries. In H. Kahne and J. Z. Giele (eds) *Women's work and women's lives: The continuing struggle worldwide*. Boulder, CO: Westview Press.

Jaccard, J. (2001). *Interaction effects in logistic regression*. Sage University Paper Series on Quantitative Applications in the Social Sciences, 07-135. Thousand Oaks, CA: SAGE.

Jacobs, J. A. and Gerson, K. (2004) *The time divide: Work, family, and gender inequality*. Cambridge, MA: Harvard University Press.

Kiernan, K. (2004). Unmarried cohabitation and parenthood in Britain and Europe. *Law & Policy 26*, 33–55.

Koons-Witt, B. (2002). The effect of gender on the decision to incarcerate before and after the introduction of sentencing guidelines. *Criminology 40*, 297–328.

Kruttschnitt, C. (1982). Women, crime and dependency: An application of the theory of law. *Criminology 19*, 495–513.

Kruttschnitt, C. (1996).Contributions of quantitative methods to the study of gender and crime, or bootstrapping our way into the theoretical thicket. *Journal of Quantitative Criminology 12*, 135–61.

Kruttschnitt, C. and Carbone-Lopez, K. (2006). Moving beyond the stereotypes: Women's subjective accounts of their violent crime. *Criminology 44*, 321–51.

Kruttschnitt, C. and Green, D. E. (1984). The sex–sanctioning issue: Is it history? *American Sociological Review 49*, 541–51.

Lappi-Seppälä, T. (1998). *Regulating the prison population: Experiences from a long-term policy in Finland*. Helsinki: National Research Institute of Legal Policy Research Communications, 38.

Lappi-Seppälä, T. (2000). *Rikosten seuraamukset* [Criminal sentencing]. Porvoo, Finland: WSLT.

Lappi-Seppälä, T. (2004). Penal policy and prison rates. Long term experiences from Finland. Unpublished manuscript, October.

Liljeroos, H. (2008). Unpublished data/personal communication. Ministry of Justice, Finland.

Maher, L. (1997). *Sexed work: Gender, race and resistance in a Brooklyn drug market*. Oxford: Clarendon Press.

Mandel, H. and Semyonov, M. (2005). Family policies, wage structures, and gender gaps: Sources of earnings inequality in 20 countries. *American Sociological Review 70*, 949–67.

Manninen, Merja (1999). Women's status in Finland. *Virtual Finland*. URL (accessed 31 December 2008): http://virtual.finland.fi/netcomm/news/showarticle. asp?intNWSAID=25736.

Messner, S. F. and Savolainen, J. (2001). Gender and the victim–offender relationship in homicide: A comparison of Finland and the United States. *International Criminal Justice Review 11*, 34–57.

Miller, J. (1998). Up it up: Gender and the accomplishment of street robbery. *Criminology 36*, 37–66.

Miller, J. and Mullins, C. W. (2006). The status of feminist theories in criminology. In F. T. Cullen, J. Wright and K. Blevin (eds) *Taking stock: The status of criminological theory. Advances in criminology theory* (Freda Adler and William Laufer, series eds), 217–50. New Brunswick, NJ: Transaction.

Miller, J. and White, N. A. (2003). Gender and adolescent relationship violence: A contextual examination. *Criminology 41*, 1501–41.

Misra, J., Moller, S. and Budig, M. J. (2007). Work–family policies and poverty for partnered and single women in Europe and North America. *Gender and Society 21*, 804–27.

Moulds, E. F. (1980). Chivalry and paternalism: Disparities of treatment in the criminal justice system. In S. K. Datesman and F. R. Scarpitti (eds) *Women, crime and justice*, 277–99. New York: Oxford University Press.

Mullins, C. W. and Wright, R. (2003). Gender, social networks, and residential burglary. *Criminology 41*, 813–40.

Nagel, I. and Hagan, J. (1983). Gender and crime: Offense patterns and criminal court sanctions. In N. Morris and M. Tonry (eds) *Crime and justice. Volume IV*, 91–144. Chicago, IL: University of Chicago Press.

Niemi, I. (1995). *Time use of women in Europe and North America.* United Nations International Research and Training Institute for the Advancement of Women to the United Nations Economic Commission for Europe. New York: United Nations.

Pridemore, W. A. and Frellich, J. D. (2005). Gender equity, traditional masculine culture, and female homicide victimization. *Journal of Criminal Justice 33*, 213–23.

Social Insurance Institution of Finland (2007). *2006 Statistical yearbook of pensioners in Finland*. Helsinki: Hakapaino.

Steffensmeier, D. (1983). Organizational properties and sex-segregation in the underworld: Building a sociological theory of sex differences in crime. *Social Forces 61*, 1010–32.

Steffensmeier, D., Kramer, J. and Streifel, C. (1993). Gender and imprisonment decisions. *Criminology 31*, 411–46.

Thorne, B. (1993). *Gender play: Girls and boys in school.* New Brunswick, NJ: Rutgers University Press.

Vieraitis, L. M., Britto, S. and Kovandzic, T.V. (2007). The impact of women's status and gender inequality on female homicide victimization rates. Evidence from U.S. counties. *Feminist Criminology* 2, 57–73.

Vieraitis, L. M. and Williams, M. R. (2002). Assessing the impact of gender inequality on female homicide victimization across U.S. cities: A racially disaggregated analysis. *Violence Against Women 8*, 35–63.

Whaley, R. B. and Messner, S. F. (2002). Gender equality and gendered homicides. *Homicide Studies 6*, 188–210.

West, C. and Fenstermaker, S. (1995). Doing difference. *Gender and Society 9*, 8–37.

Wooldredge, J. D. (1998). Analytic rigor in studies of disparities in criminal case processing. *Journal of Quantitative Criminology 14*, 155–79.

DISCUSSION QUESTIONS

1. How is the domestic position of women in Finland different from that of women in the United States?
2. Do women experience chivalrous sentencing treatment in the criminal courts in Finland?
3. What can courts in the United States learn about sentencing decision making from the Finnish legal system?

READING

As you learned in the section introduction, gender can affect the sentencing decisions of females, often resulting in preferential treatment of female offenders. However, the majority of research on the issue of chivalry is conducted at the sentencing stage, after girls may have already benefited from chivalrous treatment throughout the justice process. This reading explores whether women receive preferential treatment during the pretrial stage. The authors explore whether gender and race affect the decision to use preventive detention for offenders.

SOURCE: Freiburger, T. L., & Hilinski, C. M. (2010). The impact of race, gender, and age on the pretrial decision. *Criminal Justice Review, 35*(3), 318–334.

The Impact of Race, Gender, and Age on the Pretrial Decision

Tina L. Freiburger and Carly M. Hilinski

The majority of sentencing literature has examined the final sentencing decision (i.e., the in/out decision) and the sentence length. Few studies have examined earlier decision-making points in the judicial system, such as the pretrial release outcome. Because of this, advancements in final sentencing literature have not extended to pretrial release research. Most notably, the examination of race, gender, and age interactions has not been examined in the pretrial release research. Using the focal concerns perspective, the current research addresses this gap by examining how race, gender, and age affect defendants' odds of pretrial detention.

Instead of focusing solely on race or gender, current sentencing research has carefully examined how courtroom experiences vary across different race and gender combinations. The development of the focal concerns perspective by Steffensmeier et al. (Steffensmeier, 1980; Steffensmeier, Kramer, & Streifel, 1993; Steffensmeier, Ulmer, & Kramer, 1998) has greatly contributed to this line of sentencing research. The focal concerns perspective is comprised of the three focal concerns of blameworthiness, dangerousness, and practical constraints. Blameworthiness is largely determined by the legal factors of offense severity and prior record. Dangerousness is determined by variables such as offense type (e.g., personal, property, or drug), use of a weapon, and education and employment status of the defendant. Practical constraints consist of factors that influence a defendant's ability to serve a period of incarceration, including organizational factors such as jail space and case flow as well as individual factors such as familial responsibilities (e.g., child care duties and marital status).

According to focal concerns theory, it is through these three focal concerns that judges make their sentencing decisions. However, when judges make sentencing decisions, they must often do so with limited information and with limited time and do not have access to all of the information included in each of the three focal concerns. Thus, the demographic characteristics of an offender are often used to shape the three focal concerns. Certain demographic combinations, specifically age, gender, and race, are especially influential, as judges tend to view younger minority males as more dangerous and more blameworthy, leading to harsher sentences for these individuals (e.g., Spohn & Beichner, 2000; Steffensmeier & Demuth, 2006; Steffensmeier, Kramer, et al., 1993; Steffensmeier, Ulmer, et al., 1998).

Current research examining these interactions has supported the focal concerns perspective. Specifically, this research has found that both Black and White females are treated more leniently than males (e.g., Freiburger & Hilinski, 2009; Steffensmeier & Demuth, 2006) and that Black males are sentenced more harshly than White males (e.g., Albonetti, 1997; Steffensmeier & Demuth, 2006; Steffensmeier, Ulmer et al., 1998). Research specifically focusing on the treatment of females, however, has been more mixed, with some studies finding that Black females are treated more leniently than White females (e.g., Bickle & Peterson, 1991; Spohn & Beichner, 2000; Steffensmeier & Demuth, 2006) whereas others have found the opposite (Crawford, 2000; Steffensmeier, Ulmer, et al., 1998).

Prior sentencing studies also have found that age has varying impacts on the sentences of males and females and Black and White defendants (Steffensmeier, Ulmer, et al., 1998). The results here also are mixed, with several studies finding that young Black men are treated most harshly (Spohn & Holleran, 2000; Steffensmeier, Ulmer, et al., 1998) and others finding that middle-aged

Black men are treated most harshly (Freiburger & Hilinski, 2009; Harrington & Spohn, 2007). Despite the numerous studies examining the impact of race, gender, and age on sentencing decisions, no studies have been conducted that examined the impact of gender, race, and age interactions on early court decisions. The current research fills this gap by examining how these three factors interact to affect the pretrial release decision.

Literature Review

The pretrial release decision is a crucial point in the judicial system. A common finding in the final sentencing research is that the pretrial release status of a defendant is significantly correlated with their likelihood of incarceration; offenders who are detained have a greater chance of receiving a sentence of incarceration (see Freiburger & Hilinski, 2009; Spohn & Beichner, 2000; Steffensmeier & Demuth, 2006). There is also less scrutiny on the pretrial release decision, which allows judges a great deal of discretion. This has led some researchers to argue that it might actually be subject to more bias in judicial decision making (Hagan, 1974; Steffensmeier, 1980). Pretrial detention can further negatively affect defendants' final sentences by hindering their ability to participate in the preparation of their defense (Foote, 1954). Despite these findings illustrating the importance and significance of the pretrial decision, few studies have been conducted to examine the factors that affect this decision.

Race and Pretrial Release

The majority of research examining the effects of race on pretrial release decisions has found that White defendants receive greater leniency at this stage than Black defendants (Demuth, 2003; Katz & Spohn, 1995). Demuth (2003) and Katz and Spohn (1995) found that Black defendants were less likely to be released than White defendants. Although these studies failed to find a significant relationship between race and bail amount for both White and Black defendants, Demuth found that Black defendants were less likely to make bail. Furthermore, Demuth also found that Black defendants were significantly more likely to be ordered to detention (denied bail). No race difference was found, however, for Black and White defendants' odds of receiving a nonfinancial release (release on recognizance [ROR]) rather than bail.

Other studies examining slightly different outcome measures also have produced evidence to suggest White defendants are granted greater leniency in pretrial release. Patterson and Lynch (1991) found that non-White defendants were less likely than White defendants to receive bail amounts that were lower than the amount recommended by bail guidelines. However, they also found that White and non-White defendants were equally likely to receive bail amounts that were more than the amount recommended by bail guidelines. Albonetti (1989) did not find a direct race effect, though she did find that White defendants were less likely to be detained if they had higher levels of educational attainment and a higher income. White defendants' outcomes, however, were more negatively affected by increases in the severity of the offense.

Gender and Pretrial Release

Few studies have been conducted that examine the effect of gender on pretrial release and outcome. Overall, the studies that have examined this relationship have found that females were treated more leniently than male defendants. Daly (1987b) and Kruttschnitt and Green (1984) found that females were less likely to be detained prior to trial. Additional studies have found that females were more likely to be granted a nonfinancial release (Nagel, 1983) and be assigned lower bail amounts (Kruttschnitt, 1984). Unfortunately, no recent studies were located that focused solely on the effect of gender on pretrial release.

Race-Gender-Age Effects and Pretrial Release

Only one study has examined the interactions of race/ethnicity and gender. Demuth and Steffensmeier (2004) analyzed data from felony defendants in 75 of the most populous counties in state courts for the years 1990, 1992, 1994, and 1996. Their results indicated that race and gender significantly affected whether defendants were released prior to trial. More specifically, females were more likely to be released than males and White defendants were more likely to be released than Black and Hispanic defendants. Although females experienced leniency at every decision point (they were less likely to be detained due to failure to make bail or ordered to detention, more likely to secure nonfinancial release, and receive lower bail), the findings for race were more mixed across the different decision points. Blacks were more likely to be detained than White defendants, but Black and White defendants were equally likely to be ordered to detention (not granted bail) and be given a nonfinancial release. Demuth and Steffensmeier (2004) also found that there was no difference in the amount of bail assigned to Black and White offenders. It appeared, therefore, that the race effect was due to Black defendants' failure to post bail.

When race and gender interactions were examined, female defendants were less likely to be detained than their male counterparts across all racial and ethnic groups. The gender gap, however, was the smallest for White defendants (followed by Black defendants and Hispanic defendants).

Using categorical gender/race variables, the results further indicated that White women were the least likely to be detained, followed by Black women, Hispanic women, White men, Black men, and Hispanic men.

We were not able to locate any studies that examined the interactions of race, gender, and age on pretrial release; however, prior sentencing studies have examined the interacting effects of these factors on the final sentencing decisions. The findings of these studies are mixed. Steffensmeier et al. (1998) found that young offenders were more likely to be sentenced to incarceration, with young Black males being treated the harshest. In their examination of sentencing decisions in Chicago, Miami, and Kansas City, Spohn and Holleran (2000) found that offenders aged 20–29 received the harshest sentences. When age and race were examined, they found that young and middle-aged Black males were more likely to be incarcerated than middle-aged White defendants. In Kansas City, young Black and White males had higher odds of incarceration than middle-aged White males.

Harrington and Spohn (2007) found that Black males in the middle age (30–39) category were less likely than White males of all age groups to be sentenced to probation versus jail. When the decision to sentence an offender to prison instead of jail was examined, however, the opposite was found. White males of all ages were more likely than middle-aged Black men to receive a prison sentence. Freiburger and Hilinski (2009) found that being young benefited White males but resulted in harsher sentences for Black males. Older Black males were only granted leniency in the decision to incarcerate in jail rather than prison. For Black females, age was not a significant predictor of sentencing. Young White women, however, were treated more leniently in the decision to sentence to probation or jail.

Previous studies that have examined the influence of race and gender on pretrial release decisions have failed to consider factors that are likely to influence release decisions. When judges make pretrial release decisions, they are typically concerned with the level of risk the offender poses to the community and the likelihood that the offender will return to court for future appearances (Goldkamp & Gottfredson, 1979). Although prior record and offense severity are important factors that judges consider in this stage, other factors such as marital status, education, community ties, and employment also are used to assess these concerns (Goldkamp & Gottfredson, 1979; Nagel, 1983; Petee, 1994; Walker, 1993). In addition, the focal concerns perspective notes these factors as influential to the focal concern of dangerousness (Steffensmeier, Ulmer, et al., 1998). The only previous study that examined race and gender interactions (Demuth & Steffenmeiser, 2004) did not assess the impact of these important factors. Additionally, no studies have examined the interactions of race, gender, and age on the pretrial release decision despite the fact that judges are differently influenced by these various combinations. Thus, the current study builds on the previous research by examining the effect of race, gender, and age interactions on pretrial release outcomes while considering other factors (e.g., income, education, and marital status) that have been linked to the pretrial release decision and the focal concerns perspective.

Methods

The current study examined the effects of race, gender, and age on the pretrial detention outcomes of felony offenders in an urban county in Michigan. The data analyzed contains information collected from presentence investigation reports completed for all offenders convicted of a personal, drug, property, or public order offense in the county during 2006. The original data set contained 3,316 offenders. We removed defendants who were Hispanic or of another ethnicity ($N = 73$) from the data set because a meaningful analysis was not possible with such a small number of cases; cases that were missing important information pertaining to offense severity level and prior record level ($N = 608$) also were removed from the data set because it is necessary to include these variables in sentencing research.[1] Thus, the final data set contained 2,635 cases.

Dependent Variable

The dependent variable in the current study was a dichotomized measure of the actual pretrial outcome, with 0 representing defendants released prior to sentencing and 1 representing those detained prior to sentencing. Coding for this variable, and all independent variables, is included in Table 9.9. Although we agree with prior research that argues that the pretrial release is best assessed through the examination of both the judicial decision and the actual outcome (see arguments by Demuth, 2003 and Demuth & Steffensmeier, 2004), the current data only allows for the assessment of the actual pretrial outcome (whether the defendant was detained or released). This is considered a limitation of the current study; however, the pretrial outcome is the most telling of the decision points. It signifies the actual experience of the defendant by considering the consequences of pretrial detention (e.g., reduced ability to prepare defense and severed social ties due to incarceration). Despite this limitation, the ability to assess race, gender, and age interactions while including other facts relevant to pretrial release contributes substantially to the current literature.

Independent Variables

Several legal variables shown to be relevant in sentencing decisions were included in the analysis. The Michigan Statutory Sentencing Guideline's 6-point offense severity measure was used to control for the seriousness of the crime.[2] The state guideline 7-point measure of prior record also was used.[3] The analysis also controlled for offense type through four separate dummy variables (property, drug, personal, and public order offense), with personal crimes left out as the reference category. A dummy variable also was included for current criminal justice supervision. Those who were on probation, parole, or incarcerated at the time of the bail decision were considered under criminal justice supervision and were coded as 1.[4]

The main extralegal variables of interest (gender, race, and age) also were included in the analysis. Gender was included and coded as 0 for female and 1 for male and race was coded as 0 for White and 1 for Black. Because age was found to have a curvilinear effect on pretrial detention, it was entered into the models as three categorical variables. Similar to prior research (Freiburger & Hilinski, 2009; Harrington & Spohn, 2007), the three age categories created were 15–29, 30–39, and 40+ years, with 30–39 being left out of the analysis as the reference variable.

Several other variables that measured defendants' stability in the community also were included in the models. These variables also were important in assessing the focal concerns perspective as these factors have been theorized by Steffensmeier and colleagues (Steffensmeier, 1980; Steffensmeier, Kramer, et al., 1993; Steffensmeier, Ulmer, et al., 1998) to affect judges' perceptions of dangerousness. None of the defendants in the sample had a college education; therefore, education

Table 9.9 Description of Variables

Independent Variable	Description
Individual characteristics	
Age	Separate dummy variables for ages 15–29, 30–39, and 40+; age 30–39 is the reference category
Race	Black = 1, White = 0
Gender	Male = 1, Female = 0
Marital status	Married = 1, Not married = 0
High school (HS)	HS diploma/GED = 1, No HS diploma/GED = 0
Income over $75/month	Income over $75/month = 1, income less than $75/month = 0
Assets over $1,500	Assets over $1,500 = 1, assets < $1,500 = 0
Case characteristics	
Prior record variable (PRV)	7-category scale (1 = least serious,7 = most serious[a])
Offense variable (OV)	6-category scale (1 = least serious, 6 = most serious)
Type of conviction charge	Separate dummy variables for property offense, public order offense, personal offense, and drug offense; personal offense is the reference category
CJS supervision	CJS supervision (probation, parole, incarceration) = 1, No CJS supervision = 0
Dependent variable	
Pretrial detention	Detention = 1, Release = 0

NOTE: CJS = Criminal Justice System supervision.

a. None of the cases in the current data set had a prior record score of 7.

was entered as a dichotomous variable of high school education or General Education Diploma (GED) coded as 1 or no high school education coded 0. Marital status also was included as a dummy variable; those who were not married were coded as 0 and those who were married were coded as 1. A direct employment measure was not available; however, two income variables were recorded in the presentence investigation (PSI) reports and were included in the analysis. The first assessed whether the defendant had an income of $75 or more a month (0 = no income above $75 and 1 = income above $75). The second variable indicated whether the defendant had assets of $1,500 or more (0 = no assets totaling $1,500 and 1 = assets totaling $1,500 or more).

Results

The individual and case characteristics of the offenders included in the current research are presented in Table 9.10. The majority of both male and female offenders were 15–29 years of age. Both male and female offenders were more likely to be White, unmarried, and without a high school diploma. Further examination of the descriptive statistics reveals that over half of the females had a monthly income over $75 but less than 40% of the males earned more than $75 per month. Across both males and females, only about 15% had assets that were worth $1,500 or more. Case characteristics reveal that male offenders were charged most often with a personal offense, but female offenders were most often charged with a property offense. Males were also more likely to be under some form of criminal justice supervision at the time of the current offense. Finally, both men and women were more likely to be released prior to trial.

We estimated the effect of race, gender, and age on pretrial release using logistic regression models. First the effects of race, gender, and age were examined separately. The models were then split by gender and race; z scores also were calculated to determine whether the independent variables had a significantly different effect on the pretrial outcome for male and female and Black and White defendants. The final models contain categorical variables for gender and race and categorical variables for race, gender, and age combinations.

The logistic regression coefficients for pretrial detention are presented in Table 9.11. Four models are presented. The first model presented displays the effects of age, race, and gender without the inclusion of any other independent variables. In this model, gender, race, and both age variables are significant. The variable for gender indicates males have a significantly greater likelihood of being detained than females. Black defendants had a significantly greater likelihood of detention than White defendants. The age variables show that young defendants and older defendants were less likely to be detained than offenders 30–39. Model 2 shows the coefficients after the income variables are controlled. Although the gender and age variables remain significant race is no longer significant. When the legal variables are added in Model 3, the gender, age, and income variables remain significant. Race, however, is not significant. Therefore, it appears that the initial effect of race was due in part to differences in the financial capabilities of Black and White defendants.

The full model also is presented in Table 9.11 and contains all of the independent variables. As shown in the table, males were significantly more likely to remain detained than females; however, the coefficient for race was not significant. Young defendants and older defendants were less likely to be detained than defendants in the middle age category. Completing high school or obtaining a GED further resulted in a lower likelihood of being detained. Both income variables also were significant, indicating that defendants with an income over $75 a month and assets exceeding $1,500 were less likely to be detained.

In an attempt to garner a better understanding of the differences in the detention status of male and female defendants, we estimated split models to determine whether the same factors influenced the pretrial status of both groups. The results of this analysis are presented in Table 9.12. Examination of the female model shows that race was significant in the pretrial status for females with Black female defendants having a reduced odds of being detained compared to White females. Race did not, however, significantly affect the pretrial release status of males. The *z* score for race also was significant, indicating that the effect of race was significantly stronger for females than males. The coefficients for age indicate that young males were significantly less likely to be detained than males aged 30–39. Neither the age coefficient for females nor the *z* score for females was significant. Thus, the impact of age was not significantly different for males and females.

Further examination of the split models indicates that males and females ($b = .716$, $p < .01$) had an increased odds of detention with each increase in prior record severity. The *z* score reveals that prior record did not have a significantly greater impact on the pretrial release of males than females; however, it came very close to reaching significance. Severity of offense, conversely, was significant for males but not for females. The *z* score reveals that the difference was significant for males and females, with offense severity more strongly affecting males' pretrial detention status. Committing a property crime (compared to committing a personal crime) resulted in a decreased likelihood of being detained for females but not for males. The *z* score was significant, indicating that the differing effect was significant. The coefficient for drug offense was significant for both males and females. The *z* score also was significant, suggesting that committing a drug offense had a stronger impact for females than for males.

Discussion

The current study attempted to further the understanding of the effects of race, gender, and age on pretrial release outcomes. Given the logic of the focal concerns perspective (Steffensmeier, 1980; Steffensmeier, Kramer, et al., 1993; Steffensmeier, Ulmer, et al., 1998), it is not surprising that gender and age directly affect pretrial outcomes as females and young defendants are often viewed as less blameworthy and dangerous. The findings for race, however, were more complex. A strong race effect was found prior to entering the economic variables into the model, with Black defendants less likely to be released pretrial than Whites. Once these variables were included, however, race was no longer significant. In fact, the sign of the coefficient changed, suggesting Whites were actually more likely to be detained. Therefore, it appears that Black defendants are more likely to be detained because they do not have the financial means necessary to secure release. This indicates that Black disadvantage in the court system may not be as simple as racial bias, but instead stems from inequality and general disadvantage in society. This is especially noteworthy because prior studies on pretrial releases have not included these variables (e.g., Demuth & Steffenmeiser, 2004).

The effect of race was significant, however, when examining the sentences of men and women separately. Consistent with prior research conducted by Demuth and Steffensmeier (2004), the results also indicated that the gender gap was the smallest for White defendants. Unlike Demuth and Steffensmeier, however, White females were not most likely to be released pretrial. Instead, Black females were the least likely to be detained. This finding held across Black females of every age group. The odds of release for White women, however, were not significantly different than that of White and Black males. Although Black females were less likely to be detained than White females, the age/race/gender analysis showed that this finding was only applicable to White females aged 30–39. Younger and older White females were not significantly more likely to be detained than their Black female counterparts. When compared to males (both Black and White), however, Black females of all age groups were the least likely to be detained.

Although these findings seem inconsistent with the focal concerns perspective, it is possible that this inconsistency is actually due to the absence of practical constraint factors. Steffensmeier and colleagues (Steffensmeier, 1980; Steffensmeier, Kramer, et al., 1993; Steffensmeier, Ulmer, et al., 1998) suggest that defendants whose incarceration poses a greater practical constraint (e.g., leaving behind dependent children that will require care, need correctional facilities that are not available) will be granted leniency. It is possible, therefore, that the inclusion of family responsibility variables might account for the increased odds of Black females being released. Daly (1987a) suggests that judges are concerned with the social costs of incarcerating defendants who perform vital familial responsibilities. This is especially pertinent, given that more Black women in the criminal justice system are often single parents to dependent children (U.S. Bureau of the Census, 2000). Furthermore, prior research on the effect of gender on pretrial release has shown that the inclusion of these controls reduces the gender gap (Daly, 1987b; Kruttschnitt & Green, 1984). Unfortunately, the data used for the current study had a great deal of missing data for the measure of dependent children, making it impossible to assess this possibility.

The gender split models also show that judges give less consideration to legal factors for females

Table 9.10 Descriptive Statistics

	Total (*n* = 2635)		Males (*n* = 2187)		Females (*n* = 448)	
	n	Percentage	*n*	Percentage	*n*	Percentage
Individual characteristics						
Age						
15–29	1,397	53.0	1,184	54.1	213	47.5
30–39	597	22.7	480	21.9	117	26.1
40+	641	24.3	523	23.9	118	26.3
Race						
White	1,421	53.9	1,139	52.1	282	62.9
Black	1,214	46.1	1,048	47.9	166	37.1
Gender						
Male	2,187	83.0	—	—	—	—
Female	448	17.0	—	—	—	—
Marital status						
Married	289	11.0	237	10.8	52	11.6
Not married	2,346	89.0	1,950	89.2	396	88.4
High school (HS)						
HS diploma/GED	1,277	48.5	1,070	48.8	207	46.2
No HS diploma/GED	1,358	51.5	1,117	51.1	241	53.8
Income over $75/month						
Yes	1,081	41.0	835	38.2	246	54.9
No	1,554	59.0	1,352	61.8	202	45.1
Assets over $1500						
Yes	382	14.5	314	14.4	68	15.2
No	2,253	85.5	1,873	85.6	380	84.8

(Continued)

Table 9.10 (Continued)

	Total (n = 2635)		Males (n = 2187)		Females (n = 448)	
	n	Percentage	n	Percentage	n	Percentage
Case characteristics						
Prior record						
1	431	16.4	322	14.7	109	24.3
2	369	14.0	278	12.7	91	20.3
3	571	21.7	469	21.4	102	22.8
4	613	23.3	525	24.0	88	19.6
5	370	14.0	323	14.8	47	10.5
6	281	10.7	270	12.3	11	2.5
Offense severity						
1	1,323	50.2	1,081	49.4	242	54.0
2	812	30.8	672	30.7	140	31.3
3	298	11.3	245	11.2	53	11.8
4	98	3.7	88	4.0	10	2.2
5	70	2.7	69	3.2	1	0.2
6	34	1.3	32	1.5	2	0.4
Conviction charge						
Property offense	805	30.6	580	26.5	225	50.2
Public order offense	190	7.2	140	6.4	50	11.2
Personal offense	963	36.5	871	39.8	92	20.5
Drug offense	677	25.7	596	27.3	81	18.1
CJS supervision						
Supervision	912	34.6	774	64.6	138	30.8
No supervision	1,723	65.4	1,413	35.4	310	69.2
Pretrial detention						
Detained	960	36.4	866	39.6	94	21.0
Not detained	1,675	63.6	1,321	60.4	354	79.0

Table 9.11 Logistic Regression Estimates

	Model 1			Model 2			Model 3			Full Model		
Variable	B	SE	Exp(B)	B	SE	Exp(B)	B	SE	Exp(B)	B	SE	Exp(B)
Offender characteristics												
Age 15–29	−.467**	.102	.627	−.664**	.110	.515	−.341**	.122	.711	−.422**	.125	.656
Age 40+	−.371**	.119	.690	−.254*	.128	.776	−.259*	.139	.772	−.253**	.139	.776
Race (Black = 1)	.253**	.083	1.288	.010	.089	1.010	−.172	.102	.842	−.184	.102	.832
Gender (male = 1)	.909**	.125	2.482	.832**	.132	2.297	.533**	.146	1.705	.565**	.147	1.760
Marital status										−.270	.175	.764
High school										−.287**	.099	.750
Income over $75				−1.292**	.097	.275	−1.087**	.104	.337	−1.072**	.104	.342
Assets over $1,500				−1.255**	.164	.285	−1.056**	.176	.348	−.970**	.179	.379
Case characteristics												
Prior record							.456**	.039	1.578	.461**	.039	1.585
Offense severity							.194**	.048	1.214	.202**	.049	1.224
Property offense							−.542**	.126	.582	−.531**	.126	.588
Public order offense							−.276	.201	.758	−.266	.202	.588
Drug offense							−.530**	.134	.589	−.527**	.135	.590
CJS supervision							.733**	.103	2.082	.745**	.104	2.107
Constant	−1.116***	.138	.328	−.258*	.151	.773	−2.127**	.246	.119	−2.007**	.249	.134
Nagelkerke R^2	.046			.193			.344			.348		
Cox and Snell R^2	.034			.141			.251			.254		

Significance: $*p < .05$; $**p < .01$.

Table 9.12 Female and Male Split Models

Variable	Females			Males			Z score
	B	SE	Exp(B)	B	SE	Exp(B)	
Offender characteristics							
Age 15–29	−.072	.358	.931	−.461**	.135	.631	1.02
Age 40+	−.198	.377	.820	−.243	.151	.784	0.11
Race (Black = 1)	−.979**	.331	.376	−.090	.110	.913	2.55*
Marital status	−.854	.576	.426	−.191	.187	.826	1.09
High school	.137	.302	1.146	−.310**	.106	.733	1.40
Income over $75	−.898**	.296	.407	−1.072**	.112	.342	0.55
Assets over $1500	−.737	.589	.479	−.981**	.190	.375	0.39
Case characteristics							
Prior record	.716**	.139	2.047	.435**	.041	1.545	1.94
Offense severity	−.234	.206	.791	.234**	.051	1.264	2.21*
Property offense	1.531**	.392	.216	−.379	.135	.685	2.78*
Public order offense	−.223	.485	.800	−.553*	.237	.576	0.61
Drug offense	−1.556**	.512	.211	−.452**	.140	.636	2.08*
CJS supervision	.955*	.313	2.598	.723**	.111	2.061	0.70
Constant	−1.811*	.728	.163	−1.465	.241	.231	
Nagelkerke R^2	.423			.308			
Cox and Snell R^2	.272			.229			

$^*p < .05$; $^{**}p < .01$.

than for men. This might indicate that judges find males with more serious offenses as posing a greater risk to society. Therefore, it is possible that legal factors play less of a role in shaping the focal concerns associated with early court decisions for women as they do for men. This finding indicates a need for additional studies that closely examine the different factors that affect males' and females' sentencing decisions. It is possible that judges' focal concerns for males and females are influenced by different factors. The inconsistencies across research studies also cause questions of the ability to generalize these findings and signify a need for future research that examines the impact of race, gender, and age on sentencing decisions in other jurisdictions.

Overall, the current study has made an important contribution to the literature examining the factors affecting the pretrial decision. Most notably, it is the only study to date that has examined the effect of race, gender, and age interactions on the pretrial release outcome of defendants while also considering extralegal factors, including income, educational attainment, and marital status. The current study is limited, however, in its examination of only one jurisdiction. Although this is not uncommon in the sentencing literature, it does pose as a limitation as findings may vary across location. This study also is limited in its ability to measure the focal concern of practical constraint. It is likely that individual practical constraints (e.g., familial responsibility) as well as organization constraints (e.g., available jail space) could have an effect on pretrial detention. Additionally, these constraints may have varying effects by race, gender, and age of the defendant.

Future research should assess the impacts of race after controlling for economic factors on a more comprehensive set of dependent variables (e.g., ROR or bail, bail amount, ability to make bail). This is especially important given the finding that the race effect was eliminated once economic variables were included in the analysis. In addition, Demuth (2003) found that Black defendants were less likely to post bail than White defendants. If a more comprehensive dependent variable is assessed, it can be determined whether the Black disadvantage is due to a difference in bail amounts. In other words, are Black defendants receiving higher bail amounts or are Blacks simply more often than Whites in situations where they cannot afford to pay bail? The ability to examine a dependent variable of this nature would greatly contribute to the understanding of the effect of race on the pretrial release.

Notes

1. Significance tests performed to determine whether any differences existed between the cases excluded from the data set due to missing data and the cases included in the final analysis indicated that there were some significant differences between the two groups (presented below). Two age groups, ages 15–29 and 40+, were significantly different across the two groups; race also was significantly different across the two groups. An examination of the remaining independent variables reveals that cases excluded due to missing information were less likely to have a high school diploma or GED, less likely to have a monthly income of $75 or more, and less likely to have assets of more than $1,500. They were also less likely to be under criminal justice system supervision at the time of their arrest and more likely to be detained prior to trial. Although the missing data poses a limitation to the research, it is not unique to this study; most sentencing literature is limited in the amount of usable data. For instance, Harrington and Spohn (2007) were only able to use 59% of the cases in their original data set and Freiburger and Hilinski (2009) were only able to use 62.4% of the cases in their original data set. In the current study, nearly 80% of the cases in the original data set were able to be included in the final analysis.

2. Michigan Statutory Sentencing Guidelines assigns an offense variable (OV) to each offense. There are 19 possible offense variables that can be scored, including aggravated use of a weapon, physical or psychological injury to the victim, victim asportation or captivity, and criminal sexual penetration; the sentencing guidelines stipulate which variables will be scored based on the crime group of the current offense (e.g., crimes against a person, crimes against property, and crimes involving a controlled substance). Based on the crime group, each relevant variable is scored and then combined to create a total offense variable that ranges from 1 (least serious) to 6 (most serious; Michigan Judicial Institute, 2007). This offense variable (coded 1–6) was included in each model to control for the severity of the offense.

3. The prior record variable is a composite score based on factors such as prior adult felony and misdemeanor convictions, prior juvenile felony and misdemeanor adjudications, and the offender's relationship with the criminal justice system at the time of the current offense (i.e., whether the offender is a probationer or parolee). For each of the seven prior record variables, a numerical score is assigned. The sum of these seven scores determines the offender's prior record level, which ranges from A (least serious) to F (most serious; Michigan Judicial Institute, 2007). This variable was recoded and included in the models (coded 1–6) to control for prior record.

4. Although it is likely that those who are incarcerated are more likely to be detained pretrial than those on probation or parole, only seven offenders in the sample were actually incarcerated in jail or in prison. Due to the small number, it was impossible to meaningfully assess this difference; therefore, they were combined with those on probation and parole.

References

Albonetti, C. A. (1989). Bail and judicial discretion in the District of Columbia. *Sociology and Social Research, 74,* 40–47.

Albonetti, C. A. (1997). Sentencing under the federal sentencing guidelines: Effects of defendant characteristics, guilty pleas and departures on sentencing outcomes for drug offenses, 1991–1992. *Law and Society Review, 31,* 789–822.

Bickle, G. S., & Peterson, R. D. (1991). The impact on gender-based family roles on criminal sentencing. *Social Problems, 38,* 372–394.

Crawford, C. (2000). Gender, race, and habitual offender sentencing in Florida. *Criminology, 38,* 263–280.

Daly, K. (1987a). Structure and practice of familial-based justice in a criminal court. *Law & Society Review, 21,* 267–290.

Daly, K. (1987b). Discrimination in the criminal courts: Family, gender, and the problem of equal treatment. *Social Forces, 66,* 152–175.

Demuth, S. (2003). Racial and ethnic differences in pretrial release decisions and outcomes: A comparison of Hispanic, Black, and White felony arrestees. *Criminology, 41,* 873–907.

Demuth, S., & Steffensmeier, D. (2004). The impact of gender and race-ethnicity in the pretrial release process. *Social Problems, 51,* 222–242.

Foote, C. (1954). *Compelling appearance in court: Administration of bail in Philidelphia.* PA: Temple University Press.

Freiburger, T. L., & Hilinski, C. M. (24, February, 2009). An examination of the interactions of race and gender on sentencing decisions using a trichotomous dependent variable. *Crime & Delinquency,* Doi: 10.1177/ 0011128708330178.

Goldkamp, J. S., & Gottfredson, M. R. (1979). Bail decision making and pretrial detention: Surfacing judicial policy. *Law and Human Behavior, 3,* 227–249.

Hagan, J. (1974). Extra-legal attributes and criminal sentencing: An assessment of a sociological viewpoint. *Law and Society Review, 8,* 337–383.

Harrington, M. P., & Spohn, C. (2007). Defining sentence type: Further evidence against use of the total incarceration variable. *Journal of Research in Crime and Delinquency, 44,* 36–63.

Katz, C., & Spohn, C. (1995). The effect of race and gender on bail outcomes: Test of an interactive model. *American Journal of Criminal Justice, 19,* 161–184.

Kruttschnitt, C. (1984). Sex and criminal court dispositions: An unresolved controversy. *Journal of Research in Crime and Delinquency, 12*(3), 213–232.

Kruttschnitt, C., & Green, D. E. (1984). The sex-sanctioning issue: Is it history? *American Sociology Review, 49,* 541–551.

Mertler, C., & Vannata, R. (2002). *Advanced and multivariate statistical methods* (2nd ed.). Los Angeles: Pyrczak Publishing.

Michigan Judicial Institute. (2007). *Sentencing guidelines manual.* Retrieved from http://courts.michigan.gov/ mji/resources/sentencing-guide lines/sg.htm–srdanaly.

Nagel, I. (1983). The legal/extra-legal controversy: Judicial decision in pretrial release. *Law and Society Review, 17,* 481–515.

Paternoster, R., Brame, R., Mazerolle, P., & Piquero, A. (1998). Using the correct statistical test for the quality of regression coefficients. *Criminology, 36,* 859–866.

Patterson, E., & Lynch, M. (1991). The biases of bail: Race effects on bail decisions. In M. J. Lynch & E. Britt Patterson (Eds.), *Race and criminal justice.* NY: Harrow and Heston.

Petee, T. A. (1994). Recommended for release on recognizance: Factors affecting pretrial release recommendations. *Journal of Social Psychology, 134,* 375–382.

Spohn, C., & Beichner, D. (2000). Is preferential treatment of felony offenders a thing of the past? A multisite study of gender, race, and imprisonment. *Criminal Justice Policy Review, 11,* 149–184.

Spohn, C., & Holleran, D. (2000). The imprisonment penalty paid by young unemployed black and Hispanic male offenders. *Criminology, 38,* 281–306.

Steffensmeier, D. (1980). Assessing the impact of the women's movement on sex-based differences in the handling of adult criminal defendants. *Crime & Delinquency, 26,* 344–358.

Steffensmeier, D., & Demuth, S. (2006). Does gender modify the effects of race ethnicity on criminal sanctions? Sentences for male and female, White, Black, and Hispanic defendants. *Journal of Quantitative Criminology, 22,* 241–261.

Steffensmeier, D., Kramer, J., & Streifel, C. (1993). Gender and imprisonment decisions. *Criminology, 31,* 411–446.

Steffensmeier, D., Ulmer, J., & Kramer, J. (1998). The interaction of race, gender, and age in criminal sentencing: The punishment cost of being young, black, and male. *Criminology, 36,* 763–797.

U.S. Bureau of the Census (2003). *2000 census of the population.* Washington, DC: U.S. Government Printing Office. Available from: http://www.factfinder.census.gov

Walker, S. (1993). *Taming the system: The control of discretion in criminal justice, 1950–1990.* NY: Oxford University Press.

DISCUSSION QUESTIONS

1. What impact does gender have on the decision to release an offender during the pretrial stage?
2. What impact does race have on the decision to release an offender during the pretrial stage?
3. What impact do race *and* gender have on the decision to release an offender during the pretrial stage?

The Incarceration of Women

Section Highlights

- Historical trends in the incarceration of women
- Contemporary issues in the incarceration of women
- Gender-responsive treatment and programming for incarcerated women
- Barriers to reentry for incarcerated women

This section focuses on issues related to the supervision and incarceration of women. Drawing from historical examples of incarceration to modern-day policies, this section first looks at the treatment of women in prison and the challenges that women face. Following this discussion, this section highlights how the differential pathways of female offending affect the unique needs for women under the correctional system and presents a review of the tenets of gender-responsive programming. This section concludes with a discussion about the lives of women following incarceration and how policy decisions about offending have often succeeded in the "jailing" of women, even after their release from prison.

Historical Context of Female Prisons

Prior to the development of the all-female institution, women were housed in a separate unit within the male prison. Generally speaking, the conditions for women in these units were horrendous and were characterized by excessive use of solitary confinement and significant acts of physical and sexual abuse by both the male inmates and the male guards. Women in these facilities received few, if any, services (Freedman, 1981). At Auburn State Prison in New York, women were housed together in an attic space where they were unmonitored and received their meals from male inmates. In many cases, these men would stay longer than necessary to complete their job duties. To no surprise, there were many prison-related pregnancies that resulted from these interactions. The death of a pregnant woman named Rachel

Welch in 1825 as a result of a beating by a male guard led to significant changes in the housing of incarcerated women. In 1839, the first facility for women opened its doors. The Mount Pleasant Prison Annex was located on the grounds of Sing Sing, a male penitentiary located in Ossining, New York. While Mount Pleasant had a female warden at the facility, the oversight of the prison remained in the control of the administrators of Sing Sing, who were male and had little understanding about the nature of female criminality. Despite the intent by administrators to eliminate the abuse of women within the prison setting, the women incarcerated at Mount Pleasant continued to experience high levels of corporal punishment and abuse at the hands of the male guards.

Conditions of squalor and high levels of abuse and neglect prompted moral reformers in England and the United States to work toward improving the conditions of incarcerated women. A key figure in this crusade in the United Kingdom was Elizabeth Fry (1780–1845). Her work with the Newgate Prison in London during the early 19th century served as the inspiration for the American women's prison reform movement. Fry argued that women offenders were capable of being reformed and that it was the responsibility of women in the community to assist those who had fallen victim to a lifestyle of crime. Like Fry, many of the reformers in America throughout the 1820s and 1830s came from upper- and middle-class communities with liberal religious backgrounds (Freedman, 1981). The efforts of these reformers led to significant changes in the incarceration of women, including the development of separate institutions for women.

The Indiana Women's Prison is identified as the first stand-alone female prison in the United States. It was also the first maximum-security prison for women. At the time of its opening in 1873, IWP housed 16 women (Schadee, 2003). By 1940, 23 states had facilities designed exclusively to house female inmates. A review of facilities across the United States reveals two different models of institutions for women throughout the 20th century: reformatories and custodial institutions. The reformatory was a new concept in incarceration, as it was an institution designed with the intent to rehabilitate women. Here, women did not receive a fixed sentence length. Rather, they were sent to the reformatory for an indeterminate period of time—essentially until they were deemed to have been reformed. Women sent to the reformatories were most likely to be White, working-class women. Based on the philosophy that the reformatory was designed to "improve the moral character of women," women were sentenced for a variety of "crimes," including "lewd and lascivious conduct, fornication, serial premarital pregnancies, adultery [and] venereal disease" (Anderson, 2006, pp. 203–204). These public order offenses were based on the premise that such behaviors were "unladylike." Generally speaking, the conditions at the reformatory were superior to those found at the custodial institution. The reformatory was effective in responding to abuse of women inmates by male guards, as many of these institutions were staffed by women guards and administrations. While they were the first to provide treatment for female offenders, their rehabilitative efforts have been criticized by feminist scholars as an example of patriarchy at its finest, as women were punished for violating the socially proscribed norms of femininity. The reformatory became a place embodying attempts by society to control the autonomy of women—to punish the wayward behaviors and instill women with the appropriate morals and values of society (Kurshan, 2000).

In comparison, custodial institutions were similar in design and philosophy to male institutions. Here, women were simply warehoused, and little programming or treatment was offered to inmates. Women in custodial institutions were typically convicted on felony and property-related crimes, with a third of women convicted of violent crimes. The custodial institution was more popular with the Southern states. In cases where a state had both a reformatory and a custodial institution, the distribution of inmates was made along racial lines—custodial institutions were more likely to house women of color who were determined to have little rehabilitative potential, while reformatories housed primarily White women (Freedman, 1981). Black women were also sent to work on state-owned penal plantations under conditions that mimicked the days of slavery in the south. Women of color generally had committed less serious offenses compared to White women, and yet they were incarcerated for longer periods of time. Indeed, it was rare to see women of color convicted of moral offenses—since Black women were not held to the same standards of what was considered acceptable behavior for a lady, they were not deemed as in need of the rehabilitative tools that characterized the environments found at the reformatory (Rafter, 1985). Prison conditions for women at the custodial institution were characterized by unsanitary living environments with inadequate sewage and bathing systems, work conditions that were dominated by physical labor and corporal punishment, a lack of medical treatment for offenders, and the use of solitary confinement for women with mental health issues (Kurshan, 2000).

One of the most successful reformatories during this time frame was the Massachusetts Correctional Institution (MCI) in Framington. Opened in 1877, Framington possessed a number of unique characteristics, including an all-female staff, an inmate nursery that allowed incarcerated women to remain with their infants while they served their sentence, and an on-site hospital to address the inmates' health care needs. Additionally, several activities were provided to give women opportunities to increase their self-esteem, gain an education, and develop a positive quality of life during their sentence. While MCI Framington is the oldest running prison still in use today, it bears little resemblance to its original mission and design; the modern-day institution bears the scars of the tough-on-crime movement. Today's version of the institution has lost some of the characteristics that made Framington a unique example of the reformatory movement and now mimics the structure and design of the male prisons located in the state (Rathbone, 2005).

Today, most states have at least one facility dedicated to a growing population of female offenders. In many cases, these facilities are located in remote areas of the state, far from the cities where most of the women were arrested and where their families reside. The distance between an incarcerated woman and her family plays a significant role in the ways in which she copes with her incarceration and can affect her progress toward rehabilitation and a successful reintegration. In contrast, the sheer number of male facilities increases the probability that these men might reside in a facility closer to their home, allowing for increased frequency in visitations by family members.

Contemporary Issues for Incarcerated Women

Since the 1980s, the number of women incarcerated in the United States has multiplied at a dramatic rate. As discussed in Section VII, sentencing policies such as mandatory minimum sentences and the war on drugs have had a dramatic effect on the numbers of women in prison. These structured sentencing formats, whose intent was to reduce the levels of sentencing disparities, have only led to the increases in the numbers of women in custody. In 2008, there were more than 216,000 women incarcerated in jails and prisons in the United States[1] (West & Sabol, 2009). Table 10.1 illustrates a profile of women found in the criminal justice system today. A review of data on sentencing practices of women indicates that most women are incarcerated for nonviolent offenses. Much of the rise in female criminality is the result of minor property crimes, which reflects the economic vulnerability that women experience in society, or cases involving drug-related crimes and the public health addiction issues facing women.

While Blacks and Hispanics make up only 24% of the U.S. population, 63% of women in state prisons and 67% of women in federal prisons are Black or Hispanic, a practice that indicates that women of color are

[1]At midyear 2008, 115,779 women were incarcerated in state and federal prison facilities. The average daily population of adult females in jails in the U.S. at midyear 2008 was 99,175.

Table 10.1 Profile of Women in the Criminal Justice System

- Disproportionately women of color
- In their early to mid-thirties
- Most likely to have been convicted of a drug or drug-related offense
- Fragmented family histories, with other family members also involved with the criminal justice system
- Survivors of physical and/or sexual abuse as children and adults
- Significant substance abuse problems
- Multiple physical and mental health problems
- Unmarried mothers of minor children
- High school degree/GED, but limited vocational training and sporadic work histories

significantly overrepresented behind bars. Indeed, research indicates that Black women today are being incarcerated at a greater rate than both White females and Black males (Bush-Baskette, 1998). Poverty is also an important demographic of incarcerated women, as many women (48%) were unemployed at the time of their arrest, which affects their ability to provide a sustainable environment for themselves and their children. In addition, they tend to come from impoverished areas, which may help explain why women are typically involved in economically driven crimes such as property, prostitution, and drug-related offenses. Women also struggle with limited education and a lack of vocational training, which places them at risk for criminal behavior. The majority of women in state prisons across the United States have not completed high school and struggle with learning disabilities and literacy challenges. For example, 29% of women in custody in New York have less than a fifth-grade reading ability. Yet, many prison facilities provide limited educational and vocational training, leaving women ill prepared to successfully transition to the community following their release. For example, of the 64% of women who enter prison without a high school diploma, only 16% receive their GED and only 29% participate in any form of vocational training while they are incarcerated (Women's Prison Association [WPA], 2003, 2009a).

Physical and Mental Health Needs of Incarcerated Women

Women in custody face a variety of physical and mental health issues. In many cases, the criminal justice system is ill equipped to deal with these issues. Given the high rates of abuse and victimization these women experience throughout their lives, it is not surprising that the incarcerated female population has a high demand for mental health services. Women in prison have significantly higher rates of mental illness compared to women in the general population. Thirteen percent of women in federal facilities and 24% of women in state prisons indicate that they have been diagnosed with a mental disorder (General Accounting Office, 1999). The pains of imprisonment, including the separation from family and adapting to the prison environment, can exacerbate these conditions.

Women also face a variety of physical health needs. Women in prison are more likely to be HIV positive, presenting a unique challenge for the prison health care system. While women in the general United States population have an HIV infection rate of 0.3%, the rate of infection for women in state and federal facilities is 3.6%, a ten-fold increase. In New York state, this statistic rises to an alarming 18%, a rate sixty times that of the national infection rate. These rates are significantly higher than the rates of HIV-positive incarcerated men. Why is HIV an issue for women in prison? When we consider the lives of women prior to their incarceration, we find that these pathways are filled with experiences of abuse, which in turn places women at risk for unsafe sexual behaviors and drug use, factors that increase the potential for infection. For example, women who are HIV positive are more likely to have a history of sexual abuse, compared to women who are HIV negative (WPA, 2003). While the rates of HIV-positive women have declined since an all-time high in 1999, the rate of hepatitis C infections has increased dramatically within the incarcerated female population. Estimates indicate that between 20% and 50% of women in jails and prisons are affected by this disease. Hepatitis C is a disease that is transmitted via bodily fluids such as blood and can lead to liver damage if not diagnosed or treated. Offending women are at a high risk to contract hepatitis C given their involvement in sex and drug crimes. Few prison facilities routinely test for hepatitis C, and treatment can be expensive due to the high cost of prescriptions (Van Wormer & Bartollas, 2010).

▲ **Photo 10.1** Visitation at a women's prison. Here, a no-contact visit means that the inmate cannot touch or hug her family and friends when they come to visit. For many women, the lack of physical contact with their loved ones can contribute to the stress and loneliness of incarceration.

While women inmates have a higher need for treatment (both in terms of prevalence as well as severity of conditions) compared to male inmates, the prison system is limited in its resources and abilities to address these issues. For example, most facilities are inadequately staffed or lack the diagnostic tools needed to address women's gynecological issues. Women also have higher rates of chronic illnesses than the male population (Anderson, 2006). However, the demands for these services significantly outweigh their availability. While states such as New York indicate that more than 25% of women receive mental health treatment while they are incarcerated, the lack of accessible services ranks high on the list of inmate complaints regarding quality of life issues in prison (WPA, 2003).

While the decision in *Todaro v. Ward* (1977) mandated reforms to health care in prisons, women continue to receive fewer resources compared to the male incarcerated population (Anderson, 2006). Elaine Lord, the former superintendent of Bedford Hills Correctional Facility (a maximum-security prison for women in New York State) tells of the challenges that face a facility wherein a large percentage of the women suffer from mental health issues. She highlights how facilities struggle to provide adequate resources to address these issues and that in these instances, challenges to the court are not necessarily a bad thing, as it can force states to provide additional funds to expand the options and availability for management and treatment of these issues (Lord, 2008).

Children of Incarcerated Mothers: The Unintended Victims

Another key issue for women in prison involves the effects of incarceration on children. Children of incarcerated mothers (and fathers) deal with a variety of issues that stem from the loss of a parent, including grief, loss, sadness, detachment, and aggressive or at-risk behaviors for delinquency, and these

▲ **Photo 10.2** The rise of female incarceration has had significant impacts on the lives of incarcerees' children, who are left to grow up without their mothers. Here, children visit with their mothers at Rikers Island Prison in New York.

children are at high risk for ending up in prison themselves as adults. The location of many prisons makes it difficult for many children to retain physical ties with their mother throughout her incarceration. While more than two thirds of incarcerated mothers have children under the age of 18, only 9% of these women will ever get to be visited by their children while they are incarcerated (Van Wormer & Bartollas, 2010).

For the 5–10% of women who enter prison while pregnant, only nine states (New York, California, Illinois, Indiana, Ohio, Nebraska, South Dakota, Washington, and West Virginia) have prison nurseries, which allow for women to remain with their infant children for at least part of their sentence (WPA, 2009b). The oldest prison nursery program is located at Bedford Hills Correctional Facility in New York. Founded in 1901, this program is the largest in the country and allows for 29 mothers to reside with their infant children. Women residing in the prison nursery take classes on infant development and participate in support groups with other mothers. While most programs limit the time that a child can reside with his or her mother (generally 12–18 months), the Washington Correctional Center for Women is unique in that their prison nursery program allows for children born to incarcerated women to remain with their mothers for up to 3 years (WPA, 2009b). Other states allow for overnight visits with children, either in special family units on the prison grounds or in specialized cells within the facility. At Bedford Hills, older children can participate in programs at the facility with their mothers (Van Wormer & Bartollas, 2010). These programs help families repair and maintain ties between a mother and her child(ren) throughout her incarceration. Not only do these program help to end the cycle of incarceration, but they also assist in the reduction of recidivism once a woman is released from custody (WPA, 2009b).

While the concept of the prison nursery and programming for children of incarcerated mothers helps promote the bond between parent and child, what about those states where these types of programs are not available? What happens to these children? The majority of women in the criminal justice system are the primary custodial parents for their young children, and these women must face the issue of who will care for their children while they are incarcerated. Some may have husbands and fathers to turn to for assistance, though many will seek out extended family members, including grandparents, who will be charged with the task of raising their children. Indeed, 79% of children who have an incarcerated parent are raised by an extended family member (WPA, 2003). In cases where an extended family member in unable or unavailable to care for a woman's minor child(ren), social services will place them in foster care. When a woman faces a long term of incarceration, the Adoption and Safe Families Act of 1997 terminates the parental rights in cases where children have been in foster care for 15 months (out of the previous 22 months). Given the increases in strict sentencing practices, the effects of this law mean that the majority of incarcerated women will lose their children if a family member is unable to care for them while the mother serves her sentence (Belknap, 2007).

Gender-Responsive Programming for Women in Prison

Clearly, women have been significantly neglected by the prison system throughout history. In an effort to remedy the disparities in treatment, several court cases began to challenge the practices in women's prisons. The case of *Barefield v. Leach* (1974) was particularly important for women, as it set the standard through which the courts could measure whether women received a lower standard of treatment compared to men. Since *Barefield*, the courts have ruled that a number of policies that were biased against women were unconstitutional. For example, the case of *Glover v. Johnson* (1979) held that the state must provide the same opportunities for education, rehabilitation, and vocational training for females as provided for male offenders. *Todaro v. Ward* (1977) declared that the failure to provide access to health care for incarcerated women was a violation of the Eighth Amendment protection against cruel and unusual punishment. Cases such as *Cooper v. Morin* (1980) held that the equal protection clause prevents prison administrators from justifying the disparate treatment of women on the grounds that providing such services for women is inconvenient. Ultimately, the courts held that "males and females must be treated equally unless there is a substantial reason which requires a distinction be made" (*Canterino v. Wilson,* 1982).

While these cases began to establish a conversation on the accessibility of programming for women, these early discussions focused on the issue of parity between male and female prisoners. At the time, women comprised only about 5% of the total number of incarcerated offenders. During the 1970s, prison advocates worked toward providing women with the same opportunities for programming and treatment as men. Their efforts were relatively successful in that many gender-based policies were abolished, and new policies were put into place mandating that men and women be treated similarly (Zaitzow & Thomas, 2003). However, feminist criminologists soon discovered that parity and equality for female offenders does not necessarily mean that women require the same treatment as men (Bloom, Owen, & Covington, 2003, 2004). Indeed, research has documented that programs designed for men fail the needs of women (Belknap, 2007).

These findings led to the emergence of a new philosophy of parity for women—gender-responsive programming. What does it mean to be gender responsive in our prison environments? Research by Bloom et al. (2003, 2004) highlights how six key principles can change the way in which programs and institutions design and manage programs, develop policies, train staff, and supervise offenders. These six principles are (1) gender, (2) environment, (3) relationships, (4) services and supervision, (5) socioeconomic status, and (6) community. Together, these six principles provide guidance for the effective management of female offenders.

The first principle of gender discusses the importance for criminal justice systems and agents to recognize the role that gender plays in the offending of women and the unique treatment needs of women. As discussed in Section II, the pathways of women to crime are dramatically different from the pathways of men. Even though they may be incarcerated for similar crimes, their lives related to these offenses are dramatically different. As a result, men and women respond to treatment in different ways and have different issues to face within the context of rehabilitation. To offer the same program to men and women may not adequately address the unique needs for both populations. Given that the majority of programs have been developed about male criminality and are used for male offenders, these programs fail the unique needs of women. While the courts have held that prison officials must provide parity and equality for male and female offenders, Bloom et al. (2003, 2004) highlight that equal treatment does not necessarily mean that women require the same treatment as men.

The second principle of environment focuses on the need for officials to create a place where staff and inmates engage in practices of mutual respect and dignity. Given that many women involved in the criminal justice system come from a background of violence and abuse, it is critical that women feel safe and supported in their journey toward rehabilitation and recovery. Historically, the criminal justice system has emphasized a model of power and control, a model that limits the ability for nurturing, trust, and compassion. Yet research indicates that these elements are essential in providing effective rehabilitative environments for women. Research by Covington (1999) suggests that rehabilitative programs for women need to create an environment that is a safe place where women can share about the intimate details of their lives.

The third element of relationships refers to developing an understanding of why women commit crimes; the context of their lives prior to, during, and following incarceration; and the relationships that women build while they are incarcerated. In addition, the majority of incarcerated women attempt to sustain their relationships with family members outside the prison walls, particularly with their minor children. Given that the majority of incarcerated women present a low safety risk to the community, women should be placed in settings that are minimally restrictive, offer opportunities for programs and services, and reside in locations within reasonable proximity to their families and minor children. The concept of relationships also involves how program providers interact with and relate to their clients. Group participants need to feel supported by their treatment providers, and the providers need to be able to empower women to make positive choices about their lives (Covington, 1999).

The fourth principle identifies the need for gender-responsive programming to address the traumas that women have experienced throughout the context of their lives. As indicated throughout this text, the cycle to offending for women often begins with the experience of victimization. In addition, these victim experiences continue throughout their lives and often inform their criminal actions. Historically, treatment providers for substance abuse issues, trauma, and mental health issues have dealt with offenders on an individualized basis. Gender-responsive approaches highlight the need for program providers and institutions to address these issues as co-occurring disorders. Here, providers need to be cross-trained in these three issues in order to develop and implement effective programming options for women. In addition, community correctional settings need to acknowledge how these issues translate into challenges and barriers to success in the reentry process. This awareness can help support women in their return to the community.

The fifth principle focuses on the socioeconomic status of the majority of women in prison. Most women in prison turn to criminal activity as a survival mechanism. Earlier in this section, you learned that women in the system lack adequate educational and vocational resources to develop a sustainable life for themselves and their families and struggle with poverty, homelessness, and limited public assistance resources, particularly for drug-convicted offenders. In order to enhance the possibilities of success following their incarceration, women need to have access to opportunities to break the cycle of abuse and create positive options for their future. Without these skills and opportunities, many women will fall back into the criminal lifestyle out of economic necessity. Given that many women will reunite with their children following their release, these opportunities will not only help women make a better life for themselves, but for their children as well.

The sixth principle of community focuses on the need to develop collaborative relationships among providers in order to assist women in their transition toward independent living. Bloom et al. (2003) call for the need to develop wraparound services for women. The concept of wraparound services refers to "a holistic and culturally sensitive plan for each woman that draws on a coordinated range of services within her community" (p. 82). Examples of these services include public and mental health systems, addiction recovery, welfare, emergency shelter organizations, and educational and vocational services.

Certainly, wraparound services require a high degree of coordination between agencies and program providers. Given the multiple challenges that women face throughout their reentry process, the development of comprehensive services will help support women toward a successful transition. In addition, by having one case manager to address multiple issues, agencies can be more effective in meeting the needs of and supervising women in the community while reducing the levels of bureaucracy and "red tape" in the delivery of resources.

Table 10.2 illustrates how the principles of gender, environment, relationships, services and supervision, socioeconomic status, and community can be utilized when developing gender-responsive policies and programming. These suggestions can assist institutional administrators and program providers in developing policies and procedures that represent the realities of women's lives and reflect ways that rehabilitation efforts can be most effective for women. Within each of these topical considerations, correctional agencies should be reminded that the majority of female offenders are nonviolent in nature, are more likely to be at risk for personal injury versus harmful toward others, and are in need of services.

Reentry Issues for Incarcerated Women

With the increases in the incarceration of both men and women, it is not surprising that the numbers of people leaving the prison community have changed dramatically, creating a new population of people in need of services. By the mid-2000s, more than 650,000 people were released from prison each year (Harrison & Beck, 2005). Historically, the majority of research has focused on whether offenders will reoffend and return to prison (recidivism). Recent scholars have shifted the focus on reentry to discussions on how to successfully transition offenders back into their communities. This process can be quite traumatic, and for women, a number of issues emerge in creating a successful reentry experience.

Consider the basic needs of a woman who has just left prison. She needs housing, clothing, and food. She may be eager to reestablish relationships with friends, family members, and her children. In addition, she has obligations as part of her release—appointments with her parole officer and treatment requirements. The label of an ex-offender brings unique challenges to this process as she struggles to find employment with limited educational and vocational training (Rose, Michalsen, Wiest, & Fabian, 2008). One woman shares the struggles in meeting these demands, expressing fear and the unknown of her new life and her ability to be successful in her reentry process:

> I start my day running to drop my urine [drug testing]. Then I go see my children, show up for my training program, look for a job, go to a meeting [Alcoholics Anonymous], and show up at my part-time job. I have to take the bus everywhere, sometimes eight buses for 4 hours a day. I don't have the proper outer clothes, I don't have the money to buy lunch along the way, and everyone who works with me keeps me waiting so that I am late to my next appointment. If I fail any one of these things and my PO [probation officer] finds out, I am revoked [probation is revoked]. I am so tired that I sometimes fall asleep on my way home from work at 2:00 a.m. and that's dangerous given where I live. And then the next day I have to start over again. I don't mind being busy and working hard. . . . that's part of my recovery. But this is a situation that is setting me up to fail. I just can't keep up and I don't know where to start. (Ritchie, 2001, p. 381)

Table 10.2 Questions to Ask in Developing a Systemic Approach for Women Offenders

Operational Practices

- Are the specifics of women's behavior and circumstances addressed in written planning, policies, programs, and operational practices? For example, are policies regarding classification, property, programs, and services appropriate to the actual behavior and composition of the female population?
- Does the staff reflect the offender population in terms of gender, race/ethnicity, sexual orientation, language (bilingual), ex-offender, and recovery status? Are female role models and mentors employed to reflect the racial/ethnic and cultural backgrounds of the clients?
- Does staff training prepare workers for the importance of relationships in the lives of women offenders? Does the training provide information on the nature of women's relational context, boundaries and limit setting, communication, and child-related issues? Are staff prepared to relate to women offenders in an empathetic and professional manner?
- Are staff training in appropriate gender communication skills and in recognizing and dealing with the effects of trauma and PTSD?

Services

- Is training on women offenders provided? Is this training available in initial academy or orientation sessions? Is the training provided on an ongoing basis? Is this training mandatory for executive-level staff?
- Does the organization see women's issues as a priority? Are women's issues important enough to warrant an agency-level position to manage women's services?
- Do resource allocation, staffing, training, and budgeting consider the facts of managing women offenders?

Review of Standard Procedures

- Do classification and other assessments consider gender in classification instruments, assessment tools, and individualized treatment plans? Has the existing classification system been validated on a sample of women? Does the database system allow for separate analysis of female characteristics?
- Is information about women offenders collected, coded, monitored, and analyzed in the agency?
- Are protocols established for reporting and investigating claims of staff misconduct, with protection from retaliation ensured? Are the concepts of privacy and personal safety incorporated in daily operations and architectural design, where applicable?
- How does policy address the issue of cross-gender strip searches and pat-downs?
- Does the policy include the concept of zero tolerance for inappropriate language, touching, and other inappropriate behavior and staff sexual misconduct?

Children and Families

- How do existing programs support connections between the female offender and her children and family? How are these connections undermined by current practice? In institutional environments, what provisions are made for visiting and for other opportunities for contact with children and family?
- Are there programs and services that enhance female offenders' parenting skills and their ability to support their children following release? In community supervision settings and community treatment programs, are parenting responsibilities acknowledged through education? Through child care?

Community

- Are criminal justice services delivered in a manner that builds community trust, confidence, and partnerships?
- Do classification systems and housing configurations allow community custody placements? Are transitional programs in place that help women build long-term community support networks?
- Are professionals, providers, and community volunteer positions used to facilitate community connections? Are they used to develop partnerships between correctional agencies and community providers?

Upon release, the majority of women find themselves returning to the same communities in which they lived prior to their incarceration, where they face the same problems of poverty, addiction, and dysfunction. For those women who were able to receive some therapeutic treatment in prison, most acknowledge that these prison-based intervention programs provided few, if any, legitimate coping skills to deal with the realities of the life stressors that awaited them upon their release. Without continuing community-based resources, many women will return to the addictions and lifestyles in which they engaged prior to their incarceration. In addition, women have limited access to health care on the outside, often due to a lack of community resources, an inability to pay, or lack of knowledge about where to go to obtain assistance. Given the status of mental and physical health needs of incarcerated women, the management (or lack thereof) of chronic health problems can impede a woman's successful reentry process (Ritchie, 2001).

While women may turn to public assistance to help support their reentry transition, many come to find that these resources are either unavailable or are significantly limited. For example, the Welfare Reform Bill, signed by President Bill Clinton in 1996, not only imposed time limits on the aid that women can receive, but has significantly affected the road to success by denying services and resources for women with a criminal record, particularly in cases of women convicted on a felony drug-related charge (Hirsch, 2001). Section 115 of the welfare reform act calls for a lifetime ban on benefits such as Temporary Assistance for Needy Families (TANF) and food stamps to offenders convicted in the state or federal courts for a felony drug offense. In addition, women convicted of a drug offense are barred from living in public housing developments and, in some areas, a criminal record can limit the availability of Section 8 housing options[2] (Jacobs, 2000). Drug charges are the only offense type subjected to this ban—even convicted murderers can apply for and receive government benefits following their release (Sentencing Project, 2006). In her research on drug-convicted women and their struggles with reentry, Hirsch (2001) found that the majority of women with drug convictions were incarcerated on charges involving low levels of substances designed for personal use, not distribution. Most of these women struggled with use and addiction since adolescence and early adulthood, often in response to significant experiences with abuse and victimization, but rarely had access to treatment to address their issues. They had relatively limited educational and vocational training and faced a variety of issues such as homelessness, mental health issues, and poverty. While many of them had children whom they cared for deeply, these relationships were often strained as a result of their issues with addiction and subsequent incarceration, making their family reunification efforts a challenge. Indeed, the limits of this ban jeopardize the very efforts toward sustainable and safe housing, education, and drug treatment that are needed in order for women to successfully transition from prison. Table 10.3 presents state-level data on the implementation of the ban on welfare benefits for felony drug convictions.

How many women are affected by the lifetime bans on assistance under Section 115? Research by the Sentencing Project indicates that, as of 2006, more than 92,000 women are currently affected by the lifetime welfare ban. They also estimate that the denial of benefits places more than 135,000 children of these mothers at risk for future contact with the criminal justice system due to economic struggles. The ban also disproportionately affects women of color, with approximately 35,000 African American women and 10,000 Latina women dealing with a loss of benefits. Since its enactment in 1996, 37 states have

[2]Section 8 housing provides government subsidies for housing in nonpublic housing developments. Here, private landlords are paid the difference between the amount of rent that a tenant can afford, based on his or her available income, and the fair market value of the residence.

Table 10.3 State Implementation of Lifetime Welfare Ban

State	Denies Benefits Entirely	Partial Denial/ Term Denial	Benefits Dependent on Drug Treatment	Opted Out of Welfare Ban
Alabama	X			
Alaska	X			
Arizona	X			
Arkansas		X		
California		X		
Colorado		X		
Connecticut				X
Delaware			X	
District of Columbia				X
Florida		X		
Georgia	X			
Hawaii			X	
Idaho		X		
Illinois		X		
Indiana	X			
Iowa		X		
Kansas	X			
Kentucky			X	
Louisiana		X		
Maine				X
Maryland			X	
Massachusetts		X		
Michigan				X
Minnesota			X	
Mississippi	X			

State	Denies Benefits Entirely	Partial Denial/ Term Denial	Benefits Dependent on Drug Treatment	Opted Out of Welfare Ban
Missouri	X			
Montana	X			
Nebraska	X			
Nevada			X	
New Hampshire				X
New Jersey			X	
New Mexico				X
New York				X
North Carolina		X		
North Dakota	X			
Ohio				X
Oklahoma				X
Oregon				X
Pennsylvania				X
Rhode Island		X		
South Carolina			X	
South Dakota	X			
Tennessee			X	
Texas	X			
Utah			X	
Vermont				X
Virginia	X			
Washington			X	
West Virginia	X			
Wisconsin			X	
Wyoming				X
U.S. total	15	11	12	13

rescinded the lifetime ban on resources, either in its entirety or in part. However, 13 states have retained this ban on assistance, placing family reunification efforts between women and their children in jeopardy (Sentencing Project, 2006).

Even women without a drug conviction still face significant issues in obtaining public assistance. Federal welfare law prohibits states from providing assistance under programs such as TANF (Temporary Assistance for Needy Families), SSI (Supplementary Security Income), housing assistance, or food stamps in cases where a woman has violated a condition of her probation or parole. In many cases, this can be as simple as failing to report for a meeting with a probation officer when she has a sick child. In addition, TANF carries a 5-year lifetime limit for assistance. This lifetime limit applies to all women, not just those under the criminal justice system. In addition, the delay to receive these services ranges from 45 days to several months, a delay that significantly affects the ability of women to put a roof over their children's heads, clothes on their bodies, and food in their bellies (Jacobs, 2000). Ultimately, these reforms are a reflection of budgetary decisions that often result in the slashing of social service and government aid programs, while the budgets for criminal justice agendas such as incarceration continue to increase. These limits not only affect the women who are in the greatest need of services but their children as well, who will suffer physically, mentally, and emotionally from these economic struggles (Danner, 2003).

Despite the social stigma that comes with receiving welfare benefits, women in one study indicated that the receipt of welfare benefits represented progress toward a successful recovery and independence from reliance on friends, family, or a significant other for assistance. A failure to receive benefits could send them into a downward spiral toward homelessness, abusive relationships, and relapse. As one woman reported to Hirsch (2001),

> We still need welfare until we are strong enough to get on our feet. Trying to stay clean, trying to be responsible parents and take care of our families. We need welfare right now. If we lose it, we might be back out there selling drugs. We're trying to change our lives. Trying to stop doing wrong things. Some of us need help. Welfare helps us stay in touch with society. Trying to do what's right for us. (p. 278)

Throughout the reentry process, women also struggle with gaining access to services. Given the nature of their offenses, many women are classified as low risk, even though they may have a high level of needs. However, this classification as low risk means that they will have reduced contact with their parole/probation officer and will receive few mandates or referrals for services. Without these referrals, most women are denied access to treatment due to the limited availability of services or an inability to pay for such resources on their own. Here, women are actually at risk for recidivism, as their needs continue to be unmet. In addition, many of the therapeutic resources that are available to women fail to work within the context of their lives. For example, the majority of inpatient drug treatment programs do not provide the option for women to reside and care for their children. These programs promote sobriety first and rarely create the opportunity for family reunification until women have successfully transitioned from treatment, have obtained a job, and can provide a sustainable environment for themselves. For many women, the desire to reunite with their children is their primary focus, and the inability for women to maintain connection with their children can threaten their path toward sobriety (Jacobs, 2000).

Clearly, women who make the transition from prison or jail back to their communities must achieve stability in their lives. With multiple demands on them (compliance with the terms and conditions of their release; dealing with long-term issues such as addiction, mental health, and physical health concerns; and the need for food, clothing, and shelter), this transition is anything but easy. Here, the influence of a positive mentor can provide significant support for women as they navigate this journey.

> While it is true a woman in reentry has many tangible needs (housing, employment, family reunification, formal education), attention to intangible needs (empowerment, a sense of belonging, someone to talk to) can promote personal growth through positive reinforcement of progress, encouragement and support in the face of defeat and temptation, and a place to feel like a regular person. (WPA, 2008, p. 3)

Several key pieces of legislation have focused on the need for support and mentorship throughout the reentry process and have provided federal funding to support these networks. For example, the Ready4Work initiative (2003), the Prisoner Reentry Initiative (2005), and the Second Chance Act (2007) all acknowledged the challenges that ex-offenders face when they exit the prison environment. These initiatives help support community organizations that provide comprehensive services for ex-offenders, including case management, mentoring, and other transitional services (WPA, 2008). Given the struggles that women face as part of their journey back from incarceration, it is clear that these initiatives can provide valuable resources to assist with the reentry process.

Summary

- The first prison for women was opened in 1839 in response to the growing concerns of abuse of women in male prison facilities.
- The reformatory prison was designed to rehabilitate women from their immoral ways.
- The custodial institution offered very little in terms of rehabilitative programming for incarcerated women.
- Women of color are overrepresented in women's prisons.
- Women in custody face a variety of unique issues, many of which the prison is ill equipped to deal with.
- Some facilities have prison nursery programs, which allow mothers to remain with their infant children while incarcerated.
- Gender-responsive programming is designed to address the unique needs of female offenders.
- Upon release, many women return to the communities in which they lived prior to their incarceration, where they face issues of addiction and dysfunction in their lives.

KEY TERMS

Elizabeth Fry
Custodial institutions
Reformatory
Todaro v. Ward (1977)
Incarcerated mothers
Gender-responsive programming
Barefield v. Leach (1974)
Glover v. Johnson (1979)
Cooper v. Morin (1980)
Canterino v. Wilson (1982)
Parity
Wraparound services
Welfare Reform Bill of 1996
Reentry

DISCUSSION QUESTIONS

1. If you were to build a woman's prison that reflected gender-responsive principles, what key features would you integrate into your facility?
2. Discuss the profile for women who are incarcerated in our prison facilities. In what ways are incarcerated women different from incarcerated men?
3. What challenges do women face during their reentry process? How does the Welfare Reform Bill limit access to resources for some women following their incarceration?

WEB RESOURCES

Women's Prison Association: http://www.wpaonline.org

Our Place: DC http://www.ourplacedc.org

Hour Children: http://www.hourchildren.org

The Sentencing Project: http://www.sentencingproject.org

READING

In the section introduction, you learned about how prisons and programs for female offenders can be responsive to issues of gender and consider the unique needs of females. This reading expands on the issues that women face and discusses how the war on drugs has had significant unintended consequences for women. Here the authors highlight six key principles for consideration in developing gender-responsive programming: gender, environment, relationships, services and supervision, socioeconomic status, and community.

Women Offenders and the Gendered Effects of Public Policy

Barbara Bloom, Barbara Owen, and Stephanie Covington

Women represent a significant proportion of all offenders under criminal justice supervision in the US. Numbering over one million in 2001, female offenders make up 17% of all offenders under some form of correctional sanction. Although their numbers have grown, we maintain that public policy has ignored the context of women's lives and that women offenders have disproportionately suffered from the impact of ill-informed public policy. These policies—both within the criminal justice system and other social arenas—ignore the realities of gender. One such detrimental policy—the so-called war on drugs—has had a critical impact on the lives of women in the criminal justice system. This policy has punished women disproportionately to the harm they cause society. As the US increased the criminal penalties through mandatory sentencing and longer sentence lengths, huge increases in the imprisonment of women have been a gendered consequence of these policies. Women are most likely to be incarcerated for a drug-related crime. Nationwide, about 35% of the imprisoned women were serving a sentence for a drug related crime, with the remainder distributed somewhat equally among property and violent crime (Bureau of Justice Statistics [BJS], 1999b). This distribution differs by jurisdiction: for example, in the federal prison system, more than 80% of the female prison population in 2000 was serving a sentence for a drug-related crime. While there is some evidence that these population increases are leveling off, the US female prison population has increased from about 10,000 in 1980 to more than 96,000 in 2002 (BJS, 2003). As we discuss below, other policy changes, such as welfare reform, and public housing have combined to create a disparate impact on drug abusing women and women of color (Allard, 2002).

Attention to gender has long been absent from criminal justice policy. As Bloom and Covington (2000, p. 11) propose, an equitable system for women would be gender-responsive, defined as "creating an environment . . . that reflects an understanding of the realities of women's lives and addresses the issues of

SOURCE: Bloom, B., Owen, B., & Covington, S. (2004). Women offenders and the gendered effects of public policy. *Review of Policy Research, 21*(1), 31–48.

the women." If criminal justice policies continue to ignore these realities, the system will remain ineffective in targeting the pathways to offending that both propel women into and return them to the criminal justice system. Elsewhere, we (Bloom, Owen, & Covington, 2003) have argued that an investment in gender-responsive policy produces both short- and long-term dividends for the criminal justice system, the community, and women offenders and their families.

Acknowledging Gender Differences

Acknowledging gender requires understanding the distinction between the concepts of sex and gender. Research on the differences between women and men suggests that social and environmental factors, rather than biological determinants, account for the majority of behavioral differences between males and females. Although purely physiological differences influence some basic biological processes affecting health and medical care, and a range of reproductive issues, many of the observed behavioral differences are the result of differences in gender socialization, gender roles, gender stratification, and gender inequality. Belknap (2001) explains that sex differences are biological differences, such as those concerning reproductive organs, body size, muscle development and hormones. Gender differences are those that are ascribed by society and that relate to expected social roles (p. 11). They are neither innate nor unchangeable. These gender differences shape the reality of women's lives and the contexts in which women live.

Understanding the distinction between sex and gender informs us that most differences between men and women are societally based (gender), not biologically determined (sex). It is important to comprehend and acknowledge some of the dynamics inherent in a gendered society. The influence of the dominant culture is so pervasive that it is often unseen. One of the gender dynamics found where sexism is prevalent is that programs or policies declared "genderless" or "gender neutral" are in fact male-based (Kivel, 1992).

Race and class can also determine views of gender-appropriate roles and behavior. Differences exist among women based on race and socioeconomic status or class. Regardless of their differences, all women are expected to incorporate the gender-based norms, values, and behaviors of the dominant culture into their lives. As Kaschak (1992) said, "The most centrally meaningful principle on our culture's mattering map is gender, which intersects with other culturally and personally meaningful categories such as race, class, ethnicity, and sexual orientation. Within all of these categories, people attribute different meanings to femaleness and maleness" (p. 5). This discussion of the implications of gender within the criminal justice system is based on a simple assumption: responding to the differences between women and men in criminal behavior and to their antecedents is consistent with the goals of all correctional agencies. These goals are the same for all offenders, whether they are male or female. Across the criminal justice continuum, the goals of the system typically involve sanctioning the initial offense, controlling behavior while the offender is under its jurisdiction, and, in many cases, providing interventions, programs, and services to decrease the likelihood of future offending. At each stage in the criminal justice process, the differences between female and male offenders affect behavioral outcomes and the ability of the system to address the pathways to offending and thus achieve its goals.

Characteristics of Women in the Criminal Justice System

Understanding gender-based characteristics is also critical to gender-responsive policy. The significant increase in the number of women under criminal justice supervision has called attention to the status of women in the criminal justice system and to the particular circumstances they encounter. Current research has established that women offenders differ from their male counterparts in personal histories and pathways to crime

(Belknap, 2001). Women offenders are low-income, undereducated, and unskilled with sporadic employment histories, and they are disproportionately women of color. They are less likely than men to have committed violent offenses and more likely to have been convicted of crimes involving drugs or property. Often their property offenses are economically driven, motivated by poverty and by the abuse of alcohol and other drugs. Table 10.4 summarizes salient demographic characteristics of women in the criminal justice system.

Table 10.4 Characteristics of Women in the Criminal Justice System

	Percentage Under Community Supervision	Percentage in Jail	Percentage in Prison
Race/ethnicity			
White	62	36	33
African American	27	44	48
Hispanic	10	15	15
Median age	32	31	33
High school/GED	60	55	56
Single	42	48	47
Unemployed	—	60	62
Mother of minor children	72	70	65

Women face life circumstances that tend to be specific to their gender such as sexual abuse, sexual assault, domestic violence, and the responsibility of being the primary caretaker for dependent children. Approximately 105,000 minor children have a mother in jail and approximately 65% of women in state prisons and 59% of women in federal prisons have an average of two minor children. Women offenders reflect a population that is marginalized by race, class, and gender (Bloom, 1996). For example, African American women are overrepresented in correctional populations. While they comprise only 13% of women in the US, nearly 50% of women in prison are African American. Black women are nearly eight times more likely than white women to be incarcerated.

Eighty-five percent of women in the criminal justice system are under community supervision. In 2000, more than 900,000 women were on probation (844,697) or parole (87,063). Women represented an increasing percentage of the probation and parole populations in 2000, as compared to 1990. Women represented 22% of all probationers in 2000 (up from 18% in 1990) and 12% (up from 8% in 1990) of those on parole (BJS, 2001b).

While nearly two-thirds of women confined in jails and prisons are African American, Hispanic, or of other (non-white) ethnic origin, nearly two-thirds of those on probation are white. About 60% of women on probation have completed high school; 72% have children under 18 years of age (BJS, 2001b).

Offense Profiles

Accompanying this increase in population are several questions about women offenders. Why has women's involvement with the criminal justice system increased so dramatically? Are women committing more crimes? Are these crimes becoming more violent? The data on arrests demonstrate that the number of women under criminal justice supervision has risen disproportionately to arrest rates. For example, the total number of arrests of adult women increased by 38.2% between 1989 and 1998, while the number of women under correctional supervision increased by 71.8%. Overall, women have not become more violent as a group. In 2000, women accounted for only 17% of all arrests for violent crime. About 71% of all arrests of women were for larceny/theft or drug-related offenses (BJS, 2001b and c).

Women on probation have offense profiles that are somewhat different from those of incarcerated women. Nationwide, the majority of women on probation have been convicted of property crimes (44%). Of female probationers, 27% have been convicted of public order offenses and 19% have been convicted of drug offenses. Only 9% committed violent crimes (BJS, 2001b and c).

Data collected by the Bureau of Justice Statistics (BJS, 1999b) indicate that violent offenses are the major factor in the growth of the male prison population; however, this is not the case for women. For women, drug offenses were the largest source of growth (38% compared to 17% for males) for the female prison population. In 1998, 22% of incarcerated women had been convicted for violent offenses (BJS, 1999b). The majority of offenses committed by women in prisons and jails are nonviolent drug and property crimes.

Gender-Based Experiences

Women's most common pathways to crime are based on survival of abuse, poverty, and substance abuse. The pathway perspective (Belknap, 2001; Owen, 1998) confirms the importance of the following interconnected factors:

Family Background. Women in the criminal justice system are more likely than those in the general population to have grown up in a single-parent home. Within the incarcerated population, women are more likely than men to have had at least one incarcerated family member.

Abuse History. The prevalence of physical and sexual abuse in the childhoods and adult backgrounds of women under correctional supervision has been supported by the research literature; abuse within this segment of the population is more likely than in the general population (BJS, 1999c). In examining the abuse backgrounds of male and female probationers, the Bureau of Justice Statistics (BJS, 1999c) found a dramatic gender difference: more than 40% of the women reported having been abused at some time in their lives, compared to 9% of the men.

Substance Abuse. Women are more likely to be involved in crime if they are drug users (Merlo & Pollock, 1995). Approximately 80% of women in state prisons have substance abuse problems (Center for Substance Abuse Treatment [CSAT], 1997). About half of women offenders in state prisons had been using alcohol, drugs, or both at the time of their offense. On every measure of drug use, women offenders in state prisons reported higher usage than their male counterparts—40% of women offenders and 32% of male offenders had been under the influence of drugs when the crime occurred.

Physical Health. Women frequently enter jails and prisons in poor health, and they experience more serious health problems than do their male counterparts. This poor health is often due to poverty, poor nutrition, inadequate health care, and substance abuse (Acoca, 1998; Young, 1996). It is estimated that 20–35% of women go to prison sick call daily compared to 7–10% of men. The specific health consequences of long-term substance abuse are significant for all women, but they are particularly so for pregnant women.

Mental Health. Many women enter the criminal justice system having had prior contact with the mental health system. Women in prison have a higher incidence of mental disorders than women in the community. One-quarter of women in state prisons have been identified as having a mental illness (BJS, 2001a); the major diagnoses of mental illness are depression, post-traumatic stress disorder (PTSD), and substance abuse. Women offenders have histories of abuse associated with psychological trauma. PTSD is a psychiatric condition often seen in women who have experienced sexual abuse and other trauma.

Marital Status. Compared to the general population, women under correctional supervision are more likely to have never been married. I n 1998, nearly half of the women in jail and prison reported that they had never been married, compared to 46% in 1991 (BJS, 1994, 1999b). Forty-two percent of women on probation reported that they had never been married.

Children. Approximately 70% of all women under correctional supervision have at least one child who is under 18. Two-thirds of incarcerated women have children under the age of 18; about two-thirds of women in state prisons and half of women in federal prisons had lived with their young children prior to entering prison. It is estimated that 1.3 million minor children have a mother who is under correctional supervision and more than a quarter of a million minor children have mothers in jail or prison (BJS, 1999a).

Education and Employment. In 1998, an estimated 55% of women in local jails, 56% of women in state prisons, and 73% of women in Federal prisons had completed high school (BJS, 1999b). Approximately 40% of the women in state prisons reported that they were employed full-time at the time of their arrest. Most of the jobs held by women were low-skill and entry-level, with low pay. Women are less likely than men to have engaged in vocational training before incarceration.

In summary, a national profile of women offenders describes the following characteristics:

- disproportionately women of color
- in their early- to mid-thirties
- most likely to have been convicted of a drug or drug-related offense
- fragmented family histories, with other family members involved with the criminal justice system
- survivors of physical or sexual abuse as children and adults
- significant substance abuse problems
- multiple physical and mental health problems
- unmarried mothers of minor children
- high school degree/GED, but limited vocational training and sporadic work histories

Improving policy for women offenders begins by targeting these characteristics and their antecedents through comprehensive treatment for drug abuse and trauma recovery, education and training in job and parenting skills, and affordable and safe housing.

Theoretical Perspectives

Women in the criminal justice system come into the system in ways different from those of men. This is due partly to differences in pathways into criminality and offense patterns, and partly to the gendered effect of the war on drugs. Contemporary theorists note that most theories of crime were developed by male criminologists to explain male crime (Belknap, 2001; Chesney-Lind, 1997; Pollock, 1999). Historically, theories about women's criminality have ranged from biological to psychological and from economic to social. Social and cultural theories have been applied to men, while individual and pathological explanations have been applied to women.

Pollock (1999) found that, until recently, most criminology theory ignored the dynamics of race and class and how these factors intermix with gender to influence criminal behavior patterns (p. 8). She argues that it has been commonly believed that adding gender to these analytic variables "tended to complicate the theory and were better left out" (Pollock, 1999, p. 123). Due to this lack of attention, Belknap (2001) has called the female offender "the invisible woman."

Differences among women are also critical in providing women-sensitive policy and programs (McGee & Baker, 2003). Differences in women's pathways to the criminal justice system, women's behavior while under supervision or in custody, and the realities of women's lives in the community have significant bearing on the practices of the criminal justice system. There is significant evidence that the responses of women to community supervision, incarceration, treatment, and rehabilitation are different from those of men. Differences between men and women under community supervision and in custody have been documented in terms of the following:

- levels of violence and threats to community safety in their offense patterns
- responsibilities for children and other family members
- relationships with staff and other offenders
- vulnerability to staff misconduct and revictimization
- differences in programming and service needs while under supervision and in custody, especially in terms of physical and mental health, substance abuse, recovery from trauma, and economic/vocational skills
- differences in reentry and community integration

The Pathways Perspective

Research on women's pathways into crime indicates that gender matters significantly in shaping criminality. Steffensmeier and Allan (1998) note that the "profound differences" between the lives of women and men shape their patterns of criminal offending. Among women, the most common pathways to crime are based on survival (of abuse and poverty) and substance abuse. Belknap (2001, p. 402) has found that the pathway perspective incorporates a "whole life" perspective in the study of crime causation. Recent research establishes that because of their gender, women are at greater risk of experiencing sexual abuse, sexual assault domestic violence, and single-parent status. Pathway research has identified such key issues in producing and sustaining female criminality as histories of personal abuse, mental illness tied to early life experiences, substance abuse and addiction economic and social marginality, homelessness, and destructive relationships.

The Gendered Effects of Current Policy

While most of the policy attention has been on the impact of the war on drugs and the criminal justice system, policy changes in welfare reform, and public housing have combined to create a disparate impact on drug abusing women and women of color (Allard, 2002). Key policy areas affecting the lives of women offenders and their children include welfare benefits, drug treatment, housing, education, employment, and reunification with children.

The War on Drugs

Given the dramatic influx of women offenders into the criminal justice system, gender becomes quite important in examining the differential effect of drug policy.

As a result of the misguided drug war and its punitive consequences, women have become increasingly punished as the US continues to stiffen criminal penalties through mandatory sentencing and longer sentences. While men too have suffered as the US continues its imprisonment binge (Austin & Irwin, 2000), it is clear that women have suffered disproportionately to the harm their drug behavior represents. Inadvertently, the war on drugs became a war on women, particularly poor women and women of color. Almost a decade ago. Bloom, Chesney-Lind, and Owen (1994) suggested that women's incarceration rates were driven by more than just crime rates:

> The increasing incarceration rate for women in the State of California, then, is a direct result of short-sighted legislative responses to the problems of drugs and crime-responses shaped by the assumption that the criminals they were sending to prison were brutal males.

> Instead of a policy of last resort, imprisonment has become the first order response for a wide range of women offenders that have been disproportionately swept up in this trend. This politically motivated legislative response often ignores the fiscal or social costs of imprisonment. Thus, the legislature has missed opportunities to prevent women's crime by cutting vitally needed social service and educational programs to fund ever-increasing correctional budgets. (p. 2)

Bush-Baskette (1999) made similar observations:

> Drug use by any woman, whether she lives in suburban or urban areas, brings with it the psychological, social, and cultural experience of stigmatization that can perpetuate the continued problem of drug use. This usage and its inherent problems violate gender expectations for women in our society. Poor women who use street-level drugs experience additional societal stigma because they do not have the protective societal buffer enjoyed by women who are insulated by their families, friends, and economic status. Those who use street-level drugs are also less protected from becoming prisoners of the "war on drugs" because of their high visibility. (pp. 216–217)

The emphasis on punishment rather than treatment has brought many low-income women and women of color into the criminal justice system. Women offenders who in past decades would have been given community sanctions are now being sentenced to prison. Mandatory minimum sentencing for drug offenses has significantly increased the numbers of women in state and federal prisons. Between 1995 and 1996, female drug arrests increased by 95%, while male drug arrests increased by 55%. In 1979, approximately one in ten women in US prisons was serving a sentence for a drug conviction; in 1999, this figure was approximately one in three (BJS, 1999a).

Mauer, Potler, and Wolf (1999) measured the gender-based difference in the rates of this increase. They argued that drug policy affects women differently because women are more likely than men to commit drug offenses. In examining the overall rise in prison population between 1986 and 1995, they found that drug offenses account for about one-third of the rise in male prison population, but fully half of the increase in the female prison population. During this period, the number of women incarcerated for drug offenses rose an amazing 888%; the number of women incarcerated for other crimes rose 129%. This difference is particularly marked in states with serious penalties for drug offenses. In New York, they argue, the notorious Rockefeller drug laws account for 91% of the women's prison population increase, in California, drug offenses account for 55%, and in Minnesota, a state committed to limiting incarceration to very serious offenses, only 26%. This difference is most apparent among women of color. Compared to white women, women of color are also more likely to be arrested, convicted, and incarcerated at rates higher than their representation in the free world population (Mauer, Potler & Wolf, 1999).

Sentencing Policies

These increases have also been aggravated by mandatory minimum sentencing statutes for drug offenses that have significantly increased the numbers of women in state and federal prisons. Women offenders who would have been given community sanctions in past decades are now being sentenced to prison. Between 1995 and 1996, female drug arrests increased by 95%, while male drug arrests increased by 55%. In 1979, approximately one in ten women in US prisons was serving a sentence for a drug conviction; in 1999, this figure was approximately one in three women (BJS, 1999a).

Mandatory minimums for federal crimes, coupled with new sentencing guidelines intended to reduce racial, economic, and other disparities in sentencing males, have distinctly disadvantaged women. Twenty years ago, nearly two-thirds of the women convicted of federal felonies were granted probation; in 1991, only 28% of women were given straight probation (Raeder, 1993). Female drug couriers can receive federal mandatory sentences ranging from fifteen years to life following their first felony arrest. These gender-neutral sentencing laws fail to recognize the distinction between major players in drug organizations and minor ancillary players. According to Judge Patricia Wald (2001), "The circumstances surrounding the commission of a crime vary significantly between men and women. Yet penalties are most often based on the circumstances of crimes committed by men, creating a male norm in sentencing which makes the much-touted gender neutrality of guideline sentencing very problematical" (p. 12).

The Gendered Implications of "Three Strikes and You Are Out"

Current US prison policy is grounded in law-and-order legislative efforts to control crime, such as mandatory minimum prison sentences and increased sentence lengths. Mona Danner (1998) describes the ways in which the philosophy behind these trends in criminal justice policy has affected the lives of women, particularly the increasing penalties for drug offenses. She suggests that the consequences for women in this era of expanded punishments have been largely unexplored. In her view, public debate over "Three Strikes" and law-and-order policy ignores the reality of women's lives and that often women are forced to bear the emotional and physical brunt of these misguided policies. Danner argues that women bear these costs in three ways. First, the enormous cost of the correctional institutions needed to accommodate an increasing number of prisoners has direct implications for other social services. She cites a study by the RAND Corporation that predicts that California's Three Strikes Law will require cuts in other government services totaling 40% over eight years. She predicts that social services for the poor, especially for women and children, will be hardest hit.

Second, Danner believes that the Three Strikes laws disproportionately affect women as caregivers, through both the imprisonment of men and their own imprisonment. With nearly 1.5 million children of prisoners in the US, there are a significant number of children growing up with at least one parent incarcerated. Third, the financial and social implications for the community, as well as the individual life chances of these children, are yet another cost of Three Strikes crime control efforts.

Welfare Benefits

Section 115 of the 1996 Welfare Reform Act, "Temporary Assistance for Needy Families" (TANF), stipulates that persons convicted of a state or federal felony offense involving the use or sale of drugs are subject to a lifetime ban on receiving cash assistance and food stamps. This provision applies only to those who are convicted of a drug offense (Allard, 2002, p. 1). The lifetime welfare ban has had a disproportionate impact on African American women and Latinas with children, for several reasons. First, due to disparities in drug policies and in the enforcement of drug laws, women of color have experienced greater levels of criminal justice supervision. Second, as a result of race- and gender-based socioeconomic inequities, women of color are more susceptible to poverty and are therefore disproportionately represented in the welfare system (Allard, 2002).

Housing

Obtaining public housing may not be a viable option for women with a drug conviction. In 1996, the Federal government implemented the "One Strike Initiative" authorizing local Public Housing Authorities (PHA) to obtain from law-enforcement agencies the criminal conviction records of all adult applicants or tenants. (This policy was recently upheld by the US Supreme Court in Department of Housing and Urban Development v. Rucker et al., March 26, 2002.) Federal housing policies permit (and in some cases require) public housing authorities, Section 8 providers, and other federally assisted housing programs to deny housing to individuals who have a drug conviction or are suspected of drug involvement (Allard, 2002).

Education and Employment

As mentioned previously, a significant number of women under criminal justice system supervision have a history of low educational attainment. As of 1996, only 52% of correctional facilities for women offered postsecondary education. Access to college education was further limited when prisoners were declared ineligible for Pell Grants (Allard, 2002). Educational opportunities may also be limited by the Higher Education Act of 1998, which denies eligibility for students convicted of drug offenses. Lack of education is a key factor contributing to the underemployment and unemployment of many women in the criminal justice system.

A significant number of women under criminal justice supervision have limited employment skills and sporadic work histories, and many correctional facilities offer little in terms of gender-specific vocational training. Additionally, having a criminal record poses an additional barrier to securing employment. The transitional assistance provided through TANF and food stamps offers the financial support women need as they develop marketable employment skills and search for work that provides a living wage. Women who are denied this transitional assistance may not be able to provide shelter and food for themselves and their children while engaging in job training and placement.

Reunification With Children

The Adoption and Safe Families Act of 1997 (ASFA) mandates termination of parental rights once a child has been in foster care for fifteen or more of the past twenty-two months. While it is difficult enough for single mothers with substance abuse problems to meet ASFA requirements when they live in the community, the short deadline has particularly severe consequences for incarcerated mothers, who serve an average of eighteen months (Jacobs, 2001).

Placement of children with relatives, which would avoid the harsh ASFA mandate, is hampered by state policies that provide less financial aid to relatives who are caregivers than to non-relatives foster caregivers.

Criminal Justice System Policies

Most criminal justice policy—with few exceptions—was developed to manage the behavior of male offenders. As a result, many systems lack a written policy on the management and supervision of female offenders. In focus group interviews conducted by Bloom, Owen and Covington (2002), many managers and line staff reported that they often have to manage women offenders based on policies and procedures developed for the male offender. They also reported difficulties in modifying these policies to develop a more appropriate and effective response to women's behaviors within the correctional environment.

Gender has an undeniable effect on criminal justice processing; consider this: if gender played no role in criminal behavior and criminal justice processing, then 51.1% of those arrested, convicted, and incarcerated could be expected to be women, as that figure represents the proportion of women in the general population. Instead, men are overrepresented in most classes of criminal behavior and under all forms of correctional supervision in relation to their proportion of the general population. Gender differences have been found in all stages of criminal justice processing, including crime definition, reporting, and counting; types of crime committed; levels of harm; arrest; bail; sentencing; community supervision; incarceration; and reentry into the community (Harris, 2001). One of the most pressing policy issues affecting women involves staff sexual misconduct. In the past ten years, the problems of staff sexual misconduct have been given significant attention by the media, the public, and many correctional systems (US General Accounting Office, 1999). Yet at all levels, most criminal justice agencies have not addressed the problem through policy, training, legal penalties, or reporting/grievance procedures.

Misconduct can take many forms, including inappropriate language, verbal degradation, intrusive searches, sexual assault, unwarranted visual supervision, denying of goods and privileges, and the use or threat of force (Human Rights Watch Women's Rights Project 1996). Misconduct includes disrespectful, unduly familiar or threatening sexual comments made to inmates or parolees. Gender-neutral policies often ignore the problem of staff sexual misconduct with poor grievance procedures, inadequate investigations, and staff retaliation against inmates or parolees who "blow the whistle." Standard policies and procedures in correctional settings (e.g., searches, restraints, and isolation) can have profound effects on women with histories of trauma and abuse, and often trigger retraumatization in women who have post-traumatic stress disorder (PTSD).

Gendered Justice

A gendered policy approach calls for a new vision for the criminal justice system, one that recognizes the behavioral and social differences between female and male offenders that have specific implications for gender-responsive policy and practice. Developing gender-responsive policies, practices, programs, and services requires the incorporation of the following key findings:

- An effective system for female offenders is structured differently than a system for male offenders.
- Gender-responsive policy and practice target women's pathways to criminality by providing effective interventions that address the intersecting issues of substance abuse, trauma, mental health, and economic marginality.
- Criminal justice sanctions and interventions recognize the low risk to public safety created by the typical offenses committed by female offenders.
- Gender-responsive policy considers women's relationships, especially those with their children, and their roles in the community when delivering both sanctions and interventions.

Being gender responsive in the criminal justice system requires an acknowledgment of the realities of women's lives, including the pathways they travel to criminal offending and the relationships that shape their lives. To assist those working with women to effectively and appropriately respond to this information, Bloom and Covington (2000) developed the following definition:

> Gender-responsive means creating an environment through site selection, staff selection, program development content, and material that reflects an understanding of the realities of women's lives and addresses the issues of the participants. Gender-responsive approaches are multidimensional and are based on theoretical perspectives that acknowledge women's pathways into the criminal justice system. These approaches address social (e.g., poverty, race, class and gender inequality) and cultural factors, as well as therapeutic interventions. These interventions address issues such as abuse, violence, family relationships, substance abuse and co-occurring disorders. They provide a strength-based approach to treatment and skill building. The emphasis is on self-efficacy. (p. 11)

Guiding Principles and Strategies

Evidence drawn from a variety of disciplines and effective practice suggests that addressing the realities of women's lives through gender-responsive policy and programs is fundamental to improved outcomes at all criminal justice phases. The six guiding principles that follow are designed to address system concerns about the management operations, and treatment of women offenders in the criminal justice system.

1. *Gender:* Acknowledge that gender makes a difference

2. *Environment:* Create an environment based on safety, respect, and dignity

3. *Relationships:* Develop policies, practices, and programs that are relational and promote healthy connections to children, family, significant others, and the community
4. *Services and Supervision:* Address the issues of substance abuse, trauma, and mental health through comprehensive, integrated, culturally relevant services and appropriate supervision.
5. *Socio-economic Status:* Provide women with opportunities to improve their socioeconomic conditions.
6. *Community:* Establish a system of community supervision and reentry with comprehensive, collaborative services.

Together with the general strategies for their implementation, the guiding principles provide a blueprint for a gender-responsive approach to the development of criminal justice policy.

Developing Gender-Responsive Policy and Practice

The proposed guiding principles are intended to serve as a blueprint for the development of gender-responsive policy and practice. These principles can also provide a basis for systemwide policy and program development. Following are scenarios based on a gender-responsive model for women offenders:

- The correctional environment or setting is modified to enhance supervision and treatment.
- Classification and assessment instruments are validated on samples of women offenders.
- Policies, practices, and programs take into consideration the significance of women's relationships with their children, families, and significant others.
- Policies, practices, and programs promote services and supervision that address substance abuse, trauma, and mental health and provide culturally relevant treatment to women.
- The socioeconomic status of women offenders is addressed by services that focus on their economic and social needs.
- Partnerships are promoted among a range of organizations located within the community.

A first step in developing gender-appropriate policy and practice is to address the following questions:

- How can correctional policy address the differences in the behavior and needs of female and male offenders?
- What challenges do these gender differences create in community and institutional corrections?
- How do these differences affect correctional practice, operations, and supervision in terms of system outcomes and offender-level measures of success?
- How can policy and practice be optimized to best meet criminal justice system goals for women offenders?

Policy Considerations

As agencies and systems examine the impact of gender on their operations, policy-level changes are a primary consideration. A variety of existing policies developed by the National Institute of Corrections Intermediate Sanctions for Women Offenders Projects, the Federal Bureau of Prisons, the American Correctional Association (ACA), the Minnesota Task Force on the Female Offender, and the Florida Department of Corrections contain crucial elements of a gender-appropriate approach. Gender-responsive elements derived from this analysis are considered below.

Create Parity

As expressed in the ACA Policy Statement, "Correctional systems should be guided by the principle of parity. Female offenders must receive the equivalent range of services available to male offenders, including opportunities for individual programming and services that recognize the unique needs of this population" (American Correctional Association [ACA], 1995, p. 2). Parity differs conceptually from "equality" and stresses the importance of equivalence rather than sameness: women offenders should receive opportunities, programs, and services that are equivalent, but not identical, to those available to male offenders.

Commit to Women's Services

Executive decision-makers, administrators, and line staff must be educated about the realities of working with female offenders. Establishing mission and vision statements regarding women's issues and creating an executive-level position charged with this mission are two ways to ensure that women's issues become a priority. A focus on women is also tied to the provision of appropriate levels of resources, staffing, and training.

The National Institute of Corrections has recognized the need for gender-specific training and has sponsored a variety of initiatives designed to assist jurisdictions in addressing issues relevant to women offenders. In Florida, a staff training and development program was mandated and will be implemented for correctional officers and professionals working with female offenders in institutions and community corrections. In the Bureau of Prisons, training occurs at the local institution level. The Texas Division of Community Corrections has also created specific training for those working with female offenders in the community.

Review Standard Procedures for Their Applicability to Women Offenders

Another key element of policy for women offenders concerns a review of policies and procedures. While staff working directly with female offenders on a day-to-day basis are aware of the procedural misalignment of some procedures with the realities of women's lives, written policy often does not reflect the same understanding of these issues. As stated in the ACA policy, "Sound operating procedures that address the (female) population's needs in such areas as clothing, personal property, hygiene, exercise, recreation, and visitations with children and family" should be developed (ACA, 1995, p. 1).

Respond to Women's Pathways

Policies, programs, and services need to respond specifically to women's pathways in and out of crime and to the contexts of their lives that support criminal behavior. Procedures, programs, and services for women should be designed and implemented with these facts in mind. Both material and treatment realities of women's lives should be considered. For example, Florida's policy states that

> emphasis is placed on programs that foster personal growth, accountability, self-reliance, education, life skills, workplace skills, and the maintenance of family and community relationships to lead to successful reintegration into society and reduce recidivism. (Florida Department of Corrections, 1999, p.1)

ACA standards call for

> access to a full range of work and programs designed to expand economic and social roles for women, with an emphasis on education, career counseling and exploration of non-traditional training; relevant life skills, including parenting and social and economic assertiveness; and pre-release and work/education release programs. (ACA, 1995, p. 2)

Florida's policy states that the system must "ensure opportunities for female offenders to develop vocational and job-related skills that support their capacity for economic freedom" (Florida Department of Corrections, 1999, p. 1).

Consider Community

Given the lower risk of violence and community harm found in female criminal behavior, it is important that written policy acknowledge the actual level of risk represented by women offenders' behavior in the community and in custody. The recognition and articulation of this policy will enable the development of strong community partnerships, creating a receptive community for model reentry and transitional programs that include housing, training, education, employment, and family support services.

The ACA advocates for a range of alternatives to incarceration including pretrial and post-trial diversion, probation, restitution, treatment for substance abuse, halfway houses, and parole services. Community supervision programs need to partner with community agencies in making a wide range of services and programs available to women offenders. Community programs are better equipped than correctional agencies to respond to women's realities. After a review of its Security Designation and Custody Classification procedures, the Federal Bureau of Prisons developed additional low- and minimum-security bed space to house female offenders more appropriately and closer to their homes.

Include Children and Families

Children and family play an important role in the management of women offenders in community and custodial settings. More female than male offenders have primary responsibility for their children. However, female offenders' ties to their children are often compromised by criminal justice policy. The ACA policy states that the system should "facilitate the maintenance and strengthening of family ties, particularly between parents and children" (ACA, 1995, p. 1). In Florida, an emphasis on the relationships of women offenders with their children and other family members has potential rehabilitative effects in terms of motivation for treatment and economic responsibility (Florida Department of Corrections, 1999, p. 7).

Conclusion

This article documents the importance of understanding and acknowledging differences between female and male offenders and the impact of those differences on the development of gender-responsive policies, practices, and programs in the criminal justice system. Our analysis has found that addressing the realities of women's lives through gender-responsive policy and practice is fundamental to improved outcomes at all phases of the criminal justice system. This review maintains that consideration of women's and men's different pathways into criminality, their differential responses to custody and supervision, and their differing program requirements can result in a criminal justice system that is better equipped to respond to both male and female offenders.

References

Acoca, L. (1998). Defusing the time bomb: Understanding and meeting the growing health care needs of incarcerated women in America. *Crime & Delinquency*, 44(1), 49–70.

Allard, P (2002). *Life sentences: Denying welfare benefits to women convicted of drug offenses*. Washington, DC: The Sentencing Project.

American Correctional Association. (1995). *Public correctional policy on female offender services*. Latham, MD: Author.

Austin, J., & Irwin, J. (2000). *It's about time: America's imprisonment binge*. Belmont, CA: Wadsworth.

Belknap, J. (2001). *The invisible woman: Gender, crime, and justice*. Belmont, CA: Wadsworth.

Bloom, B. (1996). *Triple jeopardy: Race, class and gender as factors in women's imprisonment*. Riverside, CA: Department of Sociology, UC Riverside.

Bloom, B., Chesney-Lind, M., & Owen, B. (1994). *Women in California prisons: Hidden victims of the war on drugs*. San Francisco: Center on Juvenile and Criminal Justice.

Bloom, B., & Covington, S. (2000, November). *Gendered justice: Programming for women in correctional settings*. Paper presented to the American Society of Criminology, San Francisco, CA.

Bloom, B., Owen, B., & Covington (2003). *Gender-responsive strategies: Research, practice, and guiding principles for women offenders*. Washington, DC: National Institute of Corrections, US Department of Justice.

Bureau of Justice Statistics. (1994). *Women in prison*. Washington, DC: US Department of Justice.

——. (1999a). *Correctional populations in the United States, 1996.* Washington, DC: US Department of Justice.
——. (1999b). *Special report: Women offenders.* Washington, DC: US Department of Justice.
——. (1999c). *Prior abuse reported by inmates and probationers.* Washington, DC: US Department of Justice.
——. (2000). *Incarcerated parents and their children.* Washington, DC: US Department of Justice.
——. (2001a). *Mental health treatment in state prisons,* 2000. Washington, DC: US Department of Justice.
——. (2001b). *National correctional population.* Washington, DC: US Department of Justice.
——. (2001c). *Prisoners in 2000.* Washington, DC: US Department of Justice.
——. (2003). *Prison and jail inmates at midyear 2002.* Washington, DC: Bureau of Justice Statistics.
Bush-Baskette, S. (1999). The war on drugs: A war against women? In S. Cook & S. Davies (Eds.), *Harsh punishment: International experiences of women's imprisonment* (pp. 211–229). Boston: Northeastern University Press.
Center for Substance Abuse Treatment. (1997). *Substance abuse treatment for incarcerated offenders: Guide to promising practices.* Rockville, MD: Department of Health and Human Services.
Chesney-Lind, M. (1997). *The female offender: Girls, women and crime.* Thousand Oaks, CA: Sage.
Danner, M. (1998). *Three strikes and it's women who are out.* In S. Miller (Ed.) Crime Control and Women. Thousand Oaks, CA: Sage.
Florida Department of Corrections. (1999). *Operational plan for female offenders.* Tallahassee: Florida Department of Corrections.
Harris, K. (2001, May). *Women offenders in the community: Differential treatment in the justice process linked to gender.* Information session on supervision of women offenders in the community, National Institute of Corrections, Community Corrections Division, Networking Conference, Lexington, Kentucky.
Human Rights Watch Women's Rights Project. (1996). *All too familiar: Sexual abuse of women in U.S. state prisons.* New York: Ford Foundation.
Jacobs, A. (2001). Give 'em a fighting chance: Women offenders reenter society. *Criminal Justice Magazine,* 16(1), 44–47.
Kaschak, E. (1992). *Engendered lives: A new psychology of women's experience.* New York: Basic.
Kivel, P. (1992). *Men's work: Stopping the violence that tears our lives apart.* Center City, MN: Hazelden.
McGee, Z. T., & Baker, S. R. (2003). *Crime control policy and inequality among female offenders: Racial disparities in treatment among women on probation.* In R. Muraskin (Ed.). It's a crime: Women and justice (pp. 196–208). Upper Saddle River, NJ: Prentice Hall.
Merlo, A., & Pollock, J. (1995). *Women, law, and social control.* Boston: Allyn and Bacon.
Owen, B. (1998). *In the mix: Struggle and survival in a women's prison.* Albany: State University of New York Press.
Pollock, J. (1999). *Criminal women.* Cincinnati, OH: Anderson.
Raeder, M. (1993). Gender issues in the federal sentencing guidelines. *Journal of Criminal Justice,* 8(3), 20–25.
Steffensmeier, D., & Allan, E. (1998). *The nature of female offending: Patterns and explanations.* In R. T. Zaplin (Ed.). Female offenders: Critical perspectives and effective interventions (pp. 5–29). Gaithersburg, MD: Aspen.
US General Accounting Office. (1999). *Women in prison: Sexual misconduct by correctional staff.* Washington, DC:
US General Accounting Office. Wald, P. M. (2001). Why focus on women offenders? *Criminal Justice,* 16(1), 10–16.
Young, D. S. (1996). Contributing factors to poor health among incarcerated women: A conceptual model. *Affilia,* 11(4), 440–461.

DISCUSSION QUESTIONS

1. How have tough-on-crime policies such as the war on drugs become a war on women offenders?
2. What are the six guiding principles for gender-responsive programming?
3. What recommendations would you make to state legislatures and prison wardens about developing criminal justice policy that is gender responsive?

READING

In the section introduction, you learned about how the majority of incarcerated women are also mothers who are forced to leave their children either with other family members or in foster care while they serve out their sentence. Many of the women interviewed in this reading were disappointed with the lack of care they received from the courts and child protective services. In addition, their words indicate a concern for the future lives of their children and guilt that they were unable to successfully parent their kids.

Throwaway Moms

Maternal Incarceration and the Criminalization of Female Poverty

Suzanne Allen, Chris Flaherty, and Gretchen Ely

Because of the radical changes in sentencing and drug policies, the U.S. prison population has increased 500% over the past 30 years. As a result, the United States now leads the world in its rates of incarceration, with 2.1 million people currently in the nation's jails and prisons (The Sentencing Project, 2008). Female incarceration rates, in particular, are increasing at an unprecedented rate. Over the past three decades, the increase in the female prison population has continuously surpassed that of the male prison population in all 50 states, making women the fastest growing segment of the U.S. prison population (Women's Prison Association, 2006). As of June 2006, there were 203,100 women incarcerated in jails and prisons—nearly 10% of the total U.S. prison and jail population. More than 65% of these women were mothers of minor children, and 64% of them had lived with their children prior to incarceration (Women's Prison Association, 2006).

The goal of the study presented here was to gather information directly from the local (Kentucky) female prison population by means of face-to-face interviews to develop an understanding of the impact of maternal incarceration on the experience of motherhood through the eyes of the mothers themselves. To do so effectively, we first needed to understand the complex issues, programs, and policies surrounding maternal incarceration, including poverty; addiction; federal legislation, such as the Adoption and Safe Families Act (ASFA) and the War on Drugs; prison programs; child welfare practices; and systemic barriers, all of which are discussed here. Specifically, child welfare and criminal justice policies have failed to serve the needs of incarcerated women, and thus Halperin and Harris (2004) termed

SOURCE: Allen, S., Flaherty, C., & Ely, G. (2010). Throwaway moms: Maternal incarceration and the criminalization of female poverty. *Affilia, 25*(2), 160–172.

this situation "the policy vacuum." All these elements are germane to the larger issue of maternal incarceration because of the varied and complex ways in which they intersect to affect these women's situations and outcomes, as described in the narratives presented here.

This article is based on information gathered directly from the women who were interviewed as a means of creating a knowledge base of the experiences of this particular population. Therefore, it is informed through a feminist standpoint theoretical lens, which proposes that we "understand the world through the eyes and experiences of oppressed women and apply the vision and knowledge of oppressed women to social activism and social change" (Hesse-Biber & Leavy, 2007, p. 55). It is used here as a way to connect information to practice. In speaking directly to the women we interviewed, we heard their stories and learned of their experiences. From these women, we captured a snapshot of their subjective experiences as incarcerated mothers. By listening to these experiences, we can begin to understand and give voice to this long-ignored population. As Comack (1999, p. 296) stated, "the voices of women behind bars have for too long been silenced; it is time we begin to listen to what they have to say."

This project was inspired by our strong desire to hear directly from incarcerated, substance-abusing mothers. It stemmed from our concern with the implications of what was clearly a policy vacuum and the stigma attached to incarcerated women and mothers. As we explored the topic of maternal incarceration, it quickly became clear that these women, as vulnerable as they are, are often poorly served by the very system that should be helping them. "Women in prison are among the most vulnerable and marginalized members of society—women who, in other contexts, society would profess an obligation to support" (Women in Prison Project, 2006, p. 4). Hence, it became our mission to learn more, and the best way to do so, we thought, would be to hear from the women themselves.

The information gathered from speaking to these women adds to the literature in that their stories, as told by them, offer personal insights that have not otherwise been captured. Giving the women a voice makes the political personal. It is a way to identify potential areas of disconnect in policy and programming that so profoundly affect these women and their relationships with their children. It is a way to address the gender-specific programming needs of women in correctional settings and to develop appropriate programs to address the ineffectuality of "existing rehabilitative practices, which were developed for and by males [and] made available in a blanket approach to all females" (Moe & Ferraro, 2006, p. 139). Hearing directly from the women about their concrete experiences provides a foundation from which to build knowledge that accurately represents the needs of this population.

Review of the Literature

The vast majority of women who are involved in the criminal justice system are poor single mothers, most of whom are serving sentences for nonviolent drug-related offenses (Moe & Ferraro, 2006). Analyses of state and federal criminal justice statistics point to the war on drugs as the key factor not only in the increased rates of maternal incarceration but in more stringent—and longer—sentences for nonviolent drug-related crimes (Women in Prison Project, 2006). As a report of the American Civil Liberties Union (2005, p. 2) stated, "The war on drugs is having a specific, dramatic, and devastating impact on women that requires further study and attention when evaluating the success of drug policies that do far more harm than good in women's lives." Women's rates of incarceration have increased at more than double the rate of those of men over the past two decades, and women are 10% more likely than are men to be serving sentences for drug-related offenses. Much of this increase is due to the advent of crack cocaine, which has had a huge impact on low-income women and the resulting increase in nonviolent crimes that are typically associated with its use (Alleyne, 2006).

The literature has consistently documented extremely troubled and often tragic histories in the lives of these incarcerated, substance-abusing women. Some running themes that have continuously emerged on a variety of issues in these women's lives include poverty, abuse, mental health problems, and victimization or, as

Chesney-Lind and Pasko (2004) called it, their "multiple marginality." These women are more likely to live in poverty, are less likely to have been employed, and are more likely to have lower educational levels and lower household incomes than their incarcerated male counterparts (Moe & Ferraro, 2006). They also have high rates of recidivism (Alleyne, 2006; Richie, 2001). That most of these women live in high-crime neighborhoods with increased levels of homelessness to which they must return on their release from jail poses serious problems for successful reentry into the community and is a contributing factor in the women's high rates of recidivism (Alleyne, 2006). Furthermore, female drug offenders have high rates of mental health disorders and often use drugs as a means to self-medication for problems that are endemic to poverty, such as depression, anxiety, stress, trauma, and abuse (Alleyne, 2006). Most have chronic physical, emotional, and social problems as a result of their long-term drug use (Richie, 2001). Perhaps, the most widely documented characteristic is the extremely high rate of historical physical and sexual abuse—as much as 80%—among incarcerated substance-abusing women (Bush-Baskette, 2000; Inciardi, Lockwood, & Pottieger, 1993; Langan & Pelissier, 2001).

The majority of incarcerated mothers are poor (Alleyne, 2006), and many are involved with child protective services (Women in Prison Project, 2006), as were the majority of the mothers whom we interviewed. Neglect constitutes the majority of child welfare cases. According to Swift (1995), research has demonstrated that child neglect is more common among the poor than it is among the nonpoor, and poverty is often the main factor in cases of neglect. In addition, those who are accused of neglect are almost exclusively mothers, as opposed to both parents or solely fathers. "The study of child neglect is in effect the study of mothers who fail" (Swift, p. 101) and child welfare processes—indeed, society in general—reinforce this widely held assumption. Deeply embedded in our cultural psyche is the notion of the idealized mother; typically middle class, married, educated, and with access to resources. As Ferraro and Moe (2003, p. 14) stated, "The ability to mother one's children according to social expectations and personal desires depends ultimately on one's access to the resources of time, money, health, and social support."

Poor and marginalized women, such as the participants in our study, do not fit the idealized portrayal of motherhood. Consequently, they may be perceived as not only inadequate mothers but as inadequate women. They are also, by virtue of their poverty status and marginalization, the most likely to become involved in the criminal justice system and, therefore, more susceptible to having their maternal rights impinged upon. Within this long-established and widely accepted paradigm, motherhood becomes a privilege for certain women as opposed to a right for all women (Ferraro & Moe, 2003).

These women are already marginalized by their gender, class, and victimization status and the systemic barriers they consistently face. Although each has her own unique story, the one thing that incarcerated women share is their invisibility. The women have been locked away with little or no contact with the outside world. They are convicted criminals, viewed by society as social outcasts. Their multiple marginality, combined with the stigma and shame of incarceration, renders this powerless population essentially disposable in the eyes of society. They are dismissed as "throwaway moms."

Because of the rapid and unprecedented growth of this population, few procedures have been developed to address the unique issue of maternal incarceration on either the national or local level (Vera Institute of Justice, 2004). A report by the Women in Prison Project (2006) emphasized the lack of supports available to incarcerated women and their children, including visitation, parenting, substance abuse, and mental health programming; adequate legal representation; and proximity of a mother's location to that of her child. The programs that do exist, developed in response to the lack of gender-specific programs for incarcerated women, take a universal, cookie-cutter approach to programming for women in general, rather than address individual needs. No research has identified the actual needs of women who are involved in the criminal justice system, particularly mothers (Moe & Ferraro, 2006).

The implications of tougher drug-sentencing policies are further complicated by imposed time frames for reunification under ASFA. ASFA, which was passed in response to the growing number of children lingering in the foster care system, mandates that parental rights be terminated if a child has been in out-of-home care for 15 of the prior 22 months (Green, Rockhill, & Furrer, 2006). In Kentucky, the location of our study, state law (KRS 610.127) does not consider incarceration to be an exception to ASFA's federally mandated time frames. The average prison sentence for a woman is 18 months (Women's Prison Association, 2006), and the time frame for a woman to complete treatment varies from woman to woman, but is often lengthy. Thus, these ASFA time frames pose many difficult challenges for incarcerated parents, the courts, and the child welfare system (American Bar Association, 2005). Although termination of parental rights (TPR) proceedings have increased more than 100% since the enactment of ASFA, the precise number of TPR proceedings that have been filed against incarcerated women is not known (Women in Prison Project, 2006). Halperin and Harris (2004) reported that child welfare policies on children of incarcerated women have not been modified to adapt to the rapidly increasing rates of female incarceration. Furthermore, the majority of child welfare caseworkers, typically already overworked, lack the training and resources to serve incarcerated mothers adequately. This absence of a working relationship between caseworkers and imprisoned mothers puts the mothers at an obvious disadvantage with regard to the possibility of completing a case plan and, ultimately, reunifying with their children (Women in Prison Project, 2006).

It can be argued, then, that many of these women are suffering not only from radical criminal justice policies, but from inflexible child welfare policies. The experiences reported by the women in our study certainly corroborate the notion that the needs of children and mothers are often at odds with one another as a result of the intersection of policy mandates. Indeed, as is evidenced by the following interviews, the mothers' experiences indicate that child welfare protocols operate, often incorrectly, under the assumption that the welfare of the child is separate from the welfare of the mother. Although this may sometimes be the case, these women's stories indicate that more typically it is not as the women recount their experiences of feeling powerless and ensnared in and betrayed by the child welfare and criminal justice systems. Child welfare researchers have begun to argue the necessity to value mothers' subjective experiences of mothering as essential to providing good child protection practice (Davies, Krane, Collings, & Wexler, 2007).

These factors raise the issue of how child welfare policy mandates and incarceration rates interact to affect substance-addicted women and their children. Thus, the primary objective of our study was to gather data directly from incarcerated mothers to gain an understanding of the implications of incarceration, particularly on a mother's custody of and relationship to her children. In so doing, we anticipated that the information that we obtained would lead to a greater understanding of the various areas of potential disconnect between the child welfare and criminal justice systems with regard to incarcerated mothers of minor children and, in so doing, would inform policy and programming, identify systemic barriers, aid in identifying possible appropriate points of early intervention, and explore ways to extend family preservation efforts and promote reunification. Through the telling of these stories, we hope to bring attention to the plight of these and other incarcerated mothers, raise awareness of the obstacles that the women face because of their incarceration and troubled life histories, and bring to the forefront the virtually impossible odds the women face in terms of mothering their children by conveying their subjective experiences to a larger audience.

Method

Sample

Mothers of minor children, who were detained in a county jail in a midsized city in Kentucky at the time of the study and had histories of substance abuse were recruited by way of two information sessions. Only those who were serving sentences for nonviolent crimes that were related to their substance abuse were recruited to participate. The ethnicity of the sample was dependent on the ethnic makeup of the inmate population at the time of the study and did not serve as a selection or exclusion criterion. Staff of the detention center was not involved in the recruitment phase of the research to prevent any perceived coercion to participate. We fully and clearly explained to prospective participants that they would not face any negative consequences if they refused to participate for any reason or receive any special rewards or privileges. Written informed consent was obtained prior to each interview. The participants were paid $5 for participation, which was credited to their jail accounts following their interviews, typically within 2 days.

Twenty-six women who met the eligibility criteria were interviewed. The women ranged in age from 24 to 46 (mean age: 24.5, *SD* = 6.4). Of the 26, 15 were Caucasian, 9 were African American, and 2 were of other races. With regard to the educational achievement of 16 women (10 women did not answer the question), 6 had some college, 4 were high school graduates, and 6 had less than a high school education. Each woman had one to six children (mean: 2.9, *SD* =1.3).

Procedure

Qualitative interviews were conducted between May and September 2007. The face-to-face semi-structured interviews lasted from 20 min to 2 hr. The questions were designed to obtain specific information about the women's parenting, criminal, and drug-abuse histories while allowing for digressions into topic areas of unique importance to each woman. This method has become the preferred means of collecting data from marginalized and oppressed populations whose responses are not easily predicted or enumerated (Ferraro & Moe, 2003). This standpoint approach was used to capture fully the depth and uniqueness of the women's stories and to relay the women's firsthand experiences so as to bring attention to this oppressed and otherwise invisible group (Hesse-Biber & Leavy, 2007).

In the interviews, the women were encouraged to deviate from the questions to capture the full breadth of their experiences and to attempt to identify issues and experiences that would not otherwise have been explored, the majority of which were their experiences of motherhood as incarcerated and, otherwise marginalized, women. All the interviews were conducted by the first author and were audiotaped. Research procedures were approved by a local university's institutional review board.

Data Analysis

The audiotapes were transcribed and entered into ATLAS-TI qualitative data analysis software (2005). A coauthor who did not participate in the interviews open coded the qualitative responses and used a constant comparison approach (Creswell, 1998; LeCompte & Schensul, 1999) to aggregate and refine the themes. The grounded theory method (Glaser & Strauss, 1967; Strauss & Corbin, 1998) was used to generate descriptive categories. A data check was conducted by the interviewer to assess the accuracy of themes that were identified by the independent coder.

Findings

Several running themes emerged in speaking with the women. These themes were tabulated to identify the most frequently occurring ones across the set of interviews. The topics that occurred with the most frequency are explored here. It was decided that qualifying themes would be those that were present in at least 20% (5) of the interviews. These themes were parenting; drug use; involvement in the child welfare system; the revolving door of incarceration, homelessness, and recidivism; and mental health issues.

Crack cocaine was the drug of choice for 18 of the 26 women, and many of these women used alcohol and other drugs in addition to crack. Thirteen women reported a previously diagnosed mental health condition: multiple diagnoses (most often depression and anxiety, eight women; bipolar disorder, three women; and depression and anxiety without co-occurring disorders, one woman each). All 12 participants who reported a mental health diagnosis also reported having received previous mental health treatment.

Eight participants reported having their parental rights terminated for at least one child, and two had children in foster care. The majority (15) of the participants' children were currently in kinship placement, either permanently or until they could be reunified with their mothers. Seven participants reported having experienced prior homelessness and 21 had recidivated.

Nearly all the women expressed gratitude for the chance to talk about their stories so freely. They talked at great length about their children, and the majority expressed deep feelings of remorse, guilt, sadness, and love when talking about their children. Most of these dialogues were extremely emotional, and most of the women told their stories through tears and even sobs. As Joy said, "Kids are a touchy subject for all of us in here."

Parenting: The Shame of Maternal Failure

> I felt so bad about myself. I didn't feel like a good mom.... We let go of our kids because we feel it is best. (Linda, aged 47)

Perhaps, the most outstanding quality of these interviews was how deeply reflective—often philosophical—these women were about all the subjects they covered but particularly with regard to the topic of motherhood. Many of the women also struggled fiercely with their negative self-perceptions as parents. However, all the women were enthusiastic and grateful to have the opportunity to speak about their children in such an open and nonjudgmental venue. In many ways, the women expressed the same themes one would expect from any mother, including the aforementioned idealized mother. These themes included love for their children, pride in their children's accomplishments, and worry about their children's circumstances and future challenges. However, unlike idealized mothers, these poor and incarcerated mothers also expressed feelings of profound powerlessness: powerless at being separated from their children, powerless to protect their children from sharing their same fate, powerless against the child welfare system, powerless against their addictions, and powerless against the society from which they have become so disenfranchised. They were terrified that their children, too, would get caught in the devastating cycle of poverty, addiction, the criminal justice system.

Angelina, a 32-year-old White woman, was raised by her father until she was 10, at which time her father was convicted of a drug-related triple murder and put on death row, where he remains today. Her mother, a heroin addict, was incarcerated when Angelina was a child. At age 18, Angelina was arrested for the first time and ended up in the same women's prison as her mother. It was in prison that she became addicted to heroin. When her mother was released on parole, Angelina escaped to be with her and lived "on the run" with her mother for 5 years before she was caught. She said:

> It's never just about Angelina; it's about my whole damn family. I felt like I was white trash from the very beginning. My dad did a bad thing, but he's a good dad. I love my dad, and I love my mom. I understand mom more than anyone. She has drug problems, too.

Angelina is now serving a 30-year sentence as punishment for her escape. Her two children are in the permanent custody of her aunt. Angelina is in treatment for the first time through her involvement in a jail-based substance-abuse recovery program.

Virtually all the women expressed deep shame, remorse, and sadness for the mistakes they have made. Most of the women whose children were older and in kinship care, rather than foster care, had regular contact with them via telephone, letters, and visits. Some women spoke with their children every day. Several women with preschool-age and younger children described how painful it was for their children to visit because they had to visit through a glass partition; it was too difficult for them not to have physical contact, to be unable to hug and kiss their children. Some of these women made a conscious decision to forgo visits by their children, deciding that they would rather not see them at all if they were unable to express their love physically. In addition, some of them simply did not want their children to see them in jail because they were ashamed to be there.

Of the 26 women, 8 had their parental rights permanently terminated; 2 had children in foster care; and the majority, 14, had their children placed in kinship care. Most of those with children in kinship care expected to regain full or partial custody. The other two mothers' children were older than age 18.

Many of the women were imprisoned by their own guilt and remorse. As 33-year-old Lucinda put it, "Sometimes I feel like I don't deserve to be called mother. I feel like a failure, like I've failed them." Yet, despite all the challenges these mothers and their children had faced, a great number of them had extremely strong bonds, and their mutual love was evident. As 44-year-old Maggie stated, "I'm worried about my son because he laughs instead of cries. He says, 'Mama, when I play professional baseball, I'm going to buy you a house on the side of a hill, and then nobody can take you away from me again."'

Crack Cocaine: Snared in Addiction

> Crack . . . is taking a lot of women down. It's a high that you chase and never catch. (Leslie, aged 34)

The majority (75%) of the women who were interviewed were addicted to crack cocaine, and nearly half of them had been charged with possession of crack cocaine and/or crack paraphernalia. Crack has been termed the "fast-food" version of cocaine because it is inexpensive and brings with it a powerful and compelling high (Mahan, 1996). Although crack has been less popular in mainstream culture, a crack subculture is found in America's poor, struggling communities. Mahan called this subculture the "culture of powerlessness" and described it as "the epitome of poverty, ethnic segregation, and polarized gender relations" (p. 3). Those who are addicted to crack are often the poorest of the poor and subsequently the most frequently arrested, victimized, disabled, and marginalized by its use. The stigma of women who use crack and other drugs is further deepened by cultural expectations of women as nurturers and caretakers. As Campbell (2000, p. 3) stated, "When women violate gender norms by using illicit drugs, they are represented as spectacular failures—callously abandoning babies or becoming bad mothers, worse wives, or delinquent daughters."

Kearney, Murphy, Irwin, and Rosenbaum (1995) developed a grounded theory of pregnancy on crack cocaine. They found that pregnant crack-addicted mothers experienced "threatened selfhood," with selfhood consisting of self-concept and social identity. Consequently, these women sought to "evade harm," sometimes by avoiding contact with health care settings, in which their drug use might be discovered. Kearney et al. explicated a theoretical framework that describes a complex interplay of the desire both to evade harm (to self and fetus) and to face the situation. Ultimately, these two processes converge to an overarching theme of "salvaging self," in which these women sought to salvage their own lives for the sake of their children.

According to Mahan (1996), many crack addicts pay for their drugs by selling stolen goods. Indeed, 10 of the 26 women who were interviewed were serving sentences for shoplifting and/or forgery. Furthermore, the crack culture is extremely sexist, and many crack-addicted women resort to prostitution as a means of supporting their addictions (Mahan, 1996). Several of the women who were interviewed reported having prostituted themselves in the past to pay for their drugs, and three were currently serving sentences resulting from prostitution or loitering. Some women prostitute for money, others for drugs. According to Campbell (2000), women who trade sex for drugs are at the lowest end of the crack world spectrum. This fact became evident during the interviews with those women who bravely confided their experiences—women whose shame and regret were often palpable. One woman, Jane, explained that the other inmates teased her because she brushed her teeth so often. She said that she brushed her teeth incessantly—no fewer than 80 times a day—because she felt so dirty, having frequently performed oral sex while she was high for strangers in exchange for drugs and/or money.

The repeated lament of all these women was that they were painfully snared in a tangled web of addictions without the resources to help them find a way out. Over half the women who were interviewed had received treatment for their drug use in the past. Clearly, their prior treatment did not work. When asked why, many of them explained that they were unable to complete treatment for a number of reasons. This experience is not unique to this population; rather, it is typical in that the main obstacle to successful treatment for the general population is noncompletion (DeLeon, 1993). Because of the obstacles they face, though, it is particularly difficult for this population. Others stated that it was hard to maintain sobriety after being exposed again to the same environment; following treatment, their social networks—family members and friends—remained the same. Because these women have limited resources owing to their poverty status, poor employability, and lack of social networks, they end up in the same neighborhoods where drugs abound, making this a difficult cycle to break. Many of the women stated that they would want to receive treatment on release, but that money and time were against them; they could not afford the $300– $500 that the local facilities charge, and the waiting lists for a bed were often as long as 3–6 months. Those who had custody of their children also expressed frustration that there was nowhere to bring the children while participating in treatment. Many of the women expressed fear that their children would follow in their footsteps and were trying hard to prevent them from doing so. As Lisa, aged 46, stated:

> I'm just hoping he [doesn't] go down the road I went down. I'm just hoping to learn as much as I can about my addiction [through participation in the jail-based treatment program] . . . and about as much as I can about not using ever again so that I can go home and sit him down and teach him not to make the same mistakes that I made.

Child Welfare Involvement: Betrayed by the System

> I'm sitting here in jail now with my son gone. . . . I'm like, OK, I did what everybody said I should do. I think if I had more help, . . . I could have done better. He was the only thing that kept me alive. The reality of it is that my little boy is gone, and if I'm lucky I might get to see him one more time to say good-bye. (Margie, aged 24)

Margie, a 24-year-old White woman, has a 14-month-old son who is in state custody. When she was sentenced, she said she called child protective services from jail to find out if she had a caseworker and to inquire what she needed to do to get her son back. She was told that a caseworker would come to see her—that was in March. As of July, when this interview took place, she had still not heard from anyone and did not know if she had a case plan. She expressed frustration and anger with the system and was planning to go directly to child protection services on her release in 6 months to find out how to proceed. However, she did not know if by that time it would be too late, whether the ASFA clock would have already been ticking.

All the women were angry and resentful with the way they had been treated by child protection workers. The interview notes were infused with expressions of these feelings:

> There is no one out there to help you with your kids. I don't know what to do, really. There were days I wanted to get high just because I missed my kids so bad. [When they take your child away], it completely destroys everything inside of you . . . takes away your reason for trying.

Thirty-two-year-old Linda's children were taken from her 6 months before her incarceration. When her children were taken, she "fell apart" and ended up on the streets. Therefore, she did not know their status, whether there was a hearing (although she assumed there was) or if there was a case plan. She was in tears throughout the interview and angry with the system for betraying her and lying to her, as she put it.

This lack of communication between child protection services and incarcerated women is not uncommon. When a child is removed by the state, a case plan is developed that includes the parental requirements that must be fulfilled for reunification to occur. However, it is difficult, if not impossible, for incarcerated mothers to be involved in case planning because caseworkers typically fail to have contact with them (Halperin & Harris, 2004). This noncommunication between child protective services and incarcerated mothers was echoed by many of the women who were interviewed.

Several of the women reported that child protective services told them that if they did not agree to a voluntary TPR, they would never see their children again. If they were subject to an involuntary TPR, there would be no visitation, and they would never see their children again, but if they did a voluntary TPR, they could see their children a few times a year. The women reportedly felt blackmailed. They were also confused about what their rights were, and it was clear that they did not know how the system worked. The majority expressed feelings of profound powerlessness with regard to their status and their rights within the system.

The Revolving Door of Incarceration, Recidivism, and Homelessness

> The judge asked me why I broke probation, well, I can't afford $300 per month and $12 each time I have to drop [off the urine sample]. How am I supposed to come do a drug test? . . . It's like a big cycle. . . . In here at least, you know what to expect . . . you get to eat . . . got clothes to wear. . . . When you get out of here, you don't know what to expect. (Arlene, aged 42)

A remarkably high number of the women reported being homeless prior to incarceration, at a rate that is 25 times greater than that of other local citizens (Central Kentucky Housing and Homeless Initiative, 2009). According to Zlotnick, Tam, and Bradley (2007), the majority of homeless women are mothers, although many do not live with their children. Many women voluntarily opt to place their children in the custody of others to protect them from the multiple dangers and potential traumas associated with homelessness, as well as to avoid exposing them to the shelter environment, which is also often dangerous. Homeless mothers have higher rates of both substance use and mental health disorders—particularly major depression—than either the general female population or the general homeless population (Bassuk, Buckner, Perloff, & Bassuk, 1998).

The high rates of mental health disorders among homeless women contribute to a number of negative consequences, including increased recidivism and longer periods of homelessness. According to Alleyne (2006, p. 182), "Most women in prison are untreated substance abusers with high recidivism rates that correlate with greater addiction severity. Typically, each return to incarceration signifies a deeper level of addiction, with associated declines in health, employment opportunity, and social functioning."

The correlation between homelessness and reincarceration has been widely documented, and several characteristics are known to be endemic to both populations, including high rates of poverty, unemployment, substance abuse, and mental illness. These problems, combined with the continual crossover between homelessness and incarceration, result in enduring patterns of social exclusion and isolation. Given that these individuals have such high rates of substance abuse and mental illness, homeless shelters and jails have come to serve an institutional function that "effectively substitutes for more stable and appropriate housing" (Hopper, Jost, Hay, Welber, & Haugland, 1997, p. 659).

The cycle of homelessness and recidivism was glaringly evident among the women who were interviewed as well. In addition to the high rates of homelessness the women reported, 81% of these women had also recidivated. Some have had a few prior incarcerations, while others had been incarcerated 20, 30, or even 40 times in the past. Veronica, aged 36, described the cycle:

> That's the serious thing I am dealing with right now—the stress of the unknown or what will happen to me when I get out. . . . I don't have [an] address to go to. I got no family right now.

> It's just me, myself, and I. I'll walk out of here hurting with nowhere to go, . . . and that's scary. I'm hurtin' bad, and I am crying out for help, and I don't know which way to turn. I am so discombobulated, it's crazy. [Crying], I needed to know how to live without the drink and drugs [and] how to manage my money. It's just like walking all over again, feeding yourself all over again. As an adult, you have to learn everything just like a newborn baby. . . . I got caught with a five-cent piece of crack cocaine—you get clean and, at the same time, you get clean in jail, [but] you don't know what to do when you get out there. . . . You're dirty, and the only thing they do is take your kids away from you. They say they're here to help you, [but] they're not. I need help to overcome my drug addiction so I can be with my kids.

Mental Health Issues: Untreated Depression, Self-Medication

> Depression is a major problem. It is the reason why so many women are in jail. It leads to drugs and then to crime. (Carolyn, aged 40)

Of the women who were interviewed, half had an existing mental health diagnosis, and many of them had dual diagnoses in addition to their substance abuse or dependence. This situation is consistent with the findings of studies that have demonstrated the prevalence of mental health disorders among women who are involved in the criminal justice system, who are more likely to struggle with mental illness than their male counterparts (Sacks, 2004). As we stated earlier, the majority of incarcerated women have experienced past trauma and abuse, an amount reported by Green, Miranda, Daroowalla, and Siddique (2005) to be as high as 77%–90%. Psychiatric disorders, in general, are more prevalent among poor women because of the multiple stressors connected to poverty (Bassuk et al., 1998). Experiences of trauma and abuse, as well as preexisting mental health disorders, often lead to increases in substance abuse as a means of self-medication. Substance abuse, in turn, often leads to criminal behavior. The prevalence of mental health disorders among the women who were interviewed echoes that of the larger female prison population. It is endemic and, therefore, a vital area of concern because it leads to subsequent substance abuse and involvement in the criminal justice system.

Conclusion

The information garnered through these interviews revealed numerous issues that are widespread among incarcerated mothers, most of which are consistent with existing research and are documented in the literature. The women's stories were not easily quantifiable nor did this process reveal any particular construal but, rather, something much more powerful, significant, and complex. By giving voice to this invisible population through a standpoint perspective, we revealed complex stories of unfinished lives, of victimization and abuse, of poverty and exploitation, of cyclical and generational obscurity, of classism and sexism, and of stigma and shame. Perhaps, the most powerful and heartbreaking themes were those of the maternal love that these women consistently expressed for their children and the profound sense of guilt and staggering remorse they were all struggling with when they discussed the impact of their actions on their children. It became abundantly clear that their substance abuse problems and criminal justice involvement were symptomatic of extremely troubled life histories.

The extent to which these factors interfere with and disturb these women's lives was understated in prior qualitative research. For example, the degree to which incarcerated women have been involved with the child welfare system and the number of those who have had TPR proceedings filed against them have been largely undocumented. This project begins to shed light on those crucial areas. The interviews also revealed firsthand accounts of the deep layers of abuse and social problems that the women endured. Thus, our study contributes to the literature in that it explored the ways in which these issues interact to affect these women on a number of levels and how some of these issues affect each other. The following are the resulting suggested points of intervention and programmatic and policy recommendations.

All the women had negative self-perceptions as mothers, because many expressed feelings of inadequacy related to their motherhood. At the programmatic level, this finding indicates that these women could perhaps benefit from parenting classes and even mentoring programs both in jail and after their release. Mentors and advocates could also help the mothers negotiate other systems in which they and their children are involved. Another thing to be addressed is the consistent lack of successful treatment services, as reported by the women, that are geared specifically to crack cocaine addiction in the jail setting and in the larger community that address the financial barriers and obstacles related to social support, extensive waiting lists, and child care. As the literature has demonstrated and indeed as these women verified, all programming must be developed to be gender specific.

In addition, because the women expressed so much frustration and powerlessness in dealing with the systems, case advocates are needed, who can help these women navigate both the child welfare and the legal systems. Ideally, reentry programs that would implement all these elements in the form of wraparound services would be developed. These services could include treatment for substance abuse and parenting and life-skills training to prepare the women for life on the outside in an attempt to combat the high rates of recidivism reported herein. To combat some of the issues surrounding reentry and recidivism, community-based programs that help neighborhoods work with these women and connect them with needed services could be most beneficial. Community preprobation programming should include ways in which these women can explore what led them to incarceration by examining their multiple marginality, family histories, and experiences of abuse as a means of gaining a better understanding of the cycles and patterns that led them to criminality.

On the policy level, because of the growing social problem of maternal incarceration, it is essential for child welfare agencies to hire workers who work specifically with incarcerated women. In fact, the results of this and other studies suggest that specialized child welfare workers need to be trained to carry caseloads that consist only of incarcerated women, so that these women's unique needs may be addressed. Further research on the implications of the ASFA time frames, in relation to sentencing policies and family preservation and reunification, is needed. Although many may assume that these children would be better off without their mothers, this may not be the case. With proper treatment and ample opportunity to complete a case plan, these women may be able to achieve a life in which they can nurture their children—the children they so desperately love.

References

Alleyne, V. (2006). Locked up means locked out: Women, addiction and incarceration. *Women and Therapy, 29,* 181–194.

American Bar Association, Center on Children and the Law. (2005). *Parental substance abuse, child protection and ASFA: Implications for policy makers and practitioners.* Washington, DC: Author.

American Civil Liberties Union. (2005). *Caught in the net: The impact of drug policies on women and families.* New York: Author. Retrieved September 2, 2006, from http://www.aclu.org/images/asset_upload_file 393_23513.pdf

ATLAS-ti visual qualitative data analysis version 5.0.66. (2005). Berlin: ATLAS-ti Scientific Software Development.

Bassuk, E. L., Buckner, J. C., Perloff, J. N., & Bassuk, S. S. (1998). Prevalence of mental health and substance use disorders among homeless and low-income housed mothers. *American Journal of Psychiatry, 155,* 1561–1564.

Bush-Baskette, S. (2000). The war on drugs and the incarceration of mothers. *Journal of Drug Issues, 30,* 919–928.

Campbell, N. D. (2000). *Using women: Gender, drug policy, and social justice.* New York: Routledge.

Central Kentucky Housing and Homeless Initiative. (2009). *Local facts.* Lexington: Author. Retrieved July 8, 2009, from http://www.ckhhi.org/localfacts.asp

Chesney-Lind, M., & Pasko, L. (2004). *The female offender: Girls, women, and crime.* Thousand Oaks, CA: SAGE.

Comack, E. (1999). Producing feminist knowledge: Lessons from women in trouble. *Theoretical Criminology, 3,* 287–306.

Creswell, J. W. (1998). *Qualitative inquiry and research design: Choosing among five traditions.* Thousand Oaks, CA: SAGE.

Davies, L., Krane, J., Collings, S., & Wexler, S. (2007). Developing mothering narratives in child protection practice. *Journal of Social Work Practice, 21,* 23–34.

DeLeon, G. (1993). Cocaine abusers in therapeutic community treatment. In F. Tims & C. G. Leukefeld (Eds.), *Cocaine treatment: Research and clinical perspectives* (pp. 163–189). Rockville, MD: National Institute on Drug Abuse.

Ferraro, K. J., & Moe, A. M. (2003). Mothering, crime and incarceration. *Journal of Contemporary Ethnography, 32,* 9–40.

Glaser, B. G., & Strauss, A. (1967). *Discovery of grounded theory: Strategies for qualitative research.* Mill Valley, CA: Sociology Press.

Green, B., Rockhill, A., & Furrer, C. (2006). Understanding patterns of substance abuse treatment for women involved with child welfare: The influence of the adoption and safe families act (ASFA). *American Journal of Drug and Alcohol Abuse, 32,* 149–176.

Green, B. L., Miranda, J., Daroowalla, A., & Siddique, J. (2005). Trauma exposure, mental health functioning, and program needs of women in jail. *Crime & Delinquency, 51,* 133–151.

Halperin, R., & Harris, J. L. (2004). Parental rights of incarcerated mothers with children in foster care: A policy vacuum. *Feminist Studies, 30,* 339–352.

Hesse-Biber, S. N., & Leavy, P. L. (2007). *Feminist research practice: A primer.* Thousand Oaks, CA: SAGE.

Hopper, K., Jost, J., Hay, T., Welber, S., & Haugland, T. (1997). Homeless, severe mental illness, and the institutional circuit. *Psychiatric Services, 48,* 659–665.

Inciardi, J. A., Lockwood, D., & &, Pottieger (1993). *Women and crack cocaine.* New York: Macmillan.

Kearney, M. H., Murphy, S., Irwin, K., & Rosenbaum, M. (1995). Salvaging self: A grounded theory of pregnancy on crack cocaine. *Nursing Research, 44,* 208–213.

Langan, N. P., & Pelissier, B. M. M. (2001). Gender differences among prisoners in drug treatment. *Journal of Substance Abuse, 13,* 291–301.

LeCompte, M. D., & Schensul, J. J. (1999). *Analyzing and interpreting ethnographic data.* Walnut Creek, CA: SAGE.

Mahan, S. (1996). *Crack cocaine, crime, and women.* Thousand Oaks, CA: SAGE.

Moe, A. M., & Ferraro, K. J. (2006). Criminalized mothers: The value and devaluation of parenthood behind bars. *Women and Therapy, 29,* 135–164.

Richie, B. E. (2001). Challenges incarcerated women face as they return to their communities: Findings from life history interviews. *Crime and Delinquency, 47,* 368–389.

Sacks, J. Y. (2004). Women with co-occurring substance use and mental health disorders (COD) in the criminal justice system: A research review. *Behavioral Sciences and the Law, 22,* 449–466.

The Sentencing Project: Research and Advocacy for Reform. (2008). *Incarceration.* Retrieved February 3, 2008, from http://www.sentencingproject.org/template/page.cfm?id=107

Strauss, A., & Corbin, J. (1998). *Basics of qualitative research: Techniques and procedures for developing grounded theory* (2nd ed.). Thousand Oaks, CA: SAGE.

Swift, K. J. (1995). *Manufacturing "bad mothers": A critical perspective on child neglect.* Toronto: University of Toronto Press.

Vera Institute of Justice. (2004). *Hard data on hard times: An empirical analysis of maternal incarceration, foster care, and visitation.* Retrieved from http://www.prisonpolicy.org/scans/vera/245_461.pdf

Women in Prison Project of the Correctional Association of New York. (2006). *When "free" means losing your mother: The collision of child welfare and the incarceration of women in New York State.* Retrieved October 2, 2007, from http://www.correctionalassociation.org/publications/download/wipp/reports/When_Free_Rpt_Feb_2006.pdf

Women's Prison Association. (2006). *The punitiveness report—Hard hit: The growth in imprisonment of women, 1977–2004.* Retrieved October 2, 2007, from www.wpaonline.org/institute/hardhit/index.htm

Zlotnick, C., Tam, T., & Bradley, K. (2007). Impact of adult trauma on homeless mothers. *Community Mental Health Journal, 43,* 13–32.

DISCUSSION QUESTIONS

1. How does motherhood impact the struggles that incarcerated women experience?
2. How do issues such as mental illness, addiction, and homelessness impact recidivism rates of women?

READING

While most research focuses on the pains of imprisonment for women, few consider the challenges that women will face following their incarceration. This reading gives voice to women who have experienced prison and are faced with rebuilding their lives as they return to the community. These challenges are multifaceted and involve struggles with family reunification, relationships with family and friends, and reintegration as they struggle to find a place to live and employment to provide for their basic needs. For many women, they continue to struggle with issues of addiction and judgment in their communities, adding an additional layer for the challenges of integration.

Collateral Costs of Imprisonment for Women

Complications of Reintegration

Mary Dodge and Mark R. Pogrebin

Women in prison, once considered the forgotten population, have become the focus of considerable research. Incarceration rates for women have increased threefold over the past decade and created a wide range of individual and social concerns (Bloom & Chesney-Lind, 2000). This study gives voice to former women inmates who explore their experiences, feelings, and thoughts on the obstacles that they endured in prison and now face in the community. Their retrospective reflections and current accounts portray conflicted emotions about children and relationships both in and out of prison and the difficulties of community reintegration. Their narratives identify and expand on the often overlooked consequences of being an incarcerated female offender.

The stigmatization that imprisoned and paroled women experience carries great costs. The stigma (Goffman, 1963) associated with criminality becomes what Becker (1963) referred to as one's master status. Women who are labeled as criminals find confirmation of their deviant master status as they undergo the process of community reintegration with few social bonds (Braithwaite, 1989). The difficulty, if not impossibility, of attempting to disavow one's deviant label is a formidable task for many women offenders.

Once released into the community, women on parole may be treated as outcasts, excluded from the job market, and judged for their past criminal behavior. According to Braithwaite (1989), stigmatizing shaming inhibits reintegration and furthers criminal behavior. As a consequence of society's labeling and the mechanisms of self-shaming, it appears that women offenders often experience a degradation process (Garfinkel, 1956). Female inmates and parolees who have low self-esteem (Fox, 1982) and suffer from feelings of powerlessness

SOURCE: Dodge, M., & Pogrebin, M. R. (2001). Collateral costs of imprisonment for women: Complications of reintegration. *Prison Journal, 81*(1), 42–54.

and vulnerability (Bill, 1998) are likely to experience increased levels of shame in their relationships.

Punishment is compounded for many women inmates when they are separated from their children. The majority of incarcerated women are mothers—estimates range from 60% to 80% (Bloom & Steinhart, 1993; Henriques, 1996). Most women inmates were living with their children and providing the sole means of family support prior to incarceration (Baunach, 1985; Chesney-Lind, 1997; Datesman & Cales, 1983; Greenfeld & Minor-Harper, 1991; Henriques, 1982, 1996). Imprisoned mothers rank estrangement from children as their primary concern (Baunach & Murton, 1973; Glick & Neto, 1977; Henriques, 1996; Stanton, 1980; Ward & Kassebaum, 1965). Rasche (2000) noted that the harshest single aspect of being imprisoned may be the separation of mother and child. The secondary costs of imprisonment to children have been acknowledged but are largely incalculable (Henriques, 1996; McGowan & Blumenthal, 1978).

Women in prison experience an unparalleled sense of isolation. Added to the pains of women's imprisonment (Sykes, 1958) are the frustration, conflict, and guilt of being both separated from and unable to care for their children (Barry, 1987). According to Crawford (1990), as a result of imprisonment, female parents often experience feelings of despair and depression. Crawford further stated that these emotions appear to be widespread, even on the part of women inmates who believe that they were inadequate as parents when they were living with their children at home. Furthermore, anxiety arises over fear of losing custody (Bloom, 1995; Fletcher, Shaver, & Moon, 1993; Knight, 1992; Pollock-Byrne, 1990).

Divorce, another contributing factor to the loneliness of separation, is a common occurrence for imprisoned women. Rafter (1985) noted that, unlike men in prison, women are unable to count on a spouse or significant other to provide a home for their children. Because of this, female parents in prison suffer more anxiety about the type of care their children are receiving. Stanton (1980) found that a great many women prisoners report being divorced by their husbands or deserted by men with whom they lived before coming to prison. Three out of four women in prison leave children, and only 22% say that they can depend on the fathers of their children to care for them while they are incarcerated (Bloom & Steinhart, 1993). Overall, women inmates, because of their primary parental role, are not, to any great degree, receiving child care help from spouses or fathers of their children.

The obstacles imprisoned women must overcome in order to maintain a relationship with their children can be extremely frustrating (Bloom & Steinhart, 1993). Loss of contact, coupled with an inability to meet social service contract requirements resulting from a lack of visitations by the children via foster parents, places inmates at considerable risk of losing parental custody (Gabel & Johnston, 1995). Bloom and Steinhart (1993) reported results from a national study that more than 54% of the children with mothers in prison never visited them during their incarceration, despite research findings that frequent contact promotes ongoing custody and family reunification (Martin, 1997).

Reestablishing relationships and social ties often represents a barrier to successful reintegration. A majority of incarcerated female mothers expect to take responsibility for their children once they are released and rarely receive any financial or emotional support from the fathers (Prendergast, Wellisen, & Falkin, 1995). Reunification is an important although somewhat unrealistic goal for released mothers (Browne, 1998; Hairston, 1991; Harris, 1993; Henriques, 1982; Jones, 1993). If the child has been placed in foster care or state custody, it is even more difficult for a released female prisoner to show that she is able to take care of and provide for her child adequately (Pollock-Byrne, 1992). Women on parole often have to overcome many barriers in order to maintain their parental rights (Barry, 1995).

Prison is a difficult experience for most women, and the subsequent hardships that they endure upon release are no less significant. Internalized self-shame, whether derived from embarrassment or guilt, along with stigmatizing social shame from the community often constitute punishment well beyond the actual time women offenders serve and may contribute to further deviance.

Method

Qualitative data were collected from female parolees who were incarcerated at the same correctional facility located in a western U.S. state. The prison was constructed in 1968 and has a mixed classification of inmates. The prison population at the time of this study was approximately 300 women, with 61 correctional officers (37 female and 24 male). The ethnic and racial composition of the prison population was 45.6% Anglo-American, 31.5% Black, 18.4% Hispanic, 1.7% Native American, 0.4% Asian, and 2.4% unknown.

Women on parole were contacted at the time they had appointments to see their parole officers. The participants in this study were not chosen at random, but, according to a representative case sampling method, their experiences provide examples that are indicative of the issues women on parole confront (Shontz, 1965). Each person volunteered to participate and gave informed consent. A total of 54 women agreed to interviews over a 3-month period. Their ages ranged from 23 to 55 (median = 36), and their length of incarceration ranged from 1 to 12 years (median = 4.8) for all classes of offense. Seventy percent of the women interviewed were mothers.

Interviews were conducted at the parole offices in private conference rooms. Each interview lasted approximately 60 minutes and was tape recorded with the participant's consent. All women parolees were guaranteed confidentiality and told that they could choose not to tape the interview. Three women requested not to be taped, and notes were taken during those sessions. The former inmates were cooperative and seemed willing to discuss their prior prison life. We found those interviewed to be open and quite frank in relating their personal experiences, although at times the process was emotionally painful. We used a semistructured interview format, which relied on sequential probes to pursue leads provided by participants. This approach allowed the women parolees to identify and elaborate on important domains that they perceived to characterize their prison experiences retrospectively (rather than the researchers' eliciting responses to structured questions).

The interview tapes were transcribed for qualitative data analysis, which involved a search for general relationships between categories of observations using grounded theory techniques similar to those suggested by Glaser and Strauss (1967). The data were categorized into conceptual domains as portrayed by our participants. The experiences of these women may not be reflective of all women who have served time, but the narratives add depth to our understanding of the issues (Ragin, 1994; Seidman, 1991).

Findings

Separation Concerns

For female inmates in this study, being separated from their children provoked considerable stress and threatened their self-esteem. Women who violate the law are not only viewed as social outcasts but are often perceived by the community as inadequate parents. The most difficult aspect of being in prison was voiced by one respondent who seemed to portray a representative opinion for most of the women who left their children behind:

> It was so long. I missed my kids. I missed my freedom. I went to bed every night and woke up in a tiny cell. I just wish it was all a bad dream and I would wake up and I would still be there.

Often, inmates with children begin to perceive themselves as bad people, as expressed by one parent whose child has grown up not knowing her:

> Being away from my daughter affected me a lot. She is only 6, so that means that I have been in the system almost her entire life. I haven't been there for her. I feel like a horrible person because of this.

Another great concern for women in this study was the degree to which the fathers of their children took responsibility for them during the mother's incarceration. There are cases in which the husband does take responsibility for the children but leaves his imprisoned spouse for another woman. Obviously, this circumstance causes great distress for incarcerated women. There is little they can do about the situation from behind bars, and feelings of abandonment become intense. One woman stated:

> My husband chose to go to another woman. He cheated on me. It's so much to go through. You lose your husband, you lose your kids, your kid's gonna always love you, but someone else takes care of your kids, another woman, it's so much to go through. It's tragic. It's a terrible thing that you wake up and say I want to go home.

Abandonment by a husband or partner is one matter, but the additional problem of displaced children seemed insurmountable to many of these women. In the following case, one woman expressed her feelings about her husband remarrying and taking custody of her daughter. Her feeling of helplessness is apparent:

> My daughter ended up with her dad. He got married, and they took her in. He is a pretty good guy. I was upset at first when I knew he was involved with someone cause I always thought when I got out we would be together. I guess I was just young and dumb. When I first found out, I spent many nights crying over him. At first, he wrote and visited me once, but then it just stopped. Then he wrote and told me he met someone and they were getting married and were going to raise Meg. It was hard. I was so hurt. I mean I'm glad Meg is with her dad and has a family, but she is my daughter and I just wish she was with me.

As painful as having others taking one's place as the child's primary parent, nothing, it seems, can be as emotionally difficult as giving up a newborn infant while incarcerated. A respondent explained:

> The hardest part about being in prison was being away from my kids. I was pregnant when I just got in. It hurts so bad; I mean I had my daughter here. I didn't even get to hold her. I mean she was my baby. I didn't even know how she is doing or if she is alive or anything. For all I know, she could be living right by me, but I'll never know because they won't tell me, and I'll never forgive myself for getting into trouble and losing her.

In this instance, the state took custody of the newborn child and placed her in a foster home. The child was later put up for adoption. This is not an uncommon occurrence for incarcerated women.

Mothers who are in prison often find their children transferred to foster homes when there are no relatives who will be responsible for them. If multiple children are involved, they frequently are placed in different homes and separated, making it difficult for incarcerated mothers to locate them. Not being able to see one's children for a long period of time is a reality for many inmate parents. One respondent explained:

> I talked to my daughter when she was with my family and I wrote her but I never got to see her and I wasn't able to talk to her after she moved in with her dad.

Information about where their children are, who they are with, and their general welfare is not always forthcoming from state departments of social services. One parent related the difficulty she experienced:

> I wrote and stuff but they won't tell me where they are. My social worker said once I get on my feet and keep a job for 6 months we can see about visitation. What she doesn't understand is for the past year I have been trying to find a job, but no one wants to have someone who was in prison for 6 years. They [my children] are the only good thing that has ever happened to me, and I want them back. I didn't even know where they are.

In some cases, women in prison lose custody of their children. A woman related her story:

> My children, there isn't much to say. I had three boys and I lost them when I went in. I haven't seen them since I violated my probation, it's been about 5 years. I get letters from a social worker telling me how they are doing, but I can't see them or talk to them or anything. I talked to someone from social services about it, but I will never get them back. I really miss them.

Having one's children placed in foster care while incarcerated frequently is related to the financial circumstance of the female prisoner's relatives who are taking responsibility for the children. In many cases, children are being cared for by grandparents or other relatives who often cannot afford the financial burden. In these instances, family members would like to seek financial aid from the state but often are reluctant to do so. Many female prisoners do not seek government funds for relatives who are responsible for their children for fear of losing custody. This is what occurred in the following case when the inmate's mother applied for agency funding from the state to help her care for the children:

> I wanted my kids to be with my mom, but she didn't have much money, so she tried to get help and the state came in and took my kids. They helped all right. I haven't seen them since.

Problems of Reunification

The paroled women in this study had been out of prison for a period of 14 to 24 months, and many were involved in drug and alcohol rehabilitation programs. Some resided in halfway houses, whereas others were living on their own or with relatives. Most told of extreme difficulties in their attempts to regain custody of their children. A woman on parole who wishes to regain custody must meet the criteria of state social services agencies. For example, if she had an alcohol or drug abuse problem prior to her incarceration, she must show that she has actively participated in a rehabilitation program and has been off drugs and/or alcohol for a period of time. A woman on parole must show that she has sustained employment, can financially support her children, has a permanent and appropriate residence, and is no longer involved in any criminal activity. Obviously, these criteria, along with additional discretionary demands that the paroling authorities impose, present difficult obstacles to women who wish to regain custodial rights of their children.

Part of the dilemma paroled parents face is convincing child service workers that they have become responsible adults who are capable of providing adequate care for their children who remain in foster homes. Once paroled to the community, this particular parent summarized the problems she faced in proving she was a mature, responsible adult. She talked about her daughter:

> I get visits. I am trying to get her back. It is hard. The social worker had a hearing set, but I had to take a bus cause I'm not allowed to drive, but the bus never showed up, so I was late. I know that didn't look good, but I guess I'll keep trying. It's hard. I didn't even know where she is at. When I see her, we go to social services. I don't even know how she is treated or anything.

Another case clearly illustrates the conflict women on parole face between wanting their children back and not having the financial resources to adequately provide for them. One woman commented:

> I visit and we can spend the day together. It is hard cause part of me feels I should just leave her alone. She is 7 and she is doing good in school and has a lot of friends, but I just can't do it, she's my little angel and I know it might not be the best thing, but I need her.

When asked whether it was possible in the future to get custody of her daughter, the woman commented:

> No, I've tried. It's hard enough to get visits. I know I fucked up big time, but I paid the

> price and I screwed up, but now I am ready to move on.

Impediments to Reintegration

Once out of prison and on parole, women in this study reported the many difficulties they experienced in adjusting to living in the community. The one factor common to the experience of all the interviewees was the distrust community members communicated. The women constantly felt they had to prove themselves as worthy citizens to others who had knowledge of their criminal backgrounds. One respondent explained:

> I am doing very well. I have a place to live, but it's hard getting your kids back because nobody will believe that I have changed and I'm a different person now. No matter how much time we do, everyone always thinks it's like once a criminal always a criminal and that is how people see me and it's very hard to deal with.

When interacting with others in the community who have no knowledge of their past criminal background and imprisonment, the respondents reported being treated in a "normal" manner. One study participant, however, explained the change in attitudes when parents of her child's friends learned of her background:

> I became friends with some other mothers at my kid's school. They were really nice. I joined the PTA and it was going good. Then I told someone, I don't know how it came up, that I was in prison. Now, some of them won't talk to me, and they won't let their kids play with mine. So I learned my lesson. I don't really care what people think of me. Well, I kind of do, but I just don't want my kids to suffer.

The consequences of the criminal label and the stigma attached to it were experienced by another woman in a religious environment:

> It's been tough, my sister is great letting me live with her, and all at once when people find out I was in prison they look down on me. I was going to church cause I really found God and everyone was so nice. Then, someone found out I was in prison and everything changed, no one would talk to me anymore. Now I don't go, I just pray at home.

One of the biggest problems faced by the parolees was finding well-paying employment. Often, women on parole have few job skills. This, coupled with their past criminal history, leads to low-paying, dead-end employment. The negative reactions of potential employers toward their past criminal lifestyle make attaining meaningful employment with future growth potential nearly impossible for these women. A respondent said:

> I was lucky cause I had a place to live. I know a lot of people end up not having anywhere to go. When you're getting out, you are just so excited to have your freedom again. Once I got out, I couldn't find a job. It is hard. Nobody will give you a break. I could be such a good worker, but they can't see it cause I was in prison. I mean it is a lot worse in prison, and I'm glad I'm not still there, but it's been very hard for me out here.

Importance of Family Support

Close ties to families during incarceration are crucial in maintaining connections in the community. Visits from relatives, sustained correspondence, phone calls, or any type of communication serves to maintain a support system for inmates. Family contacts let the woman know that she is not forgotten and that there are people who care about her. For women returning to the community, the assistance of family is crucial to success. Family support for women on parole may mean a place to live, money for necessities, transportation, food, and a host of short-term needs until they become financially independent.

Support from relatives also enhances emotional survival. Families often provide love and a sense of caring that lifts a newly released woman's self-esteem. One example of how meaningful family support is was related by the following respondent:

> My family was great. I know that I was lucky. A lot of people in prison do not have any support, and that is what helps you get through the rough times. I don't know why my family stuck by me. My husband could have given up on me. He could have got custody of the kids and left. He must really love me, and I thank God every day that he stayed in the relationship. My family offered me the support I needed. I never would have got through it without them.

In contrast, we also found that almost half of the women had lost touch with their families. They tried several times to contact family members but never received any type of communication in return. After a while, the women stopped attempting to contact relatives: such a void of a family support system means female prisoners released to the community must function pretty much on their own. This makes for greater adjustment problems in reintegrating into the community. To illustrate the rejection by family members, one woman explained the type of response she received when she attempted to make contact while imprisoned:

> My sisters live out east and have their own lives with nice houses and kids. I am just an embarrassment to them. They won't have anything to do with me. I wrote them each a couple of times, cause when you're in a place like this, you realize how important your family really is, but they sent the letters back, and I've never heard from them.

For women without family support, being released from prison appears to be even more frightening. In these circumstances, women on parole have to become their own support system. Yet, success in the community is very much dependent on the belief that they will be accepted in society.

Discussion

Women on parole experience the pain of social and self-imposed punishment that manifests from feelings of shame or guilt connected to external and internalized norms (Cochran, Chamlin, Wood, & Sellers, 1999; Grasmick & Bursik, 1990). Although the distinction between guilt and shame is equivocal, shame is an internalized emotion that arises from public disapproval, whereas guilt is related to a specific behavior (Gehm & Scherer, 1988; Tangney, 1995). Shame for paroled women develops from being unable to live up to societal definitions of what it means to be a woman, a good parent, and a responsible citizen. Ex-offenders rarely view themselves as blameless, but continued societal alienation accentuates feelings of guilt and hinders successful reunification and reintegration. Women on parole are likely to experience "guilt with an overlay of shame" that leads to rumination and self-castigation (Tangney, 1995, p. 1142). The "bad mother" label, identified by Burkart (1973), is a painful and enduring stigma. Women in this study appear to engage in continued self-deprecation over the loss of their children, families, and relationships.

Community members often are reluctant to accept female ex-offenders and seem to engage in harsh moral judgments. Consequently, few efforts are made to reconcile the offender's presence in the community, and the person, not the deed, is labeled as bad (Braithwaite, 1989). The narratives in this research show that many of the women believe that once they are identified as a criminal, they remain a criminal in the eyes of others. Women on parole also experience disapproval from a variety of social organizations, which promotes further alienation.

Negative labels may lead to limited employment opportunities. Many parolees also lack relevant job training. Vocational education and training programs for women in most corrections facilities are limited

(Moyer, 1992). Training programs for clerical jobs, food services, and cosmetology, although cost-effective for the prison, fail to prepare women to be self-supporting upon release (Durham, 1994). The lack of job training, coupled with the label of being a female criminal, results in fewer employment opportunities.

This research represents a starting point for identifying the additional costs of imprisonment associated with displacement and the loss of significant others. The narratives, although based on women from one prison, emphasize the need for alternative sanctions, parenting programs, and community education. The collateral costs of prison and parole can be reduced by increasing opportunities that emphasize reentry into the job market, reintegration into the community, and reunification with children and families.

References

Barry, E. (1987). Imprisoned mothers face extra hardships. *National Prison Journal, 14*, 1–4.

Barry, E. (1995). Legal issues for prisoners with children. In K. Gabel & D. Johnston (Eds.), *Children of incarcerated parents* (pp. 147–156). New York: Lexington Books.

Baunach, P. J. (1985). *Mothers in prison*. New Brunswick, NJ: Transaction Books.

Baunach, P. J., & Murton, T. O. (1973). Women in prison: An awakening minority. *Crime and Corrections, 1*, 4–12.

Becker, H. (1963). *Outsiders: Studies in the sociology of deviance*. New York: Free Press.

Bill, L. (1998). The victimization and revictimization of female offenders: Prison administrators should be aware of ways in which security procedures perpetrate feelings of powerlessness among incarcerated women. *Corrections Today, 60*(7), 106–108.

Bloom, B. (1995). Public policy and the children of incarcerated parents. In K. Gabel & D. Johnston (Eds.), *Children of incarcerated parents* (pp. 271–284). New York: Lexington Books.

Bloom, B., & Chesney-Lind, M. (2000). Women in prison: Vengeful equity. In R. Muraskin (Ed.), *It's a crime: Women and justice* (pp. 183–204). Upper Saddle, NJ: Prentice Hall.

Bloom, B., & Steinhart, D. (1993). *Why punish the children? A reappraisal of the children of incarcerated mothers in America*. San Francisco: National Council on Crime and Delinquency.

Braithwaite, J. (1989). *Crime, shame and reintegration*. Cambridge, UK: Cambridge University Press.

Browne, D. C. (1998). Incarcerated mothers and parenting. *Journal of Family Violence, 4*, 211–221.

Burkart, K. (1973). *Women in prison*. Garden City, NY: Doubleday.

Chesney-Lind, M. (1997). *The female offender: Girls, women and crime*. Thousand Oaks, CA: Sage.

Cochran, J. K., Chamlin, M. B., Wood, P. B., & Sellers, C. S. (1999). Shame, embarrassment, and formal sanction threats: Extending the deterrence/rational choice model to academic dishonesty. *Sociological Inquiry, 69*, 91–105.

Crawford, J. (1990). *The female offender: What does the future hold?* Washington, DC: American Correctional Association.

Datesman, S., & Cales, G. (1983). I'm still the same mommy. *The Prison Journal, 63*, 142–154.

Durham, A. (1994). *Crisis and reform: Current issues in American punishment*. Boston: Little, Brown.

Fletcher, B., Shaver, L., & Moon, D. (1993). *Women prisoners: A forgotten population*. Westport, CT: Praeger.

Fox, J. (1982). Women in prison: A case study in the reality of stress. In R. Johnson & H. Toch (Eds.), *The pains of imprisonment* (pp. 205–220). Beverly Hills, CA: Sage.

Gabel, K., & Johnston, D. (1995). *Children of incarcerated parents*. New York: Lexington Books.

Garfinkel, H. (1956). Conditions of successful degradation ceremonies. *American Journal of Sociology, 61*, 420–424.

Gehm, T. L., & Scherer, K. R. (1988). Relating situation evaluation to emotion differentiation: Nonmetric analysis of cross-cultural questionnaire data. In K. R. Scherer (Ed.), *Facets of emotion: Recent research* (pp. 61–77). Hillsdale, NJ: Lawrence Erlbaum.

Glaser, B. G., & Strauss, A. L. (1967). *The discovery of grounded theory: Strategies for qualitative research*. London: Weidenfeld and Nicholson.

Glick, R., & Neto, V. (1977). *National study of women's correctional programs*. Washington, DC: Government Printing Office.

Goffman, E. (1963). *Stigma: Notes on the management of spoiled identity*. Englewood Cliffs, NJ: Prentice Hall.

Grasmick, H. G., & Bursik, R. J. (1990). Conscience, significant others, and rational choice: Extending the deterrence model. *Law and Society Review, 24*, 837–861.

Greenfeld, L. A., & Minor-Harper, S. (1991). *Women in prison* (Bureau of Justice Statistics special report). Washington, DC: U.S. Department of Justice.

Hairston, C. F. (1991). Mothers in jail: Parent-child separation and jail visitation. *AFFI-LIA, 6*(2), 9–27.

Harris, J. W. (1993). Comparison of stressors among female vs. male inmates. *Journal of Offender Rehabilitation, 19*, 43–56.

Henriques, Z. W. (1982). *Imprisoned mothers and their children*. Washington, DC: University Press of America.

Henriques, Z. W. (1996). Imprisoned mothers and their children: Separation-reunion syndrome dual impact. *Women & Criminal Justice, 8*, 77–95.

Jones, R. R. (1993). Coping with separation: Adaptive responses of women prisoners. *Women & Criminal Justice, 5*, 71–91.

Knight, B. (1992). Women in prison as litigants: Prospects for post prison futures. *Women & Criminal Justice, 4*, 91–116.

Martin, M. (1997). Connected mothers: A follow-up study of incarcerated women and their children. *Women & Criminal Justice, 8*, 1–23.

McGowan, B., & Blumenthal, K. (1978). *Why punish the children? A study of the children of women prisoners*. Hackensack, NJ: National Council on Crime and Delinquency.

Moyer, I. L. (1992). *The changing roles of women in the criminal justice system, offenders, victims, and professionals*. Prospect Heights, IL: Waveland Press.

Pollock-Byrne, J. M. (1990). *Women, prison, and crime*. Pacific Grove, CA: Brooks/Cole.

Pollock-Byrne, J. M. (1992). Women in prison: Why are their numbers increasing? In P. J. Benekes & A. V. Merlo (Eds.), *Corrections: Dilemmas and directions* (pp. 79–95). Cincinnati, OH: Anderson.

Prendergast, M., Wellisen, J., & Falkin, G. (1995). Assessment of and services for substance-abusing women offenders in community and correctional settings. *The Prison Journal*, *75*, 240–256.

Rafter, N. (1985). *Partial justice: Women in state prisons, 1800-1935*. Boston: Northeastern University Press.

Ragin, C. C. (1994). *Constructing social research*. Thousand Oaks, CA: Pine Forge Press.

Rasche, C. (2000). The dislike of female offenders among correctional officers. In R. Muraskin (Ed.), *It's a crime: Women and justice* (pp. 237–252). Upper Saddle, NJ: Prentice Hall.

Seidman, T. E. (1991). *Interviewing as qualitative research: A guide for researchers in education and the social sciences*. New York: Teachers College Press.

Shontz, F. (1965). *Research methods in personality*. New York: Appleton Crofts.

Stanton, A. (1980). *When mothers go to jail*. Lexington, MA: Lexington Books.

Sykes, G. M. (1958). *The society of captives: A study of a maximum security prison*. Princeton, NJ: Princeton University Press.

Tangney, J. P. (1995). Recent advances in the empirical study of shame and guilt. *American Behavioral Scientist*, *38*, 1132–1145.

Ward, D., & Kassebaum, G. (1965). *Women's prisons*. Chicago: Aldine.

DISCUSSION QUESTIONS

1. What concerns do incarcerated mothers have about their children and their lives while they are away?
2. What challenges do women face in their attempts to reunite with their children and families upon their release from prison?

Women and Work in the Criminal Justice System

Police, Courts, and Corrections

Section Highlights

- The gendered experience of women employed in the criminal justice system
- Differential challenges for women in victim services, policing, courts, and corrections
- Future directions for females working in the criminal justice system

Women in Policing

Women first began to appear in police departments in the early 20th century. However, their presence within the academy was significantly limited, as many believed that policing was a "man's job" and therefore, unsuitable as an occupation for women. Research on policing as "man's work" focuses on traditional masculine traits such as aggression, physical skill, and being tough—traits that many argued were lacking for women, making them inherently less capable of doing the job (Appier, 1998). Society assumes that police work is filled with danger, excitement, and violence, themes that are echoed in examples of popular culture, such as television and film and evening news programs that are filled with stories of dangerous offenders (Lersch & Bazley, 2012). These examples misrepresent the reality of policing, as many officers are faced with situations requiring empathy, compassion, and nurturing, traits that are often stereotypically classified as "feminine" characteristics. For example, the typical duties of an officer are not limited to the pursuit and

▲ **Photo 11.1** Captain Edyth Totten and the Women Police Reserve, New York City, 1918.

capturing of the "bad guys" and more often can include responding to victims of a crime or dealing with the welfare of children and the elderly.

An examination of the early history of women in policing indicates that the work of moral reformers was instrumental in the emergence of policewomen. During the late 19th century, women's advocacy groups were heavily involved in social issues. Examples of their efforts include the creation of a separate court for juvenile offenders as well as crime prevention outreach related to the protection of young women from immoral sexual influences. In 1910, the Los Angeles Police Department hired the first female police officer in the United States, Alice Stebbins Wells. At the time of her hiring, her job duties revolved around working with female and juvenile cases. Her philosophy as a female police officer reinforced the ideal of feminine traits of policing when she stated, "I don't want to make arrests. I want to keep people from needing to be arrested, especially young people" (Appier, 1998, p. 9). As a result of the national and international attention of her hiring, she traveled the country promoting the benefits of hiring women in municipal policing agencies. As an officer with the LAPD, Wells advocated for the protection of children and women, particularly when it came to sexual education. As part of her duties, she would inspect dance halls and movie theaters around the city. When she came into contact with girls of questionable moral status, she would lecture the girls on the dangers of premarital sex and advocate for the importance of purity.

Following in the footsteps of Alice Stebbins Wells, many women sought out positions as police officers. The hiring of women by police agencies throughout the early 20th century did not mean that these women were assigned the same duties as male police officers. Rather, policewomen were essentially social workers armed with a badge. Their duties focused on preventing crime, rather than responding to criminal activity. While hundreds of women joined the ranks of local law enforcement agencies between 1910–1920, they were a minority within the departments, and their presence was often resented by their male colleagues. In an effort to distinguish the work of policewomen, many cities created separate branches within the organization. These women's bureaus were tasked with servicing the needs of women and girls in the community. Many of these bureaus were housed outside of the walls of the city police department, in an attempt to create a more welcoming environment for citizens in need. Some scholars suggested that by making the women's bureaus look more like a home, rather than a bureaucratic institution, women and children would be more comfortable and therefore more likely to seek out the services and advice of policewomen.

The mid-20th century saw significant growth in the numbers of women in policing. In 1922, there were approximately 500 policewomen in the United States—by 1960, more than 5,600 women were employed as officers (Schulz, 1995). Throughout this time frame, the majority of these positions were limited in their duties from a traditional policing (i.e., male) model. Policewomen were not permitted to engage in the same duties as policemen, out of fear that it was too dangerous and that women would not be able to adequately serve in these positions, and most importantly, the "all boys club" that existed in most departments simply did not want women intruding on their territory. Despite these issues, women occasionally found themselves receiving expanded duties, particularly during times of war. With the decrease in manpower during World Wars I and II, many women found themselves placed in positions normally reserved for male officers, such as traffic control. In an effort to maintain adequate staffing levels during these time frames, the number of women hired within police agencies increased. However, the end of these wars saw the return of men to their policing posts and the displacement of women back to their gendered models of policing (Snow, 2010).

Like many other fields during the 1960s, the civil rights and women's movements had an effect on the presence of women in policing. Legal challenges paved the way toward gendered equality in policing by opening doors to allow women to serve in patrol capacities. In 1964, Liz Coffal and Betty Blankenship of the Indianapolis Police Department became the first women in the United States to serve as patrol officers, an assignment that was restricted to male officers throughout the country. As policewomen, Coffal and Blankenship were resented by their male colleagues, who believed that the work of a police officer was too dangerous for women. They received little training for their new positions and often had to learn things on their own. They found that dispatch often gave them the mundane and undesirable tasks, such as hospital transports. It soon became clear to Blankenship and Coffal that the likelihood of being requested for any sort of pursuit or arrest cases was slim. In an effort to gain increased experience in their position, they began to respond to calls at their own discretion. Armed with their police radio, they began to learn to interpret radio codes and would respond to cases in their vicinity. They successfully navigated calls that most male officers believed they couldn't handle. As a result of their positive performances in often tense situations, Coffal and Blankenship began to gain some respect from their male colleagues. However, they knew that any accolades could be short lived—one mistake and they ran the risk of being removed from their patrol status, and the traditional philosophy of "police work isn't for women" would be reinstated. While they eventually returned to some of the "traditional feminine roles" for women in policing, their experiences in patrol set the stage for significant changes for the futures of policewomen (Snow, 2010).

In addition to the differences in their duties, policewomen were subjected to different qualification standards for their positions. At the 1922 annual conference of the International Association of the Chiefs of Police, members suggested that policewomen should have completed college or nursing school (Snow, 2010). This standard is particularly ironic given that male officers were not required to have even a high school diploma until the 1950s and 1960s. As a result, the career path of policewomen attracted women of higher educational and intellectual standing.

Not only were policewomen limited by their role and duties within the department, they faced significant barriers in terms of the benefits and conditions of their employment. Like many other fields, policewomen were paid less for their work compared to policemen. In addition, the opportunities for advancement were significantly limited, as most departments did not allow women to participate in the exam process that would allow them to access opportunities for promotion. Generally speaking, the highest position that a policewoman could hold during this time was the commander of the women's bureau (and still, many agencies disagreed with that level of leadership, suggesting that women did not have the necessary skills or abilities to supervise officers or run a division). In some jurisdictions, women were forced to quit their positions when they got married, as many felt that women did not have enough time to care for a home, care for their husband, and fulfill their job duties. "When they marry they have to resign. You see, we might want them for some job or other when they have to be home cooking their husband's dinner. That would not be much use to us, would it?" (Snow, 2010, p. 23). In 1967, the President's Commission on Law Enforcement and the Administration of Justice advocated for expanding the numbers of policewomen and diversifying their duties. "The value should not be considered as limited to staff functions or police work with juveniles; women should also serve regularly in patrol, vice and investigative functions" (p. 125). Despite these assertions, few departments followed these recommendations, arguing that as members of a uniformed police patrol, officers required significant levels of upper-body strength in order to detain resistant offenders. In addition, many maintained the position that the job was simply too dangerous for women.

Until the 1970s, women represented only 1% of all sworn officers in the United States (Appier, 1998). However, new legislation and legal challenges in the 1960s and 1970s led to significant changes involving the presence and role of policewomen. While the Civil Rights Bill of 1964 was generally focused on eliminating racial discrimination, the word "sex" was added to the bill during the eleventh hour by House members, who hoped that this inclusion would raise objections among legislators and prohibit its passing. To the dismay of these dissenters, the bill was signed into law. In 1969, President Richard Nixon signed legislation that prohibited the use of sex as a requirement for hiring—meaning that jobs could not be restricted to men only (or vice versa). In addition, the Law Enforcement Assistance Administration (LEAA) mandated that agencies with federal funding (and police departments fell under this category) were prohibited from engaging in discriminatory hiring practices based on sex. While sex was now a protected category in terms of employment discrimination, the bill did little on its face to change the nature of women in policing, since most policewomen were employed by municipal departments (and the act exempted municipal government agencies from compliance). However, it did begin a trend within departments to introduce women into ranks that had previously been reserved exclusively for men. While several departments did take the initiative to place women into patrol positions, many of the men in these departments issued strong objections against the practice. Women in these positions often found themselves with little support from their colleagues. Eight years later, in 1972, the Civil Rights Act was amended to extend employment protections to state and municipal government agencies, which opened the door to allow women to apply to police departments as police officers without restrictions. While these changes increased the number of positions available to women (and to minorities), they also shifted the roles of policing away from the social service orientation that had been historically characteristic for women in policing. Their jobs now included the duties of crime fighting and the maintenance of order and public safety, just like their male counterparts (Schulz, 1995).

While women today are able to serve in the same positions within a department as their male colleagues, research indicates that women in policing employ different tools and tactics in their daily experiences as an officer compared to their male counterparts. The National Center for Women and Policing indicates that women officers are typically not involved in cases of police brutality and corruption. In contrast, male officers are 8½ times more likely than female officers to be accused of excessive force. Their research also indicates that women are more likely to be successful at verbally diffusing situations and have significantly fewer citizen complaints than men (Lonsway et al., 2002). These communication skills come in handy, as women officers appear to be able to connect with citizens in the community easier than male officers (Harrington & Lonsway, 2004).

By 2002, women made up 12.7% of all sworn officers in large agencies in the United States. Sadly, little has changed since 1990, which reported that only 9% of officers were women. It is particularly interesting to note that women are more likely to be found in municipal (14.2%) and county agencies (13.9%), compared to state agencies (5.9%; Lonsway et al., 2001). While court rulings in the 1970s opened the possibilities for promotion for policewomen, few women have successfully navigated their way to the top position of police chief. In 2009, there were 212 women serving in the top-ranking position in their departments

(O'Connor, 2012). Most of these women served in small communities or led agencies with a specific focus, such as university police divisions or transit authorities (Schulz, 2003). Within metropolitan agencies (more than 100 sworn officers), only 7.3% of the top-level positions and 9.6% of supervisory positions are held by women. Additionally, women of color make up only 4.8% of sworn officers, and minority women are even less likely to appear in upper-level management positions, with only 1.6% of top-level positions and 3.1% of supervisory roles fulfilled by a woman of color. The situation is even bleaker for small and rural agencies, where only 3.4% of the top-level positions are staffed by women.

Why is the representation of women so low in the policing field? While legal challenges have mandated equal access to employment and promotion opportunities for women in policing, research indicates that the overemphasis on the physical fitness skills component of the hiring process excludes a number of qualified women from employment (Lonsway et al., 2002). The physical fitness tests have been criticized as being used as a tool to exclude women from policing, despite evidence that it is not the physical abilities of officers that are most desirable. Rather, it is their communication skills which are the best asset for the job requirements. The number of pushups that a woman can complete compared to a man says little about how well each will complete their job duties. Yet standards such as these are used as evidence to suggest that women are inferior to their male colleagues. Women who are able to achieve the male standard of physical fitness are viewed more favorably by male colleagues, compared to women who satisfy only the basic requirements for their gender (Schulze, 2012). While historical perspectives have suggested that women lacked the physical prowess to manage the job duties of an officer, research by Rabe-Hemp suggests that these physical differences actually helped maintain peace when dealing with offenders. "A man, whether they don't want to hurt you or they have that innate respect that they can't hit a woman or something, I don't know what it is, or that motherly instinct, I don't know. But generally, men for the most part don't want to fight with females" (Rabe-Hemp, 2009, p. 121).

The emergence of community policing philosophies in the 1990s provided a new shift in police culture, one that increased the number of women working in the field. The values of community policing emphasize relationships, care, and communication between officers and citizens. It allows officers to develop rapport with members of their community and respond to their concerns. Effective community policing strategies have led to improved relationships and respect of officers by residents. Research indicates that policewomen have been particularly successful within models of community policing due to their enhanced problem-solving skills through communication (Lersch & Bazley, 2012).

▲ **Photo 11.2** Legal challenges throughout the 20th century have increased the number of women and minorities employed within law enforcement.

Given the historical context of women in policing, it is not surprising that attributes such as compassion, fear, or anything else that is considered "feminine" are often looked down upon, particularly by male officers. Given this masculine subculture that exists in policing, how does this affect women who are employed within the agency? What does it mean to be a woman in law enforcement? Are female police officers viewed as POLICEwomen or policeWOMEN? Cara Rabe-Hemp (2009) explores the issue of gender identity for women in policing, and finds that female officers do believe that their gender affects the way in which they function as officers. Like similar research, policewomen agree that they are more likely to rely on their interpersonal and communication skills within the context of their daily work and are less likely to "jump" to physical interventions. They acknowledged that their gender was an asset when dealing with certain populations, such as children and women who had been victimized.

While research indicates that the use of gender is interwoven through the identities of women as police officers, policewomen acknowledge that they balance their skills of communication and care in a way to protect the masculine identity of law enforcement (Rabe-Hemp, 2009).

Women continue to face a number of barriers in policing. One area that has revealed some interesting findings related to the issues of gender and job performance in policing is the topic of pregnancy. Indeed, given the physical challenges of their job duties, women in advanced stages of pregnancy may be physically unable to fulfill the contemporary job duties of policing, which could potentially present a danger to not only their own life, but also that of their unborn child. While laws such as the Pregnancy Discrimination Act prohibit the discrimination of women during pregnancy and have improved the conditions of employment for gestational women, they do not solve the issues for expecting women within law enforcement. In addition, laws such as the Family Medical Leave Act protect the employment status of women while on leave (up to 12 weeks, without compensation, a feature that often requires women to exhaust their sick and vacation reserves in order to maintain financial support for their family). However, these laws do little to respond to the needs of women who are able to work in a reduced capacity prior to the birth of their child. Given that many women may choose to have children, policies need to be constructed that not only reflect the needs of officers during pregnancy, but also accommodate the needs of families for both male and female familied officers.

A review of 203 departments in the United States found that very few departments have written policies on family leave for its officers. Only 11% of departments had any sort of clear policy on family leave, and 5.4% of departments had a policy that specifies the availability of limited-duty assignments for gestational women. While a lack of policy does not indicate that there is a negative belief system by the agencies against motherhood, it does mean that there may be little consistency on how cases are handled. Most departments focus on family-based policies, which allow for the equal treatment of mothers and fathers, rather than focus specifically on the needs of pregnant women and postpartum motherhood (Schulze, 2008). While it is important that departments accommodate the needs of pregnant officers, this can be a delicate issue. Many women express frustration that, upon informing their superiors that they are expecting, they run the risk of being removed from their position, even though they remain (depending on the stage of their pregnancy) able to adequately fulfill the requirements of their job (Kruger, 2006).

Despite the significant advances that women in policing have made over the past century, research is mixed on whether the contemporary situation is improving for women in law enforcement. While legal challenges have required equal access to employment and promotion within law enforcement, research indicates that many women continue to be "passed over" for positions that were ultimately filled by a male officer. In many cases, women felt that they continually had to prove themselves to their male colleagues, regardless of the number of years that they had spent within an organization. This experience was particularly prevalent when women moved to a new position (Rabe-Hemp, 2012).

While women in policing report experiences of discrimination and harassment within their agencies, they also acknowledge that the culture of policing has become more accepting to women throughout their careers. However, these ideals of peace were not easily won and required daily support and maintenance by the women. Research by Rabe-Hemp (2008) identifies three ways in which policewomen gained acceptance within the masculine culture of policing: (1) experiences in violent confrontations requiring the use of force, (2) achieving a high rank within the department structure, and (3) distinguishing themselves as different from their male counterparts in terms of their skills and experience. Female police officers acknowledge that the acceptance in the male-dominated police culture often comes with significant costs to their personal life and ideals. In many cases, policewomen talk about putting up with disrespect and harassment in order to achieve their goals. For others, they renegotiate their original goals and settle for "second best."

Female Correctional Officers

Correctional officers are a central component of the correctional system. Responsible for the security of the correctional institution and the safety of the inmates housed within its walls, correctional officers are involved with every aspect of the inmate life. Indeed, correctional officers play an important part in the lives of the inmates as a result of their constant interaction. Contrary to other work assignments within the criminal justice field, the position of the correctional officer is integrated into every aspect of the daily life of prisoners. Duties of the correctional officer range from enforcing the rules and regulations of the facility, to responding to inmate needs, to diffusing inmate conflicts and supervising the daily movement and activities of the inmate (Britton, 2003).

Historically, the workforce of corrections has been largely male and White, regardless of the race or gender of the offender. As discussed in Section IX, the treatment of female offenders by male guards during the early days of the prison led to significant acts of neglect and abuse of the women. These acts of abuse resulted in the hiring of women matrons to supervise the incarcerated female population. However, these early positions differed significantly from the duties of the male officers, and their opportunities to work outside of this population were significantly limited. For those women who were successful in gaining employment in a male institution, their job duties were significantly limited. In particular, prison policies did not allow female correctional officers to work in direct supervision roles with male offenders. Like the realm of policing, correctional masculine culture believed that to assign a woman to supervise male inmates was too dangerous. In male facilities, female guards were restricted to positions that had little to no inmate contact, such as administrative positions, entry and control booths, and general surveillance (Britton, 2003).

Despite the increased access to employment opportunities for women, many female guards resented these restrictions on their job duties and filed suit with the courts, alleging that the restriction of duties based on gender constituted sex discrimination. While many cases alleged that the restriction of female guards from male units was done to maintain issues of privacy for the offenders, the courts rejected the majority of these arguments. In *Griffin v. Michigan Department of Corrections* (654 F. Supp. 690, 1982), the Court held that inmates do not possess any rights to be protected against being viewed in stages of undress or naked by a correctional officer, regardless of gender. In addition, the Court held that the positive aspects of rehabilitation of an offender outweighed any potential risks of assault for female correctional officers, and therefore they should not be barred from working with a male incarcerated population. Other cases, such as *Grummett v. Rushen* (779 F2d 491, 1985), have concluded that the pat-down search of a male inmate (including their groin area) does not violate one's Fourth Amendment protection against unreasonable search and seizure. However, the Courts have held that the inverse gender relationship can be considered offensive. In *Jordan v. Gardner* (986 F.2d 1137, 1992), the Court found that pat-down policies designed to control the introduction of contraband into a facility could be viewed as unconstitutional if conducted by male staff members against female inmates. Here, the court held that a cross-sex search could amount to a deliberate indifference with the potential for psychological pain (under the Eighth Amendment) given the high likelihood of a female offender's history of physical and sexual abuse.

As a result of equal employment opportunity legislation, the doors of prison employment have been opened for women to serve as correctional officers. Today, women are increasingly involved in all areas of supervision of both male and female offenders. Many women choose corrections as a career out of interest in the rehabilitation interests, as well as a perception that such a career provides job security (Hurst & Hurst, 1997). According to the 2007 Directory of Adult and Juvenile Correctional Departments, Institutions and Agencies and Probation and Parole Authorities, women made up 37% of correctional officers in state adult facilities and 51% of juvenile correctional officers (American Correctional Association, 2007). Within these facilities, both men and women are assigned to same-sex as well as cross-sex supervision positions. In addition, more women are working as correctional officers in exclusively male facilities, where they constitute 24.5% of the correctional personnel in these institutions (DiMarino, 2009). Yet research indicates that gender can affect how officers approach their position, regardless of the sex of the inmate. For women involved in the supervision of male inmates, their philosophies differed significantly from that of male officers. While male officers functioned within a paramilitary role and were ready to use force if necessary, women saw their role as mentors and mothers, and they focused on the rehabilitation of the inmates (Britton, 2003).

Despite significant backlash and criticism, research indicates that the integration of women into the correctional field has significant benefits for prison culture. First, female correctional officers are less likely to be victimized by inmates. This finding goes against some of the traditional concerns that women would be at risk for harm if they were involved in the direct supervision of male offenders. Second, women officers are more likely to use their communication skills, rather than physical uses of force, to obtain compliance from inmates. Finally, female officers indicate a greater level of satisfaction from their work, compared to male officers (Tewksbury & Collins, 2006).

While women have made significant career gains as correctional officers, women still struggle in this masculine, male-dominated environment. Research indicates that women as officers are frequent targets of sexual harassment (Chapman, 2009). The good ol' boy network remains quite pervasive in many facilities. Many women in leadership positions face significant challenges in navigating this culture. Women discuss how gender changes the way in which they do their jobs. "Men will perceive being assertive as a good quality in a guy, [but for women] they will still say, 'oh she's such a bitch.' So you need to couch what you're saying a little differently so as not to offend these poor guys over here" (Greer, 2008, p. 5).

Like correctional officers, inmates also have conflicting perceptions of women working in the correctional field. Studies indicate that upon their first interactions, male inmates draw on stereotypical assumptions regarding female officers. Yet, women in these positions possess the unique opportunity to offer a positive image of women (Cheeseman & Worley, 2006). Despite these gains, women who work in this field indicate that they experience persistent occurrences of sexual harassment by inmates. Yet, these experiences do not affect their levels of satisfaction with their jobs—indeed many accept that incidents of sexual harassment come with the territory of being a woman working in a male-dominated arena (Chapman, 2009).

How does gender affect the perceptions of work in a correctional setting? Like other criminal justice occupations, how do female correctional officers "do" gender in the context of their job duties? As an officer working in a juvenile facility in the 1990s, I vividly recall many of the ways in which I was aware of my status as a woman and how my gender affected the ways in which I interacted with my fellow staff and inmates. In some avenues, gender was very much a part of my identity as a correctional officer. I would utilize traits and characteristics that many scholars identify as "feminine traits"—communication and care for the juveniles under my custody. I also instituted a system of mutual respect between the wards and myself. Some of my colleagues who emphasized control and order may have looked down upon these perspectives as "soft," but it was how I did my job. At the same time, I became very aware of my physical status as a woman, particularly when working with male offenders. I noticed that I began to dress down and wear baggier clothing (we did not have uniforms) to hide my figure. I would also appear very simple in terms of my hairstyle and makeup. In essence, I was limiting many of the physical displays of my gender.

Research on women in corrections echoes many of these themes that I experienced on the job. Women working in the correctional field are more likely to emphasize the "social worker" aspects of the job, compared to their male counterparts (Stohr, Mays, Lovrich, & Gallegos, 1996). Here, women use their gender to their advantage—by drawing upon their communication skills, they are able to diffuse potentially dangerous situations before violence ensues. However, it is important to find balance between the feminine traits and masculine traits—too much communication between staff and inmates can be viewed by some as a negative thing, out of fear that staff will grow too close to an inmate and risk being taken advantage of (Britton, 2003). Given the increase of the prison population and the opportunities for employment, it is important for facilities to recognize the strengths and weaknesses for women who work in this field and their relationships with the incarcerated population.

Much of the research on women working in corrections deals with the issues of job satisfaction, stress, and burnout. While several scholars have indicated that women in corrections experience higher degrees of stress than male officers, not all have come to this conclusion. While certainly the correctional setting is a high-stress work environment, research by Griffin (2006) indicates that the levels of stress do not differ by gender. In contrast, Tewksbury and Collins (2006) find evidence that female correctional officers do indicate higher levels of job-related stress than male correctional officers. Two major themes of stress for women include concern over issues of safety and of "bringing their work home." In addition, research by Lambert, Altheimer, and Hogan (2010) indicates that conflicts between work and family had a significant effect on the levels of job-related stress and job satisfaction for women, while conflict and ambiguity in the role played by an officer changed the levels of job satisfaction for men. Of particular interest are the findings reported by Tewksbury and Collins (2006), that women cite that the major source of their job-related stress comes from their interactions with their coworkers, whereas male officers report that their job stress comes from working with inmates. In addition, men and women deal with their stress differently. While women are more likely to seek out social support as a mechanism for dealing with stress, men are more likely to engage in problem-solving skills to resolve their source of stress (Hurst & Hurst, 1997).

Another major theme within the literature involves gender differences in inmate supervision preferences. Many years ago, when I worked as a line staff member at a juvenile detention center, several of my fellow female staff members would express disdain when they were assigned to work the female unit. Many of my colleagues would comment that the girls were much more difficult to work with than the boys, were more dramatic, manipulative, needy, emotional, and time consuming. Research by Pollock-Byrne (1986) provides details on why male and female correctional officers believe that working with women is less desirable. While both male and female staff believe that women inmates are more demanding, defiant, and complaining, male officers also express concerns about being blamed for accusations of inappropriate behaviors by female inmates. Female officers express that they would prefer to work with male inmates because they feel that they are more likely to be respected and appreciated by the male inmates. Belief systems such as these have a significant impact on perceptions of working with female offenders and translate into a belief that working with women is an undesirable assignment (Rasche, 2012). Research indicates that among both male and female correctional officers (and regardless of rank), there appears to be a male inmate preference, despite the increased risks for violence.

Staff members who are assigned to work with women against their preference can do a significant disservice to programs that employ a gender-responsive approach. Research by Rasche (2012) finds that the only population that does not express a preference toward the male inmate in terms of job assignment is the experienced female line staff members who are employed at all-female facilities. Interestingly, while correctional officers indicated that female offenders were more needy than male offenders, they attributed these differences to the differences between men and women generally in society. Indeed, as previous sections of this book have indicated, incarcerated women do have a variety of unique needs that the correctional system must respond to, including gynecological health needs, increased likelihood of mental health issues, and the strain of being separated from their children. Yet Gaarder and Belknap (2004) found that detention workers received little to no training on how to work with female offenders and on how, as a population, they present a variety of needs that differ from those of the male population. This lack of specialized training led to the (mis)perceptions that girls were more difficult to work with. Just as programming needs to be gender responsive, so does the training of officers who supervise female offenders.

Women and the Law

The Feminization of the Legal Field

The first woman was admitted to the American Bar Association in 1918. In the time since the integration of women into the legal field, several decades of discrimination have faced those women who made a career out of the practice of law. While times have improved significantly for women in this field, they still endure several challenges based on gender. Today, women have reached near parity in terms of law school enrollment and faculty positions. Women make up almost half of all students enrolled in law school (47.1% in 2009–2010; American Bar Association [ABA], 2009), and several high-ranking law schools have clinics and journals relating to gender issues, including Columbia's Sexuality and Gender Law Clinic, and the *Duke Journal of Gender, Law and Policy.* Women also make up a significant presence among law school faculty. According to the American Bar Association (2010), women make up 54.6% of the tenured, tenure-track, or visiting full-time faculty at law schools across the nation. In addition, more women are finding their way into the top administrative positions within these schools. While women are less likely to hold the highest office (26.9% of dean positions are staffed by women), they are more likely than men to hold the office of associate or vice dean (60.6%), and assistant dean/director (69.7%).

Despite the increased presence of women within the legal academy, research indicates that women employed in the legal field continue to face challenges and adversity based on their gender, such as differences in pay, denial for promotion and partnership, and the challenges of balancing family life with the demands of a legal career. For example, women continue to earn less than their male counterparts; sources indicate that women in the legal field earn approximately 81% of male salaries (Redhage, 2009). Contributing to this pay gap is the finding that women are less likely to make partner than men (Noonan, Corcoran, & Courant, 2008).

Like many of the fields within the criminal justice system, women in the legal profession also face challenges with balancing the needs of their career with the demands of motherhood and family life. Within the corporate model of the legal system, the emphasis on billable hours requires attorneys to work long hours, both during the week and weekend. The demands of this type of position often conflict with family duties.

For many women, this conflict results in women either delaying starting a family or choosing their career over motherhood. Others choose to leave their position prior to making partner or leave their positions for one that is less stressful and has greater flexibility.

While firms may offer opportunities for part-time work, research indicates that few women take advantage of these opportunities out of fear that it would damage the potential for career advancement. For those women who did choose these career trajectories, research indicates that these positions did not necessarily involve compensatory reductions in workload, forcing many to bring their work home with them, work for which they did not receive compensation. In addition, the women believed that the reduction in time spent in the office could ultimately affect their chances for promotion and earning potential as well as create a negative assumption regarding their work ethic and level of commitment by their colleagues (Bacik & Drew, 2006).

As women in private law practice become discouraged regarding their likelihood of achieving partner status, many make the decision to leave their positions. Indeed, men are more likely to become partners than women and also earn significantly higher salaries. While the decisions to get married, have children, and take time away from their jobs, or reduce their employment status to part-time, do not have a significant effect for men or women in their likelihood to leave private practice, these variables are associated with levels of satisfaction surrounding the balance of work and family needs. Here, satisfaction is the key, not their decisions regarding family status (Noonan & Corcoran, 2004).

The majority of research on women in the legal profession lacks any discussion of how race and ethnicity interact with gender for women of color. Indeed, 67% of lawyers are White males, 25% are White females, and 8% are people of color. Research by Garcia-Lopez (2008) finds that race and ethnicity have significant effects on the gendered nature of legal work. Generally speaking, men were more likely to be assigned high-profile cases, whereas women were assigned cases related to educational and other social issues. In addition, one respondent indicated that White women and women of other minority groups were more likely to be viewed as "good attorneys," while Chicanas are less likely to be viewed as valuable professionals in their field. Here, women of color are put in a position wherein they need to constantly prove themselves to their colleagues—"They just didn't appreciate me; (they) didn't think I was capable" (p. 598). In addition, Chicana women were more likely than White women to be overburdened with larger, lower-profile caseloads. They also felt as though they were the key representatives and spokespersons for their racial-ethnic group. "It's like they expect you to answer for the entire Latino population; like you should know everything there is to know about Latinos" (p. 601). Unlike other racial/ethnic and gender groups, Chicana women attorneys did not define their success by financial achievements. Rather, social justice and helping people in their community play a key function in their concept of success and happiness with their lives and careers (Garcia-Lopez, 2008).

Scholars debate whether or not women can achieve equality in the legal profession. Some suggest that as older (and mostly male) partners retire, younger attorneys will be more likely to include a greater representation of women, given the increase in the number of women who attend and graduate from law school (French, 2000). Others argue that this theory neglects the fact that any change in the culture of the law firm will be slow in coming, due to the small numbers of women who choose to work within these types of positions and are successful on the partnership track (Reichman & Sterling, 2001).

Women and the Judiciary

In the judiciary, the representation of women has grown substantially. Sandra Day O'Connor was the first woman appointed to the U.S. Supreme Court in 1981. At that time, there were few women in high-ranking judicial positions at the state and federal level. O'Connor began her tenure on the court as a conservative voice,

▲ **Photo 11.3** The four women of the U.S. Supreme Court. From left to right: Sandra Day O'Connor, Sandra Sotomayor, Ruth Bader Ginsburg, and Elena Kagan.

yet later became the swing vote in many cases with some of the liberal judges. O'Connor remained the lone woman on the court until 1993, when Clinton appointed a second woman to the court (Ruth Bader Ginsburg). While O'Connor retired in 2005 to spend time with her ailing husband, Ginsburg remains on the court today. Recently, Ginsburg has been joined by two additional female justices: Sonya Sotomayor (2009) and Elena Kagan (2010). Their appointments mark a shift in the judiciary of the highest court in the land. Sotomayor is the first woman of color, a Latina, to serve on the Supreme Court, and the inclusion of Kagan creates a historical first, as this is the first time in history that three women have served simultaneously on the Court.

In terms of the appointment of women to the judiciary, the appointments made by President Bill Clinton in the 1990s made a significant impact on the judiciary, as he appointed more women to judicial positions at the federal appellate level than any other president (Goldman, 1995, in Palmer, 2001). By 2000, 25% of state Supreme Court justices were women. Research indicates that despite these increases, women often serve as the token female on these courts, limiting the future possibilities of more women on the bench in these regions. Indeed, the majority of these positions held by women are in the courts of general jurisdiction, not the upper ranking positions. For example, in California, 92% of women judges serve in courts at the trial level (Williams, 2007).

What factors affect the appointment of women to the judiciary? Research by Williams (2007) indicates that more women receive a judicial appointment as a result of a nonpartisan election, compared to partisan elections. Liberal states are more likely to have women in judicial positions, compared to conservative states. In addition, the presence of female attorneys in the state also increases the representation of women as judges in the trial courts. At the appellate level, three variables affect the representation of women in these positions: (1) as more seats are generally available on the appellate bench, the representation of women at this level increases; (2) as the number of female attorneys in a state increases, so does the number of women judges at the appellate level; (3) states that use the merit selection process to fill seats have an increased number of women on the bench, compared to those states that rely on a partisan election to fill these positions.

Does being female affect the way in which judges make decisions? In a study involving hypothetical vignettes, the findings indicated several areas where gender differences existed among judges who participated in the survey. In most of the scenarios, the female judges imposed longer sentences in cases of simple assault and were less likely to award civil damages for these cases. However, when damages were to be awarded, female judges awarded significantly higher monetary levels compared to male judges (Coontz, 2000). When reviewing outcomes in real-life cases, research by Steffensmeier and Hebert (1999) finds that women judges tend to be harsher in their sentencing decisions compared to their male counterparts. Controlling for offender characteristics, the presence of a woman on the bench increases both the likelihood of prison time for offenders (10%) and the length of their sentences (+5 months longer). In addition, property offenders and repeat offenders are the ones most likely to bear the brunt of this increased severity when facing a female judge. Similar research on gender differences in sentencing by Songer, Davis, and Haire (1994) indicates that male and female judges do not differ in judicial decision making in federal cases involving obscenity charges or criminal search and seizure cases, but female judges were significantly more likely to decide in favor of the victim in cases of employment discrimination. At the state Supreme Court level, research indicates that not only do women tend to vote more liberally in death penalty and obscenity cases, but also the presence of a woman on the court increases the likelihood that the male judges will vote in a liberal fashion (Songer & Crews-Meyer, 2000).

Conclusions

Despite the gains of women in criminal justice occupations, they continue to confront a glass ceiling in terms of equal representation, compensation, and opportunity within the field. Women who work in these fields become a symbol for all things gender related in these male-dominated fields. At the heart of the research for each of these fields, two major themes emerge—(1) how gender can affect the way in which women who work in these fields satisfy the demands of their positions, and (2) how gender affects the experiences that they have within their jobs. These factors are multiplied for women of color, whereby race serves as yet another variable through which discrimination can occur. For some of the most masculine positions, such as policing and corrections, women must fight against firmly held beliefs that such jobs are inappropriate for women. While equal employment opportunity legislation has opened the doors for access for women in these traditionally male-dominated fields, women still face an uphill battle as they have been denied opportunities and promotion throughout history. In occupations such as attorneys and judges, while the proportion of women in these fields has significantly increased, women still remain underrepresented at the upper levels. Despite these struggles, women remain an important presence in these fields with significant contributions that need to be encouraged and acknowledged, particularly for future generations of women in these fields.

Summary

- Traditional male occupations such as policing and corrections historically excluded employment options for women on the grounds that the work was too dangerous.
- Early policewomen were involved in crime prevention efforts, primarily with juvenile and female populations.
- While equal opportunity legislation may have opened access for women in policing and corrections, institutional cultures and standards continued to create barriers for women in these occupations for entry and advancement.
- Women police and correctional officers use different tools and techniques in their daily experiences in their positions, compared to male officers.
- Few women have successfully navigated to the top levels of their fields in law enforcement and the legal arena.
- As correctional officers, women are subjected to issues with job satisfaction, stress, and burnout.
- Many correctional officers prefer not to work with female inmates, as they assume that women are more difficult to work with.
- Women in the legal field struggle with balancing work demands with family life. These struggles can affect the advancement of women in their field.
- While the number of women in the judiciary has increased, the majority of these positions are at the lower levels.

KEY TERMS

Alice Stebbins Wells

Betty Blankenship

Liz Coffal

POLICEwomen vs. policeWOMEN

Community policing

Good ol' boy network

Griffin v. Michigan Department of Corrections

Grummett v. Rushen

Jordan v. Gardner

Male inmate preference

Work-family balance

"Mommy track"

DISCUSSION QUESTIONS

1. Based on the research, how do women "do" gender within traditional "male-dominated" criminal justice occupations?
2. What challenges do women who work in criminal justice occupations face that their male counterparts do not?
3. What suggestions can be made to improve the status of women within criminal justice occupations?

WEB RESOURCES

National Center for Women and Policing: http://www.womenandpolicing.org

American Correctional Association: http://www.aca.org

Association of Women Executives in Corrections: http://www.awec.us

National Association of Women Lawyers: http://www.nawl.org

National Association of Women Judges: http://www.nawj.org

International Association of Women Judges: http://www.iawj.org

READING

In the section introduction, you learned about the challenges that women in policing face as a result of their gender. In this reading, Dr. Rabe-Hemp explores the experiences of policewomen to assess where they are confronted with issues as a result of their gender and how they cope with these barriers in the workplace. Her research indicates that officers do experience acts of sexual harassment on the job. Understanding these issues can help departments assess the culture of policing for women and raise awareness in departments regarding the challenges of policing for women.

Survival in an "All Boys Club"

Policewomen and Their Fight for Acceptance

Cara Rabe-Hemp

Introduction

Since the 1972 Civil Rights Act opened the door to patrol for women and made it illegal for police agencies to utilize gender as a barrier to employment and promotion, female police officers have struggled through both internal and external obstacles to gain equality in the police role. The ability of women to integrate into the masculine culture of policing has been hotly debated since this time. In early research, police officers, administrators, and even police officers' wives questioned female officers' abilities to maintain the authority and strength necessary to the police role (Bartlett and Rosenblum, 1977; Bloch and Anderson, 1974; California Highway Patrol, 1976; Sherman, 1975). Given that policemen did not trust the ability of female officers to perform police duties, the resultant resistance to women integrating into the male culture comes at little surprise (Brown and Sargent, 1995; Burligame and Baro, 2005; Franklin, 2005; Garcia, 2003; Hunt, 1990; Martin, 1980; Paoline, 2003; Waddington, 1999). Despite this resistance, between 1978 and 1986, the proportion of women in policing doubled, from 4.2 to 8.8 percent of municipal officers.

Little is known about the ensuing cohort of women who joined the policing occupation amidst the rise of community policing, consent decrees, and affirmative action suits (Martin, 1991). Their struggles to integrate into police agencies has been exemplified in the past 30 years of research underscoring continued resistance and sexual harassment from colleagues, a glass ceiling to promotion, and gender-specific obstacles with childcare and pregnancy (Belknap and Shelley, 1992; Franklin, 2005; Heidensohn, 1992; Garcia, 2003; Martin, 1980; Miller, 1999; Pike, 1985; Price, 1985; Remmington, 1983; Schulz, 2004).

Less yet is known about the acceptance of female officers in medium- and small-town police agencies.

SOURCE: Rabe-Hemp, C. (2008). Survival in an "all boys club": Policewomen and their fight for acceptance. *Policing: An International Journal of Police Strategies and Management, 31*(2), 251–270.

Previous studies of female officer integration have taken place in large, urban departments (Burligame and Baro, 2005; Garcia, 2003; Lonsway, 2003; Martin, 1980; Pogrebin et al., 2000; Sims et al., 2003). While this research is important, it does not represent the experiences of women in small and medium, municipal, county, and state departments (Weisheit et al., 2006). A recent report by the National Center for Women and Policing (2000) found women in small and rural agencies were severely under-represented and disproportionately limited to the lower tiers of the rank structure, suggesting an even greater potential for isolation and discrimination.

In spite of the obstacles presented to women in policing, female officers comprise 17 percent of the nation's urban municipal police agencies and 6 percent of non-urban municipal agencies suggesting female officers do find measures to overcome these obstacles and succeed in police employment (Bureau of Justice Statistics, 2006). These "success stories," female officers who have succeeded in an "all boys club" despite early and continued resistance from colleagues, isolation in the organizational culture, and inability to climb the proverbial administration ladder (Hunt, 1990; Martin, 1980; Price, 1985; Remmington, 1983) necessitate a discussion of the occupational experiences of female officers, the importance of determining the coping mechanisms female officers utilize to overcome adversity confronted in police work, and the legacy they leave in their respective departments.

Theoretical Framework

Although female officers are still tokens or numerical oddities in many police agencies, recent research has suggested the theory of tokenism alone may not fully describe the experiences of female officers (Franklin, 2005; Greene and del Carmen, 2002). Instead, research must examine how the expectations and stereotypes of police, gender, and work amalgamate to form the guidelines for behavior for female police officers to understand their occupational experiences.

The occupation of policing is masculine by its nature, as law enforcement is typically associated with aggressive behavior, physical strength, and solidarity. Expectations for police based on a crime-fighting image of ruggedness, toughness, and masculinity (Bittner, 1970; Koenig, 1978; Young, 1991) in concert with expectations of femininity of understanding, sensitivity, and softness (Werner and LaRussa, 1985) create an incongruent role for female officers. Martin (1999) argue female police officers have the choice of either maintaining their gender identity as police women or the police identity as police women. Women who attempt to meet the crime-fighting image of police may be negatively labeled as "butch or dyke" (Pike, 1985, p. 264), but female officers who do not attempt to meet this perceived ideal may risk being defined as weak or "pansy police" (Miller, 1999, p. 70). Neither solution allows them to be fully integrated or accepted in police work, as bona fide or "real" police officers (Martin, 1990; Remmington, 1983).

The token status of women in police organizations may exacerbate the expectations and stereotypes already associated with female officers. When a group makes up less than 15 percent of an organization, they perceive themselves to be highly visible, attracting disproportionate attention to themselves, being perceived as "in but not of" the organization (Belknap and Shelley, 1992; Kanter, 1977). In organizations with less female representation, the behaviors of female officers are more salient and draw more attention as novelties than those of male officers, which is especially relevant in non-urban agencies where women are un-common.

Female officers negotiate the roles associated with their gender and work through a backdrop of hegemonic masculinity that reinforces the power of men both on the cultural and collective levels. Hegemonic masculinity at work in police agencies is maintained through: authority heterosexism, subordination of women, and the ability to display force (Connell, 1995; Messerschmidt, 1993). Through academy and field training experiences, female officers are acculturated to the hyper-masculine values and attitudes of policing which restrict women's behavior to feminine tasks, mostly involving women and children (Garcia, 2003). By adopting the roles for "women's work," female officers maintain the hegemonic masculinity operating in the police occupation. Those who dare to challenge these roles may be isolated, harassed, and put both physically and emotionally at risk (Hunt, 1990; Martin, 1990). The accounts of female officers' great efforts to manage their status as policewomen and survive in an "all boys club" are the basis for this research.

The Police Culture and Female Officers' Integration

Although women had been active in police work as prison matrons since the 1800s, it was not until female officers began encroaching on the stereotypically male task of patrol that the debate of the proper role of women in policing commenced. Concern over female officers' abilities to maintain the authority and strength necessary to the police role provided a backdrop for the continued resistance to women in the policing culture, including daily harassment and sexism (Burligame and Baro, 2005; Franklin, 2005; Garcia, 2003; Hunt, 1990; Martin, 1980).

Police Culture

Existing literature on police culture defends and explains the theories that drive resistance to female officers (Brown and Sargent, 1995; Paoline, 2003; Waddington, 1999). There is wide speculation on the nature and purpose of the police subculture. Some theorists posit the solitary, masculine police subculture is a dated vestige of the past, and modern police organizations have many different subcultures, each vying for resources (Wood et al., 2004). Despite the fact that women represent almost 48 percent of the nation's workforce, no police organization in the USA has reported female employment equal to that percentage, suggesting if women do possess a unique subculture, it would be subordinate in strength and power to the male subculture (National Center for Women and Policing, 2001; US Census Bureau, 1999). The police subculture has also been described as a manifestation of the nature of police work (i.e. stress, shift work, danger), as well as a social structure which exists purposely and specifically to oppress female officers (Brown and Sargent, 1995; Franklin, 2005). There is consensus that the sovereign police culture is a distinctive occupational subculture that celebrates masculine values which engender particular views of women, the nature of policing, and the roles for which men and women officers are believed to be most suitable (Dick and Jankowicz, 2001). The intrusion of women into the police culture has the potential to change these norms, values, and customs and hence has been met with great resistance.

Male Officer Resistance

Male police officers' resistance to female officers has been amply documented (Haarr, 1997; Herrington, 2002; Hunt, 1990; Martin, 1980). Martin (1980, p. 290) found, "most women officers have experienced both sex discrimination and sexual harassment"... and frequently these behaviors were "blatant, malicious, widespread, organized, and involved supervisors; occasionally it was life-threatening". Hunt (1990) examined the underlying logic of police sexism among police when she researched the hesitation of male police to accept women into the rank-and-file and concluded, "sexism ... is a deep structure which is articulated in every aspect of the police world" (Hunt, 1990, p. 26). More recently, Haarr (1997) found women are maintained outside of the informal police structure and continue to face sexual harassment, sexism, and discrimination.

Glass Ceiling

In addition to resistance from colleagues, women face a glass ceiling in regards to promotion. There are few supervisory positions available in municipal departments and the progress up the proverbial ladder is slow. More than half the agencies surveyed in the Status of Women in Policing (2001) reported no women in top command or supervisory positions. Martin (1980) argues that a lack of variety of female role models in higher ranks is a major obstacle to younger female officers. Without the guidance of seasoned veterans to teach rookies the ropes, female officers are excluded from informal networks which are essential to the police culture (Martin, 1980; Wexler and Logan, 1983). The lack of female role models also may foster feelings of isolation in policing organizations.

Current Status of Women in Policing

Despite these obstacles, female officers have made significant gains in the police occupation in the past 30 years. Within large police agencies, sworn women currently hold approximately 10 percent of supervisory positions (National Center for Women and Policing, 2001), suggesting improvement in many, but not all police agencies. Urban municipal police departments report dramatic increases in female representation in the past decade, topping out at 25 percent sworn (Los Angeles, 19 percent; Chicago, 22 percent; Philadelphia, 24 percent; Detroit, 25 percent; see National Center for Women and Policing, 2001).

Rural and small town agencies have not maintained the same growth. Women police represent only 6 percent sworn in municipal departments in 2003, up from 4.5 percent in 1993 (Bureau of Justice Statistics, 1996, 2006). A recent report written regarding women in small and rural communities found women were severely under-represented in small and rural police agencies, and disproportionately represented in the lower tiers of the rank structure (National Center for Women and Policing, 2000), suggesting the promotion disparities are even wider for women in small and rural agencies than they are in larger and more urban departments. The potential for isolation and discrimination may be greater in these departments due to their token status.

The rise of community policing has provided female officers the opportunity to adopt skills that have not previously been associated with the traditional crime fighter image of the police. Instead, descriptions of the role and responsibilities of the police officer in community policing have marked similarities to policewomen in the late nineteenth and early twentieth centuries (Miller, 1999). The same feminine gendered traits of communication, trust, and relationship-building that maintained women outside of the crime fighter model are now being embraced by community policing (Miller, 1999). During the rise of community policing, female representation in policing doubled, suggesting new opportunities created by community policing for women officers and the progress of affirmative action programs in many police departments. The increased representation of women in policing provides a greater opportunity for women to support one another, to define women's roles in policing, and to impact the police culture and the way police "do justice" (Menkel-Meadow, 1989; MacKinnon, 1989; Scarborough and Collins, 2002; West and Zimmerman, 1987).

Purpose of the Study

The purpose of this study is to explore the experiences of policewomen in a variety of police agencies to determine the extent to which female officers have and continue to face resistance and obstacles to police work, to examine the coping mechanisms female officers utilize to overcome impediments to police employment, and to establish common themes in female officers' success stories of acceptance and integration into police agencies.

Methods

Participants

The sample consists of 24 female officers, each with varying police experience ranging from 10 to 30 years. In an attempt to determine the onset and desistance of female officers' experiences with resistance and acceptance, only tenured policewomen were interviewed for inclusion into the study (served at least ten years; mean years served were 21.4). Most of the respondents were white (one minority officer). The sample included nine municipal officers, three county officers, six campus police officers, and six state police, all from a Midwestern state. About 11 female officers in this study represent departments with over 100 sworn members; seven female officers serve in departments with more than 50 and less than 100 sworn members; and six women serve in departments with fewer than 50 members sworn. Overall, 12 of the 24 women interviewed were administrators: one chief, one colonel, four captains/commanders, two lieutenants, three sergeants, and one corporal and 12 were line officers: four detectives and eight patrol officers. Policewomen interviewed and included in the sample represented 12 unique state (one), county (three), municipal (six), and campus (two) agencies.

Data Collection

The most difficult aspects of researching female officers, especially those in rural, small and medium agencies, is finding them and making initial contact. To encourage honesty and frankness by female officers, department administrators were not contacted for lists of possible interviewees. Instead, due to the rarity of the population being examined[1], potential participants were identified by "snowball sampling." A few known female officers helped compile a short list of other female officers; these officers, in turn, provided referrals to other female officers. The sampling frame of 26 female officers was completed when no new names were produced. Initial contact was made by e-mail, telephone, or personal contact. Only two of the 26 female officers identified declined participation.

The interview protocol consisted of open-ended questions designed to examine female officers' perceived acceptance and integration in their departments, difficulties associated with the police roles and organization, and coping mechanisms utilized to overcome resistance. Since the study concerns women's subjective perceptions and assessments of their experiences as female police, the questions allowed the respondents to elaborate on their experiences and the events that shaped their tenures. Interviews were conducted at coffee shops, respondents' offices, or other places convenient for the officers. The interview length varied from 35 minutes to two and half hours but generally lasted an hour. The project and procedures were approved by the researcher's institutional review board.

Analysis

During each interview, notes were taken, and each interview was taped with the subject's consent and then typed verbatim for qualitative data analysis. The auditor conducted all of the interviews and transcribing was completed by graduate assistants. Owing to the risk of exposure, it might seem female officers may be less than candid with a researcher, although my general impression is a good and honest rapport was maintained throughout the interviews. Most female officers were happy to talk about their experiences as policewomen and were open about both their struggles and accomplishments throughout their careers.

Each transcript was analyzed for emerging themes, concerns, and phrases that had been presented by the participants. These were coded using an opened-ended approach. After the first reading, tags corresponding to relevant research issues were placed on the transcriptions by hand. The issues covered by the tags were very broad in nature (i.e., acceptance, resistance, coping mechanisms, and success) and left a great deal of scope for establishing variability in participants' responses that were later organized into categories that represented the participants' distinctions. The themes within each tagged topic were refined and reorganized through multiple re-readings of the transcripts. In the last step, representative quotes were pulled to illustrate the major themes reported by the participants (Padgett, 1998; Tutty et al, 1996).

Results

Despite experiences of sexual harassment, discrimination, and disrespect, almost all the participants reported they had achieved acceptance in their current agencies and the organizational culture of policing had improved for women since they had started. They acknowledged hegemonic masculinity, inherent in the police culture, was established through training academy and field training experiences and maintained through isolation, intimidation, and resistance. Female officers described the process by which acceptance was nurtured, negotiated, and maintained on a daily basis, rather than achieved. Throughout this negotiation process, women varied in the roles they adopted to achieve their desired goals. For many female officers this meant doing "women's work" and avoiding promotion opportunities. For others who defied the system and achieved high ranks, these opportunities often came at a significant cost.

Resistance to Police Work

Participants were asked several questions regarding resistance from and acceptance by colleagues and administrators. Most reported experiences with sexual harassment, discrimination, or disrespect starting early in field training and police academy experiences and continuing through promotions and job assignments:

> *Captain:* So this instructor goes up and starts writing down clearly bar names on the board right, and I am going huh, they are going to tell us make sure to stay out of those places. He starts talking to the guys. He starts pointing to these places . . . He starts pointing out all of these places and telling the guys. This is a good place to get laid. This is not a good place to get laid. I swear to God, I swear to God. And then to top matters off, I am going are you, I think I must have had a dropped jaw, a completely dropped jaw expression on my face. He looks at me and he goes Honey, if you have problems getting laid give me a call, in front of the whole class, the whole class.
>
> *Sergeant:* I remember sitting at the academy for some kind of training and I can remember sitting there with some old goat and he was talking about, "I can't believe you know we have all these women here. Some day we are going to have to deal with all this menopausal stuff."
>
> *Captain:* I got into a car with an FTO [Field training officer] that told me, because of my, you're a woman you don't belong on the f-ing job [spelling in original]. You won't make it through my f-ing FTO program. Go to work, I am not even here. That was my FTO program.
>
> *Sergeant:* I was with a training officer one night afternoon. It was Easter . . . and I was like, what are we going to find to eat today. He was like, let's go to the Y, and I am thinking, are they having some kind of family dinner? He was like, no, the Y, get it? I was like no, no. He's like, the Y between your legs.

Most officers' experiences, while not as extreme, defined and maintained policing as man's work and women as outsiders in the police culture. The forms of resistance varied significantly for each officer depending on type of department, height of the bureaucracy, reputation of token officers who came before and after the participants, and the participants' behaviors and attitude. Obstacles included a hostile work environment maintained through teasing, joking, isolation, physical assaults, denying "plum" work assignments and backup, or being forced to do "women's work." The length of the perceived resistance varied from 2 to 23 years and the pervasiveness varied from one "old goat" to the entire department. The saliency of being female was constantly displayed and maintained through a variety of mechanisms. Even after succeeding in institute and field training experiences, women were reminded of their token status through harassing comments, discrimination, and being "babysat":

> *Detective:* They told me, I can't get rid of you, but we don't have to let you take a good man's place.
>
> *Patrol:* The Chief had said he would never hire a female. So, he would show up off duty on my calls. At first this irritated me, I don't need a baby sitter. If I can't do the job like a man, I will be the first out.
>
> *Patrol:* One Sergeant in particular forbid me when he was working for me to work one of the rougher areas of the city.

Questions posed specifically to learn about resistance female officers faced yielded answers that defined how hegemonic masculinity in police cultures was maintained through isolation of female officers, confirming past research that suggested male officers work individually and collectively to keep women outside of the police subculture (Brown and Sargent, 1995; Franklin, 2005):

> *Captain:* Well we were all pretty much separated from each other. At that time they didn't want us taking vacations at the same time, didn't want us really on the same shift, they want us kind of spread out.
>
> *Chief:* I think at one time it was referred to as the Estrogen Mafia . . . I was head of investigations. I had a female sergeant and a female detective along me in there but it was called the Estrogen Mafia.
>
> *Chief:* The guys are extremely scared, that is probably not the word, very nervous, or it bothers them when, you know, they have a lot of you women.
>
> *Commander:* Of course then you know if there is more than 2 women in a conversation you know three makes a conspiracy.

Many women in the sample had not taken part in the promotion process and defined promotion as "too much of a pain in the ass" to attempt. Even women who had achieved promotion earlier in their police careers suggested little interest in existing promotional experiences. Commonly cited reasons for not attempting promotion included: being happy with the current shift they were working or current assignment; family and childcare conflicts; and not wanting to fight the seniority system. Others may have simply been avoiding the glass ceiling and inherent repercussions experienced in previous attempts with promotion:

> *Captain:* And I wasn't getting those positions even though I had a really good work record and I had really good relationships with my co-workers and peers. One of the guys that got a position that I didn't get was a complete idiot. So I started to say, you know that the writing is on the wall. They are not going to let me go into what I want to do, there is still some stigma going on here. There was still some issues with where they wanted females to go.
>
> *Chief:* He said "Rick's a boy", I said yeah I have been doing this job. He said I can't promote you over a man.
>
> *Sergeant:* When I got promoted, I remember a gruff old Master Sergeant down there that he felt one of the other guys should have gotten promoted here. So he was disappointed that one of the other people that he wanted to get promoted, didn't get promoted. And I can just remember he questioned every decision I wanted to make and you could just tell he just did not, he did not like I was in that position.
>
> *Commander:* Always a bridesmaid, never a bride. And I think part of that was my mindset and part of it was the department's mindset. I would settle for the second position instead of the top position. And I am not sure why I did that.

Coping Mechanisms

One way of coping with the stress and resistance to police work is for officers to accept segregation into the feminine, paperwork-dominated aspects of the job, dealing with children, female victims, or administrative duties (Garcia, 2003). In an organizational culture which prizes law enforcement above all other police roles, maintaining women in administrative and service positions is one way of defining "women's work" as inferior. Many women experienced the administrative push for these "girl jobs":

> *Detective:* But you know [it's women's work] when you get called back from your task force assignment because there is a sexual abuse case going on and the guys can't handle it themselves in investigations.
>
> *Detective:* When I was in patrol however I would get assault calls all the time. That would make me so mad because it would not be my

area and they would do whatever they could because no one wanted to work them so they work in she really wants to talk to a female.

Detective: I would get called along to go on raids, if they knew there would be children there. Once, we are getting ready for a raid and I looked around and I was the only person there that did not have kids. What was I doing there?

Patrol: I've never had a problem. I think because of my attitude. I don't try to compete with the men. I try to do the best I can do and I try to work with them. By compete I mean I don't butt heads with them.

Other female officers attempted to better themselves through education to be more competitive for potential and future police positions as a way of overcoming obstacles in police work. Almost all of the women interviewed had attained Bachelor's degrees, which may not be unusual for police recruits today, but was rare for police recruits of the 1980s (Carter et al., 1989; Herrington, 2002). Many women suggested their degrees helped them move up the proverbial slow administrative ladder, which may explain the high degree of administrators in the sample:

Colonel: If you think women have to be twice as good just to be considered average, then don't spend your energy complaining about it just spend your energy being twice as good.

Women who are drawn to policing as an occupation may not enter police work as a means of displaying feminine characteristics (Garcia, 2003; Rabe-Hemp, 2006). Many female officers interviewed had chosen dangerous, masculine assignments historically reserved for men exclusively. Female officers who pursued assignments as firearms instructors, undercover narcotics officers, terrorism task force officers, or patrol officers usually reported individual esteem and prestige were associated with these "plum" law enforcement assignments. Being female in these highly masculine and preferred police roles was especially arduous for women, but for many these assignments were the most rewarding. Owing to the threat of their intruding into the "male domain" of police work, these female officers also risked feeling particularly isolated by the departmental culture:

Patrol: I am an outcast.

Detective: Accepted professionally, feared by peers.

Commander: And so you are isolated on the job and now you, you go home and your social life and you are isolated even more, I think we lose a lot of people that way and especially women.

Many female officers suggested their families, significant others, and children helped them cope with stresses of police work and acted as a buffer to the resistance experienced in the police organization. Balancing the needs of children with police work was reportedly difficult for the few women in the sample who did have children (11 of 24 had children), especially in light of their department's lack of formal pregnancy policy. Only four of the 12 agencies represented in this study had formal policies according to the participants[2]:

Corporal: That's why I was Sergeant for a year and a half on midnights after I got done with my degree I said, "I don't want to be on midnights for the next 20 years. I have got a 5-year-old at home." So I gave up my Sergeant stripes and went back to Corporal and now I am on days. And that was a personal choice, what was best for my family.

Chief: We said okay we are considering having families. Could we be assigned to desk duty during our pregnancy and our job. And his answer was no, I would have to fire you if you couldn't go on and do your job.

Patrol: Opportunities that I might be interested in probably didn't fit in with my schedule or you know what is going on in my personal life [with children].

Most of the women interviewed who were married, were married to other police officers, which reportedly help them gain acceptance in police agencies and cope with the stresses of police work. In addition, being sexually unavailable to the male officers, either through marriage or sexual orientation, was often mentioned by the participants as easing the organizational tensions in police work (Hunt, 1990; Martin, 1980; Rohrbaugh, 1976). By comparison, women without families reported feeling acutely isolated by the police culture, many due to relationships with co-workers that had ended:

Detective: In the last three years, I have become pretty isolated from the department. I pick and choose my friends wisely because I have gotten burnt. I wish I would have learned . . . not to date police officers in general but I didn't learn that until too late I think.

Detective: I was married to someone here and all of our friends were here and when we got divorced, I really needed some outside influence.

Other women coped with the resistance faced in police work by connecting with others outside of policing. Women, who defined their work as separate from their social lives, reported having more friends outside of police work than inside and appeared to be less concerned with the values and norms of the police culture. These women worried less about being accepted and often reported they had chosen over time to create an identity outside the police role. Past literature on the police culture asserts male officers find solidarity through formal social and elite group membership which segregates them from society and also binds them together (Bittner, 1974; Dick and Jankowicz, 2001). This membership gives officers an identity and a community. Female officers who worked outside of this membership, found community in other places:

Patrol: I have a partner at home who is very understanding and I can talk to her.

Patrol: Sometimes I think you have to talk about if something really happens here that really stresses you out you need to talk to somebody. And it is nice to have a husband who is a police officer because he understands exactly what is going on.

Acceptance

The glorification of violence, so important to the masculine culture of policing, dominated female participants' discussions of acceptance (Connell, 1987; Messerschmidt, 1993; Martin, 1980). Female officers who felt accepted discussed three mechanisms they utilized to gain status:

1. through a violent show of force;
2. through achieving a rank that demanded respect; or
3. by being unique or different from male officers.

When asked about a moment when they knew they had achieved acceptance by their peers and colleagues, many retold "war stories" about pursuits, physical confrontations, and fights, in which the female officer proved her ability to maintain the power and authority necessary to the law enforcement function of policing:

Captain: My first fight, yeah. Before that no one would really speak to me . . . Once they saw I was going to jump in and help somebody and actually win the fight then it was, you know, oh we will slap you on the back but we are going to keep an eye on you, but you are okay as far as we are concerned.

Patrol: When the toughest guy in the team came up and slapped me on the back and said good job bad boy.

Patrol: Probably the most acceptance was when I was almost killed in a hotel room. I had to use my gun on a 17 year old. I think, you hate to say it that way, but I think officers felt, she can do the job. She did prove that she made it out of this hotel room alive and protected herself. So, I think you can get accepted in a way like that.

Other officers highlighted their rank, and the respect associated with rank in the police culture. These officers defined their success in terms of the norms and culture of policing, although their means of achieving high ranks may have been due to their unique gendered status in policing:

Captain: There's a certain amount of courtesy that is prefunctionary because I wear this particular rank.

Colonel: I had a lot of opportunities because being a woman in [names agency], back then I was very unique. So I think I had the opportunity to do a lot of things. Opportunities the others may not have had. In part because the department was showcasing me. Ya know, I looked good in a uniform, but sometimes you would get frustrated because they would not take you seriously.

Captain: I am a female in a male dominated organization. I'm not particularly tall. I'm not particularly short. I'm not particularly fat. I'm not particularly thin. So what sets me apart? My gender does.... You can make it an advantage if you use that visibility to your future. Now, there is risk involved. Okay. So, when I screw up people are going to remember well that you know so and so woman you know the one.

Despite the proverbial glass ceiling in regarding promotions in police occupations, over half of the tenured (those with ten or more years experience) participants interviewed were administrators with high ranks in their respective departments, suggesting female officers are overcoming resistance and thriving in police work. The hard-fought rank for female officers was believed to be the greatest potential for bringing about change in police work:

Colonel: You gotta play the game if you will, by the rules that are there to get into positions in order to change it.... It's growing, is starting to be established, you know women helping women, but the good old boy network has been around for so much longer.

Female officers who felt accepted in their respective departments, suggested it was a fluid, dynamic status that must be negotiated on a day-by-day basis to be maintained. Acceptance was not perceived as something that was achieved and then cached. Continuously negotiating their status in the organizational culture meant maintaining high levels of self-awareness and monitoring:

Captain: Acceptance on a police department when you are female is something that you never completely obtain but you, you kind of constantly strive towards to some extent.

Chief: I do feel that you have to constantly go back out and nurture that acceptance.

Sergeant: You must prove yourself everyday.

For these reasons, women reported maintaining higher expectations and more fear of cohorts of female officers currently being mentored and guided into policing. A misstep by another female officer in the agency could put all female officers at risk for chastisement:

Detective: When I first came on the department I was in the police academy and our eighth week on we had a shooting at our academy and it was a female officer who shot another, she shot another classmate. And that really shaped my values about what I thought about women in police work. And even though I didn't want to, I was so hard on the other females because I have such a high expectation of myself and I want that to be in other people also. So you know, when she shot him all the guys were now looking at every other woman in that academy class going, is she going to shoot me next.

Commander: I think the hardest obstacle is that you know when you are you know I will never forget this experience when I remember I met my female counterpart up there who was on the road. There were only two of us there at the time. And one of the male officers that I went to introduce myself and I stuck my hand out and said, I'm [officer's name] and she turned her back and walked away.

Almost all of the female officers interviewed reported the organizational culture of policing has become more accepting of women during their tenures. Despite early resistance, most women reported after ten or fewer years on, they felt accepted by their respective agencies. Examples of the acceptance women felt in the police culture included: less overt discrimination and harassment exemplified by women being able to achieve high ranks, and less visible signs of change, such as uniforms and vests designed specifically for women. Some female officers believed the discrimination was simply less overt, but still present:

Detective: Everything is not on a level playing field. It's not equal.

Chief: You know I have never been invited to go play golf.... I've never been invited on the fishing trips and I have never been invited to the ballgame.

Other female officers were more optimistic about the current state of policing and the opportunities for women, citing significant and long lasting changes in their departments:

Detective: A lot of that old timer, machismo, you know being herewith you know, when it was just a good old boys club, that's, those guys are pretty much gone now.

Captain: Lots of sexist [behavior] went on down there. It doesn't go on down there now though. I can tell you that right now. That crap is long gone.

Commander: I had Sergeants sit there and tell me walking into headquarters when I was a trooper, you don't belong here. It is a job for the white man. I don't believe that happens anymore.

Detective: I think now it is easy for women.... It just seems like everyone is much more accepting.

Sergeant: It's [the organizational culture] much better. Oh my goodness, compared to when I came on 22 years ago to now. I mean it is much more accepting to have a female out here as it was when I came on 22 years ago.

Detective: Women find it easier today entering, but she will still have to prove herself, but the men do too.

Discussion

The purpose of this study was to describe the resistance and obstacles women continued to face in policing, the coping mechanisms they utilized to overcome those obstacles, and their ability to integrate into an "all boys club." To that end, female police officers were interviewed to allow the women to articulate their own experiences. In examining female officers' status in policing, much has changed and much has stayed the same since the early research of the 1970s and 1980s.

Female officers are achieving acceptance and high ranks in police agencies, accomplishing new "firsts" every day, which suggests a significant mark away from the prior decade's experiences of resistance that were blatant, malicious, widespread, organized, and often life-threatening in nature (Martin, 1980; Remmington, 1981). In addition, women interviewed reported the evolution of the police culture in their respective departments included more "enlightened" men, and the weakened "boys club" attitude present in earlier years (Barker, 1999; Hunt, 1990; Martin, 1980; Young, 1991). Clearly the vestiges of the "old boys club," reported in the early police subculture literature, must be updated to incorporate recent studies that underscore the variety of value systems at work in the police culture (Barker, 1999; Herbert, 1998; Wood et al, 2004).

Female officers interviewed generally defined their acceptance into the police culture in one of three ways: through achieving rank, through completing some tough, manful act, or through being different or unique to the typical male police role. The first two solutions required female officers to adopt the values and norms of violence, aggression, danger, solidarity, and courageousness, typical of the masculine police culture. The last mechanism to gain acceptance was risky for women, because it encouraged them to highlight their token status in the department to exaggerate their skills as being polarized to their male officers, and to become highly visible in the process. Women who challenged the norms and values inherent in the police culture also assumed the greatest risk of resistance, harassment, and isolation (Garcia, 2003; Hunt, 1990). For many women, this risk appeared to be worthwhile since those who had attempted to market themselves as unique from their male counterparts, were most successful in gaining promotion.

Promotion was not reported by all officers as a sound definition of success. Many women repeated the same issues with promotion men have cited for decades: promotion meant loss of current shift or family and lifestyle changes; while others feared the risk of isolation and harassment associated with promotion. Kanter (1977) maintains when in relatively powerless and low-mobility situations within an organization, employees tend to lower their expectations and develop different work patterns from those employees who have more power and greater opportunities for advancement. This creates a self-fulfilling prophecy by which female officers have less confidence in their own work performance, avoiding high-profile positions, confirming peer and supervisor suspicion that female officers are less than adequate to the crime-fighting tasks. It was unclear if women had consciously adopted lower expectations for themselves in these roles or had decided the convenience of their current shift and position was worth the lower rank. A second explanation argues family and child-care issues play a larger role in women's decisions to forego promotional opportunities than they do for men, reflected in a preference to stay in their current assignment and job shift (Whetstone and Wilson, 1999). In American society, women are consistently the primary caregiver for children and the home, making employment with shift work especially difficult (Business Week, 2005). Police agencies have not kept pace with the private sector in implementing more "family-friendly" policies such as maternity/paternity leave, flex time, and in-house day care options.

Promotion also appears to be a divergent issue in the variety of agencies represented. This study included an inordinately high number of female police officers who had participated in the promotion experience at some stage of their tenures (12 of the 24 are administrators versus the national average of 10 percent; see the National Center for Women and Policing, 2001). One explanation for these findings is the unusually high-education levels of women in the sample. Virtually all of the policewomen interviewed hold Bachelors degrees and many masters and doctorates, which make them more competitive in promotion bids. Owing to the tenure requirement for inclusion, another possible explanation is policewomen chosen for the sample should be higher ranked simply due to the seniority system utilized for promotion in police agencies. Consistent with prior research, the height of the bureaucracy, the size of the agency, and the administrative leadership all significantly altered female officers' promotion experiences (Burligame and Baro, 2005; Weisheit et al, 2006). Despite past literature that argues women in large agencies have more promotional opportunities, the height of the bureaucratic structure may provide an alternate explanation. Women serving in agencies with more elaborate bureaucratic structures appeared to have more promotional opportunities (i.e., State police and campus police agencies versus municipal and county agencies), despite agency size. Agencies with higher percentages of sworn female officers and female administrative leadership also provided female officers with greater promotional opportunities (Martin, 1980).

Despite the preceding positive changes, obstacles and resistance to women integrating into police work are still many (Brown, 2000; Pogrebin et al., 2000). While the duration and the form varied widely, all female officers interviewed identified personal instances of sexual harassment, discrimination, or disrespect that impeded their successes and acceptance in police work. Surprisingly, most participants suggested the serious obstacles they faced, such as physical assault and obstacles to promotion, had occurred early in their careers and had desisted after several years experience; often to re-surface following agency or officer changes. Rather than being a level of achievement or an obstacle overcome, female officers reported acceptance by the police culture was constantly negotiated and re-negotiated throughout their careers, underscoring the need for additional research which analyzes the continuum of female officers' careers to determine the pattern of the onset and desistance of resistance. This process compels female officers to be in a constant state of self-awareness which has implications for officer health and well being. Perpetual behavior awareness and regulation may explain the higher than average rates of stress, burnout, and turnover reported by female officers (Hear, 1997; Kirschman, 1997; Peak, 2003).

In keeping with past literature, training academy and field experiences appear to be the first line of defense to stop "the female invasion" (Haarr, 2005; Herrington, 2002; Pike, 1985; Prokos and Padavic, 2002). Research suggests police employment standards are designed to keep out candidates who do not exhibit the macho traits of toughness, confidence, bravery, emotional detachment, and aggressiveness (Crank, 1998; Garcia, 2003; Martin, 1980; Prokos and Padavic 2002). Women, by their gender identification, often fail to meet these macho expectations associated with the image of a "crime-fighter." Field and academy training appear to be the first of many social institutions to inculcate and perpetuate women as "outsiders" in police work. Male solidarity and power is enhanced by sex separation (Franklin, 2005). This research extends past literature by demonstrating women's status as outsiders within the police culture is exacerbated by the antagonism female officers faced when attempting to form a work identity for support and morale.

Fear of a subculture of female officers or "estrogen mafia," suggested even within organizations with multiple subcultures premised by current research, women who actively compete for organizational resources such as assignments and promotion are faced with resistance (Herbert, 1998; Hunt, 1990). Officers maintained outside the main subculture of the department face additional isolation if also prohibited from finding community in an alternate group. This may also explain why many of the female officers interviewed reported finding identity and social support outside the police role, avoiding the separation, distrust, and opposition to the general population usually associated with the police subculture (Heidensohn, 1992; Herbert, 1998; Paoline, 2003). It is also plausible the cohort of medium- and small-town officers are usually more connected to their communities than those in urban locales and are less likely to view them in terms of opposition (Weisheit et al., 2006). Female officers struggle to integrate into the police subculture and limited adoption of the resultant us vs. them attitude, may explain female officers' astonishingly low rates of police force and misconduct and have important implications for police accountability (Lonsway et al., 2002; Rabe-Hemp, 2006).

Limitations

These findings should be evaluated in light of the study's methodological limitations. First, caution must be used in generalizing from these findings. The results are based on 24 interviews collected in a snowball sample from 12 agencies in the Midwest. Although snowball sampling is an efficient method for gaining access to difficult populations, this sample cannot be assumed to represent the larger population, which may limit the generalizability of the results. On the other hand, the rarity of non-urban research on female police officers makes these data quite rich. Additionally, the data are self-reported. While there is not obvious motivation for officers to be untruthful, self-reported data are, by its nature, subjective. Despite the limitations, the present findings are still compelling. Exploring women's accounts of their experiences in police work through in-depth interviews in multiple agencies provides a more comprehensive discussion of their integration into police work, than the often utilized survey in a very few agencies.

Implications

Beyond adding to the literature regarding female police officers' experiences, this research has practical and theoretical implications for police agencies and the communities they serve. Police departments nationwide are having difficulties maintaining recruitment standards and full rosters (Egan, 2005; Raymond et al., 2005). The more financially secure departments have even begun offering "signing bonuses," while others have simply been cutting back on services provided to their communities (Egan, 2005; Raymond et al., 2005). The current study can provide insight into the obstacles women face when joining police agencies and methods to alleviating these impediments, revealing an under-recruited population for police agencies nationwide.

Conclusion

In conclusion, the voice of this cohort of women and their legacy is heartening. These tenured officers have the potential through their administrative policies and statures to exact change in how police "do" justice and gender. Female officers' integration into policing has forced police officials to rethink traditional practices in selection, training, and performance evaluation, and has improved the quality of police services for the community (Miller, 1999; Sherman, 1975). The potential for enduring change in the field of policing is great as women continue to make great strides in achieving high ranks, breaking down assignments barriers, and ensuring just opportunities for future generations of policewomen.

Notes

1. The National Center of Policing reports municipal agencies typically have between 6 and 17 percent female officers. However, due to the tenure requirement for this sample, the population of possible officers was significantly smaller. This issue is exacerbated by the fact that the majority of the agencies in this sample had fewer than 75 officers, and less than 10 percent female, some as low as 2 percent. For these reasons, snowball sampling was the only way to access this under-utilized population. In addition, most previous research exploring female officers' experiences come from one agency, so comparisons based on department type and culture is difficult. This research fills that research gap as women from 12 diverse agencies were interviewed.

2. The governor of the state recently signed legislation requiring police departments to maintain a pregnancy policy.

References

Barker, J. (1999), Danger, Duty and Disillusionment: The Worldview of Los Angeles Police Officer, Waveland Press, Prospect Heights, IL.

Bartlett, H.W. and Rosenblum, A. (1977), Policewomen Effectiveness, Denver Civil Service Commission Denver, CO.

Belknap, J., and Shelley, J.K. (1992), "The new lone ranger policewomen on patrol", American Journal of Police, Vol. XII, pp. 47-76.

Bittner, E. (1970), The Functions of Police in Modern Society: A Review of the Background Factors. Current Practices, and Possible Role Models, National Institute of Mental Health, Bethesda, MD.

Bittner, E. (1974), "Esprit de corps and the code of secrecy", in Goldsmith, J. and Goldsmith, S.S. (Eds), The Police Community: Dimensions of an Occupational Subculture, Palisades Publishers, Pacific Palisades, CA pp. 237-46.

Bloch, P. and Anderson, D. (1974), Policewomen on Patrol, Police Foundation, Washington, DC.

Brown, J. (2000), "Occupational culture as a factor in the stress experiences of police officers", in Leishman, F., Loveday, B. and Savage, S. (Eds), Core Issues in Policing, Pearson Education, Harlow, pp. 249-63.

Brown, J. and Sargent. S. (199S), "Policewomen and firearms in British police service", Policing, Vol. 18, pp. 1-16.

Bureau of Justice Statistics (1996), Local Police Departments, 1993, Law Enforcement Management and Administrative Statistics, Washington, DC.

Bureau of Justice Statistics (2006), Local Police Departments, 2003, Law Enforcement Management and Administrative Statistics, Washington, DC

Burligame, D. and Baro, A.L. (2005), "Women's representation and status in law enforcement: does CALEA involvement make a difference", Criminal Justice Policy Review, Vol. 16 No. 4, pp. 391-411.

Business Week (2005), "Working mom tearing down office walls", Business Week, available at: www.businessweek.com/bwdaily/dnflash/may 2005/nf2005054_3294_db_083.htm (accessed August 28,2007).

California Highway Patrol (1976), Women Traffic Officer Project: Final Report California Highway Patrol, Sacramento, CA

Carter, D.L, Sapp, A. and Stephens, D.W. (1989), The State of Police Education, Police Executive Research Forum. Washington, DC.

Connell, R.W. (1987), Gender and Power, Stanford University Press, Palo Alto, CA

Connell, R.W. (1995), Masculinities, University of California Press, Berkeley, CA.

Crank, J.P. (1998), Understanding the Police Culture, Anderson, Cincinnati, OH.

Dick. P. and Jankowicz, D. (2001), "A social constructionist account of police culture and it influence on the representation and progression of female officers". Policing. Vol. 24, pp. 181-99.

Egan, T. (2005), "Police forces, their ranks thin, offer bonuses, bounties and more". New York Times, 28 December, p. Al.

Franklin, C. (200S), "Male peer support and the police culture: understanding the resistance and opposition of women in policing". Women & Criminal Justice, Vol. 16 No. 3, pp. 1-25.

Garcia, V. (2003), "Difference in the police department: women, policing, and doing gender. Journal of Contemporary Criminal Justice, Vol. 19 No. 3, pp. 330-44.

Greene, H. and del Carmen, A. (2002), "Female police officers in Texas: perceptions of colleagues and stress". Policing: An International Journal of Police Strategies & Management, Vol. 25 No. 2, pp. 385-98.

Haarr, R. (1997), "Patterns of interaction in a police bureau: race and gender barriers to integration". Justice Quarterly, Vol. 14, pp. 53-85.

Haarr, R. (2005), "Factors influencing the decision of police recruits to drop out of police work". Police Quarterly, Vol. 8 No. 4, pp. 431-53.

Heidensohn, F. (1992), Women in Control? The Role of Women in Law Enforcement, Oxford University Press, London.

Herbert S. (1998), "The police subculture reconsidered", Criminology, Vol. 36, pp. 343-69.

Herrington, P. (2002), "Advice to women beginning a career in policing". Women & Criminal Justice, Vol. 14 No. 1, pp. 1-13.

Hunt J. (1990), "The logic underlying police sexism". Women & Criminal Justice, Vol. 1 No. 2, pp. 3-30.

Kanter, R. (1977), Men and Women of the Corporation, Basic Books, New York, NY.

Kirschman, E. (1997), I Love a Cop: What Families Need to Know, The Guilford Press, New York. NY.

Koenig, E.J. (1978), "An overview of attitudes toward women in law enforcement". Public Administration Review, Vol. 38 No. 3, pp. 267-76.

Lonsway, K.A. (2003), "Tearing down the wall: problems with consistency, validity, and adverse impact of physical agility testing in police selection", Police Quarterly, Vol. 6 No. 3, pp. 237-77.

Lonsway, K.A, Wood. M. and Spillar, K (2002), "Officer gender and excessive force". Law & Order, Vol. 50 No. 12, pp. 60-6.

MacKinnon, C. (1989), Towards a Feminist Theory of the State, Harvard University Press, Cambridge, MA.

Martin, S. (1980), Breaking and Entering: Police Women on Patrol, University of California Press, Berkley, CA.

Martin, S. (1990), On the Move: The Status of Women in Policing, Police Foundation, Washington, DC.

Martin, S. (1991), "The effectiveness of affirmative action: the case of women in policing", Justice Quarterly, Vol. 8, pp. 489-504.

Martin, S. (1999), "Police force or police service? Gender and emotional labor", Annals of the American Academy of Political and Social Sciences, Vol. 561, pp. 111-26.

Menkelf-Meadow, C. (1989), "Exploring a research agenda of the feminization of the legal profession: theories of gender and social change". Law & Social Inquiry, Vol. 14, pp. 289-314.

Messerschmidt, J, (1993), Masculinities and Crime: Critique and Reconceptualization of Theory, Rowman and Littlefield, Lanham, MD.

Miller, S. (1999), Gender and Community Policing: Walking the Talk, Northeastern University Press, Boston, MA.

National Center for Women and Policing (2000), Status of Small and Rural Police Agencies, NCWP, Los Angeles, CA.

National Center for Women and Policing (2001), Equality Denied: The Status of Women in Policing. NCWP, Los Angeles, CA.

Padgett, D.K. (1998), Qualitative Methods in Social Work: Challenges and Rewards, Sage, Thousand Oaks, CA.

Paoline, E. III (2003), "Taking stock: toward a richer understanding of police culture", Journal of Criminal Justice, Vol. 31, pp. 199-214.

Peak, K.L (2003), Policing America: Methods, Issues and Challenges, 4th ed., Prentice-Hall, Upper Saddle River, NJ.

Pike, E.L. (1985), "Women in police academy training: some aspects of organizational response", in Moyer, I.L. (Ed.), The Changing Roles of Women in the Criminal Justice System: Offenders, Victims, and Professionals, Waveland Press, Prospect Heights, IL, pp. 250-70.

Pogrebin, M., Dodge, M. and Chatham, H. (2000), "Reflections of African-American women on their careers in urban policing: their experiences of racial and sexual discrimination". International Journal of the Sociology of Law, Vol. 28 No. 4, pp. 311-26.

Price, B.R. (1985), "Sexual integration in American law enforcement", in Hefferman, W.C. and Stroup, T. (Eds), Police Ethics, John Jay Press, New York, NY, pp. 205-14.

Prokos, A. and Padavic, I. (2002), "There outta be a law against bitches: masculinity lessons in police academy training". Gender, Work and Organization, Vol. 9 No. 4, pp. 439-59.

Rabe-Hemp, C. (2006), "The effect of officer gender in police-citizen interactions", paper presented at the Annual Meetings of the American Society of Criminology, Toronto.

Raymond, B., Hickman, L.J., Miller, L. and Wong J.S. (2005), Police Personnel Challenges after September 11: Anticipating Expanded Duties and a Changing Labor Pool, RAND Corporation, Washington, DC.

Remmington, P. (1981), Policing: The Occupation and Introduction of Female Officers, University Press of America, Washington, DC.

Remmington, P. (1983), "Women in police: integration or separation?" Qualitative Sociology, Vol. 6 No. 2. pp. 118-35.

Rohrbaugh, J.B. (1976), "Women in the workplace", in Skolnick, J. and Currie, E. (Eds), Crisis in American Institutions, Little, Brown, Boston, MA, pp. 215-30.

Searoborough, K, and Collins, P. (2002), Women in Public and Private Law Enforcement, Butterworth-Heinemann, Boston, MA.

Schulz, D.M. (2004), Breaking the Brass Ceiling: Women Police Chiefs and their Paths to the Top, Praeger. Westport. CT.

Sherman, L.J. (1975), "An evaluation of policewomen on patrol in a suburban police department", Journal of Police Science and Administration, Vol. 3 No. 4, pp. 434-8.

Sims, B., Scarborough, K. and Ahmad J. (2003), "The relationship between police officers attitudes toward women and perceptions of police models". Police Quarterly, Vol. 6 No. 3, pp. 278-97.

Tutty, LM, Rothery, MA. and Grinnell, RM (1996), Qualitative Research for Social Worker, Allyn & Bacon, Boston, MA.

United States Census Bureau (1999), Population, US Census Bureau, Washington. DC.

Waddington, P.A.J. (1999), "Police (canteen) subculture: an appreciation", British Journal of Criminology, Vol. 39, pp. 287-309.

Weisheit, RA, Falcone, D. and Wells, E.L (2006), Crime and Policing in Rural and Smalltown America, Waveland Press, Long Grove, IL.

Werner, P.D. and LaRussa, G.W. (1985), "Persistence and change in sex role stereotypes". Sex Roles, Vol. 12 pp. 1089-100.

West, C. and Zimmerman, DH. (1987), "Doing gender, Gender & Society, Vol. 1 No. 2. pp. 125-51.

Wexler, J. and Logan, D. (1983), "Sources of stress among police women police officers". Journal of Police Science and Administration, Vol. 11, pp. 46-53.

Whetstone, T.S. and Wilson D. (1999), "Dilemmas confronting female police officers promotional candidates: glass ceilings, disenfranchisement. or satisfaction?". International Journal of Police Science and Management, Vol. 2 No. 2, pp. 128-43.

Wood, R.L, Davis, M. and Rouse, A. (2004), "Diving into quicksand: program implementation and police subcultures", in Skogan, W. (Ed.), Community Policing: Can it Work?, Wadsworth/Thomson Learning, Belmont CA, pp. 136-62.

Young, M. (1991), An Inside Job: Policing and Police Culture in Britain, Clarendon Press, Oxford.

DISCUSSION QUESTIONS

1. What challenges do female police officers face within the context of their job that are related to their gender?
2. What coping mechanisms do policewomen employ in their struggle for acceptance and support within their job?
3. How does the masculine policing culture affect the experience of female police officers?

READING

Does gender affect the sentencing practices of judges? Scholars argue that gender provides a different lens through which women make decisions. In this study, Phyllis Coontz uses hypothetical case examples in a survey administered to trial court judges in the state of Pennsylvania. The hypothetical cases involve four different situations: (1) a homicide case where the defendant argued that she killed her former boyfriend in self-defense, (2) a personal injury case involving an automobile accident that left the plaintiff paralyzed from the waist down, (3) an alimony case following a divorce, and (4) a case of simple assault where two people got into a fight over a bet on a basketball game. Her results indicate that in some instances, male and female judges decide these cases differently, though not always in the ways in which many people would expect.

SOURCE: Coontz, P. (2000). Gender and judicial decisions: Do female judges decide cases differently than male judges? *Gender Issues, 18*(4), 59–73.

Gender and Judicial Decisions

Do Female Judges Decide Cases Differently Than Male Judges?

Phyllis Coontz

Introduction

When asked whether women judges decide cases differently by virtue of being female, Minnesota Supreme Court Justice Jeanne Coyne replied that, in her experience, "a wise old man and a wise old woman reach the same conclusion" (as quoted by Martin, 1993:126). Legal scholars and researchers continue to debate women's impact on the judiciary and whether women judges decide cases differently than men judges. Since the principle of equal treatment is central to our concept of justice, knowing whether women judges decide cases differently from men judges has profound implications for equal treatment under the law.

Feminist scholars argue that women do bring a different perspective to the law and it matters because they seek different outcomes from legal processes than do their male colleagues (Goldstein, 1992; MacKinnon, 1989; Menkel-Meadow, 1990; Scales-Trent, 1986). Conventional law, feminist scholars assert, embraces a male perspective that tends to emphasize separation, individual rights, and abstract rules. In contrast, a feminist jurisprudence examines how women's experience affects perspectives on the law itself and legal processes. The recent Supreme Court decision in Davis vs. Monroe County Board of Education illustrates this point. The day after the decision, the headline on the front page of the New York Times read: "A Conservative Voice, and Clearly a Woman's" (New York Times, May 26, 1999). The case in question dealt with sexual harassment in schools and in its decision the court was sharply divided by a 5 to 4 vote. The real news in the decision was that Justice Sandra Day O'Connor broke rank with her dissenting conservative colleagues (all male) and her own pattern of siding with states' sovereignty over Federal intervention. The dissenters maintained that the ruling would "teach little Johnny a perverse lesson in Federalism," to which Justice O'Connor replied that the decision "assures that little Mary may attend class." Justice O'Connor's decision may well have been influenced by her own experiences of differential treatment in the pursuit of a legal career.

Much of the foundation for the "different voice" argument comes from the work of Carol Gilligan who argued that women and men resolve moral problems differently (1982). Women tend to define themselves through their connections with others and perceive morality in terms of these interconnections. Men tend to define themselves in terms of individual achievement and autonomy, and stratify morality on a hierarchy that is based on levels of abstract rights and principles. In research drawing upon Gilligan's work, Suzanna Sherry developed a model of a feminine jurisprudence (one that emphasizes connection, subjectivity, and responsibility) and used it to examine the decision-making of Supreme Court Justice Sandra Day O'Connor. Sherry concluded that Justice O'Connor's opinions reflected concern for the rights of individuals as members of communities rather than as autonomous independent beings (1986:543-615).

Women in the Legal Profession

We know from statewide task force reports on gender bias in the profession from every state and a growing body of empirical evidence that women's experiences in the legal profession have been different. The most transparent example of this is the historical exclusion of women in the legal profession. Until the passage of Title VII, women's representation in the legal profession never exceeded 3 percent. Today women make up approximately 28 percent of those in the legal profession and comprise over half of all law students. According to Rosenberg et al. (1990), 77 percent of all women currently practicing law entered the profession after 1970.

Despite gains in the number of women in the profession, the empirical evidence shows that they lag behind their male colleagues on every indicator of success (Chambers, 1989; Coontz, 1995; Curran, 1986; Erlanger, 1980; Fuchs-Epstein, 1981, 1990; Hagan et al., 1988; Hagan, 1990; Kanter, 1977; and Liefland, 1986). Not only is there a gap in the career trajectories of women and men, but it also appears that the gap widens as more women enter the profession (Hagan, 1990:840). In fact the career gap is greater between women and men in the legal profession than it is between women and men in the general labor force.

Notwithstanding the fact that women were barred from the profession until 1869 and that the most significant gains for women have been made since the passage of Title VII, we see that the number of women on the bench has not kept pace with the number of women in the profession. This suggests that the barriers that have worked to exclude women from the legal profession may be greater for women trying to enter the judiciary. Looking at the historical record, we see that although women were appointed to limited jurisdiction courts in the nineteenth century, it was not until 1921 that the first woman judge was elected to a general jurisdiction court.[1] By 1940, only twenty-one states had women judges and by 1980 sixteen states had no female judges on courts of general jurisdiction. Today every state has women on state courts, but women make up a paltry 7.2 percent (873) of the 12,093 law-trained judges on full time state courts (State Justice Institute, 1996). If we examine women's presence on state-by-state basis, we see that some states have more women judges than others. For example, Alaska leads all fifty states with women accounting for 21.1 percent of their judges while Virginia falls at the bottom with women accounting for a meager 2.3 percent of their judges. There are still twenty-five states with no women on the State Courts of Last Resort (State Supreme Courts), twenty-two states have just one woman serving at this level, and three have women on their highest court for a national average of 8.3 percent. At the federal level, we see that women comprise only 7.4 percent of the 753 full-time federal judges, 9.5 percent of the 168 U.S. Circuit Court of Appeals justices and 6.9 percent of the 576 U.S. District Court judges. Of the nine United States Supreme Court justices, two are female. These statistics underscore a history of exclusion of women from the profession and the judiciary.

Research on Gender Differences in the Judiciary

The empirical evidence exploring whether women and men judges approach legal issues differently is both limited and inconclusive. In an examination of the voting behavior of state Supreme Court justices on women's issues, criminal rights, and economic liberties, Allen and Wall (1993, 1990) found little evidence to support the "different voice" claim. They did find that women justices are more likely to be the most outspoken advocate for women's issues on their court; to occupy positions at the extreme liberal or conservative wings of their court; and to voice extreme and dissenting views in criminal and economic cases. Allen and Wall suggest that the reason for this may be that women justices may have personality traits that are uncharacteristic of the general group of women from which they are drawn—they may be more like men supreme court justices in their thinking than they are like women judges in general. In a study of gender and voting behavior on U.S. courts of appeals, Davis et al. (1993) tested whether women voted differently from their male counterparts in ways that would reflect a tendency to emphasize rights of inclusion rather than individual rights (as suggested by Gilligan's work), and found only partial support for the notion of a "different voice" for women judges. In another study of state Supreme Court judges, Gryski et al. (1986) found that courts with women were more likely to produce pro-female decisions. However, in two studies of criminal court judges, Kritzer and Uhlman (1977) and Gruhl et al. (1981) found no significant differences between women and men judges, although male judges were more likely to give lesser sentences to female defendants. Walker and Barrow (1985) found that female and male judges differed on issues of personal liberties with male judges more likely to support a liberal position than female judges.

There are several possible explanations for disparities in the empirical research. First, as was pointed out earlier, the number of women judges generally is quite small. Most women judges are concentrated on state trial courts with only token representation on higher-level courts, yet most of the research examining gender differences has been done with judges on higher level courts. Higher level judges may not be representative of the groups from which they are drawn, thus it is possible that women judges on higher level courts may be unique from women judges on state courts and far less inclined to break new legal ground. A second possibility is that among the few studies which have been done, most have analyzed published cases and have focused on criminal courts. Family court may be more relevant than criminal court for examining gender differences since more women are affected by the decisions that are made in family court, e.g., divorce, alimony, custody, and child support, than they are by criminal court cases. And although a published case stands as a judge's decision, it is impossible to tell whether another judge deciding the same case would reach the same decision.

A third possibility is that as a group, judges are elites and gaining access to them is difficult. Elites are highly visible, but judges are more so because the judiciary is a public office and political. Like other elite groups, judges are less inclined than nonelites to participate in research examining their conduct. None of the systematic research to date has sampled judges directly about decision-making processes.[2] Another possible reason is methodological. Most studies have focussed on judicial voting behavior as recorded in case decisions and as noted earlier recorded cases do not provide information about how another judge reviewing the same case would decide. In addition, published case decisions are proxies for the decision-making process and do not contain information about what led to final decisions. It is not at all clear from published opinions what factors may have influenced the actual decision.

Finally, there is the inadequacy of the theoretical frameworks available for understanding and interpreting judicial behavior. While the tradition of the judiciary stresses impartiality, objectivity, and disinterestedness, there are no conceptual frameworks for reconciling the gap between legal theory and legal practice. The fact that women and minorities have historically been barred from the profession raises several relevant questions about their integration in the profession and how they reconcile their personal experiences within a theoretical tradition of impartiality and objectivity. Some feminist legal scholars have focused on the legal legacy of "coverture" and "protectionism" to show how the law has restricted women's rights and economic and social opportunities (Rhode, 1990; West, 1987; Binion, 1991; Bender, 1988). While these analyses have been very instrumental for challenging traditional definitions of equality, they still beg the question of how we might interpret judicial behavior. Other feminist scholars have drawn attention to patriarchy as the root of disenfranchisement for women and minorities and some scholars have argued that the law itself is patriarchal (Smart, 1989; Scales-Trent, 1989; Menkel-Minow, 1990; MacKinnon, 1974; Bartlett, 1990).

The question of women's impact on the judiciary is particularly salient and timely, especially as more and more women seek judicial positions. To advance our understanding of women's impact on the judiciary, clearly more research (and research that attempts to overcome the weaknesses cited earlier) is needed. The analyses reported here represent one such effort to add to our knowledge base about women's impact on the judiciary. The question I explore in this analysis is whether the gender of either the judge and/or the plaintiff/defendant makes a difference in the outcome of a case. The underlying logic for the study is that ideally nonlegal (personal and demographic) factors should play no role in judicial decisions. I hypothesize that when judges' decisions are based on legal factors there will be little variance among judges. Conversely if nonlegal factors influence decisions, we should expect variation in judge's decisions.

Methodology

The analyses presented here are derived from a statewide survey of trial court (and law-trained) judges in Pennsylvania. The survey was part of a larger study conducted by the Pennsylvania Conference of State Trial Judges and the Pennsylvania Bar Association Joint Task Force to Ensure Gender Fairness in the Courts. The aim of the study was to examine the effect that nonlegal (such as gender and race) factors have on judicial decisions. Four hypothetical cases consisting of a self-defense homicide case, a personal injury case, an alimony case, and an assault case were developed.[3] Two versions of each case were created in which the personal/demographic (nonlegal) characteristics of the plaintiff/defendant were mixed while the facts of the case were held constant. One version presented plaintiffs/defendants with one set of characteristics while the other version presented plaintiffs/defendants with altered characteristics. In both versions, the facts of the hypothetical situations were the same.

The versions were alternated (Version 1, Version 2, Version 1, Version 2, and so forth) and sent to the entire population of trial court judges in Pennsylvania ($N = 366$). A total of 183 surveys of Version 1 were sent out and 183 surveys of Version 2 were sent out. A total of 195 surveys were returned (a 53% response rate which we considered to be exceptionally good since judges are a difficult population to sample and no follow-up efforts were used)[4] and were roughly evenly divided between the two versions ($N = 96$ and $N = 99$). In addition to responses to the four hypothetical cases, demographic information on the judges was gathered.

Findings

All of the female law trained judges ($N = 28$) responded (representing 14% of all respondents) while 86 percent of the respondents were male ($N = 167$). For the entire sample, 92 percent were white, 6 percent were African American, and 2 percent classified themselves as "other." The average age was 53. Twenty-nine percent of the judges reported having been on the court for less than five years, 28 percent between six and ten years, 20 percent between eleven to fifteen years, and 23 percent for more than fifteen years. As a way of preserving anonymity, judges were not asked to identify the county in which they presided, however, they were asked to report the number of associate judges in their county. Thirty-two percent reported sitting in small counties (5 or less judges), twenty-five percent indicated they sat in medium-sized counties (6 to 10 judges), and forty-one percent reported sitting in larger counties (more than 10 judges). The overwhelming majority (89%) reported being married, 3 percent divorced, 3 percent separated, 2 percent not married and living alone, 1 percent not married, but living with someone, and 1 percent widowed. Sixty-one percent indicated that they did not have minor children living at home while 38 percent did.

Comparing female and male judges, the average responding female judge was white, 46 years old, married with minor children, and had been a judge for less than 6 years. The average responding male judge was white, 54 years old, married, with no minor children living at home, and had been a judge for more than 10 years.

T-tests[5] were used to examine the effect of nonlegal characteristics of either litigants or judges on judicial decisions. For the litigant analyses, Version 1 and Version 2 of the survey were compared. For the judge's analysis, female and male judges were grouped and compared. Below is a brief description of the 4 vignettes.

Situation 1—Self Defense/Homicide

This vignette presented an abuse/self-defense claim in a homicide case. The defendant had previously obtained a temporary protection from abuse order from her boyfriend. One Saturday, the boyfriend followed her to a bar and threatened to kill her. The boyfriend left then broke into the defendant's apartment, frightening the baby-sitter and the defendant's children. He told the baby-sitter to tell the defendant that he would be back later. When the defendant returned and learned of the incident, she reported it immediately to the police. Approximately five hours later, the boyfriend returned and assaulted the defendant. She then stabbed and killed the boyfriend. In Version 1, the defendant was named Patricia Lawson and in Version 2 the defendant was named Shanika Washington.[6] Judges were asked whether the defendant had grounds for self-defense, whether a judge and whether a jury would find the defendant guilty of homicide, and if so, what the length of sentence would be (assuming a standard sentencing range from 60 to 120 months per mandatory sentencing guidelines).

The only statistically significant difference (though small at the .10 confidence level) was for length of sentence and here we see that the range is wider for Patricia Lawson (46 to 180 months) than Shanika Washington (45 to 120 months). The only apparent explanation is that race (albeit implied) figured in the decision-making process. While not statistically significant, it is nevertheless interesting to note that on the question regarding whether a jury would find the defendant guilty, more judges reported Patricia Lawson would more likely be found guilty by a jury than Shanika Washington.

Table 11.1 shows the comparison between Version 1 and 2 for vignette one.

Table 11.1 Vignette One

	Yes	No	F-Statistic	S
Sufficient Grounds for Self-Defense?				
Combined Response		93%	5%	
Version 1	92%	4%		
Version 2	93%	4%		
Would Jury Find Guilty of Homicide?				
Combined Response		12%	75%	
Version 1	14%	79%	2.613	.108
Version 2	11%	71%		
Would Judge Find Guilty of Homicide?				
Combined Response		14%	76%	
Version 1	12%	82%		
Version 2	15%	17%		

Length of Sentence[7]	Mean	Std. Dev.	Max.	F Statistic	S
Combined Response	46 mos.	28.00	180 mos.		
Version 1	46 mos.	31.10	180 mos.		
Version 2	45 mos.	25.33	120 mos.	3.304	.071*

*Significant at the .10 level.

Situation 2—Personal Injury Case

This vignette presents a personal injury case where an automobile accident left the plaintiff paralyzed from the waist down and unable to return to work or perform routine domestic/family activities. The plaintiff sued seeking a higher award than what a jury had given (seeking $2.2 million—$650,000 for pain and suffering and $1,550,000 for loss of earnings and medical expenses). In this case the gender was varied—in Version 1 the plaintiff was named Bob Kramer and in Version 2 the plaintiff was named Pam Foster.

There was a $45,258.22 difference in the mean amount of lost earnings awards between vignettes. While this difference is sizable, t-test results indicate the difference is not statistically significant.

Situation 3—Alimony Case

In this vignette a husband files for divorce from his wife. In Version 1 the couple is named Sheila and Michael Arnold and in Version 2 the couple is named Jose and Maria Garcia. The wife is seeking alimony. She is 56 years old, overweight, and was forced to stop working the previous year in the family owned hardware store due to high blood pressure and hypertension. Judges were asked whether they would award alimony and if so, the amount and the length of the award. None of the differences between versions for this case were significant.

Situation 4—Simple Assault

Vignette 4 deals with a civil and criminal assault case. The parties involved in Version 1 the defendant was named Sally Squires while the victim was named Joe DeLuca and in Version 2 the defendant was named Joe Lanza and the victim was named Sally Squires. The defendant and victim had a physical altercation following a dispute over a basketball game bet. The defendant won the bet, but refused to pay.

The victim demanded payment and threw a beer in the defendant's face. The defendant got angry and retaliated by throwing a beer glass at the victim. A scuffle ensued wherein the defendant and victim sustained injury. Judges were asked whether they would find the defendant guilty of simple assault, if so the length of sentence they would impose (assuming a prior record score of 5 and noting that simple assault is a misdemeanor with an offense gravity score of 3 carrying a sentencing range of 6 to 12 months), whether they would award the victim damages in a civil case, and the amount of that award. Table 11.2 shows the results of the t-test analyses.

Here we see that gender appears to make a difference. Judges would be less likely to find Sally Squires guilty of assault than they would Joe Lanza. Similarly, judges would impose a higher civil damage award for a male defendant.

The earlier analyses show that nonlegal factors entered into the decisions of some judges in vignettes one and four. Does the gender of the judge make a difference in the outcome of a case? T-test results show that the judge's gender was a significant factor in at least one of the decision points in each of the four vignettes. Table 11.3 shows which decision points were significant in all four vignettes.

In the comparison between Version 1 and Version 2, i.e., examining litigant nonlegal factors, *t*-test analyses showed little variance. There were a total of fifteen decisions involved in the four vignettes, and in eleven of the fifteen, there were no statistically significant differences. It can be concluded that litigant factors were not involved in 73 percent of the responding judges' decisions.

However, when the gender of the judge is considered, we see that in almost half of the decisions (7 out of 15) statistically significant differences were found. Briefly these differences were:

- Female judges were more likely than male judges to find a male defendant guilty of assault than a female defendant, under the same factual circumstances.

Table 11.2 Vignette Four

	Yes	No	SD	F Statistic	S
Would Judge Find Defendant Guilty of Assault?					
Combined Response		39%	61%		
Version 1	30%	68%	16.54	.000*	
Version 2	46%	54%			
If Guilty, the Length of Sentence Imposed					
Combined Response		5 mos.			
Version 1	5 mos.				
Version 2	5 mos.				
Likelihood of Civil Damages Awarded					
Combined Response		13%	85%		
Version 1	10%	86%			
Version 2	15%	85%			
Amount of Civil Damage Award	**Mean**		**SD**	**F-Statistic**	**S**
Combined Response			$440.98		
Version 1	$256.97		$1,725		
Version 2	$598.46		$1,197	7.087	.008*

*Significant at the .10 level of confidence.

- Male judges were half as likely as female judges to find a female defendant claiming self-defense guilty of homicide than female judges, under the same factual circumstances.
- Male judges' personal injury awards were over twice as large as female judges, under the same factual circumstances.
- Male judges were almost half as likely as female judges to believe a jury would find a female defendant claiming self-defense guilty of homicide, under the same factual circumstances.
- Male judges were not unanimous, as were women judges, in awarding alimony, under the same factual circumstances.
- Male judges imposed shorter sentences for simple assault than female judges, under the same factual circumstances.
- Male judges were more likely to award civil damages for simple assault, under the same factual circumstances.
- Male judges awarded one-third the damages awarded by female judges for simple assault, under the same factual circumstances.

Table 11.3 Comparison of Female and Male Judges' Decisions

	Yes	No	F Statistic	S
Situation 1				
Would Jury Find Guilty?				
Female Judges	24%	76%	4.057	.046*
Male Judges	13%	86%		
Would Judge Find Guilty?				
Female Judges	27%	73%	4.927	.028*
Male Judges	13%	86%		
Situation 2				
Amount Awarded	**Mean**			
Female Judges	$107,150.64		3.287	.072*
Male Judges	$232,711.65			
Situation 3				
Award Wife Alimony	**Yes**	**No**		
Female Judges	100%	0%	3.952	.048*
Male Judges	97%	3%		
Situation 4				
Length of Sentence 0–6 Months				
Female Judges	61%		9.871	.001*
Male Judges	77%			
Likely Civil Award				
Female Judges	22%	78%	7.968	.005*
Male Judges	11%	89%		
Amount of Award				
Female Judges	$955		8.873	.003*
Male Judges	$353			

*Significant at the .10 level of confidence.

To rule out the possibility that experience on the job or judicial culture (the size of the county) contributed to the observed differences between male and female judges, an ANOVA adding these two factors was performed. The results were not significant, strongly suggesting that number of years on the bench and/or judicial culture did not neutralize the effect of gender. Given the statistical maxim that there is greater variance within groups than between groups, variance within the female judge group and within the male judge group was examined. While some variance within each group was present, it was not statistically significant. This suggests that as a group, female judges decide certain aspects of some cases differently than male judges.

Discussion

Two caveats need to be made before discussing the results. The first is that although the entire population of female judges participated in this study, there are still too few women judges relative to male judges. The lack of a comparable number of female judges is not a methodological problem, but rather a professional problem that must be addressed within the profession itself. While it is true that women have made inroads on the judiciary, their representation is smaller than the overall percentage of women in the profession and is minuscule relative to the number of men on the bench. It has been argued that the barriers women face when considering the judiciary are directly related to the level of gender bias in the profession, particularly since the gatekeepers to the judiciary are predominantly white males who are inclined to nominate and support candidates most like themselves in terms of background and experience—that is, nominate and support other males (Epstein, 1983; Kanter, 1977; and Martin and Jurik, 1996). This is a matter that most appropriately should be addressed by the profession.

The second caveat is that the analyses are based on self-reported decisions to a set of hypothetical cases. It is always possible that actual decisions would differ from reported decisions, and the only way to reconcile this possibility is to do research on the decision-making process of judges using a variety of research methods. Nevertheless, self-report data are valuable in their own right for what they tell us about the decision-making process which is not available from analyzing published cases. While hypothetical cases are only proxies for real cases, they provide valuable information not present in published cases. That is they contain a specific set of legal and nonlegal factors that may influence decision making. Another advantage hypothetical cases have over published cases is that they allow comparison among multiple judges on a single case. In published cases there is no way of knowing whether other judges would reach the same decision.

Nevertheless, the analyses reveal two interesting patterns. The first pattern relates to litigant characteristics and the finding that litigant characteristics did not influence judges' decisions as a whole. This is reassuring. One possible explanation for the congruence among judges has to do with the effectiveness of legal training in preparing judges for the bench. A legal education emphasizes legal principles and focuses on the factual matters of a case.

The second pattern, showing differences between female and male judges, is more difficult to explain because this pattern reflects differences in judges' interpretations and may reflect the relevance that judges' lived experiences have on their interpretations of factual information. The survey did not ask judges to identify the legal principles they applied, but rather presented judges with concrete situations and asked them to indicate how they would decide. The application of legal principles to concrete situations is always interpretive, and while legal education can train judges to focus on the factual matters of cases, the meaning that factual matters have for judges is an interpretive social process and this is precisely where a judge's experiences could have bearing on the decision making process. We, of course, expect judges to set aside personal viewpoints when deciding cases, yet beneath the robe of justice is an individual whose perceptions of the world have been influenced by their experiences in it.

The second pattern provides some support for Carol Gilligan's (1982) research, particularly with respect to her views about justice and fairness. Since gender, race, ethnicity, and socioeconomic status affect people's perceptions of fairness and justice, it is conceivable that female and male judges attach different weights to the factual aspects of identical situations. Until we explore what these differences are and acknowledge that they exist, we cannot begin to discuss what these differences may mean in terms of the behavior of judges.

Legal and judicial systems are historically male institutions—basically derived by men based on male behavioral standards. It was not until 1920 that women were permitted to practice law in every state. Admission barriers and restrictions kept women's numbers in law schools low until the mid 1970s. Despite these and other struggles, the presence of women throughout the legal profession has had an impact on the substantive and procedural aspects of law. There have been legal actions and lawsuits initiated because of women. Issues of date rape, spousal rape, sexual harassment, and other issues involving inequality have become part of the public debate because of women's perspectives.

Finally, this study adds to our understanding of impact that gender has on judicial decision-making in several important ways. Unlike previous research, this study sampled law-trained state trial judges. While the sample may not be representative of all male judges, the entire population of female judges responded, which means that the voices of female judges sitting on the bench in Pennsylvania are heard in these findings and that they are different from the voices of their male colleagues. This is important if for no other reason than the fact that state courts are where the largest concentration of women judges is found and where judicial decisions have the broadest reach. By comparing judges' decisions on the same case, this study provides information about how different judges with different backgrounds and experiences interpret the same factual matters. The study did not explore the rationale underlying judges' decisions, consequently we can only speculate about the reasons for specific decisions. Such questions are left for future research. At a minimum, however, these results underscore the value of legal training for achieving fairness in the court system and provide empirical evidence that female and male judges do bring different interpretations to the same situation.

Notes

1. Florence Allen became the first woman judge elected to a general jurisdiction court in the Court of Common Pleas in Cuyahoga County, Ohio.

2. Many state-wide task force efforts have gathered information from judges, but this has not been done systematically because of sampling problems.

3. Members of the Joint Task Force included judges and practicing attorneys. In developing the hypothetical cases, the Joint Task Force drew upon their broad experience on the bench and in the courtroom.

4. Members of the Joint Task Force believed that judges would be less likely to respond to the survey if they could be identified from the survey. In addition to being encouraged to participate in the research project by a cover letter from the Chief Trial Justice, it was decided that no follow-up attempts would be made.

5. A significance level of .10 was used for these analyses since the entire population was sampled.

6. The members of the Task Force participated in naming the plaintiff/defendant presented in each vignette.

7. Judges were asked to fill in actual length of sentences (which ranged from 0 to 200 months), these were then collapsed and coded into 5 distinct categories: 0–30 months = 1, 31 to 60 months = 2, 61 to 90 months = 3, 91 to 120 months = 4, and everything above 120 months = 5.

References

Allan, David and Wall, Diane. 1987. "The Behavior of Women State Supreme Court Justices: Are They Tokens or Outsiders?" *Justice System Journal* 12:232-244.

Bartlett, Katherine. 1987. "Jurisprudence and Gender." University of Chicago Law Review, 55: 1-72.

Bender, Leslie. 1988. "A Lawyer's Primer on Feminist Theory and Tort." *Journal of Legal Education*, 38: 3-37.

Binion, Gayle. 1993. "The Nature of Feminist Jurisprudence." *Judicature* 77(3).

Carp, Robert and Stidham, Ronald. 1990. Judicial Process in America. Washington, D.C.: Congressional Quarterly Press.

Chambers, David. 1989. "Accommodation and Satisfaction: Women and Men Lawyers and the Balance of Work and Family." *Law and Social Inquiry*

Coontz, Phyllis. 1995. "Gender Bias in the Legal Profession: Women 'See' It, Men Don't." *Women and Politics* 15:1-22.

Curran, Barbara. 1986. "American Lawyers in the 1980s: A Profession in Transition." *Law and Society Review* 20:19-52.

Davis, Sue. 1993. "Do Women Judges Speak in 'A Different Voice?' Carol Gilligan, Feminist Legal Theory and the Ninth Circuit." *Wisconsin Women's Legal Journal* December 8: 143-173.

Epstein, Cynthia. 1988. *Deceptive Distinctions: Sex, Gender, and the Social Order*. New Haven: Yale University Press.

Gilligan, Carol 1982. *In A Different Voice*. Cambridge: Harvard University Press.

Gruhl, J., Spohn C., and Welch, S. 1981. "Women as Policy Makers: The Case of Trial Judges" *American Journal of Political Science* 25:308-322.

Gryski, G. et al. 1986. "Models of State High Court Decision Making in Sex Discrimination Cases." *Journal of Politics* 48: 143-155.

Hagan, J., Huxter, M. and Parker, P. 1988. "Class Structure and Legal Practice: Inequality and Mobility Among Toronto Lawyers." *Law & Society Review* 22:9-55.

Hagan, John. 1990. "The Gender Stratification of Income Inequality Among Lawyers." *Social Forces* 68:836-855.

Kanter, Rosabeth. 1977. *Men and Women in Corporations*. New York: Basic Books.

Kritzer, H. and Uhlman, T. 1977. "Sisterhood in the Courtroom: Sex of Judge and Defendant in Criminal Case Disposition." *Social Science Quarterly* 14:77-88.

Liefland, Linda. 1986. "Career Patterns of Male and Female Lawyers." *Buffalo Law Review* 36:601–631.

MacKinnon, K. 1989. *Toward a Feminist Theory of the State*. Cambridge, MA: Harvard University Press.

Martin, Elaine 1993. "The Representative Role of Women Judges." *Judicature* 7:166-173.

Martin, Susan and Jurik, Nancy. 1995. *Doing Justice, Doing Gender*. Thousand Oaks, CA: Sage Publications.

Menkel-Minow, Carrie. 1990. *Making all the Difference: Inclusion, Exclusion, and American Law*. Cornell University Press.

New York Times, May 26, 1999.

Sherry, S. 1986. "Civic Virtue and the Feminine Voice in Constitutional Adjudication." *VA Law Review* 72.

Sacles-Trent, Judy. 1989. "Black Women and the Constitution: Finding Our Place, Asserting Our Rights." *Harvard Civil Rights-Civil Liberties Law Review* 24.

Smart, Carol. 1976. *Women, Crime and Criminology: A Feminist Critique*. London: Routledge & Kegan Paul.

State Justice Institute, National Center for State Courts, personal correspondence, October 10, 1995.

Walker, Thomas, and Barrow, Deborah. 1985. "The Diversification of the Federal Bench: Policy and Process Ramifications. *Judicial Policy* 47:596-617.

DISCUSSION QUESTIONS

1. What impact do legal and nonlegal factors have on judicial decision making?
2. Under which types of cases do female judges appear to engage in lenient decision-making practices?
3. Under what circumstances do female judges appear to hand down decisions that are more severe than their male counterparts?

READING

Working in the prison environment is a stressful and hazardous occupation. Gender can often add to the stress levels of correctional officers. This article explores the different types of stress that correctional officers face on the job and investigates the role of gender in how correctional officers handle stress in the workplace.

SOURCE: Griffin, M. L. (2006). Gender and stress: A comparative assessment of sources of stress among correctional officers. *Journal of Contemporary Criminal Justice, 22*(1), 4–25.

Gender and Stress

A Comparative Assessment of Sources of Stress Among Correctional Officers

Marie L. Griffin

In most workplace settings, stress is a daily occurrence. Studies spanning a myriad of occupations and organizations have identified the sources of job-related stress as well as the often negative behavioral outcomes associated with high levels of stress (Grossi, Keil, & Vito, 1996; Lambert, 2004; Lambert, Edwards, Camp, & Saylor, 2005; Moon & Maxwell, 2004; Robinson, Porporino, & Simourd, 1997; Slate, Vogel, & Johnson, 2001; Van Voorhis, Cullen, Link, & Wolfe, 1991; Walters, 1996). Similar to employees in other large organizations, correctional officers encounter multiple sources of stress. Prior studies have shown that relationships with coworkers and supervisors, as well as one's perception of the broader organization, were sources of work-related stress (Brodsky, 1982; Cullen, Link, Wolfe, & Frank, 1985; Lindquist & Whitehead, 1986; Poole & Regoli, 1980; Triplett, Mullings, & Scarborough, 1996). More recent studies have examined the pressures associated with work-family conflict among correctional officers (Lambert, Hogan, & Barton, 2003, 2004). In addition, the prison environment possesses rather unique work-related stressors, especially for those employees who are most directly responsible for maintaining safety and security within prison facilities (Cullen et al., 1985; Finn, 1999; Grossi & Berg, 1991). As noted by Armstrong and Griffin (2004), "few other institutions are charged with the central task of supervising and securing an unwilling and potentially violent population" (p. 577). Line officers, in particular, face stressful conditions as they negotiate often-routine work-related tasks within an environment of unpredictable inmate violence. For correctional officers, "safety is a particularly salient issue" (Triplett et al., 1996, p. 303).

The correctional context also is unique given its historical resistance to the integration of women and minorities and its highly masculinized orientation (Britton, 1997). Studies have shown that within correctional organizations, employee traits such as physical strength and a willingness to use force were viewed as essential job skills and were assumed to be masculine in nature (Belknap, 1991, 2001; Britton, 1997; Hemmens, Stohr, Schoeler, & Miller, 2002; Jurik, 1985; Owen, 1985; Pogrebin & Poole, 1997, 1998; Pollock-Byrne, 1986; Zimmer, 1986). Multiple studies have illustrated the nature and extent of resistance encountered by women from male officers who questioned the legitimacy of a woman's place (outside of clerical duties) in the prison organization (Belknap, 1991; Britton, 2003; Hemmens et al., 2002; Jurik, 1985, 1988; Owen, 1985; Pogrebin & Poole, 1997, 1998; Pollock-Byrne, 1986; Savicki, Cooley, & Gjesvold, 2003; Stohr, Mays, Beck, & Kelley, 1998; Zimmer, 1986). Specifically, studies have shown that female officers entering the prison workplace experienced harassment and discrimination at the hands of colleagues and supervisors (Belknap, 1991; Pogrebin & Poole, 1997, 1998; Savicki et al., 2003; Stohr et al., 1998). Such factors may very well function as a background stressor, exacerbating existing relationships between officers' perceptions of the organizational environment and increased levels of stress (Savicki et al., 2003).

This present study examined the mediating effect of gender on the influence of multiple work environment factors on correctional line officers' reported stress. Specifically, this study assessed the influence of officer perceptions of the broader organization (organizational support for the individual employee, and organizational support for equal treatment policies), of more immediate interactions (quality of supervision, coworker support, and management of inmates), and of work-family conflict on expressed levels of stress and how these sources of stress varied by gender. The importance of this study is threefold. First, high levels of stress among correctional officers have been well documented, as have strategies and programs to address this issue (U.S. Department of Justice, 2000). The U.S. Department of Labor's Occupational Outlook Handbook (U.S. Department of Labor, 2004/2005) highlighted the stressful and hazardous nature of the correctional environment. A number of studies have described varied negative outcomes at both the individual and organizational level, associated with increased stress among correctional officers, including both increased health problems and turnover rates and decreased levels of job satisfaction and organizational commitment (Adwell & Miller, 1985; Armstrong & Griffin, 2004; Blau, Light, & Chamlin, 1986; Cheek & Miller, 1983; Dowden & Tellier, 2004; Lambert, 2004; Robinson etal., 1997; Van Voorhis et al., 1991). Second, research examining stress among correctional officers has highlighted multiple themes regarding the sources of stress. This study attempted to further clarify this notion by identifying and assessing four broad sources of stress among correctional officers, including organizational policies and practices, interpersonal relationships, safety, and balance between work and personal life. Finally, although several studies compared stress levels by gender and include gender as a control variable when examining job-related stress or other organizational outcomes, few examined the way in which antecedents of stress varied by gender (Triplett, Mullings, & Scarborough, 1999). With an increasing number of women entering the field of corrections, it is essential to explore further the mediating effect of gender on the relationship between the work environment and work-related stress.

Literature Review

Generally, work-related stress has been characterized within the corrections literature as the individual response to strain within the work environment (Armstrong & Griffin, 2004; Triplett et al., 1999). Work-related stress has been measured in terms of "psychological work-related discomfort or anxiety," as well as physiological or health-related issues (Dowden & Tellier, 2004, p. 33). Studies have examined the relationship between stress and a variety of personal characteristics and work environment variables. The vast majority of studies have provided inconsistent evidence as to the influence of demographic variables such as age, race, and education level on work-related stress (Auerbach, Quick, & Pegg, 2003; Blau et al., 1986; Dowden & Tellier, 2004; Mitchell, MacKenzie, Styve, & Gover, 2000). Indeed, Brodsky (1982) argued that sources of stress "cannot be pinned to the characteristics of individual employees alone, but must also be seen as part and parcel of the structure and culture of the correctional institution and its role in American society" (p. 83). As such, studies have assessed the influence of multiple factors on stress levels among correctional officers, including officer perceptions of institutional policies and practices, officer interactions with coworkers and inmates, and officer concerns regarding the overlap between work and family responsibilities.

Gender and Stress in the Correctional Setting

For the purpose of this study, it is important to review more closely the extant literature on gender and stress in the correctional environment. Because of the gendered nature of the prison environment, studies have examined the way in which men and women experience differently the role of correctional officer, as well as the stresses associated with this role. Several of those studies found that female correctional officers reported significantly higher levels of job stress (Cullen et al., 1985; Mitchell et al., 2000; Zupan, 1986). In a study by Wright and Saylor (1991), for instance, female correctional officers expressed higher levels of work-related stress, but the magnitude of the difference was small. Other studies observed no significant relationship between gender and job stress (Griffin, 2001b; Grossi & Berg, 1991; Slate, 1993; Triplett et al., 1996; Walters, 1992). Still others noted that gender was not a strong predictor of stress among correctional officers (Fry & Glaser, 1987; Jurik & Halemba, 1984; Lancefield,

Lennings, & Thomas, 1997). In a similar manner, Carlson, Anson, and Thomas (2003) found that women correctional officers reported higher levels of personal accomplishment that they argued provided little support for the "stress-gender connection" (p. 284). In the only meta-analysis of work-related stress among correctional officers, Dowden and Teller (2004) failed to find evidence to suggest that gender was a significant predictor of stress among correctional officers. They argued, however, for continued research on gender and stress, suggesting that future studies that separate "both male and female correctional officers might facilitate the discovery of differential specific predictors of work stress" (p. 42).

Officer Perceptions of Organizational Support, Policies, and Practices

Relatively few studies examined the relationship between officers' perceptions of organizational support and work-related stress. Studies showed increased perceptions of organizational support by correctional officers to be a predictor of various work place behaviors, including decreased stress levels (Armstrong & Griffin, 2004; Auerbach et al., 2003; Cheek & Miller, 1983; Poole & Regoli, 1980). Such studies indicated that officers who expressed concern regarding the nature and extent of support from their organization or agency reported higher levels of stress. For example, Armstrong and Griffin (2004) found that the more an officer "perceived that their organization valued their work and input, the less stress they experienced in the workplace" (p. 587). Few studies, however, examined the manner in which the relationship between perceived organizational support and stress varied by gender.

Studies also examined the impact of employee perceptions of organizational support for affirmative action and other types of policies and practices that promote equal treatment in the workplace on employee attitudes and work outcomes (e.g., loyalty, job stress, and job satisfaction) (Griffin, Armstrong, & Hepburn, in press; Parker, Baltes, & Christiansen, 1997). These studies reflected the concern that White male employees, in particular, interpreted policies regarding equal treatment as unnecessary, unfair, and a promotion of the advancement of unqualified personnel, thus leading to increased negative work attitudes and behaviors. In general, these studies found that perceived organizational support for affirmative action policies was not negatively related to work attitudes among White men as had been hypothesized. When examining the correctional setting, Griffin et al. (in press) found that White male officers reported a greater perception of organizational support for equal treatment policies than any other group of officers (White males, White females, minority men, minority women), and this increased perception of organizational support for equal treatment significantly lowered job stress among all groups of correctional officers. These findings appear contrary to the frequently held belief that organizational support for fair and equal treatment would result in a backlash effect from those who would not appear to benefit from these efforts. Especially given the highly gendered nature of the prison organization, the current study sought to further explore officer perceptions of organizational support for equity policies and their role as a source of stress for correctional officers.

Officer Perceptions of Interpersonal Relationships

Research outside the correctional setting suggested that perceived job characteristics had a greater impact on employee attitudes than perceptions of the organization (Hall, Goodale, Robinowitz, & Morgan, 1978). Research examining officer perceptions of quality of supervision found support for the notion that more immediate social contacts have a greater influence on work outcomes—in this case, job satisfaction—than broader organizational concerns (Griffin, 2001a). Studies have highlighted the significance of interpersonal relationships as a source of stress among correctional officers (Armstrong & Griffin, 2004; Lindquist & Whitehead, 1986; Liou, 1995; Triplett et al., 1996; U.S. Department of Justice, 2000; Wright & Saylor, 1991). Lindquist and Whitehead (1986) found poor supervisory practices correlated with increased levels of reported stress among correctional officers, and officer trust in supervisors also was negatively correlated with work stress (Liou, 1995). Interestingly, in their study of South Korean correctional officers, Moon and Maxwell (2004) failed to find a significant relationship between negative interactions with supervisors and increased levels of stress, similar to Armstrong and Griffin (2004) in their assessment of both treatment and correctional staff.

Regarding gender and perceptions of supervision, findings from studies outside the correctional work setting suggested that the quality of supervision was more important to women's professional lives than men's professional lives (Ehly & Reimers, 1988). Studies also have shown that men and women interacted differently with their supervisors and that supervisory style preference varied by gender (Comer, Jolson, Dubinsky, & Yammarino, 1995; Goodyear, 1990; Miller, 1992). In addition, the superior and subordinate relationship was found to exert a significantly robust influence on women's occupational behaviors (Smith, Smits, & Hoy, 1998). Fewer studies, however, have examined the gendered nature of supervisory interaction and assessment among correctional officers. In her examination of race and gender in the prison setting, Britton (1997) found that compared to White men, White female correctional officers reported a more positive evaluation of supervisory personnel, leading to increased levels of job satisfaction. Wright and Saylor (1991), however, found gender unrelated to officers' perceptions of the quality of supervision.

Coworker support within the correctional setting also has been examined as an important factor influencing job stress among correctional officers. Studies suggested that employees who were able to employ coping mechanisms, such as coworker support, were more effective at addressing job-related stress (Triplett et al., 1996). Researchers found that low levels of social support in the workplace led to increased levels of stress among correctional officers (Armstrong & Griffin, 2004; Holahan & Moos, 1982). Cullen et al. (1985), however, found little evidence to support the relationship between coworker support and stress. At the same time, qualitative studies have illustrated at some length the significance of coworker support (or lack of) for women working in corrections (Belknap, 1991; Jurik, 1985; Owen, 1985; Pogrebin & Poole, 1997, 1998; Pollock-Byrne, 1986; Savicki et al., 2003; Stohr et al., 1998; Zimmer, 1986). For example, Pogrebin and Poole (1998) argued that male officer resistance to women working as detention officers exacerbated job stress among female officers. Women deputies lack of acceptance in this traditional male fraternity "denies them a critical organization coping mechanism to mitigate the impact of work-related stress" (Pogrebin & Poole, 1998, p. 114).

Officer Perceptions of Safety in the Work Environment

According to the U.S. Department of Justice (2000), "except for police officers, the number of workplace non-fatal violent incidents is higher per 1,000 employees for correctional officers than any other profession" (p. 14). Given the undoubtedly dangerous nature of the prison work environment, a significant body of both quantitative and qualitative research has examined officer safety concerns and the effect of such concerns on organizational outcomes, particularly job stress (Brodsky, 1982; Cullen et al., 1985; Finn, 1999; Guenther & Guenther, 1974; Lindquist & Whitehead, 1986; Lombardo, 1989; Triplett et al, 1996; Stalgaitis, Meyer, & Krisak, 1982; Veneziano, 1984; Wright & Saylor, 1991). Studies found safety concerns among officers to have significant effect on stress levels, with officers who perceived their working environment to be unsafe or dangerous expressing increased levels of job stress and health-related problems (Armstrong & Griffin, 2004; Brodsky, 1982; Triplett et al., 1996, 1999). Other research noted that although heightened safety concerns increased stress levels among officers, organizational matters proved to be more robust predictors of stress (Auerbach et al., 2003; Veneziano, 1984). In one of the few studies to assess gender differences in perceptions of safety concerns among corrections officers, Wright and Saylor (1991) found that women reported feeling less safe than men; yet compared to men, these female correctional officers perceived the work environment as safer. Similarly, a study of detention officers found that although no gender differences existed in reported levels of fear of victimization, such safety concerns significantly influenced job satisfaction among female but not male officers (Griffin, 2001a).

Officer Perceptions of Work and Personal Life Balance

A significant body of research outside of corrections has explored the connection between work and an employee's social or family life and the manner by which stress is transferred from one sphere to the other (Barnett, Marshall, & Sayer, 1992; Bolino, & Turnley, 2005; Greenhaus & Beutell, 1985). Researchers found a gender effect in the nature and extent of work-home conflict, suggesting that women tended to emphasize their role within the family in a way that men did not, with women more likely to take on primary responsibility of balancing work and family obligations (Bolino & Turnley, 2005; Gutek, Searle, & Klepa, 1991; Pleck, 1977; Parasuraman & Greenhaus, 1993). Studies have shown that women were more likely than men to adjust their jobs around their family responsibilities (Gerson, 1985; Reskin & Padavic, 1994) and were more likely than men to temporarily leave the workforce or reduce work hours to care for others (Gerstel & McGonagle, 1999; Sandberg, 1999; Sandberg & Cornfield, 2000). Bolino and Turnley (2005) found that when men and women engaged in high levels of individual initiative within their workplace (i.e., going above and beyond what is minimally expected by the organization), women, as compared to men, experience higher levels of work-family conflict.

Fewer studies examined the conflict between family and work as a source of stress among correctional officers (Triplett et al., 1999; Lambert et al., 2003, 2004). According to Lambert et al. (2004), "work-family conflict occurs when the two primary focuses in a person's life (i.e., work and family/social) are incompatible and, therefore, cause conflict that leads to spillover into both the work and familial/social milieus" (p. 148). This study found the issue of competing work and family obligations to be a stressor for all correctional staff and especially significant for correctional line officers. Lambert et al. (2004) also found a gender effect, with female correctional officers reporting higher levels of time-based work-on-family conflict than their male colleagues.[1] Triplett et al. (1999) found work-home conflict (measured as conflict between work and personal roles) to contribute significantly to increased stress levels at work among correctional officers. When examined by gender, however, this type of work-home conflict remained a significant predictor of work-related stress only for women. Triplett et al. (1999) argued that this finding suggests "that women are still dealing with conflict engendered by a culture that holds one set of role expectations for women at home and another set for them at work" (p. 384).

Research Problem

According to Savicki et al. (2003), "although women tend to perform as well in correctional employment setting as do men, women's sources of stress in the workplace are quite different than men's sources of stress" (p. 603). Given the highly masculinized work environment that characterizes the prison setting, some scholars have argued that female correctional officers experience stress as a result of their need to adapt to this predominately male work environment (Jurik, 1985). Studies have shown that in addition to the normal pressures associated with working as a correctional officer, female officers must navigate a complex work environment made more difficult because of negative expectations, informal stereotypes, denial of informal training and socialization, and resistance from inmates and fellow officers (Britton, 2003; Jurik, 1985; Hemmens et al., 2002; Huckabee, 1992; Pogrebin & Poole, 1998; Savicki et al., 2003; Wright & Saylor, 1991; Zimmer, 1986). This study examined multiple sources of stress and assessed their influence on male and female officers' reported levels of job stress.

Hypothesis 1: The mean level of stress reported by officers does not differ significantly by gender.

Hypothesis 2: Officer perceptions of the organizational environment have a greater influence on reported stress than individual characteristics of the officer.

Hypothesis 3: The sources of officer-reported stress differ by gender.

Hypothesis 4: Officers who report increased perceptions of quality of supervision, coworker support, work safety, and organizational support for employees, and decreased perceptions of work-family conflict report lower levels of stress.

Hypothesis 5: Organizational support for equal treatment policies has a differential effect on the reported level of stress by gender, with increased perceptions of organizational support for equal treatment policies increasing male officer stress while decreasing stress among female officers.

Research Method

Participants

The entire population of employees in all 10 adult state prisons in a southwestern state received a self-administered Quality of Work Life survey. The survey was part of the Department of Corrections' effort to assess employee perceptions of the prison work environment. Along with the questionnaire, employees received a cover letter explaining the purpose of the survey and an envelope, which facilitated the anonymous return of the questionnaire to the Departmental Research Unit. Employees were allowed to complete the survey while on duty. Of the 9,457 staff actively employed by the department, 5,540 individuals (58.6%) returned a usable questionnaire.[2] These analyses are based on data from respondents (N = 2,576) who classified themselves as correctional line officers (non treatment staff).

Descriptive statistics for these officers are presented by gender in Table 11.4. More than three quarters of the participants were men. Among the male officers, almost two thirds were White. On average, male correctional officers were approximately 35 years old, with more than 75% reporting high school as their highest educational level. More than 65% of the male officers reported having been employed with the organization 4 years or fewer at the time of the survey. In comparison, 67% of female officers were White, with an average age of 35 years. Approximately 75% of female officers were employed by this organization for 4 years or fewer. Almost 24% of female officers reported some post-high school education experience.

Measures

Dependent Variables

The dependent variable job stress assessed the extent to which staff perceived feelings of tension and anger as a result of their work. Job stress was operationalized by a 5-item Likert-type scale (.73) and was based on items previously used by Crank, Regoli, Hewitt, and Culbertson (1995).[3]

Independent Variables

Six scales were used to measure individual perceptions of work environment factors that prior research has suggested act as sources of stress for correctional officers: quality of supervision, coworker support, safety, organizational support for employee, organizational support for equal treatment policies, and work-family conflict. Scale items are listed in the appendix.

A 7-item scale measured the employee's perception of the quality of supervision (.82). The items for this scale were based on ones used previously by Armstrong and Griffin (2004), Griffin (2001a, 2001b), and Saylor (1984) and included items such as "I often receive feedback on my performance from my supervisor" and "My supervisor is knowledgeable and competent."

Coworker support was measured using a 6-item scale that assessed the extent to which individuals believed they had established a positive social support network within their work environment (.75). Based on scales previously used by Haines, Hurlbert, and Zimmer (1991), this scale included such statements

Table 11.4 Descriptive Statistics for Individual Demographic Characteristics, Work Environment Variables, and the Dependent Variable

	Male Correctional Officers (n = 1,940)			Female Correctional Officers (n = 636)		
Variable	**%**	**M**	**SD**	**%**	**M**	**SD**
Individual characteristics						
Age (years)		34.5	10.3		34.5	8.9
Race						
1 = White	60.5			67.0		
0 = Not White	38.7			32.2		
Missing	0.8			0.8		
Tenure (years employed)						
1 = 5 or more	34.7			25.5		
0 = 4 or less	65.2			74.4		
Missing	0.1			0.2		
Education						
0 = High school	73.8			74.5		
1 = More than high school	24.7			23.6		
Missing	1.4			1.9		

	No. of Items	**Range**	**M**	**SD**	**α**
Environmental scales					
Quality of supervision	7	7-35	24.0	5.1	.82
Coworker support	6	6-30	21.7	3.5	.75
Safety	4	4-20	13.2	2.8	.60
Organizational support for employee	3	3-15	7.8	2.8	.80
Organization support for equal treatment policies	7	7-35	25.0	4.5	.74
Work-family conflict	6	6-30	14.6	4.5	.72
Dependent variable scale					
Job stress	4	4-20	11.9	3.4	.73

as "The people I work with are helpful to me in getting my job done" and "Coworkers criticize my work to others."

A 4-item safety scale assessed officers' perceptions of workplace safety (.60). This scale was informed by those previously used by Hepburn and Crepin (1984), Logan (1993), and Griffin (2001a, 2001b) and includes such items as "I feel safe when working among the inmates" and "I received the kind of training I need to keep myself safe while working here."

A 3-item scale measured an individual's perception of organizational support (.80) and is based on a scale previously used by Eisenberger, Huntington, Hutchinson, and Sowa (1986; see also Armstrong & Griffin, 2004; Griffin, 1999, 2001a, 2001b, 2002). The scale includes items such as "The department takes pride in my accomplishments at work" and "Even if I did the best job possible, the department probably would not notice."

Organizational support for equal treatment policies was measured by a 7-item Likert-type scale (alpha = .74). This scale, previously used by Griffin, Armstrong, and Hepburn (in press), measured officer perceptions of organizational efforts to promote equal treatment using policies and procedures to target universal issues of cultural diversity, unfair treatment of women and minorities, and equal access to merit increases and promotion.

Correctional officers' perceptions of work-family conflict were measured using a 6-item Likert-type scale (alpha = .72). This locally constructed scale assessed the extent to which officers experienced conflict, or spillover, between work and domestic obligations and included items such as "Often when I am out with family or friends, I still am thinking about work" and "What happens at work affects my relationship with my spouse or partner."

In addition to these work environment scales, four individual-level variables were included in this analysis as control variables. These variables were race, age, tenure or length of employment at the institution, and education. Race was measured as a binary variable (1 = White; 0 = non-White).[4] Age of the employee was measured as a continuous variable. Tenure was measured as a dichotomous variable so that 0 represented 4 or fewer years of employment, and 1 represented 5 or more years with the institution. Finally, an officer's level of education was measured as a dichotomous variable so that 0 represented a high school degree and 1 represented any educational level beyond high school.

Findings

Table 11.5 presents results of a bivariate comparison between male and female correctional officers on the self-reported measures of the dependent variable, job stress, and the six work environment independent variables. A difference of means test indicated a significant difference in the mean level of some but not all variables of interest. Male and female officers reported similar levels of job stress, perceptions regarding the quality of supervision, and coworker support. A significant difference was found by gender in officers' mean level of perceptions of safety, organizational support for employees, organizational support for equal treatment policies, and work-family conflict. Compared to female officers, male officers reported significantly higher levels of both perceived organizational support for equal treatment policies and concern regarding work-family conflict. Female officers, on the other hand, reported a significantly greater concern about safety issues and a greater level of perceived organizational support for employees.

To assess the influence of multiple sources of job-related stress on correctional officers, and the extent to which they vary by gender, two multivariate ordinary least squares (OLS) regression models were used. The first model regressed individual level and work environment variables on reported levels of job-related stress for male officers. The second model assessed the influence of these same variables on reported levels of job-related stress among female officers. Standardized beta coefficients resulting from the two models are presented in Table 11.6.

Both OLS regression equations that incorporated individual level and work environment variables

Table 11.5 Means and Standard Deviations of Organizational Variables by Gender

	Male		Female		
Work Environment Variable	M	SD	M	SD	*t* **Value**
Job stress	14.83	4.51	14.05	4.39	−1.48
Quality of supervision	24.00	5.06	24.31	5.27	1.23
Coworker support	21.68	3.52	21.90	3.60	1.32
Safety	13.07	2.84	13.44	2.65	2.02*
Organizational support for employee	7.70	2.72	8.04	2.89	2.59**
Organizational support for equal treatment policies	25.30	4.89	24.23	4.35	−4.69**
Work-family conflict	14.83	4.51	14.05	4.39	−3.67**

*$p < .05$. **$p < .01$.

explained a sizeable proportion of the variance in both male and female correctional officers' reported job stress ($R^2 = .38, p < .01$; $R^2 = .41, p < .01$, respectively). Individual level variables had no significant effect on male officers' reported levels of job stress. Among female officers, however, both age ($\beta = -.10, p < .05$) and tenure ($\beta = .09, p < .05$) exerted significant influence on job stress. Women who were younger and women who had been employed longer expressed higher levels of job stress.

When examining the influence of work environment variables on job stress, male and female officers shared many of the same stressors and were influenced by these stressors in a similar manner. For male officers, all work environment variables included in the model significantly influenced job stress. Positive perceptions regarding the quality of supervision ($\beta = -.11, p < .01$), coworker support ($\beta = -.13, p < .01$), safety in the workplace ($\beta = -.20, p < .01$), and the organization's support for employees ($\beta = -.16, p < .01$) significantly reduced stress among male officers. Among male officers, increased perceptions that the organization supported policies promoting equal treatment significantly increased job stress ($\beta = .06, p < .05$). Finally, increased concerns regarding work-family conflict exerted the strongest effect on male officers' job-related stress ($\beta = .35, p < .01$).

Among female correctional officers, only four of the six work environment variables had a significant influence on job stress. Similar to male officers, increased positive perceptions of coworkers ($\beta = -.14, p < .01$), of safety in the workplace ($\beta = -.16, p < .01$), and of the organization's support for employees ($\beta = -.16, p < .01$) significantly reduced stress among female officers. Work-family conflict also proved to have the greatest effect on women's stress levels, with increased concerns regarding a lack of balance between work and personal obligations leading to increased levels of job stress ($\beta = .35, p < .01$). Unlike their male colleagues, stress levels of female correctional officers were not influenced significantly by perceptions of the quality of supervision or by the organization's support for equal treatment policies.

Table 11.6 Summary of Multiple Regression Equations by Gender

	Males		Females	
	β	SE	β	SE
Individual-level variables				
Age	.03	.01	–.10*	.02
Race	.02	.15	–.00	.27
Tenure	.03	.17	.09*	.31
Education	–.01	.16	–.01	.28
Work environment variables				
Quality of supervision	–.11**	.02	–.05	.03
Coworker support	–.13**	.02	–.14**	.04
Safety	–.20**	.03	–.16**	.06
Organizational support for employee	–.16**	.03	–.14**	.06
Organization support for equal treatment policies	.06*	.02	–.04	.03
Work-family conflict	.35**	.02	.35**	.03
Adjusted R^2	.38	.41		
F ratio	95.85**	34.17**		

*$p < .05$. **$p < .01$.

Discussion

Given the complex and unique context of the correctional setting, correctional officers encounter a variety of stressors. Prior studies have suggested a rather ambiguous relationship between gender and stress, yet it is critical to assess the differential influence of gender on stress in an organizational environment as gendered as the correctional workplace. Organizational studies outside the area of corrections have pointed to the continued existence of sex stereotyping in the workplace (Leuptow, Garovich-Szabo, & Lueptow, 2001), an especially salient factor for women engaging in nontraditional careers (Heilman, Block, & Lucas, 1992; Jackson, Esses, & Burris, 2001). Taking into consideration the multiple factors representing the structure and culture of the correctional environment as well as the relevant background stressors female correctional officers face in this nontraditional occupation, this study sought to explore perceptions of the organizational environment, the influence of such perceptions on reported levels of officer stress, and the way in which sources of stress varied by gender.

Supporting the first hypothesis, no significant difference was found in the mean level of job stress between male and female correctional officers. This adds to the existing literature that has failed to find a consistent difference between male and female officers' stress levels (Griffin, 2001b; Grossi & Berg, 1991; Slate, 1993; Triplett et al., 1996; Walters, 1992; Wright & Saylor, 1991). Differences by gender were found, however, in the mean level of several other variables of interest, including perceptions of safety, organizational support for employees, organizational support for equal treatment policies, and work-family conflict. Compared to male officers, female officers reported significantly greater concerns regarding safety and perceptions of organizational support for employees. Male officers reported significantly higher perceptions of organizational support for equal treatment policies and greater concerns regarding work-family conflict compared to their female colleagues. This past finding regarding work-family conflict is particularly interesting given the oft-cited belief that women face comparatively higher levels of work-family conflict because of family roles and gendered socialization (Camp, 1994; Lambert et al., 2005; Lambert et al., 2004; Samak, 2003).

Individual-Level Variables

This study also hypothesized that officer perceptions of the organizational environment would have a greater influence on reported stress than individual characteristics of the officer. In general, this second hypothesis was supported; personal characteristics of the correctional officers exerted little influence on levels of reported work stress, especially among male officers. A gender difference was found, however, with increased levels of stress reported by women who were younger and women who had been employed longer. Yet the magnitude of the effect was minimal when compared to work environment variables. This finding does suggest a need to examine more closely the relationship between age, tenure, and important organizational behaviors, especially among women.

Work Environment Variables

The findings regarding the influence of officers' perceptions of work environment variables on stress were noteworthy given their similarity by gender, thus providing only partial support for the third and fourth hypotheses. Stress levels among male and female officers were similarly influenced in manner and magnitude by coworker support, safety concerns, organizational support for the employee, and work-family conflict. This is comparable to studies in the policing literature that found similarity in workplace stressors by gender (Morash & Haarr, 1995). Concerns regarding the quality of supervision significantly influenced only male officers' reported level of stress, with male officers who reported a poor supervisory relationship expressing increased levels of stress. This finding adds some clarity to existing corrections literature and is important in that few studies have examined the gendered influence of supervision on stress. It remains uncertain, however, as to why quality of supervision appears to be related to stress for male but not female officers, especially given the qualitative data that have described resistance on the part of both coworkers and supervisors to women entering corrections. Pogrebin and Poole (1998), in particular, outlined the ways in which female officers "routinely experience derision, hostility, and exclusion from male supervisors" (p. 111) and examined in some detail the manner in which supervisors' paternalistic attitudes and exclusionary practices limited women's ability to fully participate in their work environment. In addition, prior studies assessing other workplace outcomes, such as job satisfaction, have found a significant relationship between gender and supervisory practices (Britton, 1997; Griffin, 2001a). Clearly, given the research outside the correctional organization that has noted the significant impact of gender on the superior and subordinate relationship (Smith et al., 1998), future research is needed to clarify this relationship.

Perhaps the most intriguing findings are those related to the effects of stress of officer concerns regarding the spillover between work and family obligations. As noted previously, there was a significant difference in the mean level of reported work-family conflict by gender, with male officers reporting higher levels of work-family conflict than female officers. In addition, work-family conflict proved to have the greatest impact on stress for both male and female officers. As a source of stress, studies have suggested that the conflict between work and family obligations will have a greater impact on women because of gender role expectations and socialization as was found by Triplett et al. (1999). The present study found, however, that correctional officers' concerns regarding the extent to which their work life negatively affects their private life acted as a major source of stress, regardless of gender.[5] In their discussion of multiple forms of work-family conflict and their influence on correctional personnel (staff and officers), Lambert et al. (2004) argued that it is the nature of the job that accounts for higher levels of conflict for those who work as line officers. They noted that

> officers rarely have input into scheduling, and they often are required to give far advance notice for a request of days off, even if the days off are for unexpected family events. Further, shift work is prescribed, without the ability to alter start or finish times. Moreover, budget cuts and understaffing add to the dilemma by forcing officers to work mandatory overtime. (p. 164)

It would appear that regardless of gender, correctional officers experience this inability to manage the spillover from work to their personal lives in a similar manner.

Finally, it was hypothesized that male and female correctional officers' perceptions that the organization promoted policies that supported equal treatment would have a differential effect on the reported level of stress by gender. In this study, male officers were more likely than female officers to believe that their organization supported policies promoting equitable treatment and diversity in the workplace. This supports prior research that found White men were more likely than women or minorities to believe that their organization supported affirmative action policies (Griffin et al., in press; Kossek & Zonia, 1993; Parker et al., 1997). Some support was found for the backlash theory in that male officers who believed that the organization supported policies that promoted equitable treatment based on race and gender reported increased levels of stress. Although few studies have explored this issue within the correctional context, Camp, Steiger, Wright, Saylor, and Gilman (1997) found that White male correctional officers tended to have exaggerated perceptions of promotional opportunities available to Black male correctional officers. Studies of noncorrectional organizations noted that differences in employees' perception of equal opportunity policies often reflect anticipated effects of such policies on an individual's self-interest (Summers, 1995; Veilleux & Tougas, 1989). That such perceptions act as a source of job-related stress for male officers suggests a concern among male officers that policies promoting equitable treatment may act to unfairly advantage particular groups of employees to the detriment of others. Such policies also may act as a stressor for male correctional officers in that officers may feel it necessary to assess more closely interactions or behaviors that may be deemed as inappropriate or violating organizational policies. This is a significant finding in that prison organizations historically have offered considerable resistance to the integration of women and minority officers (Britton, 1997) and where today, the majority of correctional officers and supervisory personnel remains male.

Limitations and Implications

The findings of this study and subsequent interpretations are subject to limitations. First, perhaps other measures of work-family conflict (see Lambert et al., 2004) and the inclusion of variables that more directly measure the nature and extent of family responsibilities (e.g., marital status, number of dependents) could assess more accurately the gendered nature of work-family conflict as a contributor to increased stress levels among correctional officers. Regarding officer perceptions of equitable treatment policies, these data do not include objective measures of variables that may be related to perceptions of fair treatment and equal opportunity within the organization. Information on the number of minority hires or recent promotions (see Camp et al., 1997) or a measure of racial diversity within the work group (see Camp, Saylor, & Wright, 2001) would add considerably to an understanding of the complex dynamics of individual perceptions of equitable treatment policies.

This research does contribute significantly to the extant literature on stress and corrections. Importantly, this study found more similarity than difference among male and female officers in terms of organizational stressors. Irrespective of gender, support from coworkers and the organization, safe work conditions, and the need for balance between work and personal life are significant contributors to stress among correctional officers. It is difficult, however, to reconcile these findings with prior studies that have noted highly divergent work experiences among men and women employed in prisons and jails. As such, these findings highlight the complexity of the correctional environment and the need to continue to explore the gendered work experiences of correctional officers. Indeed, the finding that male officers reported increased stress as a result of organizational policies aimed at promoting an equitable work environment suggests the need for organizations to revisit the manner in which rank and file employees perceive such policies. Organizational climate or culture can be difficult to change, particularly in correctional organizations where the social setting is one in which "ultramasculine sex-role stereotypes are promoted" (Lutze & Murphy, 1999, p. 727). As correctional administrators assess stress within their own organizations and implement strategies to promote positive work behaviors and increase retention, they must continue to attend to the gendered nature of this work environment.

Appendix: Work Environment Scale Items

Job Stress

When I'm at work, I often feel tense or uptight.

I usually feel that I am under a lot of pressure when I am at work.

There are a lot of things about my job that can make me pretty upset.

My work environment allows me to be attentive, yet relaxed and at ease.*

Quality of Supervision

I often receive feedback on my performance from my supervisor.

On my job, I know what my supervisor expects of me.

My supervisor asks my opinion when a work-related problem arises.

I am free to disagree with my supervisor.

I can tell my supervisor when things are wrong.

My supervisor respects my work.

My supervisor is knowledgeable and competent.

Coworker Support

I usually get along very well with my coworkers.

The people I work with are helpful to me in getting my job done.

I know I can get help from my coworkers when I need it.

Coworkers criticize my work to others.*

The people I work with are competent.

My coworkers respect my work and abilities.

Safety

I feel safe when working among the inmates.

I received the kind of training I need to keep myself safe while working here.

I have the equipment I need to keep staff from getting hurt by inmates.

I have the back-up support I need if things get rough.

Organizational Support for Employees

The department takes pride in my accomplishments at work.

Even if I did the best job possible, the department probably would not notice.*

The department values my input.

Perceived Organizational Support for Equal Treatment Policies

Policies and procedures here discourage sexual harassment.

Policies and procedures here promote cultural diversity among ADC employees in terms of race, ethnicity, and gender.

Women and minorities have equal access to merit recognition and promotion opportunities (the same as men and nonminorities).

People here are treated fairly regardless of their race or gender.

Policies and procedures here create standards so that decisions are fair and consistent.

Allegations of sexual harassment are taken seriously by management.

Anyone who treats women or minorities unfairly will receive meaningful sanctions.

Work-Family Conflict

Often, when I am out with family or friends, I still am thinking about work.

What happens at work affects my relationship with my spouse or partner.

Whenever I make plans to do something special with family or friends, it seems like I have to change my plans just to get the work done here.

It seems that there has been so much work to be done lately that I just don't have much time for a personal life.

I take work home or stay to finish up work, even if not specifically asked to do so.

I postpone my vacation or day off, in spite of personal inconvenience, to meet the needs of the organization.

*Reverse coded.

Notes

1. Lambert et al. (2004) suggested caution when interpreting this finding, noting that there were only 20 (of the approximately 136 correctional officers represented in the sample) female correctional officers in this study.

2. Some significant differences were found when comparing the demographic characteristics, job classification, or institutional location of those who responded to the survey with those who did not respond. Employees who were female, older, held longer tenure, held higher level (i.e., warden, deputy warden, associate deputy warden), or held security oriented positions were more likely to respond. Response rates varied between 43.9% and 75.3% at the 10 institutions, with an average response rate of 58.4%.

3. With the exception of the Equality in the Work Environment scale, all scales used in this study were informed by scales from previous studies. Some items were altered to incorporate the name of the local agency or to update wording. All items were measured on a five point scale (1 = strongly agree and 5 = strongly disagree) with recoding of items such that higher numeric values represented a higher level of the variable measured (e.g., higher level of stress, more positive attitude toward the quality of supervision, etc.). Confirmatory factor analyses and reliability analyses verified the integrity of all scales. Only items that loaded on a single factor were used to construct each scale.

4. The variable race was dichotomized. The following is a breakdown of race or ethnicity for all correctional officers: White (62.6%), Hispanic (24.6%), African American (5.7%), Other (3.2%), Native American (2.5%), Asian (1.4%).

5. The scale used in these analyses represented a more global measurement of work-family conflict that assessed the impact of increased work obligations on the extent and quality of time spent with family and friends. Triplett et al. (1999) used a measure of work-home conflict that reflected the role conflict that arises when an officer moves from the role of officer to that of parent, partner, and so forth.

References

Adwell, S. T., & Miller, L. E. (1985). Occuapational burnout. *Corrections Today, 47,* 70–72.

Armstrong, G. S., & Griffin, M. L. (2004). Does the job matter? Comparing correlates of stress among treatment and correctional staff in prisons. *Journal of Criminal Justice, 32,* 577–592.

Auerbach, S. M., Quick, B. G., & Pegg, R O. (2003). General job stress and job-specific stress in juvenile correctional officers. *Journal of Criminal Justice, 31,* 25–36.

Barnett, R., Marshall, N., & Sayer, A. (1992). Positive-spillover effects from job to home: A closer look. *Women and Health, 19,* 13–41.

Belknap, J. (1991). Women in conflict: An analysis of women correctional officers. *Women and Criminal Justice, 2,* 89–115.

Belknap, J. (2001). *The invisible woman: Gender, crime and justice* (2nd ed.). Belmont, CA: Wadsworth.

Blau, J. R., Light, S. C., & Chamlin, M. (1986). Individual and contextual effects on stress and job satisfaction. *Work and Occupations, 13,* 131–156.

Bolino, M. C., & Turnley, W. H. (2005). The personal costs of citizenship behavior: The relationship between individual initiative and role overload, job stress, and work-family conflict. *Journal of Applied Psychology, 90,* 740–748.

Britton, D. M. (1997). Perceptions of the work environment among correctional officers: Do race and sex matter? *Criminology, 35,* 85–105.

Britton, D. M. (2003). *At work in the iron cage: The prison as gendered organization.* New York: New York University Press.

Brodsky, C. M. (1982). Work stress in correctional institutions. *Journal of Prison and Jail Health, 2,* 74–102.

Camp, S. (1994). Assessing the effects of organizational commitment and job satisfaction on turnover: An event history approach. *Prison Journal, 74,* 279–305.

Camp, S. D., Saylor, W. G., & Wright, K. N. (2001). Racial diversity of correctional workers and inmates: Organizational commitment, teamwork, and workers' efficacy in prisons. *Justice Quarterly, 18,* 411–427.

Camp, S. D., Steiger, T. L., Wright, K. N., Saylor, W. G., & Oilman, E. (1997). Affirmative action and the "level playing field": Comparing perceptions of own and minority job advancement opportunities. *The Prison Journal, 77,* 313–334.

Carlson, J. R., Anson, R. H., & Thomas, G. (2003). Correctional officer burnout and stress: Does gender matter? *The Prison Journal, 83,* 277–288.

Cheek, F. E., & Miller, M. D. (1983). The experience of stress for correction officers: A double-bind theory of correctional stress. *Journal of Criminal Justice, 11,* 105–120.

Comer, L. B., Jolson, M. A., Dubinsky, A. J., & Yammarino, F. J. (1995). When the sales manager is a woman: An exploration into the relationship between salesperson's gender and their responses to leadership style. *Journal of Personal Selling and Sales Management, 15,* 17–32.

Crank, J., Regoli, R., Hewitt, J., & Culbertson, R. (1995). Institutional and organizational antecedents of role stress, work alienation and anomie among police executives. *Criminal Justice and Behavior, 22,* 152–171.

Cullen, F. T., Link, B. G., Wolfe, N. T, & Frank, J. (1985). The social dimensions of correctional stress. *Justice Quarterly, 2,* 505–533.

Dowden, C., & Tellier, C. (2004). Predicting work-related stress in correctional officers: A meta-analysis. *Journal of Criminal Justice, 32,* 31–47.

Ehly, S., & Reimers, T. M. (1988). Gender differences in job site perceptions. *Special Services in the Schools, 4,* 131–144.

Eisenberger, R., Huntington, R., Hutchinson, S., & Sowa, D. (1986). Perceived organizational support. *Journal of Applied Psychology, 71,* 500–507.

Finn, P. (1999). Correctional officer stress: A cause for concern and additional help. *Federal Probation, 62,* 65–74.

Fry, L. J., & Glaser, D. (1987). Gender differences in work adjustment of prison employees. *Journal of Offender Counseling, Services and Rehabilitation, 12,* 39–52.

Gerson, K. (1985). *Hard choices: How women decide about work, career, and motherhood.* Berkeley: University of California Press.

Gerstel, N., & McGonagle, K. (1999). Job leaves and the limits of the Family and Medical Leave Act. *Work and Occupations, 26,* 510–534.

Goodyear, R. K. (1990). Gender configurations in supervisory dyads: Their relation to supervisee influence strategies and to skill evaluations of the supervisee. *Clinical Supervisory, 8,* 67–79.

Greenhaus, I., & Beutell, M. (1985). Sources of conflict between work and family roles. *Academy of Management Review, 10,* 76–88.

Griffin, M. L. (1999). The influence of organizational climate on detention officers' readiness to use force in a county jail. *Criminal Justice Review, 24,* 1–26.

Griffin, M. L. (2001a). Job satisfaction among detention officers: Assessing the relative contribution of organizational climate variables. *Journal of Criminal Justice, 29,* 219–232.

Griffin, M. L. (2001b). *The use of force by detention officers.* New York: LFB Scholarly Publishing.

Griffin, M. L. (2002). The influence of professional orientation on detention officers' attitudes toward the use of force. *Criminal Justice and Behavior, 29,* 250–277.

Griffin, M. L., Armstrong, G. S., & Hepburn, I. R. (in press). Correctional officers' perceptions of equitable treatment in the "masculinized" prison environment. *Criminal Justice Review,* 30.

Grossi, E., & Berg, B. (1991). Stress and job dissatisfaction among correctional officers: An unexpected finding. *International Journal of Offender Therapy and Comparative Criminology, 35,* 73–81.

Grossi, E., Keil, T, & Vito, G. (1996). Surviving "the joint": Mitigating factors of correctional officer stress. *Journal of Crime and Justice, 19,* 103–120.

Guenther, A., & Guenther, M. (1974). Screws vs. thugs. *Society, 12,* 42–50.

Gutek, B., Searle, S., & Klepa, L. (1991). Rational versus gender role expectations for work-family conflict. *Journal of Applied Psychology, 76,* 560–568.

Haines, V., Hurlbert, I., & Zimmer, C. (1991). Occupational stress, social support and the buffer hypothesis. *Work and Occupations, 18,* 212–235.

Hall, D. T., Goodale, I. G., Rabinowitz, S., & Morgan, M. A. (1978). Effects of top-down departmental and job change upon perceived employee behavior and job attitudes: A natural field experiment. *Journal of Applied Psychology, 63,* 62–72.

Heilman, M. E., Block, C. I., & Lucas, I. A. (1992). Presumed incompetent? Stigmatization and affirmative action efforts. *Journal of Applied Psychology, 77,* 536–544.

Hemmens, C., Stohr, M., Schoeler, M., & Miller, B. (2002). One step up, two steps back: The progression of perceptions of women's work in prisons and jails. *Journal of Criminal Justice, 30,* 473–489.

Hepburn, J. R., & Crepin, A. E. (1984). Relationship strategies in a coercive institution: A study of dependence among prison guards. *Journal of Social and Personal Relationships, 1,* 139–157.

Holahan, C. J., & Moos, R. H. (1982). Social support and adjustment: Predictive benefits of social climate indices. *American Journal of Community Psychology, 10,* 403–415.

Huckabee, R. D. (1992). Stress in corrections: An overview of the issues. *Journal of Criminal Justice, 20,* 479–486.

Jackson, L. M., Esses, V. M., & Burris, C. T. (2001). Contemporary sexism and discrimination: The importance of respect for men and women. *Personality and Social Psychology Bulletin, 27,* 48–61.

Jurik, N. (1985). An officer and a lady: Organizational barriers to women working as correctional officers in men's prisons. *Social Problems, 32,* 375–388.

Jurik, N. (1988). Striking a balance: female correctional officers, gender role stereotypes, and male prisons. *Sociological Inquiry, 58*(3), 291–305.

Jurik, N. C, & Halemba, G. J. (1984). Gender, working conditions and the job satisfaction of women in a nontraditional occupation: Female correctional officers in men's prisons. *Sociological Quarterly, 25,* 551–556.

Kossek, E. E., & Zonia, S. C. (1993). Assessing diversity climate: A field study of reactions to employer efforts to promote diversity. *Journal of Organizational Behavior, 14,* 61–81.

Lambert, E. (2004). The impact of job characteristics on correctional staff members. *The Prison Journal, 84,* 208–227.

Lambert, E., Edwards, C, Camp, S., & Saylor, W. (2005). Here today, gone tomorrow, back again the next day: Antecedents of correctional absenteeism. *Journal of Criminal Justice, 33,* 165–175.

Lambert, E., Hogan, N., & Barton, S. (2003). The impact of work-family conflict on correctional staff job satisfaction. *American Journal of Criminal Justice, 27,* 35–51.

Lambert, E., Hogan, N., & Barton, S. (2004). The nature of work-family conflict among correctional staff: An exploratory examination. *Criminal Justice Review, 29,* 145–172.

Lancefield, K., Lennings, C. J., & Thomas, D. (1997). Management style and its effect on prison officers' stress. *International Journal of Stress Management, 4,* 205–219.

Lindquist, C. A., & Whitehead, J. T. (1986). Burnout, job stress, and job satisfaction among Southern correctional officers: Perception and causal factors. *Journal of Offender Counseling, Services, and Rehabilitation, 10,* 5–26.

Liou, K. T. (1995). Role stress and job stress among detention care workers. *Criminal Justice and Behavior, 22,* 425–436.

Logan, C. (1993). *Criminal justice performance measures for prisons.* Washington, DC: National Institute of Justice, U.S. Department of Justice.

Lombardo, L. X. (1989). *Guards imprisoned: Correctional officers at work.* New York: Elsevier.

Lueptow, L. B., Garovich-Szabo, L., & Lueptow, M. B. (2001). Social change and the persistence of sex typing: 1974-1997. *Social Forces, 80,* 1–36.

Lutze, E, & Murphy, D. (1999). Ultramasculine prison environments and inmates' adjustment: It's time to move beyond the "boys will be boys" paradigm. *Justice Quarterly, 16,* 709–733.

Miller, J. (1992). Gender and supervision: The legitimation of authority in relationship to task. *Sociological Perspectives, 35,* 137–162.

Mitchell, O., MacKenzie, D. L., Styve, G. J., & Gover, A. R. (2000). The impact of individual, organizational, and environmental attributes on voluntary turnover among juvenile correctional staff members. *Justice Quarterly, 17,* 333–358.

Moon, B., & Maxwell, S. (2004). The sources and consequences of corrections officers' stress: A South Korean example. *Journal of Criminal Justice, 32,* 359–370.

Morash, M., & Haarr, R. (1995). Gender, workplace problems, and stress in policing. *Justice Quarterly, 12,* 113–140.

Owen, B. (1985). Race and gender relations among prison workers. *Crime & Delinquency, 31,* 147–159.

Parasuraman, S., & Greenhaus, J. (1993). Personal portrait: The life-style of the woman manager. In E. A. Fagenson (Ed.), *Women in management: Trends, issues, and challenges in managerial diversity. Women and work: A research and policy series* (Vol. 4, pp. 186–211). Thousand Oaks, CA: Sage.

Parker, C. P., Baltes, B. B., & Christiansen, N. D. (1997). Support for affirmative action, justice perceptions, and work attitudes: A study of gender and racial-ethnic group differences. *Journal of Applied Psychology, 82,* 376–389.

Pleck, I. (1977). The work-family role system. *Social Problems, 24,* 417–427.

Pogrebin, M. R., & Poole, E. D. (1997). The sexualized work environment: A look at women jail officers. *The Prison Journal,* 77, 41–57.

Pogrebin, M. R., & Poole, E. D. (1998). Women deputies and jail work. *Journal of Contemporary Criminal Justice, 14,* 117–134.

Poole, E. D., & Regoli, R. M. (1980). Role stress, custody orientation, and disciplinary actions: A study of prison guards. *Criminology, 18,* 215–226.

Pollock-Byrne, J. M. (1986). *Sex and supervision: Guarding male and female inmates.* New York: Greenwood.

Reskin, B., & Padavic, I. (1994). *Women and men at work.* Thousand Oaks, CA: Pine Forge.

Robinson, D., Porporino, F. I., & Simourd, L. (1997). The influence of educational attainment on the attitudes and job performance of correctional officers. *Crime & Delinquency, 43,* 60–77.

Samak, Q. (2003). *Correctional officers of CSC and their working conditions: A questionnaire-based study.* Ottawa, Ontario, Canada: Prevention Group, Labour Relations Department, Correctional Service of Canada.

Sandberg, I. C. (1999). The effects of family obligations and workplace resources on men's and women's use of family leave. *Research in the Sociology of Work, 7,* 261–281.

Sandberg, I. C., & Cornfield, D. B. (2000). Returning to work: The impact of gender, family, and work on terminating a family or medical leave. In T. L. Parcel & D. B. Cornfield (Eds.), *Work & family: Research informing policy* (pp. 161–184). Thousand Oaks, CA: Sage.

Savicki, V., Cooley, E., & Gjvesvold, I. (2003). Harassment as a predictor of job burnout in correctional officers. *Criminal Justice and Behavior, 30,* 602–619.

Saylor, W. G. (1984). *Surveying prison environments.* Washington, DC: Federal Bureau of Prisons, Office of Research.

Slate, R. N. (1993). *Stress levels and thoughts of quitting of correctional personnel: Do perceptions of participatory management make a difference?* Unpublished doctoral dissertation, Claremont Graduate School, Claremont, CA.

Slate, R. N., Vogel, R. E., & Johnson, W. W. (2001). To quit or not to quit: Perceptions of participation in correction decision making and the impact of organizational stress. *Corrections Management Quarterly, 5,* 68–78.

Smith, P. L., Smits, S., & Hoy, F. (1998). Employee work attitudes: The subtle influence of gender. *Human Relations, 51,* 649–667.

Stalgaitis, S., Meyer, A., & Krisak, I. (1982). A social learning theory model for reduction of correctional officer stress. *Federal Probation, 3,* 33–41.

Stohr, M. K., Mays, L. G., Beck, A. C., & Kelley, T. (1998). Sexual harassment in women's jails. *Journal of Contemporary Criminal Justice, 14,* 135–155.

Summers, R. I. (1995). Attitudes toward different methods of affirmative action. *Journal of Applied Social Psychology, 25,* 1090–1104.

Triplett, R., Mullings, I. L., & Scarborough, K. E. (1996). Work-related stress and coping among correctional officers: Implications from organizational literature. *Journal of Criminal Justice, 24,* 291–308.

Triplett, R., Mullings, I. L., & Scarborough, K. E. (1999). Examining the effect of work-home conflict on work-related stress among correctional officers. *Journal of Criminal Justice, 27,* 371–384.

U.S. Department of Justice. (2000). *Addressing correctional officer stress programs and strategies: Issues and practices* (NCI No. 183474). Washington, DC: U.S. Government Printing Office.

U.S. Department of Labor, Bureau of Labor Statistics. (2004/2005). *Occupational outlook handbook 2004-2005.* Retrieved June 2005 from http://www.bls.gov/oco/ocosl56.htm

Van Voorhis, P., Cullen, F. T., Link, B. G., & Wolfe, N. T. (1991). The impact of race and gender on correctional officers' orientation to the integrated environment. *Journal of Research in Crime and Delinquency, 28,* 472–500.

Veilleux, F., & Tougas, F. (1989). Male acceptance of affirmative action programs for women: The results of altruistic or egoistic motives? *International Journal of Psychology, 24,* 485–496.

Veneziano, C. (1984). Occupational stress and the line correctional officer. *Southern Journal of Criminal Justice, 8,* 214–231.

Walters, S. (1992). Attitudinal and demographic differences between male and female corrections officers. *Journal of Offender Rehabilitation, 18,* 173–189.

Walters, S. (1996). The determinants of job satisfaction among Canadian and American correctional officers. *Journal of Crime and Justice, 19,* 145–158.

Wright, K., & Saylor, W. (1991). Male and female employees' perceptions of prison work: Is there a difference. *Justice Quarterly, 8,* 505–524.

Zimmer, L. E. (1986). *Women guarding men.* Chicago: University of Chicago Press.

Zupan, L. (1986). Gender related differences in correctional officers' perceptions and attitudes. *Journal of Criminal Justice, 14,* 349–361.

DISCUSSION QUESTIONS

1. What variables provide the greatest levels of stress for correctional officers?
2. How does this experience of stress differ between male and female correctional officers?

Glossary

Abusive incident: the second phase of Lenore Walker's cycle of violence, in which the batterer becomes highly abusive and engages in physical and/or sexual violence to control his victim.

Acquaintance rape: the victim knows the perpetrator; it usually accounts for the majority of rape and sexual assault cases.

Adler, Freda: her works were inspired by the emancipation of women that resulted from the effects of the second wave of feminism. Adler suggested that women's rates of violent crime would increase.

Affiliate: a person who has a direct or indirect role within the gang. An example: a female who married a gang member but is no longer active in criminal activity.

Age-of-consent campaign: was designed to protect young women from men who preyed on the innocence of girls by raising the age of sexual consent to 16 or 18 in all states by 1920.

Altruistic filicide: the mother believes that it is in the best interest of the child to be dead and that the mother is doing a good thing by killing the child.

***Barefield v. Leach* (1974):** set the standard through which the courts could measure whether women received a lower standard of treatment compared to men.

Battered woman syndrome: developed by Lenore Walker (1979); the most recognized explanation of the consequences of intimate partner abuse for victims; has been introduced as evidence to explain the actions of women on trial for killing their abusers.

Battered women's movement: Shelters and counseling programs were established throughout the United States to help women in need as a result of the feminist movements in the 1960s and 1970s. It led to systemic changes in how the police and courts handled cases of domestic violence.

Biological theories of crime: led to early psychological theories of female criminality and to the interpretation that the root for all causes of crime was related to a women's sexuality.

Blankenship, Betty: one of the first women in the United States to serve as a patrol officer; she worked for the Indianapolis Police Department in 1964. She helped set the stage for significant changes for the futures of policewomen.

Bootstrapping: modern-day practice of institutionalizing girls for status offenses.

Brothels: legal businesses where people could go to engage in sexual acts for money with prostitutes.

***Canterino v. Wilson* (1982):** held that males and females must all be treated equally unless there is a substantial reason that requires a distinction to be made.

Chivalry: women receive preferential treatment by the justice system.

Coffal, Liz: one of the first women in the United States to serve as a patrol officer for the Indianapolis Police Department in 1964. She helped set the stage for significant changes for the future of policewomen.

Community policing: a policing strategy that is based on the idea that the community is extremely important in achieving shared goals. It emphasizes community support from its members, which can help reduce crime and fear.

***Cooper v. Morin* (1980):** the equal protection clause prevents prison administrators from justifying the disparate treatment of women on the grounds that providing such services for women is inconvenient for the institution.

Custodial institutions: similar to male institutions, women were warehoused and little programming or treatment was offered to the inmates.

Cyberstalking: incidents that create fear in the lives of its victims; the anonymity that is involved creates opportunities for offenders to control, dominate, and manipulate their victims.

Cycle of victimization and offending: explains how young girls often run away from home in an attempt to escape from an abusive situation.

Cycle of violence: conceptualized by Lenore Walker in 1979 to help explain how perpetrators of intimate partner violence maintain control over their victims over time. The cycle is made up of three distinct time frames: tension building, abusive incident, and the honeymoon period.

Dating violence: violence that occurs between two people who are unmarried; teenagers are seen as the most at-risk population.

Decriminalization: allowed brothel owners to have legal sites of business, which created a tax base and revenue for the government, as well as enabled labor laws to be enacted to provide safe working conditions for prostitutes.

Discretionary arrest: police officers have the option to arrest or not arrest the offender based on their free choice within the context of their professional judgment.

Drug-facilitated rape: an unwanted sexual act following the deliberate intoxication of a victim.

Emancipation/Liberation theory: leads to an increased participation of women in criminal activities; however true this theory may be, it may not indicate that women are more compelled to actually engage in crime.

Emotional battering/abuse: one of the most damaging types of abuse. It robs the victim of self-esteem. The perpetrator seeks to control the victim by derogatory name calling; limiting the victim from being social, whether in the workplace, in school, or with family and friends; controlling all the finances; and limiting access to information regarding the money.

Evil woman hypothesis: women are punished not only for violating the law, but also for breaking the socialized norms of gender-role expectations.

Ex-intimate stalkers: perpetrators who stalk following the rejecting or severing of a relationship.

Extralegal factors: can include the type of attorney (private or public defender), which can significantly affect the likelihood of pretrial release for women.

Fear of victimization: a gendered experience where women experience higher rates of fear of crime compared to males. This idea is based on the distorted portrayal of the criminal justice system by the media.

Federal Interstate Stalking Law: passed in 1996 and amended in 2000, it restricted the use of mail or electronic communications for the purposes of stalking and harassments.

Felony Drug Provision of the Welfare Reform Act of 1996: bans women with a felony drug conviction from collecting welfare benefits and food stamps.

Femicide: the killing of women based on gender discrimination. The murders often involve sexual torture and body mutilation.

Feminist criminology: developed as a reaction against traditional criminology, which failed to address women and girls in research. It reflects several of the themes of gender roles and socialization that resulted from the second wave of feminism.

Feminist pathways perspective: provides some of the best understanding of how women find themselves stuck in a cycle that begins with victimization and leads to offending.

Feminist research methods: largely qualitative in nature and allows for emotions and values to be present as part of the research process.

Filicide: the homicide of children older than one year of age by their parent.

Formal processing: a petition is filed requesting a court hearing, which can initiate the designation of being labeled as a delinquent.

Fry, Elizabeth (1780–1845): a key figure in the crusade to improve the conditions of incarcerated women in the United Kingdom, and an inspiration for the American women's prison reform movement.

Gender-neutral laws: have significantly increased the number of women serving time in U.S. prisons, because they assume that men and women have an equal position in society, and women's needs are not considered when making sentencing decisions.

Gender-responsive programming: designed to address the unique needs of the female offender; includes six key principles that provide guidance for effective management: gender, environment, relationships, services and supervision, socioeconomic status, and community.

Gender-specific programming: must be able to address the wide variety of needs of the delinquent girl. Efforts by Congress have been made to allocate the resources necessary for analyzing, planning, and implementing these services.

Gendered justice (also referred to as injustice): the discrimination of individuals based on their gender. This idea is often seen in the criminal justice system where females' needs and unique experiences go unmet due to the fact that the theories of offending have come from the male perspective.

***Glover v. Johnson* (1979):** holds that the state must provide the same opportunities for education, rehabilitation, and vocational training for female offenders as those provided for male offenders.

Good ol' boy network: a social network of people who provide access and grant favors to each other. It is usually made up of elite White males, and they tend to exclude other members of their community.

***Griffin v. Michigan Department of Corrections* (1982):** held that inmates do not possess any rights to be protected against being viewed in stages of undress or naked by a correctional officer, regardless of gender.

***Grummett v. Rushen* (1984):** the pat-down search of a male inmate (including the groin area) does not violate one's fourth amendment protection against unreasonable search and seizure.

Harassment: acts which are indicative of stalking behaviors, but do not ignite feelings of fear in the victim.

Hollywood stalkers: the stalking offender experiences delusions of love toward the victim.

Honeymoon period: the third phase of Lenore Walker's cycle of violence; the batterer becomes very apologetic and is often loving and promises to change. This phase has a tendency to not last and can sometimes become nonexistent.

Honor-based violence: murders that are executed by a male family member and are a response to a belief that the woman has offended a family's honor and has brought shame to the family.

Hostile work environment: involves significant sexual advances or derogatory comments by the victim's superior or colleagues. Leads to a very uncomfortable work environment and fear because the victims feel that they have no other options.

Human trafficking: the exploitation and forced labor of individuals for the purposes of prostitution, domestic servitude, and other forms of involuntary servitude in agricultural and factory industries.

Immigrant victims of intimate partner abuse: do not report abuse or crimes nor do they seek out assistance for fear that they will be deported. In many cultural settings, these victims have come to accept the abuse.

Incapacitated rape: an unwanted sexual act that occurs after a victim voluntarily consumes drugs or alcohol.

Incarcerated mothers: has a significant effect on children. The geographical location of the prison and sentencing determine whether mothers can have ties with their children; in many cases, the children are either cared for by family members or they will be placed in foster care.

Independent female gangs: the absence of a male gang hierarchy; they tend to experience high levels of violence as a result of selling drugs and their interactions on the streets with other girls.

Infanticide: acts where a parent kills his or her child within the first year.

Informal processing: sanctions that the youth participates in on a voluntary basis, such as community service, victim restitution, mediation, and voluntary supervision.

Intimate partner abuse: any form of abuse between individuals who have or have had an intimate relationship.

Jail the offender vs. protect the victim: prioritization is given to the prosecution of offenders over the needs of the victims; however, these models are widely criticized due to their limitations and inability to deter individuals from participating in the offenses.

***Jordan v. Gardner* (1993):** the pat-down policy designed to control the introduction of contraband into the facility could be viewed as unconstitutional if conducted by male staff members against female inmates.

Jumped in (Walking the line): gang initiation process for girls in which they are subjected to assault by their fellow gang members.

Just world hypothesis: society has a need to believe that people deserve whatever comes to them, and this paradigm is linked to patterns of victim blaming.

Juvenile delinquency: young children and adolescents that are repeatedly committing crimes.

Juvenile Justice and Delinquency Prevention (JJDP) Act of 1974: provides funding for state and local governments to help decrease the number of juvenile delinquents and to help provide community and rehabilitative programs to offenders.

Karo-kari: literally "black woman/black man," this is a form of premeditated killing of both male and female adulterers that is a part of Pakistan's cultural tradition.

Las muertas de Juarez/Dead women of Juarez: the "club" of the women who are killed and tortured in Ciudad Juarez.

Learned helplessness: victims may believe that their batterer is exempt from laws against battering and that their status as a victim is unworthy.

Legal factors: have an impact on the decision-making process for both males and females in different ways. They vary from jurisdiction to jurisdiction and they can range from criminal history to offense severity.

Legalization: allowing brothels to register as businesses, authorities are also able to gain control of the industry by mandating public health and safety screenings for sex workers.

Liberation hypothesis: female participation in gangs is increasing, which is related to an increase in opportunities to participate in their traditionally male domains of crime.

Lifestyle theory: developed to explore the risks of victimization from personal crimes, and seeks to relate the patterns of one's everyday activities to the potential for victimization.

Lombroso, Cesare, and William Ferrero: the first criminologists to investigate the nature of the female offender, they worked together to publish *The Female Offender* in 1985.

Machista: also referred to as chauvinistic, a belief in which masculinity is praised and seen as superior.

Male inmate preference: women inmates are perceived as more demanding, defiant, and just hard to work with, so male and female officers would much rather work with male inmates.

Mandatory arrest: surfaced during the 1980s and 1990s with the intention to stop domestic violence by deterring offenders. It clarified the roles of police officers when dealing with domestic violence calls and removed the responsibility of arrest from the victim.

Maquiladores: assembly plants or factories that reside in another country, they are responsible for manufacturing and assembling parts and then shipping the products to the originating country.

Masked criminality of women: Otto Pollak's theory, which suggested that women gain power by deceiving men through sexual play-acting, faked sexual responses, and menstruation.

Mendelsohn's, Benjamin, six categories of victims: distinguished based on the responsibility of the victim and the degree to which victims have the power to make decisions that can alter their likelihood of victimization.

Minneapolis Domestic Violence Experiment: helped show the decrease in recidivism rates when an actual arrest was made in misdemeanor domestic violence incidents, in comparison to when a police officer just counseled the aggressor.

Mixed-gender gangs: gang comprising both male and female members. For females, it can often serve as a protective factor but it also places the girls at risk of rape and sexual assault by their male counterparts.

"Mommy track": a work-related path that allows mothers time off, giving them fewer opportunities for advancement in their careers.

National Crime Victimization Survey (NCVS): gathers additional data about crimes to help fill in the gap between reported and unreported crime, also known as the dark figure of crime.

National Incident Based Reporting System (NIBRS): an incident-based reporting system that was created to provide a more comprehensive understanding of crime in the United States.

Neonaticide: an act of homicide of an infant during the first 24 hours after the birth of the child.

Net-widening: alternatives are provided as a means to deter the offenders from the system, but in reality the number of offenders in the courts increases.

1992 Reauthorization of the Juvenile Justice and Delinquency Prevention (JJDP) Act: acknowledged the need to provide gender-specific services to address the unique needs of female offenders.

No-drop policies: developed in response to a victim's lack of participation in the prosecution of her batterer; these policies have led to the disempowering of victims.

Parens patriae: originated in the English Chancery Courts; gives the state custody of children in cases where the child had no parents, or the parents were deemed unfit care providers.

Parity: providing the same opportunities for programming and treatment to men and women.

Perpetrators of sexual harassment: divided into several different typologies: persistent harasser, malicious harasser, exploitative harasser, vulnerable harasser, power player, opportunist, mother figure, and comedian.

Physical battering/abuse: can sometimes lead to murder; it is very hard to research and measure because it tends to happen behind closed doors and the victims are usually too afraid to report it.

POLICEwomen vs. policeWOMEN: the gender identity debate, in which female officers believe that their gender does in fact affect the way in which they function as officers.

Pollak, Otto: wrote *The Criminality of Women* in 1961 to further explain his belief that crime data sources failed to reflect the true extent of female crime.

Postpartum syndrome: a severe case of mental illness; makes the mother unaware of her actions or unable to appreciate the wrongfulness of her behaviors.

Posttraumatic stress disorder (PTSD): having high levels of stress or anxiety that are related to traumatizing events.

Predatory stalkers: the stalking perpetrator engages in behaviors with the intent to harm the victim.

Prostitution recovery programs: programs that help recovering prostitutes by building their self-esteem and addressing the issues that have caused drug addictions and the overall lifestyle.

Pulling a train (sexed in): example of the gang initiation process, which requires sexual assault by multiple male members.

Quid pro quo: literally, "this for that." A claim of sexual harassment in the housing arena; can include cases whereby a tenant is evicted for refusing the sexual advances of a landlord or property manager.

Rape: the act of forced sexual intercourse, which involves penile-vaginal penetration; the laws regarding this act tend to vary from state to state.

Rape myths: false beliefs that are seen as justifiable causes for sexual aggression against women. The acceptance by society is a contributing factor in the practice of victim blaming.

Reentry: the transition from an incarcerated setting to the community; usually involves meetings with parole officers who provide referrals to receive treatment; unsuccessful reentry often leads to recidivism.

Reformatory: a new concept that saw incarceration as an institution designed with the intent to rehabilitate women from their immoral ways.

Resiliency: also known as protective factors, can enable female victims and female offenders to succeed in light of risk factors for delinquency.

Restraining orders: available in every jurisdiction, designed to provide the victim with the opportunity to separate from the batterer and prohibit the batterer from contacting the victim.

Risk factors for female delinquency: a poor family relationship, a history of abuse, poor school performance, negative peer relationships, and issues with substance abuse.

Romantic stalkers: the perpetrators stalk in an effort to establish a romantic relationship.

Routine activities theory: created to discuss the risk of victimization in property crimes. It suggests that the likelihood of a criminal act or the likelihood of victimization occurs when you have an offender, a potential victim, and the absence of a guardian that would deter said offender from making contact with the victim.

Safe Haven Law: policies that permit parents to surrender their newborn children, without question or judgment, to safe locations such as hospitals, police, and fire stations. The children are then placed into the custody of child protective services and placed up for adoption.

Same-sex sexual assault: oftentimes male-on-male assault, due to the limited research on woman-on-woman sexual violence.

Secondary victimization: the idea that victims become more traumatized after the primary victimization. It can stem from victim blaming or from the process of collecting evidence (physical or testimonial).

Sentencing guidelines: created in conjunction with the Sentencing Reform Act of 1984, the only factors to be considered in imposing a sentence were offense committed, the presence of aggravating or mitigating circumstances, and the criminal history of the offender.

Sexual assault: used to identify forms of sexual victimization that are not included under the definition of rape. Laws expanded the definitions to include sodomy, forced oral copulation, and unwanted touching of a sexual nature.

Sexual harassment: a social problem that can be presented in many ways, such as verbal comment with a sexual reference, inappropriate physical touching, derogatory looks or gestures, sharing of sexually based jokes or stories, displaying of sexually graphic pictures or writings, or even comments about a person's clothing. It tends to happen in the workplace, education, housing, and public spaces.

Sexual slavery: the use of lies, deceit, and kidnapping tactics to force people to participate in sexual acts.

Shadow of sexual assault: women experience a greater fear of crime because they believe that any crime could ultimately become sexually based, and thus links to causes of rape or sexual assault.

Simon, Rita: hypothesized that women would make up a greater proportion of property crimes as a result of their "liberation" from traditional gender roles and restrictions.

Social injury hypothesis: girls in gangs experience higher levels of risk, danger, and injury compared to their male counterparts.

Spousal rape: involves emotional coercion or physical force against a spouse to achieve nonconsensual sexual intercourse; it can often lead to domestic violence.

Stalking: a course of conduct directed at a specific person that would cause a reasonable person to feel fear.

Stalking and intimate partner abuse: very common for the victims of domestic violence. The degrees to which victims are stalked vary depending on the levels of emotional or physical abuse that they have experienced.

Status offenses: acts that are only illegal if committed by juveniles, such as underage consumption of alcohol, running away from home, truancy, and curfew violations.

Stranger rape: the perpetrator is unknown to the victim and is usually associated with a lack of safety, such as walking home at night or not locking the doors.

Street prostitution: an illegal form of prostitution that takes place in public places.

Tension building: the first phase of Lenore Walker's cycle of violence; the batterer increases control over a victim. It is usually characterized by poor communication skills between the partners.

***Todaro v. Ward* (1977):** mandated reforms to provide health care in prisons; failure to do so was a violation of the Eighth Amendment protection against cruel and unusual punishment.

Trafficking Victims Protection Act of 2000: designed to punish traffickers, protect victims, and facilitate prevention efforts in the community to fight against human trafficking.

Uniform Crime Reports (UCR): crime statistics that are compiled and then published in an annual report by the Federal Bureau of Investigation (FBI).

Unintended consequences: very significant in the war on drugs; the loss of a mother deeply affects the children, usually bringing in social services intervention.

U.S. Equal Employment Opportunity Commission (EEOC): represents cases of workplace sexual harassment and discrimination.

Vice Queens: a female auxiliary gang that was an affiliate of the Vice Lords (a black conflict gang in Chicago during the 1960s); they lacked cohesion within the group and had no formal leadership structure.

Victim blaming: shifting the blame of rape from the offender to the victim; by doing so the confrontation of the realities of rape and sexual assault in society are avoided.

Violence Against Women Act (VAWA): passed in 1994; provided funding for battered women's shelters and outreach education, funding for domestic violence training for police and court personnel, and the opportunity for victims to sue for civil damages as a result of violent acts perpetrated against them.

von Hentig's, Hans, 13 categories of victims: focuses on how personal factors such as biological, social, and psychological characteristics influence risk factors for victimization.

Welfare Reform Bill of 1996: signed by President Bill Clinton in 1996, imposed time limits on the aid that women can receive and denies services and resources for women with a criminal record.

Wells, Alice Stebbins: the first police officer hired in the United States by the Los Angeles Police Department in 1920, she advocated for the protection of children and women, especially when it came to sexual education.

"Westernized": to participate in modern-day activities such as wearing jeans, listening to music, and developing friendships.

Work-family balance: a challenge that affects women in every profession, especially in careers that require the work to be done at home or on the weekends. The stress and issues regarding this often put females on the mommy track, or in some cases the females decide to not start a family.

Workplace sexual harassment: linked with job-related stress, decreased personal mental health, poor job performance, and general feelings of job dissatisfaction.

Wraparound services: holistic and culturally sensitive plans for each woman that draw on a coordinated range of services within her community, such as public and mental health systems, addiction recovery, welfare, emergency shelter organizations, and educational and vocational services.

Zero-tolerance policies: used as a deterrent. They prevent women with felony drug convictions from receiving assistance such as public housing and federal financial aid to attend college.

Credits and Sources

Section I. Women and Crime: An Introduction

Photo 1.1: The National Museum of American History, Smithsonian Institution

Photo 1.2: Stacy L. Mallicoat. Printed with permission.

Section III. Women and Victimization: Rape and Sexual Assault

Photo 3.1: Carl Gehring/Corbis

Photo 3.2: Stacy L. Mallicoat. Printed with permission.

Table 3.2: http://www.rainn.org, http://www.womenshealth.gov and National Drug Intelligence Center

Figure 3.1: The Rape and Incest National Network—http://www.rainn.org

Case Study Box: http://today.duke.edu/showcase/lacrosseincident/ and http://www.huffingtonpost.com/2011/04/19/crystal-mangum-duke-lacro_n_850910.html

Section IV. Women and Victimization: Intimate Partner Abuse

Photo 4.1: Can Stock Photo Inc.

Photo 4.2: IStock

Case Study Box: Warner, J. (2011, April 6). Molly Midyette, a mother sentenced to sixteen years for the death of her son, speaks out. *Denver Westword News*. Retrieved April 15, 2011, from http://www.westword.com

Photo 4.3: Stacy L. Mallicoat. Printed with permission.

Section V. Women and Victimization: Stalking and Sexual Harassment

Photo 5.1: Comstock/Thinkstock

Table 5.1: Adapted from Baum, K., Catalano, S., Rand, M., & Rose, K. (2009). *Stalking victimization in the United States*. U.S. Department of Justice, Bureau of Justice Statistics. Retrieved March 1, 2010, from http://ojp.usdoj.gov/content/pub/pdf/svus.pdf

Table 5.2: Logan, T. K., Shannon, L., Cole, J., & Swanberg, J. (2007). Partner stalking and implications for women's employment. *Journal of Interpersonal Violence, 22*(3), 268–291.

Section VI. International Issues for Women and Crime

Photo 6.1: Katie Orlinsky/Corbis

Section VII. Girls and Juvenile Delinquency

Table 7.1: Snyder, H. N., & Sickmund, M. (2006). *Juvenile offenders and victims: 2006 national report.* National Center for Juvenile Justice. Office of Juvenile Justice and Delinquency Prevention. Retrieved December 1, 2010, from http://www.ojjdp.gov/ojstatbb/nr2006/

Photo 7.1: ©iStockphoto.com/zorani

Section VIII. Female Offenders and Their Crimes

Photo 8.1: STR/Reuters/Corbis

Section IX. Processing and Sentencing of Female Offenders

Photo 9.1: Stockbyte/Thinkstock

Section X. The Incarceration of Women

Photo 10.1: Comstock/Thinkstock Images

Photo 10.2: © Viviane Moos/CORBIS

Table 10.1: Bloom, B., Owen, B., & Covington, S. (2003). *Gender responsive strategies: Research, practice and guiding principles for women offenders.* National Institute of Corrections. U.S. Department of Justice. Retrieved January 13, 2011, from http://nicic.gov/pubs/2003/018017.pdf

Table 10.2: Bloom, B., Owen, B., & Covington, S. (2003). *Gender responsive strategies: Research, practice and guiding principles for women offenders.* National Institute of Corrections. U.S. Department of Justice. Retrieved January 13, 2011, from http://nicic.gov/pubs/2003/018017.pdf

Table 10.3: Sentencing Project. (n.d.) *Life sentences: Denying welfare benefits to women convicted of drug offenses.* Retrieved December 28, 2010, from http://www.sentencingproject.org/doc/publications/women_smy_lifesentences.pdf

Section XI. Women and Work in the Criminal Justice System: Police, Courts, and Corrections

Photo 11.1: Library of Congress (http://www.loc.gov/pictures/item/2006675559/)

Photo 11.2: Darrin Klimek/Digital Vision/ThinkStock

Photo 11.3: Steve Petteway, Collection of the Supreme Court of the United States

References

Abe-Kim, J., Takeuchi, D. T., Hong, S., Zane, N., Sue, S., & Spencer, M. S. (2007). Use of mental health-related services among immigrant and U.S.-born Asian Americans: Results from the National Latino and Asian American Study. *American Journal of Public Health, 97*(1), 91–98.

Acoca, L., & Dedel, K. (1998). *Identifying the needs of young women in the juvenile justice system.* San Francisco: National Council on Crime and Delinquency.

Adler, F. (1975). *Sisters in crime: The rise of the new female criminal.* New York: McGraw-Hill.

Agosin, M. (2006). *Secrets in the sand: The young women of Ciudad Juarez.* Buffalo, NY: White Pine Press.

Alabama Coalition Against Domestic Violence (ACADV). (n.d.) *Dating violence.* Retrieved February 5, 2010, from http://www.acadv.org/dating.html

Albonetti, C. A. (1986). Criminality, prosecutorial screening, and uncertainty: Toward a theory of discretionary decision making in felony case processing. *Criminology,* 24(4), 623–645.

Alder, C. M. (1998). Passionate and willful girls: Confronting practices. *Women and Criminal Justice, 9*(4), 81–101.

Alexy, E. M., Burgess, A. W., Baker, T., & Smoyak, S. A. (2005). Perceptions of cyberstalking among college students. *Brief Treatment and Crisis Intervention, 5*(3), 279–289.

Althaus, D. (2010, April 18). Juarez's women in power struggle. *Houston Chronicle.* Retrieved April 20, 2010, from http://www.chron.com/disp/story.mpl/world/6964738.html

American Bar Association (ABA). (2009). *Enrollment and degrees awarded 1963–2009 academic years.* Retrieved February 20, 2011, from http://www.americanbar.org/content/dam/aba/migrated/legaled/statistics/charts/stats_1.authcheckdam.pdf

American Bar Association. (2010). *Law school staff by gender and ethnicity.* Retrieved February 20, 2011, from http://www.americanbar.org/content/dam/aba/migrated/legaled/statistics/charts/facultyinformationbygender.authcheckdam.pdf

American Correctional Association. (2007). *Directory of adult and juvenile correctional departments, institutions and agencies and probation and parole authorities.* Alexandria, VA: Author.

Amnesty International. (1999). *Pakistan: Violence against women in the name of honor.* Retrieved June 1, 2010, from http://www.amnesty.org/en/library/asset/ASA33/017/1999/en/5d9201f4-e0f2-11dd-be39-2d4003be4450/asa330171999en.pdf

Anderson, T. L. (2006). Issues facing women prisoners in the early twenty-first century. In C. Renzetti, L. Goodstein, & S. L. Miller (Eds.), *Rethinking gender, crime and justice.* Los Angeles: Roxbury.

Appier, J. (1998). *Policing women: The sexual politics of law enforcement and the LAPD.* Philadelphia: Temple University Press.

Arin, C. (2001). Femicide in the name of honor in Turkey. *Violence Against Women, 7*(7), 821–825.

Bachman, R., Zaykowski, H., Lanier, C., Poteyva, M., & Kallmyer, R. (2010). Estimating the magnitude of rape and sexual assault against American Indian and Alaska Native (AIAN) women. *The Australian and New Zealand Journal of Criminology, 43*(2), 199–222.

Bacik, I., & Drew, E. (2006). Struggling with juggling: Gender and work/life balance in the legal professions. *Women's Studies International Forum, 29,* 136–146.

Ball, J. D., & Bostaph, L. G. (2009). He versus she: A gender-specific analysis of legal and extralegal effects on pretrial release for felony defendants. *Women and Criminal Justice, 19*(2), 95–119.

Barata, P. C., & Schneider, F. (2004). Battered women add their voices to the debate about the merits of mandatory arrest. *Women's Studies Quarterly, 32*(3/4), 148–163.

Baum, K., Catalano, S., Rand, M., & Rose, K. (2009). *Stalking victimization in the United States.* U.S. Department of Justice, Bureau of Justice Statistics. Retrieved March 1, 2010, from http://ojp.usdoj.gov/content/pub/pdf/svus.pdf

Beger, R. R., & Hoffman, H. (1998). Role of gender in detention dispositioning of juvenile probation violators. *Journal of Crime and Justice, 21,* 173–188.

Belknap, J. (2007). *The invisible woman: Gender, crime and justice* (3rd ed.). Belmont, CA: Wadsworth.

Belknap, J., Dunn, M., & Holsinger, K. (1997). *Moving toward juvenile justice and youth serving systems that address the distinct experience of the adolescent female* (Report to the Governor). Columbus, OH: Office of Criminal Justice Services.

Belknap, J., & Holsinger, K. (1998). An overview of delinquent girls: How theory and practice failed and the need for innovative change. In R. Zaplin (Ed.), *Female crime and delinquency: Critical perspectives and effective interventions.* Gaithersburg, MD: Aspen.

Belknap, J., & Holsinger, K. (2006). The gendered nature of risk factors for delinquency. *Feminist Criminology, 1*(1), 48–71.

Bennice, J. A., & Resick, P. A. (2003). Marital rape history, research and practice. *Trauma Violence Abuse, 4*(3), 228–246.

Bent-Goodley, T. B. (2004). Perceptions of domestic violence: A dialogue with African American women. *Health and Social Work, 29*(4), 307–316.

Bernard, T. J. (1992). *The cycle of juvenile justice*. New York: Oxford University Press.

Bernhard, L. A. (2000). Physical and sexual violence experienced by lesbian and heterosexual women. *Violence Against Women, 6*(1), 68–79.

Beynon, C. M., McVeigh, C., McVeigh, J., Leavy, C., & Bellis, M. A. (2008). The involvement of drugs and alcohol in drug-facilitated sexual assault: A systematic review of the evidence. *Trauma Violence Abuse, 9,* 178–188.

Block, C. R. (2003). How can practitioners help an abused woman lower her risk of death. *NIJ Journal, 250,* 4–7. Retrieved February 5, 2010, at http://www.ncjrs.gov/pdffiles1/jr000250c.pdf

Bloom, B., Owen, B., & Covington, S. (2003). *Gender responsive strategies: Research, practice and guiding principles for women offenders.* National Institute of Corrections. U.S. Department of Justice. Retrieved January 13, 2011, from http://nicic.gov/pubs/2003/018017.pdf

Bloom, B., Owen, B., & Covington, S. (2004). Women offenders and the gendered effects of public policy. *Policy Research, 21*(1), 31–48.

Bloom, B., Owen, B., Deschenes, E. P., & Rosenbaum, J. (2002a). Improving juvenile justice for females: A statewide assessment in California. *Crime and Delinquency, 48*(4), 526–552.

Bloom, B., Owen, B., Deschenes, E. P., & Rosenbaum, J. (2002b). Moving toward justice for female offenders in the new millennium: Modeling gender-specific policies and programs. *Journal of Contemporary Criminal Justice, 18*(1), 37–56.

Bodinger-deUriarte, C., & Austin, G. (1991). *Substance abuse among adolescent females* (Prevention research update no. 9). Washington, DC: Western Center for Drug Free Schools and Communities, Department of Education.

Boykins, A. D., Alvanzo, A. A. H., Carson, S., Forte, J., Leisey, M., & Plichta, S. B. (2010). Minority women victims of recent sexual violence: Disparities in incident history. *Journal of Women's Health, 19*(3), 453–461.

Brents, B. G., & Hausbeck, K. (2005). Violence and legalized brothel prostitution in Nevada: Examining safety, risk and prostitution policy. *Journal of Interpersonal Violence, 20*(3), 270–295.

Britton, D. M. (2003). *At work in the iron cage: The prison as a gendered organization*. New York: New York University Press.

Brown, L. M., Chesney-Lind, M., & Stein, N. (2007). Patriarchy matters: Towards a gendered theory of teen violence and victimization. *Violence Against Women, 13*(12), 1249–1273.

Bryant-Davis, T., Chung, H., & Tillman, S. (2009). From the margins to the center: Ethnic minority women and the mental health effects of sexual assault. *Trauma, Violence and Abuse, 10*(4), 330–357.

Bui, H. (2007). The limitations of current approaches to domestic violence. In R. Muraskin (Ed.), *It's a crime* (4th ed.). Upper Saddle River, NJ: Pearson Prentice Hall.

Bui, H., & Morash, M. (2008). Immigration, masculinity and intimate partner violence from the standpoint of domestic violence service providers and Vietnamese-origin women. *Feminist Criminology, 3*(3), 191–215.

Bureau of Justice Statistics. (2006). *Intimate partner violence.* Office of Justice Programs. Retrieved February 5, 2010, from http://bjs.ojp.usdoj.gov/content/pub/press/ipvpr.cfm

Burgess-Jackson, K. (1999). A history of rape law. In K. Burgess-Jackson (Ed.), *A most detestable crime: New philosophical essays on rape.* New York: Oxford University Press.

Bush-Baskette, S. (1998). The war on drugs as a war on Black women. In S. Miller (Ed.), *Crime control and women*. Thousand Oaks, CA: Sage.

Bush-Baskette, S. R. (2000). The war on drugs and the incarceration of mothers. *Journal of Drug Issues, 30,* 919–928.

Calderon Gamboa, J. (2007). Seeking integral reparations for the murders and disappearance of women in Ciudad Juarez: A gender and cultural perspective. *Human Rights Brief, 14,* 31.

California State Legislature Assembly Bill 1825. (2007). Retrieved March 30, 2010, from http://library.findlaw.com/2005/Feb/6/133651.html

Campbell, A. (1984). *The girls in the gang*. New Brunswick: Rutgers University Press.

Campbell, A. (1995). Female participation in gangs. In M. W. Klein, C. L. Maxson, & J. Miller (Eds.), *The modern gang reader* (pp. 70–77). Los Angeles: Roxbury.

Campbell, J. C., Webster, D., Koziol-McLain, J., Block, C. R., Campbell, D., Curry, M. A. et al. (2003). Assessing risk factors for intimate partner homicide. *NIJ Journal, 250,* 14-19. Accessed February 5, 2010 at http://www.ncjrs.gov/pdffiles1/jr000250e.pdf

Campbell, N. D. (2000). *Using women: Gender, drug policy and social justice*. New York: Routledge.

Carr, N. T., Hudson, K., Hanks, R. S., & Hunt, A. N. (2008). Gender effects along the juvenile justice system: Evidence of a gendered organization. *Feminist Criminology, 3*(1), 25–43.

Catalano, S. (2007). *Intimate partner violence in the United States.* Bureau of Justice Statistics. U.S. Department of Justice: Washington D.C. Retrieved February 5, 2010, from http://bjs.ojp.usdoj.gov/content/pub/pdf/IPAus.pdf

Catalano, S. M. (2006). *Criminal victimization 2005.* Bureau of Justice Statistics. U.S. Department of Justice. Retrieved September 14, 2009, from http://bjs.ojp.usdoj.gov/content/pub/pdf/cv05.pdf

Centers for Disease Control and Prevention (CDC). (2003). *Costs of intimate partner violence against women in the United States: 2003.* Atlanta, GA: National Centers for Injury Prevention and Control.

Chamberlain, L. (2007). 2 cities and 4 bridges where commerce flows. *New York Times.* Retrieved April 20, 2010, from http://www.nytimes.com/2007/03/28/realestate/commercial/28juarez.html?_r=2&scp=1&sq=Where+commerce+flows&st=nyt

Chapman, S. B.. (2009). *Inmate-perpetrated harassment: Exploring the gender-specific experience of female correctional officers.* Unpublished dissertation, City University of New York.

Cheeseman, K. A., & Worley, R. (2006). Women on the wing: Inmate perceptions about female correctional office job competency in a southern prison system. *Southwest Journal of Criminal Justice, 3*(2), 86–102.

Chesler, P. (2010). Worldwide trends in honor killings. *Middle East Quarterly, 17*(2), 3–11. Retrieved May 12, 2010, from http://www.meforum.org

Chesney-Lind, M. (1973). Judicial enforcement of the female sex role. *Issues in Criminology, 8,* 51–70.

Chesney-Lind, M. (1997). *The female offender: Girls, women and crime.* Thousand Oaks, CA: Sage.

Chesney-Lind, M., & Shelden, R. G. (1998). *Girls, delinquency and juvenile justice.* Belmont, CA: West/Wadsworth.

Chesney-Lind, M., & Shelden, R. (2004). *Girls, delinquency, and juvenile justice* (3rd ed.). Belmont, CA: Wadsworth.

Chiricos, T., Padgett, K., & Gertz, M. (2000). Fear, TV news, and the reality of crime. *Criminology, 38,* 755–786.

Cohen, A. K., & Felson, M. (1979). Social change and crime rate trends: A routine activities approach. *American Sociological Review, 44,* 588–608.

Cook, J. A., & Fonow, M. M. (1986). Knowledge and women's interests: Issues of epistemology and methodology in feminist sociological research. *Sociological Inquiry, 56,* 2–29.

Coontz, P. (2000). Gender and judicial decisions: Do female judges decide cases differently than male judges? *Gender Issues, 18*(4), 59–73.

Covington, S. (1999). *Helping women recover: A program for treating substance abuse.* San Francisco: Jossey-Bass.

Cox, L., & Speziale, B. (2009). Survivors of stalking: Their voices and lived experiences. *Affilia: Journal of Women and Social Work, 24*(1), 5–18.

Crandall, M., Senturia, K., Sullivan, M., & Shiu-Thornton, S. (2005). No way out: Russian-speaking women's experiences with domestic violence. *Journal of Interpersonal Violence, 20*(8), 941–958.

Craven, D. (1997) *Sex differences in violent victimization: 1994.* Retrieved September 14, 2009, from http://bjs.ojp.usdoj.gov/content/pub/pdf/SDVV.PDF

Crosnoe, R., Erickson, K. G., & Dornbusch, S. M. (2002) Protective functions of family relationships and school factors on the deviant behavior of adolescent boys and girls. *Youth and Society, 33*(4), 515–544.

Curry, G. D., Ball, R. A., & Fox, R. J. (1994). *Gang crime and law enforcement recordkeeping. Research in brief.* Washington, DC: U.S. Department of Justice, Office of Justice Programs, National Institute of Justice. Accessed September 28, 2010, from http://www.ncjrs.gov/txtfiles/gcrime.txt

Daigle, L. E., Cullen, F. T., & Wright, J. P. (2007). Gender differences in the predictors of juvenile delinquency: Assessing the generality-specificity debate. *Youth Violence and Juvenile Justice, 5*(3), 254–286.

Dalla, R. L. (2000). Exposing the "Pretty Woman" myth: A qualitative examination of the lives of female streetwalking prostitutes. *Journal of Sex Research, 37*(4), 344–353.

Dalla, R. L., Xia, Y., & Kennedy, H. (2003). "You just give them what they want and pray they don't kill you": Street-level workers' reports of victimization, personal resources and coping strategies. *Violence Against Women, 9*(11), 1367–1394.

Daly, K. (1994). *Gender, crime and punishment.* New Haven: Yale University Press.

Danner, M. J. E. (2003). Three strikes and it's women who are out: The hidden consequences for women of criminal justice policy reforms. In R. Muraskin (Ed.), *It's a crime: Women and justice* (2nd ed.). Upper Saddle River, NJ: Prentice-Hall.

Davies, K., Block, C. R., & Campbell, J. (2007). Seeking help from the police: Battered women's decisions and experiences. *Criminal Justice Studies, 20,* 15–41.

Davis, C. P. (2007). At risk girls and delinquency: Career pathways. *Crime and Delinquency, 53*(3), 408–435.

Davis, K. E., Coker, A. L., & Sanderson, M. (2002). Physical and mental health effects of being stalked for men and women. *Violence and Victims, 17*(4), 429–443.

De Coster, S., Estes, S., & Mueller, C. W. (1999). Routine activities and sexual harassment in the workplace. *Work and Occupations, 26,* 21–49.

De Groof, S. (2008). And my mama said: The (relative) parental influence on fear of crime among adolescent girls and boys. *Youth and Society, 39*(3), 267–293.

de Vaus, D., & Wise, S. (1996). The fear of attack: Parents' concerns for the safety of their children. *Family Matters, 43,* 34–38.

Demuth, S., & Steffensmeier, D. (2004). The impact of gender and race-ethnicity in the pretrial release process. *Social Problems, 51*(2), 222–242.

Dietz, N. A., & Martin, P. Y. (2007). Women who are stalked: Questioning the fear standard. *Violence Against Women, 13*(7), 750–776.

DiMarino, F. (2009). *Women as corrections professionals.* Corrections.com. Retrieved February 28, 2011, from http://www.corrections.com/articles/21703-women-as-corrections-professionals

Diviney, C. L., Parekh, A., & Olson, L. M. (2009). Outcomes of civil protective orders: Results from one state. *Journal of Interpersonal Violence, 24*(7), 1209–1221.

Dobash, R., & Dobash, R. E. (1992). *Women, violence and social change.* New York: Routledge.

Dowler, K. (2003). Media consumption and public attitudes toward crime and justice: The relationship between fear of crime, punitive attitudes and perceived police effectiveness. *Journal of Criminal Justice and Popular Culture, 10,* 109–126.

Dutton, M. A., Orloff, L. E., & Hass, G. A. (2000). Characteristics of help seeking behaviors, resources and service needs of battered immigrant Latinas: Legal and policy implications. *Georgetown Journal on Poverty, Law and Policy, 7,* 245–305.

Dziech, B. W., & Weiner, L. (1990). *The lecherous professor: Sexual harassment on campus.* Chicago: University of Illinois Press.

Eghigian, M., & Kirby, K. (2006). Girls in gangs: On the rise in America. *Corrections Today, 68*(2), 48–50.

Esbensen, F.-A., Deschenes, E. P., & Winfree, L. T., Jr. (1999). Differences between gang girls and gang boys: Results from a multisite survey. *Youth and Society, 31*(1), 27–53.

Esfandiari, G. (2006). *Rights watchdog alarmed at continuing "honor killings" in Afghanistan.* Retrieved May 5, 2010, from http://www.rawa.org/honorkilling.htm

Farley, M. (2004). "Bad for the body, bad for the heart": Prostitution harms women even if legalized or decriminalized. *Violence Against Women, 10*(10), 1087–1125.

Farley, M., & Barkin, H. (1998). Prostitution, violence and post-traumatic stress disorder. *Woman and Health, 27*(3), 37–49.

Farley, M., & Kelly, V. (2000). Prostitution: A critical review of the medical and social sciences literature. *Women and Criminal Justice, 11*(4), 29–64.

Farnworth, M., & Teske, R. H. C. (1995). Gender differences in felony court processing. *Women and Criminal Justice, 6,* 23–44.

Fattah, E. A., & Sacco, V. F. (1989). *Crime and victimization of the elderly.* New York: Springer-Verlag.

Feld, B. C. (2009). Violent girls or relabeled status offenders? An alternative interpretation of the data. *Crime and Delinquency, 55*(2), 241–265.

Felix, Q. (2004, July 22). *Honour killings and "karo-kari" in Pakistan.* Retrieved May 12, 2010, from http://www.asianews.it/index.php?l=en&art=1187

Fisher, B. S., Cullen, F. T., & Turner, M. G. (2000). *The sexual victimization of college women.* Washington, DC: National Institute of Justice.

Fisher, B. S., Daigle, L. E., & Cullen, F. T. (2010). What distinguishes single from recurrent sexual victims? The role of lifestyle-routine activities and first incident characteristics. *Justice Quarterly, 27*(1), 102–129.

Fisher, B. S., Daigle, L. E., Cullen, F. T., & Turner, M. G. (2003). Reporting sexual victimization to the police and others: Results from a national-level study of college women. *Criminal Justice and Behavior, 30*(1), 6–38.

Fisher, B. S., & May, D. (2009). College students' crime-related fears on campus: Are fear-provoking cues gendered? *Journal of Contemporary Criminal Justice, 25*(3), 300–321.

Fisher, B. S., & Sloan, J. J. (2003). Unraveling the fear of victimization among college women: Is the "shadow of sexual assault hypothesis" supported? *Justice Quarterly, 20,* 633–659.

Fishman, L. (1995). The Vice-Queens: An ethnographic study of Black female gang behavior. In M. W. Klein, C. L. Maxon, & J. Miller (Eds.), *The modern gang reader* (pp. 83–91). Los Angeles: Roxbury.

Fitzgerald, L. F. (1993). Sexual harassment: Violence against women in the workplace. *American Psychologist, 48*(10), 1070–1076.

Fleisher, M. S., & Krienert, J. L. (2004). Life-course events, social networks, and the emergence of violence among female gang members. *Journal of Community Psychology, 32*(5), 607–622.

Fleury-Steiner, R., Bybee, D., Sullivan, C. M., Belknap, J., & Melton, H. C. (2006). Contextual factors impacting battered women's intentions to reuse the criminal legal system. *Journal of Community Psychology, 34*(3), 327–342.

Franiuk, R., Seefelt, J. L., Cepress, S. L., & Vandello, J. A. (2008). Prevalence and effects of rape myths in print journalism: The Kobe Bryant case. *Violence Against Women, 14*(3), 287–309.

Freedman, E. B. (1981). *Their sisters' keepers: Women prison report in America, 1830–1930.* Ann Arbor: University of Michigan Press.

Freiburger, T. L., & Hilinski, C. M. (2009, February 24). An examination of the interactions of race and gender on sentencing decisions using a trichotomous dependent variable. *Crime and Delinquency.*

Freiburger, T. L., & Hilinski, C. M. (2010). The impact of race, gender and age on the pretrial decision. *Criminal Justice Review, 35*(3), 318–334.

French, S. (2000). Of problems, pitfalls and possibilities: A comprehensive look at female attorneys and law firm partnership. *Women's Rights Law Reporter, 21*(3), 189–216.

Gaarder, E., & Belknap, J. (2002). Tenuous borders: Girls transferred to adult court. *Criminology, 40*(3), 481–518.

Gaarder, E., Rodriguez, N., & Zatz, M. S. (2004). Cries, liars and manipulators: Probation officers' views of girls. *Justice Quarterly, 21*(3), 547–578.

Garcia-Lopez, G. (2008). Nunca te toman en cuenta (They never take you into account): The challenges of inclusion and strategies for success of Chicana attorneys. *Gender and Society, 22*(5), 590–612.

Gavazzi, S. M., Yarcheck, C. M., & Lim, J.-Y. (2005). Ethnicity, gender, and global risk indicators in the lives of status offenders coming to the attention of the juvenile court. *International Journal of Offender Therapy and Comparative Criminology, 49*(6), 696–710.

General Accounting Office (GAO). (1999). *Women in prison: Issues and challenges confronting U.S. correctional systems.* Washington, DC: U.S. Department of Justice.

Gerbner, G., & Gross, L. (1980). The violent face of television and its lesson. In E. Palmer & A. Dorr (Eds.), *Children and the faces of television: Teaching, violence, selling.* New York: Academic Press.

Gidycz, C. A., Orchowski, L. M., King, C. R., & Rich, C. L. (2008). Sexual victimization and health-risk behaviors: A prospective analysis of college women. *Journal of Interpersonal Violence, 23*(6), 744–763.

Gilbert, E. (2001). Women, race and criminal justice processing. In C. Renzetti & L. Goodstein (Eds.), *Women, crime and criminal justice: Original feminist readings.* Los Angeles: Roxbury.

Gillum, T. L. (2008). Community response and needs of African American female survivors of domestic violence. *Journal of Interpersonal Violence, 23*(1), 39–57.

Girard, A. L., & Senn, C. Y. (2008). The role of the new "date rape drugs" in attributions about date rape. *Journal of Interpersonal Violence, 23*(1), 3–20.

Girls Incorporated. (1996). *Prevention and parity: Girls in juvenile justice.* Indianapolis, IN: Girls Incorporated National Resource Center & Office of Juvenile Justice and Delinquency Prevention.

Girshick, L. B. (2002). No sugar, no spice: Reflections on research on woman-to-woman sexual violence. *Violence Against Women, 8*(12), 1500–1520.

Goddard, C., & Bedi, G. (2010). Intimate partner violence and child abuse: A child-centered perspective. *Child Abuse Review, 19,* 5–20.

Goodkind, S. (2005). Gender specific services in the juvenile justice system: A critical examination. *Affilia, 20*(1), 52–70.

Gormley, P. (2007). The historical role and views towards victims and the evolution of prosecution policies in domestic violence. In R. Muraskin (Ed.), *It's a crime* (4th ed.). Upper Saddle River, NJ: Pearson Prentice Hall.

Greenfeld, L. A. (1997). *Sex offenses and offenders: An analysis of data on rape and sexual assault.* Bureau of Justice Statistics, U.S. Department of Justice. Retrieved February 5, 2010, from http://bjs.ojp.usdoj.gov/content/pub/pdf/SOO.PDF

Greenfeld, L. A., & Snell, T. L. (2000). *Women offenders.* Washington, DC: Bureau of Justice Statistics. Retrieved October 15, 2010, from http://bjs.ojp.usdoj.gov/content/pub/pdf/wo.pdf

Greer, K. (2008). *When women hold the keys: Gender, leadership and correctional policy.* Management and Training Institute. Retrieved February 28, 2011, from http://nicic.gov/Library/023347

Griffin, M. L. (2006). Gender and stress: A comparative assessment of sources of stress among correctional officers. *Journal of Contemporary Criminal Justice, 22*(1), 4–25.

Griffin, T., & Wooldredge, J. (2006). Sex-based disparities in felony dispositions before versus after sentencing reform in Ohio. *Criminology, 44,* 893–923.

Guevara, L., Herz, D., & Spohn, C. (2008). Race, gender and legal counsel: Differential outcomes in two juvenile courts. *Youth Violence and Juvenile Justice, 6*(1), 83–104.

Harrington, P., & Lonsway, K. A. (2004). Current barriers and future promise for women in policing. In B. R. Price & N. J. Sokoloff (Eds.), *The criminal justice system and women: Offenders, prisoners, victims and workers* (3rd ed.). Boston: McGraw Hill.

Harrison, P. M., & Beck, A. J. (2005). *Prisoners in 2005.* Bureau of Justice Statistics. Retrieved July 31, 2011, from http://bjs.ojp.usdoj.gov/content/pub/pdf/p05.pdf

Hassouneh, D., & Glass, N. (2008). The influence of gender-role stereotyping on female same sex intimate partner violence. *Violence Against Women: An International and Interdisciplinary Journal, 14*(3), 310–325.

Haynes, D. F. (2004). Used, abused, arrested and deported: Extending immigration benefits to protect the victims of trafficking and to secure the prosecution of traffickers. *Human Rights Quarterly, 26,* 221–272.

Heidensohn, F. M. (1985). *Women and crime: The life of the female offender.* New York: New York University Press.

Hessy-Biber, S. N. (2004). *Feminist perspectives on social research.* New York: Oxford University Press.

Hindelang, M. J., Gottfredson, M. R., & Garafalo, J. (1978). *Victims of personal crime: An empirical foundation for a theory of personal victimization.* Cambridge, MA: Ballinger.

Hirsch, A. E. (2001). Bringing back shame: Women, welfare reform and criminal justice. In P. J. Schram & B. Koons-Witt (Eds.), *Gendered (in) justice: Theory and practice in feminist criminology* (pp. 270–286). Long Grove, IL: Waveland Press.

Hirschi, T. (1969). *Causes of delinquency.* Berkeley: University of California Press.

Hunt, G., & Joe-Laidler, K. (2001). Situations of violence in the lives of girl gang members. *Health Care for Women International, 22,* 363–384.

Hurst, T. E., & Hurst, M. M. (1997). Gender differences in mediation of severe occupational stress among correctional officers. *American Journal of Criminal Justice, 22*(1), 121–137.

Illinois Coalition Against Sexual Assault (ICASA). (n.d.). *Emotional and physical effects of sexual assault.* Accessed February 5, 2010, from www.icasa.org/docs/emotional_&_physical_effects_-_DRAFT-4.doc

Inciardi, J. A., Lockwood, D., & Pottiger, A. E. (1993). *Women and crack-cocaine.* Toronto: Maxwell Macmillian.

Ingram, E. M. (2007). A comparison of help seeking between Latino and non-Latino victims of intimate partner violence. *Violence Against Women, 13*(2), 159–171.

Inter-American Commission on Human Rights. (2003, March 7). *The situation of the rights of women in Ciudad Juarez, Mexico: The right to be free from violence and discrimination.* Retrieved April 20, 2010, from http://www.cidh.org/annualrep/2002eng/chap.vi.juarez.htm

Jacobs, A. (2000). *Give 'em a fighting chance: The challenges for women offenders trying to succeed in the community.* Retrieved July 31, 2011, from http://www.wpaonline.org/pdf/WPA_FightingChance.pdf

Joe, K., & Chesney-Lind, M. (1995). Just every mother's angel: An analysis of gender and ethnic variation in youth gang membership. *Gender and Society, 9*(4), 408–430.

Joe-Laidler, K., & Hunt, G. (2001). Accomplishing femininity among girls in the gang. *British Journal of Criminology, 41,* 656–678.

Kardam, F. (2005). *The dynamics of honor killings in Turkey.* United Nations Development Programme Population Association, United Nations Population Fund. Retrieved June 6, 2010, from www.stop-stoning.org/files/676_filename_honourkillings.pdf

Katz, C. M., & Spohn, C. (1995). The effect of race and gender and bail outcomes: A test of an interactive model. *American Journal of Criminal Justice, 19,* 161–184.

Kaukinen, C. (2004). Status compatibility, physical violence and emotional abuse in intimate relationships. *Journal of Marriage and Family, 66,* 452–471.

Kaukinen, C., & DeMaris, A. (2009). Sexual assault and current mental health: The role of help seeking and police response. *Violence Against Women, 15*(11), 1331–1357.

Kilpatrick, D. G., Resnick, H. S., Ruggiero, K. J., Conoscenti, L. M., & McCauley, J. (2007). *Drug facilitated, incapacitated and forcible rape: A national study.* U.S. Department of Justice. Retrieved October 1, 2009, from http://ovc.ncjrs.gov/ncvrw2009/pdf/crime_clock_hr.pdf

Klein, A. R. (2004). *The criminal justice response to domestic violence.* Belmont, CA: Wadsworth Thomson Learning.

Knoll, C., & Sickmund, M. (2010). *Delinquency cases in juvenile court, 2007.* Office of Justice Programs. Office of Juvenile Justice and Delinquency Prevention. Retrieved December 1, 2010, from http://www.ncjrs.gov/pdffiles1/ojjdp/230168.pdf

Koons-Witt, B. (2006). Decision to incarcerate before and after the introduction of sentencing guidelines. *Criminology, 40*(2), 297–328.

Kraaij, V., Arensman, E., Garnefski, N., & Kremers, I. (2007). The role of cognitive coping in female victims of stalking. *Journal of Interpersonal Violence, 22*(12), 1603–1612.

Kruger, K. J. (2006). Pregnancy and policing: Are they compatible? *bePress Legal Series* Paper 1357. Retrieved February 19, 2011, from http://law.bepress.com/cgi/viewcontent.cgi?article=6262&context=expresso

Kurshan, N. (2000). *Women and imprisonment in the United States: History and current reality.* Retrieved January 3, 2011, from http://www.prisonactivist.org/archive/women/women-and-imprisonment.html

Kyckelhahn, T., Beck, A. J., & Cohen, T. H. (2009). *Characteristics of suspected human trafficking incidents, 2007-08.* Bureau of Justice Statistics, U.S. Department of Justice. Retrieved June 10, 2010, from http://content.news14.com/human_trafficking.pdf

Lambert, E., Altheimer, I., & Hogan, N. L. (2010). An exploratory examination of a gendered model of the effects of role stressors. *Women and Criminal Justice, 20*(3), 193–217.

Lane, J., Gover, A. R., & Dahod, S. (2009). Fear of violent crime among men and women on campus: The impact of perceived risk and fear of sexual assault. *Violence and Victims, 24*(2), 172–192.

Lauritsen, J. L., Heimer, K., & Lynch, J. P. (2009). Trends in the gender gap in violent offending: New evidence from the National Crime Victimization Survey. *Criminology, 47*(2), 361–399.

Lee, R. K. (1998). Romantic and electronic stalking in a college context. *William and Mary Journal of Women and the Law, 4,* 373–466.

Leiber, M. J., Brubacker, S. J., & Fox, K. C. (2009). A closer look at the individual and joint effects of gender and race on juvenile justice decision making. *Feminist Criminology, 4*(4), 333–358.

Leonard, E. B. (1982). *Women, crime and society*. New York: Longman.

Leonard, E. D. (2001). Convicted survivors: Comparing and describing California's battered women inmates. *The Prison Journal, 81*(1), 73–86.

Lerner, M. J. (1980). *The belief in a just world: A fundamental delusion*. New York: Plenum Press.

Lersch, K. M., & Bazley, T. (2012). A paler shade of blue? Women and the police subculture. In R. Muraskin (Ed.), *Women and justice: It's a crime* (5th ed., pp. 514–526). Upper Saddle River, NJ: Prentice-Hall.

Lipsky, S., Caetano, R., Field, C. A., & Larkin, G. L. (2006). The role of intimate partner violence, race and ethnicity in help-seeking behaviors. *Ethnicity and Health,* 11(1), 81–100.

Littleton, H., Breitkopf, C. R., & Berenson, A. (2008). Beyond the campus: Unacknowledged rape among low-income women. *Violence Against Women, 14*(3), 269–286.

Logan, T. K., & Cole, J. (2007). The impact of partner stalking on mental health and protective order outcomes over time. *Violence and Victims, 22*(5), 546–562.

Logan, T. K., Evans, L., Stevenson, E., & Jordan, C. E. (2005). Barriers to services for rural and urban survivors of rape. *Journal of Interpersonal Violence, 20*(5), 591–616.

Logan, T. K., Shannon, L., & Cole, J. (2007). Stalking victimization in the context of intimate partner violence. *Violence and Victims, 22*(6), 669–683.

Logan, T. K., Shannon, L., Cole, J., & Swanberg, J. (2007). Partner stalking and implications for women's employment. *Journal of Interpersonal Violence, 22*(3), 268–291.

Lombroso, C., & Ferrero, W. (1895). *The female offender*. New York: Barnes and Company.

Lonsway, K., Carrington, S., Aguirre, P., Wood, M., Moore, M., Harrington, P., et al. (2001). *Equality denied: The status of women in policing: 2001.* The National Center for Women and Policing. Retrieved February 19, 2011, from http://www.womenandpolicing.org/PDF/2002_Status_Report.pdf

Lonsway, K. A., & Fitzgerald, L. F. (1994). Rape myths: In review. *Psychology of Women Quarterly, 18,* 133–164.

Lonsway, K., Wood, M., Fickling, M., De Leon, A., Moore, M., Harrington, P., et al. (2002). *Men, women and police excessive force: A tale of two genders: A content analysis of civil liability cases, sustained allegations and citizen complaints.* The National Center for Women and Policing. Retrieved February 19, 2011, from http://www.womenandpolicing.org/PDF/2002_Excessive_Force.pdf

Lord, E. A. (2008). The challenges of mentally ill female offenders in prison. *Criminal Justice and Behavior, 35*(8), 928–942.

Lucero, M. A., Middleton, K. L., Finch, W. A., & Valentine, S. R. (2003). An empirical investigation of sexual harassers: Towards a perpetrator typology. *Human Relations, 56*(12), 1461–1483.

MacKinnon, C. (1979). *Sexual harassment of working women: A case of sex discrimination*. New Haven: Yale University Press.

Maguire, B. (1988). Image vs. reality: An analysis of prime-time television crime and police programs. *Crime and Justice, 11,* 165–188.

Maher, L. (1996). Hidden in the light: Occupational norms among crack-using street level sex workers. *Journal of Drug issues, 26,* 143–173.

Maher, L. (2004). A reserve army: Women and the drug market. In B. Price & N. Sokoloff (Eds.), *The criminal justice system and women: Offenders, prisoners, victims and workers* (3rd ed., pp. 127–146). New York: McGraw-Hill.

Mahoney, J. L., Cairns, B. D., & Farmer, T. W. (2003). Promoting interpersonal competence and educational success through extracurricular activity participation. *Journal of Educational Psychology, 95*(2), 409–418.

Mallicoat, S. L. (2006, August). *Mary Magdalene Project: Kester program evaluation.* Presented to the Program Committee of the Mary Magdalene Project, Van Nuys, CA.

Mallicoat, S. L. (2007). Gendered justice: Attributional differences between males and females in the juvenile courts. *Feminist Criminology, 2*(1), 4–30.

Mallicoat, S. L. (2011). Lives in transition: A needs assessment of women exiting from prostitution. In R. Muraskin (Ed.), *It's a crime: Women and justice* (4th ed., pp. 241–255). Upper Saddle River, NJ: Prentice-Hall.

Martin, E. K., Taft, C. T., & Resick, P. A. (2007). A review of marital rape. *Aggression and Violent Behavior, 12,* 329–347.

Martin, L. (1991). *A report on the glass ceiling commission.* Washington, DC: U.S. Department of Labor.

Masood, S. (2004, October 27). Pakistan tries to curb "honor killings." *New York Times.* Retrieved May 12, 2010, from http://www.nytimes.com/2004/10/27/international/asia/27stan.html?_r=1&oref=login

Matsueda, R. (1992). Reflected appraisals, parental labeling and delinquency: Specifying a symbolic interactionist theory. *American Journal of Sociology.*

Mayell, H. (2002, February 12). Thousands of women killed for family honor. *National Geographic.* Retrieved May 5, 2010, from http://news.nationalgeographic.com/news/2002/02/0212_020212_honorkilling_2.html

McCartan, L. M., & Gunnison, E. (2010). Individual and relationship factors that differentiate female offenders with and without a sexual abuse history. *Journal of Interpersonal Violence, 25*(8), 1449–1469.

McKnight, L. R., & Loper, A. B. (2002). The effect of risk and resilience factors on the prediction of delinquency in adolescent girls. *School Psychology International, 23*(2), 186–198.

McLaughlin, H., Uggen, C., & Blackstone, A. (2008). Social class and workplace harassment during the transition to adulthood. *New Directions for Child and Adolescent Development, 119,* 85–98.

McLaughlin, H., Uggen, C., & Blackstone, A. (2009). *A longitudinal analysis of gender, power and sexual harassment in young adulthood.* Presented at the 104th Annual Meeting of the American Sociological Association, San Francisco, CA.

Melton, H. C. (2007). Predicting the occurrence of stalking in relationships characterized by domestic violence. *Journal of Interpersonal Violence, 22*(1), 3–25.

Mendelsohn, B. (1963). The origin and doctrine of victimology. *Excerpta Criminologica,* 3, 239–244.

Merolla, D. (2008). The war on drugs and the gender gap in arrests: A critical perspective. *Critical Sociology, 34,* 25.

Meyer, C. L., & Oberman, M. (2001). *Mothers who kill their children: Understanding the acts of moms from Susan Smith to the "Prom Mom."* New York: University Press.

Millar, G. M., Stermac, L., & Addison, M. (2002). Immediate and delayed treatment seeking among adult sexual assault victims. *Women & Health, 35*(1), 53–63.

Miller, J. (1994). Race, gender and juvenile justice: An examination of disposition decision-making for delinquent girls. In M. Schwartz & D. Milovanivoc (Eds.), *Race, gender and class in criminology: The intersection* (pp. 219–246). New York: Garland.

Miller, J. (1998). Gender and victimization risk among young women in gangs. *Journal of Research in Crime and Delinquency, 35,* 429–453.

Miller, J. (2000). *One of the guys: Girls, gangs and gender.* Oxford, UK: Oxford University Press.

Miller, L. J. (2003). Denial of pregnancy. In M. G. Spinelli (Ed.), *Infanticide: Psychosocial and legal perspectives on mothers who kill* (pp. 81–104). Washington, DC: American Psychiatric Publishing.

Miller, S., Loeber, R., & Hipwell, A. (2009). Peer deviance, parenting and disruptive behavior among young girls. *Journal of Abnormal Child Psychology, 37*(2), 139–152.

Miller, S. L., & Peterson, E. S. L. (2007). The impact of law enforcement policies on victims of intimate partner violence. In R. Muraskin (Ed.), *It's a crime* (4th ed.). Upper Saddle River, NJ: Pearson Prentice Hall.

Ministry of Labour in cooperation with the Ministry of Justice and the Ministry of Health and Social Affairs, Government of Sweden. (1998). *Fact sheet.* Secretariat for Information and Communication, Ministry of Labour. Retrieved October 2, 2010, from http://www.bayswan.org/swed/swedishprost1999.html

Moe, A. M. (2009). Battered women, children, and the end of abusive relationships. *Afilia: Journal of Women and Social Work, 24*(3), 244–256.

Moffitt, T. E., Caspi, A., Rutter, M., & Silva, P. A. (2001). *Sex differences in antisocial behavior: Conduct disorder, delinquency and violence in the Denedin Longitudinal Study.* New York: Cambridge University Press.

Molidor, C. E. (1996). Female gang members: A profile of aggression and victimization. *Social Work, 41*(3), 251–257.

Moore, J. W. (1991). *Going down to the barrio: Homeboys and homegirls in change.* Philadelphia: Temple University Press.

Moore, J., & Terrett, C. P. (1998). *Highlights of the 1996 National Youth Gang Survey. Fact sheet.* Washington, DC: U.S. Department of Justice, Office of Justice Programs, Office of Juvenile Justice and Delinquency Prevention.

Mustaine, E. E., & Tewksbury, R. (2002). Sexual assault of college women: A feminist interpretation of a routine activities analysis. *Criminal Justice Review, 27*(1), 89–123.

Nagel, I. H., & Hagan, J. (1983). Gender and crime: Offense patterns and criminal court sanctions. In M. Tonry & N. Morris (Eds.), *Crime and justice* (Vol. 4). Chicago: University of Chicago Press.

Nagel, I. H., & Johnson, B. L. (2004). The role of gender in a structured sentencing system: Equal treatment, policy choices and the sentencing of female offenders. In P. Schram & B. Koons-Witt (Eds.), *Gendered (in)justice: Theory and practice in feminist criminology.* Long Grove, IL: Waveland Press.

National Coalition Against Domestic Violence (NCADV). (n.d.). Retrieved February 5, 2010, from www.ncadv.org

National Drug Intelligence Center. (n.d.). *Drug-facilitated sexual assault fast facts.* Accessed March 9, 2010, from http://www.justice.gov/ndic/pubs8/8872/8872p.pdf

National Public Radio. (n.d.). *Timeline: America's war on drugs.* Retrieved July 31, 2011, from http://www.npr.org/templates/story/story.php?storyId=9252490

National Youth Gang Center. (2009). *National Youth Gang Survey analysis.* Retrieved September 25, 2010, from http://www.nationalgangcenter.gov/Survey-Analysis

Newton, M. (2003). *Ciudad Juarez: The serial killer's playground.* Retrieved April 20, 2010, from http://www.trutv.com/library/crime/serial_killers/predators/ciudad_juarez/index.html

Nixon, K., Tutty, L., Downe, P., Gorkoff, K., & Ursel, J. (2002). The everyday occurrence: Violence in the lives of girls exploited through prostitution. *Violence Against Women, 8*(9), 1016–1043.

Nokomis Foundation. (2002). *We can do better: Helping prostituted women and girls in Grand Rapids make healthy choices: A prostitution round table report to the community.* Retrieved September 3, 2008, from http://www.nokomisfoundation.org/documents/WeCanDoBetter.pdf

Noonan, M. C., & Corcoran, M. E. (2008). The mommy track and partnership: Temporary delay or dead end? *Annals of the American Academy of Political and Social Science, 596,* 130–150.

Noonan, M. C., Corcoran, M. E., & Courant, P. N. (2008). Is the partnership gap closing for women? Cohort differences in the sex gap in partnership chances. *Social Science Research, 37,* 156–179.

Norton-Hawk, M. (2004). A comparison of pimp and non-pimp controlled women. *Violence Against Women, 10*(2), 189–194.

Nurge, D. (2003). Liberating yet limiting: The paradox of female gang membership. In L. Kontos, D. Brotherton, & L. Barrios (Eds.), *Gangs and society: Alternative perspectives.* New York: Columbia University Press.

O'Connor, M. L. (2012). Early policing in the United States: "Help wanted—women need not apply! In R. Muraskin (Ed.), *Women and justice: It's a crime* (5th ed., pp. 487–499). Upper Saddle River, NJ: Prentice-Hall.

Odem, M. E. (1995). *Delinquent daughters: Protecting and policing adolescent female sexuality in the United States: 1885–1920.* Chapel Hill: University of North Carolina Press.

Office on Violence Against Women (OVW). (n.d.). U.S. Department of Justice. Retrieved February 5, 2010, from http://www.ovw.usdoj.gov

Owen, B., & Bloom B. (1998). *Modeling gender-specific services in juvenile justice: Final report to the office of criminal justice planning.* Sacramento, CA: OCJP.

Oxman-Martinez, J. (2003). *Human smuggling and trafficking: Achieving the goals of the UN protocols?* Retrieved July 15, 2010, from http://www1.maxwell.syr.edu/uploadedFiles/campbell/events/Oxman Martinez%20oped.pdf

Pakes, F. (2005). Penalizeation and retreat: The changing face of Dutch criminal justice. *Criminal Justice, 5*(2), 145–161.

Palmer, B. (2001). Women in the American judiciary: Their influence and impact. *Women and Politics, 23*(3), 91–101.

Patel, S., & Gadit, A. M. (2008). Karo-kari: A form of honour killing in Pakistan. *Transcultural Psychiatry, 45*(4), 683–694.

Pathe, M., & Mullen, P. E. (1997). The impact of stalkers on their victims. *British Journal of Psychiatry, 170,* 12–17.

Platt, A. M. (1969). *The child savers.* Chicago: University of Chicago Press.

Pollak, O. (1950). *The criminality of women.* Westport, CT: Greenwood Press.

Pollak, O. (1961). *The criminality of women.* New York: A. S. Barnes.

Pollock-Byrne, J. M. (1986). *Sex and supervision: Guarding male and female inmates.* New York, NY: Greenwood Press.

Portillo, L. (Producer & Director). (2001). *Senorita extravida: Missing young women* [Film]. New Jersey: Xochitl Films.

Potter, H. (2007a). Battered Black women's use of religious services and spirituality for assistance in leaving abusive relationships. *Violence Against Women, 13*(3), 262–284.

Potter, H. (2007b). *Battle cries: Understanding and confronting intimate partner abuse against African-American women.* New York: New York University Press.

President's Commission on Law Enforcement and the Administration of Justice. (1967). *The challenge of crime in a free society.* Washington DC: U.S. Government Printing Office.

Proano-Raps, T. C., & Meyer, C. L. (2003). Postpartum syndrome and the legal system. In R. Muraskin (Ed.), *It's a crime: Women and justice* (3rd ed., pp. 53–76). Upper Saddle River, NJ: Prentice-Hall.

Rabe-Hemp, C. E. (2008). Survival in an "all boys club": Policewomen and their fight for acceptance. *Policing: An International Journal of Police Strategies and Management, 31*(2), 251–270.

Rabe-Hemp, C. E. (2009). POLICEwomen or PoliceWOMEN? Doing gender and police work. *Feminist Criminology, 4*(2), 114–129.

Rabe-Hemp, C. (2012). The career trajectories of female police executives. In R. Muraskin (Ed.), *Women and justice: It's a crime* (5th ed., pp. 527–543). Upper Saddle River, NJ: Prentice-Hall.

Raeder, M. S. (1995). *The forgotten offender: The effect of the sentencing guidelines and mandatory minimums on women and their children.* 8 Fed. Sent. R. 157.

Rafferty, Y. (2007). Children for sale: Child trafficking in Southeast Asia. *Child Abuse Review, 16,* 410–422.

Rafter, N. H. (1985). *Partial justice: Women in state prisons 1800–1935.* Boston: New England University Press.

Rand, M. R. (2008). *Criminal victimization 2007.* Bureau of Justice Statistics, U.S. Department of Justice. Retrieved September 14, 2009, from http://bjs.ojp.usdoj.gov/content/pub/pdf/cv07.pdf

Raphael, J. (2000). *Saving Bernice: Battered women, welfare and poverty.* Boston: Northeastern University Press.

Raphael, J. (2004). *Listening to Olivia: Violence, poverty and prostitution.* Boston: Northeastern University Press.

Raphael, J., & Shapiro, D. L. (2004). Violence in indoor and outdoor venues. *Violence Against Women, 10*(2), 126–139.

Rasche, C. E. (2012). The dislike of female offenders among correctional officers: A need for specialized training. In R. Muraskin (Ed.), *Women and justice: It's a crime* (5th ed., pp. 544–562). Upper Saddle River, NJ: Prentice-Hall.

Rathbone, C. (2005). *A world apart: Women, prison and life behind bars.* New York: Random House.

Ravensberg, V., & Miller, C. (2003). Stalking among young adults: A review of the preliminary research. *Aggression and Violent Behavior, 9,* 455–469.

Raymond, J. G. (2004). Prostitution on demand: Legalizing the buyers as sexual consumers. *Violence Against Women, 10*(10), 1156–1186.

Redhage, J. (2009, August 24). Gender gap in legal pay the widest of any profession. *Daily Journal.* Retrieved October 10, 2011 at http://www.balanomics.net/press/8.pdf

Reed, M. E., Collinsworth, L. L., & Fitzgerald, L. F. (2005). There's no place like home: Sexual harassment of low income women in housing. *Psychology, Public Policy and Law, 11*(3), 439–462.

Reichman, N. J., & Sterling, J. S. (2001). Recasting the brass ring: Deconstructing and reconstructing workplace opportunities for women lawyers. *Capital University Law Review, 29,* 923–977.

Reinharz, S. (1992). *Feminist methods in social research.* New York: Oxford University Press.

Rennison, C. M. (2009). A new look at the gender gap in offending. *Women and Criminal Justice, 19,* 171–190.

Resnick, H. S., Holmes, M. M., Kilpatrick, D. G., Clum, G., Acierno, R., Best, C. L., et al. (2000). Predictors of post-rape medical care in a national sample of women. *American Journal of Preventive Medicine, 19,* 214–219.

Resnick, P. J. (1970). Murder of the newborn: A psychiatric review of neonaticide. *American Journal of Psychiatry, 126,* 1414–1420.

Revolutionary Worker Online. (2002, September 15) *The disappearing women of Juarez.* Retrieved April 20, 2010, from http://revcom.us/a/v24/1161-1170/1166/juarez.htm

Rickert, V. I., Wiemann, C. M., & Vaughan, R. D. (2005). Disclosure of date/acquaintance rape: Who reports and when. *Journal of Pediatric and Adolescent Gynecology, 18*(1), 17–24.

Ritchie, B. E. (2001). Challenges incarcerated women face as they return to their communities: Findings from life history interviews. *Crime and Delinquency, 47*(3), 368–389.

Roberts, C. A. (1999). Drug use among inner-city African American women: The process of managing loss. *Qualitative Health Research, 9*(5), 620–638.

Rodriguez, S. F., Curry, T. R., & Lee, G. (2006). Gender differences in criminal sentencing: Do effects vary across violent, property and drug offenses. *Social Science Quarterly, 87*(2), 318–339.

Romero-Daza, N., Weeks, M., & Singer, M. (2003). "Nobody gives a damn if I live or die": Violence, drugs, and street-level prostitution in inner city Hartford, Connecticut. *Medical Anthropology, 22,* 233–259.

Rose, D. R., Michalsen, V., Wiest, D. R., & Fabian, A. (2008). *Women, re-entry and everyday life: Time to work?* Women's Prison Association. Retrieved January 20, 2011, from http://www.wpaonline.org/resources/publications.htm

Rosenbaum, A. (2009). Batterer intervention programs: A report from the field. *Violence and Victims, 24*(6), 757–770.

Rosenbaum, J. L. (1989). Family dysfunction and female delinquency. *Crime and Delinquency, 35,* 31–44.

Rospenda, K. M., Richman, J. A., Ehmke, J. L. Z., & Zlatoper, K. W. (2005). Is workplace harassment hazardous to your health? *Journal of Business and Psychology, 20*(1), 95–110.

Sampson, R. (2006). *Acquaintance rape of college students.* Office of Community Oriented Policing Services, U.S. Department of Justice. Retrieved October 1, 2009, from www.cops.usdoj.gov

Sampson, R. J., & Laub, J. (1993). *Crime in the making: Pathways and turning points through life.* Cambridge, MA: Harvard University Press.

Saulters-Tubbs, C. (1993). Prosecutorial and judicial treatment of female offenders. *Federal Probation,* 37–42.

Schadee, J. (2003). Passport to healthy families. *Corrections Today, 65*(3), 64.

Schalet, A., Hunt, G., & Joe-Laidler, K. (2003). Respectability and autonomy: The articulation and meaning of sexuality among girls in the gang. *Journal of Contemporary Ethnography, 32*(1), 108–143.

Schoot, E., & Goswami, S. (2001). *Prostitution: A violent reality of homelessness.* Chicago: Chicago Coalition for the Homeless.

Schuller, R. A., & Rzepa, S. (2002). Expert testimony pertaining to battered woman syndrome: Its impact on jurors' decisions. *Law and Human Behavior, 26*(6), 655–673.

Schulz, D. M. (1995). *From social worker to crime fighter: Women in United States municipal policing.* Westport, CT: Praeger.

Schulz, D. M. (2003). Women police chiefs: A statistical profile. *Police Quarterly, 6*(3), 330–345.

Schulze, C. (2008). *Maternity leave policy in U.S. police departments and school districts: The impact of descriptive and social group representation in a context of gendered institutions.* Unpublished doctoral dissertation, University of New Orleans.

Schulze, C. (2012). The policies of United States police departments: Equal access, equal treatment. In R. Muraskin (Ed.), *Women and justice: It's a crime* (5th ed., pp. 500–513). Upper Saddle River, NJ: Prentice-Hall.

Schwartz, M. D., & DeKeseredy, W. S. (2008). Interpersonal violence against women: The role of men. *Journal of Contemporary Criminal Justice, 24*(2), 178–185.

Schwartz, M. D., DeKeseredy, W. S., Tait, D., & Alvi, S. (2001). Male peer support and a feminist routine activities theory: Understanding sexual assault on the college campus. *Justice Quarterly, 18*(3), 623–649.

Scott-Ham, M., & Burton, F. (2005). Toxicological findings in cases of alleged drug-facilitated sexual assault in the United Kingdom over a 3-year period. *Journal of Clinical Forensic Medicine, 12,* 175–186.

Sellers, C., & Bromley, M. (1996). Violent behavior in college student dating relationships. *Journal of Contemporary Criminal Justice, 12*(1), 1–27.

Sentencing Project. (2006). *Life sentences: Denying welfare benefits to women convicted of drug offenses.* Retrieved December 28, 2010, from http://www.sentencingproject.org/doc/publications/women_smy_lifesentences.pdf

Sexual Harassment Support. (n.d.). *A support community for anyone who has experienced sexual harassment.* Retrieved April 2, 2010, from http://www.sexualharassmentsupport.org

Shannon-Lewy, C., & Dull, V. T. (2005). The response of Christian clergy to domestic violence: Help or hindrance? *Aggression and Violent Behavior, 10*(6), 647–659.

Shelden, R. G. (1981). Sex discrimination in the juvenile justice system: Memphis, Tennessee, 1900–1917. In M. Q. Warren (Ed.), *Comparing male and female offenders.* Beverly Hills, CA: Sage.

Sherman, L. W., & Berk, R. A. (1984). *The Minneapolis Domestic Violence Experiment. Police Foundation reports.* Retrieved February 5, 2010, from http://www.policefoundation.org/pdf/minneapolisdve.pdf

Silverman, J. G., Raj, A., Mucci, L. A., & Hathaway, J. E. (2001). Dating violence against adolescent girls and associated substance use, unhealthy weight control, sexual risk behavior, pregnancy and suicidality. *Journal of American Medical Association, 285*(5), 572–579.

Simkhada, P. (2008). Life histories and survival strategies amongst sexually trafficked girls in Nepal. *Children and Society, 22,* 235–248.

Simmons, W. P. (2006). Remedies for the women of Ciudad Juarez through the Inter-American Court of Human Rights. *Northwestern Journal of International Human Rights, 4*(3).

Simon, R. (1975). *Women and crime.* Lexington, MA: D. C. Heath.

Sipe, S. R., Johnson, C. D., & Fisher, D. K. (2009). University students' perceptions of sexual harassment in the workplace. *Equal Opportunities International, 28*(4), 336–350.

Smith, A. (2000). It's my decision, isn't it? A research note on battered women's perceptions of mandatory intervention laws. *Violence Against Women, 6*(12), 1384–1402.

Smith, E. L., & Farole, D. J., Jr. (2009). *Profile of intimate partner violence cases in large urban counties. Bureau of Justice Statistics, U.S. Department of Justice.* Retrieved February 5, 2010, at http://bjs.ojp.usdoj.gov/content/pub/pdf/pipvcluc.pdf

Snow, R. L. (2010). *Policewomen who made history: Breaking through the ranks.* Lanham, MD: Rowman and Littlefield.

Snyder, H. N., & Sickmund, M. (2006). *Juvenile offenders and victims: 2006 national report.* National Center for Juvenile Justice. Office of Juvenile Justice and Delinquency Prevention. Retrieved December 1, 2010, from http://www.ojjdp.gov/ojstatbb/nr2006/

Sokoloff, N. J. (2004). Domestic violence at the crossroads: Violence against poor women and women of color. *Women Studies Quarterly, 32*(3/4), 139–147.

Songer, D. R., & Crews-Meyer, K. A. (2000). Does judge gender matter? Decision making in state Supreme Courts. *Social Science Quarterly, 8*(3), 750–762.

Songer, D. R., Davis, S., & Haire, S. (1994). A reappraisal of diversification in the federal courts: Gender effects in the court of appeals. *Journal of Politics, 56*(2), 425–439.

Spencer, G. C. (2004). Her body is a battlefield: The applicability of the Alien Tort Statute to corporate human rights abuses in Juarez, Mexico. *Gonzaga Law Review, 4,* 503–533.

Spinelli, M. G. (2001). A systematic investigation of 16 cases of neonaticide. *American Journal of Psychiatry, 158,* 811–813.

Spitzberg, B. H., & Cupach, W. R. (2003). What mad pursuit? Obsessive relational intrusion and stalking related phenomena. *Aggression and Violent Behavior, 8,* 345–375.

Spohn, C., & Beichner, D. (2000). Is preferential treatment of female offenders a thing of the past? A multisite study of gender, race, and imprisonment. *Criminal Justice Policy Review, 11*(2), 149–184.

Spohn, C., Gruhl, J., & Welch, S. (1987). The impact of the ethnicity and gender of defendants on prosecutors' decisions to reject or dismiss felony charges. *Criminology, 25,* 175–191.

Spohn, C., Welch, S., & Gruhl, J. (1985). Women defendants in court: The interaction between sex and race in convicting and sentencing. *Social Science Quarterly, 66,* 178–185.

Stangle, H. L. (2008). Murderous Madonna: Femininity, violence, and the myth of postpartum mental disorder in cases of maternal infanticide and filicide. *William and Mary Law Review, 50,* 699–734.

Starzynski, L. L., Ullman, S. E., Townsend, S. M., Long, L. M., & Long, S. M. (2007). What factors predict women's disclosure of sexual assault to mental health professionals? *Journal of Community Psychology, 35*(5), 619–638.

Stattin, H., & Magnusson, D. (1990). *Pubertal maturation in female development, vol. 2.* Hillsdale, NJ: Erlbaum.

Steffensmeier, D., & Allan, E. (1996). Gender and crime: Toward a gendered theory of female offending. *American Review of Sociology, 22,* 459–487.

Steffensmeier, D., & Hebert, C. (1999). Women and men policymakers: Does the judge's gender affect the sentencing of criminal defendants? *Social Forces, 77*(3), 1163–1196.

Steffensmeier, D., Kramer, J., & Streifel, C. (1993). Gender and imprisonment decisions. *Criminology, 31,* 411–446.

Steffensmeier, D., Schwartz, J., Zhong, H., & Ackerman, J. (2005). An assessment of recent trends in girls' violence using diverse longitudinal sources: Is the gender gap closing? *Criminology, 43,* 355–405.

Steffensmeier, D., & Ulmer, J. (2005). *Confessions of a dying thief: Understanding criminal careers and illegal enterprise.* New York: Aldine Transaction.

Steffensmeier, D., Zhong, H., Ackerman, J., Schwartz, J., & Agha, S. (2006). Gender gap trends for violent crimes: A UCR-NCVS comparison. *Feminist Criminology, 1*(1): 72–98.

Stevenson, T., & Love, C. (1999). *Her story of domestic violence: A timeline of the battered women's movement.* Safework: California's Domestic Violence Resource. Retrieved February 5, 2010, from http://www.mincava.umn.edu/documents/herstory/herstory.html

Stohr, M. K., Mays, G. L., Lovrich, N. P., & Gallegos, A. M. (1996). *Partial perceptions: Gender, job enrichment and job satisfaction among correctional officers in women's jails.* Presented at the Annual Meeting of the Academy of Criminal Justice Sciences, Las Vegas, Nevada.

Sugar, N. F., Fine, D. N., & Eckert, L. O. (2004). Physical injury after sexual assault: Findings of a large case series. *American Journal of Obstetrics and Gynecology, 190*(1), 71–76.

Sullivan, M., Senturia, K., Negash, T., Shiu-Thornton, S., & Giday, B. (2005). For us it's like living in the dark: Ethiopian women's experiences with domestic violence. *Journal of Interpersonal Violence, 20*(8), 922–940.

Sutton, J. R. (1988). *Stubborn children: Controlling delinquency in the United States, 1640–1981.* Berkeley: University of California Press.

Svensson, R. (2003). Gender differences in adolescent drug use. *Youth and Society, 34,* 300–329.

Tester, G. (2008). An intersectional analysis of sexual harassment in housing. *Gender and Society, 22*(3), 349–366.

Tewksbury, R., & Collins, S. C. (2006). Aggression levels among correctional officers. *The Prison Journal, 86*(3), 327–343.

Tillman, S., Bryant-Davis, T., Smith, K., & Marks, A. (2010). Shattering silence: Exploring barriers to disclosure for African American sexual assault survivors. *Trauma, Violence, Abuse, 11*(2), 59–70.

Tjaden, P., & Thoennes, N. (2006). *Extent, nature and consequences of rape victimizations: Findings from the National Violence Against Women Survey.* National Institute of Justice. U.S. Department of Justice. Retrieved February 5, 2010, from http://www.ncjrs.gov/pdffiles1/nij/210346.pdf

Tolman, D. L. (1994). Doing desire: Adolescent girls' struggles for/with sexuality. *Gender and Society, 8*(3), 324–342.

Triplett, R., & Myers, L. B. (1995). Evaluating contextual patterns of delinquency: Gender based differences. *Justice Quarterly, 12*(1), 59–84.

U.S. Bureau of the Census. (2000). *The American Indian and Alaska native population 2000. Census 2000 brief.* Washington, DC: U.S. Department of Commerce, Bureau of the Census.

U.S. Department of Housing (HUD). (2008, November 17). *Questions and answers on sexual harassment under the Fair Housing Act.* Retrieved March 25, 2010, from http://www.hud.gov/content/releases/q-and-a-111708.pdf

U.S. Department of State. (2008). *Trafficking in persons report 2008.* Retrieved July 19, 2011, from http://www.state.gov/g/tip/rls/tiprpt/2008/

U.S. Department of Justice. (2003). *Criminal victimization, 2003.* Washington, DC: U.S. Department of Justice.

U.S. Department of State. (2009). *Trafficking in persons (TIP) report 2009.* Retrieved June 17, 2010, from http://www.state.gov/documents/organization/123360.pdf

U.S. Department of State. (2010). *Trafficking in persons (TIP) report 2010.* Retrieved March 27, 2011, from http://www.state.gov/g/tip/rls/tiprpt/2010/

U.S. Equal Employment Opportunity Commission (EEOC). (n.d.). *Sexual harassment.* Retrieved March 3, 2010, from http://www.eeoc.gov/laws/types/sexual_harassment.cfm and http://www.eeoc.gov/eeoc/statistics/enforcement/sexual_harassment.cfm

U.S. Merit Systems Protection Board. (1995). *Sexual harassment in the federal workforce: Trends, progress, and continuing challenges.*

Retrieved March 20, 2010, from http://www.mspb.gov/netsearch/viewdocs.aspx?docnumber=253661&version=253948&application=ACROBAT

Valera, R. J., Sawyer, R. G., & Schiraldi, G. R. (2000). Violence and post traumatic stress disorder in a sample of inner city street prostitutes. *American Journal of Health Studies, 16*(3), 149–155.

Van Wormer, K. S., & Bartollas, C. (2010). *Women and the criminal justice system.* Boston: Allyn and Bacon.

von Hentig, H. (1948). *The criminal and his victim: Studies in the sociobiology of crime*. New Haven, CT: Yale University Press.

Wagenaar, H. (2006). Democracy and prostitution: Deliberating the legalization of brothels in the Netherlands. *Administration and Society, 38*(2), 198–235.

Walker, L. E. (1979). *The battered woman*. New York: Harper and Row.

Wang, M.-C., Horne, S. G., Levitt, H. M., & Klesges, L. M. (2009). Christian women in IPV relationships: An exploratory study of religious factors. *Journal of Psychology and Christianity, 28*(3), 224–235.

Warr, M. (1984). Fear of victimization: Why are women and the elderly more afraid? *Social Science Quarterly, 65,* 681–702.

Warr, M. (1985). Fear of rape among urban women. *Social Problems, 32,* 238–250.

Warshaw, R. (1994). *I never called it rape: The Ms. report on recognizing, fighting and surviving date & acquaintance rape*. New York: Harper Perennial.

Washington, P. A. (2001). Disclosure patterns of Black female sexual assault survivors. *Violence Against Women,* 7(11), 1254–1283.

West, C. M. (2004). Black women and intimate partner violence: New directions for research. *Journal of Interpersonal Violence, 19*(12), 1487–1493.

West, D. A., & Lichtenstein, B. (2006). Andrea Yates and the criminalization of the filicidal maternal body. *Feminist Criminology, 1*(3), 173–187.

West, H. C., & Sabol, W. J. (2009). *Prison inmates at midyear 2008: Statistical tables.* Bureau of Justice Statistics. Retrieved July 31, 2011, from bjs.ojp.usdoj.gov/content/pub/pdf/pim08st.pdf

Westervelt, S. D., & Cook, K. J. (2007). Feminist research methods in theory and action: Learning from death row exonerees. In S. Miller (Ed.), *Criminal justice research and practice: Diverse voices from the field.* Boston: University Press of New England.

Widom, C. S. (1989). The cycle of violence. *Science, 244,* 160–166.

Widom, C. S. (1991). Childhood victimization: Risk factor for delinquency. In M. E. Colten & S. Gore (Eds.), *Adolescent stress: Causes and consequences.* New York: Aldine De Gruyter.

Williams, M. (2007). Women's representation on state trial and appellate courts. *Social Science Quarterly, 88*(5), 1192–1204.

Women's Health. (2004). *UMHS Women's Health Program.* Retrieved February 5, 2010, from http://www.med.umich.edu/whp/newsletters/summer04/p03-dating.html

Women's Prison Association. (2003). *WPA focus on women and justice: A portrait of women in prison.* Retrieved January 3, 2011, from http://www.wpaonline.org/pdf/Focus_December 2003.pdf

Women's Prison Association. (2008). *Mentoring women in reentry.* Retrieved January 21, 2011, from http://www.wpaonline.org

Women's Prison Association. (2009a). *Quick facts: Women and criminal justice 2009.* Retrieved January 5, 2011, from http://www.wpaonline.org

Women's Prison Association. (2009b). *Mothers, infants and imprisonment: A national look at prison nurseries and community-based alternatives.* Retrieved January 5, 2011, from http://www.wpaoline.com

Wooldridge, J., & Griffin, T. (2005). Displaced discretion under Ohio sentencing guidelines. *Journal of Criminal Justice, 33,* 301–316.

Yacoubian, G. S., Urbach, B. J., Larsen, K. L., Johnson, R. J., & Peters, R. J. (2000). A comparison of drug use between prostitutes and other female arrestees. *Journal of Alcohol and Drug Education, 46*(2), 12–26.

Yahne, C. E., Miller, W. R., Irvin-Vitela, L., & Tonigan, J. S. (2002). Magdalena Pilot Project: Motivational outreach to substance abusing women street sex workers. *Journal of Substance Abuse Treatment, 23*(1), 49–53.

Young, Vernetta. D. (1992). Fear of victimization and victimization rates among women: A paradox? *Justice Quarterly, 9,* 419–442.

Zaitzow, B. H., & Thomas, J. (2003). *Women in prison: Gender and social control.* Boulder, CO: Lynne Rienner Publishers.

Author Index

Subject Index

Note: f = figures; t = tables.

About the Author

Stacy L. Mallicoat is Associate Professor of Criminal Justice in the Division of Politics, Administration and Justice at California State University, Fullerton. Her teaching and research interests include issues of women and crime, juvenile delinquency, and capital punishment. Her work has appeared in edited books and various journals, including *Feminist Criminology*, *The Journal of Ethnicity and Crime, Journal of Criminal Justice,* and *The Southwest Journal of Criminal Justice.*